# Graduate Texts in Physics

T0171916

## Graduate Texts in Physics

Graduate Texts in Physics publishes core learning/teaching material for graduate- and advanced-level undergraduate courses on topics of current and emerging fields within physics, both pure and applied. These textbooks serve students at the MS- or PhD-level and their instructors as comprehensive sources of principles, definitions, derivations, experiments and applications (as relevant) for their mastery and teaching, respectively. International in scope and relevance, the textbooks correspond to course syllabi sufficiently to serve as required reading. Their didactic style, comprehensiveness and coverage of fundamental material also make them suitable as introductions or references for scientists entering, or requiring timely knowledge of, a research field.

More information about this series at http://www.springer.com/series/8431

Edouard B. Manoukian

# Quantum Field Theory I

Foundations and Abelian and Non-Abelian
Gauge Theories

 Springer

Edouard B. Manoukian
The Institute for Fundamental Study
Naresuan University
Phitsanulok, Thailand

ISSN 1868-4513        ISSN 1868-4521   (electronic)
Graduate Texts in Physics
ISBN 978-3-319-80922-9      ISBN 978-3-319-30939-2   (eBook)
DOI 10.1007/978-3-319-30939-2

Printed on acid-free paper

This Springer imprint is published by Springer Nature
The registered company is Springer International Publishing AG Switzerland

# Preface to Volume I

This textbook is based on lectures given in quantum field theory (QFT) over the years to graduate students in theoretical and experimental physics. The writing of the book spread over three continents: North America (Canada), Europe (Ireland), and Asia (Thailand). QFT was born about 90 years ago, when quantum mechanics met relativity, and is still going strong. The book covers, pedagogically, the wide spectrum of developments in QFT emphasizing, however, those parts which are reasonably well understood and for which satisfactory theoretical descriptions have been given.

The legendary Richard Feynman *in* his 1958 Cornell, 1959–1960 Cal Tech lectures on QFT of fundamental processes, the first statement he makes, the very first one, is that the *lectures cover all of physics*.[1] One quickly understands what Feynman meant by covering all of physics. The role of fundamental physics is to describe the basic interactions of Nature and *QFT, par excellence, is supposed to do just that*. Feynman's statement is obviously more relevant today than it was then, since the recent common goal is to provide a unified description of *all* the fundamental interactions in nature.

The book requires as background a good knowledge of quantum mechanics, including rudiments of the Dirac equation, as well as elements of the Klein-Gordon equation, and the reader would benefit much by reading relevant sections of my earlier book: *Quantum Theory: A Wide Spectrum (2006), Springer* in this respect.

This book differs from QFT books that have appeared in recent years[2] in several respects and, in particular, it offers something new in its approach to the subject, and the reader has plenty of opportunity to be exposed to many topics not covered, or

---

[1] R. P. Feynman, The Theory of Fundamental Processes, The Benjamin/Cummings Publishing Co., Menlo Park, California. 6th Printing (1982), page 1.

[2] Some of the fine books that I am familiar with are: L. H. Ryder, Quantum Field Theory; S. Weinberg, The Quantum Theory of Fields I (1995) & II (1996), Cambridge: Cambridge University Press; M. Peskin and D. V. Schroeder, An Introduction to Quantum Field Theory, New York: Westview Press (1995); B. DeWitt, The Global Approach to Quantum Field Theory, Oxford: Oxford University Press (2014).

just touched upon, in standard references. Some notable differences are seen, partly, from unique features in the following material included in ours:

- The very elegant functional *differential* approach of Schwinger, referred to as the quantum dynamical (action) principle, and its underlying theory are used systematically in generating the so-called vacuum-to-vacuum transition amplitude of both abelian and non-abelian gauge theories, in addition to the well-known functional integral approach of Feynman, referred to as the path-integral approach, which are simply related by functional Fourier transforms and delta functionals.

- Transition amplitudes are readily extracted by a direct expansion of the vacuum-to-vacuum transition amplitude in terms of a unitarity sum, which is most closely related to actual experimental setups with particles emitted and detected prior and after a given process and thus represent the underlying physics in the clearest possible way.

- Particular emphasis is put on the concept of a quantum field and its particle content, both physically and technically, as providing an appropriate description of physical processes at sufficiently high energies, for which relativity becomes the indispensable language to do physics and explains the exchange that takes place between energy and matter, allowing the creation of an unlimited number of particles such that the number of particles need not be conserved, and for which a variable number of particles may be created or destroyed. Moreover, quantum mechanics implies that a wavefunction renormalization arises in QFT field independent of any perturbation theory – a point not sufficiently emphasized in the literature.

- The rationale of the stationary action principle and emergence of field equations, via field variations of transformation functions and generators of field variations. The introduction of such generators lead, self consistently, to the field equations. Such questions are addressed as: "Why is the variation of the action, *within* the boundaries of transformation functions, set equal to *zero* which eventually leads to the Euler-Lagrange equations?", "How does the Lagrangian density appear in the formalism?" "What is the significance in commuting/anti-commuting field components within the interaction Lagrangian density in a theory involving field operators?" These are some of the questions many students seem to worry about.

- A panorama of all the fields encountered in present high-energy physics, together with the details of the underlying derivations are given.

- Schwinger's point splitting method of currents is developed systematically in studying abelian and non-abelian gauge theories anomalies. Moreover, an explicit experimental test of the presence of an anomaly is shown by an example.

- Derivation of the Spin & Statistics connection and CPT symmetry, emphasizing for the latter that the invariance of the action under CPT transformation is not sufficient for CPT symmetry, but one has also to consider the roles of incoming and outgoing particles.

- The fine-structure effective coupling $\alpha \simeq 1/128$ at high energy corresponding to the mass of the neutral $Z^0$ vector boson based on all the charged leptons and all those contributing quarks of the three generations.

- Emphasis is put on renormalization theory, including its underlying general subtractions scheme, often neglected in treatments of QFT.

- Elementary derivation of Faddeev-Popov factors directly from the functional differential formalism, with constraints, and their *modifications*, and how they may even arise in some abelian gauge theories.

- A fairly detailed presentation is given of "deep inelastic" experiments as a fundamental application of quantum chromodynamics.

- Schwinger line integrals, origin of Wilson loops, lattices, and quark confinement.

- Neutrino oscillations,[3] neutrino masses, neutrino mass differences, and the "seesaw mechanism."

- QCD jets and parton splitting, including gluon splitting to gluons.

- Equal importance is put on both abelian and non-abelian gauge theories, witnessing the wealth of information also stored in the abelian case.[4]

- A most important, fairly detailed, and semi-technical introductory chapter is given which traces the development of QFT since its birth in 1926 without tears, in abelian and non-abelian gauge theories, including aspects of quantum gravity, as well as examining the impact of supersymmetry, string theory, and the development of the theory of renormalization, as a *pedagogical* strategy for the reader to be able to master the basic ideas of the subject at the outset before they are encountered in glorious technical details later.

- *Solutions* to all the problems are given right at the end of the book.

With the mathematical rigor that renormalization has met over the years and the reasonable agreement between gauge theories and experiments, the underlying theories are in pretty good shape. This volume is organized as follows. The first introductory chapter traces the subject of QFT since its birth, elaborating on many of its important developments which are conveniently described in a fairly simple language and will be quite useful for understanding the underlying technical details of the theory covered in later chapters including those in Volume II. A preliminary chapter follows which includes the study of symmetry transformations in the quantum world, as well as of intricacies of functional differentiation and functional integration which are of great importance in field theory. Chapter 3 deals with quantum field theory methods of spin 1/2 culminating in the study of anomalies in the quantum world. The latter refers to the fact that a conservation law

---

[3]It is rather interesting to point out that the theory of neutrino oscillations was written up in this book much earlier than the 2015 Nobel Prize in Physics was announced on neutrino oscillations.

[4]With the development of non-abelian gauge theories, unfortunately, it seems that some students are not even exposed to such derivations as of the "Lamb shift" and of the "anomalous magnetic moment of the electron" in QED.

in classical physics does not necessarily hold in the quantum world. Chapter 4, a critical one, deals with the concept of a quantum field, the Poincaré algebra, and particle states. Particular attention is given to the stationary action principle as well as in developing the solutions of QFT via the quantum dynamical principle. This chapter includes the two celebrated theorems dealing with CPT symmetry and of the Spin & Statistics connection. A detailed section is involved with the basic quantum fields one encounters in present day high-energy/elementary-particle physics and should provide a useful reference source for the reader. Chapter 5 treats abelian gauge theories (QED, scalar boson electrodynamics) in quite details and includes, in particular, the derivations of two of the celebrated results of QED which are the anomalous magnetic moment of the electron and the Lamb shift. Chapter 6 is involved with non-abelian gauge theories (electroweak, QCD, Grand unification).[5] Such important topics are included as "asymptotic freedom," "deep inelastic" scattering, QCD jets, parton splittings, neutrino oscillations, the "seesaw mechanism" and neutrino masses, Schwinger-line integrals, Wilson loops, lattices, and quark confinement. Unification of coupling parameters of the electroweak theory and of QCD are also studied, as well as of spontaneous symmetry breaking in both abelian and non-abelian gauge theories, and of renormalizability aspects of both gauge theories, emphasizing the so-called BRS transformations for the latter. We make it a point, pedagogically, to derive things in detail, and some of such details are *relegated* to appendices at the end of the respective chapters with the main results *given* in the sections in question. Five general appendices, at the end of this volume, cover some additional important topics and/or technical details. In particular, I have included an appendix covering some aspects of the general theory of renormalization and its underlying subtractions scheme itself which is often neglected in books on QFT. Fortunately, my earlier book, with *proofs* not just words, devoted completely to renormalization theory – *Renormalization (1983), Academic Press* – may be consulted for more details. The problems given at the end of the chapters form an integral part of the book, and many developments in the text depend on the problems and may include, in turn, additional material. They should be attempted by every serious student. *Solutions* to all the problems are given right at the end of the book for the convenience of the reader. The introductory chapter together with the introductions to each chapter provide the motivation and the *pedagogical* means to handle the technicalities that follow them in the texts.

I hope this book will be useful for a wide range of readers. In particular, I hope that physics graduate students, not only in quantum field theory and high-energy physics, but also in other areas of specializations will also benefit from it as, according to my experience, they seem to have been left out of this fundamental area of physics, as well as instructors and researchers in theoretical physics. The content of this volume may be covered in one-year (two semesters) quantum field theory courses.

---

[5]QED and QCD stand, respectively, for quantum electrodynamics and quantum chromodynamics.

In Volume II, the reader is introduced to quantum gravity, supersymmetry, and string theory,[6] which although may, to some extent, be independently read by a reader with a good background in field theory, the present volume sets up the language, the notation, provides additional background for introducing these topics, and will certainly make it much easier for the reader to follow. In this two-volume set, aiming for completeness in covering the basics of the subject, I have included topics from the so-called conventional field theory (the classics) to ones from the modern or the new physics which I believe that every serious graduate student studying quantum field theory should be exposed to.

Without further ado, and with all due respect to the legendary song writer Cole Porter, let us find out "what is this thing called QFT?"

Edouard B. Manoukian

---

[6]Entitled: Quantum Field Theory II: *Introductions to Quantum Gravity, Supersymmetry, and String Theory*" (2016), Springer.

# Acknowledgements

In the beginning of it all, I was introduced to the theoretical aspects of quantum field theory by Theodore Morris and Harry C. S. Lam, both from McGill and to its mathematical intricacies by Eduard Prugovečki from the University of Toronto. I am eternally grateful to them. Over the years, I was fortunate enough to attend a few lectures by Julian Schwinger and benefited much from his writings as well. Attending a lecture by Schwinger was quite an event. His unique elegant, incisive, physically clear approach and, to top it off, short derivations were impressive. When I was a graduate student, I would constantly hear that Schwinger "does no mistakes." It took me years and years to understand what that meant. My understanding of this is because he had developed such a powerful formalism to do field theory that, unlike some other approaches, everything in the theory came out automatically and readily without the need to worry about multiplicative factors in computations, such as $2\pi$'s and other numerical factors, and, on top of this, is relatively easy to apply. Needless to say this has much influenced my own approach to the subject. He had one of the greatest minds in theoretical physics of our time.

I want to take this opportunity as well to thank Steven Weinberg, the late Abdus Salam, Raymond Streater, and Eberhard Zeidler for the keen interest they have shown in my work on renormalization theory.

I acknowledge with thanks the support I received from several colleagues, while visiting their research establishments for extended periods, for doing my own thing and writing up the initial notes on this project. These include Yasushi Takahashi and Anton Z. Capri from the University of Alberta, Lochlainn O'Raifeartaigh and John Lewis from the Dublin Institute for Advanced Studies, and Jiri Patera and Pavel Winternitz from the University of Montreal. For the final developments of the project, I would like to thank Sujin Jinahyon, the President of Naresuan University, Burin Gumjudpai, Seckson Sukhasena, Suchittra Sa-Nguansin, and Jiraphorn Chomdaeng of the university's Institute for Fundamental Study for encouragement, as well as Ahpisit Ungkitchanukit and Chai-Hok Eab from Chulalongkorn University.

I am also indebted to many of my former graduate students, who are now established physicists in their own rights, particularly to Chaiyapoj Muthaporn,

Nattapong Yongram, Siri Sirininlakul, Tukkamon Vijaktanawudhi (aka Kanchana Limboonsong), Prasopchai Viriyasrisuwattana, and Seckson Sukhasena, who through their many questions, several discussions, and collaborations have been very helpful in my way of analyzing this subject.

Although I have typed the entire manuscripts myself, and drew the figures as well, Chaiyapoj Muthaporn prepared the LATEX input files. Without his constant help in LATEX, this work would never have been completed. I applaud him, thank him and will always remember how helpful he was. My special thanks also go to Nattapong Yongram for downloading the endless number of papers I needed to complete the project.

I was fortunate and proud to have been associated with the wonderful editorial team of Maria Bellantone and Mieke van der Fluit of Springer. I would like to express my deepest gratitude to them for their excellent guidance, caring, patience, and hard work in making this project possible and move forward toward its completion. I have exchanged more emails with Mieke than with anybody else on the globe. This has led to such an enjoyable association that I will always cherish.

This project would not have been possible without the patience, encouragement, and understanding of my wife Tuenjai. To my parents, who are both gone, this work is affectionately dedicated.

# Contents

# Notation and Data

○ Latin indices $i, j, k, \ldots$ are generally taken to run over 1,2,3, while the Greek indices $\mu, \nu, \ldots$ over $0, 1, 2, 3$ in 4D. Variations do occur when there are many different types of indices to be used, and the meanings should be evident from the presentations.

○ The Minkowski metric $\eta_{\mu\nu}$ is defined by $[\eta_{\mu\nu}] = \text{diag}[-1, 1, 1, 1] = [\eta^{\mu\nu}]$ in 4D.

○ Unless otherwise stated, the fundamental constants $\hbar$, c are set equal to one.

○ The gamma matrices satisfy the anti-commutation relations $\{\gamma^\mu, \gamma^\nu\} = -2\,\eta^{\mu\nu}$.

○ The Dirac, the Majorana, and the chiral representations of the $\gamma^\mu$ matrices are defined in Appendix I at the end of the book.

○ The charge conjugation matrix is defined by $\mathscr{C} = i\gamma^2\gamma^0$.

○ $\overline{\psi} = \psi^\dagger\gamma^0$, $\overline{u} = u^\dagger\gamma^0$, $\overline{v} = v^\dagger\gamma^0$. A Hermitian conjugate of a matrix $M$ is denoted by $M^\dagger$, while its complex conjugate is denoted by $M^*$.

○ The step function is denoted by $\theta(x)$ which is equal to 1 for $x > 0$, and 0 for $x < 0$.

○ The symbol $\varepsilon$ is used in dimensional regularization (see Appendix III). $\epsilon$ is used in defining the boundary condition in the denominator of a propagator $(Q^2 + m^2 - i\epsilon)$ and should not be confused with $\varepsilon$ used in dimensional regularization. We may also use either one when dealing with an infinitesimal quantity, in general, with $\epsilon$ more frequently, and this should be self-evident from the underlying context.

○ For units and experimental data, see the compilation of the "Particle Data Group": Beringer et al. [1] and Olive et al. [2]. The following (some obviously approximate) numerical values should, however, be noted:

$$
\begin{aligned}
1 \text{ MeV} &= 10^6 \text{ eV} \\
1 \text{ GeV} &= 10^3 \text{ MeV} \\
10^3 \text{ GeV} &= 1 \text{ TeV} \\
1 \text{ erg} &= 10^{-7} \text{ J}
\end{aligned}
$$

$$1\,\text{J} = 6.242 \times 10^9 \text{ GeV}$$
$$c = 2.99792458 \times 10^{10} \text{ cm/s (exact)}$$
$$\hbar = 1.055 \times 10^{-34} \text{ J s}$$
$$\hbar c = 197.33 \text{ MeV fm}$$
$$1 \text{ fm} = 10^{-13} \text{ cm}$$

(Masses) $M_p = 938.3\,\text{MeV}/c^2$, $M_n = 939.6\,\text{MeV}/c^2$,

$M_W = 80.4 \text{ GeV}/c^2$, $M_Z = 91.2\,\text{GeV}/c^2$,

$m_e = 0.511\,\text{MeV}/c^2$, $m_\mu = 105.66 \text{ MeV}/c^2$, $m_\tau = 1777\,\text{MeV}/c^2$.

Mass of $\nu_e < 2\,\text{eV}/c^2$, Mass of $\nu_\mu < 0.19\,\text{MeV}/c^2$, Mass of $\nu_\tau < 18.2\,\text{MeV}/c^2$,

Mass of the neutral Higgs $H^0 \approx 125.5\,\text{GeV}/c^2$.

For approximate mass values of some of the quarks taken, see Table 5.1 in Sect. 5.19.2. For more precise range of values, see Olive et al. [2].

(Newton's gravitational constant) $G_N = 6.709 \times 10^{-39}\,\hbar\,c^5/\text{GeV}^2$.

(Fermi weak interaction constant) $G_F = 1.666 \times 10^{-5}\,\hbar^3\,c^3/\text{GeV}^2$.

Planck mass $\sqrt{\hbar c/G_N} \approx 1.221 \times 10^{19}\,\text{GeV}/c^2$,

Planck length $\sqrt{\hbar\,G_N/c^3} \approx 1.616 \times 10^{-33}$ cm.

Fine structure constant $\alpha = 1/137.04$ at $Q^2 = 0$, and $\approx 1/128$ at $Q^2 \approx M_Z^2$.

For the weak-mixing angle $\theta_W$, $\sin^2\theta_W \approx 0.232$, at $Q^2 \approx M_Z^2$.

$\alpha/\sin^2\theta_W \approx 0.034$, at $Q^2 \approx M_Z^2$.

Strong coupling constant $\alpha_s \approx 0.119$, at $Q^2 \approx M_Z^2$.

# References

1. Beringer, J., et al. (2012). Particle data group. *Physical Review D, 86*, 010001.
2. Olive, K. A., et al. (2014). Particle data group. *Chinese Physics C, 38*, 090001.

# Chapter 1
# Introduction

Donkey Electron, Bare Electron, Electroweak Frog, God Particle, "Colored" Quarks
and Gluons, Asymptotic Freedom, Beyond Resonances into the Deep Inelastic
Region, Partons, QCD Jets, Confined Quarks, Bekenstein – Hawking Entropy
of a Black Hole, Sparticles, Strings, Branes, Various Dimensions and even Quanta
of Geometry, AdS/CFT Correspondence and Holographic Principle, CPT, and
Spin & Statistics

The major theme of quantum field theory is the development of a unified theory
that may be used to describe nature from microscopic to cosmological distances.
Quantum field theory was born 90 years ago, when quantum theory met relativity,
and has captured the hearts of the brightest theoretical physicists in the world. It is
still going strong. It has gone through various stages, met various obstacles on the
way, and has been struggling to provide us with a coherent description of nature
in spite of the "patchwork" of seemingly different approaches that have appeared
during the last 40 years or so, but still all, with the common goal of unification.

As mentioned in our Preface, Feynman, in his 1958 Cornell, 1959–1960 Cal
Tech, lectures on the quantum field theory of fundamental processes, the first
statement he makes, the very first one, is that the lectures will cover *all* of physics
[76, p. 1]. One quickly understands what Feynman meant by covering all of physics.
After all, the role of fundamental physics is to describe the basic interactions we
have in nature and quantum field theory is supposed to do just that. Feynman's
statement is obviously more relevant today than it was then, since the recent
common goal is to provide a unified description of *all* the fundamental forces in
nature. With this in mind, let us trace the development of this very rich subject from
the past to the present, and see what the theory has been telling us all these years.

When the energy and momentum of a quantum particle are large enough, one is
confronted with the requirement of developing a formalism, as imposed by nature,
which extends quantum theory to the relativistic regime. A relativistic theory, as
a result of the exchange that takes place between energy and matter, allows the
creation of an unlimited number of particles and the number of particles in a given
physical process need not be conserved. An appropriate description of such physical
processes for which a variable number of particles may be created or destroyed, in
the quantum world, is provided by the very rich concept of a quantum field. For
example, photon emissions and absorptions, in a given process, are explained by the
introduction of the electromagnetic quantum field. The theory which emerges from
extending quantum physics to the relativistic regime is called "Relativistic Quantum

© Springer International Publishing Switzerland 2016
E.B. Manoukian, *Quantum Field Theory I*, Graduate Texts in Physics,
DOI 10.1007/978-3-319-30939-2_1

Field Theory" or just *"Quantum Field Theory"*. Quantum Electrodynamics is an example of a quantum field theory and is the most precise theory devised by man when confronted with experiments. The essence of special relativity is that all inertial frames are completely equivalent in explaining a physical theory as one inertial frame cannot be distinguished from another. This invariance property of physical theories in all inertial frames, as required by special relativity, as well as by the many symmetries one may impose on such theories, are readily implemented in the theory of quantum fields. The implementation of symmetries and describing their roles in the explanation of observed phenomena has played a key role in elementary particle physics.

Of course it took years before the appropriate language of quantum field theory, described concisely above, by marrying quantum theory and relativity, was spelled out and applied consistently to physical processes in the quantum world in the relativistic regime. An appropriate place to start in history is when Dirac [47–49] developed his relativistic equation of spin 1/2, from which one learns quite a bit about the subsequent development of the subject as a multi-particle theory. We will then step back a year or two, and then move again forward in time to connect the dots between the various stages of the underlying exciting developments. His relativistic equation, which incorporated the spin of the electron, predicted the existence of negative energy states with negative mass, with energies going down to $-\infty$, implying the instability of the corresponding systems. For example, an electron in the ground-state energy of an atom would spontaneously decay to such lower and lower negative energy states emitting radiation of arbitrary large energies leading eventually to the collapse of the atom with the release of an infinite amount of energy. Historically, a relativistic equation for spin 0, was developed earlier by Klein and Gordon in 1926,[1] referred to as the Klein-Gordon equation, which also shared this problem, but unlike Dirac's theory it led to negative probabilities as well. Dirac being aware of the negative probabilities encountered in the theory of the latter authors, was able to remedy this problem in his equation. To resolve the dilemma of negative energies, Dirac, in 1930,[2] assumed that a priori all the negative energy states are filled with electrons in accord to the Pauli exclusion principle, giving rise to the so-called Dirac sea or the Dirac vacuum, so that no transitions to such states are possible, thus ensuring the stability of the atom.

The consequences of the assumption made by Dirac above were many. A negative energy electron in the Dirac sea, may absorb radiation of sufficient energy so as to overcome an energy gap arising from the level $-mc^2$ to $+mc^2$, where $m$ is the mass of an electron, thus making such a negative energy electron jump to a positive energy state, leaving behind a surplus of positive energy and a surplus of

---

[1]Klein [128] and Gordon [101].
[2]Dirac [50, 51].

positive charge $+|e|$ *relative* to the Dirac sea. This has led Dirac eventually,[3] in 1931 [52], to interpret the "hole" left behind by the transition of the negative energy electron to a positive energy state, as a particle that has the same mass as the electron but of opposite charge. It is interesting to note that George Gamow referred[4] to Dirac's predicted particle as a *"donkey electron"*, because it would move in the opposite direction of an appropriate applied force. The physics community found it difficult to accept Dirac's prediction until Anderson[5] discovered this particle (the positron $e^+$), who apparently was not aware of Dirac's prediction at the time of the discovery.[6] With the positron now identified, the above argument just given has provided an explanation of the so-called *pair production* $\gamma \rightarrow e^+ e^-$ by a photon (in the vicinity of a nucleus).[7] Conversely, if a "hole" is created in the vacuum, then an electron may make a transition to such a state releasing radiation giving rise to the phenomenon of *pair annihilation*. A Pair created, as described above, in the vicinity of a positively charged nucleus, would lead to a partial screening of the charge of the nucleus as the electron within the pair would be attracted by the nucleus and the positively charged one would be repelled. Accordingly, an electron, in the atom, at sufficiently large distances from the nucleus would then see a smaller charge on the nucleus than an electron nearby (such as one in an $s$-state). This leads to the concept of *vacuum polarization*, and also to the concept of *charge renormalization* as a result of the partial charge screening mentioned above.

The Dirac equation is Lorentz covariant, that is, it has the same form in every inertial frame with its variables being simply relabeled reflecting the variables used in the new inertial frame. It predicted, approximately, the gyromagnetic ratio $g = 2$ of the electron, the fine-structure of the atom, and eventually anti-matter was discovered such as antiprotons.[8] It was thus tremendously successful. Apparently,[9] Dirac himself remarked in one of his talks that *his equation was more intelligent than its author*.[10]

Thus the synthesis of relativity and quantum physics, led to the discovery of the antiparticle. The Dirac equation which was initially considered to describe a single particle necessarily led to a *multi*-particle theory, and a single particle description in the relativistic regime turned out to be not complete. A formalism which would naturally describe creation and annihilation of particles and take into account this

---

[3]Dirac [50, 51] assumed that the particle is the proton as the positron was not discovered yet at that time. Apart from the large mass difference between the proton and the electron, there were other inconsistencies with such an assumption.

[4]Weisskopf [242].

[5]Anderson [5, 6].

[6]Weisskopf [242].

[7]The presence of the nucleus is to conserve energy and momentum.

[8]Chamberlain et al. [30].

[9]Weisskopf [242].

[10]For a systematic treatment of the intricacies of Dirac's theory and of the quantum description of relativistic particles, in general, see Manoukian [151], Chapter 16.

multi-particle aspect became necessary. The so-called "hole" theory although it gave insight into the nature of fundamental processes involving quantum particles in the relativistic regime, and concepts such as vacuum polarization, turned out to be also not complete. For example, in the "hole" theory, the number of electrons minus the number of positrons, created is conserved by the simultaneous creation of a "hole" for every electron ejected from the Dirac Sea. In nature, there are processes, where just an electron or just a positron is created while conserving charge of course. Examples of such processes are $\beta^-$ decay: $n \rightarrow p + e^- + \tilde{\nu}_e$, muon decay: $\mu^- \rightarrow e^- + \tilde{\nu}_e + \nu_\mu$, and $\beta^+$ decay: $p \rightarrow n + e^+ + \nu_e$, for a bound proton in a nucleus for the latter process. Finally, Dirac's argument of a sea of negatively charged bosons did not work with the Klein-Gordon equation because of the very nature of the Bose statistics of the particles. A new description to meet all of the above challenges including the creation and annihilation of particles, mentioned above, was necessary.

After the conceptual framework of quantum mechanics was developed, Born, Heisenberg, and Jordan in 1926 [26], applied quantum mechanical methods to the electromagnetic field, now, giving rise to a system with an infinite degrees of freedom, and described as a set of independent harmonic oscillators of various frequencies. Then Dirac in 1927 [46], prior to the development of his relativistic spin 1/2 equation, also extended quantum mechanical methods to the electromagnetic field now with the latter field treated as an operator, and provided a theoretical description of how photons emerge in the quantization of the electromagnetic field. This paper is considered to mark the birthdate of "Quantum Electrodynamics", a name coined by Dirac himself, and provided a prototype for the introduction of field operators for other particles with spin, such as for spin 1/2, where in the latter case commutators in the theory are replaced by anti-commutators [125, 126] for the fermion field.

The first comprehensive treatment of a general quantum field theory, involving Lagrangians, as in modern treatments, was given by Heisenberg and Pauli in 1929, 1930 [116, 117], where canonical quantization procedures were applied directly to the fields themselves. A classic review of the state of affairs of quantum electrodynamics in 1932 [68] was given by Fermi. The problem of negative energy solutions was resolved and its equivalence to the Dirac "hole" theory was demonstrated by Fock in 1933 [83], and Furry and Oppenheimer in 1934 [90], where the (Dirac) field operator and its adjoint were expanded in terms of appropriate creation and annihilation operators for the electron and positron, thus providing a unified description for the particle and its antiparticle. The method had a direct generalization to bosons. The old "hole" theory became unnecessary and obsolete.[11] The problem of negative energy solutions was also resolved for spin 0 bosons by

---

[11]As a young post-doctoral fellow, I remember attending Schwinger's lecture tracing the Development of Quantum Electrodynamics in "The Physicist's Conception of Nature" [202], making the statement, regarding the "hole" theory, that it is now best regarded as an historical curiosity, and forgotten.

similar methods by Pauli and Weisskopf in 1934 [170]. The fields thus introduced from these endeavors have become operators for creation and annihilation of particles and antiparticles, rather than probability amplitudes.[12]

The explanation that interactions are generated by the exchange of quanta was clear in the classic work of Bethe and Fermi in 1932 [18]. For example, charged particles, as sources of the electromagnetic field, influence other charged particles via these electromagnetic fields. Fields as operators of creation and destruction of particles, and the association of particles with forces is a natural consequence of field theory. The same idea was used by Yukawa in 1935 [249], to infer that a massive scalar particle is exchanged in describing the strong interaction (as understood in those days), with the particle necessarily being massive to account for the short range nature of the strong force unlike the electromagnetic one which is involved with the massless photon describing an interaction of infinite range. The mass $\mu$ of the particle may be estimated from the expression $\mu \approx \hbar/Rc$, obtained formally from the uncertainty principle, where $R$ denotes the size of the proton, i.e., $R = 1\,\text{fm} = 10^{-13}\,\text{cm}$. In natural units, i.e., for $\hbar = 1, c = 1$, $1\,\text{fm} \approx 1/(200\,\text{MeV})$. This gives $\mu \approx 200\,\text{MeV}$. Such a particle (the pion) was subsequently discovered by the C. F. Powell group in 1947 [136].

As early as 1930s, infinities appeared in explicit computations in quantum electrodynamics by Oppenheimer [168], working within an atom, by Waller [233, 234], and by Weisskopf [239]. The nature of these divergences, arising in these computations, came from integrations that one had to carry out over energies of photons exchanged in describing the interaction of the combined system of electrons and the electromagnetic field to arbitrary high-energies. By formally restricting the energies of photons exchanged, as just described, to be less than, say, $\kappa$, Weisskopf, in his calculations, has shown [239, 240], within the full quantum electrodynamics, that the divergences encountered in the self-energy acquired by the electron from its interaction with the electromagnetic field is of the logarithmic[13] type $\sim \ln(\kappa/mc^2)$, improving the preliminary calculations done earlier, particularly, by Waller, mentioned above. That such divergences, referred to as "ultraviolet divergence",[14] are encountered in quantum field theory should

---

[12]It is important to note, however, that the matrix elements of these field operators between particle states and the vacuum naturally lead to amplitudes of particles creation by the fields and to the concept of *wavefunction* renormalization (see Sect. 4.1) independently of any perturbation theories.

[13]The corresponding expression occurs with higher powers of the logarithm for higher orders in the fine-structure constant $e^2/4\pi\hbar c$.

[14]That is, divergences arising from the high-energy behavior of a theory. Another type of divergence, of different nature occurring in the low energy region, referred to as the "infrared catastrophe", was encountered in the evaluation of the probability that a photon be emitted in a collision of a charged particle. In computations of the scattering of charged particles, due to the zero mass nature of photons, their simultaneous emissions in arbitrary, actually infinite, in number must necessary be taken into account for a complete treatment. By doing so finite expressions for the probabilities in question were obtained [22].

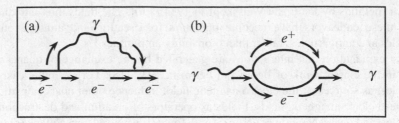

**Fig. 1.1** Processes leading to an electron self-energy correction, and vacuum polarization, respectively

be of no surprise as one is assuming that our theories are valid up to infinite energies![15]

The $2S_{1/2}$, $2P_{1/2}$ states of the Hydrogen atom are degenerate in Dirac's theory. In 1947 [134], Lamb and Retherford, however, were able to measure the energy difference between these states, referred to as the "Lamb Shift", using then newly developed microwave methods with great accuracy. Bethe [17] then made a successful attempt to compute this energy difference by setting an upper limit for the energy of photon exchanged in describing the electromagnetic interaction of the order of the rest energy of the electron $mc^2$, above which relativistic effects take place, relying on the assumption that the electron in the atom is non-relativistic, and, in the process, took into consideration of the mass shift[16] of the electron. He obtained a shift of the order of 1000 megacycles which was in pretty good agreement with the Lamb-Retherford experiment.

Very accurate computations were then made, within the full relativistic quantum electrodynamics, and positron theory. Notably, Schwinger[17] in 1948 [192], computed the magnetic moment of the electron modifying the gyromagnetic ratio, $g = 2$ in the Dirac theory, to $2(1 + \alpha/2\pi)$, to lowest order in the fine-structure constant. The computation of the Lamb-Shift was also carried out in a precise manner by Kroll and Lamb [133], and, for example, by French and Weisskopf [85], and Fukuda et al. [88, 89].

State of affairs changed quite a bit. It became clear that an electron is accompanied by an electromagnetic field which in turn tends to alter the nature of the electron that one was initially aiming to describe. The electron $e^-$, being a charged particle, produces an electromagnetic field ($\gamma$). This field, in turn, interacts back with the electron as shown below in Fig. 1.1a. Similarly, the electromagnetic field ($\gamma$) may lead to the creation of an electron-positron pair $e^+ e^-$, which in turn annihilate each other re-producing an electromagnetic field, a process referred to as vacuum-polarization, shown in part (b). Because of these processes, the parameters initially

---

[15]See also the discussion in Sect. 5.19.

[16]See also the important contribution to this by Kramers [130]. This reference also includes contributions of his earlier work.

[17]See also Appendix B of Schwinger [193].

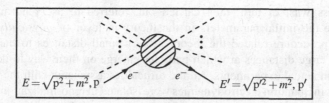

**Fig. 1.2** As a result of the self-energy correction in Fig. 1.1a, where an electron emits and re-absorbs a photon, the mass parameter $m_0$, one initially starts with, does not represent the physical mass of the electron determined in the lab. Here this is emphasized by the energy dependence on the physical mass $m$ of an electron in a scattering process. The *dashed lines* represent additional particles participating in the process

appearing in the theory, such as mass, say, $m_0$, vis-à-vis Fig. 1.1a, and the electron charge, say, $e_0$, vis-à-vis Fig. 1.1b, that were associated with the electron one starts with, are not the parameters actually measured in the lab. For example, the energy of a scattered electron of momentum **p**, in a collision process, turned up to be not equal to $\sqrt{\mathbf{p}^2 + m_0^2}$ but rather to $\sqrt{\mathbf{p}^2 + m^2}$, self-consistently,[18] with $m$ identified with the actual, i.e., tabulated, mass of the electron, and $m \neq m_0$, with a scattering process shown in Fig. 1.2, where the dashed lines represent other particles (such as $\gamma, e^-, e^+$), where the total charge as well as the total energy and momentum are conserved in the scattering process.

Similarly, the potential energy between two widely separated electrons, by a distance $r$, turned up to be not $e_0^2/4\pi r$ but rather $e^2/4\pi r$, with $e^2 \neq e_0^2$, where e is identified with the charge, i.e., the tabulated charge, of the electron. As we will see later, the physical parameters are related to the initial ones by scaling factors, referred to as mass and charge renormalization constants, respectively. An electron parametrized by the couple $(m_0, e_0)$, is referred to as a *bare electron* as it corresponds to measurements of its properties by going down to "zero" distances all the way into the "core" of the electron – a process that is unattainable experimentally. On the other hand, the physical parameters $(m, e)$, correspond to measurements made on the electron from sufficiently large distances.

One thus, in turn, may generate parameters, corresponding to a wide spectrum of scales running from the very small to the very large. Here one already notices that in quantum field theory, one encounters so-called effective parameters which are functions of different scales (or energies). Functions of these effective parameters turn out to satisfy invariance properties under scale transformations, thus introducing a concept referred to as the renormalization group. Clearly, due to the screening effect via vacuum polarization of $e^+e^-$ pairs creation, as discussed earlier, the magnitude of the physical charge is smaller than the magnitude of the bare charge.

---

[18]An arbitrary number of photons of vanishingly small energies are understood to be attached to the external electron lines, as discussed in Footnote 14 when dealing with infrared divergence problems.

A process was, in turn, then carried out, referred to as "renormalization", to eliminate the initial parameters in the theory in favor of *physically* observed ones. This procedure related the theory at very small distances to the theory at sufficiently large distances at which particles emerge on their way to detectors as it happens in actual experiments. All the difficulties associated with the ultraviolet divergences in quantum electrodynamics were isolated in renormalization constants, such as the ones discussed above, and one was then able to eliminate them in carrying out physical applications giving rise to completely finite results. This basic idea of the renormalization procedure was clearly spelled out in the work of Schwinger, Feynman, and Tomonaga.[19] The renormalization group,[20] mentioned above, describes the connection of renormalization to scale transformations, and relates, in general, the underlying physics at different energy scales.

In classic papers, Dyson [59, 60] has shown not only the equivalence of the Schwinger, Feynman, and Tomonaga approaches,[21] and the finiteness of the so-called renormalized quantum electrodynamics, but also developed a formalism for computations that may be readily applied to other interacting quantum field theories. Theories that are consistently finite when all the different parameters appearing initially in the theory are eliminated in favor of the physically observed ones, which are finite in number, are said to be renormalizable. Dyson's work, had set up:

*renormalizability as a condition for generating field theory interactions.*

In units of $\hbar = 1$, $c = 1$, $[\text{Mass}] = [\text{Length}]^{-1}$. Roughly speaking, in a renormalizable theory, no coupling constants can have the dimensions of negative powers of mass. (Because of dimensional reasons, we note, in particular, that one cannot have too many derivatives of the fields, describing interactions, as every derivative necessitates involving a coupling of dimensionality reduced by one in units of mass.)

The photon as the agent for transmitting the interaction between charged particles, is described by a vector – the vector potential. In quantum electrodynamics, as a theory of the interaction of photons and electrons, for example, the photon is coupled locally to the electromagnetic current. The latter is also a vector, and the interaction is described by their scalar product (in Minkowski space) ensuring the relativistic invariance of the underlying theory. To lowest order in the charge e of the electron $e$, this coupling may be represented by the diagram Fig. 1.3a. On the other hand, for a spin 0 charged boson $\varphi$, say, one encounters two such diagrams, each shown to lowest order in the charge e in Fig. 1.3b, where we note that in the second diagram in the latter part, two photons emerge locally from the same point.

---

[19]This is well described in their Nobel lectures: Schwinger [201], Feynman [75], Tomonaga [225], as well as in the collection of papers in Schwinger [198, 201].

[20]Stueckelberg and Peterman [209], Gell-Mann and Low [93], Bogoliubov and Shirkov [25], Ovsyannikov [169], Callan [28, 29], Symanzik [213–215], Weinberg [237], and 't Hooft [218].

[21]The best sources for these approaches are their Nobel Lectures: Schwinger [201], Feynman [75], Tomonaga [225], as well as Schwinger [198].

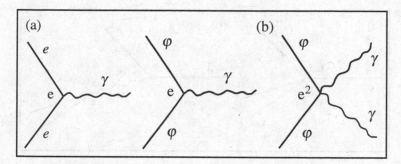

**Fig. 1.3** Local couplings for photon emission by an electron, and by a spin 0 charged particle described by the field $\varphi$, respectively

Quantum Electrodynamics, was not only the theory of interest. There was also the weak interaction. The preliminary theory of weak interaction dates back to Fermi [69, 70]. Based on weak processes such as $\beta^-$ decay: $n \rightarrow p + e^- + \tilde{\nu}_e$, he postulated that the weak interactions may be described by local four-point interactions involving a universal coupling parameter $G_F$. The four particles of the process just mentioned, interact locally at a point with a zero range interaction. The Fermi theory was in good agreement in predicting the energy distribution of the electron. For dimensional reasons, however, the dimensions of the coupling constant $G_F$ involved in the theory has the dimensions of $[\text{Mass}]^{-2}$, giving rise to a non-renormalizable theory.[22] In analogy to quantum electrodynamics, the situation with this type of interaction may be somehow improved by introducing, in the process, a vector Boson[23] $W^-$ which mediates an interaction[24] between the two pairs (so-called currents), $(n, p)$ and $(e^-, \tilde{\nu}_e)$, with both necessarily described by entities carrying (Lorentz) vector indices, to ensure the invariance of the underlying description. Moreover, in units of $\hbar = 1, c = 1$, a *dimensionless* coupling $g$ is introduced. The Fermi interaction and its modification are shown, respectively in parts Fig. 1.4a, b.

In order that the process in diagram given in part Fig. 1.4b, be consistent with the "short-range" nature of the Fermi interaction, described by the diagram on the left, the vector particle $W^-$ must not only be massive but its mass, $M_W$ must be quite large. This is because the propagator of a massive vector particle of mass $M_W$, which mediates an interaction between two spacetime points $x$ and $x'$, as we will discuss below, behaves like $\eta_{\mu\nu}\delta^{(4)}(x - x')/M_W^2$ for a large mass, signifying

---

[22]It is interesting to point out as one goes to higher and higher orders in the Fermi coupling constant $G_F$, the divergences increase (Sect. 6.14) without any bound and the theory becomes uncontrollable.

[23] A quantum relativistic treatment of a problem, implies that a theory involving the $W^-$ particle, must also include its antiparticle $W^+$, having the same mass as of $W^-$.

[24]Such a suggestion was made, e.g., by Klein [129].

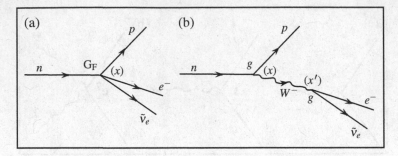

**Fig. 1.4** (a) The old Fermi theory with a coupling $G_F$ is replaced by one in (b) where the interaction is mediated by a vector boson with a dimensionless coupling $g$

necessarily a vanishingly small range of the interaction.[25] Upon comparison of both diagrams, one may then infer that

$$G_F \approx \frac{g^2}{M_W^2}. \tag{1.1}$$

Evidently, the Fourier transform of the propagator in the energy-momentum description, at energies much less than $M_W$ is, due to the $\delta^{(4)}(x-x')$ function given above, simply $\eta_{\mu\nu}/M_W^2$, and (1.1) may be obtained from a low-energy limit.

With some minimal effort, the reader will understand better the above two limits and some of the difficulties encountered with a massive vector boson, in general, if, at this stage, we write down explicitly its propagator between two spacetime points $x, x'$ in describing an interaction carried by the exchange of such a particle which is denoted by[26]:

$$\Delta_+^{\mu\nu}(x - x') = \int \frac{(dk)}{(2\pi)^4} e^{i k_\nu(x^\nu - x'^\nu)} \Delta_+^{\mu\nu}(k), \quad (dk) = dk^0 dk^1 dk^1 dk^3, \tag{1.2}$$

$$\Delta_+^{\mu\nu}(k) = \frac{1}{(k^2 + M_W^2 - i0)} \left( \eta^{\mu\nu} + \frac{k^\mu k^\nu}{M_W^2} \right), \tag{1.3}$$

---

[25]Here $\eta_{\mu\nu}$ is the Minkowski metric.

[26]This expression will be derived in Sect. 4.7. For a so-called virtual particle $k^2 = \mathbf{k}^2 - (k^0)^2 \neq -M_W^2$. The $-i0$ in the denominator in (1.3) just specifies the boundary condition on how the $k^0$ integration is to be carried out. These things will be discussed in detail later on and are not needed here.

where $k^0$ is its energy, and $\mathbf{k} = (k^1, k^2, k^3)$ its momentum. Formally for $M_W^2 \to \infty$, $\triangle_+^{\mu\nu}(k) \to \eta^{\mu\nu}/M_W^2$, leading from (1.2) to

$$\triangle_+^{\mu\nu}(x - x') \to \frac{\eta^{\mu\nu}}{M_W^2} \int \frac{(\mathrm{d}k)}{(2\pi)^4} \, \mathrm{e}^{\mathrm{i}k_\nu(x^\nu - x'^\nu)} = \frac{\eta^{\mu\nu}}{M_W^2} \delta^4(x - x'), \tag{1.4}$$

signalling, in a limiting sense, a short range interaction for a heavy-mass particle. On the other hand, for $|k^\nu| \ll M_W$ for each component, one has

$$\triangle_+^{\mu\nu}(k) \approx \frac{\eta^{\mu\nu}}{M_W^2}, \tag{1.5}$$

in the energy-momentum description.

Although the introduction of the intermediate boson $W$ improves somehow the divergence problem, it is still problematic. The reason is not difficult to understand. In the energy-momentum description, the propagator of a massive vector particle, as given in (1.3), has the following behavior at high energies and momenta

$$\triangle_+^{\mu\nu}(k) \to \frac{1}{k^2} \frac{k^\mu k^\nu}{M_W^2}, \tag{1.6}$$

providing *no damping* in such a limit. Moreover, as one goes to higher orders in perturbation theory the number of integration variables, over energy and momenta arising in the theory, increase, and the divergences in turn increase without bound and the theory becomes uncontrollable.[27] On the other hand, an inherited property of quantum electrodynamics is *gauge symmetry* due to the masslessness of the photon. In the present context of ultraviolet divergences, the photon propagator has a very welcome vanishing property at high energies. This gauge symmetry as well as the related *massless* aspect of the photon, which are key ingredients in the self consistency of quantum electrodynamics, turned out to provide a *guiding principle* in developing the so-called electroweak theory.

In 1956 [138], an important observation was made by Lee and Yang that parity P is violated in the weak interactions. Here we recall that, given a process, its parity transformed (mirror) version, is obtained by reversing the directions of the space variables.[28] This has led Gershtein and Zel'dovich [95], Feynman and Gell-Mann [77], Sudarshan and Marshak [210], and Sakurai [178], to express the currents

---

[27] The damping provided by the propagators of a massless vector particle, a spin 1/2 particle, and a spin 0 particle, for example, in the ultraviolet region vanish like 1/energy$^2$, 1/energy, 1/energy$^2$, respectively.

[28] It was later observed that the product of charge conjugation, where a particle is replaced by its antiparticle, and parity transformation "CP", is also not conserved in a decay mode of $K$ mesons at a small level [38, 39, 82]. As the product "CPT", of charge conjugation, parity transformation, and time reversal "T", is believed to be conserved, the violation of time reversal also follows. For a test of such a violation see CPLEAR/Collaboration [36].

constructed out of the pairs of fields: $(n, p), (e^-, \tilde{\nu}_e), \ldots$ in the Fermi theory to reflect, in particular, this property dictated by nature. The various currents were eventually expressed and conveniently parametrized in such a way that the theory was described by the universal coupling parameter $G_F$. The construction of such fundamental currents together with idea of intermediate vector bosons exchanges to describe the weak interaction led eventually to its modern version.

Quantum Electrodynamics may be considered to arise from local gauge invariance in which the electron field is subjected to a local phase transformation $e^{i\vartheta(x)}$. The underlying group of transformations is denoted by $U(1)$ involving simply the identity as the *single* generator of transformations with which the photon is associated as the single gauge field. In 1954 [247], Yang and Mills, and Shaw in 1955 [203], generalized the just mentioned abelian gauge group of phase transformations, encountered in quantum electrodynamics, to a non-abelian[29] gauge theory, described by the group $SU(2)$,[30] and turned out to be a key ingredient in the development of the modern theory of weak interactions. This necessarily required the introduction, in addition to the charged bosons $W^\pm$, a neutral one. What distinguishes a non-abelian gauge theory from an abelian one, is that in the former theory, direct interactions occur between gauge fields, carrying specific quantum numbers, unlike in the latter, as the gauge field – the photon – being uncharged.

As early as 1956, Schwinger believed that the weak and electromagnetic interactions should be combined into a gauge theory [97, 159, 199]. Here we may pose to note that both in electrodynamics and in the modified Fermi theory, interactions are mediated by vector particles. They are both described by universal dimensionless coupling constants e, and by $g$ (see (1.1)) in the intermediate vector boson description, respectively. In a unified description of electromagnetism and the weak interaction, one expects these couplings to be comparable, i.e.,

$$g^2 \approx e^2 = 4\pi\alpha, \qquad \text{where} \quad \alpha \approx \frac{1}{137}, \quad G_F \approx 1.166 \times 10^{-5}/(\text{GeV})^2. \qquad (1.7)$$

in units $\hbar = 1, c = 1$. From (1.1), we may then estimate the mass of the $W$ bosons to be

$$M_W \approx \sqrt{\frac{4\pi\alpha}{G_F}} \approx 90 \text{ GeV}/c^2, \qquad (1.8)$$

---

[29]Non-abelian refers to the fact that the generators do not commute. In contrast a $U(1)$ gauge theory, such as quantum electrodynamics, is an abelian one.

[30]$SU(2)$ consists of $2 \times 2$ unitary matrices of determinant one. (The letter S in the group stands for the special property of determinant one.) It involves three generators, with which are associated three gauge fields. This will be studied in detail in Sect. 6.1.

re-inserting the constant c for convenience, in good agreement with the observed mass. We may also estimate the range of the weak interaction to be

$$R_W \approx \frac{\hbar c}{M_W c^2} \approx 2.2 \times 10^{-16} \text{cm}. \tag{1.9}$$

Glashow, a former graduate student of Schwinger, eventually realized [96][31] the important fact that the larger group SU(2) × U(1), is needed to include also electrodynamics within the context of a Yang-Mills-Shaw theory. A major problem remained: the local gauge symmetry required that the gauge fields associated with the group must, a priori, be *massless* in the initial formulation of the theory.

The problem of the masslessness of the vector bosons was solved by Weinberg [236, 238] and Salam [182, 183],[32] by making use of a process,[33] referred to as spontaneous symmetry breaking, where a scalar field interacting with the vector bosons, whose expectation value in the vacuum state is non zero, leads to the generation of masses to them.[34] This is referred to as the Higgs[35] mechanism, in which the group SU(2) × U(1) is spontaneously broken to the group U(1) with the latter associated with the photon, and, in the process, the other bosons, called $W^{\pm}, Z^0$, acquiring masses, thanks to the Higgs boson, and renormalizability may be achieved. The latter particle has been also called the "God Particle".[36] The mere existence of a neutral vector boson $Z^0$ implies the existence of a weak interaction component in the theory without a charge transfer, the so-called neutral currents. A typical process involving the neutral $Z^0$ boson exchange is in $\tilde{\nu}_{\mu} + e^- \rightarrow \tilde{\nu}_{\mu} + e^-$ shown in Fig. 1.5 not involving the muon itself. Neutral currents[37] have been observed,[38] and all the vector bosons have been observed[39] as well. It turned out that the theory with spontaneous symmetry did not spoil the renormalizability of

---

[31] See also Salam and Ward [189].

[32] See also Salam and Ward [187–189] and Salam [181].

[33] Some key papers showing how spontaneous symmetry breaking using spin 0 field may generate masses for vector bosons are: Englert and Brout [63], Englert et al. [64], Guralnik et al. [110], and Kibble [127].

[34] Apparently the Legendary Victor Weisskopf was not impressed by this way of generating masses. In his CERN publication [241], on page 7, 11th line from below, he says that this is an awkward way to explain masses and that he believes that Nature should be more inventive, but experiments may prove him wrong.

[35] Higgs [119–121]. This work followed earlier work of Schwinger [200], where he shows, by the exactly solvable quantum electrodynamics in two dimensions, that gauge invariance does not prevent the gauge field to acquire mass dynamically, as well as of the subsequent work of Anderson [7] in condensed matter physics.

[36] This name was given by Lederman and Teresi [137].

[37] Neutral current couplings also appear in Bludman's [23] pioneering work on an SU(2) gauge theory of weak interactions but did not include electromagnetic interactions.

[38] Hasert et al. [112, 113] and Benvenuti et al. [15].

[39] See, e.g., C. Rubbia's Nobel Lecture [176].

**Fig. 1.5** A process involving
the exchange of the neutral
vector boson $Z^0$

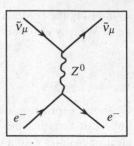

the resulting theory with massive vector bosons. Proofs of renormalizability were given by 't Hooft [216, 217].[40] It seems that Sydney Coleman used to say that 't Hooft's proof has turned the Weinberg-Salam frog into an enchanted prince.[41] The "Electroweak Theory" turned up to be quite a successful theory.[42]

Another interaction which was also developed in the "image" of quantum electrodynamics was quantum chromodynamics, as a theory of strong interactions based, however, on the non-abelian gauge symmetry group SU(3). Here one notes that a typical way to probe the internal structure of the proton is through electron-proton scattering. The composite nature of the proton, as having an underlying structure, becomes evident when one compares the differential cross sections for elastic electron-proton scattering with the proton described as having a finite extension to the one described as a point-like particle. With a one photon exchange description, the form factors in the differential cross section are seen to vanish rapidly for large momentum transfer (squared) $Q^2$ of the photon imparted to the proton. As $Q^2$ is increased further one reaches the so-called resonance region,[43] beyond which, one moves into a deep inelastic region, where experimentally the reaction changes "character", and the corresponding structure functions of the differential cross section have approximate scaling properties (Sect. 6.9), instead of the vanishing properties encountered with elastic form factors, the process of which is depicted in Fig. 1.6. Such properties indicate the presence of approximately free point-like structures within the proton referred as partons, which consist of quarks, gluons together with those emitted[44] from their scattering reactions. This led to the development of the so-called parton model,[45] as a first approximation, in which these point-like particles within the proton are free and the virtual photon interacts

---

[40]See also 't Hooft and Veltman [221], Lee and Zinn-Justin [139–142], and Becchi et al. [13].

[41]See Salam [183], p. 529.

[42]The basic idea of the renormalizability of the theory rests on the fact that renormalizability may be established for the theory with completely massless vector bosons, as in QED, one may then invoke gauge symmetry to infer that the theory is also renormalizable for massive vector bosons via spontaneous symmetry breaking.

[43]A typical resonance is $\Delta^+$, of mass 1.232 GeV, consisting of a proton $p$ and a $\pi^0$ meson.

[44]See, e.g., Fig. 6.7c.

[45]Feynman [73, 74] and Bjorken and Pachos [21].

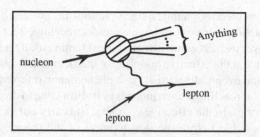

**Fig. 1.6** In the process, "Anything" denotes anything that may be created in the process consistent with the underlying conservation laws. The wavy line denotes a neutral particle ($\gamma$, $Z^0$,...) of large momentum transfer

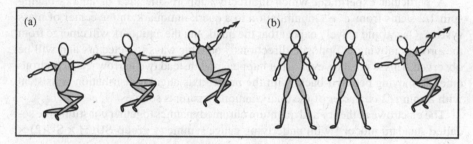

**Fig. 1.7** (a) If interactions between quarks may be represented, as an analogy, by people holding hands, then pulling one person would drag everybody else along. In the parton model, the situation is represented as in part (b) rather than in part (a)

with each of its charged constituents independently,[46] instead of interacting with the proton as a whole.

The non-abelian gauge symmetry group SU(3), is needed to accommodate quarks and gluons, involving eight generators with which the gluons are associated. Here, in particular, a quantum number referred to as "color" (three of them)[47] is assigned to the quarks. One of the many reasons for this is that the spin 3/2 particle $\Delta^{++}$, which is described in terms of three identical quarks (the so-called u quarks) as a low lying state with no orbital angular momentum between the quarks, behaves as a symmetric state under the exchange of two of its quarks and would violate the Spin & Statistics connection without this additional quantum number. The color degrees of freedom are not observed in the hadronic states themselves and the latter behave as scalars, that is they are color singlets, under SU(3) transformations. As the group SU(3) involves "color" transformations within each quark flavor, it may be denoted by SU(3)$_{\text{color}}$ or just by SU(3)$_{\text{C}}$. The gluons also carry "color" and direct gluon-gluon interactions then necessarily occur, unlike the situation with

---

[46]For an analogy to this, see part (b) of Fig. 1.7b.

[47]Greenberg [104], Han and Nambu [111], Nambu [163], Greenberg and Zwanziger [105], Gell-Mann [92], and Fritzsch and Gell-Mann [87].

photons in quantum electrodynamics since photons do not carry a charge. These gluon-gluon interactions turn out to have an anti-screening effect on a source field which dominate over the screening effect of quark/antiquark interactions leading to the interesting fact that the effective coupling of quark interactions becomes smaller at high-energies, and eventually vanish[48] – a phenomenon referred to as asymptotic freedom. This has far reaching consequences as it allows one to develop perturbation theory at high energies, in the effective coupling, and carry out various applications which were not possible before the development of the theory, and is consistent with deep-inelastic experiments of leptons with nucleons, with the latter described by point-like objects which, at high energies, scatter almost like free particles,[49] as mentioned above, the process of which is shown Fig. 1.6.

A particular experiment which indirectly supports the idea of quarks having spin 1/2 stems from $e^- e^+$ annihilation to a quark-antiquark, in the center of mass system. One would naïvely expect that the quark and the antiquark will emerge from the process moving in opposite directions[50] on their ways to detectors and will be observed. This is not, however, what happens and instead two narrow jets of hadrons emerge, moving back-yo-back, with the net jet-axis angular distribution consistent with a spin 1/2 character of the quark/antiquark parents sources.

The electroweak theory and quantum chromodynamics together constitute the so-called standard model[51] with underlying gauge symmetry group $SU(3) \times SU(2) \times U(1)$.

The effective coupling in quantum chromodynamics is expected to become larger at large distances increasing with no bound providing a strong confining force of quarks and gluons restricting them within hadrons – a phenomenon that is sometimes referred to as *infrared slavery*.[52]

From our discussion of quantum electrodynamics, we recall that the effective coupling of a U(1) gauge theory, as an abelian gauge theory, increases with energies. On the other hand, the asymptotic free nature of non-abelian gauge theories, imply that the effective couplings associated with the groups SU(3), SU(2), decrease with energy. Due to the smallness of the U(1) coupling in comparison to the other two at the present low energies, this gives one the hope that at sufficient high energies these three couplings merge together and the underlying theory would be described by one single force. A theory which attempts to unify the electroweak and the strong interactions is called a *grand unified theory*. Such theories have been developed[53] and the couplings seem to run to merge roughly

---

[48]This was discovered by Gross and Wilczek [106] and Politzer [172]. See also Vanyashin and Terentyev [229] for preliminary work on vector bosons.

[49]Chromodynamics means Colordynamics, and the name Quantum Chromodynamics is attributed to Gell-Mann, see, e.g., Marciano and Pagels [158].

[50]See, e.g., Fig. 6.7d, and Sect. 6.10.3

[51]The name "Standard Model" is usually attributed to Weinberg.

[52]Unfortunately, no complete proof of this is available.

[53]See, e.g., Georgi et al. [94] for pioneering work. See also Beringer et al. [16] and Olive et al. [167].

somewhere around $10^{15}$–$10^{16}$ GeV. This, in turn, gives the hope of the development of a more fundamental theory in which gravitation, which should be effective at energy scale of the order of Planck energy scale $\sqrt{\hbar c / G_N} \simeq 10^{19}$ GeV,[54] or even at a lower energy scale, where $G_N$ denotes Newton's gravitational constant, is unified with the electroweak and strong forces. If the standard model is the low energy of such a fundamental theory, then the basic question arises as to what amounts for the enormous difference between the energy scale of such a fundamental theory ($\sim 10^{16}$–$10^{19}$ GeV) and the defining energy scale of the standard model ($\sim 300$ GeV)? This has been termed as the *hierarchy* problem which will be discussed again later. We will see in Vol. II, in particular, that the above mentioned couplings seem to be unified at a higher energy scale of the order $10^{16}$ GeV, when supersymmetry is taken into account, getting it closer to the energy scale at which gravitation may play an important role.

One may generalize the symmetry group of the standard model, and consider transformations which include transformations between quarks and leptons, leading to a larger group such as, for example, to the SU(5) group, or a larger group, which include SU(3) × SU(2) × U(1). The advantage of having one larger group is that one would have only one coupling parameter and the standard model would be recovered by spontaneous symmetry breaking at lower energies. This opens the way to the realization of processes in which baryon number is not conserved, with a baryon, for example, decaying into leptons and bosons. The experimental[55] bound on lifetime of proton decay seems to be $> 10^{33}$ years and is much larger than the age of the universe which is about 13.8 billion years.[56] Such a rare event even if it occurs once will give some support of such grand unified theories.

We have covered quite a large territory and before continuing this presentation, we pose for a moment, at this appropriate stage, to discuss three aspects of importance that are generally expected in order to carry out reliable computations in perturbative quantum field theory. These are:

1. The development of a powerful and simple formalism for doing this.
2. To show how the renormalization process is to be carried, and establish that the resulting expressions are finite to any order of perturbation theory.
3. The physical interpretation will be completed if through the process of renormalization, the initial experimentally unattainable parameters in the theory are eliminated in favor of physically observed ones, which are finite in number, and are generally determined experimentally as discussed earlier in a self consistent manner.

---

[54]The Planck energy (mass) will be introduced in detail later.

[55]See, e.g., Olive et al. [167].

[56]A decay of the proton may have a disastrous effect in the stability of matter over anti-matter itself in the universe. See, however, the discussion given later on the dominance of matter in the visible universe.

Perturbatively renormalizable theories are distinguished from the non-renormalizable ones by involving only a finite number of parameters in the theory that are
determined experimentally.

We discuss each of these in turn.

1. A powerful formalism is the *Path Integral* one, pioneered by Feynman,[57] defining
   a generating functional for so-called Green functions from which physical
   amplitudes may be extracted, and has the general structure $\int d\mu[\chi] e^{i\text{Action}}$. Here
   $d\mu[\chi]$ defines a measure of integration over classical fields as the counterparts
   of the quantum fields of the theory.[58] "Action" denotes the classical action. In
   the simplest case, the measure $d\mu[\chi]$ takes the form $\Pi_x d\chi(x)$ as a product of
   all spacetime points. In gauge theories, due to constraints, the determination of
   the measure of integrations requires special techniques[59] and takes on a much
   more complicated expression and was successfully carried out by Faddeev and
   Popov in 1967.[60] The path integral expression as it stands, involves continual
   integrations to be carried out.

   An equally powerful and quite an elegant formalism is due to Schwinger,[61]
   referred to as the *Action Principle* or the *Quantum Dynamical Principle*. For
   quantum field theory computations, the latter gives the variation of the so-called
   vacuum-to-vacuum transition amplitude (a generating functional): $\delta\langle 0_+|0_-\rangle$ as
   any of the parameters of the theory are made to vary. The latter is then expressed
   as a differential operator acting on a simple generating functional expressed
   in closed form. This formalism involves only functional differentiations to be
   carried out, no functional integrations are necessary, and hence is relatively
   easier to apply than the path integral. We will learn later, for example, that the
   path integral may be simply obtained from the quantum dynamical principle
   by a functional Fourier transform thus involving functional integrals. Again the
   application of the quantum dynamical principle to the quantization of gauge
   theories with underlying constraints require special techniques and it was carried
   out in Manoukian [150].

   All the fundamental interactions in nature are presently described theoretically
   by gauge theories, involving *constraints*, and the two approaches of their
   quantization discussed above will be both treated in this book for pedagogical
   reasons and are developed [62]

---

[57] See Feynman and Hibbs [78] for the standard pedagogical treatment. See also Feynman [75].

[58] We use a general notation $\chi(x)$ for the fields as functions of spacetime variable and suppress all
indices that they may carry to simplify the notation. These fields may include so-called Grassmann
fields.

[59] Feynman [72], DeWitt [43, 44], and Faddeev and Popov [66].

[60] *Op. cit.*

[61] Schwinger [194–197, 201].

[62] There is also the canonical formalism, see, e.g., Mohapatra [161, 162] and Utiyama and
Sakamoto [228].

via the
Path Integral [66],
or via the
Action Principle (Quantum Dynamical Principle) [150].

2. Historically, Abdus Salam, was the first "architect" of a general theory of renormalization. In 1951, he carried out a systematic study[63] of renormalization [180], introduced and sketched a subtraction scheme in a general form. Surprisingly, this classic paper was not carefully reexamined until much later. This was eventually done in 1976 [147] by Manoukian, and inspired by Salam's work, a subtraction scheme was developed and brought to a mathematically consistent form, and the finiteness of the subtracted, i.e., renormalized, theory was proved by the author[64] to *any* order of perturbation theory.[65] by using, in the process, a power counting theorem established by Weinberg [235] for integrals of a special class of functions, thus completing the Dyson-Salam program. The subtraction was carried out directly in momentum space and no cut-offs were introduced.

Shortly after the appearance of Salam's work, two other "architects" of a general theory of renormalization theory, Bogoliubov and Parasiuk, in a classic paper in 1957 [24], also developed a subtraction scheme. In 1966 [118], Hepp gave a complete proof of the finiteness of the Bogoliubov-Parasiuk to any order of perturbation theory, by using in the intermediate stages ultraviolet cut-offs, and in 1969 [251], Zimmermann formulated their scheme in momentum space, without cut-offs, and provided a complete proof of finiteness as well, thus completing the Bogoliubov-Parasiuk program. This scheme is popularly known as the BPHZ scheme.

The equivalence of the Bogoliubov-Parasiuk scheme, in the Zimmermann form, and our scheme was then proved by Manoukian,[66] after some systematic cancelations in the subtractions. This equivalence theorem[67] unifies the two monumental approaches of renormalization.[68]

---

[63] Salam [180], see also Salam [179].

[64] Manoukian [149].

[65] For a pedagogical treatment of all these studies, see my book "Renormalization" [149]. This also includes references to several of my earlier papers on the subject as well as many results related to renormalization theory.

[66] See Manoukian [149] *op. cit.*

[67] This result has been also referred to as "Manoukian's Equivalence Principle", Zeidler [250], p. 972. See also Streater [207].

[68] I was pleased to see that our equivalence theorem has been also considered, by completely different methods, by Figueroa and Gracia-Bondia [80]. For other earlier, and recent, but different, approaches to renormalization theory, see, e.g., Epstein and Glaser [65], Kreimer [131, 132], Connès and Kreimer [34, 35], and Figueroa and Gracia-Bondia [79, 81]. See also Landsman [135] and Aschenbrenner [11].

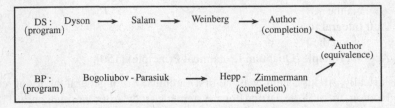

**Fig. 1.8** Developments of the general theory of renormalization from the DS and BP programs. The intricacies of this layout also appear in Zeidler [250], pp. 972–975. Regarding the author's work shown in the above layout and of his completion of the renormalization program stemming out of Salam's, Streater [207] writes: *"It is the end of a long chapter in the history of physics"*

The development of the general theory of renormalization from the DS and BP programs may be then summarized as given in Fig. 1.8[69].

3. The physical interpretation of the theory is completed by showing that the subtractions of renormalization are implemented by counterterms in the theory which have the same structures as the original terms in the theory (i.e., in the Lagrangian density),[70] thus establishing the self-consistency involved in the elimination of the initial parameters in the theory in favor of physically observed ones. As mentioned above, for a theory to be renormalizable, i.e., involving only a finite number of parameters determined, in general, experimentally, the counterterms of the theory must be finite in number as well.

All particles due to their energy content experience the gravitational attraction. Einstein's theory of gravitation, also referred to as general relativity (GR), is described by a second rank tensor with the energy-momentum tensor of matter as its source from which the energy density of matter may be defined. It may not be described just by a scalar or just by a vector field as they are inconsistent with experiment. It is easy to see that due to the fact that masses have the same signs (positive) a theory based on a vector field alone will lead to a repulsive rather an attractive gravitational force.[71] GR theory predictions are well supported experimentally in our solar system.

The key observation, referred to as the principle of equivalence, of Einstein is that at any given point in space and any given time, one may consider a frame in which gravity is wiped out at the point in question. For example, in simple Newtonian gravitational physics, a test particle placed at a given point inside a freely falling elevator on its way to the Earth, remains at rest, inside the elevator, for a very short time, depending on the accuracy being sought, and, depending on its position relative to the center of the Earth, eventually moves, in general, from its original position in a given instant. Einstein's principle of equivalence applies only locally

---

[69]The layout in Fig. 1.8 is based on Manoukian [149], and it also appears in Zeidler [250], p. 974. See also Streater [207] and Figueroa and Gracia-Bondia [80].

[70]For a detailed study of this see, Manoukian ([148]; Appendix, p. 183 in [149]).

[71] Attempts have been made to include such fields as well for generalizations of Einstein's theory, but we will not go into it here.

at a given point and at a given time. At the point in question, in the particular frame in consideration, gravity is wiped out and special relativity survives. The reconciliation between special relativity and Newton's theory of gravitation, then readily leads to GR, where gravity is accounted for by the curvature of spacetime and its departure from the flat spacetime of special relativity one has started out with upon application of the principle of equivalence. By doing this, one is able to enmesh non-gravitational laws with gravity via this principle.

Quantum gravity (QG) is needed in early cosmology, black hole physics, and, in general, to deal with singularities that arise in a classical treatment. QG must also address the problem of the background geometry. A common interest in fundamental physics is to provide a unified description of nature which is applicable from microscopic to cosmological distances. A fundamental constant of unit of length that is expected to be relevant to this end is the Planck length as well as the Planck mass. Out of the fundamental constants of quantum physics $\hbar$, of relativity c, and the Newtonian gravitational one $G_N$, we may define a unit of length and mass, the Planck length and Planck mass, respectively, relevant in quantum gravity, through the following

$$\ell_P = \sqrt{\frac{\hbar G_N}{c^3}} \simeq 1.616 \times 10^{-33}\,\text{cm}, \quad m_P = \sqrt{\frac{\hbar c}{G_N}} \simeq 1.221 \times 10^{19}\,\text{GeV}/c^2.$$

$$(1.10)$$

In units $\hbar = 1$, c $= 1$, dimensions of physical quantities may be then expressed in powers of mass ([Energy] = [Mass], [Length] = [Mass]$^{-1}$ = [Time], . . .), and, as gravitation has a universal coupling to all forms of energy, one may hope that it may be implemented within a unified theory of the four fundamental interactions, with the Planck mass providing a universal mass scale. Unfortunately, it is difficult experimentally to investigate the quantum properties of spacetime as one would be working at very small distances.

GR predicts the existence of Black holes. Here it is worth recalling of the detection ("Observational waves from a binary black hole merger", Phys. Rev. Lett. 116, 061102 (1–16) (2016)) by B. P. Abbott et al. of gravitational waves from the merger of two black holes 1.3 billion light-years from the Earth. Recall that a black hole (BH) is a region of space into which matter has collapsed and out of which light may not escape. It partitions space into an inner region which is bounded by a surface, referred to as the event horizon which acts as a one way surface for light going in but not coming out. The sun's radius is much larger than the critical radius of a BH which is about 2.5 km to be a black hole for the sun. We will see that for a spherically symmetric BH of mass $M$, the radius of the horizon is given by $R_{BH} = 2G_N M/c^2$.[72]

---

[72]This may be roughly inferred from Newton's theory of gravitation from which the escape speed of a particle in the gravitational field of a spherically symmetric massive body of mass $M$, at a distance $r$, is obtained from the inequality $v^2/2 - G_N M/r < 0$, and by formally replacing $v$ by

One may argue that the Planck length may set a lower limit spatial cut-off. The following formal and rough estimates are interesting. Suppose that by means of a high energetic particle of energy $E$, $\langle E^2 \rangle \sim \langle p^2 \rangle c^2$, with $\langle p^2 \rangle$ very large, one is interested in measuring a field within an interval of size $\delta$ around a given point in space. Such form of energy acts as an effective gravitational mass $M \sim \sqrt{\langle E^2 \rangle / c^4}$ which, in turn distorts space around it. The radius of the event horizon of such a gravitation mass $M$ is given by $r_{BH} = 2 G_N M / c^2$. Clearly we must have $\delta > r_{BH}$, otherwise the region of size $\delta$ that we wanted to locate the point in question will be hidden beyond a BH horizon, and localization fails. Also:

$\langle p^2 \rangle \geq \langle (p - \langle p \rangle)^2 \rangle \geq \hbar^2 / 4 \delta^2$. Hence $M \geq \hbar / 2 c \delta$,

$$\delta > \frac{2 G_N M}{c^2} \geq \frac{\hbar G_N}{c^3 \delta},$$

which gives $\delta > r_{BH} = \sqrt{\hbar G_N / c^3} = \ell_P$.

Interesting investigations by Hawking[73] have shown that a BH is not really a black body, it is a thermodynamic object, it radiates and has a temperature associated with it.[74] In Chapter 7 in Vol. II , we will see, considering a spherically symmetric BH, that its temperature is given by[75]

$$T_{BH} = \frac{\hbar c^3}{8 \pi G_N M k_B}. \tag{1.11}$$

where $k_B$ is the Boltzmann constant. Note that a very massive black hole is cold.

Recall that entropy $S$ represents a measure of the amount of disorder with information encoded in it, and invoking the thermodynamic interpretation of a BH, we may write

$$\frac{\partial S}{\partial (M c^2)} = \frac{1}{T}, \tag{1.12}$$

which upon integration with boundary condition that for $M \to 0$, $S \to 0$, gives the celebrated result

$$S_{BH} = \frac{c^3 k_B}{4 \hbar G_N} A = k_B \frac{A}{4 \ell_P^2}, \qquad A = 4\pi \left( \frac{2 G_N M}{c^2} \right)^2 \tag{1.13}$$

---

the ultimate speed c to obtain for the critical radius $R_{critical} = 2 G_N M / c^2$ such that for $r < R_{critical}$ a particle cannot escape.

[73]Hawking [114, 115].

[74]Particle emission from a BH is formally explained through virtual pairs of particles created near the horizon with one particle falling into the BH while the other becoming free outside the horizon.

[75]A pedestrian approach in determining the temperature is the following. By comparing the expression of energy expressed in terms of the wavelength of radiation $\lambda$: $E = hc/\lambda$, with the expression $E = k_B T$, gives $T = hc/k_B \lambda$. On dimensional grounds $\lambda \sim 2 G_N M / c^2$, which gives $T \sim \pi \hbar c^3 / G_N M k_B$. This is the expression given for the temperature up to a proportionality constant.

referred to as the Bekenstein-Hawking Entropy formula[76] of a BH. This relation relates quantum gravity to information theory. This result is expected to hold in any consistent formulation of quantum gravity, and shows that a BH has entropy unlike what would naïvely expect from a BH with the horizon as a one way classical surface through which information is lost to an external observer. The proportionality of the entropy to the area rather than to the volume of a BH horizon should be noted. It also encompasses Hawking's theorem of increase of the area with time with increase of entropy. We will discuss the Bekenstein-Hawking Entropy formula below in conjunction with loop quantum gravity and string theory.

Now we turn back to the geometrical description of gravitation given earlier, and introduce a gravitational field to account for the departure of the curved spacetime metric from that of the Minkowski one to make contact with the approaches of conventional field theories, dealing now with a field permeating an interaction between all dynamical fields. The quantum particle associated with the gravitational field, the so-called graviton, emerges by considering the small fluctuation of the metric, associated with curved spacetime of GR about the Minkowski metric, as the limit of the full metric, where the gravitational field becomes weaker and the particle becomes identified. This allows us to determine the graviton propagator in the same way one obtains, for example, the photon propagator in QED, and eventually carry out a perturbation theory as a first attempt to develop a quantum theory of gravitation.

In units of $\hbar = 1, c = 1$, Newtons gravitational constant $G_N$, in 4 dimensional spacetime, has the dimenionality $[G_N] = [\text{Mass}]^{-2}$, which is a dead give away of the non-renormalizabilty of a quantum theory of gravitation based on GR. The non-renormalizability of the theory is easier to understand by noting that the divergences, in general, tend to increase as we go to higher orders in the gravitational coupling constant without a bound, implying the need of an infinite of parameters need to be fixed experimentally[77] and hence is not of any practical value. Some theories which are generalizations of GR, involving higher order derivatives, turn out to be renormalizable[78] but violate, in a perturbative setting, the very sacred principle of positivity condition of quantum theory. Unfortunately, such a theory involves ghosts in a perturbative treatment, due to the rapid damping of the propagator at high energies faster than $1/\text{energy}^2$, and gives rise, in turn, to negative probabilities.[79]

One is led to believe that Einstein's general relativity is a low energy effective theory as the low energy limit of a more complicated theory, and as such it provides a reliable description of gravitation at low energies. Moreover, one may argue that the non-renormalizability of a quantum theory based on GR is due to the fact that one is trying to use it at energies which are far beyond its range of validity. As a

---

[76]Bekenstein [14].

[77]Manoukian [149] and Anselmi [9].

[78]Stelle [204].

[79]Unitarity (positivity) of such theories in a non-perturbative setting has been elaborated upon by Tomboulis [224].

matter of fact the derivatives occurring in the action, in a momentum description via Fourier transforms, may be considered to be small at sufficiently low energies. In view of applications in the low energy regime, one then tries to separate low energy effects from high energy ones even if the theory has unfavorable ultraviolet behavior such as in quantum gravity.[80] Applications of such an approach have been carried out in the literature as just cited, and, for example, the modification of Newton's gravitational potential at long distances has been determined to have the structure

$$U(r) = -\frac{G_N m_1 m_2}{r}\left[1 + \alpha\frac{G_N(m_1 + m_2)}{c^2 r} + \beta\frac{G_N \hbar}{c^3 r^2}\right], \qquad (1.14)$$

for the interaction of two spin 0 particles of masses $m_1$ and $m_2$. Here $\alpha$, $\beta$, are dimensionless constants,[81] and the third term represents a quantum correction being proportional to $\hbar$.

Conventional quantum field theory is usually formulated in a fixed, i.e., in, a priori, given background geometry such as the Minkowski one. This is unlike the formalism of "Loop Quantum Gravity" (LQG) also called "Quantum Field Theory of Geometry". The situation that we will encounter in this approach is of a *quantum field theory in three dimensional space*, which is a non-perturbative background independent formulation of quantum gravity. The latter means that no specific assumption is made about the underlying geometric structure and, interestingly enough, the latter rather emerges from the theory. Here by setting up an eigenvalue equation of, say, an area operator, in a quantum setting, one will encounter a granular structure of three-dimensional space yielding a discrete spectrum for area measurements with the smallest possible having a non-zero value given to be of the order of the Planck length squared: $\hbar G_N/c^3 \sim 10^{-66}\,\mathrm{cm}^2$.[82] The emergence of space in terms of "quanta of geometry", providing a granular structure of space, is a major and beautiful prediction of the theory. The 3 dimensional space is generated by a so-called time slicing procedure of spacetime carried out by Arnowitt, Deser and Misner.[83] The basic field variables in the theory is a gravitational "electric" field, which determines the geometry of such a 3 dimensional space and naturally emerges from the definition of the area of a surface in such a space, and its canonical conjugate variable is referred to as the connection. By imposing equal time commutation relation of these two canonically conjugate field variables, the quantum version of the theory arises, and the fundamental problem of the quantization of geometry follows. The basic idea goes to Penrose [171] whose interest was to construct the concept of space from combining angular momenta. It is also interesting that the proportionality of entropy and the surface area of the BH horizon in the Bekenstein-Hawking Entropy formula has been derived in loop quantum gravity.[84]

---

[80]Donoghue [55–57] and Bjerrum-Bohr et al. [19], Bjerrum-Bohr et al. [20].

[81]Recent recorded values are $\alpha = 3$, and $\beta = 41/10\pi$ Bjerrum-Bohr et al. [19].

[82]Rovelli and Smolin [174], Ashtekar and Lewandoski [12], and Rovelli and Vidotto [175].

[83]Arnowitt et al. [10].

[84]See, e.g., Meissner [160] and Ansari [8].

Supersymmetry is now over 40 years old. Supersymmetry provides a symmetry between fermions and bosons. Borrowing a statement made by Dirac, speaking of theories, in general, it is *a theory with mathematical beauty*.[85] The name "Supersymmetry" for this symmetry is attributed to Salam and Strathdee as it seemed to have first appeared in the title of one of their papers [184]. An abbreviated name for it is SUSY, as some refer to this symmetry. The latter is not only beautiful but is also full of thought-provoking surprises. To every degree of freedom associated with a particle in the standard model, in a supersymmetric version, there corresponds a degree of freedom associated with a partner, referred to as a sparticle, with the same mass and with opposite statistics to the particle.[86] Unlike other discoveries, supersymmetry was not, a priori, invented under pressure set by experiments and was a highly intellectual achievement. Theoretically, however, it quickly turned out to be quite important in further developments of quantum field theory. For one thing, in a supersymmetric extention of the standard model, the electroweak and strong effective couplings do merge at energies about $10^{16}$ GeV, signalling the possibility that these interactions are different manifestations of a single force in support of a grand unified theory of the fundamental interactions. Also gravitational effects are expected to be important at the quantum level at the Planck energy of the order $10^{19}$ GeV, or possibly at even lower energies, giving the hope of having a unified theory of the four fundamental interactions at high energies. Supersymmetry leads to the unification of coupling constants. SUSY also tends to "soften" divergences of a theory in the sense that divergent contributions originating from fermions tend to cancel those divergent contributions originating from bosons due to their different statistics.

One of the important roles that supersymmetry may play in a supersymmetric extension of the standard problem is in the so-called *hierarchy* problem. The basic idea of a facet of this is the following. A fundamental energy scale arises in the standard model from the vacuum expectation value of the Higgs boson which sets the scale for the masses in the theory, such as for the masses of the vector bosons. It turns out that the masses imparted to the initially massless vector bosons, for example, via the Higgs mechanism, using the parameters in the Lagrangian density are in very good agreement with experimental results. On the other hand, if one introduces a large energy scale cut off $\kappa \sim 10^{15}$ GeV, of the order of a grand unified energy scale, or the Planck energy scale $10^{19}$ GeV, at which gravitation may play a significant role, to compute the shift of the squared-mass of the Higgs boson, as a scalar particle, due to the dynamics (referred to as radiative corrections), it turns out to be quadratic[87] in $\kappa$, which is quite large for such a large cut-off. This requires that the bare mass squared of the Higgs boson to be correspondingly large to cancel

---

[85]Here we recall the well known statement of Dirac, that a theory with mathematical beauty is more likely to be correct than an ugly one that fits some experimental data [53].

[86]This is such that the total number of fermion degrees of freedom is equal to the total bosonic ones.

[87]See, e.g., Veltman [230].

such a quadratic dependence on $\kappa$ and obtain a physical mass of the Higgs boson of the order of magnitude of the minute energy[88] ($\sim 10^2$ GeV), in comparison, characterizing the standard model, and this seems quite unnatural for the cancelation of such huge quantities.[89] This unnatural cancelation of enormously large numbers has been termed a facet of the hierarchy problem. Supersymmetric theories have, in general, the tendency to cancel out such quadratic divergences, up to possibly of divergences of logarithmic type which are tolerable, thus protecting a scalar particle from acquiring such a large bare mass. So supersymmetry may have an important role to play here.

SUSY relates fermions to bosons, and vice versa, and hence a generator is required which is of fermionic type, that is, it carries a spinor index as in the Dirac field[90] to carry out a transformation fermion $\leftrightarrow$ boson. Since the spins of fermions and bosons are different, this necessarily means that such a generator does *not* commute with the angular (spin) momentum operator as supersymmetry unites particles of the same mass and different spins into multiplets. Bosons and fermions have, in general, different masses, which means that SUSY is to be spontaneously broken if such a symmetry is to have anything to do with nature. If supersymmetry breaking sets at such an energy scale as 1 TeV or so, then some of the lowest mass superpartners may be hopefully discovered.[91]

Of particular interest was also the development of the superspace concept as an extension of the Minkowski one, where one includes an additional degree of freedom usually denoted by[92] $\theta = (\theta_a)$ to the space coordinates $(t, \mathbf{x})$, which turns out to be quite convenient in defining and setting up SUSY invariant integrals such as the action of a dynamical system.[93] To describe dynamics this, in turn, necessitates to introduce superfields of different types[94] as functions of these variables.

The extension of the algebra of the Poincaré group to a superalgebra was first carried out by Gol'fand and Likhtman in [100] to construct supersymmetric field theory models, and with the implementation of spontaneous symmetry breaking

---

[88] Aad et al. [1] and Chatrchyan et al. [31].

[89] As mentioned earlier, the question, in turn, arises as to what amounts for the enormous difference between the energy scale of grand unification and the energy scale that characterizes the standard model.

[90] This point is of importance because an earlier attempt by Coleman and Mandula [33] to enlarge the Poincaré group did not work. They considered only so-called "Bose" generators (that is tensors, and not spinors) in their analysis.

[91] Perhaps an optimist would argue that since antiparticles corresponding to given particles were discovered, the discovery of superpartners associated with given particles would not be out of the question either. The underlying symmetries involved in these two cases are, however, quite of different nature.

[92] This is called a Grassmann variable.

[93] See Salam and Strathdee [185, 186].

[94] Details on superfields will be given in Vol. II. The *explicit* expression of the pure vector superfield has been recently obtained in Manoukian [155].

by Volkov and Akulov in [232]. In [243], Wess and Zumino also,[95] independently, developed supersymmetric models in 4 dimensions, and this work has led to an avalanche of papers on the subject and to a rapid development of the theory. In particular, supersymmetric extensions of the standard model were developed,[96] supergravity, as a supersymmetric extension of gravitational theory, was also developed.[97] Unfortunately, things do not seem to be much better for supergravity, as far as its renormalizability is concerned.[98]

Now we come to String Theory. String Theory is a theory which attempts to provide a unified description of all the fundamental interactions in Nature and, in particular, give rise to a consistent theory of quantum gravity. A string is a fundamental one dimensional extended object, and if it has to do with quantum gravity, it is, say, of the order of the Plank length $\ell_P = \sqrt{G_N \hbar / c^3} \sim 10^{-33}$ cm, involving the three fundamental constants: Newton's gravitational constant $G_N$, the quantum unit of action $\hbar$, and the speed of light $c$. Since no experiments can probe distances of the order of the Planck length, such a string in present day experiments is considered to be point-like. When a string, whether closed or open, moves in spacetime, it sweeps out a two dimensional surface referred to as a worldsheet. String Theory is a *quantum field theory which operates on such a two dimensional worldsheet*. This, as we will see, has remarkable consequences in spacetime itself, albeit in higher dimensions. Particles are identified as vibrational modes of strings, and a single vibrating string may describe several particles depending on its vibrational modes. Strings describing bosonic particles are referred to as a bosonic strings, while those involving fermionic ones as well are referred to as superstrings. The remarkable thing is that the particles needed to describe the dynamics of elementary particles *arise* naturally in the mass spectra of oscillating strings, and are not, a priori, assumed to exist or put in by hand in the underlying theories. The *dimensionality* of the spacetime in which the strings live are *predicted* by the underlying theory as well and are necessarily of higher dimensions than four for consistency with Lorentz invariance of spacetime at the quantum level, consisting of a dimensionality of 26 for the bosonic strings and a spacetime dimensionality of 10 for the superstrings. The extra dimensions are expected to curl up into a space that is too small to be detectable with present available energies. For example the surface of a hollow extended cylinder with circular base is two dimensional, with one dimension along the cylinder, and another one encountered as one moves on its circumference. If the radius of the base of the cylinder is relatively small, the cylinder will appear as one dimensional when viewed from a large distance (low energies). Accordingly, the extra dimensions in string theory are expected to be

---

[95]These basic papers, together with other key ones, are conveniently collected in Ferrara [71].

[96]See Fayet [67], Dimopoulos and Georgi [45].

[97]See Freedman, van Nieuwenhuizen and Ferrara [84], Deser and Zumino [42].

[98]See, e.g., Deser et al. [41], Deser [40], Stelle [205, 206], and Howe and Stelle [124].

small and methods, referred to as compactifications,[99] have been developed to deal with them thus ensuring that the "observable" dimensionality of spacetime is four.

Superstring theories involve fermions and are thus relevant to the real world, but there are, however, several superstring theories. Also unlike the loop quantum gravity, which provides a background independent formulation of spacetime with the latter emerging from the theory itself, as discussed earlier, the strings in string theories are assumed to move in a pre-determined spacetime, and thus spacetime plays a passive role in them.[100] A theory, referred to as M-Theory,[101] based on non-perturbative methods, is envisaged to unify the existing superstrings theories into one single theory, instead of several ones, and be related to them by various limiting and/or transformation rules, referred to as dualities,[102] and is of 11 dimensional spacetime. M-Theory is believed to be approximated by 11 dimensional supergravity,[103] and the spacetime structure is envisaged to emerge from the theory as well. Bosonic strings involve tachyonic states. This is unlike the situation in superstring theories in which supersymmetry plays a key role in their definitions, and a process referred to as a GSO projection method, ensuring the equality of the degrees of freedom of bosonic and fermionic states, as required by supersymmetry, in turn implies that no tachyonic states appear in the theory.[104]

String theory was accidentally discovered through work carried out by Veneziano in 1968 when he attempted to write down consistent explicit expressions of meson-meson scattering amplitudes in strong interactions physics.[105] This was, of course before the discovery of QCD. With the many excited states of mesons and baryons (resonances), it was observed experimentally that there exists a linear relationship between spin $J$ and the mass $M$ squared of a resonance given by a linear relationship

$$\text{with a universal slope :} \quad \frac{dJ}{dM^2} = \alpha', \quad \alpha' \cong 1\,\text{GeV}^{-2}, \tag{1.15}$$

defining so-called Regge trajectories. Veneziano postulated and wrote down a scattering amplitude of meson – meson scattering: $p_1(m_1) + p_2(m_2) \rightarrow p_3(m_3) + p_4(m_4)$, which, in particular, showed that the amplitude involves the exchange of an infinite number of particles (corresponding to arbitrary integer spins). This is unlike the situation in conventional field theory as QED or the standard model, where they involve the exchange of a finite number of particles to any given order. String theory shares this property of the Veneziano amplitude. As a matter of fact the Veneziano

---

[99] An idea used by Kaluza and Klein in their attempt to unify gravity and electromagnetism in a 5 dimensional generalization of general relativity.

[100] See also Horowitz [122].

[101] Townsend [227], Witten [244], and Duff [58].

[102] Duff [58] and Schwarz [191].

[103] Cremmer et al. [37].

[104] The GSO method of projection was proposed in Gliozzi et al. [98, 99].

[105] Veneziano [231], see also Lovelace and Squires [145] and Di Vecchia [54].

amplitude may be derived from string theory. Nambu [164], Nielsen [166] and Susskind [211] have shown that the famous expression of the amplitude postulated by Veneziano may be interpreted as a quantum theory of scattering of relativistic strings. Although, a priori, this was assumed to describe a strong interaction process, Yoneya [248], and Scherk and Schwarz [190] made use of the fact that string theory (involving closed strings) contains a spin 2 massless state, which was identified with the elusive graviton, in addition to a whole spectrum of other excitation modes, to propose that string theory provides a framework for the unification of general relativity and quantum mechanics. As early as 1971, Neveu and Schwarz [165], and Raymond [173] included fermions in their analyses, which eventually led to the notion of superstrings, and during a short period of time, several types[106] of superstrings were introduced in the literature.

Due to the assumed non-zero extensions of strings, it is hoped that they provide, naturally, an ultraviolet cut-off $\Lambda \sim (\ell_P)^{-1}$ and render all processes involving strings ultraviolet finite. This is unlike conventional quantum field theory interactions where all the quantum fields are multiplied locally at the same spacetime points, like multiplying distributions at the same point, and are, in this sense, quite troublesome.

In string theory, two strings with given vibrational modes, identifying two given particles, may combine forming one string with an arbitrary number of different vibrational modes associated with a myriad number of particles, defining generalized 3-vertices. The combined string may again split into two strings with associated vibrational modes, identified appropriately with two more particles, describing a scattering process of 2 particles $\rightarrow$ 2 particles. Thus interactions involve string worldsheets of various topologies arise.

Other extended objects are also encountered in string theory called branes which, in general, are of higher spatial dimensions than one, with the string defined as a one dimensional brane. For example, an open string, satisfying a particular boundary condition, referred to as a Dirichlet boundary condition, specifies a hypersurface, referred to as a D brane, on which the end points of the open string reside. On the other hand, the graviton corresponds to a vibrational mode of closed strings, and since the latter, having no ends, may not be restricted to a brane and moves away from it. This might explain the weakness of the gravitational field, if our universe is a 3 dimensional brane embedded in a higher dimensional spacetime. Massless particles encountered in string theory are really the physically relevant ones because of the large unit of mass $(\ell_P)^{-1} \sim 10^{19}\,\text{GeV}$ in attributing masses to the spectrum of massive particles.[107] As we will see a massless particle may acquire mass if,

---

[106]Green and Schwarz [102, 103] and Gross et al. [107, 108].

[107]A systematic analysis of all the massless field excitations encountered in both bosonic and superstrings are investigated in Manoukian [152–154], in their respective higher dimensional spacetimes, and include the determinations of the degrees of freedom associated with them. Note that in four dimensional spacetime the number of degrees of freedom (spin states) of non-scalar fields is always two. This is not true in higher dimensional spacetime. For example, the degrees of freedom associated with a massless vector particle is 8 in 10 dimensions, while for the graviton is

for example, the end points of the open string are attached to two different branes, instead of a single brane.

We will learn the remarkable facts that Einstein's general relativity as well as Yang-Mills field theory may be obtained from string theory.

Interesting high energy scattering amplitudes have been computed in string theory over the years,[108] which provide a hint that space may not be probed beyond the Planck length – a result shared with "loop quantum gravity". It is worth mentioning that the Bekenstein-Hawking Entropy relation has been also derived in string theory.[109]

In recent years much work has been done, which is worth mentioning here but rather briefly, indicating that general relationships may exist between field theories and string theories, and consequently considerable attention was given trying to make such a statement more and more precise, with the ultimate hope of providing, in turn, a consistent and acceptable quantum theory of gravitation relevant to our world but much work still remains to be done. In particular, much study has been made to study the equivalence relation between certain four dimensional gauge theories and superstring theories, referred to as the AdS/CFT correspondence, where AdS space stands for anti-de-Sitter space, and CFT stands for conformal field theory.[110] Such correspondences have been also referred to as Gauge/Gravity duality, as well as Maldacena duality, a duality which was first proposed by Maldacena.[111] Without going into technical details, the aim of this work is to show, for example, the existence of an equivalence relation between a certain supersymmetric $SU(N)$ Yang-Mills field theory in 4 dimensional Minkowski spacetime, and a superstring theory in a 5 dimensional AdS space, having one additional dimension to the Minkowski one, and with the 5 dimensions of the AdS space supplemented by 5 extra dimensions defined by a five-sphere, making up the 10 dimensions of superstrings mentioned earlier. The interest in this work is that it deals with a connection between string theory (involving gravity) and

---

35, as shown later in Chapter 3 of Vol. II. In 4 dimensions, their degrees of freedom are, of course, two.

[108] See, e.g., Amati et al. [3, 4] and 't Hooft [219].

[109] See, e.g., Strominger and Fava [208] and Horowitz et al. [123].

[110] AdS space and CFT symmetry may be introduced as follows. AdS space, in $D$ dimensions, may be defined in terms of coordinates $z = (z^0, z^1, \ldots, z^{D-1}, z^D)$ satisfying a quadratic equation $\sum_{k=1}^{D-1}(z^k)^2 - (z^0)^2 - (z^D)^2 = -R^2$, for a given constant $R^2$, embedded in a $(D+1)$ dimensional space with interval squared $ds^2 = \sum_{j=1}^{D-1} dz^{j\,2} - dz^{0\,2} - dz^{D\,2}$. On the other hand a $D-$Sphere is defined in terms of coordinates $y^1, \ldots, y^{D+1}$ satisfying a quadratic equation $\sum_{j=1}^{D+1}(y^j)^2 = \rho^2$ for a given constant $\rho$. The conformal group, as applied in 4 dimensional Minkowski spacetime, is defined by a scale transformation $x^\mu \rightarrow \lambda\, x^\mu$, and a so-called special (conformal) transformation

$$\frac{x'^\mu}{x'^2} = \frac{x^\mu}{x^2} + a^\mu,$$

for a constant 4-vector $a^\mu$, in addition to the Poincaré ones.

[111] Maldacena [146]. See also Witten [245], Gubser et al. [109], and Aharoni et al. [2].

supersymmetric gauge theories. This brings us into contact with the holographic principle, in analogy to holography in capturing 3 dimensional images of objects on a 2 dimensional (holographic) plate,[112] showing that an equivalence relation exists between the 3 and the 2 dimensional set-ups. The 4 dimensional quantum field theory is like a hologram capturing information about the higher dimensional quantum gravity theory. In this case the SU(N) theory provides a holographic description of gravitational field. This is in analogy to black hole entropy with its encoded information being proportional to the area rather than to the volume of the region enclosed by the horizon. Perhaps holography is a basic property of string theory and one expects that much has to be done before developing a realistic quantum gravity, and in turn provide a background independent formulation for string theory. The holographic principle was first proposed by 't Hooft.[113]

We close this chapter by commenting on two symmetries which seem to be observed in Nature, that is of the CPT symmetry and of the Spin & Statistics connection and of their relevance to our own existence. We will see how these symmetries arise from quantum field theory in Sect. 4.10 and Sect. 4.5, respectively.

CPT taken in any order, seems to be an observed symmetry in Nature, where C stands for charge conjugation with which particles are replaced by their antiparticles and vice versa, P represents space reflection, while T denotes time reversal.

Local Lorentz invariant quantum field theories preserve (Sect. 4.10) the CPT symmetry. Experimentally, symmetry violations are well known to occur when one restricts to one or to the product of two transformations in CPT in dealing with some fundamental processes. For example the violation of parity was already established in 1957[114] as well as the violation of charge symmetry.[115] Later, in 1964 CP violation, and hence also of T, was observed in neutral Kaon decays.[116] The CP transformation and C, provide the fundamental relations between matter and anti-matter. The question then arises as to why we observe, apart in accelerator experiments, only one form (matter) than the other form in the visible universe – a key criterion for our own *existence*. If an equal amount of matter and anti-matter was produced at some stage then why, our visible universe is matter dominated. Sakharov in 1967 [177] proposed that a key reason for this is CP violation. In more details to explain this asymmetry, he proposed that (1) baryon number is not conserved. (This is supported by recent grand unified field theories,) (2) CP and C are violated, (3) the universe has gone through a phase of extremely rapid expansion to avoid the pairing of matter and anti-matter. The violation of such symmetries, at

---

[112]Recall that the *two* dimensional holographic plate which registers the interference of reflected light off an object and an unperturbed Laser beam stores information of the shape of the *three* dimensional object. As one shines a Laser beam on it an image of the three dimensional object emerges.

[113]'t Hooft [220], see also especially Thorn [223], as well as the analysis with further interpretations by Susskind [212]. See also Bousso [27].

[114]Wu et al. [246], Garwin et al. [91], and Friedman and Telegdi [86].

[115]Garwin et al. [91].

[116]Christenson et al. [32].

the microscopic level, and their consequences on a macroscopic scale is certainly intriguing.

Clearly, the "Spin & Statistics" connection, of which the Pauli exclusion principle is a special case applicable to spin $1/2$ particles, is important not only in physics but in all of the sciences, and is relevant to our own existence. For one thing, the periodic table of elements in chemistry is based on the exclusion principle. In simplest terms, the upshot of this is that half-odd-integer spin fields are quantized by anti-commutators, while integer spins fields are quantized by commutators. This result is of utmost significance for our existence. As a matter of fact the Pauli exclusion principle is not only sufficient for the stability of matter[117] in our world but it is also necessary.[118] In the problem of stability of neutral matter, with a finite number of electrons per atom, but involving several nuclei, and correspondingly a large number of electrons $N$, the stability of matter, based on the Pauli exclusion principle, or instability of so-called "bosonic matter", in which the exclusion principle is abolished, rests rather on the following. For "bosonic matter", the ground state energy $E_N \sim -N^\alpha$, with $\alpha > 1$,[119] where $(N + N)$ denotes the number of the negatively charged particles plus an equal number of positively charged particles. This behavior for "bosonic matter" is unlike that of matter, with the exclusion principle, for which $\alpha = 1$.[120] A power law behavior with $\alpha > 1$ implies instability as the formation of a single system consisting of $(2N + 2N)$ particles is favored over two separate systems brought together each consisting of $(N + N)$ particles, and the energy released upon collapse of the two systems into one, being proportional to $[(2N)^\alpha - 2(N)^\alpha]$, will be overwhelmingly large for realistic large $N$, e.g., $N \sim 10^{23}$. Dyson [61], has estimated that without the exclusion principle, the assembly of two macroscopic objects would release energy comparable to that of an atomic bomb, and such "matter" in bulk would collapse into a condensed high-density phase and our world will cease to exist.[121] Ordinary matter, due to the exclusion principle, occupies a very large volume. This point was emphasized by Ehrenfest in a discussion with Pauli in 1931[122] on the occasion of the Lorentz medal to this effect: "We take a piece of metal, or a stone. When we think about it, we are astonished that this quantity of matter should occupy so large a volume". He went on by stating that the exclusion principle is the reason: "Answer: only the Pauli principle, no two electrons in the same state". In this regard,

---

[117]For a pedagogical treatment of the problem of "stability of matter" and related intricacies, see Manoukian [151], Chapter 14.

[118]Lieb and Thirring [144] and Thirring [222].

[119]Dyson [61], Lieb [143], and Manoukian and Muthaporn [156].

[120]Lieb and Thirring [144] and Thirring [222].

[121]In the Preface of Tomonaga's book on spin [226], one reads: "The existence of spin, and the statistics associated with it, is the most subtle and ingenious design of Nature – without it the whole universe would collapse".

[122]See Ehrenfest [62].

a rigorous treatment[123] shows that the extension of matter radially grows not any slower than $N^{1/3}$ for large $N$. No wonder why matter occupies so large a volume. The importance of the "Spin & Statistics" connection and the role it plays in our world cannot be overemphasized. Needless to say, no quantum field theory treatment is complete without the CPT Theorem and the Spin & Statistics Connection.

The present volume deals with the foundations of quantum field theory and with the intricacies of abelian and non-abelian gauge theories. Volume II deals with quantum gravity, supersymmetry, and string theory.

# References

1. Aad, G. et al. (2012). Observation of a new particle in the search for the Standard Model Higgs Boson with the ATLAS detector at the LHC. *Physics Letters, B716*, 1–29.
2. Aharoni, O. et al. (2008). $N = 6$ superconformal Chern-Simons matter theories, M2-branes and their gravity duals. *JHEP, 0810*, 091.
3. Amati, D., Ciafaloni, M., & Veneziano, G. (1987). Superstring collisions at planckian energies. *Physics Letters, B197*, 81–88.
4. Amati, D., Ciafaloni, M., & Veneziano, G. (1988). Classical and Quantum effects from Planckian energy superstring collisions. *International Journal of Modern Physics, 3*, 1615–1661.
5. Anderson, C. D. (1932). The apparent existence of easily deflectable positives. *Science, 76*, 238–239.
6. Anderson, C. D. (1933). The positive electron. *Physical Review, 43*, 491–494.
7. Anderson, P. W. (1963). Plasmons, gauge invariance, and mass. *Physical Review, 130*, 439–442.
8. Ansari, M. H. (2008). Generic degeneracy and entropy in loop quantum gravity. *Nuclear Physics, B795*, 635–644.
9. Anselmi, D. (2003). Absence of higher derivatives in the renormalization of propagators in quantum field theory with infinitely many couplings. *Class. Quantum Grav., 20*, 2344–2378.
10. Arnowitt, R. S., Deser, S., & Misner, W. (2008). The dynamics of general relativity. *General Relativity and Gravitation, 40*, 1997–2027. Reprinted from *Gravitation: An Introduction to current research* (Chap.7), Edited by L. Witten. John Wiley & Sons Inc., New York/London, 1962.
11. Aschenbrenner, M. (1996). A decoupling theorem for the BPHZL-scheme. *Annals of Physics, 250*, 320–351.
12. Ashtekar, A., & Lewandoski, J. (1997). Quantum theory of gravity I: Area operators. *Classical Quantum Gravity, 14*, A55–A81.
13. Becchi, C., Rouet, A., & Stora, R. (1976). Renormalization of gauge theories. *Annals of Physics, 98*, 287–321.
14. Bekenstein, J. D. (1973). Black holes and entropy. *Physical Review, D7*, 2333–2346.
15. Benvenuti, A. et al. (1974). Observation of Muonless neutrino-induced inelastic interactions. *Physical Review Letters, 32*, 800–803.
16. Beringer, J. et al. (2012). Particle data group. *Physical Review D, 86*, 010001.
17. Bethe, H. (1947). The Electromagnetic shift of energy levels. *Physical Review, 72*, 339–341.
18. Bethe, H., & Fermi, E. (1932). Über die Wechselwirkung von Zwei Elektronen. *Zeitschrift fur Physik, 77*, 296–306.

---

[123]Manoukian and Sirininlakul [157].

19. Bjerrum-Bohr, N. E. J., Donoghue, J. F., & Holstein, B. R. (2003a). Quantum gravitational corrections to the non-relativistic scattering potential of two mesons. *Physical Review, D 67*, 084033. Erratum: *ibid., D71*, 069903 (2005).
20. Bjerrum-Bohr, N. E. J., Donoghue, J. F., & Holstein, B. R. (2003b). Quantum corrections to the schwarzchild and kerr metrics. *Physical Review, 68*, 084005–084021.
21. Bjorken, J. D., & Pachos, E. A. (1969). Inelastic electron-proton and y-proton scattering and the structure of the nucleon. *Physical Review, 185*, 1975–1982.
22. Bloch, F., & Nordsieck, A. (1937). Notes on the radiation field of the electron. *Physical Review, 52*, 54–59.
23. Bludman, S. (1958). On the universal fermi interaction. *Nuovo Cimento, 9*, 433–445.
24. Bogoliubov, N. N., & Parasiuk, O. S. (1957). On the multiplication of propagators in quantum field theory. *Acta Physics Mathematics, 97*, 227–266. Original German Title: Über die Multiplikation der Kausalfunctionen in der Quantentheorie der Felder.
25. Bogoliubov, N. N., & Shirkov, D. V. (1959). *Introduction to the theory of quantized fields*. Interscience, New York.
26. Born, M., Heisenberg, W., & Jordan, P. (1926). Zur Quantenmechanik III. *Zeitschrift fur Physik, 35*, 557–615. Reprinted in *Sources of quantum mechanics*, (ed. B. L. vander Waerden), Dover Publications, New York (1968).
27. Bousso, R. (2002). The holographic principle. *Reviews of Modern Physics, 74*, 825–874.
28. Callan, C. G. (1970). Broken scale invariance in scalar field theory. *Physical Review, D2*, 1541–1547.
29. Callan, C. G. (1972). Broken scale invariance and asymptotic behavior. *Physical Review, D5*, 3202–3210.
30. Chamberlain, O., Segrè, E., Wiegand, C. and Ypsilantis, T. (1955). Observation of antiprotons. *Physical Review, 100*, 947–950.
31. Chatrchyan, S. et al. (2012). Observation of a New Boson at Mass 125 GeV with the CMS experiment at LHC. *Physics Letters, B716*, 30–61.
32. Christenson, J. H. et al. (1964). Evidence for the $2\pi$ decay of the $K_2^0$ meson. *Physical Review Letters, 13*, 138–140.
33. Coleman, S., & Mandula, J. (1967). All possible symmetries of the S matrix. *Physical Review, 150*, 1251–1256.
34. Connes, A., & Kreimer, D. (1998). Hopf algebras, renormalization and noncommutative geometry. *Communications in Mathematical Physics, 119*, 203–242.
35. Connes, A., & Kreimer, D. (2000). Renormalization in quantum field theory and the Riemann-Hilbert problem I: The Hopf algebra structure and the main theorem. *Communications in Mathematical Physics, 210*, 249–273.
36. CPLEAR/Collaboration (2000). T violation and CPT tests at CPLEAR, symmetries in subatomic physics. In: *3rd International Symposium. AIP Conference Proceedings* (Vol. 539, pp. 187–196), Adelaide (Australia).
37. Cremmer, E., Julia, B., & Scherk, J. (1978). Supergravity theory in eleven-dimensions. *Physics Letters, B76*, 409–412.
38. Cronin, J. W. (1981). CP symmetry violation: The search of its origin. *Reviews of Modern Physics, 53*, 373–383.
39. Cronin, J. W., & Fitch, V. L. (1964). Evidence for the $2\pi$ decay of the $K_2^{\ 0}$ meson. *Physical Review Letters, 13*, 138–140.
40. Deser, S. (2000). Infinities in quantum gravities. *Annalen der Physik, 9*, 299–306.
41. Deser, S., Kay, J. H., & Stelle, K. S. (1977). Renormalizability properties of supergravity. *Physical Review Letters, 38*, 527–530.
42. Deser, S., & Zumino, B. (1976). Consistent supergravity. *Physics Letters, 62B*, 335–337.
43. DeWitt, B. (1964). Theory for radiative corrections for non-Abelian gauge fields. *Physical Review Letters, 12*, 742–746.
44. DeWitt, B. (1967a). Quantum theory of gravity. II. The manifestly covariant theory. *Physical Review, 162*, 1195–1239.

45. Dimopoulos, S., & Georgi, H. (1981). Softly broken supersymmetry and SU(5). *Nuclear Physics, B193*, 150–162.
46. Dirac, P. A. M. (1927). The quantum theory of the emssion and absorption of radiation. *Proceedings of the Royal Society of London, A114*, 243–265.
47. Dirac, P. A. M. (1928a). The quantum theory of the electron, I. *Proceedings of the Royal Society of London, A114*, 610–624.
48. Dirac, P. A. M. (1928b). The quantum theory of the electron, II. *Proceedings of the Royal Society of London, A117*, 610–624.
49. Dirac, P. A. M. (1928c). Über die Quantentheorie des Elektrons. *Physikalishce Zeitschrift, 29*, 561–563.
50. Dirac, P. A. M. (1930a). A theory of electrons and protons. *Proceedings of the Royal Society of London, A126*, 360–365.
51. Dirac, P. A. M. (1930b). On the annihilation of electrons and protons. *Proc. Cambridge Phil. Soc., 26*, 361–375.
52. Dirac, P. A. M. (1931). Quantized singularities in the electromagnetic field. *Proceedings of the Royal Society of London, A133*, 60–72.
53. Dirac, P. A. M. (1970). Can equations of motion be used in high-energy physics? *Physics Today, 23*, 29. April, issue (4).
54. Di Vecchia, P. (2008). The birth of string theory. In M. Gasperini & J. Maharana (Eds.), *String theory and fundamental interactions: Gabriele Veneziano and theoretical physics: Historical and contemporary perspectives*. (Lecture notes in physics, vol. 737, pp. 59–118). Berlin/New York: Springer.
55. Donoghue, J. F. (1994a). Leading correction to the newtonian potential. *Physical Review Letters, 72*, 2996.
56. Donoghue, J. F. (1994b). General relativity as an effective field theory, the leading corrections. *Physical Review, D50*, 3874–3888. (gr-qg/9405057).
57. Donoghue, J. F. (1997). In Fernando, C., & Herrero, M.-J. (Eds.), *Advanced school on effective theories*. World Scientific, Spain. UMHEP - 424, gr-qc/9512024.
58. Duff, M. J. (1996). M-theory (The theory formerly known as superstrings). *International Journal of Modern Physics, A11*, 5623–5642, hep-th/9608117.
59. Dyson, F. J. (1949a). The radiation theories of Tomonaga, Schwinger and Feynman. *Physical Review, 75*, 486–502. Reprinted in Schwinger (1958a).
60. Dyson, F. J. (1949b). The S-Matrix in quantum electrodynamics. *Physical Review, 75*, 1736–1755. Reprinted in Schwinger (1958a).
61. Dyson, F. J. (1967). Ground-state energy of a finite system of charged particles. *Journal of Mathematics and Physics, 8*(8), 1538–1545.
62. Ehrenfest, P. (1959). Ansprache zur Verleihung der Lorentzmedaille an Professor Wolfgang Pauli am 31 Oktober 1931. (Address on award of Lorentz medal to Professor Wolfgang Pauli on 31 October 1931). In M. J. Klein (Ed.), *Paul Ehrenfest: Collected scientific papers* (p. 617). Amsterdam: North-Holland. [The address appeared originally in P. Ehrenfest (1931). *Versl. Akad. Amsterdam, 40*, 121–126.].
63. Englert, F., & Brout, R. (1964). Broken symmetry and the mass of gauge vector bosons. *Physical Review Letters, 13*, 321–323.
64. Englert, F., Brout, R., & Thiry, M. F. (1966). Vector mesons in presence of broken symmetry. *Nuovo Cimento, A43*, 244–257.
65. Epstein, H., & Glaser, V. (1973). The role of locality in perturbation theory. *Annales de l'Institute Henri Poincaré, A19*, 211–295.
66. Faddeev, L. D., & Popov, V. N. (1967). Feynman diagrams for the Yang-Mills field. *Physics Letters, B25*, 29–30.
67. Fayet, P. (1977). Spontaneously broken supersymmetric theories of weak, electromagnetic, and strong interactions. *Physics Letters, 69B*, 489–494.
68. Fermi, E. (1932). Quantum theory of radiation. *Reviews of Modern Physics, 4*, 87–132.
69. Fermi, E. (1934a). Tentativo di una teoria dei raggi $\beta$. *Nuovo Cimento, 11*, 1–19.
70. Fermi, E. (1934b). Versuch einer Theorie der $\beta$ - Strahlen. *Zeitschrift fur Physik, 88*, 161–171.

71. Ferrara, S. (Ed.). (1987). *Supersymmetry* (Vol. 1 & 2). New York: Elsevier.
72. Feynman, R. P. (1963). Quantum theory of gravitation. *Acta Phys. Polon., 24*, 697–722.
73. Feynman, R. P. (1969a). The behavior of hadron collisions at extreme energies. In *Proceedings of the 3rd topical conference on high energy collisions*, Stony Brook. New York: Gordon & Breach.
74. Feynman, R. P. (1969b). Very high-energy collisions of hadrons. *Physical Review Letters, 23*, 1415–1417.
75. Feynman, R. P. (1972). The development of the space-time view of quantum electrodynamics. In *Nobel Lectures, Physics 1963–1970*, 11 Dec 1965. Amsterdam: Elsevier.
76. Feynman, R. P. (1982). *The theory of fundamental processes*, 6th printing. The Menlo Park: Benjamin/Cummings.
77. Feynman, R. P., & Gell-Mann, M. (1958). Theory of fermi interaction. *Physical Review, 109*, 193–198.
78. Feynman, R. P., & Hibbs, A. R. (1965). *Quantum mechanics and path integrals*. New York: McGraw-Hill.
79. Figueroa, H., & Gracia-Bondia, J. M. (2001). On the antipode of Kreimer's Hopf algebra. *Modern Physics Letters, A16*, 1427–1434. hep–th/9912170v2.
80. Figueroa, H., & Gracia-Bondia, J. M. (2004). The uses of Connes and Kreimer's algebraic formulation of renormalization. *International Journal of Modern Physics, A19*, 2739–2754. hep–th/0301015v2.
81. Figueroa, H., & Gracia-Bondia, J. M. (2005). Combinatorial Hopf algebras in quantum field theory. I. *Reviews in Mathematical Physics, 17*, 881–961. hep–th/0408145v2.
82. Fitch, V. L. (1981). The discovery of charge conjugation parity asymmetry. *Reviews of Modern Physics, 53*, 367–371.
83. Fock, V. (1933). *C. R. Leningrad (N.S.) no. 6*, pp. 267–271.
84. Freedman, D. Z., van Nieuwenhuizen, P., & Ferrara, S. (1976). Progress toward a theory of supergravity. *Physical Review, B13*, 3214–3218.
85. French, J. B., & Weisskopf, V. F. (1949). The electromagnetic shift of energy levels. *Physical Review, 75*, 1240–1248.
86. Friedman, J. I., & Telegdi, V. L. (1957). Nuclear emulsion evidence for parity nonconservation in the decay chain $\pi^+ \rightarrow \mu^+ \rightarrow e^+$. *Physical Review, 105*, 1681–1682.
87. Fritzsch, H., & Gell-Mann, M. (1972). Quatks and what else? In J. D. Jackson, & A. Roberts (Eds.), *Proceedings of the XVI International Conference on High Energy Physics* (Vol. 2). Chicago: Chicago University Press.
88. Fukuda, H., Miyamoto, Y., & Tomonaga, S. (1949a). A self consistent method in the quantum field theory. II-1. *Progress of Theoretical Physics, 4*, 47–59.
89. Fukuda, H., Miyamoto, Y., & Tomonaga, S. (1949b). A self consistent method in the quantum field theory. II-2. *Progress of Theoretical Physics, 4*, 121–129.
90. Furry, W. H., & Oppenheimer, J. R. (1934). On the theory of the electron and positive. *Physical Review, 45*, 245–262.
91. Garwin, R. L. et al. (1957). Observations of the failure of conservation of parity and charge conjugation in meson decays: The magnetic moment of the free muon. *Physical Review, 105*, 1415–1417.
92. Gell-Mann, M. (1972). Quarks. *Acta Physica Austriaca Supplement IX, 9*, 733–761.
93. Gell-Mann, M., & Low, F. E. (1954). Quantum electrodynamics at small distances. *Physical Review, 95*, 1300–1312.
94. Georgi, H., Quinn, H. R., & Weinberg, S. (1974). Hierarchy of interactions in unified gauge theories. *Physical Review Lett, 33*, 451–454.
95. Gershtein, S. S., & Zel'dovich, Y. B. (1956). On corrections from mesons to the theory of $\beta$-decay. *Soviet Physics JETP, 2*, 576. Original Russian version: *Zhurnal Experimental'noi i Teoreticheskoi Fiziki, 29*, 698 (1955)
96. Glashow, S. L. (1961). Partial symmetries of weak interactions. *Nuclear Physics, 22*, 579–588.

97. Glashow, S. L. (1980). Towards a unified theory: Threads in a tapestry. *Reviews of Modern Physics, 52*, 539–543.

98. Gliozzi, F., Scherk, J., & Olive, D. (1976). Supergravity and the spinor dual model. *Physics Letters, 65B*, 282–286.

99. Gliozzi, F., Scherk, J., & Olive, D. (1977). *Supersymmetry, supergravity theories and the dual Spinorl model. Nuclear Physics, B22*, 253–290.

100. Gol'fand, A., & Likhtman, E. P. (1971). *Extension of the Poincaré Group Generators and Violation of P Invariance. JETP Letters, 13*, 323–326. Reprinted in Ferrara (1987).

101. Gordon, W. (1926). Der Compton Effect nach der Schrödingerschen Theorie. *Zeitschrift fur Physik, 40*, 117–133.

102. Green, M. B., & Schwarz, J. H. (1981). Supersymmetrical dual string theory. *Nuclear Physics, B181*, 502–530.

103. Green, M. B., & Schwarz, J. H. (1982). Supersymmetrical string theories. *Physics Letters, 109B*, 444–448.

104. Greenberg, O. W. (1964). Spin and unitary spin independence in a paraquark model of baryons and mesons. *Physical Review Letters, 13*, 598–602.

105. Greenberg, O. W., & Zwanziger, D. (1966). Saturation in triplet models of hadrons. *Physical Review, 150*, 1177–1180.

106. Gross, D., & Wilczek, F. (1973). Ultraviolet behavior of non-Abelian gauge theories. *Physical Review Letters, 30*, 1342–1346.

107. Gross, D. J., Harvey, J. A., Martinec, E. J., & Rhom, R. (1985a). Heterotic string theory (I). The free hetrotic string. *Nuclear Physics, B256*, 253–284.

108. Gross, D. J., Harvey, J. A., Martinec, E. J., Rohm, R. (1985b). The heterotic string. *Physical Review Letters, 54*, 502–505.

109. Gubser, S. S., Klebanov, I. R., & Polyakov, A. M. (1998). Gauge theory correlations from non-critical string theory. *Physics Letters, B428*, 105–114. (hep-th/9802150).

110. Guralnik, G. S., Hagen, C. R., & Kibble, T. W. B. (1964). Global conservation laws and massless particles. *Physical Review Letters, 13*, 585–587.

111. Han, M. Y., & Nambu, Y. (1965). Three-triplet model with double SU(3) symmetry. *Physical Review, 139*, B1006–B1010.

112. Hasert, F. J. et al. (1973). Observation of neutrino-like interactions without muon-electron in the Gargamellr neutrino experiment. *Physics Letters, B 46*, 138–140.

113. Hasert, F. J. et al. (1973). Search for elastic muon-neutrino electron scattering. *Physics Letters, B 46*, 121–124.

114. Hawking, S. W. (1974). Black hole explosions? *Nature, 248*, 230–231.

115. Hawking, S. W. (1975). Particle creation by Black Holes. *Communications in Mathematical Physics , 43*, 199–220.

116. Heisenberg, W., & Pauli, W. (1929). Zur Quantenelektrodynamic der Wellenfelder, I. *Zeitschrift fur Physik, 56*, 1–61.

117. Heisenberg, W., & Pauli, W. (1930). Zur Quantenelektrodynamic der Wellenfelder, II. *Zeitschrift fur Physik, 59*, 168–190.

118. Hepp, K. (1966). Proof of the Bogoliubov-Parasiuk theorem of renormalization. *Communications in Mathematical Physics, 2*, 301–326.

119. Higgs, P. W. (1964a). Broken symmetries, massles particles and gauge fields. *Physics Letters, 12*, 132–133.

120. Higgs, P. W. (1964b). Broken symmetries and the masses of Gauge Bosons. *Physical Review Letters, 13*, 508–509.

121. Higgs, P. W. (1966). Spontaneous symmetry breaking without massless particles. *Physical Review, 145*, 1156–1163.

122. Horowitz, G. T. (2005). Spacetime in string theory. *New Journal of Physics, 7*, 201 (1–13).

123. Horowitz, G., Lowe, D. A., & Maldacena, J. (1996). Statistical entropy of non-extremal four dimensional black holes and U-duality. *Physical Review Letters, 77*, 430–433.

124. Howe, P. S., & Stelle, K. S. (2003). Supersymmetry counterterms revisited. *Physics Letters, B554*, 190–196. hep-th/0211279v1.

125. Jordan, P., & Pauli, W. (1928). Zur Quantenelektrodynamic Ladungfreier Felder. *Zeitschrift fur Physik, 47*, 151–173.
126. Jordan, P., & Wigner, E. (1928). Über das Paulische Äquivalenzverbot. *Zeitschrift fur Physik, 47*, 631–651. Reprinted in Schwinger (1958a).
127. Kibble, T. W. B. (1968). Symmetry breaking in non-abelian gauge theories. *Physical Review, 155*, 1554–1561.
128. Klein, O. (1926). Quantentheorie und Fünfdimensionale Relativitätstheorie. *Zeitschrift fur Physik, 37*, 895–906.
129. Klein, O. (1948). Mesons and nuclei. *Nature, 161*, 897–899.
130. Kramers, H. A. (1948). Non-relativistic quantum-electrodynamics and correspondence principle. In *Rapport et Discussions du 8e Conseil de Physique Solvay 1948* (pp. 241–265). R. Stoop, Bruxelles, 1950.
131. Kreimer, D. (1999). On the Hopf algebra structure of perturbative quantum field theory. *Advances in Theoretical and Mathematical Physics, 2*, 303–334.
132. Kreimer, D. (2003). New mathematical structures in renormalizable quantum field theories. *Annals of Physics, 303*, 179–202.
133. Kroll, N. M., & Lamb, W. E. (1949). On the self-energy of a bound electron. *Physical Review, 75*, 388–398. Reprinted in Schwinger (1958a).
134. Lamb, W. E., Jr., & Retherford, R. C. (1947). Fine structure of the hydrogen atom by a microwave method. *Zeitschrift fur Physik, 72*, 241–243. Reprinted in Schwinger (1958a).
135. Landsman, N. P. (1989). Large-mass and high-temperature behaviour in perturbative quantum field theory. *Communications in Mathematical Physics, 125*, 643–660.
136. Lattes, C. M. G. et al. (1947). Processes involving charged mesons. *Nature, 159*, 694–697.
137. Lederman, L., & Teresi, D. (2006). *The god particle: If the universe is the answer, what is the question?* New York: Mariner Books.
138. Lee, T. D., & Yang, C. N. (1956). Question of parity conservation in weak interactions. *Physical Review, 104*, 254–258. See also ibid. *106*, 1671 (1957).
139. Lee, B., & Zinn-Justin, J. (1972a). Spontaneously Broken Gauge symmetries. I. Preliminaries. *Physical Review, D5*, 3121–3137.
140. Lee, B., & Zinn-Justin, J. (1972b). Spontaneously Broken Gauge symmetries. II. Perturbation theory and renormalization. *Physical Review, D5*, 3137–3155.
141. Lee, B., & Zinn-Justin, J. (1972c). Spontaneously Broken Gauge symmetries. III. Equivalence. *Physical Review, D5*, 3155–3160.
142. Lee, B., & Zinn-Justin, J. (1973). Spontaneously Broken Gauge symmetries. IV. general Gauge formulation. *Physical Review, D7*, 1049–1056.
143. Lieb, E. H. (1979). The $N^{5/3}$ law for bosons. *Physics Letters, A70*(2), 71–73. Reprinted in Thirring (2005).
144. Lieb, E. H., & Thirring, W. E. (1975). Bound for the kinetic energy of fermions which proves the stability of matter. *Physical Review Letters, 35*(16), 687–689. [Erratum *ibid., 35*(16), 1116 (1975).] Reprinted in Thirring (2005).
145. Lovelace, C., & Squires, E. (1970). Veneziano theory. *Proceedings of the Royal Society of London, A318*, 321–353.
146. Maldacena, J. (1998). The large $N$ limit of superconformal theories and gravitation. *Advances in Theoretical and Mathematical Physics, 2*, 231–252. (hep-th/9711200).
147. Manoukian, E. B. (1976). Generalization and improvement of the Dyson-Salam renormalization scheme and equivalence with other schemes. *Physical Review, D14*, 966–971. *ibid., 2202* (E).
148. Manoukian, E. B. (1979). Subtractions vs counterterms. *Nuovo Cimento, 53A*, 345–358.
149. Manoukian, E. B. (1983a). *Renormalization*. New York/London/Paris: Academic.
150. Manoukian, E. B. (1986a). Action principle and quantization of gauge fields. *Physical Review, D34*, 3739–3749.
151. Manoukian, E. B. (2006). *Quantum theory: A wide spectrum*. Dordrecht: Springer.
152. Manoukian, E. B. (2012a). All the fundamental massless bosonic fields in bosonic string theory. *Fortschritte der Physik, 60*, 329–336.

153. Manoukian, E. B. (2012b). All the fundamental bosonic massless fields in superstring theory. *Fortschritte der Physik, 60*, 337–344.
154. Manoukian, E. B. (2012c). All the fundamental massless fermion fields in supersring theory: A rigorous analysis. *Journal of Modern Physics, 3*, 1027–1030.
155. Manoukian, E. B. (2012d). The explicit pure vector superfield in Gauge theories. *Journal of Modern Physics, 3*, 682–685.
156. Manoukian, E. B., & Muthaporn, C. (2003). $N^{5/3}$ Law for bosons for arbitrary large $N$. *Progress of Theoretical Physics, 110*(2), 385–391.
157. Manoukian, E. B., & Sirininlakul, S. (2005). High-density limit and inflation of matter. *Physical Review Letters, 95*, 190402: 1–3.
158. Marciano, W., & Pagels, H. (1978). Quantum chromodynamics. *Physics Reports, C36*, 137–276.
159. Martin, P. C., & Glashow, S. L. (2008). Julian Schwinger 1918–1994: A biographical memoir. *National Academy of Sciences*, Washington, DC, Copyright 2008.
160. Meissner, K. (2004). Black-Hole entropy in loop quantum gravity. *Classical Quantum Gravity, 21*, 5245–5251.
161. Mohapatra, R. N. (1971). Feynman rules for the Yang-Mills field: A canonical quantization approach. I, II. *Physical Review, D4*, 378–392, 1007–1017.
162. Mohapatra, R. N. (1972). Feynman rules for the Yang-Mills field: A canonical quantization approach. III. *Physical Review, D4*, 2215–2220.
163. Nambu, Y. (1966). A systematic of hadrons in subnuclear physics. In A. de Shalit, H. Feshback, & L. van Hove (Eds.), *Preludes in theoretical physics in Honor of V. F. Weisskopf* (p. 133). Amsterdam: North-Holland.
164. Nambu, Y. (1969). In: *Proceedings of the Internatinal Conference on Symmetries and Quark Models* (p. 269), Wayne State University. New York: Gordon and Breach, 1970.
165. Neveu, A., & Schwarz, J. H. (1971a). Factorizable dual model of pions. *Nuclear Physics, B31*, 86–112.
166. Nielsen, H. (1970). *Internatinal Conference on High Energy Physics*, Kiev Conference, Kiev.
167. Olive, K. A. et al. (2014). Particle data group. *Chinese Physics C, 38*, 090001.
168. Oppenheimer, J. R. (1930). Note on the theory of the interaction of field and matter. *Physical Review, 35*, 461–477.
169. Ovsyannikov, L. V. (1956). General solution to renormalization group equations. *Doklady Akademii Nauk SSSR, 109*, 1112–1115.
170. Pauli, W., & Weisskopf, V. (1934). Über die Quantisierung der Skalaren Relativistischen Wellengleichung. *Helvetica Physica Acta, 7*, 709–731.
171. Penrose, R. (1971). In T. Bastin (Ed.), *Quantum theory and beyond* (pp. 151–180). Cambridge: Cambridge University Press.
172. Politzer, H. D. (1973). Reliable perturbative results for strong interactions. *Physical Review Letters, 30*, 1346–1349.
173. Raymond, P. (1971). Dual theory for free fermions. *Physical Review, D3*, 2415–2418.
174. Rovelli, C., & Smolin, L. (1995). Discreteness of area and volume in quantum gravity. *Nuclear Physics, B442*, 593–622.
175. Rovelli, C., & Vidotto, F. (2015). *Covariant loop quantum gravity*. Cambridge: Cambridge University Press.
176. Rubbia, C. (1984). Experimental observation of the intermediate vecor bosons, $W^+$, $W^-$ and $Z^0$. *Nobel Lecture, 8 December* (pp. 240–287).
177. Sakharov, A. D. (1967). Violation of CP invariance, C asymmetry, and Baryon asymmetry of the universe. *Soviet Physics – JETP Letters, 5*, 24–27.
178. Sakurai, J. J. (1958). Mass reversal and weak interactions. *Nuovo Cimento, 7*, 649–660.
179. Salam, A. (1951a). Overlapping divergences and the S-matrix. *Physical Review, 82*, 217–227.
180. Salam, A. (1951b). Divergent integrals in renormalizable field theories. *Physical Review, 84*, 426–431.
181. Salam, A. (1962). Renormalizability of gauge theories. *Phys. Rev, 127*, 331–334.

182. Salam, A. (1968). Weak and electromagnetic interactions. In N. Svartholm (Ed.), *Elementary Particle Theory, Proceedings of the 8th Nobel Symposium*, Almqvist and Wiksell, Stockholm.

183. Salam, A. (1980). Grand unification and fundamental forces. *Reviews of Modern Physics, 52*, 525–538.

184. Salam, A., & Strathdee, J. (1974a). Supersymmetry and non-abelian gauges. *Physics Letters, 51B*, 353–355. Reprinted in Ferrara (1987).

185. Salam, A., & Strathdee, J. (1974b). Supergauge Transformations. *Nuclear Physics, B76*, 477–482.

186. Salam, A., & Strathdee, S. (1975b). Feynman rules for superfields. *Nuclear Physics, B86*, 142–152.

187. Salam, A., & Ward, J. (1959). Weak and electromagnetic interactions. *Nuovo Cimento, 11*, 568–577.

188. Salam, A., & Ward, J. (1961). On a gauge theory of elementary interactions. *Nuovo Cimento, 19*, 165–170.

189. Salam, A., & Ward, J. (1964). Electromagnetic and weak interactions. *Physics Letters, 13*, 168–170.

190. Scherk, J., & Schwarz, J. H. (1974). Dual models for non-hadrons. *Nuclear Physics, B81*, 118–144.

191. Schwarz, J. H. (1997). Lectures on superstrings and m-theory. *Nuclear Physics Supp., B55*, 1–32, hep–th/9607201.

192. Schwinger, J. (1948). On Quantum-electrodynamics and the magnetic moment of the electron. *Physical Review, 73*, 416.

193. Schwinger, J. (1951). On gauge invariance and vacuum polarization. *Physical Review, 82*, 664–679.

194. Schwinger, J. (1951a). On the Green's functions of quantized fields. I. *Proceedings of the National academy of Sciences of the United States of America, 37*, 452–455.

195. Schwinger, J. (1951b). The theory of quantized fields. I. *Physical Review, 82*, 914–927.

196. Schwinger, J. (1953). The theory of quantized fields. II, III. *Physical Review, 91*, 713–728, 728–740.

197. Schwinger, J. (1954). The theory of quantized fields. V. *Physical Review, 93*, 615–628.

198. Schwinger, J. (Editor) (1958a). *Selected Papers on Quantum Electrodynamics*. New York: Dover.

199. Schwinger, J. (1958b). A theory of fundamental interactions. *Annals of Physics, 2*, 404–434.

200. Schwinger, J. (1962a). Gauge invariance and mass. II. *Physical Review, 128*, 2425–2429.

201. Schwinger, J. (1972). Relativistic quantum field theory. In Nobel Lectures, Physics 1963–1970, 11 Dec 1965. Amsterdam: Elsevier.

202. Schwinger, J. (1973a). A report on quantum electrodynamics. In J. Mehra (Ed.), *The physicist's conception of nature*. Dordrecht-Holland: D. Reidel Publishing Company.

203. Shaw, R. (1955). *The problem of particle types and other contributions to the theory of elementary particles*. Ph.D. Thesis, Cambridge University.

204. Stelle, K. S. (1977). Renormalization of higher-derivative quantum gravity. *Physical Review, D16*, 953–969.

205. Stelle, K. S. (2001). Revisiting supergravity and super Yang-Mills renormalization. In J. Lukierski & J. Rembielinski (Eds.), *Proceedings of the 37th Karpacz Winter School of Theoretical Physics*, Feb 2001. hep-th/0203015v1.

206. Stelle, K. S. (2012). String theory, unification and quantum gravity. In *6th Aegean Summer School, "Quantum Gravity and Quantum Cosmology"*, 12–17 Sept 2011, Chora, Naxos Island. hep-th/1203.4689v1.

207. Streater, R. F. (1985). Review of renormalization by E. B. Manoukian. *Bulletin of London Mathematical Society, 17*, 509–510.

208. Strominger, A., & Fava, G. (1996). Microscopic origin of the 'Bekenstein-Hawking Entropy'. *Physics Letters, B379*, 99–104.

209. Stueckelberg, E. C. G., & Peterman, A. (1953). La Normalisation des Constantes dans la Théorie des Quanta. *Helvetica Physica Acta, 26*, 499–520.

210. Sudarshan, E. C. G., & Marshak, R. (1958). Chirality invariance and the universal fermi interaction. *Physical Review, 109*, 1860–1862.
211. Susskind, L. (1970). Dual symmetric theory of hadrons. I. *Nuovo Cimento, 69A*, 457–496.
212. Susskind, L. (1995). The world as a hologram. *Journal of Mathematics and Physics, 36*, 6377–6396.
213. Symanzik, K. (1970). Small distance behaviour in field theory and power counting. *Communications in Mathematical Physics, 18*, 227–246.
214. Symanzik, K. (1971). Small distance behavior in field theory. In G. Höhler (Ed.), *Springer tracts in modern physics* (Vol. 57). New York: Springer.
215. symanzik, k. (1971). small distance behaviour analysis in field theory and Wilson expansions. *Communications in Mathematical Physics, 23*, 49–86.
216. 't Hooft, G. (1971a). Renormalizable of massless Yang-Mills fields. *Nuclear Physics, B33*, 173–199.
217. 't Hooft, G. (1971b). Renormalizable Lagrangians for massive Yang-Mills fields. *Nuclear Physics, B35*, 167–188.
218. 't Hooft, G. (1973). Dimensional regularization and the renormalization group. *Nuclear Physics, B61*, 455–468.
219. 't Hooft, G. (1987). Can spacetime be probed below string size? *Physics Letters, B198*, 61–63.
220. 't Hooft, G. (1995). Black holes and the dimensional reduction of spacetime. In Lindström, U. (Ed.), *The Oskar Klein Centenary, 19–21 Sept 1994, Stockholm, (1994)*, World Scientific. See also: "Dimensional Reduction in Quantum gravity", Utrecht preprint THU-93/26 (gr-qg/9310026).
221. 't Hooft, G., & Veltman, M. J. G. (1972). Regularization and Renormalization of gauge fields. *Nuclear Physics, B44*, 189–213.
222. Thirring, W. E. (Ed.). (2005). *The stability of matter: From atoms to stars: Selecta of Elliott H. Lieb* (4th ed.). Berlin: Springer.
223. Thorn, C. B. (1991). Reformulating string theory with the $1/N$ expansion. In *International A. D. Sakharov Conference on Physics*, Moscow (pp. 447–454). (hep-th/9405069).
224. Tomboulis, E. T. (1984). Unitarity in higher-derivative quantum gravity. *Physical Review Letters, 52*, 1173–1176.
225. Tomonaga, S. (1972). Development of quantum electrodynamics: Personal recollections. In *Nobel Lectures, Physics 1963–1970*, 6 May 1966, Amsterdam: Elsevier Publishing Company.
226. Tomonaga, S.-I. (1997). *The story of spin*. Chicago: University of Chicago Press. Translated by T. Oka.
227. Townsend, P. K. (1995). The eleven-dimensional supermembrane revisited. *Physics Letters, B350*, 184–187, hep-th/9501068.
228. Utiyama, R., & Sakamoto, J. (1976). Canonical quantization of non-abelian gauge fields. *Progress of Theoretical Physics, 55*, 1631–1648.
229. Vanyashin, V. S., & Terentyev, M. V. (1965). *Soviet Physics JETP, 21*, 375–380. Original Russian Version: *Zhurnal Experimental'noi i Teoreticheskoi Fiziki, 48*, 565 (1965)
230. Veltman, M. J. G. (1981). The infrared-ultraviolet connection. *Acta Physica Polonica, B12*, 437.
231. Veneziano, G. (1968). Construction of a crossing-symmetric regge-behaved amplitude for linearly rising trajectories. *Nuovo Cimento, 57A*, 190–197.
232. Volkov, D. V., & Akulov, V. P. (1973). Is the neutrino a goldstone particle. *Physics Letters, 46B*, 109–110. Reprinted in Ferrara (1987).
233. Waller, I. (1930a). Die Streuung von Strahlung Durch Gebundene und Freie Elektronen Nach der Diracshen Relativistischen Mechanik. *Zeitschrift fur Physik, 61*, 837–851.
234. Waller, I. (1930b). Bemerkung über die Rolle der Eigenenergie des Elektrons in der Quantentheorie der Strahlung. *Zeitschrift fur Physik, 62*, 673–676.
235. Weinberg, S. (1960). High-energy behavior in quantum field theory. *Physical Review, 118*, 838–849.
236. Weinberg, S. (1967). A model of leptons. *Physical Review Letters, 19*, 1264–1266.

237. Weinberg, S. (1973). New approach to renormalization group. *Physical Review, D8*, 3497–3509.
238. Weinberg, S. (1980). Conceptual foundations of the unified theory of weak and electromagnetic interactions. *Reviews of Modern Physics, 52*, 515–523.
239. Weisskopf, V. F. (1934). Über die Selbstenergie des Elektrons. *Zeitschrift fur Physik, 89*, 27–39.
240. Weisskopf, V. F. (1939). On the self-energy of the electromagnetic field of the electron. *Physical Review, 56*, 72–85.
241. Weisskopf, V. F. (1979). Personal impressions of recent trends in particle physics. CERN Ref.Th. 2732, CERN, Geneva, Aug 1979. [Also presented at the 17 International School of Subnuclear Physics "Ettore Majorana", Erice, 31 July-11 August, 1979. pp.1–9, (ed. A. Zichichi), Plenum, New York,1982].
242. Weisskopf, V. F. (1980). Growing up with field theory, and recent trends in particle physics. "The 1979 Bernard Gregory Lectures at CERN", 29 pages, CERN, Geneva.
243. Wess, J., & Zumino, B. (1974a). Supergauge transformations in four dimensions. *Nuclear Physics, B70*, 39–50. Reprinted in Ferrara (1987).
244. Witten, E. (1995). String theory dynamics in various dimensions. *Nuclear Physics, B443*, 85–126, hep–th/9503124.
245. Witten, E. (1998). Anti-de-Sitter space and holography. *Advances in Theoretical and Mathematical Physics, 2*, 253–291. (hep-th/9802150).
246. Wu, C. S. et al. (1957). Experimental tests of parity conservation in beta decay. *Physical Review, 105*, 1413–1415.
247. Yang, C. N., & Mills, R. L. (1954). Conservation of isotopic spin and isotopic gauge invariance. *Physical Review, 96*, 191–195.
248. Yoneya, T. (1974). Connection of dual models to electrodynamics and gravidynamics. *Progress of Theoretical Physics, 51*, 1907–1920.
249. Yukawa, H. (1935). On the interaction of elementary particles. I. *Proceedings of the Physical Mathematical Society Japan, 17*, 48–57. Reprinted in "Foundations of Nuclear Physics", (ed. R. T. Beyer), Dover Publications, Inc. New York (1949).
250. Zeidler, E. (2009). *Quantum field theory II: Quantum electrodynamics*. Berlin: Springer.
251. Zimmermann, W. (1969). Convergence of Bogoliubov's method of renormalization in momentum space. *Communications in Mathematical Physics, 15*, 208–234.

# Recommended Reading

1. Becker, K, Becker, M., & Schwarz, J. H. (2006). *String theory and M-theory: A modern introduction*. Cambridge: Cambridge University Press.
2. Davies, P. (Ed.). (1989). *The new physics*. Cambridge: Cambridge University Press.
3. DeWitt, B. (2014). *The global approach to quantum field theory*. Oxford: Oxford University Press.
4. Manoukian, E. B. (1983). *Renormalization*. New York/London/Paris: Academic Press.
5. Manoukian, E. B. (2006). *Quantum theory: A wide spectrum*. Dordrecht: Springer.
6. Manoukian, E. B. (2016). *Quantum field theory II: Introductions to quantum gravity, supersymmetry, and string theory*. Dordrecht: Springer.
7. Martin, P. C., & Glashow, S. L. (2008). Julian Schwinger 1918–1994": A Bio-graphical memoir. *National Academy of Sciences*, Washington, DC, Copyright 2008.
8. Oriti, D. (Ed.), (2009). *Approaches to quantum gravity*. Cambridge: Cambridge University Press.
9. Schweber, S. S. (2008). Quantum field theory: From QED to the standard model In M. Jo Nye (Ed.), *The Cambridge history of science. The modern physical and mathematical sciences* (Vol. 5, pp. 375–393). Cambridge: Cambridge University Press.

10. Streater, R. F. (1985). Review of Renormalization by E. B. Manoukian. *Bulletin of London Mathematical Society, 17*, 509–510.
11. Weinberg. S. (2000). *The quantum theory of fields. III: supersymmetry*. Cambridge: Cambridge University Press.
12. Weisskopf, V. F. (1980). Growing up with field theory, and recent trends in particle physics. The 1979 Bernard Gregory Lectures at CERN, 29 pages. "Personal Impressions of Recent Trends in Particle Physics". CERN Ref. Th. 2732 (1979). CERN, Geneva.
13. Zeidler, E. (2009). *Quantum field theory II: Quantum electrodynamics*. Berlin: Springer. pp. 972–975.

# Chapter 2
# Preliminaries

This preliminary chapter deals with basic tools needed for dealing with quantum field theory. It begins with Wigner's theory of symmetry transformations by showing how symmetry is implemented in the quantum world via unitary or anti-unitary operators. This subject matter will be also important in deriving commutation (anti-commutation) rules between symmetry generators such as in developing the Poincaré algebra of spacetime transformations including those in supersymmetric field theories which also involve fermion-boson exchanges. Some properties of the Dirac equation and related aspects are summarized in Appendix I, at the end of the book, for the convenience of the reader. Various representations of the gamma matrices are spelled out, however, in the present chapter which turn out to be important in modern field theory, and the concept of a Majorana spinor is also introduced. Special emphasis is put in the remaining sections on functional differentiations and functional integrations involving Grassmann variables as well. Functional Fourier transforms are introduced which clearly show the intimate connection that there exists between the functional differential formalism pioneered by Julian Schwinger and the functional integral formalism pioneered by Richard Feynman. The last section on delta functionals, makes this connection even more transparent and shows the simplicity of the functional formalism in general, whether it is in differential or integral forms. It is, however, relatively easier to functionally differentiate than to deal with continual functional integrals. Both formalisms are used in this book for greater flexibility and for a better understanding of quantum field theory.

© Springer International Publishing Switzerland 2016
E.B. Manoukian, *Quantum Field Theory I*, Graduate Texts in Physics,
DOI 10.1007/978-3-319-30939-2_2

## 2.1  Wigner's Symmetry Transformations in the Quantum World

Invariance of physical laws under some given transformations leads to conservation laws and the underlying transformations are referred to as symmetry transformations. For example, invariance under time translation (e.g., by setting one's clocks back by a certain amount) or under space translation (by shifting the origin of one's coordinate system) lead, respectively, to energy and momentum conservations. Invoking invariance properties in developing a dynamical theory, conveniently narrows down one's choices in providing the final stages of a theory. Obviously, not all transformations of a given physical system are symmetry transformations. But invoking the invariance of a system under a set of given transformations may provide the starting point in describing a dynamical theory, and one may then consistently modify the theory to take into account any symmetry breaking. One may also have symmetry breaking spontaneously which will be discussed later.

A celebrated analysis of Eugene Wigner in the thirties[1] originating on symmetry under rotations in space, spells out the nature of the transformations implemented on elements of a Hilbert space. This result is of central importance for doing quantum physics[2] and quantum field theory.

The physical situation that presents itself is the following. One prepares a system in a state $|\psi\rangle$. The question then arises as to what is the probability of finding the system in some state $|\phi\rangle$ if $|\psi\rangle$ is what one initially has?.[3] The latter is given by $|\langle\phi|\psi\rangle|^2$.

If $|\psi'\rangle$, $|\phi'\rangle$ denote the states $|\psi\rangle$, $|\phi\rangle$, resulting from a symmetry transformation, then invariance of the above probability under such a transformation means that

$$|\langle\phi'|\psi'\rangle|^2 = |\langle\phi|\psi\rangle|^2, \tag{2.1.1}$$

and $\{|\psi'\rangle, |\phi'\rangle\}$ provide an equivalent description as $\{|\psi\rangle, |\phi\rangle\}$. One may scale $|\psi'\rangle, |\phi'\rangle, |\psi\rangle, |\phi\rangle$ by arbitrary phase factors without changing the relevant probabilities given in (2.1.1). Although such overall phase factors are not important, the relative phases occurring when considering addition of such states are physically relevant with far reaching consequences.[4]

---

[1] See, e.g., his book Wigner [3]. See also Wigner [4].

[2] For intricacies of Wigner's Symmetry Transformations in quantum mechanics, see, e.g., Manoukian [1], pp. 55–65.

[3] More generally one may also have initially a mixture described by a density operator.

[4] One should also consider unit rays in this discussion and in the subsequent definitions but we will not go into these points here.

To understand Wigner's Theorem, one needs to define the following operators:

1. An operator $L$ is called linear or anti-linear, if for any $|\psi\rangle, |\phi\rangle, [a\,|\psi\rangle + b\,|\phi\rangle]$, where $a, b$ are, in general, complex numbers, then

$$L[a\,|\psi\rangle + b\,|\phi\rangle] = aL|\psi\rangle + bL|\phi\rangle, \qquad (2.1.2)$$

or

$$L[a\,|\psi\rangle + b\,|\phi\rangle] = a^*L|\psi\rangle + b^*L|\phi\rangle, \qquad (2.1.3)$$

respectively, with * denoting complex conjugation.

2. A linear or anti-linear operator $U$ is called unitary or anti-unitary, if

$$\langle U\phi | U\psi \rangle = \langle \phi | \psi \rangle, \qquad (2.1.4)$$

or

$$\langle U\phi | U\psi \rangle = \langle \phi | \psi \rangle^* = \langle \psi | \phi \rangle, \qquad (2.1.5)$$

respectively.

## 2.1.1 Wigner's Symmetry Transformations

Under a symmetry transformation, there exists a unitary or anti-unitary operator $U$ such that (2.1.2)/(2.1.4) or (2.1.3)/(2.1.5) hold true, respectively, with

$$|\psi'\rangle = U\,|\psi\rangle, \quad |\phi'\rangle = U\,|\phi\rangle, \qquad (2.1.6)$$

$$|[a\psi + b\phi]'\rangle = U[a\,|\psi\rangle + b\,|\phi\rangle]. \qquad (2.1.7)$$

Continuous transformations, for which one may also consider infinitesimal ones close to the identity operation, such as space or time translations and rotations, are implemented by unitary operators since the identity operator $I$ itself is trivially a unitary one. An infinitesimal change of a parameter $\delta\xi$, under consideration, is implemented by a given operator $G$, referred to as the generator of symmetry transformation, and the corresponding unitary operator, for such an infinitesimal transformation, would take the form

$$U = I + i\,\delta\xi G. \qquad (2.1.8)$$

The unitarity condition $U^\dagger U = I = UU^\dagger$ implies that $\delta\xi\, G$ is a self-adjoint operator[5]:

$$(\delta\xi\, G)^\dagger = \delta\xi\, G. \tag{2.1.9}$$

For example, for an infinitesimal space translation $\delta\mathbf{x}$, via the momentum operator $\mathbf{P}$ associated with the system in consideration, the corresponding unitary operator is given by

$$U|_{\text{space}} = I + i\,\delta\mathbf{x}\cdot\mathbf{P}. \tag{2.1.10}$$

For infinitesimal time translation $\delta\tau$, via the Hamiltonian $H$ of the system, one has

$$U|_{\text{time}} = I - i\delta\tau H. \tag{2.1.11}$$

Moreover for a rotation of a coordinate system by an infinitesimal angle $\delta\xi$ about a unit vector $\mathbf{n}$, via the angular operator $\mathbf{J}$,

$$U|_{\text{rotation}} = I + i\,\delta\xi\mathbf{n}\cdot\mathbf{J}. \tag{2.1.12}$$

The latter may be more conveniently rewritten by introducing parameters $\delta\omega^{ij}$, and rewrite $J^i$ as

$$J^i = \frac{1}{2}\varepsilon^{ijk}J^{jk}, \qquad \delta\omega^{ij} = \varepsilon^{ijk}n^k\delta\xi, \qquad i,j,k = 1,2,3, \tag{2.1.13}$$

where a summation over repeated indices is assumed, and $\varepsilon^{ijk}$ is totally anti-symmetric, with $\varepsilon^{123} = +1$. We may then rewrite (2.1.12) as

$$U|_{\text{rotation}} = I + \frac{i}{2}\delta\omega^{ij}J^{ij}. \tag{2.1.14}$$

The generators $H$, $\mathbf{P}$, and $\mathbf{J}$ of the transformations are self-adjoint.

In a relativistic setting, in which this book is based, these spacetime transformations may be combined by introducing, in the process, additional operators to this set of operators, denoted by $J^{0i}$, $i = 1,2,3$, which impart a frame with an infinitesimal velocity change $\delta\mathbf{v}$. The unitary operator corresponding to these infinitesimal spacetime transformations then takes the elegant form

$$U = 1 + i\left(\delta b_\mu P^\mu + \frac{1}{2}\delta\omega_{\mu\nu}J^{\mu\nu}\right), \qquad \delta\omega_{\mu\nu} = -\delta\omega_{\nu\mu}, \qquad \mu,\nu = 0,1,2,3, \tag{2.1.15}$$

[5]One may conveniently, in general, absorb $\delta\xi$ in $G$.

with[6] $P^0 = H$,

$$\delta b_\mu = (-\delta\tau, \delta\mathbf{x}), \quad \delta\omega^{ij} = \varepsilon^{ijk}n^k\delta\xi, \quad \delta\omega_{0i} = \delta v^i, \quad \delta\omega_{00} = 0. \quad (2.1.16)$$

First we have to learn how the labels attached to an event in different, so-called, inertial frames are related. This will be taken up in the next section. The transformation rules relating this different labeling, are referred to as Lorentz transformations in Minkowski spacetime. On the other hand, the genertors $P^\mu, J^{\mu\nu}$ satisfy basic commutations relations and form an algebra in the sense that the commutator of two generators is equal to a linear combination of the generators in question. The underlying algebra is called the Poincaré algebra and will be studied in Sect. 4.2.

In incorporating supersymmetry into physics, with an inherited symmetry existing between fermions and bosons, one introduces a generator[7] of fermionic type, that is, it carries a spinor index, referred to as a supercharge operator, and the Poncaré algebra is extended to a larger one including anti-commutation relations as well. This algebra is referred to as Super-Poincaré algebra.[8]

Typical discrete transformations, are provided by space P or time T reflections (parity and time reversal) or charge conjugation C, for which a particle is replaced by its antiparticle. A key criterion for finding out if T, for example, should be unitary or anti-unitary is to eliminate the choice which would lead to the inconsistent result that a Hamiltonian is unbounded from below. This is inferred as follows. Under an infinitesimal time translation $\delta\tau$, via the Hamiltonian $H$ of the system under consideration, followed by a time-reversal applied to it is equivalent to a time reversal followed by a time translation, i.e., we have the equality

$$[1 - i\,\delta\tau H]\,T = T\,[1 + i\,\delta\tau(H + \varphi)], \quad (2.1.17)$$

up to a phase factor $\varphi$. If T were unitary, this gives $HT = -T(H + \varphi)$. For a state $|\eta\rangle$ with positive and arbitrary large energy $E$, there would correspond a state $T|\eta\rangle$ with energy $-(E+\varphi)$ and the Hamiltonian would be unbounded from below. That is, T is to be implemented by an anti-unitary operator which, in the process of applying time reversal, it would complex conjugate the i factor on the right-hand side of (2.1.17). A similar analysis applied to P leads one to infer that it is to be implemented by a unitary operator. On the other hand, C, unlike P, and T, is not involved with space and time reflections, and may be implemented by a unitary operator. Accordingly, the product CPT, in turn, is implemented by an anti-unitary

---

[6]With the Minkowski metric adopted in this book $[\eta_{\mu\nu}] = \text{diag}[-1, 1, 1, 1]$, $b^i = b_i$, $b^0 = -b_0$.

[7]One may also introduce more than one such operator.

[8]This will be studied in the accompanying book Manoukian [2]: Quantum Field Theory II: Introductions to Quantum Gravity, Supersymmetry, and string Theory. Minkowski spacetime is, in turn, extended to what has been called superspace.

operator. The corresponding symmetry, embodied in the so-called  CPT  Theorem, will be dealt with in Chap. 4.[9]

## 2.2  Minkowski Spacetime: Common Arena of Elementary Particles

A fundamental principle which goes to the heart of special relativity is the equivalence of inertial frames in describing physical laws. By this it is meant, in particular, that dynamical equations take the same form in such frames up to a mere relabeling of the variables of the underlying theories. Because of such relabeling of the variables, these equations are said to transform covariantly as one goes from one inertial frame to another and the corresponding rules of transformations are, in general, called Lorentz transformations.

An event labeled by $x = (x^0, \mathbf{x})$, where $x^0 = t$ (time), in one frame, say, F will be labeled by $x' = (x'^0, \mathbf{x}')$ in some other frame, say, F$'$. The transformation rules which connect this different labeling of the same event are the so-called Lorentz transformations. One inertial frame F$'$ may be moving with a uniform velocity with respect to another frame F, as determined in F, with a possible orientation of the cartesian space coordinate axes of F$'$ as also determined in F at some initial time $x^0 = 0$. If an $x^\mu = 0$ reading in F corresponds to an $x'^\mu = 0$ reading in F$'$, then the Lorentz transformations are called homogeneous ones. Otherwise, they are called inhomogeneous. In the former case, the origins of the space coordinate axes set up in F and F$'$ coincide at time readings $x^0 = 0$, $x'^0 = 0$ by observers located at the corresponding origins of these respective coordinate systems.

We use the notation $(x^\mu) = (x^0, \mathbf{x}) \equiv x$, $\mu = 0, 1, 2, 3$, and our Minkowski metric is defined by $[\eta_{\mu\nu}] = \mathrm{diag}\,[-1, 1, 1, 1]$, $[\eta^{\mu\nu}] = \mathrm{diag}\,[-1, 1, 1, 1]$. Authors who have also used this signature for the metric include, Julian Schwinger, Steven Weinberg,.... Others may use the minus of ours. Surprisingly, Paul Dirac has used both signatures at different times of his career. It doesn't matter which one to use. It is easy to keep track of the relative minus sign. In our notation, $x_0 = -x^0$, $x_i = x^i$ for $i = 1, 2, 3$, $x^\mu = \eta^{\mu\nu}x_\nu$, $x_\mu = \eta_{\mu\nu}x^\nu$, $[\eta^{\mu\nu}\eta_{\nu\sigma}] = [\delta^\mu_{\ \sigma}] = \mathrm{diag}\,[1, 1, 1, 1]$.

Under a homogeneous Lorentz transformation,

$$x'^\mu = \Lambda^\mu_{\ \nu} x^\nu, \tag{2.2.1}$$

---

[9]The reader may wish to consult Manoukian [1], pp. 55–65, 112–115, where additional details, and proofs, are spelled out in quantum theory.

where $\Lambda^{\mu}{}_{\nu}$ is independent of $x$, $x'$. Before we spell out the structure of the matrix $(\Lambda^{\mu}{}_{\nu})$, we note that

$$\frac{\partial x'^{\mu}}{\partial x^{\nu}} = \Lambda^{\mu}{}_{\nu}, \tag{2.2.2}$$

and hence from the chain rule

$$\partial_{\nu} = \Lambda^{\mu}{}_{\nu}\,\partial'_{\mu}, \qquad \left(\partial_{\nu} \equiv \frac{\partial}{\partial x^{\nu}}\right). \tag{2.2.3}$$

Quite generally, we have the Lorentz invariant property

$$(x'^{\mu} - y'^{\mu})\,\eta_{\mu\nu}\,(x'^{\nu} - y'^{\nu}) = (x^{\mu} - y^{\mu})\,\eta_{\mu\nu}\,(x^{\nu} - y^{\nu}), \tag{2.2.4}$$

from which one may infer that

$$\Lambda^{\mu}{}_{\rho}\,\eta_{\mu\nu}\,\Lambda^{\nu}{}_{\lambda} = \eta_{\rho\lambda}, \quad (\Lambda^{-1})^{\rho\nu} = \Lambda^{\nu\rho}, \tag{2.2.5}$$

$$\partial'_{\mu} = \Lambda_{\mu}{}^{\nu}\,\partial_{\nu}, \qquad \square = \square', \qquad (\square \equiv \partial^{\mu}\partial_{\mu}). \tag{2.2.6}$$

The Poincaré (inhomogeneous Lorentz) transformations in Minkowski spacetime $x \rightarrow x'$ are defined by

$$x'^{\mu} = \Lambda^{\mu}{}_{\nu}\,x^{\nu} - b^{\mu}. \tag{2.2.7}$$

Consistency of such a transformation requires that its structure remains the same under subsequent transformations leading to group properties spelled out below. For a subsequent transformation $x' \rightarrow x''$, we may use (2.2.7) to write

$$x''^{\lambda} = \Lambda'^{\lambda}{}_{\mu}\,x'^{\mu} - b'^{\lambda}, \tag{2.2.8}$$

or

$$x''^{\lambda} = \left(\Lambda'^{\lambda}{}_{\mu}\,\Lambda^{\mu}{}_{\nu}\right)x^{\nu} - \left(\Lambda'^{\lambda}{}_{\mu}\,b^{\mu} + b'^{\lambda}\right). \tag{2.2.9}$$

Thus a Poincaré transformation may be specified by a pair $(\Lambda, b)$, with $\Lambda = (\Lambda^{\mu}{}_{\nu})$, $b = (b^{\nu})$, satisfying the group properties:

1. Group multiplication: $(\Lambda', b')(\Lambda, b) = (\Lambda'\Lambda, \Lambda'b + b')$.
2. Identity $(I, 0)$: $(I, 0)(\Lambda, b) = (\Lambda, b)$.
3. Inverse $(\Lambda, b)^{-1} = (\Lambda^{-1}, -\Lambda^{-1}b)$: $(\Lambda, b)^{-1}(\Lambda, b) = (I, 0)$.
4. Associativity Rule: $(\Lambda_3, b_3)[(\Lambda_2, b_2)(\Lambda_1, b_1)] = [(\Lambda_3, b_3)(\Lambda_2, b_2)](\Lambda_1, b_1)$.

To spell out the general structure of the matrices $\Lambda^{\mu}{}_{\nu}$, we first consider spatial rotations of coordinate systems. A point $\mathbf{x}$ specified in a given 3D coordinate system will be read as $\mathbf{x}'$ in a coordinate system obtained from the first by a c.c.w rotation by an angle $\xi$ about a unit vector $\mathbf{n}$,

$$x'^{i} = R^{ij}x^{j}, \qquad R^{ij}R^{ik} = \delta^{jk}, \qquad i,j,k = 1,2,3, \tag{2.2.10}$$

with a summation over repeated indices understood. Here $R^{ik}$ are the matrix elements of the rotation matrix with

$$R^{ik} = \delta^{ik} - \varepsilon^{ijk}n^{j}\sin\xi + (\delta^{ik} - n^{i}n^{k})(\cos\xi - 1), \tag{2.2.11}$$

$\varepsilon^{ijk}$ is totally anti-symmetric with $\varepsilon^{123} = +1$.

If for a given unit vector $\mathbf{n}$, we use the notation $\mathbf{x}' = \mathbf{x}(\xi)$, then from (2.2.10), (2.2.11), it is worth noting that

$$\frac{\mathrm{d}}{\mathrm{d}\xi}\mathbf{x}(\xi) = -\mathbf{n} \times \mathbf{x}(\xi), \qquad \mathbf{x}(0) = \mathbf{x}. \tag{2.2.12}$$

The general structure of the matrices $\Lambda^{\mu}{}_{\nu}$, involving such rotations and so-called Lorentz boosts is given by

$$\Lambda^{i}{}_{j} = \quad \Lambda^{ij} = R^{ij} + (\varrho - 1)R^{ik}\frac{\beta^{k}\beta^{j}}{\beta^{2}}, \tag{2.2.13}$$

$$\Lambda^{0}{}_{0} = -\Lambda^{00} = \varrho \equiv (1 - \beta^{2})^{-1/2}, \tag{2.2.14}$$

$$\Lambda^{0}{}_{i} = \quad \Lambda^{0i} = -\varrho\beta^{i}, \tag{2.2.15}$$

$$\Lambda^{i}{}_{0} = -\Lambda^{i0} = -\varrho R^{ij}\beta^{j}, \tag{2.2.16}$$

Using rather a standard notation for $\boldsymbol{\beta}$, its physical interpretation is as follows. From the second identity in (2.2.5) it follows from the transformation law $x'^{\mu} = \Lambda^{\mu}{}_{\nu}x^{\nu}$ in (2.2.1) that $x^{\nu} = \Lambda_{\mu}{}^{\nu}x'^{\mu}$. The origin of the spacial coordinate system of the frame $F'$ is specified by $x'^{j} = 0, j = 1,2,3$. From this and (2.2.14), (2.2.15), we have $x^{0} = \varrho x'^{0}$, $x^{i} = \varrho\beta^{i}x'^{0}$. That is, $\boldsymbol{\beta}$ denotes the velocity with which the origin of the spatial coordinate system of $F'$ moves relative to the corresponding origin of F. Now put a (massive $m \neq 0$) particle at the origin of the spatial coordinate system of F, the so-called rest frame of the particle, i.e., with energy-momentum $p^{\mu} = (m, \mathbf{0})$. Then if no initial relative rotation of the spatial coordinate systems of the frames is involved, the velocity of the particle in $F'$ (the "laboratory" frame) would be simply $-\boldsymbol{\beta}$, i.e., of momentum $\mathbf{p}' = -m\varrho\boldsymbol{\beta}$. With a relative rotation of the coordinate systems, the relation $p'^{j} = \Lambda^{j}{}_{0}m$ gives $p'^{j} = R^{jk}(-m\varrho\beta^{k})$ showing simply a c.w. rotation of the earlier expression.

Equally important are the transformations (2.2.7), (2.2.13), (2.2.14), (2.2.15), (2.2.16) for infinitesimal changes $\delta\mathbf{b}$, $\delta\xi$, $\delta\boldsymbol{\beta}$. To first order,

$$\delta x^{\mu} = x^{\mu} - x'^{\mu} = \delta b^{\mu} - \delta\omega^{\mu}{}_{\nu}x^{\nu}, \tag{2.2.17}$$

where

$$\Lambda^{\mu}{}_{\nu} = \delta^{\mu}{}_{\nu} + \delta\omega^{\mu}{}_{\nu}, \qquad \delta\omega^{\mu\nu} = -\delta\omega^{\nu\mu}, \tag{2.2.18}$$

$$\delta\omega^{ij} = \varepsilon^{ijk} n^k \delta\xi, \tag{2.2.19}$$

$$\delta\omega^{i}{}_{0} = -\delta\beta^{i}, \tag{2.2.20}$$

$$\delta\omega^{0}{}_{i} = -\delta\beta^{i}, \tag{2.2.21}$$

$$\delta\omega^{0}{}_{0} = 0. \tag{2.2.22}$$

and $i, j = 1, 2, 3$; $\mu, \nu = 0, 1, 2, 3$.

When the parameters $\delta\mathbf{b}$, $\delta\xi$, $\delta\boldsymbol{\beta}$ are led continuously to go to zero, the induced transformations in the underlying vector space of particle states go over to the identity. The latter being unitary, we may refer to Wigner's Theorem (Sect. 2.1) to infer that an induced transformation is represented by a unitary rather than by an anti-unitary operator. Such transformations, in turn, generate an algebra of the corresponding generators referred to as the Poincaré algebra. The corresponding analysis will be carried out in Sect. 4.2.

Before closing this section, we note that the general structure in (2.2.5), rewritten in matrix form reads

$$\Lambda^{\mathsf{T}} \eta \Lambda = \eta, \tag{2.2.23}$$

and since $\det \eta = -1$, we may infer that for the transformations in (2.2.1) det $\Lambda = +1$. It is not $-1$ as this transformation includes neither time nor space reflections. From this and (2.2.2) we learn that the Jacobian of the transformation $x' \to x$ is given by $\det \Lambda = 1$, establishing the Lorentz invariance of the volume element in Minkowski spacetime:

$$(\mathrm{d}x) = (\mathrm{d}x'), \qquad \text{where } (\mathrm{d}x) \equiv \mathrm{d}x^0 \, \mathrm{d}x^1 \, \mathrm{d}x^2 \, \mathrm{d}x^3. \tag{2.2.24}$$

This together with the definition of a Lorentz scalar $\Phi(x)$ by the condition

$$\Phi'(x') = \Phi(x), \tag{2.2.25}$$

under the above Lorentz transformations, guarantees the invariance of integrals of the form

$$\mathscr{A} = \int (\mathrm{d}x) \, \Phi(x), \tag{2.2.26}$$

such as the action integral, and lead to the development of Lorentz invariant theories.

## 2.3   Representations of the Dirac Gamma Matrices; Majorana Spinors

The Dirac formalism for the description of the relativistic electron in a quantum setting via the Dirac equation $(\gamma^\mu \partial_\mu/i + m)\psi = 0$, is summarized in Appendix I at the end of the book. The Dirac representation of the gamma $\gamma^\mu$ matrices, in particular, are defined by

$$\gamma^0 = \begin{pmatrix} I & 0 \\ 0 & -I \end{pmatrix}, \quad \boldsymbol{\gamma} = \begin{pmatrix} \mathbf{0} & \boldsymbol{\sigma} \\ -\boldsymbol{\sigma} & \mathbf{0} \end{pmatrix}, \quad \gamma^5 \equiv i\gamma^0\gamma^1\gamma^2\gamma^3 = \begin{pmatrix} 0 & I \\ I & 0 \end{pmatrix}, \qquad (2.3.1)$$

satisfying the anti-commutation relations: $\{\gamma^\mu, \gamma^\nu\} = -2\eta^{\mu\nu}$. Various representations of the gamma matrices, satisfying the same anti-commutation relations, may be similarly defined. We here introduce two other representations which turn up to be of quite importance in modern field theory. They arise as follows.

The unitary matrix

$$G = \frac{1}{\sqrt{2}} \begin{pmatrix} I & I \\ -I & I \end{pmatrix}, \quad G^{-1} = G^\dagger = \frac{1}{\sqrt{2}} \begin{pmatrix} I & -I \\ I & I \end{pmatrix}, \qquad (2.3.2)$$

introduces via the transformation $G\gamma^\mu G^{-1}$ the following representation of gamma matrices, known as the chiral representation, in which $\gamma^5$ is diagonal

$$\gamma^0 = \begin{pmatrix} 0 & -I \\ -I & 0 \end{pmatrix}, \quad \boldsymbol{\gamma} = \begin{pmatrix} \mathbf{0} & \boldsymbol{\sigma} \\ -\boldsymbol{\sigma} & \mathbf{0} \end{pmatrix}, \quad \gamma^5 = \begin{pmatrix} I & 0 \\ 0 & -I \end{pmatrix}, \qquad (2.3.3)$$

where for simplicity we have used the same notation for the resulting $\gamma^\mu$ matrices. Note that $\boldsymbol{\gamma}$ coincides with the one in the Dirac representation. The chiral representation is important in dealing with massless particles and in the study of supersymmetry. While the unitary matrix

$$G = G^{-1} = G^\dagger = \frac{1}{\sqrt{2}} \begin{pmatrix} I & \sigma^2 \\ \sigma^2 & -I \end{pmatrix}, \qquad (2.3.4)$$

leads to the so-called Majorana representation

$$\gamma^0 = \begin{pmatrix} 0 & \sigma^2 \\ \sigma^2 & 0 \end{pmatrix}, \quad \gamma^5 = \begin{pmatrix} \sigma^2 & 0 \\ 0 & -\sigma^2 \end{pmatrix}, \qquad (2.3.5)$$

$$\gamma^1 = \begin{pmatrix} i\sigma^3 & 0 \\ 0 & i\sigma^3 \end{pmatrix}, \quad \gamma^2 = \begin{pmatrix} 0 & -\sigma^2 \\ \sigma^2 & 0 \end{pmatrix}, \quad \gamma^3 = \begin{pmatrix} -i\sigma^1 & 0 \\ 0 & -i\sigma^1 \end{pmatrix}, \qquad (2.3.6)$$

in which $\left(\gamma^{\mu}/i\right)^{*} = \left(\gamma^{\mu}/i\right)$, making the Dirac operator $\left(\gamma^{\mu}\partial_{\mu}/i + m\right)$ *real*. This representation is particularly convenient in analyzing general results in field theory such as the Spin & Statistics Connection .

If $\psi$ satisfies the Dirac equation in an external electromagnetic field $eA^{\mu}(x)$ of charge e then $\psi^{\mathscr{C}} = \mathscr{C}\overline{\psi}^{\mathsf{T}}$, where $\overline{\psi} = \psi^{\dagger}\gamma^{0}$, and $\mathscr{C}$ is the charge conjugation matrix defined by $\mathscr{C} = i\gamma^{2}\gamma^{0}$, satisfies the Dirac equation with $eA^{\mu}(x)$ replaced by $-eA^{\mu}(x)$ i.e., with the sign of the charge e reversed. (See Eq. (I.3) in Appendix I at the end of the book). In the Dirac and chiral representations the charge conjugation matrix are given by $\left((\sigma^{2})^{\mathsf{T}} = -\sigma^{2}\right)$

$$
\mathscr{C}_{\text{Dirac}} = \begin{pmatrix} 0 & -i\sigma^{2} \\ -i\sigma^{2} & 0 \end{pmatrix}, \quad \mathscr{C}_{\text{chiral}} = \begin{pmatrix} -i\sigma^{2} & 0 \\ 0 & i\sigma^{2} \end{pmatrix}. \tag{2.3.7}
$$

We note that $\gamma^{5}\mathscr{C}$, $\gamma^{5}\gamma^{\mu}\mathscr{C}$, together with $\mathscr{C}$, are anti-symmetric matrices.

Some properties of the charge conjugation matrix $\mathscr{C} = i\gamma^{2}\gamma^{0}$, in general, are: ($\sharp$ stands for any of the operations $(.)^{\dagger}, (.)^{-1}, (.)^{\mathsf{T}}$)

$$
\mathscr{C}^{\sharp} = -\mathscr{C}, \quad \mathscr{C}^{-1}\gamma^{\mu}\mathscr{C} = -(\gamma^{\mu})^{\mathsf{T}}, \quad [\mathscr{C}, \gamma^{5}] = 0, \quad \mathscr{C}^{-1}\gamma^{5}\gamma^{\mu}\mathscr{C} = (\gamma^{5}\gamma^{\mu})^{\mathsf{T}}, \tag{2.3.8}
$$

$$
\mathscr{C}^{-1}[\gamma^{\mu}, \gamma^{\nu}]\mathscr{C} = -([\gamma^{\mu}, \gamma^{\nu}])^{\mathsf{T}}, \quad \mathscr{C}^{-1}\gamma^{5}[\gamma^{\mu}, \gamma^{\nu}]\mathscr{C} = -\left(\gamma^{5}[\gamma^{\mu}, \gamma^{\nu}]\right)^{\mathsf{T}}. \tag{2.3.9}
$$

Once a charge matrix $\mathscr{C}$ has been defined, one may define a Majorana spinor $\theta$, not to be confused with the Majorana representation of the gamma matrices, by the condition $\theta = \theta^{\mathscr{C}}$, that is

$$
\theta = \mathscr{C}\overline{\theta}^{\mathsf{T}}, \qquad \overline{\theta} = -\theta^{\mathsf{T}}\mathscr{C}^{-1} = -(\mathscr{C}\theta)^{\mathsf{T}}. \tag{2.3.10}
$$

From these definitions, one may infer the general structure of a Majorana spinor, for example, in the *chiral* representation, to be of the form:

$$
\theta = \begin{pmatrix} \theta_{1} \\ \theta_{2} \\ \theta_{3} \\ \theta_{4} \end{pmatrix} = \begin{pmatrix} \theta_{4}^{*} \\ -\theta_{3}^{*} \\ \theta_{3} \\ \theta_{4} \end{pmatrix}. \tag{2.3.11}
$$

## 2.4   Differentiation and Integration with Respect to Grassmann Variables

Consider $n$ real Grassmann variables $\rho_1, \ldots \rho_n$, i.e., variables that satisfy $\{\rho_i, \rho_k\} = 0$, pairwise, which, in particular, implies that for any $k$: $\rho_k^2 = 0$.

The left-hand derivative with respect to a variable $\rho_j$ of any product $\rho_{i_1} \ldots \rho_{i_k}$ is defined by

$$\frac{\partial}{\partial \rho_j} \rho_{i_1} \ldots \rho_{i_k} = (-1)^{\delta_j} \rho_{i_1} \ldots \widehat{\rho}_j \ldots \rho_{i_k}, \qquad (2.4.1)$$

where $\delta_j$ denotes the position of $\rho_j$ from the left $-1$, in the product $\rho_{i_1} \ldots \rho_{i_k}$, *and* $\rho_j$ is omitted on the right-hand side of the equation as indicated by the "hat" sign on it. One may also define a right-hand derivative, taking a derivative from the right, with a corresponding rule.

In particular note that,

$$\frac{\partial}{\partial \rho_k} \rho_i \rho_j = \delta_{ik} \rho_j - \delta_{jk} \rho_i, \qquad (2.4.2)$$

and

$$\left\{ \frac{\partial}{\partial \rho_j}, \rho_k \right\} = \delta_{jk}. \qquad (2.4.3)$$

One may define the differential operator $\mathrm{d} = \sum_{k=1}^{n} \mathrm{d}\rho_k \partial / \partial \rho_k$ satisfying the rule

$$\mathrm{d}(\rho_i \rho_j) = \mathrm{d}\rho_i \, \rho_j + \rho_i \, \mathrm{d}\rho_j. \qquad (2.4.4)$$

On the other hand by using the rule in (2.4.2) in applying the differential operator $\mathrm{d}$ gives

$$\mathrm{d}(\rho_i \rho_j) = \mathrm{d}\rho_i \, \rho_j - \mathrm{d}\rho_j \, \rho_i, \qquad (2.4.5)$$

which upon comparison with (2.4.4) gives

$$\mathrm{d}\rho_j \, \rho_i = -\rho_i \, \mathrm{d}\rho_j. \qquad (2.4.6)$$

To extend the above rules for differentiations with respect to complex variables, one may proceed as follows. Given two real Grassmann variables $\rho_R, \rho_I$ satisfying $\{\rho_R, \rho_I\} = 0$, one may define a complex Grassmann variable $\rho = \rho_R + \mathrm{i}\,\rho_I$. Using the notation $\rho^*$ for the complex conjugate of $\rho$, the following anti-commutation rules emerge

$$\{\rho, \rho\} = 0, \quad \{\rho, \rho^*\} = 0, \quad \{\rho^*, \rho^*\} = 0. \qquad (2.4.7)$$

Imposing a reality restriction on the product

$$\rho^* \rho = -\mathrm{i}\,\rho_{\mathrm{I}}\,\rho_{\mathrm{R}},\tag{2.4.8}$$

i.e., $(\rho^* \rho)^* = (\rho^* \rho)$, leads to the following definition

$$(\rho_{\mathrm{I}}\,\rho_{\mathrm{R}})^* = -\rho_{\mathrm{I}}\,\rho_{\mathrm{R}} \quad \text{or} \quad (\rho_{\mathrm{I}}\,\rho_{\mathrm{R}})^* = \rho_{\mathrm{R}}\,\rho_{\mathrm{I}},\tag{2.4.9}$$

showing that *complex conjugation* of the product of two real Grassmann variables *reverses* their order. This in turn implies that for complex (or real) Grassmann variables $\rho_1, \ldots, \rho_n$.

$$(\rho_1 \ldots \rho_n)^* = \rho_n^* \ldots \rho_1^*.\tag{2.4.10}$$

One may also define two sets of derivatives with respect to variables $\rho_i$, $\rho_i^*$, as before, and note that

$$\left\{ \frac{\partial}{\partial \rho_i}, \frac{\partial}{\partial \rho_j} \right\} = 0, \quad \left\{ \frac{\partial}{\partial \rho_i}, \frac{\partial}{\partial \rho_j^*} \right\} = 0.\tag{2.4.11}$$

We now consider integrations with respect to real Grassmann variables. Due to the property $\rho_i^2 = 0$, one has to investigate only the meanings of the following two integrals, for a given $i$,

$$\int \mathrm{d}\rho_i, \quad \int \mathrm{d}\rho_i\,\rho_i.\tag{2.4.12}$$

Imposing translational invariance of these integrals under $\rho_i \to \rho_i + \alpha_i = \rho_i'$, with $\alpha_i$ anti-commuting with $\rho_i$, simply means that

$$\int \mathrm{d}\rho_i' = \int \mathrm{d}\rho_i, \quad \int \mathrm{d}\rho_i'\,\rho_i' = \int \mathrm{d}\rho_i\,\rho_i.\tag{2.4.13}$$

In particular, translation invariance gives

$$\int \mathrm{d}\rho_i'\,\rho_i' = \int \mathrm{d}\rho_i\,\rho_i + \left( \int \mathrm{d}\rho_i \right) \alpha_i,\tag{2.4.14}$$

for arbitrary $\alpha_i$. We may thus infer that

$$\int \mathrm{d}\rho_i = 0.\tag{2.4.15}$$

The second integral in (2.4.12) may be normalized as

$$\int d\rho_i\, \rho_i = 1, \tag{2.4.16}$$

for any $i$, a normalization condition that will be adopted in this book.

For any function $f(\rho_i) = \alpha_0 + c_1\,\rho_i$, where $c_1$ is a c-number, and $\alpha_0$ anti-commutes with $\rho_i$, one clearly has

$$\int d\rho_i\, (\rho_i - \beta_i)f(\rho_i) = f(\beta_i). \tag{2.4.17}$$

This allows one to introduce a Dirac delta given by

$$\delta(\rho_i - \beta_i) = (\rho_i - \beta_i), \tag{2.4.18}$$

being a Grassmann variable itself. That is $d\rho_i\, \delta(\rho_i - \beta_i)$ commutes with Grassmann variables.

Before introducing integrations over complex Grassmann variables, consider making a change of integration variables

$$\rho_i = C_{ij}\,\xi_j, \tag{2.4.19}$$

in an integral, where a summation over $j$ understood, and the $C_{ij}$ are c-numbers, $i,j = 1,\ldots,n$. We are thus led to consider, in general, the integral

$$\int d\rho_1 \ldots d\rho_n\, \rho_1 \ldots \rho_n = \int d\xi_1 \ldots d\xi_n\, J\, C_{1i_1}\xi_{i_1} \ldots C_{ni_n}\xi_{i_n}, \tag{2.4.20}$$

where $J$ is the Jacobian of the transformation to be determined. Using the anti-commutativity property of the $\xi_j$, we may write

$$\xi_{i_1} \ldots \xi_{i_n} = \varepsilon^{\,i_1 \ldots i_n}\,\xi_1 \ldots \xi_n, \tag{2.4.21}$$

where $\varepsilon^{\,i_1 \ldots i_n}$ is the Levi-Civita symbol totally anti-symmetric with $\varepsilon^{\,1 \ldots n} = +1$. Using the definition of a determinant of the matrix $[\,C_{ij}\,]$

$$C_{1i_1} \ldots C_{ni_n}\, \varepsilon^{\,i_1 \ldots i_n} = \det C, \tag{2.4.22}$$

we immediately obtain

$$J = 1/\det C, \tag{2.4.23}$$

being the *inverse* of $\det C$.

As before, out of two real Grassmann variables $\rho_R$, $\rho_I$, we may define a complex one $\rho = \rho_R + i\,\rho_I$. Now consider the transformation $\rho_R$, $\rho_I \rightarrow \rho$, $\rho^*$ defined by

$$\begin{pmatrix} \rho_R \\ \rho_I \end{pmatrix} = \frac{1}{2} \begin{pmatrix} 1 & 1 \\ i & -i \end{pmatrix} \begin{pmatrix} \rho^* \\ \rho \end{pmatrix}, \tag{2.4.24}$$

with the Jacobian of the transformation given by

$$J = \left( \det \frac{1}{2} \begin{pmatrix} 1 & 1 \\ i & -i \end{pmatrix} \right)^{-1} = 2\,i. \tag{2.4.25}$$

Using the fact that $\rho_I \rho_R = \rho\rho^*/2i$, we may infer that

$$1 = \int d\rho_R\, d\rho_I\, \rho_I\, \rho_R = \int d\rho^* d\rho\, (2i)\, (\rho\rho^*/2i), \tag{2.4.26}$$

or

$$\int d\rho^* d\rho\, \rho\, \rho^* = 1. \tag{2.4.27}$$

Hence for consistency with the above integral we may set

$$\int d\rho\, \rho = 1, \qquad \int d\rho^*\, \rho^* = 1. \tag{2.4.28}$$

For integrations over $n$ complex Grassmann variables $\rho_1, \ldots, \rho_n$, consider the integral

$$I = \int \frac{d\rho_1^* d\rho_1}{i} \cdots \frac{d\rho_n^* d\rho_n}{i} \exp\left[ -i\rho_i^* A_{ij}\, \rho_j \right] \tag{2.4.29}$$

where the $A_{ij}$ are c-numbers. The commutativity of $\rho_i^* A_{i1}\rho_1$, for example, with all the Grassmann variables, and so on, allows us to rewrite the integral $I$ as (since $(\rho_1^*)^2 = 0, \ldots, (\rho_n^*)^2 = 0$)

$$I = \int d\rho_1^* \left( -\rho_1^* A_{1j_1} \rho_{j_1} \right) d\rho_1 \cdots d\rho_n^* \left( -\rho_n^* A_{nj_n} \rho_{j_n} \right) d\rho_n$$

$$= \int d\rho_1 \rho_{j_1} \cdots d\rho_n \rho_{j_n}\, A_{1j_1} \cdots A_{nj_n}$$

$$= \int d\rho_1 \rho_1 \cdots d\rho_n \rho_n\, \varepsilon^{j_1 \cdots j_n} A_{1j_1} \cdots A_{nj_n}, \tag{2.4.30}$$

where we have used (2.4.21). From (2.4.28) and the definition of a determinant (2.4.22), the following expression for the integral emerges

$$\int \frac{d\rho_1^* d\rho_1}{i} \cdots \frac{d\rho_n^* d\rho_n}{i} \, \exp[-i \, \rho_i^* A_{ij} \, \rho_j] = \det A. \tag{2.4.31}$$

For $\det A \neq 0$, the above integral may be used to give another useful integral. To this end, we note the following. We may call the variables $\rho_i^*$ in (2.4.31) anything we like and use any notation for them. In the language of calculus they are dummy variables. We may go even further by noting that we are free to make completely different change of variables for the $\rho_j$ and the former variables. Taking these two points into account, we make the substitutions $\rho_j \to \rho_j - (A^{-1}\eta)_j$ and $\rho_i^* \to \overline{\rho}_i - (\overline{\eta}A^{-1})_i$ to obtain directly from (2.4.31)

$$\int \frac{d\overline{\rho}_1 d\rho_1}{i} \cdots \frac{d\overline{\rho}_n d\rho_n}{i} \, \exp[i \, (-\overline{\rho}_i A_{ij} \, \rho_j + \overline{\rho}_i \eta_i + \overline{\eta}_i \rho_i)] = (\det A) \exp[i \, \overline{\eta}_i A_{ij}^{-1} \, \eta_j]. \tag{2.4.32}$$

This integral is to be compared with the corresponding one for integrations over commuting complex variables

$$\int \frac{dz_1^* dz_1}{2\pi/i} \cdots \frac{dz_n^* dz_n}{2\pi/i} \, \exp[i \, (-z_i^* A_{ij} z_j + z_i^* K_i + K_i^* z_i)] = \left(\frac{1}{\det A}\right) \exp\left(i \, K_i^* A_{ij}^{-1} K_j\right). \tag{2.4.33}$$

For integrations over real commuting variables one also has

$$\int \frac{dy_1}{\sqrt{2\pi/i}} \cdots \frac{dy_n}{\sqrt{2\pi/i}} \, \exp\left[i \left(-\frac{1}{2} y_i A_{ij} y_j + K_i y_i\right)\right] = \left(\frac{1}{\sqrt{\det A}}\right) \exp\left[\frac{i}{2} K_i A_{ij}^{-1} K_j\right]. \tag{2.4.34}$$

## 2.5  Fourier Transforms Involving Grassmann Variables

The study of Fourier transform theory of Grassmann variables rests on the following two integrals. The first one is

$$\int d\overline{\eta}_1 d\eta_1 \ldots d\overline{\eta}_n d\eta_n \, \exp i \left[\overline{\eta}(\rho - \alpha) + (\overline{\rho} - \overline{\alpha})\eta\right]$$

$$= (\rho_1 - \alpha_1)(\overline{\rho}_1 - \overline{\alpha}_1) \ldots (\rho_n - \alpha_n)(\overline{\rho}_n - \overline{\alpha}_n), \tag{2.5.1}$$

which, by now, should be straightforward to verify (see Problem 2.6). In the above equation, we have used the standard notation $\overline{\eta}(\rho - \alpha) = \sum_j \overline{\eta}_j (\rho_j - \alpha_j)$. The second one is

$$\int d\overline{\rho}_1 \, d\rho_1 \, \ldots \, d\overline{\rho}_n \, d\rho_n \, (\overline{\rho}_1)^{\kappa_1} (\rho_1)^{\varepsilon_1} \ldots (\overline{\rho}_n)^{\kappa_n} (\rho_n)^{\varepsilon_n} \, (\rho_1 - \alpha_1) \times$$

$$\times (\overline{\rho}_1 - \overline{\alpha}_1) \ldots (\rho_n - \alpha_n)(\overline{\rho}_n - \overline{\alpha}_n) = (\overline{\alpha}_1)^{\kappa_1} (\alpha_1)^{\varepsilon_1} \ldots (\overline{\alpha}_n)^{\kappa_n} (\alpha_n)^{\varepsilon_n}, \qquad (2.5.2)$$

where the $\kappa_j$, $\varepsilon_j$ are either 0 or 1, and a factor such as $(0)^0$, (e.g., when $\alpha_j = 0, \varepsilon_j = 0$), is to be replaced by 1 on the right-hand side of the equation.

The second one seems more difficult to verify but easily follows by first writing

$$(\overline{\rho})^{\kappa} = \delta_{\kappa 0} + \overline{\rho} \, \delta_{\kappa 1}, \qquad (\rho)^{\varepsilon} = \delta_{\varepsilon 0} + \rho \, \delta_{\varepsilon 1}. \qquad (2.5.3)$$

The integral on the left-hand side of (2.5.2) then becomes

$$\int d\overline{\rho}_1 (\delta_{\kappa_1 0} - \overline{\rho}_1 \delta_{\kappa_1 1}) \, (\rho_1 - \alpha_1) \, (\overline{\rho}_1 - \overline{\alpha}_1) \, d\rho_1 \, (\delta_{\varepsilon_1 0} + \rho \delta_{\varepsilon_1 1}) \times$$

$$\cdots \times d\overline{\rho}_n (\delta_{\kappa_n 0} - \overline{\rho}_n \delta_{\kappa_n 1})(\rho_n - \alpha_n)(\overline{\rho}_n - \overline{\alpha}_n) \, d\rho_n \, (\delta_{\varepsilon_n 0} + \rho \delta_{\varepsilon_n 1}). \qquad (2.5.4)$$

Therefore it remains to face the following integral which works out as follows:

$$\int d\overline{\rho} \, (\delta_{\kappa 0} - \overline{\rho} \delta_{\kappa 1})(\rho - \alpha)(\overline{\rho} - \overline{\alpha}) \, d\rho \, (\delta_{\varepsilon 0} + \rho \delta_{\varepsilon 1})$$

$$= \int d\overline{\rho} \, (\delta_{\kappa 0} - \overline{\rho} \delta_{\kappa 1})(\overline{\rho} - \overline{\alpha}) \, d\rho \, (\rho - \alpha)(\delta_{\varepsilon 0} + \rho \delta_{\varepsilon 1})$$

$$= \int d\overline{\rho} \, \overline{\rho} \, (\delta_{\kappa 0} + \overline{\alpha} \delta_{\kappa 1}) \, d\rho \, \rho \, (\delta_{\varepsilon 0} + \alpha \delta_{\varepsilon 1}) = (\overline{\alpha})^{\kappa} (\alpha)^{\varepsilon}, \qquad (2.5.5)$$

which leads from (2.5.4) to the expression on the right-hand side of (2.5.2). Now we are ready to introduce Fourier transforms involving Grassmann variables.

Given a function $F[\overline{\rho}, \rho]$ which is a linear combination of terms as $[(\overline{\rho}_1)^{\kappa_1} (\rho_1)^{\varepsilon_1} \ldots (\overline{\rho}_n)^{\kappa_n} (\rho_n)^{\varepsilon_n}]$, we define its Fourier transform: $(\overline{\rho}, \rho) \rightarrow (\overline{\eta}, \eta)$ by

$$\tilde{F}[\overline{\eta}, \eta] = \int \frac{d\overline{\rho}_1 \, d\rho_1}{i} \ldots \frac{d\overline{\rho}_n \, d\rho_n}{i} F[\overline{\rho}, \rho] \exp[i(\overline{\rho} \eta + \overline{\eta} \rho)]. \qquad (2.5.6)$$

To find the inverse Fourier transform: $(\overline{\eta}, \eta) \rightarrow (\overline{\rho}, \rho)$, we multiply the above integral by

$$i \, d\overline{\eta}_1 \, d\eta_1 \ldots i \, d\overline{\eta}_n \, d\eta_n \, \exp[-i(\overline{\eta} \alpha + \overline{\alpha} \eta)],$$

which commutes with everything, and integrate using (2.5.1) and then (2.5.2). The right-hand becomes[10]

$$\int d\overline{\rho}_1 d\rho_1 \ldots d\overline{\rho}_n d\rho_n \, F[\overline{\rho}, \rho] \int d\overline{\eta}_1 d\eta_1 \ldots d\overline{\eta}_n \, d\eta_n \exp i \, [\, \overline{\eta}(\rho - \alpha) + (\overline{\rho} - \overline{\alpha})\eta \,]$$

$$= \int d\overline{\rho}_1 d\rho_1 \ldots d\overline{\rho}_n d\rho_n \, F[\overline{\rho}, \rho] \, (\rho_1 - \alpha_1)(\overline{\rho}_1 - \overline{\alpha}_1) \ldots (\rho_n - \alpha_n)(\overline{\rho}_n - \overline{\alpha}_n) = F[\overline{\alpha}, \alpha].$$

$$(2.5.7)$$

By equating this expression with the resulting one on the left-hand side of the equation in question, the following expression emerges for the inverse Fourier transform

$$F[\overline{\rho}, \rho] = \int i \, d\overline{\eta}_1 d\eta_1 \ldots i \, d\overline{\eta}_n d\eta_n \, \tilde{F}[\overline{\eta}, \eta] \exp [-i(\overline{\eta}\,\rho + \overline{\rho}\,\eta)], \qquad (2.5.8)$$

written in terms of variables $(\overline{\rho}, \rho)$.

For integration over commuting complex variables, with $z = (z_1, \ldots, z_n)^\mathsf{T}$, we define a Fourier transform by

$$\tilde{F}[K^\dagger, K] = \int \Big( \prod_j^n \frac{dz_j^* \, dz_j}{2\pi/i} \Big) \, F[z^\dagger, z] \, \exp[i\,(z^\dagger K + K^\dagger z)], \qquad (2.5.9)$$

then for the inverse Fourier transform, we have

$$F[z^\dagger, z] = \int \Big( \prod_j^n \frac{dK_j^* \, dK_j}{2\pi \, i} \Big) \, \tilde{F}[K^\dagger, K] \, \exp[-i\,(z^\dagger K + K^\dagger z)]. \qquad (2.5.10)$$

For integrations over commuting real variables, we define a Fourier transform as

$$\tilde{F}[K] = \int \Big( \prod_j^n \frac{d\phi_j}{\sqrt{2\pi/i}} \Big) \, F[\phi] \, \exp[i\,K^\mathsf{T}\phi], \qquad (2.5.11)$$

and the inverse Fourier transform is then given by

$$F[\phi] = \int \Big( \prod_j^n \frac{dK_j}{\sqrt{2\pi \, i}} \Big) \, \tilde{F}[K] \, \exp[-i\,K^\mathsf{T}\phi]. \qquad (2.5.12)$$

---

[10]Note that the factors $(\rho_j - \alpha_j)(\overline{\rho}_j - \overline{\alpha}_j)$ commute with everything.

## 2.6  Functional Differentiation and Integration; Functional Fourier Transforms

We begin by extending the concept of differentiation with respect to Grassmann variables $\eta_1, \ldots, \eta_n, \overline{\eta}_1, \ldots, \overline{\eta}_n$ to a continuum limit as $n \to \infty$. Such an extension for commuting variables will then become straightforward to carry out. By a continuum limit as $n \to \infty$, it is meant that one introduces Grassmann type functions $\eta_a(x)$, $\overline{\eta}_a(x)$ of a continuous variable $x$, such as specifying spacetime points, moreover they may also depend, in general, on discrete parameters, such as a spinor index. By Grassmann type functions, one also means that these functions all anti-commute for all $x$ and all $a$.

The simple rules of differentiations taken from the left with respect to a discrete set of Grassmann variables, defined earlier in Sect. 2.4, such as

$$\frac{\partial}{\partial \eta_i}\, \eta_j = \delta_{ji}, \qquad \frac{\partial}{\partial \overline{\eta}_i}\, \overline{\eta}_j = \delta_{ji}, \qquad \frac{\partial}{\partial \overline{\eta}_i}\, \eta_j = 0, \qquad \frac{\partial}{\partial \eta_i}\, \overline{\eta}_j = 0, \tag{2.6.1}$$

$$\frac{\partial}{\partial \eta_i}\, \eta_j \overline{\eta}_k \overline{\eta}_m \eta_s = \delta_{ji}\, \overline{\eta}_k \overline{\eta}_m \eta_s - \eta_j \overline{\eta}_k \overline{\eta}_m \delta_{si}, \tag{2.6.2}$$

become replaced, in this limit, by[11]

$$\frac{\delta}{\delta \eta_a(x)}\, \eta_b(x') = \delta_{ba}\delta(x'-x), \qquad \frac{\delta}{\delta \overline{\eta}_a(x)}\, \overline{\eta}_b(x') = \delta_{ba}\delta(x'-x), \tag{2.6.3}$$

$$\frac{\delta}{\delta \eta_a(x)}\, \eta_b(x')\, \overline{\eta}_c(y')\, \overline{\eta}_d(y'')\, \eta_e(x'') = \delta_{ba}\, \delta(x'-x)\, \overline{\eta}_c(y')\, \overline{\eta}_d(y'')\, \eta_e(x'')$$
$$- \eta_b(x')\, \overline{\eta}_c(y')\, \overline{\eta}_d(y'')\, \delta_{ea}\, \delta(x''-x), \tag{2.6.4}$$

involving Dirac deltas instead of just Kronecker deltas. One uses the notation $\delta/\delta \eta_a(x)$ etc, for the functional derivatives , i.e., for derivatives with respect to functions.

These functions $\eta_a(x)$, $\overline{\eta}_a(x)$ together with the functional derivatives with respect to them satisfy the anti-commutation rules:

$$\{\eta_a(x), \eta_b(x')\} = 0, \quad \{\overline{\eta}_a(x), \overline{\eta}_b(x')\} = 0, \quad \{\eta_a(x), \overline{\eta}_b(x')\} = 0, \tag{2.6.5}$$

$$\left\{\frac{\delta}{\delta \eta_a(x)}, \frac{\delta}{\delta \eta_b(x')}\right\} = 0, \quad \left\{\frac{\delta}{\delta \overline{\eta}_a(x)}, \frac{\delta}{\delta \overline{\eta}_b(x')}\right\} = 0, \quad \left\{\frac{\delta}{\delta \eta_a(x)}, \frac{\delta}{\delta \overline{\eta}_b(x')}\right\} = 0, \tag{2.6.6}$$

---

[11]We interchangeably use the notations $\delta(x-x')$ and $\delta^{(D)}(x-x')$ where $D$ is the dimensionality of spacetime.

$$\left\{ \frac{\delta}{\delta\eta_a(x)}, \eta_b(x') \right\} = \delta_{ba}\,\delta(x'-x), \quad \left\{ \frac{\delta}{\delta\overline{\eta}_a(x)}, \overline{\eta}_b(x') \right\} = \delta_{ba}\,\delta(x'-x). \quad (2.6.7)$$

$$\left\{ \frac{\delta}{\delta\eta_a(x)}, \overline{\eta}_b(x') \right\} = 0, \quad \left\{ \frac{\delta}{\delta\overline{\eta}_a(x)}, \eta_b(x') \right\} = 0. \quad (2.6.8)$$

Important functionals, i.e., functions of functions, that occur repeatedly in field theory are the exponentials of linear and/or bilinear forms such as:

$$\exp[i\,(\overline{\eta}\rho + \overline{\rho}\eta)], \quad \text{where} \quad \overline{\eta}\rho \equiv \sum_a \int (dx)\,\overline{\eta}_a(x)\,\rho_a(x), \quad (2.6.9)$$

$\rho_a(x)$, $\overline{\rho}_a(x)$ anti-commute with $\eta_b(x')$, $\overline{\eta}_b(x')$ as well as with the functional derivatives with respect to the latter, and

$$\exp[i\,\overline{\eta}A\eta], \quad \text{where} \quad \overline{\eta}A\eta \equiv \sum_{a,b} \int (dx)\,(dx')\,\overline{\eta}_a(x)A_{ab}(x,x')\eta_b(x'), \quad (2.6.10)$$

$\eta_a(x)$, $\overline{\eta}_a(x)$ *commute* with $A_{bc}(x',x'')$. Note that the above two functionals being even in anti-commuting functions commute with everything.

Typical functional derivatives of such functionals are

$$\frac{\delta}{\delta\eta_a(x)}\exp[i\,(\overline{\eta}\rho + \overline{\rho}\eta)] = -i\,\overline{\rho}_a(x)\,\exp[i\,(\overline{\eta}\rho + \overline{\rho}\eta)], \quad (2.6.11)$$

$$\frac{\delta}{\delta\overline{\eta}_a(x)}\exp[i\,(\overline{\eta}\rho + \overline{\rho}\eta)] = +i\,\rho_a(x)\,\exp[i\,(\overline{\eta}\rho + \overline{\rho}\eta)], \quad (2.6.12)$$

note the minus sign on the right-hand side of the first equation,

$$\frac{\delta}{\delta\overline{\eta}_a(x)}\exp[i\,\overline{\eta}A\eta] = i\sum_b \int (dx')A_{ab}(x,x')\eta_b(x')\,\exp[i\,\overline{\eta}A\eta], \quad (2.6.13)$$

$$\frac{\delta}{\delta\eta_a(x)}\exp[i\,\overline{\eta}A\eta] = -i\sum_c \int (dx')\overline{\eta}_c(x')A_{ca}(x',x)\,\exp[i\,\overline{\eta}A\eta], \quad (2.6.14)$$

$$\frac{\delta}{\delta\eta_c(y)}\frac{\delta}{\delta\overline{\eta}_a(x)}\exp[i\,\overline{\eta}A\eta] = i\,A(x,y)_{ac}\,\exp[i\,\overline{\eta}A\eta] + \exp[i\,\overline{\eta}A\eta]\times$$

$$\times \left( \sum_d \int (dx'')\overline{\eta}_d(x'')A_{dc}(x'',y) \right)\left( \sum_b \int (dx')A_{ab}(x,x')\eta_b(x') \right), \quad (2.6.15)$$

$$\frac{\delta}{\delta\overline{\eta}_a(x)}\frac{\delta}{\delta\eta_b(x')}\exp[i\,\overline{\eta}A\eta] = -\frac{\delta}{\delta\eta_b(x')}\frac{\delta}{\delta\overline{\eta}_a(x)}\exp[i\,\overline{\eta}A\eta]. \quad (2.6.16)$$

A point which is perhaps not sufficiently emphasized is that a functional derivative may be also applied in a simple manner to a derivative such as shown below

$$\partial_\mu \eta_a(x) = \partial_\mu \int (\mathrm{d}x') \, \delta(x - x') \eta_a(x') = \int (\mathrm{d}x') \, \eta_a(x') \partial_\mu \delta(x - x'),$$

$$\frac{\delta}{\delta \eta_b(x'')} \, \partial_\mu \eta_a(x) = \int (\mathrm{d}x') \, \delta_{ab} \, \delta(x' - x'') \, \partial_\mu \delta(x - x') = \delta_{ab} \, \partial_\mu \delta(x - x'').$$

$$(2.6.17)$$

We may take over the integral in (2.4.32) in a continuum limit, and use the notation $\prod_{x\,a} (\mathrm{d}\overline{\rho}_a(x) \, \mathrm{d}\rho_a(x)/i) \equiv (\mathscr{D}\overline{\rho}\,\mathscr{D}\rho)$, which is even in anti-commuting variables, to obtain for a matrix $M = [M_{ab}(x, x')]$:

$$\exp[i\,\overline{\eta}\,M^{-1}\eta] = (\det M)^{-1} \int (\mathscr{D}\overline{\rho}\,\mathscr{D}\rho) \exp[i\,(-\overline{\rho}M\rho + \overline{\rho}\eta + \overline{\eta}\rho)], \quad (2.6.18)$$

where we note that

$$\det M = \int (\mathscr{D}\overline{\rho}\,\mathscr{D}\rho) \exp[i\,(-\overline{\rho}M\rho + \overline{\rho}\eta + \overline{\eta}\rho)]\Big|_{\eta=0,\overline{\eta}=0}. \quad (2.6.19)$$

Here one may use a matrix notation for $M_{ab}(x, x') \equiv \langle x | M_{ab} | x' \rangle$, $\langle x | I_{ab} | x' \rangle = \delta_{ab} \, \delta(x - x')$, and note that

$$\sum_b \int \langle x | M_{ab}^{-1} | x' \rangle (\mathrm{d}x') \langle x' | M_{bc} | x'' \rangle = \delta_{ac} \, \delta(x - x''). \quad (2.6.20)$$

To see how these integrals work, consider, for example, the Dirac operator in the presence of an external electromagnetic field $A_\mu(x)$ (see Eq. (I.2) in Appendix I at the end of the book)

$$M = \left[\gamma^\mu \left(\frac{\partial_\mu}{i} - eA_\mu\right) + m\right] \equiv M(e), \quad (2.6.21)$$

$$\langle x | M(e) | x' \rangle = \left[\gamma^\mu \left(\frac{\partial_\mu}{i} - eA_\mu(x)\right) + m\right] \delta(x - x'), \quad (2.6.22)$$

corresponding to the interaction of an electron with an external electromagnetic potential. The following notation is often used for the matrix

$$M(0) = \left(\gamma^\mu \frac{\partial_\mu}{i} + m\right) \equiv S_+^{-1}. \quad (2.6.23)$$

Subject to an appropriate boundary condition, we will learn in Sect. 3.1 how to invert the matrix $S_+^{-1}$, thus defining the matrix $S_+$. We may conveniently write $M(e) = (S_+^{-1} - e\,\gamma A)$, and note that

$$\det M(e) = \exp[\operatorname{Tr}\ln M(e)] = \exp[\operatorname{Tr}\ln(S_+^{-1} - e\,\gamma A)]. \tag{2.6.24}$$

In particular,

$$\frac{\det M(e)}{\det M(0)} = \exp[\operatorname{Tr}\ln(I - e\,S_+\gamma A)], \tag{2.6.25}$$

and the expression in the exponential, on the right-hand side may be formally expanded as follows

$$\operatorname{Tr}\ln(I - e\,S_+\gamma A) = -\sum_{n\geqslant 1}\frac{(e)^n}{n}\int(dx_1)\ldots(dx_n)A_{\mu_1}(x_1)\ldots A_{\mu_n}(x_n)\times$$

$$\times\operatorname{Tr}\Big[S_+(x_1,x_2)\gamma^{\mu_2}S_+(x_2,x_3)\gamma^{\mu_3}\ldots S_+(x_n,x_1)\gamma^{\mu_1}\Big],\quad (dx)\equiv dx^0\,dx^1\,dx^2\,dx^3, \tag{2.6.26}$$

where the trace operation is over the gamma matrices. Such expressions will be studied in detail later and applications will be given.

Using the notation $M(e) \equiv [S_+^A]^{-1}$, for the matrix depending on the electromagnetic potential, and taking into account the expression in (2.6.19), we may equivalently rewrite (2.6.18) as

$$\exp[i\,\bar{\eta}\,S_+^A\,\eta] = \frac{Z[\bar{\eta},\eta]}{Z[0,0]}. \tag{2.6.27}$$

$$Z[\bar{\eta},\eta] = \int\mathscr{D}\bar{\rho}\,\mathscr{D}\rho\,\exp i\Big[-\bar{\rho}\Big(\gamma^\mu\Big(\frac{\partial_\mu}{i} - eA_\mu\Big) + m\Big)\rho + \bar{\rho}\eta + \bar{\eta}\rho\Big]. \tag{2.6.28}$$

We note that this defines a *functional Fourier transform* of

$$\exp\Big[-i\,\bar{\rho}\Big(\gamma^\mu\Big(\frac{\partial_\mu}{i} - eA_\mu\Big) + m\Big)\rho\Big], \tag{2.6.29}$$

giving the Fourier transform from $(\bar{\rho},\rho)$ to $(\bar{\eta},\eta)$ variables We will recognize the expression in the exponential in (2.6.29), multiplying the i factor, as the Dirac Lagrangian in the presence of an external electromagnetic field, with $(\bar{\eta},\eta)$ in (2.6.28) representing external sources coupled to the Dirac field. $Z[\bar{\eta},\eta]/Z[0,0]$

acts as a generating functional for determining basic components of an underlying theory as we will see this on several occasions later on in the book. In particular by functional differentiating (2.6.27) with respect to $\bar{\eta}$ and then by $\eta$, and setting them equal to zero, gives[12] $i\,S_+^A$, with $S_+^A$ interpreted as the propagator of a charged particle of charge e in the presence of an external electromagnetic field $A_\mu(x)$. $S_+^A$ may be then analyzed depending on how complicated the external field is. This will be dealt with in the next chapter.

By using the notation $\prod_x d\phi^*(x)\,d\phi(x)/(2\pi/i) \equiv (\mathscr{D}\phi^*\mathscr{D}\phi)$, for complex functions $\phi^*(x)$, $\phi(x)$, of a real variable, the integral (2.4.33) in the continuum limit may be written for a matrix $M$

$$\frac{1}{\det M}\,\exp[\,i\,K^*M^{-1}K] = \int (\mathscr{D}\phi^*\mathscr{D}\phi)\exp[\,i\,(-\phi^*M\phi + \phi^*K + K^*\phi)].$$

$$(2.6.30)$$

On the other-hand for a real function $\phi(x)$ of a real variable, the integral in (2.4.34) in the continuum limit, with the notation $\prod_x d\phi(x)/\sqrt{2\pi/i} \equiv (\mathscr{D}\phi)$, takes the form

$$\frac{1}{\sqrt{\det M}}\,\exp\left[\frac{i}{2}KM^{-1}K\right] = \int(\mathscr{D}\phi)\exp\left[i\left(-\frac{1}{2}\phi M\phi + \phi K\right)\right]. \qquad (2.6.31)$$

Functionals $F[\delta/\delta K, \delta/\delta K^*]$ and $F[\delta/\delta K]$ may be applied, respectively, to (2.6.30) and (2.6.31) to generate more complicated functional integrals.

The Functional Fourier transforms in the continuum case may be taken over from (2.5.6), (2.5.8), (2.5.9), (2.5.10), (2.5.11), (2.5.12). For anti-commuting fields we define

$$\tilde{F}[\bar{\eta}, \eta] = \int (\mathscr{D}\bar{\rho}\,\mathscr{D}\rho)\,F[\,\bar{\rho}, \rho]\exp[\,i\,(\bar{\rho}\,\eta + \bar{\eta}\,\rho)], \qquad (2.6.32)$$

and the inverse Functional Fourier transform of $\tilde{F}[\bar{\eta}, \eta]$ is then given by

$$F[\,\bar{\rho}, \rho] = \int(\tilde{\mathscr{D}}\bar{\eta}\,\tilde{\mathscr{D}}\eta)\,\tilde{F}[\bar{\eta}, \eta]\exp[-i\,(\bar{\rho}\,\eta + \bar{\eta}\,\rho)], \qquad (2.6.33)$$

using the notation $\prod_{xa}\left(i\,d\bar{\eta}_a(x)\,d\eta_a(x)\right) = (\tilde{\mathscr{D}}\bar{\eta}\,\tilde{\mathscr{D}}\eta)$.

For complex fields, the Functional Fourier transform reads (see (2.5.9), (2.5.10))

$$\tilde{F}[K^*, K] = \int (\mathscr{D}\phi^*\mathscr{D}\phi)\,\tilde{F}[\phi^*, \phi]\exp[\,i\,(\phi^*K + K^*\phi)]. \qquad (2.6.34)$$

---

[12]See Problem 2.8.

where $\prod_x \big( (d\phi^*(x)\, d\phi(x))/(2\pi/\mathrm{i}) \big) = (\mathscr{D}\phi^* \mathscr{D}\phi)$, and for the inverse Fourier transform one has

$$F[\phi^*, \phi] \;=\; \int (\tilde{\mathscr{D}}K^* \tilde{\mathscr{D}}K)\, \tilde{F}[K^*, K]\, \exp[-\mathrm{i}\,(K^*\phi + \phi^*K)], \qquad (2.6.35)$$

where now $\prod_x \big( (dK^*(x)\, dK(x)(/(2\pi\,\mathrm{i})) \big) = (\tilde{\mathscr{D}}K^* \tilde{\mathscr{D}}K)$.

For real fields, the Functional Fourier transform reads (see (2.5.11), (2.5.12))

$$\tilde{F}[K] \;=\; \int (\mathscr{D}\phi)\, F[\phi]\, \exp[\mathrm{i}K\phi], \qquad (2.6.36)$$

where $\prod_x \big( d\phi(x)/\sqrt{(2\pi/\mathrm{i})} \big) = (\mathscr{D}\phi)$, and the Fourier transform of $\tilde{F}[K]$ is given by

$$F[\phi] \;=\; \int (\tilde{\mathscr{D}}K)\, F[K]\, \exp[-\mathrm{i}\,K\phi], \qquad (2.6.37)$$

where $\prod_x \big( dK(x)/\sqrt{(2\pi\,\mathrm{i})} \big) = (\tilde{\mathscr{D}}K)$.

## 2.7  Delta Functionals

Consider a real function $K(x)$ and define the functional

$$G = -\frac{1}{2} \int (dx')(dx'')\, (-\mathrm{i})\frac{\delta}{\delta K(x')}\, M(x', x'')\, (-\mathrm{i})\frac{\delta}{\delta K(x'')}, \qquad (2.7.1)$$

where $M(x', x'') = M(x'', x')$. Then we have the commutation relation

$$\mathrm{i}\,[G, K(x)] \;=\; \mathrm{i} \int (dx')\, M(x, x')\frac{\delta}{\delta K(x')}. \qquad (2.7.2)$$

By using the relation $MM^{-1} = I$, we obtain

$$\Big( \mathrm{i}\,[G, K(x)] + K(x) \Big) \exp\!\Big[ \frac{\mathrm{i}}{2} K M^{-1} K \Big] = 0. \qquad (2.7.3)$$

Since $[G, [G, K(x)]] = 0$, we may use the identity

$$\mathrm{e}^A B\, \mathrm{e}^{-A} = [A, B] + B, \qquad \text{for } [A, [A, B]] = 0, \qquad (2.7.4)$$

with $B = K$, $A = iG$, to infer, upon multiplying (2.7.3), in the process, by $\exp(-iG)$, that

$$K(x)\left(\exp(-iG)\exp\left[\frac{i}{2}KM^{-1}K\right]\right) = 0, \qquad (2.7.5)$$

for all $x$.

Therefore for every point $x$ we may introduce a delta function $\delta(K(x))$ and introduce the delta functional

$$\delta(K) = \prod_x \delta\big(K(x)\big), \qquad (2.7.6)$$

and use the fact that $xf(x) = 0$ implies that $f(x)$ is a delta function $\delta(x)$, up to a multiplicative constant, to infer that the expression between the round brackets in (2.7.5) is equal to $\delta(K)$, up to a proportionality constant, since it is valid for all $x$, i.e.,

$$\exp\left[\frac{i}{2}KM^{-1}K\right] = C\exp(iG)\,\delta(K), \qquad (2.7.7)$$

where $C$ is a $K$-independent multiplicative factor, and $G$ is defined in (2.7.1).

Equivalence of the above formulation with the Functional Fourier transform method developed in the previous section may be established by writing

$$\delta(K) = \prod_x \int_{-\infty}^{\infty} \frac{d\phi(x)}{2\pi}\exp\left[iK(x)\phi(x)\right], \qquad (2.7.8)$$

which leads from (2.7.7) to

$$\exp\left[\frac{i}{2}KM^{-1}K\right] = C\left(\prod_x \int_{-\infty}^{\infty} \frac{d\phi(x)}{2\pi}\right)\exp\left[i\left(-\frac{1}{2}\phi M\phi + K(x)\phi(x)\right)\right]. \qquad (2.7.9)$$

Since the left-hand side is unity for $K = 0$, we obtain for $1/C$ the functional integral

$$\frac{1}{C} = \left(\prod_x \int_{-\infty}^{\infty} \frac{d\phi(x)}{2\pi}\right)\exp\left[-\frac{i}{2}\phi M\phi\right]. \qquad (2.7.10)$$

Obviously (2.7.9) coincides with that given in (2.6.31) since the $(2\pi)$ factors in (2.7.9) cancel out anyway in determining $\exp\left[iKM^{-1}K/2\right]$. The interest in the expression in (2.7.7) is that one may also apply to it an arbitrary functional $F[\delta/\delta K]$ which, as we will see later, essentially provides the solution of field theory.

For complex functions $K(x)$, $K^*(x)$ of a real variable $x$, the same analysis as above involving both functions, with

$$G = -\int (dx')(dx'')\,(-i)\frac{\delta}{\delta K(x)}\,M(x',x'')\,(-i)\frac{\delta}{\delta K^*(x)}, \qquad (2.7.11)$$

gives

$$\exp[\,i\,K^*\,M^{-1}K] = C\exp(i\,G)\,\delta(K^*,K), \qquad (2.7.12)$$

up to a $(K^*,K)$-independent multiplicative factor $C$, and $\delta(K^*,K)$ may be taken as the product $\prod_x \delta(K_R(x))\,\delta(K_I(x))$ or equivalently as

$$\delta(K^*,K) = \prod_x \int \frac{d\phi^*(x)\,d\phi(x)}{i\,(2\pi)^2}\,\exp\!\big[\,i\,\big(K^*(x)\phi(x)+\phi^*(x)K(x)\big)\big]. \qquad (2.7.13)$$

For anti-commuting $\eta,\overline{\eta}$, paying special attention to this property, the same analysis as above gives

$$\exp[\,i\,\overline{\eta}M^{-1}\eta] = C\exp(i\,G)\,\delta(\overline{\eta},\eta), \qquad (2.7.14)$$

up to a $(\overline{\eta},\eta)$-independent multiplicative factor $C$, with

$$G = -\int (dx')(dx'')\,(i)\frac{\delta}{\delta\eta_a(x')}\,M_{ab}(x',x'')\,(-i)\frac{\delta}{\delta\overline{\eta}_b(x'')}, \qquad (2.7.15)$$

and $\delta(\overline{\eta},\eta)$, absorbing any $\pi$ factors in $C$, may be taken as

$$\delta(\overline{\eta},\eta) = \prod_x \int d\overline{\rho}(x)\,d\rho(x)\,\exp i\,[\,\overline{\eta}(x)\rho(x)+\overline{\rho}(x)\eta(x)\,], \qquad (2.7.16)$$

with the functional differentiations on the right-hand side of (2.7.14) to be carried out before carrying out the integrations in (2.7.16).

## Problems

**2.1** Consider the matrix $G = I\cos\theta + \boldsymbol{\gamma}\cdot\mathbf{n}\sin\theta$, $G^\dagger = G^{-1} = I\cos\theta - \boldsymbol{\gamma}\cdot\mathbf{n}\sin\theta$, where $\mathbf{n} = \mathbf{a}/|\mathbf{a}|$, for some non-zero dimensionless 3-vector $\mathbf{a}$, and

$$\cos\theta = \Big[\big(\sqrt{\mathbf{a}^2+1}+1\big)\big/\big(2\sqrt{\mathbf{a}^2+1}\big)\Big]^{1/2}.$$

Show that $G\gamma^0(\boldsymbol{\gamma}\cdot\mathbf{a}+1)G^{-1} = \gamma^0\sqrt{\mathbf{a}^2+1}$.

**2.2** The Dirac Hamiltonian is given by $H = \gamma^0[\boldsymbol{\gamma} \cdot \mathbf{p} + m]$ in the momentum description. Use the result in Problem 2.1 to diagonalize the Hamiltonian in the Dirac representation.

**2.3** Evaluate the integral over the Grassmann variable $\rho_j$ in (2.4.17).

**2.4** As a curious property of integration over a Grassmann variable, consider again, as in the previous problem, the function: $f(\rho) = \alpha_0 + c_1\rho$ of a single variable. Show that $\int d\rho f(\rho) = (\partial/\partial\rho)f(\rho)$ implying an equivalence relationship between integration and differentiation.

**2.5** Use the property of complex conjugation of the product $d\rho_R \, \rho_R$ of a real Grassmann variable $\rho_R$ to infer that

(i) $(d\rho_R)^* = -d\rho_R$. From this conclude that for a complex Grassmann variable $\rho$, $(d\rho)^* = -d\rho^*$.

(ii) $\left(\int d\rho\, \rho\right)^* = \int d\rho^*\, \rho^*$. This justifies of assignining the same real numerical values for both integrals: $\int d\rho\, \rho$, and $\int d\rho^*\, \rho^*$ over complex Grassmann variables.

**2.6** Verify the validity of the integral in (2.5.1).

**2.7** Verify the functional differentiations carried out in (2.6.15).

**2.8** Evaluate the functional integral in (2.6.27) to the leading order in the external potential as a functional of $S_+$. Then use this expression to determine $S_+^A$ to this leading order.

# References

1. Manoukian, E. B. (2006). *Quantum theory: A wide spectrum*. Dordrecht: Springer.
2. Manoukian, E. B. (2016). *Quantum field theory II: Introductions to quantum gravity, supersymmetry, and string theory*. Dordrecht: Springer.
3. Wigner, E. P. (1959). *Group theory, and its applications to the quantum mechanics of atomic spectra*. New York: Academic.
4. Wigner, E. P. (1963). Invariant quantum mechanical equations of motion. In: *Theoretical physics* (p. 64). Vienna: International Atomic Energy Agency.

# Recommended Reading

1. Manoukian, E. B. (2006). *Quantum theory: A wide spectrum*. Dordrecht: Springer.

# Chapter 3
# Quantum Field Theory Methods of Spin 1/2

With the many-particle aspect encountered by emerging relativity and quantum physics and the necessity of describing their creation and annihilation in arbitrary numbers, one introduces the very rich concept of a quantum field. As a first encounter with this concept, we investigate, in this chapter, the very significance of a quantum field for the very simple system described by the celebrated and well known Dirac equation as given in Eq. (I.1) in Appendix I, as well as of the Dirac equation in the presence of a general external electromagnetic potential in Eq. (I.2).

The quantum field inevitably leads to the concept of a propagator as describing the transfer of energy and momentum between emission sources and particle detectors or between particles, in general. In this particular situation involved with the Dirac field, this necessarily leads one to investigate and understand how to invert the Dirac operator $(\gamma \partial/i + m)$, based on physical grounds, and define carefully the boundary conditions involved in doing this (Sect. 3.1).

The experimental set-up where a particle may be emitted or detected before or after having participated in some physical process, respectively, as well as of the underlying physical process itself are quite appropriately described by introducing the so-called vacuum-to-vacuum transition amplitude (Sects. 3.1, 3.2, 3.3, 3.4, 3.5, 3.6, 3.7, and 3.8). The latter amplitude leads quite naturally to the determination of transition amplitudes of physical processes. In particular, the positron is readily re discovered by extracting the amplitude by such a method for the system described by the Dirac equation in an external electromagnetic potential probe (Sect. 3.3).

Various applications will be given including to the Coulomb scattering of relativistic electrons, derivation of the Euler-Heisenberg effective Lagrangian, as an example of a modification of Maxwell's equations, as well the decay of the vacuum and of the underlying Schwinger effect of pair productions by a constant electric field (Sect. 3.8). Particular emphasis is put on deriving the spin & statistics connection of the Dirac quantum field, and of the generation of gauge invariant currents (Sects. 3.6, 3.9, and 3.10) based on Schwinger's elegant point splitting method. Finally it is noted that a conservation law that may hold classically does

© Springer International Publishing Switzerland 2016     73
E.B. Manoukian, *Quantum Field Theory I*, Graduate Texts in Physics,
DOI 10.1007/978-3-319-30939-2_3

not necessarily hold quantum mechanically and lead to modifications of the theory in the quantum world. The breakdown of such conservation laws lead to calculable anomalies and are treated in Sect. 3.9 and 3.10, and we provide as well of a concrete physical example where such an anomaly is actually verified experimentally.

## 3.1  Dirac Quantum Field, Propagator and Energy-Momentum Transfer: Schwinger-Feynman Boundary Condition

In this section, we investigate the very significance of a quantum field for the very simple system described by the Dirac equation in (I.1). To this end, one may consider the matrix element of the Dirac field $\psi(x)$, treated as a quantum field, between the no-particle state (vacuum) and a single particle state of a particle of momentum-spin $(\mathbf{p}, \sigma)$ as follows $\langle \text{vac} \mid \psi(x) \mid \mathbf{p}\sigma \rangle$.

The particle may have been produced by some external emission source, and a much more transparent treatment for understanding the meaning of such a matrix element is to introduce a source term $\eta(x)$ on the right-hand side of the Dirac equation to reflect this fact:

$$\left( \frac{\gamma \partial}{\mathrm{i}} + m \right) \psi(x) = \eta(x). \tag{3.1.1}$$

The presence of the external source $\eta(x)$ makes the *physical* interpretation of the problem at hand clearer, as it plays the role of an emitter and detector of particles. The source is assumed to be of compact support in time, i.e., it vanishes except within a given interval or given sets of intervals in time. Physically it means outside the interval(s) the source ceases to operate, i.e., it is switched off, and is set equal to zero.[1] The explicit structure of $\eta(x)$ is *unimportant*. It does not appear in the expressions of transition amplitudes of quantum particles. The introduction of such sources facilitate tremendously the computation of transition amplitudes, and hence their inclusion in the analysis of the underlying physical problems is of quite practical value. $\eta(x)$, by definition of an external source, is not a quantum field. It is not, however, a classical one either as such, but a Grassmann variable. This latter point does not concern us yet at this stage and we will come back to it later.

Let $\mid 0_-\rangle$ designates the vacuum state before the source $\eta(x')$ is switched on. After the source is eventually switched off, the latter may have created some particles. This means that if $\mid 0_+\rangle$ designates the vacuum state after the source is switched-off, then quantum mechanics says that $\langle 0_+ \mid 0_- \rangle$ is not necessarily one or just a phase factor as the system may be in some other state involving some

---

[1] It is important to emphasize, that the source need not vanish abruptly at any point and may be taken to be continuous at a point where it vanishes (see Problem 3.1).

particles. This amplitude corresponds to the case where we begin with the vacuum before the source $\eta(x')$ is switched on, and end up with the vacuum after the source is switched off. Here we have conveniently used the argument $x'$ for the source.

We consider the matrix element $\langle 0_+ | \psi(x) | 0_- \rangle$, and solve (3.1.1) for the latter. We will then investigate the meaning of this matrix element later.

By carrying out a 4D Fourier transform of (3.1.1) and inverting, in the process, the matrix $(\gamma \partial/i + m)$, we obtain for the matrix element in question

$$\frac{\langle 0_+ | \psi(x) | 0_- \rangle}{\langle 0_+ | 0_- \rangle} = \int (dx') \int \frac{(dp)}{(2\pi)^4} e^{ip(x-x')} \frac{(-\gamma p + m)}{(p^2 + m^2)} \eta(x'), \qquad (3.1.2)$$

$$(dx) = dx^1 dx^2 dx^3 dx^4, \quad p(x - x') = \mathbf{p} \cdot (\mathbf{x} - \mathbf{x}') - p^0(x^0 - x^0), \qquad (3.1.3)$$

where we have used the relation $(-\gamma p + m)(\gamma p + m) = (p^2 + m^2)$.

Regarding (3.1.2), we are faced with the interpretation of the pole that may arise from the vanishing of the denominator $(p^2 + m^2)$, whose physical significance and the evaluation of the expression on the right-hand side of (3.1.2) provides the first step for carrying out computations in quantum field theory.

The denominator in (3.1.2) may be rewritten as

$$(p^2 + m^2) = -(p^0 - E)(p^0 + E), \qquad E = \sqrt{\mathbf{p}^2 + m^2}. \qquad (3.1.4)$$

If $x^0$ is such that $x^0 = x_2^0 > x'^0$, as shown in Fig. 3.1, for *all* $x'^0$ contributing to the $x'^0$-integral in (3.1.2), that is, if the source is switched off before the time $x^0$, we have the physical situation of a causal arrangement ($x^0 > x'^0$), where energy is being transferred from the source to the point $x_2$, with the source playing the role of an emitter. Thus we have to pick up the pole $p^0 = +E$ in (3.1.2) corresponding to an energy gain at the point in question. To do this, we make a transition to a complex $p^0$-plane. Let $\operatorname{Im} p^0$ denote the imaginary part of $p^0$. For the case considered above, $x^0 - x'^0 \equiv T > 0$, and the real part of $(-ip^0 T)$ is given by $\operatorname{Re}(-ip^0 T) = (\operatorname{Im} p^0)T$ (Fig. 3.2).

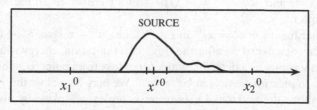

**Fig. 3.1** $x_1^0$, $x_2^0$ are, respectively, two arbitrary points before the source is switched on and after it is switched off at which $\langle 0_+ | \psi(x) | 0_- \rangle$ is examined

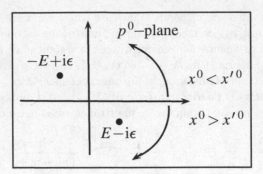

**Fig. 3.2** Figure showing the direction of the contour that is to be taken in the complex $p^0$–plane when applying the residue theorem, taken c.w. below the horizontal axis for $x^0 > x'^0$, and c.c.w. above it for $x^0 < x'^0$, generating, in turn, two *semi-circles* of infinite radii, one below and one above the *horizontal axis*, respectively

To pick up the pole at $p^0 = +E$ we displace $E$ by $-i\epsilon$, $\epsilon \to 0$, in (3.1.4), i.e., we make the replacement

$$(p^2 + m^2) \to -[p^0 - (E - i\epsilon)][p^0 + (E - i\epsilon)] = (p^2 + m^2 - i\epsilon). \qquad (3.1.5)$$

We close the $p^0$–contour c.w. from below the horizontal axis, for which $\text{Im} \, p^0 < 0$, ensuring that the infinite semi-circle part of the contour in the lower plane gives no contribution. The latter is as a consequence of the fact that $\exp[(\text{Im} \, p^0)T] \to 0$ for $(\text{Im} \, p^0)T \to -\infty$, $T > 0$, and, in the process pick up the pole $p^0 = +E$, by the application of the Cauchy Theorem. We note that for any real number $a > 0$, $a\epsilon$ is equivalent to $\epsilon$ in complex integration for $\epsilon \to +0$. An application of the residue theorem gives

$$\frac{\langle 0_+ | \psi(x) | 0_- \rangle}{\langle 0_+ | 0_- \rangle} = i \int (dx') \int \frac{d^3\mathbf{p}}{(2\pi)^3 2p^0} \, e^{ip(x-x')} \, (-\gamma p + m) \, \eta(x'), \qquad (3.1.6)$$

where $x^0 > x'^0$, and now $p^0 = +\sqrt{\mathbf{p}^2 + m^2}$. Here we recall the source $\eta(x')$ acts as an emitter.

On the other hand for $x^0 = x_1^0$ in Fig. 3.1, i.e., $x^0 - x'^0 \equiv T' < 0$, meaning that the source is switched on after the time $x^0$ in question, energy is absorbed by the source, now acting as a detector, with energy loss from point $x_1$. That is, in this case, we have to pick up the pole at $p^0 = -E$. We may then close the $p^0$–contour c.c.w. from above the horizontal axis in the complex $p^0$–plane by noting, in the process, that $\exp[(\text{Im} \, p^0)T'] \to 0$, for $\text{Im} \, p^0 \to +\infty$, $T' < 0$, to obtain by the residue theorem

$$\frac{\langle 0_+ | \psi(x) | 0_- \rangle}{\langle 0_+ | 0_- \rangle} = i \int (dx') \int \frac{d^3\mathbf{p}}{(2\pi)^3 2p^0} \, e^{-ip(x-x')} \, (\gamma p + m) \, \eta(x'), \qquad (3.1.7)$$

where $x^0 < x'^0$, $p^0 = +\sqrt{\mathbf{p}^2 + m^2}$ (not minus), and we made a change of variables of integration $\mathbf{p} \to -\mathbf{p}$ to finally write down the above integral.

Accordingly, and quite generally, from (3.1.5), (3.1.6), (3.1.7) one may rewrite the expression in (3.1.2) in a unified manner as

$$\frac{\langle 0_+ | \psi(x) | 0_- \rangle}{\langle 0_+ | 0_- \rangle} = \int (dx') \, S_+(x - x') \, \eta(x'), \tag{3.1.8}$$

where $S_+(x - x')$ is referred to as the Dirac (fermion) propagator given by

$$S_+(x - x') = \int \frac{(dp)}{(2\pi)^4} \, e^{i p (x - x')} \frac{(-\gamma p + m)}{(p^2 + m^2 - i\epsilon)}, \tag{3.1.9}$$

responsible for the transfer of energy (and momentum) as discussed above. The $i\epsilon$ term, with $\epsilon \to +0$, in the denominator, dictated by the physics of the situation, specifies the (Schwinger-Feynman) boundary condition for the proper integration in the complex $p^0$-plane. Also with $p^0 = +\sqrt{\mathbf{p}^2 + m^2}$,

$$S_+(x - x') = i \int \frac{d^3\mathbf{p}}{(2\pi)^3 2p^0} \, e^{i p (x - x')} (-\gamma p + m), \quad \text{for } x^0 > x'^0, \tag{3.1.10}$$

$$S_+(x - x') = i \int \frac{d^3\mathbf{p}}{(2\pi)^3 2p^0} \, e^{-i p (x - x')} (\gamma p + m), \quad \text{for } x^0 < x'^0, \tag{3.1.11}$$

$$\left( \frac{\gamma \partial}{i} + m \right) S_+(x - x') = \delta^{(4)}(x - x'), \tag{3.1.12}$$

as follows from (3.1.9). The propagator also satisfies

$$S_+(x' - x) \left( -\frac{\gamma \overleftarrow{\partial}}{i} + m \right) = \delta^{(4)}(x' - x). \tag{3.1.13}$$

The field equation for $\overline{\psi}(x) \equiv \psi^\dagger(x)\gamma^0$, in the presence of the external source, is given by

$$\overline{\psi}(x) \left( -\frac{\gamma \overleftarrow{\partial}}{i} + m \right) = \overline{\eta}(x). \tag{3.1.14}$$

The vacuum expectation value of $\overline{\psi}(x)$ is then easily worked out to be

$$\frac{\langle 0_+ | \overline{\psi}(x) | 0_- \rangle}{\langle 0_+ | 0_- \rangle} = \int (dx') \, \overline{\eta}(x') \, S_+(x' - x), \tag{3.1.15}$$

and from (3.1.10), (3.1.11),

$$\frac{\langle 0_+ | \overline{\psi}(x) | 0_- \rangle}{\langle 0_+ | 0_- \rangle} = i \int (dx') \int \frac{d^3 \mathbf{p}}{(2\pi)^3 2p^0} e^{ip(x'-x)} \overline{\eta}(x')(-\gamma p + m), \qquad (3.1.16)$$

if $x^0 < x'^0$, i.e., if the source $\overline{\eta}(x')$ is switched on after the time $x^0$,

$$\frac{\langle 0_+ | \overline{\psi}(x) | 0_- \rangle}{\langle 0_+ | 0_- \rangle} = i \int (dx') \int \frac{d^3 \mathbf{p}}{(2\pi)^3 2p^0} e^{-ip(x'-x)} \overline{\eta}(x')(\gamma p + m), \qquad (3.1.17)$$

if $x^0 > x'^0$, i.e., if the source $\overline{\eta}(x')$ is switched off before the time $x^0$.

Thus the Schwinger-Feynman boundary condition, based on physical grounds, amounts in replacing the *mass* $m$ *by* $m - i\epsilon$ in the denominator in defining the propagator. Needless to say for $m = 0$ the $-i\epsilon$ should survive.

We note that due to the equality

$$(dp)\, \delta(p^2 + m^2)\, \theta(p^0) = \frac{d^3 \mathbf{p}}{2p^0}\, dp^0\, \delta(p^0 - \sqrt{\mathbf{p}^2 + m^2}), \qquad (3.1.18)$$

where $\theta(p^0) = 1$ for $p^0 > 0$, $\theta(p^0) = 0$ for $p^0 < 0$, and due to the fact that a Lorentz transformation does not change the sign of $p^0$, we may infer the Lorentz invariance of the measure on the right-hand side of (3.1.18) as well.

We are now ready to investigate, in the next section, the role of $\psi(x)$ as a quantum field and the particle content of the theory.

## 3.2   The Dirac Quantum Field Concept, Particle Content, and C, P, T Tansformations

For the interpretation of the role played by the quantum field $\psi(x)$ for the simple system described by the Dirac equation, we consider the expression for the matrix element $\langle 0_+ | \psi(x) | 0_- \rangle / \langle 0_+ | 0_- \rangle$ in (3.1.6), that is for $x^0 > x'^0$ in reference to Fig. 3.1, where the source acts as an emitter.

With the functional time dependence of $\eta(x')$ being of compact support in time explicitly *absorbed* in $\eta(x')$, we may carry out the $x'$-integral in (3.1.6) over all $x'$ to obtain

$$\frac{\langle 0_+ | \psi(x) | 0_- \rangle}{\langle 0_+ | 0_- \rangle} = \int \frac{d^3 \mathbf{p}}{(2\pi)^3 2p^0} e^{ipx} (-\gamma p + m)\, i\, \eta(p), \qquad (3.2.1)$$

$$\eta(p) = \int (dx')\, e^{-ipx'} \eta(x'). \qquad (3.2.2)$$

Using the expression for the projection operator in Eq. (I.21) in Appendix I over spin states, we rewrite (3.2.1) as

$$\frac{\langle 0_+ | \psi(x) | 0_- \rangle}{\langle 0_+ | 0_- \rangle} = \int \sum_\sigma \frac{m}{p^0} \frac{\mathrm{d}^3 \mathbf{p}}{(2\pi)^3} \mathrm{e}^{ipx} u(\mathbf{p}, \sigma) \langle \mathbf{p}\sigma | 0_- \rangle. \tag{3.2.3}$$

where

$$\langle \mathbf{p}\sigma | 0_- \rangle \equiv [i\, \overline{u}(\mathbf{p}, \sigma)\, \eta(p)], \tag{3.2.4}$$

and as it will be evident below, and as shown in the next section, $\langle \mathbf{p}\sigma | 0_- \rangle$ represents an amplitude that the source, as an emitter, emits a particle of momentum-spin $(\mathbf{p}, \sigma)$ which persists after the source is switched off.

Since we have the vacuum state on the left-hand side of (3.2.3), we reach the inevitable conclusion that $\psi(x)$ has annihilated the particle in question to end up in the vacuum.

At this stage, it is instructive to introduce the normalization condition of a particle state,[2] as well as of the resolution of the identity of the momenta and spin of such a particle defined, in turn, by

$$\langle \mathbf{p}'\sigma' | \mathbf{p}\sigma \rangle = (2\pi)^3 \frac{p^0}{m} \delta_{\sigma'\sigma} \delta^3(\mathbf{p}' - \mathbf{p}), \quad I = \int \sum_\sigma \frac{m}{p^0} \frac{\mathrm{d}^3 \mathbf{p}}{(2\pi)^3} | \mathbf{p}\sigma \rangle \langle \mathbf{p}\sigma |, \tag{3.2.5}$$

to rewrite (3.2.3) as

$$\frac{\langle 0_+ | \psi(x) | 0_- \rangle}{\langle 0_+ | 0_- \rangle} = \int \sum_\sigma \frac{\langle 0_+ | \psi(x) | \mathbf{p}\sigma \rangle}{\langle 0_+ | 0_- \rangle} \frac{m}{p^0} \frac{\mathrm{d}^3 \mathbf{p}}{(2\pi)^3} \langle \mathbf{p}\sigma | 0_- \rangle. \tag{3.2.6}$$

and infer from (3.2.3) that

$$\frac{\langle 0_+ | \psi(x) | \mathbf{p}\sigma \rangle}{\langle 0_+ | 0_- \rangle} = \mathrm{e}^{ipx} u(\mathbf{p}, \sigma). \tag{3.2.7}$$

A similar analysis as above dealing with $\langle 0_+ | \overline{\psi}(x) | 0_- \rangle$ in (3.1.16) for the causal arrangement, now, with $x^0 < x'^0$, relative to the source $\overline{\eta}(x')$, in reference to Fig. 3.1, leads to the identification

$$[i\, \overline{\eta}(p)\, u(\mathbf{p}, \sigma)] \equiv \langle 0_+ | \mathbf{p}\sigma \rangle, \tag{3.2.8}$$

---

[2] Particle states and the Poincaré Algebra will be considered in Sect. 4.2.

with the latter representing an amplitude that the source $\overline{\eta}$, as a detector, absorbs a particle, and finally obtain

$$\frac{\langle \mathbf{p}\,\sigma \mid \overline{\psi}(x) \mid 0_- \rangle}{\langle 0_+ \mid 0_- \rangle} = e^{-ipx}\,\overline{u}(\mathbf{p}, \sigma). \tag{3.2.9}$$

Thus we reach the inevitable conclusion that $\overline{\psi}(x)$ has created a particle since we began with the vacuum and ended up having a particle detected by the source.

It remains to consider the situation for $\langle 0_+ \mid \psi(x) \mid 0_- \rangle$ with $x^0 < x'^0$ relative to $\eta(x')$, and $\langle 0_+ \mid \overline{\psi}(x) \mid 0_- \rangle$ with $x^0 > x'^0$, relative to $\overline{\eta}(x')$. This will be done in the next section.

In the remaining part of this section and in view of applications in the following sections, we consider the Dirac equation in an external electromagnetic field $A_\mu$ in the presence of an external source $\eta$. This is given by

$$\left[ \gamma^\mu \left( \frac{\partial_\mu}{i} - e A_\mu(x) \right) + m \right] \psi(x) = \eta(x), \tag{3.2.10}$$

with e denoting the charge. We define it to carry its own sign, i.e., $e = -|e|$ for an electron.

In analogy to (3.1.12), we introduce the propagator $S^A_+(x, x')$, now in the presence of an external electromagnetic field, satisfying

$$\left[ \gamma^\mu \left( \frac{\partial_\mu}{i} - e A_\mu(x) \right) + m \right] S^A_+(x, x') = \delta^{(4)}(x - x'). \tag{3.2.11}$$

The reason as to why we have written the functional dependence of $S^A_+$ as $(x, x')$ rather than $(x - x')$ is that for $A_\mu(x) \neq 0$ translational invariance is generally broken. One may convert (3.2.11) to the following integral equation

$$S^A_+(x, x') = S_+(x-x') + e \int (dx'')\, S_+(x-x'')\gamma^\mu A_\mu(x'') S^A_+(x'', x'). \tag{3.2.12}$$

This is easily verified by applying the uperator $(\gamma \partial/i + m)$ to (3.2.12), use (3.1.12) and finally integrate over a delta function to see that (3.2.11) is indeed satisfied.

The expression for $\langle 0_+ \mid \psi(x) \mid 0_- \rangle_A / \langle 0_+ \mid 0_- \rangle_A$, now in the presence of the external electromagnetic field $A_\mu(x)$, follows from (3.2.10) and takes the form

$$\langle \psi(x) \rangle_A \equiv \frac{\langle 0_+ \mid \psi(x) \mid 0_- \rangle_A}{\langle 0_+ \mid 0_- \rangle_A} = \int (dx')\, S^A_+(x, x')\, \eta(x'). \tag{3.2.13}$$

We consider the following physical situation where $\eta(x')$ is so chosen that $x'^0$ lies sufficiently in the remote past and $x^0$ lies sufficiently forward in the future, with $A_\mu(x'')$ in (3.2.12) being effective only in the region, $x'^0 < x''^0 < x^0$. In

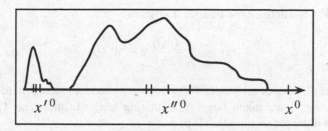

**Fig. 3.3** $\eta(x')$ is so chosen that $x'^0$ lies sufficiently in the remote past and $x^0$ lies sufficiently forward in the future, with $A_\mu(x'')$ being effective only in the region $x'^0 < x''^0 < x^0$

practice, one may choose $x'^0 \ll x''^0 \ll x^0$, if necessary, depending on $A_\mu(x'')$, in a limiting sense for the time variables $x'^0$, $x^0$ (Fig. 3.3).

For a sufficiently weak external field $A_\mu$, (3.2.12) becomes

$$S_+^A(x, x') \simeq S_+(x-x') + e \int (dx'') \, S_+(x-x'') \gamma^\mu A_\mu(x'') S_+(x''-x'). \quad (3.2.14)$$

The condition of a weak external field will be relaxed in Sect. 3.6/Appendix IV and an all order treatment will be given.

We may carry out the integrations in (3.2.13), (3.2.14) to obtain[3]

$$\langle \psi(x) \rangle_A \simeq \int \sum_{\sigma\sigma'} \frac{m}{p^0} \frac{d^3\mathbf{p}}{(2\pi)^3} \frac{m}{p'^0} \frac{d^3\mathbf{p}'}{(2\pi)^3} \, e^{ip'x} u(\mathbf{p}',\sigma') \langle \mathbf{p}'\sigma' \, | \, \mathbf{p}\sigma \rangle_A \, [\, i\, \bar{u}(\mathbf{p},\sigma)\eta(p)\,],$$

$$(3.2.15)$$

where

$$\langle \mathbf{p}'\sigma' \, | \, \mathbf{p}\sigma \rangle_A = \left[ (2\pi)^3 \frac{p^0}{m} \delta_{\sigma'\sigma} \delta^3(\mathbf{p}'-\mathbf{p}) + i\, e\, \bar{u}(\mathbf{p}',\sigma') \gamma^\mu A_\mu(p'-p) u(\mathbf{p},\sigma) \right],$$

$$(3.2.16)$$

which coincides with (3.2.5) for $e = 0$. In the process of deriving (3.2.15), (3.2.16), we have, by using (3.1.10), (I.21), conveniently rewritten the first term $S_+(x-x')$ on the right-hand side of (3.2.14) as

$$S_+(x-x') = i \int \sum_{\sigma\sigma'} \frac{m}{p^0} \frac{d^3\mathbf{p}}{(2\pi)^3} \frac{m}{p'^0} \frac{d^3\mathbf{p}'}{(2\pi)^3} \, e^{i(p'x-px')} \times$$

$$\times u(\mathbf{p}',\sigma') \, [\, (2\pi)^3 \frac{p^0}{m} \delta_{\sigma\sigma'} \delta^3(\mathbf{p}'-\mathbf{p}) \,] \, \bar{u}(\mathbf{p},\sigma), \quad (3.2.17)$$

---

[3]See Problem 3.3.

for $x^0 > x'^0$, and defined the Fourier transform

$$A_\mu(x) = \int \frac{(dQ)}{(2\pi)^4} e^{iQx} A_\mu(Q). \tag{3.2.18}$$

We note that (3.2.16) is also valid for a time-independent external field $A_\mu(\mathbf{x})$ as long as $x^0$, $-x'^0$ are taken large in a limiting sense. In this case $A_\mu(p' - p)$ in (3.2.16) will be proportional to $\delta(p'^0 - p^0)$.

In the causal arrangement discussed above, the quantum field $\psi(x)$, in a source free region, destructs the particle after it has scattered off the external field $A_\mu$, with the transition amplitude for the process corresponding to $\mathbf{p}\sigma \to \mathbf{p}'\sigma'$, say, for $\mathbf{p} \neq \mathbf{p}'$, given from (3.2.16) to be

$$\langle \mathbf{p}'\sigma' | \mathbf{p}\sigma \rangle_A = [i e \bar{u}(\mathbf{p}', \sigma')\gamma^\mu A_\mu(p' - p)u(\mathbf{p}, \sigma)]. \tag{3.2.19}$$

The external electromagnetic field $A_\mu(\mathbf{x})$ may in turn create particles. This will be taken up in Sect. 3.8.

The structure given in (3.2.16) is interesting as, in particular, it provides us with the value of the magnetic moment of the charged particle (electron). To this end, we may use the equation $(\gamma p + m)u(\mathbf{p}, \sigma) = 0$ (see (I.10)), to write $u(\mathbf{p}, \sigma) = -(\gamma p/m) u(\mathbf{p}, \sigma)$, multiply the latter from the left by $\gamma^\mu$ to derive the identity

$$\gamma^\mu u(\mathbf{p}, \sigma) = \frac{p^\mu}{m} u(\mathbf{p}, \sigma) - p_\nu \frac{[\gamma^\mu, \gamma^\nu]}{2m} u(\mathbf{p}, \sigma), \tag{3.2.20}$$

and similarly

$$\bar{u}(\mathbf{p}', \sigma')\gamma^\mu = \bar{u}(\mathbf{p}', \sigma')\frac{p'^\mu}{m} + \bar{u}(\mathbf{p}', \sigma')p'_\nu \frac{[\gamma^\mu, \gamma^\nu]}{2m}. \tag{3.2.21}$$

By multiplying (3.2.20) from the left by $\bar{u}(\mathbf{p}', \sigma')$ and (3.2.21) from the right by $u(\mathbf{p}, \sigma)$, and adding the resulting two equations lead to the following decomposition, referred to as the Gordon decomposition,

$$e[\bar{u}(\mathbf{p}', \sigma')\gamma^\mu u(\mathbf{p}, \sigma)]A_\mu(Q)$$

$$= e\bar{u}(\mathbf{p}', \sigma')\left[ \frac{(p^\mu + p'^\mu)}{2m} + \frac{[\gamma^\mu, \gamma^\nu]}{4m} Q_\nu \right]u(\mathbf{p}, \sigma)A_\mu(Q)$$

$$= \bar{u}(\mathbf{p}', \sigma')\left[ e\frac{(p^\mu + p'^\mu)}{2m} A_\mu(Q) + \frac{i e}{2m}(Q_\mu A_\nu(Q) - Q_\nu A_\mu(Q))\frac{i}{4}[\gamma^\mu, \gamma^\nu] \right]u(\mathbf{p}, \sigma), \tag{3.2.22}$$

and $Q^\nu = p'^\nu - p^\nu$ is the (four) momentum transfer. In reference to the second term in the second equality on the right-hand side of (3.2.22), note that the specific contribution over the sum of the indices $\mu$, $\nu$, for $\mu = i$, $\nu = j$, is given by

$$\frac{\mathrm{i}\,e}{2m}\, \varepsilon^{\,ijk}\left(Q_i A_j(Q) - Q_j A_i(Q)\right)\left[\bar{u}(\mathbf{p}',\sigma')\,S^k\,u(\mathbf{p},\sigma)\right], \quad i,j,k = 1,2,3,$$

(3.2.23)

with a sum over repeated indices understood, where $S^k = \mathrm{i}\varepsilon^{s\ell k}[\gamma^s, \gamma^\ell]/8$ is the spin matrix in (I.15), and in writing (3.2.23), we have used, in the process, the identity

$$\varepsilon^{\,ijk}\varepsilon^{\,s\ell k} = \delta^{is}\delta^{j\ell} - \delta^{i\ell}\delta^{js}.$$

(3.2.24)

We note that $\mathrm{i}\varepsilon^{\,ijk}\left(Q_i A_j(q) - Q_j A_i(Q)\right)/2$ denotes the Fourier transform of the magnetic field

$$B^k(Q) = \frac{\mathrm{i}}{2}\,\varepsilon^{\,ijk}\left(Q_i A_j(Q) - Q_j A_i(Q)\right)$$

(3.2.25)

and we may rewrite (3.2.23) as

$$\left[\bar{u}(\mathbf{p}',\sigma')\,\boldsymbol{\mu}\,u(\mathbf{p},\sigma)\right]\cdot\mathbf{B}(Q), \qquad \boldsymbol{\mu} = \frac{g\,e}{2m}\,\mathbf{S},$$

(3.2.26)

where $\boldsymbol{\mu}$ is the magnetic dipole moment of the charged particle (electron) with the $g$-factor equal to 2. The modification of this value due to so-called radiative corrections in QED will be considered later in Sect. 5.11.2.

### 3.2.1 Charge Conjugation (C), Parity Transformation (T), and Time Reversal (T) of the Dirac Quantum Field

In the remaining part of this section, we investigate the transformation rules of the Dirac quantum field under charge conjugation (C), parity transformation (P), and time reversal (T).

In the presence of an external electromagnetic field, the Dirac equation reads from (3.2.10), with $\eta = 0$,

$$\left[\gamma^\mu\left(\frac{\partial_\mu}{\mathrm{i}} - e A_\mu(x)\right) + m\right]\psi(x) = 0,$$

(3.2.27)

while from Eq. (I.3), $\mathscr{C}\,\overline{\psi}^{\top} \equiv \psi^{\mathscr{C}}$ satisfies the same equation with sign of the charge e reversed, i.e.,

$$\left[\gamma^\mu\left(\frac{\partial_\mu}{\mathrm{i}} + e A_\mu(x)\right) + m\right]\psi^{\mathscr{C}}(x) = 0,$$

(3.2.28)

where $\mathscr{C} = i\gamma^2\gamma^0$ is the charge conjugation matrix defined in Eq. (I.3). Hence we may define the charge conjugation of a Dirac quantum field by

$$C\,\psi(x)\,C^{-1} = \mathscr{C}\,\overline{\psi}^{\mathsf{T}}(x), \tag{3.2.29}$$

up to a phase factor.

On the other hand, according to Eq. (I.18), $\gamma^0\psi(x')$, with $x = (x^0, -\mathbf{x})$, satisfies the same Dirac equation $(\gamma^\mu\partial_\mu/i + m)\psi(x) = 0$, i.e.,

$$\left(\frac{\gamma^\mu\partial_\mu}{i} + m\right)\gamma^0\psi(x') = 0, \qquad x' = (x^0, -\mathbf{x}). \tag{3.2.30}$$

Hence we may infer that the parity transformation of the Dirac quantum field may be defined by

$$P\,\psi(x)\,P^{-1} = \gamma^0\psi(x'), \tag{3.2.31}$$

up to a phase factor.

For time reversal, we note that by setting $x'' = (-x^0, \mathbf{x})$, that we may rewrite the Dirac equation $(\gamma^\mu\partial_\mu/i + m)\psi(x) = 0$ as

$$\left[-\frac{\gamma^0\partial_0''}{i} + \frac{\boldsymbol{\gamma}\cdot\boldsymbol{\partial}}{i} + m\right]\psi(x) = 0. \tag{3.2.32}$$

Upon multiplying the latter equation by $\gamma^5\mathscr{C}$, making the substitution $x'' \leftrightarrow x$, and using the facts that

$$[\gamma^5\mathscr{C}, \gamma^0] = 0, \quad [\gamma^5\mathscr{C}, \gamma^2] = 0, \quad \{\gamma^5\mathscr{C}, \gamma^1\} = 0, \quad \{\gamma^5\mathscr{C}, \gamma^3\} = 0, \tag{3.2.33}$$

as is easily verified, we may rewrite (3.2.32) as

$$\left[+\frac{\gamma^2\partial_2}{i} - \frac{\gamma^0\partial_0}{i} - \frac{\gamma^1\partial_1}{i} - \frac{\gamma^3\partial_3}{i} + m\right]\gamma^5\mathscr{C}\,\psi(x'') = 0. \tag{3.2.34}$$

Finally we make use of the property of T, as an anti-unitary operator, that it complex conjugates, and note that, e.g., in the Dirac representation, $(\gamma^2)^* = -\gamma^2$, and $\gamma^0, \gamma^1, \gamma^3$ are real, to obtain upon multiplying (3.2.34) from the left by $T^{-1}$, and from the right by T, the equation

$$\left(\frac{\gamma^\mu\partial_\mu}{i} + m\right)T^{-1}\gamma^5\mathscr{C}\,\psi(x'')T = 0, \tag{3.2.35}$$

which by comparing it with the Dirac equation, gives the following transformation rule for time reversal of the Dirac quantum field

$$\mathrm{T}\,\psi(x)\,\mathrm{T}^{-1} = \gamma^5 \mathscr{C}\,\psi(x''), \quad x'' = (-x^0, \mathbf{x}), \tag{3.2.36}$$

up to a phase factor.

## 3.3   Re-Discovering the Positron and Eventual Discovery of Anti-Matter

We have seen in Sect. 3.2, (3.2.1),(3.2.6), that the vacuum expectation value of the Dirac quantum field $\psi(x)$, in the presence of an external source $\eta(x)$, i.e., satisfying (3.1.1), is given by

$$\frac{\langle\,0_+ |\,\psi(x)\,|\,0_-\rangle}{\langle\,0_+ |\,0_-\rangle} = i\int (dx')\int\frac{d^3\mathbf{p}}{(2\pi)^3 2p^0}\,e^{ip\,(x-x')}\,(-\gamma p + m)\,\eta(x')$$

$$= \int\sum_\sigma\frac{m}{p^0}\frac{d^3\mathbf{p}}{(2\pi)^3}\,\frac{\langle\,0_+ |\,\psi(x)\,|\,\mathbf{p}\sigma\rangle}{\langle\,0_+ |\,0_-\rangle}\,\langle\mathbf{p}\sigma\,|\,0_-\rangle, \tag{3.3.1}$$

when the source $\eta(x')$ is in operation, i.e., non-zero, and then switched off sometime *before* the time $x^0$, i.e., $x^0 > x'^0$ in reference to Fig. 3.1, for all $x'^0$ contributing to the above integral. Here $\langle\,\mathbf{p}\sigma\,|\,0_-\rangle$ denotes an amplitude that the source emits a particle of momentum-spin $(\mathbf{p}, \sigma)$, energy $p^0 = +\sqrt{\mathbf{p}^2 + m^2}$ (see (3.2.4) and later on in this section),

$$\frac{\langle\,0_+ |\,\psi(x)\,|\,\mathbf{p}\sigma\rangle}{\langle\,0_+ |\,0_-\rangle} = e^{ipx}\,u(\mathbf{p},\sigma), \qquad p^0 = +\sqrt{\mathbf{p}^2 + m^2}, \tag{3.3.2}$$

leading to the inevitable conclusion that $\psi(x)$ annihilates such a particle to end up with the vacuum. The so-called wave-function of the particle is given by (3.3.2). Similarly, when the source $\overline{\eta}(x')$ is switched on *after* the time $x^0$, i.e., $x'^0 > x^0$ (see (3.1.16), (3.2.8), (3.2.9)),

$$\frac{\langle\,0_+ |\,\overline{\psi}(x)\,|\,0_-\rangle}{\langle\,0_+ |\,0_-\rangle} = i\int (dx')\int\frac{d^3\mathbf{p}}{(2\pi)^3 2p^0}\,\overline{\eta}(x')\,(-\gamma p + m)\,e^{ip\,(x'-x)}$$

$$= \int\sum_\sigma\frac{m}{p^0}\frac{d^3\mathbf{p}}{(2\pi)^3}\,\langle\,0_+ |\,\mathbf{p}\sigma\rangle\,\frac{\langle\,\mathbf{p}\sigma\,|\,\overline{\psi}(x)\,|\,0_-\rangle}{\langle\,0_+ |\,0_-\rangle}, \tag{3.3.3}$$

with

$$\frac{\langle\,\mathbf{p}\,\sigma\,|\,\overline{\psi}(x)\,|\,0_-\rangle}{\langle\,0_+\,|\,0_-\rangle} = e^{-ipx}\,\overline{u}(\mathbf{p},\sigma), \qquad p^0 = +\sqrt{\mathbf{p}^2 + m^2}, \qquad (3.3.4)$$

$$\langle\,0_+\,|\,\mathbf{p}\,\sigma\,\rangle = [\,i\,\overline{\eta}(p)\,u(\mathbf{p},\sigma)\,], \qquad\qquad (3.3.5)$$

representing an amplitude that the source $\overline{\eta}(x')$ has absorbed a particle of momentum-spin $(\mathbf{p}, \sigma)$, thus playing the role of a detector, and leading to the inevitable conclusion that $\overline{\psi}(x)$ creates such a particle which eventually goes to the detector.

We have also considered the Dirac equation in an external electromagnetic field in (3.2.10) under the same condition as in (3.3.1), with the source switched off sometime before the time $x^0$, and we have identified in (3.2.16), (3.2.10), the charge of the particle to be $e$.

What is the role of $\overline{\psi}(x)$ if the source is nonzero and is finally switched off sometime *before* the time $x^0$, i.e., $x^0 > x'^0$ instead of the situation encountered in (3.3.3).

In this case, we may use the expression in (3.1.17) and the orthogonal projection in Eq. (I.22), to write

$$\frac{\langle\,0_+\,|\,\overline{\psi}(x)\,|\,0_-\rangle}{\langle\,0_+\,|\,0_-\rangle} = i\int(dx')\int\frac{d^3\mathbf{p}}{(2\pi)^3 2p^0}\,e^{-ip(x'-x)}\,\overline{\eta}(x')(\gamma p + m)$$

$$= \int\sum_\sigma\frac{m}{p^0}\frac{d^3\mathbf{p}}{(2\pi)^3}\,e^{ipx}\,\overline{v}(\mathbf{p},\sigma)\,[-i\,\overline{\eta}(-p)\,v(\mathbf{p},\sigma)]. \qquad (3.3.6)$$

Upon comparing this equation with (3.3.1), (3.3.2), for the *same* causal arrangement where now the source $\overline{\eta}(x')$ is also switched off before the time $x^0$, we may infer that $\overline{\psi}(x)$ annihilates a particle with wave-function

$$\frac{\langle\,0_+\,|\,\overline{\psi}(x)\,|\,\mathbf{p}\,\sigma\,\rangle}{\langle\,0_+\,|\,0_-\rangle} = e^{ipx}\,\overline{v}(\mathbf{p},\sigma), \qquad p^0 = +\sqrt{\mathbf{p}^2 + m^2}, \qquad (3.3.7)$$

and

$$[-i\,\overline{\eta}(-p)\,v(\mathbf{p},\sigma)] \equiv \langle\,\mathbf{p}\,\sigma\,|\,0_-\rangle, \qquad\qquad (3.3.8)$$

represents an amplitude that the source $\overline{\eta}(x')$ creates a particle of momentum-spin $(\mathbf{p}, \sigma)$.

What is the nature of this particle? To answer this question, we introduce an external electromagnetic potential probe $\delta A_\mu(x)$ off which a particle, after being created, may, for example, be scattered. Before doing this, we rewrite (3.3.6) as

$$\frac{\langle\, 0_+ \mid \overline{\psi}(x) \mid 0_-\rangle}{\langle\, 0_+ \mid 0_-\rangle} = \int \sum_{\sigma,\sigma'} \frac{m}{p^0}\frac{d^3\mathbf{p}}{(2\pi)^3}\frac{m}{p'^0}\frac{d^3\mathbf{p}'}{(2\pi)^3}$$

$$\times \frac{\langle\, 0_+ \mid \overline{\psi}(x) \mid \mathbf{p}'\sigma'\rangle}{\langle\, 0_+ \mid 0_-\rangle}\langle\, \mathbf{p}'\sigma' \mid \mathbf{p}\sigma\rangle\langle\, \mathbf{p}\sigma \mid 0_-\rangle. \qquad (3.3.9)$$

where $\langle\, \mathbf{p}'\sigma' \mid \mathbf{p}\sigma\rangle$ is defined in (3.2.5).

In the presence of the external probe, the Dirac equation for $\overline{\psi}(x)$ reads

$$\overline{\psi}(x)\left[-\gamma^\mu\left(\frac{\overleftarrow{\partial_\mu}}{i} + e\,\delta A_\mu(x)\right) + m\right] = \overline{\eta}(x). \qquad (3.3.10)$$

This simply amounts in the modification of the matrix element of $\overline{\psi}(x)$ between the vacuum states, in the absence of the probe from

$$\frac{\langle\, 0_+ \mid \overline{\psi}(x) \mid 0_-\rangle}{\langle\, 0_+ \mid 0_-\rangle} = \int(dx')\,\overline{\eta}(x')\,S_+(x',x), \qquad (3.3.11)$$

where

$$S_+(x',x)\left(-\gamma^\mu\frac{\overleftarrow{\partial_\mu}}{i} + m\right) = \delta^{(4)}(x'-x), \qquad (3.3.12)$$

to

$$\frac{\langle\, 0_+ \mid \overline{\psi}(x) \mid 0_-\rangle_{\delta A}}{\langle\, 0_+ \mid 0_-\rangle_{\delta A}} \simeq \int(dx')\,\overline{\eta}(x')$$

$$\times \left[S_+(x'-x) + e\int(dx'')\,S_+(x'-x'')\,\gamma^\mu\delta A_\mu(x'')\,S_+(x''-x)\right], \qquad (3.3.13)$$

in the presence of the external probe as a perturbation of the system.

Repeating the same analysis as the one leading to (3.2.15), (3.2.16), as before, gives[4]

$$\frac{\langle\, 0_+ \mid \overline{\psi}(x) \mid 0_-\rangle_{\delta A}}{\langle\, 0_+ \mid 0_-\rangle_{\delta A}} = \int \sum_{\sigma\sigma'} \frac{m}{p^0}\frac{d^3\mathbf{p}}{(2\pi)^3}\frac{m}{p^0}\frac{d^3\mathbf{p}}{(2\pi)^3}\,e^{ip'x}\,\overline{v}(\mathbf{p}',\sigma')$$

$$\times \langle\mathbf{p}'\sigma' \mid \mathbf{p}\sigma\rangle_{\delta A}\,[-i\,\overline{\eta}(-p)\,v(\mathbf{p},\sigma)], \qquad (3.3.14)$$

$$\langle\mathbf{p}'\sigma' \mid \mathbf{p}\sigma\rangle_{\delta A} = \left[(2\pi)^3\frac{p^0}{m}\,\delta_{\sigma'\sigma}\,\delta^3(\mathbf{p}'-\mathbf{p}) - i\,e\,\overline{v}(\mathbf{p},\sigma)\gamma^\mu\delta A_\mu(p'-p)\,v(\mathbf{p}',\sigma')\right]. \qquad (3.3.15)$$

---

[4] See Problem 3.5.

Upon comparing this equation with (3.2.16), and taking into account (3.3.6), (3.3.7), (3.3.8), we may infer that $\overline{\psi}(x)$ annihilates a particle of mass $m$ and charge $-e$, i.e., of charge $+|e|$. For such a particle of momentum $\mathbf{p}$, its energy is given by $p^0 = +\sqrt{\mathbf{p}^2 + m^2}$, and its wave-function is given in (3.3.7). Also $\langle\,\mathbf{p}\sigma\,|\,0_-\rangle$, as defined in (3.3.8), represents the amplitude that the source $\overline{\eta}(x)$ creates such a particle. This is the anti-electron, the positron.

As mentioned in the Introductory Chapter of the book, prior to its experimental discovery of the positron, it was referred to as the donkey electron by George Gamow because it would move in the opposite direction to that of an electron in an applied field.

It is important to note that $\overline{v}(\mathbf{p}, \sigma)$, corresponding to the positron in its initial state, is to the left of $\gamma^\mu$, and $v(\mathbf{p}', \sigma')$, corresponding to the its final state, is on its right-hand side.

Under space reflection, we have from below Eq. (I.18), that

$$u(\mathbf{p}, \sigma) \;\rightarrow\; \gamma^0 u(\mathbf{p}, \sigma) \;=\; +u(-\mathbf{p}, \sigma),$$

$$\overline{v}(\mathbf{p}, \sigma) \;\rightarrow\; \overline{v}(\mathbf{p}, \sigma)\gamma^0 \;=\; v^\dagger(\mathbf{p}, \sigma) \;=\; -\overline{v}(-\mathbf{p}, \sigma),$$

and hence the electron and positron have opposite intrinsic parities.

Finally we note that for the source switched on after $x^0$, i.e., $x'^0 > x^0$,

$$\frac{\langle\,0_+|\,\psi(x)\,|\,0_-\rangle}{\langle\,0_+|\,0_-\rangle} = \mathrm{i}\int(\mathrm{d}x')\int \frac{\mathrm{d}^3\mathbf{p}}{(2\pi)^3 2p^0}\, \mathrm{e}^{-ip\,(x-x')}(\gamma p + m)\,\eta(x'),$$

$$= \int\sum_\sigma \frac{m}{p^0}\frac{\mathrm{d}^3\mathbf{p}}{(2\pi)^3}\,[-\mathrm{i}\,\overline{v}(\mathbf{p}, \sigma)\,\eta(-p)]\,\mathrm{e}^{-ipx}\,v(\mathbf{p}, \sigma), \qquad (3.3.16)$$

where we have used Eq. (II.21). Now

$$\frac{\langle\,\mathbf{p}\sigma\,|\,\psi(x)\,|\,0_-\rangle}{\langle\,0_+|\,0_-\rangle} \;=\; \mathrm{e}^{-ipx}\,v(\mathbf{p}, \sigma), \qquad p^0 = +\sqrt{\mathbf{p}^2 + m^2}, \qquad (3.3.17)$$

and $\psi(x)$ creates a positron,

$$[-\mathrm{i}\,\overline{v}(\mathbf{p}, \sigma)\,\eta(-p)] \;=\; \langle\,0_+|\,\mathbf{p}\,\sigma\rangle, \qquad (3.3.18)$$

denotes an amplitude that the source $\eta(x)$ absorbs a positron with momentum-spin $(\mathbf{p}, \sigma)$ and hence it plays the role of a detector.

Clearly to distinguish between the electron and positron we may rewrite the states as $|\, \mathbf{p}\sigma, \varepsilon\,\rangle$, with $\varepsilon = \mp$ for the electron/positron, respectively, with the conventions

$$\langle\, \mathbf{p}'\sigma', \varepsilon'\, |\, \mathbf{p}\sigma, \varepsilon\,\rangle = (2\pi)^3 \frac{p^0}{m} \delta_{\sigma'\sigma}\, \delta_{\varepsilon'\varepsilon} \delta^3(\mathbf{p}' - \mathbf{p}), \qquad (3.3.19)$$

and

$$\frac{\langle\, 0_+ |\, \psi(x)\, |\, \mathbf{p}\sigma, -\rangle}{\langle\, 0_+ |\, 0_-\rangle} = \mathrm{e}^{ipx}\, u(\mathbf{p}, \sigma), \qquad (3.3.20)$$

$$\frac{\langle\, \mathbf{p}\sigma, + |\, \psi(x)\, |\, 0_-\rangle}{\langle\, 0_+ |\, 0_-\rangle} = \mathrm{e}^{-ipx}\, v(\mathbf{p}, \sigma), \qquad (3.3.21)$$

with $\psi(x)$ annihilating/creating an electron/positron, respectively,

$$\frac{\langle\, \mathbf{p}\sigma, - |\, \overline{\psi}(x)\, |\, 0_-\rangle}{\langle\, 0_+ |\, 0_-\rangle} = \mathrm{e}^{-ipx}\, \overline{u}(\mathbf{p}, \sigma), \qquad (3.3.22)$$

$$\frac{\langle\, 0_+ |\, \overline{\psi}(x)\, |\, \mathbf{p}\sigma, +\rangle}{\langle\, 0_+ |\, 0_-\rangle} = \mathrm{e}^{ipx}\, \overline{v}(\mathbf{p}, \sigma), \qquad (3.3.23)$$

with $\overline{\psi}(x)$ creating/annihilating an electron/positron, respectively.

Clearly, this was the beginning of the discovery of anti-matter, in general, thanks to the Dirac equation. It is interesting to remember (see the Introductory Chapter of the book) that, in one of his talks, Dirac remarked that his equation was more intelligent than its author.

With the above analysis now carried out, it will be quite evident in the sequel, that the computations of transition amplitudes of processes in quantum field theory become more manageable.

The amplitudes for detection of a particle/antiparticle and emission of particle/antiparicle are spelled out below using the sign of the charge $\varepsilon = \mp$.[5] To do this we obtain, in the process, the expression for the vacuum-to-vacuum transition amplitude $\langle\, 0_+ |\, 0_-\rangle$ for the Dirac equation in the presence of an external source.

### 3.3.1   $\langle\, 0_+ |\, 0_-\rangle$ *for the Dirac Equation*

For the Dirac equation (3.1.1) in the presence of an external source,

$$\left(\frac{\gamma\partial}{i} + m\right)\psi(x) = \eta(x), \qquad (3.3.24)$$

---

[5]When it is evident we suppress the parameter $\varepsilon$ in amplitudes.

we have seen in (3.1.8) that

$$\langle\, 0_+ |\, \psi(x) \,| \, 0_- \rangle = \int (\mathrm{d}x')\, S_+(x - x')\, \eta(x')\, \langle\, 0_+ \,| \, 0_- \rangle, \qquad (3.3.25)$$

where $S_+(x - x')$ is given in (3.1.9).

In Sect. 4.6, on the Quantum Dynamical Principle, we will learn that the matrix element $\langle\, 0_+ \,|\, \psi(x) \,|\, 0_- \rangle$, for the Dirac equation in (3.3.24), may be generated by functionally differentiating $\langle\, 0_+ \,|\, 0_- \rangle$ with respect to the external source $\overline{\eta}(x)$. More precisely,

$$(-\,\mathrm{i})\,\frac{\delta}{\delta\overline{\eta}(x)}\,\langle\, 0_+ \,|\, 0_- \rangle = \langle\, 0_+ |\, \psi(x) \,|\, 0_- \rangle, \qquad (3.3.26)$$

which from (3.3.25) leads to

$$(-\,\mathrm{i})\,\frac{\delta}{\delta\overline{\eta}(x)}\,\langle\, 0_+ \,|\, 0_- \rangle = \int (\mathrm{d}x')\, S_+(x - x')\, \eta(x')\, \langle\, 0_+ \,|\, 0_- \rangle. \qquad (3.3.27)$$

Upon integration this gives

$$\langle\, 0_+ \,|\, 0_- \rangle = \exp\Big[\, \mathrm{i} \int (\mathrm{d}x)\, (\mathrm{d}x')\, \overline{\eta}(x) S_+(x - x')\eta(x') \Big], \qquad (3.3.28)$$

which should be compared with (2.6.13)/(2.6.10) and obviously verifies (3.3.27). We have normalized $\langle\, 0_+ \,|\, 0_- \rangle$ to unity for $\eta(x) = 0$, $\overline{\eta}(x) = 0$, i.e., in the absence of the external source, dispensing with any phase factor, signalling the fact that the vacuum stays the same and nothing happens in the latter case.

Consider a source function $\eta(x)$ given by (Fig. 3.4)

$$\eta(x) = \eta_1(x) + \eta_2(x), \qquad (3.3.29)$$

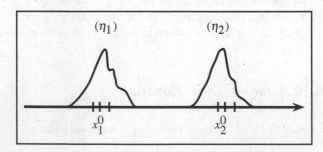

**Fig. 3.4** A causal arrangement with the source $\eta_2$ switched on after the source $\eta_1$ is switched off

with a causal arrangement such that $\eta_2(x)$ is switched on after $\eta_1(x)$ is switched off. We will learn in Sect. 3.5, in the study of the Spin & Statistics Connection for the Dirac field, that $\eta(x)$, $\overline{\eta}(x)$ are Grassmann variables.

Thus we may write

$$\langle \, 0_+ \mid 0_- \rangle^{\eta} = \langle \, 0_+ \mid 0_- \rangle^{\eta_2} \exp \mathrm{i} W_{21} \langle \, 0_+ \mid 0_- \rangle^{\eta_1}, \qquad (3.3.30)$$

where

$$\mathrm{i} W_{21} = \mathrm{i} \int (\mathrm{d}x_1)(\mathrm{d}x_2)\overline{\eta}_2(x_2)S_+(x_2 - x_1)\eta_1(x_1)$$

$$+ \mathrm{i} \int (\mathrm{d}x_1)(\mathrm{d}x_2)\overline{\eta}_1(x_1)S_+(x_1 - x_2)\eta_2(x_2). \qquad (3.3.31)$$

In (3.3.30) we have emphasized the source-dependence by rewriting, for example, (3.3.28) as $\langle \, 0_+ \mid 0_- \rangle^{\eta}$.

Using the corresponding two expressions for $S_+(x_2 - x_1)$, $S_+(x_1 - x_2)$, in (3.1.10), (3.1.11), respectively, and the projection operations in (I.21), (I.22), we obtain

$$\mathrm{i} W_{21} = \int \sum_{\sigma} \frac{m}{p^0} \frac{\mathrm{d}^3 \mathbf{p}}{(2\pi)^3} \left[ \mathrm{i}\,\overline{\eta}_2(p)\,u(\mathbf{p}, \sigma) \right] \left[ \mathrm{i}\,\overline{u}(\mathbf{p}, \sigma)\eta_1(p) \right]$$

$$+ \int \sum_{\sigma} \frac{m}{p^0} \frac{\mathrm{d}^3 \mathbf{p}}{(2\pi)^3} \left[ -\mathrm{i}\,\overline{v}(\mathbf{p}, \sigma)\,\eta_2(-p) \right] \left[ -\mathrm{i}\,\overline{\eta}_1(-p)v(\mathbf{p}, \sigma) \right]. \qquad (3.3.32)$$

In writing the second integral, we have used the anti-commutativity of $\overline{\eta}_1(-p)$, $\eta_2(-p)$, to move $\eta_2(-p)$ to the left of $\overline{\eta}_1(-p)$, thus picking another minus sign.

For bookkeeping purposes, we use an ingenious notation due to Schwinger, which facilitates further the analysis as well as the physical interpretation of the underlying theory. We write

$$\sqrt{2m\mathrm{d}\omega_p}\,\overline{\eta}(p)u(\mathbf{p}, \sigma) = \eta^*_{\mathbf{p}\sigma -}, \qquad \sqrt{2m\mathrm{d}\omega_p}\,(-)\overline{v}(\mathbf{p}, \sigma)\eta(-p) = \eta^*_{\mathbf{p}\sigma +}$$

$$\qquad (3.3.33)$$

$$\sqrt{2m\mathrm{d}\omega_p}\,\overline{u}(\mathbf{p}, \sigma)\eta(p) = \eta_{\mathbf{p}\sigma -}, \qquad \sqrt{2m\mathrm{d}\omega_p}\,\overline{\eta}(-p)(-)v(\mathbf{p}, \sigma) = \eta_{\mathbf{p}\sigma +}$$

providing *four* different entries, where

$$\mathrm{d}\omega_p = \frac{\mathrm{d}^3 \mathbf{p}}{2p^0(2\pi)^3}. \qquad (3.3.34)$$

Don't let the notation in (3.3.33) scare you. It is for bookkeeping purposes, and it provides a simple worry free formalism for getting correct numerical factors such as $(2\pi)^3$ and so on.

Let $r = (\mathbf{p}, \sigma, \epsilon)$, $\epsilon = \mp$, corresponding to all possible values that the variables take in a convenient discrete values notation. This allows us to rewrite $i\,W_{21}$ as a formal sum

$$i\,W_{21} = \sum_{(r)} [\, i\,\eta_{r2}^* \, i\,\eta_{r1}\,]. \tag{3.3.35}$$

One may also rewrite $\langle\, 0_+ \mid 0_- \rangle^\eta$ in the form of a unitarity sum

$$\langle\, 0_+ \mid 0_- \rangle^\eta = \sum_n \sum_{n_1+n_2+\ldots=n} \langle\, 0_+ \mid n; n_1, n_2, \ldots \rangle^{\eta_2} \langle\, n; n_1, n_2, \ldots \mid 0_- \rangle^{\eta_1}, \tag{3.3.36}$$

with $\langle\, n; n_1, n_2, \ldots \mid 0_- \rangle^{\eta_1}$ denoting the amplitude that the source $\eta_1$ has emitted $n$ particles, $n_1$ of which have the value $r_1$, and so on. The Fermi character of the anti-commuting source requires that $n_1 = 0$ or 1, and so on. Similarly, $\langle\, 0_+ \mid n; n_1, n_2, \ldots \rangle^{\eta_2}$, denotes the amplitude that the source $\eta_2^*$ has absorbed (detected) $n$ particles, and so on.

Upon expanding the exponential of (3.3.35), using in the process (3.3.30), and comparing with (3.3.36), we obtain for any given source $\eta$:

$$\langle\, n; n_1, n_2, \ldots \mid 0_- \rangle^\eta = \langle\, 0_+ \mid 0_- \rangle^\eta \frac{(i\eta_{r_1})^{n_1}}{\sqrt{n_1!}} \frac{(i\eta_{r_2})^{n_2}}{\sqrt{n_2!}} \cdots, \tag{3.3.37}$$

$$\langle\, 0_+ \mid n; n_1, n_2, \ldots \rangle^\eta = \langle\, 0_+ \mid 0_- \rangle^\eta \cdots \frac{(i\eta_{r_2}^*)^{n_2}}{\sqrt{n_2!}} \frac{(i\eta_{r_1}^*)^{n_1}}{\sqrt{n_1!}}, \tag{3.3.38}$$

where the $n_i = 0$ or 1.

Physically, we note that $\langle\, 0_+ \mid 0_- \rangle$ corresponds to processes where arbitrary number of particles are emitted and re-absorbed through the time of operation of $(\eta, \overline{\eta})$, with no particles (antiparticles) present after the sources cease to operate, and before they begin to operate. Accordingly, in scattering theory, we define amplitudes for emission of a given particle (antiparticle) by a given source, that escapes this parent source as a free particle, by dividing the corresponding above expression by $\langle\, 0_+ \mid 0_- \rangle$. Thus one eliminates particles absorbed and then re-absorbed that do not participate in dynamical scattering process of interest. Similarly, the amplitude of absorption of a given particle (antiparticle), way after the particle (antiparticle) emerges, say, from a scattering process, is defined by dividing the corresponding above expression by $\langle\, 0_+ \mid 0_- \rangle$. Accordingly, we introduce the amplitudes of either absorption or emission of a particle (antiparticle) by a given source as follows:

$$\sqrt{2m\,d\omega_p}\,[\,i\,\overline{\eta}(p)\,u(\mathbf{p},\sigma)\,] = \langle\,0_+\,|\,\mathbf{p}\,\sigma, -\rangle, \tag{3.3.39}$$

$$\sqrt{2m\,d\omega_p}\,[\,-i\,\overline{v}(\mathbf{p},\sigma)\,\eta(-p)\,] = \langle\,0_+\,|\,\mathbf{p}\,\sigma, +\rangle, \tag{3.3.40}$$

$$\sqrt{2m\,d\omega_p}\,[\,i\,\overline{u}(\mathbf{p},\sigma)\,\eta(p)\,] = \langle\,\mathbf{p}\,\sigma, -\,|\,0_-\rangle, \tag{3.3.41}$$

$$\sqrt{2m\,d\omega_p}\,[\,-i\,\overline{\eta}(-p)\,v(\mathbf{p},\sigma)\,] = \langle\,\mathbf{p}\,\sigma, +\,|\,0_-\rangle, \tag{3.3.42}$$

where $\langle\,0_+\,|\,\mathbf{p}\,\sigma, -\rangle$ denotes the amplitude that a source $\overline{\eta}$ absorbs a particle, thus playing the role of a detector, and $\langle\,0_+\,|\,\mathbf{p}\,\sigma, +\rangle$, denotes the amplitude that $\eta$ also absorbs an antiparticle. Given that $\langle\,0_+\,|\,\mathbf{p}\,\sigma, -\rangle$ denotes the amplitude for the event just mentioned, i.e., that $\overline{\eta}$ absorbs a given particle (electron), the analysis leading to (3.3.18), shows that $\langle\,0_+\,|\,\mathbf{p}\,\sigma, +\rangle$ necessarily represents the amplitude that $\eta$ absorbs an *anti*particle (the positron). Similarly, $\langle\,\mathbf{p}\,\sigma, -\,|\,0_-\rangle$ represents the amplitude that $\eta$ emits a particle, and $\langle\,\mathbf{p}\,\sigma, +\,|\,0_-\rangle$ represents the amplitude that $\overline{\eta}$ emits an antiparticle.

It is advisable to keep the factors $\sqrt{2m\,d\omega_p}$ in the amplitudes in applications in order not to waste time guessing numerical multiplicative factors in physical applications.

The moral of the above analysis is this. For a general interacting theory, we obtain an expression for $\langle\,0_+\,|\,0_-\rangle$ as in (3.3.30), where now the expression corresponding to $W_{21}$ may be expanded in powers of the source functions. Transition amplitudes for arbitrary processes are then extracted, from a unitarity sum, and are given by the coefficients of these source functions written as in (3.3.39), (3.3.40), (3.3.41), and (3.3.42), with the sources appropriately arranged causally to reflect the actual process where the particles in question are emitted and detected by these sources. The procedure will be spelled out in detail in Sect. 5.8, (5.8.14), (5.8.15), (5.8.16), (5.8.17), (5.8.18), (5.8.19), (5.8.20), and (5.8.21). Applications of these will be given when considering interacting theories. See also (3.2.16), (3.2.19), (3.2.26), (3.3.15), (3.4.1) and Problem 3.6 for simple demonstrations. More involved non-linear theories will be dealt with later.

## 3.4 Coulomb Scattering of Relativistic Electrons

We have seen in (3.2.19) that, to the leading order in an external electromagenetic potential $A_\mu(x)$, the transition amplitude of an electron scattering off such a field, with $\mathbf{p}\,\sigma \rightarrow \mathbf{p}'\sigma'$, for $\mathbf{p}' \neq \mathbf{p}$, is given by

$$\langle\mathbf{p}'\sigma'\,|\,\mathbf{p}\,\sigma\rangle_A = i\,e\,\overline{u}(\mathbf{p}',\sigma')\,\gamma^\mu A_\mu(p'-p)\,u(\mathbf{p},\sigma), \tag{3.4.1}$$

where

$$A_\mu(Q) = \int (\mathrm{d}x)\, \mathrm{e}^{-iQx} A_\mu(x). \tag{3.4.2}$$

For Coulomb scattering ($A_0 = -\phi$)

$$(eA_\mu(x)) = \left(\frac{Ze^2}{4\pi|\mathbf{x}|}, \mathbf{0}\right), \qquad eA_\mu(Q) = [-2\pi\,\delta(Q^0)\,\eta_{\mu 0}]\,\frac{Ze^2}{|\mathbf{Q}|^2}, \tag{3.4.3}$$

$\mathbf{Q} = \mathbf{p}' - \mathbf{p}$, and where $Z|e|$ is the charge of the scattering center. Also

$$\frac{1}{|\mathbf{p} - \mathbf{p}'|^2} = \frac{1}{4|\mathbf{p}|^2 \sin^2\frac{\theta}{2}}, \qquad \mathbf{p}\cdot\mathbf{p}' = |\mathbf{p}|\,|\mathbf{p}'|\cos\theta. \tag{3.4.4}$$

Using the formal Fermi rule $[2\pi\,\delta(p^0 - p'^0)]^2 \to [2\pi\,\delta(p^0 - p'^0)]\int \mathrm{d}x^0$, we find for the transition probability for the process in question from (3.4.1), per unit time, the expression

$$\frac{m}{p^0}\,\frac{\mathrm{d}^3\mathbf{p}}{(2\pi)^3}\,\frac{m}{p'^0}\,\frac{\mathrm{d}^3\mathbf{p}'}{(2\pi)^3}\,\frac{[2\pi\,\delta(p'^0 - p^0)\,Z^2\mathrm{e}^4]}{16|\mathbf{p}|^4\,\sin^4\frac{\theta}{2}}\,|\,\bar{u}(\mathbf{p}',\sigma')\gamma^0 u(\mathbf{p},\sigma)\,|^2. \tag{3.4.5}$$

Averaging over the initial electron spin projections $\sigma/2 = \pm 1/2$, and summing over the final ones, we obtain for the last term in the above equation, for $p^0 = p'^{0},$[6]

$$\frac{1}{2}\sum_{\sigma\sigma'}|\,\bar{u}(\mathbf{p}',\sigma')\gamma^0 u(\mathbf{p},\sigma)\,|^2 = \frac{1}{8m^2}\,\mathrm{Tr}\,[\,(-\gamma p + m)\gamma^0(-\gamma p' + m)\gamma^0\,]$$

$$= \frac{1}{m^2}\left((p^0)^2 \cos^2\frac{\theta}{2} + m^2\sin^2\frac{\theta}{2}\right), \tag{3.4.6}$$

where in writing the first equality, we have used Eq. (I.21) for the projection operation, and in the second equality we have used the properties of the gamma matrices (see Appendix I).

Since the initial velocity of the electron may be written as $\mathbf{v} = \mathbf{p}/p^0$, the incident flux is given by $(\mathrm{d}^3\mathbf{p}/(2\pi)^3)(|\mathbf{p}|/p^0)$ in a unit of volume. Using the relations

$$\mathrm{d}^3\mathbf{p}' = \mathrm{d}\Omega\,|\mathbf{p}'|^2\mathrm{d}|\mathbf{p}'|, \quad \delta(p'^0 - p^0) = \frac{p'^0}{|\mathbf{p}'|}\delta(|\mathbf{p}'| - |\mathbf{p}|), \quad |\mathbf{p}'|^2 = (p'^0)^2 - m^2, \tag{3.4.7}$$

---

[6] See Problem 3.7.

we have for the differential cross section of the process

$$\frac{d\sigma}{d\Omega} = \int_0^\infty \frac{d|\mathbf{p}'|}{(2\pi)^3} \frac{\delta(\mathbf{p}' - \mathbf{p})\, 2\pi\, Z^2 e^4}{16\, |\mathbf{p}|^4 \sin^4\frac{\theta}{2}} \left[ (p^0)^2 \cos^2\frac{\theta}{2} + m^2 \sin^2\frac{\theta}{2} \right], \qquad (3.4.8)$$

which integrates out to

$$\frac{d\sigma}{d\Omega} = \frac{Z^2 \alpha^2}{4 \sin^4\frac{\theta}{2}} \left( \frac{p^0}{|\mathbf{p}|^2} \right)^2 \left[ \cos^2\frac{\theta}{2} + \frac{m^2}{(p^0)^2} \sin^2\frac{\theta}{2} \right], \qquad (3.4.9)$$

and

$$\alpha = \frac{e^2}{4\pi}, \qquad (3.4.10)$$

defines the fine-structure constant. The expression in (3.4.9) is known as the Mott differential cross section. It may be equivalently rewritten as

$$\left.\frac{d\sigma}{d\Omega}\right|_{\text{Mott}} = \frac{Z^2 \alpha^2}{4\, \mathbf{p}^2 \beta^2 \sin^4\frac{\theta}{2}} \left[ 1 - \beta^2 \sin^2\frac{\theta}{2} \right], \qquad (3.4.11)$$

where we have used

$$p^0 = \frac{m}{\sqrt{1 - \beta^2}}, \qquad |\mathbf{p}| = \frac{m\beta}{\sqrt{1 - \beta^2}}. \qquad (3.4.12)$$

with $\beta$ denoting the speed of the electron ($c = 1$).

We note that since $d\Omega = 2\pi \sin\theta\, d\theta$, we encounter a divergence in the total cross section for $\theta \simeq 0$, corresponding to the forward direction $\mathbf{p} \simeq \mathbf{p}'$. Also note that (3.4.11) develops a singularity at low energies $|\mathbf{p}| \simeq 0$, referred to as an infra-red divergence. We will re-consider these points together with the so-called radiative corrections to this scattering process within QED later on in Sect. 5.12.

The familiar Rutherford formula follows directly from (3.4.9) by considering the non-relativistic (NR) limit

$$\left( \frac{d\sigma}{d\Omega} \right)_{\text{NR}} = \frac{Z^2 m^2 (e^2/4\pi)^2}{4|\mathbf{p}|^4 \sin^4\frac{\theta}{2}}, \qquad (3.4.13)$$

as expected.

## 3.5   Spin & Statistics and the Dirac Quantum Field; Anti-Commutativity Properties Derived

Every student who has ever heard of spin 1/2 is aware that the corresponding field theory should be quantized with anti-commutations relations. This is a special case of the Spin & Statistics Connection which is the subject matter of Sect. 4.5. As we will see, this property of Spin 1/2 is already inherited in the Dirac equation itself, and establishing the underlying statistics of the Dirac quantum field is straightforward and follows by a simple application of the main result obtained in Sect. 4.5, and it is conveniently treated here at this stage. We work in the Majorana representation of the gamma matrices (see (2.3.5), (2.3.6)).

One of the most important facts of the Majorana representation is that $(\gamma^\mu/i)^* = (\gamma^\mu/i)$, i.e., the matrices $\gamma^\mu/i$ are real. (See Sect. 2.3).

Define the $8 \times 8$ matrices

$$\Gamma^\mu = \begin{pmatrix} \gamma^\mu & 0 \\ 0 & \gamma^\mu \end{pmatrix}, \tag{3.5.1}$$

and write the Dirac quantum field in terms of a real and imaginary parts as

$$\psi_a = \frac{1}{\sqrt{2}} \left( \psi_a^1 + i\, \psi_a^2 \right), \qquad a = 1, 2, 3, 4. \tag{3.5.2}$$

We introduce the 8 component real field

$$\chi^\mathsf{T} = \left( \psi_1^1 \ \psi_2^1 \ \psi_3^1 \ \psi_4^1 \ \psi_1^2 \ \psi_2^2 \ \psi_3^2 \ \psi_4^2 \right). \tag{3.5.3}$$

Similarly, we write the external source as

$$\eta_a = \frac{1}{\sqrt{2}} \left( \eta_a^1 + i\, \eta_a^2 \right), \qquad a = 1, 2, 3, 4, \tag{3.5.4}$$

$$\rho^\mathsf{T} = \left( \eta_1^1 \ \eta_2^1 \ \eta_3^1 \ \eta_4^1 \ \eta_1^2 \ \eta_2^2 \ \eta_3^2 \ \eta_4^2 \right). \tag{3.5.5}$$

The Dirac equation then takes the form

$$\left( \Gamma^\mu \frac{\partial_\mu}{i} + m \right) \chi = \rho, \qquad (\Gamma^\mu)^* = -\Gamma^\mu, \tag{3.5.6}$$

and note the *complex conjugate* property of $\Gamma^\mu$ just stated. This is a *first* order equation. In this reformulation, the following replacements are then made

$$\psi_a^\dagger [\gamma^0]_{ab} \psi_b \rightarrow \chi_A [\Gamma^0]_{AB} \chi_B, \tag{3.5.7}$$

$$\psi_a^\dagger [\gamma^0 \gamma^\mu]_{ab} \psi_b \rightarrow \chi_A [\Gamma^0 \Gamma^\mu]_{AB} \chi_B, \tag{3.5.8}$$

$a, b = 1, 2, 3, 4; \ A, B = 1, \ldots, 8$.

We note that the matrices, $Q^\mu \equiv i\Gamma^0 \Gamma^\mu$, and $\Gamma^0$ satisfy the following key properties (the i factor in $Q^\mu$ is chosen for convenience)

$$(Q^\mu)^\dagger = -Q^\mu \qquad (Q^\mu)^\mathsf{T} = +Q^\mu, \qquad (\Gamma^0)^\mathsf{T} = -\Gamma^0. \tag{3.5.9}$$

Note the signs on the right-hand sides of these equations. The underlying statistics would be different if the signs of the last two equalities were reversed. According to the underlying theory of the Spin & Statistics Connection (Sect. 4.5), these inherited signs in the Dirac equation imply that the field components $\chi_A$ satisfy the following *anti*-commutation relations for $x'^0 = x^0$:

$$(Q^0)_{AB} \{\chi_C(x), \chi_B(x')\} = i\delta_{CA}\delta^3(\mathbf{x} - \mathbf{x}'), \qquad \text{where} \quad Q^0 = iI, \tag{3.5.10}$$

and, the anti-symmetry property of $\Gamma^0$, in particular, implies that $\rho$ is a Grassmann variable and anti-commutes with $\chi$.

From the definition of $\chi$ in terms of its components in (3.5.3), these anti-commutations, in turn, lead to the anti-commutation relations of the Dirac field components for $x'^0 = x^0$:

$$\{\psi_a(x), \psi_b^\dagger(x')\} = \delta_{ab}\delta^3(\mathbf{x} - \mathbf{x}'), \quad \{\psi_a(x), \psi_b(x')\} = 0, \quad \{\psi_a^\dagger(x), \psi_b^\dagger(x')\} = 0. \tag{3.5.11}$$

One may, of course, use Pauli's Fundamental Theorem (see, e.g., [6]), to infer that the anti-commutation relations in (3.5.11) are valid in other representations, by the application of the necessary $G$ matrices as follows: $\psi \rightarrow G \psi G^\dagger$ (see (2.3.2), (2.3.4), (I.1)) to the *quantum* field. The Grassmann variable property of $\rho$ and its anti-commutativity with $\chi$, in turn, imply the Grassmann variable property of $\eta$, $\overline{\eta}$, and their anti-commutativity with the Dirac field and its adjoint.

In the absence of external sources, we may also use the general expression for $\psi(x)$ in Problem 3.8, with the coefficients $a(\mathbf{p}, \sigma)$, $b^\dagger(\mathbf{p}, \sigma)$ promoted to quantum variables, and the first anti-commutation relation in (3.5.11) to find the following anti-commutation relation valid at *all* times (see Problem 3.8)

$$\{\psi_a(x), \overline{\psi}_b(x')\} = \left(-\frac{\gamma\partial}{i} + m\right)_{ab} \frac{1}{i} \Delta(x - x'), \tag{3.5.12}$$

where

$$\Delta(x - x') = \int \frac{d^3\mathbf{p}}{(2\pi)^3} \, e^{i\mathbf{p}\cdot(\mathbf{x}-\mathbf{x}')} \, \frac{\sin p^0(x^0 - x'^0)}{p^0}, \tag{3.5.13}$$

and reduces to the previous relation for $x'^0 = x^0$, upon multiplying it by $\gamma^0_{bc}$.

## 3.6   Electromagnetic Current, Gauge Invariance and $\langle\, 0_+ \,|\, 0_- \rangle$ with External Electromagnetic Field

As a preparation for the main study of this section, re-consider the Dirac equation in the presence of an external source $\eta(x)$

$$\left( \frac{\gamma \partial}{i} + m \right) \psi(x) = \eta(x). \tag{3.6.1}$$

As we have seen in Sect. 3.1, (3.1.8), the matrix element $\langle\, 0_+ \,|\, \psi(x) \,|\, 0_- \rangle$, is given by

$$\langle\, 0_+ \,|\, \psi(x) \,|\, 0_- \rangle = \int (dx') \, S_+(x - x') \eta(x') \langle\, 0_+ \,|\, 0_- \rangle, \tag{3.6.2}$$

where $S_+(x - x')$ is the fermion propagator in (3.1.9). As discussed at the of Sect. 3.3, we will learn from a direct application of the Quantum Dynamical Principle in Sect. 4.6, that the matrix element $\langle\, 0_+ \,|\, \psi(x) \,|\, 0_- \rangle$, may be generated by a functional differentiation of $\langle\, 0_+ \,|\, 0_- \rangle$ with respect to $\overline{\eta}(x)$ as follows

$$(-i)\frac{\delta}{\delta\overline{\eta}(x)}\langle\, 0_+ | 0_- \rangle = \langle\, 0_+ \,|\, \psi(x) \,|\, 0_- \rangle, \tag{3.6.3}$$

which upon comparison with (3.6.2) gives

$$(-i)\frac{\delta}{\delta\overline{\eta}(x)}\langle\, 0_+ | 0_- \rangle = \left( \int (dx') \, S_+(x - x')\eta(x') \right) \langle\, 0_+ | 0_- \rangle. \tag{3.6.4}$$

The solution is given by

$$\langle\, 0_+ | 0_- \rangle = \exp\left[ i \int (dx)(dx') \, \overline{\eta}(x) \, S_+(x - x')\eta(x') \right]. \tag{3.6.5}$$

This is normalized to unity for $\eta(x) = 0$, $\overline{\eta}(x) = 0$, in which case $\langle\, 0_+ \,|\, 0_- \rangle \rightarrow \langle 0|0 \rangle$, where $|\, 0 \rangle$ is the so-called Fock vacuum state , i.e., the vacuum state in the absence of the external source (in the absence of any interaction).

In the presence of an external electromagnetic field, the Dirac equation reads

$$\left[\gamma^\mu\left(\frac{\partial_\mu}{i} - eA_\mu(x)\right) + m\right]\psi(x) = \eta(x). \tag{3.6.6}$$

The purpose of this section is to derive the expression for $\langle\,0_+\,|\,0_-\rangle_A$, for this system. As will be seen throughout this book, the vacuum persistence amplitude $\langle\,0_+\,|\,0_-\rangle_A$, contains much useful information and gives rise to the actual solution of an underlying quantum field theoretical problem.

There are several reasons why we consider this system at this stage. It provides a simple interacting system involving fermions. Also it provides an excellent training for the direct application of the rules of the Quantum Dynamical Principle developed in Sect. 4.6, and to appreciate its power in applications and witness the simplicity of its underlying rules before getting involved in its intricate details. Before spelling out these rules, as we apply them to the system described by (3.6.6), we introduce the electromagnetic current of the system.

To the above end, we write the equations for $\psi$, and $\overline{\psi}$, in terms of components, as follows

$$(\partial_\mu\gamma^\mu\psi)_a - i(\gamma^\mu\psi)_a eA_\mu + im\psi_a = i\eta_a, \tag{3.6.7}$$

$$-(\partial_\mu\overline{\psi}\gamma^\mu)_a - i(\overline{\psi}\gamma^\mu)_a eA_\mu + im\overline{\psi}_a = i\overline{\eta}_a. \tag{3.6.8}$$

The combination $[\overline{\psi}_a\times(3.6.7) - (3.6.8)\times\psi_a - (3.6.7)\times\overline{\psi}_a + \psi_a\times(3.6.8)]$, gives

$$\partial_\mu j^\mu(x) = ie\left(\overline{\psi}(x)\eta(x) - \overline{\eta}(x)\psi(x)\right), \quad j^\mu(x) = \frac{e}{2}[\overline{\psi}_a(x), (\gamma^\mu\psi)_a(x)], \tag{3.6.9}$$

where $j^\mu(x)$ defines the electromagnetic and is conserved in the absence of the external source. In writing (3.6.9), we have used the Grassmann variable property of $\eta$, $\overline{\eta}$, that the latter anti-commute with the Dirac field and its adjoint (Sect. 3.5).

The commutator in the definition of the current is of significance, as under charge conjugation (see Sect. 3.2), it changes sign, as an electromagnetic current should. Recall that under charge conjugation, the Dirac quantum field transforms as (see (3.2.29))

$$\psi \to \mathscr{C}\overline{\psi}^\mathsf{T}, \qquad \overline{\psi} \to \psi^\mathsf{T}\mathscr{C}, \qquad \overline{\psi}_a(\gamma^\mu\psi)_a \leftrightarrow (\gamma^\mu\psi)_a\overline{\psi}_a. \tag{3.6.10}$$

where $\mathscr{C}$ is the charge conjugation matrix, defined in (I.3), satisfying (2.3.8), (2.3.9). Hence

$$j^\mu(x) \to -j^\mu(x), \tag{3.6.11}$$

under charge conjugation. It is common in the literature to write the current naïvely as $e\overline{\psi}\gamma^\mu\psi$ with the understanding that it should be defined as a commutator as given above in (3.6.9) to ensure the correct transformation given in (3.6.11).

We now spell out some basic facts that we will learn in applying the Quantum Dynamical Principle (Sect. 4.6) to the system described by (3.6.6):

♦ **Lessons from the Quantum Dynamical Principle:**

$$(-\,\mathrm{i})\frac{\delta}{\delta\overline{\eta}(x)}\langle\,0_+\mid 0_-\rangle = \langle\,0_+\mid\psi(x)\mid 0_-\rangle, \tag{3.6.12}$$

$$(\mathrm{i})\frac{\delta}{\delta\eta(x)}\langle\,0_+\mid 0_-\rangle = \langle\,0_+\mid\overline{\psi}(x)\mid 0_-\rangle, \tag{3.6.13}$$

$$e\frac{\partial}{\partial e}\langle\,0_+\mid 0_-\rangle = \mathrm{i}\int(dx)\,\langle\,0_+\mid j^\mu(x)\mid 0_-\rangle\,A_\mu(x), \tag{3.6.14}$$

$$(-\,\mathrm{i})\frac{\delta}{\delta\overline{\eta}(x)}(\mathrm{i})\frac{\delta}{\delta\eta(x')}\langle\,0_+\mid 0_-\rangle = \langle\,0_+\mid\big(\psi(x)\overline{\psi}(x')\big)_+\mid 0_-\rangle, \tag{3.6.15}$$

where $(\ldots)_+$ defines the time-ordered product of Fermi fields

$$\big(\psi_a(x)\overline{\psi}_b(x')\big)_+ = \theta(x^0-x'^0)\,\psi_a(x)\overline{\psi}_b(x') - \theta(x'^0-x^0)\,\overline{\psi}_b(x')\psi_a(x), \tag{3.6.16}$$

and $\theta(x^0) = 1$ for $x^0 > 0$, $\theta(x^0) = 0$ for $x^0 < 0$. Note the minus sign between the two terms on the right-hand side of the equation that arises due to the Fermi character of the Dirac field (Sect. 3.5). The anti-commutativity of Fermi fields *within* a time-ordered product is directly inferred from the Grassmann variable character of the sources and the functional *derivative* operations with respect to them in (3.6.15). ♦

Using (3.6.12), and recalling (3.2.13), we may write

$$(-\,\mathrm{i})\frac{\delta}{\delta\overline{\eta}(x)}\langle\,0_+\mid 0_-\rangle = \langle\,0_+\mid\psi(x)\mid 0_-\rangle = \int(dx')\,S_+^A(x,x')\,\eta(x')\langle\,0_+\mid 0_-\rangle, \tag{3.6.17}$$

where we have omitted the "subscript" $A$ in $\langle\,0_+\mid(.)\mid 0_-\rangle_A$ for simplicity of the notation, and (see (3.2.11))

$$\left[\gamma^\mu\left(\frac{\partial_\mu}{\mathrm{i}} - eA_\mu(x)\right) + m\right]S_+^A(x,x') = \delta^{(4)}(x-x'). \tag{3.6.18}$$

As in (3.6.3), (3.6.4), (3.6.5), we may integrate (3.6.17), to obtain

$$\langle\,0_+\mid 0_-\rangle = \langle\,0_+\mid 0_-\rangle^{(e)}\exp\left[\mathrm{i}\int(dx)(dx')\overline{\eta}(x)S_+^A(x,x')\eta(x')\right], \tag{3.6.19}$$

*where* the amplitude

$$\langle\, 0_+ \mid 0_-\rangle^{(e)} = \langle\, 0_+ \mid 0_-\rangle \mid_{\eta=0,\bar\eta=0}, \tag{3.6.20}$$

and *depends* on $eA_\mu$.

Now we use (3.6.15) and (3.6.19), to conclude that for $\eta = 0$, $\bar\eta = 0$,

$$\frac{\langle\, 0_+ \mid \big(\psi_a(x)\overline\psi_b(x')\big)_+ \mid 0_-\rangle^{(e)}}{\langle\, 0_+ \mid 0_-\rangle^{(e)}} = -\,\mathrm{i}\,(S_+^A)_{ab}(x,x'). \tag{3.6.21}$$

where $\psi(x)$ (similarly $\overline\psi$) within this time-ordered product satisfies the Dirac equation (3.6.6) in the external electromagnetic field, in the absence of external source $\eta$ and, as already indicated in (3.6.20), $\langle\, 0_+ \mid 0_-\rangle^{(e)}$ denotes the vacuum persistence amplitude in the absence of $\eta$ (i.e., for $\eta = 0$, $\bar\eta = 0$).

We already know from the anti-commutation relation of the Dirac field and its adjoint in (3.5.12), singularities may develop in defining an expression like $[\overline\psi_a(x), (\gamma^\mu \psi)_a(x')]$ at coincident points $x = x'$. Accordingly, one may formally define the current[7] as a limit of $\epsilon \to 0$

$$j^\mu(x,\epsilon) = \frac{e}{2}\left(\overline\psi_a\Big(x+\frac{\epsilon}{2}\Big)(\gamma^\mu\psi)_a\Big(x-\frac{\epsilon}{2}\Big) - (\gamma^\mu\psi)_a\Big(x-\frac{\epsilon}{2}\Big)\overline\psi_a\Big(x+\frac{\epsilon}{2}\Big)\right), \tag{3.6.22}$$

Obviously the right-hand side of (3.6.22) reduces to the commutator in (3.6.9) in the just mentioned limit. By initially choosing the arguments of $\overline\psi(x)$, $\psi(x')$ in (3.6.22) at non-coincident points, however, one destroys gauge invariance as we now discuss. This, in turn, will allow one to define a gauge invariant current at the outset.

To ensure the generation of a gauge invariant electromagnetic current, we proceed as follows. Under a gauge transformation of the electromagnetic potential, $A_\mu \to A_\mu + \partial_\mu\lambda(x)$, where $\lambda(x)$ is arbitrary, the field equation (3.6.6) remains invariant provided the spinor $\psi(x)$ transforms simultaneously as $\psi(x) \to \exp\mathrm{i}\,[e\lambda(x)] \times \psi(x)$, and, in turn, $\overline\psi_a(x)\psi_b(x')$ transforms as

$$\overline\psi_a(x)\psi_b(x') \to \overline\psi_a(x)\psi_b(x')\exp\big[-\mathrm{i}e\big(\lambda(x) - \lambda(x')\big)\big]. \tag{3.6.23}$$

Accordingly, one may define the electromagnetic current as the limit of[8]

$$j^\mu(x) = j^\mu(x,\epsilon)\,\exp\Big[\mathrm{i}\,e\int_{x-}^{x+} \mathrm{d}\xi^\mu\, A_\mu(\xi)\Big], \quad \epsilon \to 0, \tag{3.6.24}$$

---

[7]In order to distinguish between an external, i.e., c-number, current and a quantum current, as given in (3.6.22), we use the notation with small $j^\mu$ for the latter, and capital $J^\mu$ for the former in this book.

[8]See Problem 3.10 where the line integral over $\xi$ is carried out showing its dependence on $\epsilon^\mu$, and for more details see also Appendix IV at the end of the book.

where $x_\pm = x \pm \epsilon/2$, and is gauge invariant at the outset. This method is known as Schwinger's point splitting method. The exponential of the line-integral involving the vector potential is referred to as the Schwinger line-integral.

The point of interest is the matrix element of the current between the vacuum states in (3.6.14). From (3.6.21), (3.6.23) we may, in a limiting sense, write

$$\frac{\langle\, 0_+ \mid j^\mu(x) \mid 0_-\rangle^{(e)}}{\langle\, 0_+ \mid 0_-\rangle^{(e)}} = i\,e\,\mathrm{Tr}\,\mathrm{Av}\left( [\,S_+^A(x_-,x_+)\gamma^{\,\mu}\,]\,\exp\!\big[\,i\,e\int_{x_-}^{x_+}\!d\xi^\mu A_\mu(\xi)\big]\right),$$
$$(3.6.25)$$

where Av stands for an average over $\epsilon^0 > 0$ and $\epsilon^0 < 0$, with $\epsilon$ in a space-like direction. Here Tr denotes a trace over the spinor indices of gamma matrices.

To emphasize the dependence of $S_+^A(x,x')$ on the combination $eA_\mu$, it is convenient to use the notation $S_+(x,x';eA)$ for the former. With the boundary condition, $\langle\, 0_+ \mid 0_-\rangle^{(e)} = 1$, for $e = 0$, we may use (3.6.14) to integrate over e to obtain

$$\langle\, 0_+ \mid 0_-\rangle^{(e)} = \exp\,[\,i\,W\,],\qquad\qquad(3.6.26)$$

$$i\,W = -\int_0^e de'\,(dx)\left(\mathrm{Tr}\,[\,S_+(x_-,x_+;e'A)\,\gamma A\,]\,\exp\!\big[\,i\,e\int_{x_-}^{x_+}\!d\xi^\mu A_\mu(\xi)\big]\right),$$
$$(3.6.27)$$

and an average taken of the limits $\epsilon^0 \to +0$, $\epsilon^0 \to -0$ is understood.

In Appendix IV, at the end of the book, $i\,W$ is determined to arbitrary orders in $eA_\mu$ in the limit $\epsilon \to 0$. It is shown that the $\epsilon \to 0$ limits give rise to a modification from the naïve expression only up to fourth order in the charge e.

In particular, to second order in $eA_\mu$ (IV.27), in the just mentioned appendix, gives

$$i\,W\Big|_{2\text{nd order}} = -\frac{e^2}{2}\int \frac{(dQ_1)}{(2\pi)^4}\frac{(dQ_1)}{(2\pi)^4}(2\pi)^4\,\delta^{(4)}(Q_1 + Q_2)$$
$$\times\, A_{\mu_1}(Q_1)\,A_{\mu_2}(Q_2)\,L^{\mu_1\mu_2}(Q_1,Q_2),\qquad(3.6.28)$$

$$L^{\mu_1\mu_2}(Q,-Q) = \int \frac{(dp)}{(2\pi)^4}\left(\mathrm{Tr}\,\big[\,\gamma^{\,\mu_1}S_+(p-\tfrac{Q}{2})\gamma^{\,\mu_2}S_+(p+\tfrac{Q}{2})\,\big]\right.$$
$$\left. +\, Q^{\nu_1}Q^{\nu_2}\frac{1}{24}\frac{\partial}{\partial p^{\nu_1}}\frac{\partial}{\partial p^{\nu_2}}\frac{\partial}{\partial p_{\mu_2}}\mathrm{Tr}\,[\,\gamma^{\,\mu_1}S_+(p)\,] + \frac{\partial}{\partial p_{\mu_2}}\mathrm{Tr}\,[\,\gamma^{\,\mu_1}S_+(p)\,]\right),$$
$$(3.6.29)$$

which with (3.6.27) providing a contribution to the vacuum-to-vacuum transition amplitude in (3.6.26), in the presence of the external electromagnetic field $A_\mu$.

The following tensor is referred to as the vacuum polarization tensor, to lowest order in e,

$$\Pi^{\mu\nu}(Q) = -i\,e^2\, L^{\mu_1\mu_2}(Q, -Q), \tag{3.6.30}$$

and will find applications in Chap. 5. It is evaluated in Appendix A of this chapter, and the following expression emerges for it

$$\Pi^{\mu\nu}(Q) = (\eta^{\mu\nu}Q^2 - Q^\mu Q^\nu)\,\Pi(Q^2), \tag{3.6.31}$$

$$\Pi(Q^2) = \frac{e^2}{12\,\pi^2}\left(\left[\frac{1}{i\pi^2}\int\frac{(dp)}{[p^2+m^2]^2} + \frac{1}{2}\right] - 6\int_0^1 dz\, z(1-z)\ln\left[1 + \frac{Q^2}{m^2}z(1-z)\right]\right). \tag{3.6.32}$$

This expression will be used in Sect. 5.10.2 in studying the photon propagator. The presence of the last two terms in (3.6.29) ensure the gauge invariance of the formulation and leads necessarily to a transverse expression for $\Pi^{\mu\nu}(Q)$, in response to a gauge transformation $A_\mu(Q) \to A_\mu(Q) + Q_\mu\lambda(Q)$ in (3.6.28),

$$Q_\mu\Pi^{\mu\nu}(Q) = 0, \qquad \Pi^{\mu\nu}(Q)\,Q_\nu = 0. \tag{3.6.33}$$

In the next section, we also provide another expression for ⟨0₊ | 0₋⟩⁽ᵉ⁾, for special external electromagnetic potentials $A_\mu$, which turns out to be useful for applications as to the one given in Sect. 3.8 in pair production by a constant electric field.

## 3.7 ⟨0₊|0₋⟩⁽ᵉ⁾ in the Presence of a Constant $F^{\mu\nu}$ Field and Effective Action

An expression for ⟨0₊ | 0₋⟩⁽ᵉ⁾ is readily obtained for the Dirac equation in the presence of a classical constant $F^{\mu\nu}$ Field, or effectively constant over sufficiently extended regions of spacetime. An interesting application of this is that the quantum nature of the fermion field adopted here leads to a modification of the Maxwell Lagrangian density giving rise to an effective one involving higher powers of the Faraday tensor $F^{\mu\nu}$ and of its dual to be defined shortly. This will be the subject matter of this section. Another interesting application is that it provides an explicit expression for the decay of the system initially in the vacuum state leading to $e^+e^-$ pair creation by a constant electric field. This will be developed in the next section.

Consider the response of $\langle 0_+ \mid 0_- \rangle^{(e)}$ to variations of $[eA_\mu(x)]$, which from (3.6.14) is given by

$$\delta\langle 0_+ \mid 0_- \rangle^{(e)} = i \int (dx) \langle 0_+ \mid \frac{1}{2} [\overline{\psi}_a(x), (\gamma^\mu \psi)_a(x)] \mid 0_- \rangle \delta[e A_\mu(x)], \quad (3.7.1)$$

where the current is defined in (3.6.9) by $j^\mu = e [\overline{\psi}_a(x), (\gamma^\mu \psi)_a(x)]/2$.

At this stage it is convenient to introduce a matrix notation in spacetime as well. For the propagator $S^A(x, x') \equiv S(x, x'; eA)$, we write

$$S_+^A(x, x') = \langle x \mid S_+^A \mid x' \rangle, \qquad \langle x \mid 1 \mid x' \rangle = \delta^{(4)}(x - x'). \quad (3.7.2)$$

The differential equation satisfied by $S_+(x, x'; eA)$ in (3.6.18) may be rewritten in matrix form as

$$[\gamma\Pi + m] S_+^A = 1, \quad S_+^A = \frac{1}{\gamma\Pi + m}, \quad \Pi_\mu = p_\mu - eA_\mu, \quad p_\mu = \frac{\partial_\mu}{i}. \quad (3.7.3)$$

From the definition $\langle 0_+ \mid [\overline{\psi}_a(x), (\gamma^\mu \psi)_a(x)]/2 \mid 0_- \rangle^{(e)} = i S_+^A \langle 0_+ \mid 0_- \rangle^{(e)}$, in (3.6.21), (3.6.22), and the fact that $\delta[e\gamma A] = -\delta[\gamma\Pi]$, (3.7.1), in matrix form, becomes

$$\delta\langle 0_+ \mid 0_- \rangle^{(e)} = \mathbf{Tr}[\delta(\gamma\Pi) S_+^A]\langle 0_+ \mid 0_- \rangle^{(e)}. \quad (3.7.4)$$

where **Tr** stands for a trace over spinor indices as well as spacetime coordinates. From (3.7.3), the propagator may be re-expressed as

$$S_+^A = (-\gamma\,\Pi + m) \frac{1}{m^2 - (\gamma\,\Pi)^2} = (-\gamma\,\Pi + m) \int_0^\infty ds \, \exp[-s(m^2 - (\gamma\,\Pi)^2)]. \quad (3.7.5)$$

Here it turns out more convenient to use this representation than the one with $-i s(m^2 - i\epsilon)$ in the exponential. We use the fact that the trace of an odd number of gamma matrices is zero. Also for a matrix $M$, $\delta e^M \neq \delta M e^M$, in general, but $\mathrm{Tr}[\delta e^M] = \mathrm{Tr}[\delta M e^M]$.[9] Finally note that for a matrix $A$: $\mathrm{Tr}[\delta e^{A^2}] = \mathrm{Tr}[(\delta A A + A \delta A) e^{A^2}] = 2\,\mathrm{Tr}[\delta A A e^{A^2}]$. Thus we may write

$$\mathbf{Tr}\left[\delta(\gamma\Pi)(-\gamma\,\Pi + m) \exp[-s(m^2 - (\gamma\,\Pi)^2)]\right]$$

$$= -\frac{1}{2} \mathbf{Tr}\left[\delta(\gamma\Pi)^2 \exp[-s(m^2 - (\gamma\,\Pi)^2)]\right], \quad (3.7.6)$$

---

[9]See Appendix B of this chapter.

from which (3.7.4) takes the convenient form

$$
\frac{\delta\langle\,0_+\,|\,0_-\rangle^{(e)}}{\langle\,0_+\,|\,0_-\rangle^{(e)}} = -\frac{1}{2}\,\mathbf{Tr}\Big[\delta\,(\gamma\varPi)^2\int_0^\infty ds\,\exp\big[-s(m^2-(\gamma\,\varPi)^2)\big]\Big]. \tag{3.7.7}
$$

Again from the property $\mathrm{Tr}\,[\delta\,e^M] = \mathrm{Tr}\,[\delta M\,e^M]$, this finally gives us the useful expression

$$
\frac{\delta\langle\,0_+\,|\,0_-\rangle^{(e)}}{\langle\,0_+\,|\,0_-\rangle^{(e)}} = -\frac{1}{2}\,\delta\Big[\,\mathbf{Tr}\int_0^\infty \frac{ds}{s}\,\exp\big[-s(m^2-(\gamma\,\varPi)^2)\big]\Big]. \tag{3.7.8}
$$

The formalism will be gauge invariant if the results obtained are expressed in terms of $F^{\mu\nu}$ and possibly in terms of its dual

$$
{}^*F^{\mu\nu} = \frac{1}{2}\,\epsilon^{\mu\nu\lambda\sigma}\,F_{\lambda\sigma}. \tag{3.7.9}
$$

Here it is important to recall the expressions for the two invariants, as the scalar and the pseudo-scalar, respectively, of the Maxwell field,

$$
-\frac{1}{4}\,F^{\mu\nu}F_{\mu\nu} = \frac{1}{2}(\mathbf{E}^2-\mathbf{B}^2), \qquad -\frac{1}{4}\,{}^*F^{\mu\nu}F_{\mu\nu} = \mathbf{E}\cdot\mathbf{B}. \tag{3.7.10}
$$

The subsequent analysis simplifies quite a bit if we first work in a frame, where $\mathbf{E}$ and $\mathbf{B}$ are parallel. In this case the second expression above would read $|\,\mathbf{E}\,||\,\mathbf{B}\,|$. The general case may be then readily inferred.

To the above end, we consider a vector potential in the form

$$
A^\mu = (0,0,x^1 B, -x^0 E), \tag{3.7.11}
$$

giving an electric and magnetic fields $E$ and $B$, respectively, in the $z$-direction. Using the identity $\gamma^\mu\gamma^\nu = -\eta^{\mu\nu} + [\gamma^\mu,\gamma^\nu]/2$, we may write

$$
m^2 - (\gamma\,\varPi)^2 = m^2 + (\varPi)^2 - \frac{[\gamma^\mu,\gamma^\nu]}{2}\frac{[\varPi_\mu,\varPi_\nu]}{2}, \qquad [\varPi_\mu,\varPi_\nu] = \mathrm{i}eF_{\mu\nu}. \tag{3.7.12}
$$

On the other hand, with

$$
F_{30} = -F_{03} = F^{03} = E, \quad F^{12} = B = -F^{21},
$$

as the only non-vanishing components, we have

$$
i s e\, \frac{[\gamma^\mu, \gamma^\nu]}{4} F_{\mu\nu} = -i s e \begin{pmatrix} 0 & \sigma^3 \\ \sigma^3 & 0 \end{pmatrix} E + s e \begin{pmatrix} \sigma^3 & 0 \\ 0 & \sigma^3 \end{pmatrix} B, \qquad (3.7.13)
$$

involving two commuting matrices.

In reference to the second term on the right-hand side of the first equality in (3.7.12), the trace Tr, over spinor indices, of the exponential of the expression in (3.7.13) is easily carried out by expanding the exponential leading to

$$
\mathrm{Tr}\left\{ \exp\left[ -i s e \begin{pmatrix} 0 & \sigma^3 \\ \sigma^3 & 0 \end{pmatrix} E + s e \begin{pmatrix} \sigma^3 & 0 \\ 0 & \sigma^3 \end{pmatrix} B \right] \right\} = 4 \cos(s|eE|)\cosh(s|eB|).
$$
$$
(3.7.14)
$$

Also

$$
\Pi^\mu \Pi_\mu = \left[ (p_1{}^2 + \left(x_1 - \frac{p_2}{eB}\right)^2 e^2 B^2 \right] + \left[ -(p_0)^2 + \left(x^0 + \frac{p_3}{eE}\right)^2 e^2 E^2 \right], \quad (3.7.15)
$$

which, with $p^0$ as the generator of translation of $x^0$ and $p_1$ as the generator of translation of $x_1$, may be rewritten as

$$
\exp\left[ -i \frac{p_2 p_1}{eB} \right] \exp\left[ -i \frac{p_3 p^0}{eE} \right]
$$
$$
\times \left[ (p_1)^2 + (x_1)^2 e^2 B^2 - (p_0)^2 + (x^0)^2 e^2 E^2 \right]
$$
$$
\times \exp\left[ i \frac{p_3 p^0}{eE} \right] \exp\left[ i \frac{p_2 p_1}{eB} \right]. \qquad (3.7.16)
$$

Accordingly, in reference to (3.7.8) we may write

$$
\mathbf{Tr}\left[ \exp\left[ -s(m^2 - (\gamma\, \Pi)^2) \right] \right] = 4\, e^{-s m^2} \cos(s|eE|)\cosh(s|eB|)\, G(s; eE, eB),
$$
$$
(3.7.17)
$$

where

$$
G(s; eE, eB) = \int (\mathrm{d}x) \Big\langle x \Big| \exp\left[ -i \frac{p_2 p_1}{eB} \right] \exp\left[ -i \frac{p_3 p^0}{eE} \right]
$$
$$
\times \exp\left( -s \left[ (p_1)^2 + (x_1)^2 e^2 B^2 - (p_0)^2 + (x^0)^2 e^2 E^2 \right] \right)
$$
$$
\times \exp\left[ i \frac{p_3 p^0}{eE} \right] \exp\left[ i \frac{p_2 p_1}{eB} \right] \Big| x \Big\rangle. \qquad (3.7.18)
$$

By using completeness relations (Fourier transforms)

$$\int \frac{(\mathrm{d}p)}{(2\pi)^4} \frac{(\mathrm{d}p')}{(2\pi)^4} \langle x \,|\, p \rangle \langle p \,|\, \cdot \,|\, p' \rangle \langle p' \,|\, x \rangle, \tag{3.7.19}$$

with $\langle x | p \rangle = \exp[\,\mathrm{i}\, xp\,]$, and noting that the expressions within the square brackets multlpying $s$ in the exponential in (3.7.18) are independent of $p_2, p_3, p'_2, p'_3$, we may use the Fourier normalization condition $\langle p_2, p_3 \,|\, p'_2, p'_3 \rangle = (2\pi)^2 \delta(p_2 - p'_2)\, \delta(p_3 - p'_3)$, we may integrate over $p_2, p_3, p'_2, p'_3$ to obtain for (3.7.18)

$$G(s\,; eE, eB) = \frac{|eE||eB|}{(2\pi)^2} \int (\mathrm{d}x) \int \frac{\mathrm{d}p^0}{2\pi} \frac{\mathrm{d}p^1}{2\pi}$$

$$\times \left\langle p^0, p^1 \left| \exp\left( -s\left[ (p_1)^2 + (x_1)^2 e^2 B^2 - (p_0)^2 + (x^0)^2 e^2 F^2 \right] \right) \right| p^0, p^1 \right\rangle \tag{3.7.20}$$

where $(x^0)^2 = -(\mathrm{d}/\mathrm{d}p^0)^2$, $(x^1)^2 = -(\mathrm{d}/\mathrm{d}p^1)^2$, in the exponential.

We may use the well known results from the one dimensional harmonic oscillator problem dealing with two independent systems. For one system, we have formally a mass $1/2\, e^2 B^2$ and an angular frequency $2\,|eB|$, and a second one with mass $1/2\, e^2 F^2$ and angular frequency $2\,|eE|/\mathrm{i}$, from which we carry out the traces $\int (\mathrm{d}p^0/2\pi)(\mathrm{d}p^1/2\pi)\langle p^0, p^1 | \cdot | p^0, p^1 \rangle$ in (3.7.20) to obtain

$$G(s\,; eE, eB) = \frac{|eE||eB|}{(2\pi)^2} \int (\mathrm{d}x) \sum_{n=0}^{\infty} \exp[2s\,\mathrm{i}\,|eE|(n+1/2)] \sum_{m=0}^{\infty} \exp[-2s|eB|(m+1/2)]$$

$$= \frac{|eE||eB|}{(2\pi)^2} \int (\mathrm{d}x)\, \frac{\mathrm{i}}{2\,\sin s|eE|}\, \frac{1}{2\,\sinh s|eB|}. \tag{3.7.21}$$

From (3.7.17), the following expression then emerges for (3.7.8)

$$\frac{\delta\langle 0_+ \,|\, 0_- \rangle^{(e)}}{\langle 0_+ \,|\, 0_- \rangle^{(e)}} = -\frac{\mathrm{i}}{8\pi^2} \int (\mathrm{d}x) \int_0^{\infty} \frac{\mathrm{d}s}{s^3}\, e^{-sm^2} \left[ \left( s|eE| \cot s|eE| \right) \left( s|eB| \coth s|eB| \right) \right]. \tag{3.7.22}$$

For a reason that will become clear shortly we normalize $\langle 0_+ \,|\, 0_- \rangle^{(e)}$ for $e = 0$, in this section, by the *phase* factor

$$\langle 0_+ \,|\, 0_- \rangle^{(e)}\big|_{e=0} = \exp\left[ \mathrm{i} \int (\mathrm{d}x)\left( -\frac{1}{4} F^{\mu\nu} F_{\mu\nu} \right) \right] = \exp\left[ \mathrm{i} \int (\mathrm{d}x) \frac{1}{2} (\mathbf{E}^2 - \mathbf{B}^2) \right], \tag{3.7.23}$$

in terms of the Maxwell action.

The power series expansions

$$x \cot x = 1 - \frac{x^2}{3} - \frac{x^4}{45} - \frac{2x^6}{945} - \cdots, \tag{3.7.24}$$

$$y \coth y = 1 + \frac{y^2}{3} - \frac{y^4}{45} + \frac{2y^6}{945} - \cdots, \tag{3.7.25}$$

lead to the expansion

$$(x \cot x)(y \coth y) = 1 - \frac{x^2 - y^2}{3} - \frac{(x^2 - y^2)^2 + 7x^2 y^2}{45} + \cdots. \tag{3.7.26}$$

For the subsequent analysis, we rewrite (3.7.22) by adding and subtracting the second order contribution $(se)^2 (E^2 - B^2)/3$ within its square brackets, giving

$$\frac{\delta \langle 0_+ \mid 0_- \rangle^{(e)}}{\langle 0_+ \mid 0_- \rangle^{(e)}} = i\kappa \int (dx) \frac{(E^2 - B^2)}{2} + i \frac{1}{8\pi^2} \int (dx) \int_0^\infty \frac{ds}{s^3} e^{-sm^2} \times$$

$$\times \left[ (s|eE| \cot s|eE|)(s|eB| \coth s|eB|) + \frac{(se)^2 (E^2 - B^2)}{3} \right], \tag{3.7.27}$$

where $\kappa$ is the divergent constant

$$\kappa = \frac{e^2}{12\pi^2} \int_{\to 0}^\infty \frac{ds}{s} e^{-sm^2}. \tag{3.7.28}$$

Using the boundary condition in (3.7.23), the functional integral of (3.7.27) gives

$$\langle 0_+ \mid 0_- \rangle^{(e)} = \exp i \left[ (1 + \kappa) \int (dx) \frac{(E^2 - B^2)}{2} \right] \exp i \left[ \frac{-1}{8\pi^2} \int (dx) \int_0^\infty \frac{ds}{s^3} e^{-sm^2} \right.$$

$$\times \left. \left\{ (s|eE| \cot s|eE|)(s|eB| \coth s|eB|) + \frac{(se)^2 (E^2 - B^2)}{3} - 1 \right\} \right], \tag{3.7.29}$$

where the expression between the curly brackets goes to zero for $e \to 0$.

The expression in the first exponential above multiplied by i is nothing but the Maxwell action scaled by the factor $1 + \kappa$, which amounts in redefining $(1 + \kappa)A_\mu(x)A_\nu(x) \to A_\mu(x)A_\nu(x)$ *provided* one also redefines, $(1+\kappa)^{-1}e^2 \to e^2$ since everything else in $\langle 0_+ \mid 0_- \rangle^{(e)}$ depends on the product $eA$, with this product remaining *invariant*, i.e., $eA \to eA$. One may use different notations for these two re-scaled quantities if one wishes. Needless to say, this process of re-parametrization is a reconciliation with the initial definition of the Maxwell Lagrangian density. It is necessary even if $\kappa$ were finite. Accordingly, the following final expression emerges

$$\langle\, 0_+ \mid 0_-\rangle^{(e)} = e^{\,i\,W^{(e)}}, \tag{3.7.30}$$

$$W^{(e)} = \int (\mathrm{d}x)\frac{(E^2 - B^2)}{2} - \left[\frac{1}{8\pi^2}\int (\mathrm{d}x)\int_0^\infty \frac{\mathrm{d}s}{s^3}\, e^{-s\,m^2}\right.$$

$$\left.\times\,\Big\{\big(s|eE|\cot s|eE|\big)\big(s|eB|\coth s|eB|\big) + \frac{(s\,e)^2(E^2 - B^2)}{3} - 1\Big\}\right]. \tag{3.7.31}$$

In this section, we consider the real part of $W^{(e)}$. The imaginary part and its physical consequences are considered in the next section. As seen from (3.7.31), the real part of $W^{(e)}$ gives rise to a modification of the Maxwell Lagrangian density which, to lowest order, is given by

$$\mathscr{L}_4 = \frac{e^4}{8\pi^2}\frac{[(E^2 - B^2)^2 + 7(EB)^2]}{45}\int_0^\infty s\,\mathrm{d}s\,e^{-s\,m^2}. \tag{3.7.32}$$

The integration is elementary, and in terms of arbitrary directions of $\mathbf{E}$, and $\mathbf{B}$, this becomes

$$\mathscr{L}_4 = \frac{2\,\alpha^2}{45\,m^4}[(\mathbf{E}^2 - \mathbf{B}^2)^2 + 7(\mathbf{E}\cdot\mathbf{B})^2], \qquad \alpha = \frac{e^2}{4\pi}, \tag{3.7.33}$$

expressed in terms of the invariants in (3.7.10). The new modified Lagrangian density is known as the Euler-Heisenberg effective Lagrangian (Euler-Heisenberg, 1936), and much additional detailed work on this was also carried out by Schwinger [7]. The technique used in this section based on a parametric representation of a Green function is referred to as the Schwinger parametric representation.

## 3.8  Pair Creation by a Constant Electric Field

The vacuum-to-vacuum transition amplitude $\langle\, 0_+ \mid 0_-\rangle^{(e)}$, for the system described by the Dirac equation in the presence of a constant $F^{\mu\nu}$ field was derived in the previous section in (3.7.30), (3.7.31). The probability that the vacuum stays the same in the presence of this external field is $|\langle\, 0_+ \mid 0_-\rangle^{(e)}|^2$. It reads

$$|\langle\, 0_+ \mid 0_-\rangle^{(e)}|^2 = e^{-2\,\mathrm{Im}\,W^{(e)}}, \tag{3.8.1}$$

and $W^{(e)}$ given in (3.7.31).

We may define the decay rate $\Gamma$ of the vacuum, that is the probability of decay of the vacuum per unit time by

$$\Gamma = 2\,\frac{\mathrm{Im}\,W^{(\mathrm{e})}}{T} \qquad (3.8.2)$$

due to the intervening $F^{\mu\nu}$ field. A decay of the vacuum, means that the system, initially in the vacuum state, emerges eventually to some other state, and is accompanied by particle production. Here $T$, in a limiting sense $T \to \infty$, denotes the time of operation of the electric field. Since this is a charge conserving process, $e^+e^-$ pairs may be created by the electric field. More precisely, we determine the decay rate per unit volume $\Gamma/V$, where $V$, in a limiting sense $V \to \infty$, denotes the volume of space where the electric field operates. Accordingly, the physical quantity of interest is

$$\frac{\Gamma}{V} = 2\,\frac{\mathrm{Im}\,W^{(\mathrm{e})}}{VT} \qquad (3.8.3)$$

with corresponding anticipated small measure of the probability, per unit volume, per unit time, for pair production.

Since $\mathrm{Re}\,\mathrm{i}\,W^{(\mathrm{e})} = -\,\mathrm{Im}\,W^{(\mathrm{e})}$, (3.7.31) leads to $(\int (\mathrm{d}x) = VT)$

$$2\,\frac{\mathrm{Im}\,W^{(\mathrm{e})}}{VT} = \frac{1}{4\pi^2}\,\mathrm{Re}\left\{\mathrm{i}\int_0^\infty \frac{\mathrm{d}s}{s^3}\,\mathrm{e}^{-sm^2}\right.$$

$$\left. \times \left[ (s|eE|\cot s|eE|)(s|eB|\coth s|eB|) + \frac{(s\mathrm{e})^2(E^2-B^2)}{3} - 1 \right] \right\}. \qquad (3.8.4)$$

For a pure electric field, i.e., $E \neq 0$, $B = 0$, this reads

$$\frac{2\,\mathrm{Im}\,W^{(\mathrm{e})}}{VT} = \frac{1}{4\pi^2}\,\mathrm{Re}\left\{\mathrm{i}\int_0^\infty \frac{\mathrm{d}s}{s^3}\,\mathrm{e}^{-sm^2}\left[ (s|eE|\cot s|eE|) + \frac{(s\mathrm{e}E)^2}{3} - 1 \right] \right\}. \qquad (3.8.5)$$

There are several ways of evaluating the integral. One way is given in Problem 3.12. An easier way is to note that $\cot(s|eE|)$ has singularities at $s = n\pi/|eE|$ coming from the zeros of the sine function. At all other points the integral is real and hence do not contribute after applying the $\mathrm{Re} \times (\mathrm{i})$ operation. Note that $s = 0$ is not a singularity of the integrand, as is easily verified by considering the $s \to 0$ limit of the integrand, hence $n = 1, 2, \ldots$. Near a singularity $s \simeq n\pi/|eE|$, the integrand multiplying the $\exp(-sm^2)$ factor behaves like

$$\frac{1}{s^3}\left[ (s|eE|\cot s|eE|) + \frac{(s\mathrm{e}E)^2}{3} - 1 \right] \simeq \frac{e^2E^2}{n^2\pi^2}\,\frac{1}{\left(s - \frac{n\pi}{|eE|}\right)}. \qquad (3.8.6)$$

The replacement $m^2 \rightarrow m^2 - i\epsilon$, is equivalent to rewriting $s - i\epsilon$ instead of $s$, since $m^2$ as well as $s$ are non-negative. Setting $s - i\epsilon = u$ and using the fact that

$$\text{Re}\left\{ i\frac{1}{\left(u - \frac{n\pi}{|eE|} + i\epsilon\right)}\right\} = \pi\,\delta\left(u - \frac{n\pi}{|eE|}\right), \tag{3.8.7}$$

leads to a decay rate per unit time, per unit volume, corresponding to a measure of pair production in (3.8.5), given by[10]

$$\frac{2\,\text{Im}\,W}{VT} = \frac{\alpha E^2}{\pi^2}\sum_{n=1}^{\infty}\frac{\exp\left[-n\pi m^2/|eE|\right]}{n^2}. \tag{3.8.8}$$

The phenomenon described by the above equation is often referred to as the Schwinger effect. For $|eE| \ll m^2\,\pi$, the expression on the right-hand side of (3.8.8) reduces to $(\alpha E^2/\pi^2)\exp(-\pi m^2/|eE|)$ which is extremely small in practice, difficult to observe. It is easily verified that $2\,\text{Im}\,W/VT = 0$ for a pure magnetic field, i.e., $B \neq 0$, $E = 0$, as the integral (3.8.4) is real with *no* singularities on the real axis, and, as expected, no creation of such pairs is possible.

## 3.9 Fermions and Anomalies in Field Theory: Abelian Case

A classical symmetry does not necessarily hold true quantum mechanically. This is well illustrated by considering the divergence of the so-called axial vector current defined by $j^{\mu 5}(x) = \frac{1}{2}[\overline{\psi}_a(x), (\gamma^\mu\gamma^5\psi)_a(x)]$. This symmetry breaking is due to the fact that the fermions give rise to closed loops, represented by the trace of products of Dirac propagators, referred to as radiative corrections, which are absent classically.

We will learn, in particular, that for a massless fermion that although classically the current $j^{\mu 5}(x)$ is conserved, i.e., $\partial_\mu j^{\mu 5}(x) = 0$, quantum mechanically it is not $\partial_\mu j^{\mu 5}(x) \neq 0$, and the right-hand of the latter equation is referred to as an anomaly. What is most interesting is that although the current $j^{\mu 5}(x)$ leads to divergent integrals, its derivatives gives rise to a finite expression for anomaly. Moreover, we will see that the presence of the anomaly, a purely quantum mechanical effect, may be tested experimentally. We first derive the explicit expression of the just mentioned anomaly and then show how the anomaly is tested experimentally.

---

[10]For an equivalent contour-integration derivation in the $s$-complex plane of this result see Problem 3.12.

### 3.9.1  Derivation of the Anomaly

Consider first a global phase transformation of the Dirac field $\psi(x) \rightarrow e^{\lambda \gamma^5} \psi(x)$ $= \psi'(x)$, where $\lambda$ is a constant, in the Dirac equation (3.6.6) in the presence of an external electromagnetic $A_\mu(x)$ field (in the absence of external sources $\eta, \overline{\eta}$).

Upon multiplying the Dirac equation from the left by $\exp(-\lambda \gamma^5)$, we obtain

$$\left[ \gamma^\mu \left( \frac{\partial_\mu}{i} - eA_\mu(x) \right) + m \exp\left[-2\lambda\gamma^5\right] \right] \psi'(x) = 0, \tag{3.9.1}$$

where we have used the relation $\{\gamma^5, \gamma^\mu\} = 0$. That is, the equation does not remain invariant for $m \neq 0$.

In connection to the above observation, consider the divergence of the axial vector current. To this end, multiply the Dirac equation for $\psi$ from the left by $\gamma^5$ and the Dirac equation for $\overline{\psi}$ (see, e.g., (3.6.8) for $\overline{\eta} = 0$) from the right also by $\gamma^5$. The resulting equations may be conveniently written as

$$\gamma^\mu \gamma^5 \partial_\mu \psi(x) = i \left( e \gamma^\mu \gamma^5 A_\mu(x) + m\gamma^5 \right) \psi(x), \tag{3.9.2}$$

$$\partial_\mu \overline{\psi}(x) \gamma^\mu \gamma^5 = -i \overline{\psi}(x) \left( e \gamma^\mu \gamma^5 A_\mu(x) - m\gamma^5 \right), \tag{3.9.3}$$

which naïvely lead to $\partial_\mu j^{\mu 5}(x) = 2 i m j^5(x)$, where

$$j^5(x) = \frac{1}{2} [\overline{\psi}_a(x), (\gamma^5 \psi)_a(x)], \quad j^{\mu 5}(x) = \frac{1}{2} [\overline{\psi}_a(x), (\gamma^\mu \gamma^5 \psi(x))_a].$$

For $m = 0$ the above, formally, leads to a seemingly conservation law $\partial_\mu j^{\mu 5}(x) = 0$. We will see that due to the singularity of the current $j^{\mu 5}(x)$, defined at coincident spacetime points of the fields, its divergence is modified giving rise to an anomaly and, for $m = 0$, it reads

$$\partial_\mu j^{\mu 5}(x) = -\frac{e^2}{16\pi^2} \varepsilon^{\mu\nu\alpha\beta} F_{\mu\nu} F_{\alpha\beta}. \tag{3.9.4}$$

The singularity of the axial vector current arises due to the quantum nature of the fermion field. To show this, we use Schwinger's point splitting method to generate a gauge invariant expression for $j^{\mu 5}(x)$.[11]

To the above end, consider the derivative of the product $(x_\pm = x \pm \epsilon/2)$

$$P^{\mu 5}(x, \epsilon) = \frac{1}{2} [\overline{\psi}_a(x_+), (\gamma^\mu \gamma^5 \psi)_a(x_-)] \exp\left[ i e \int_{x_-}^{x_+} d\xi^\nu A_\nu(\xi) \right], \tag{3.9.5}$$

---

[11] See (3.6.24) as well as Appendix IV at the end of the book which will be quite useful to follow the present section.

An application of (3.9.2), (3.9.3) gives for $m = 0$,

$$\partial_\mu \left[ \overline{\psi}_a(x_+), (\gamma^\mu \gamma^5 \psi)_a(x_-) \right] = -\mathrm{i}\, e \left[ \overline{\psi}_a(x_+), (\gamma^\mu \gamma^5 \psi)_a(x_-) \right]$$
$$\times \left[ A_\mu(x_+) - A_\mu(x_-) \right]. \tag{3.9.6}$$

and

$$A_\mu(x_+) - A_\mu(x_-) = \epsilon^\nu \partial_\nu A_\mu(x) + \mathcal{O}(\epsilon^3). \tag{3.9.7}$$

On the other hand, upon setting $\xi^\nu = x^\nu + (\epsilon^\nu/2)\lambda,\ -1 \leq \lambda \leq 1$, this leads to

$$\partial_\mu \exp\left[ \mathrm{i}\, e \int_{x_-}^{x_+} \mathrm{d}\xi^\nu A_\nu(\xi) \right] = \mathrm{i}\, e[\epsilon^\nu \partial_\mu A_\nu(x) + \mathcal{O}(\epsilon^3)] \exp\left[ \mathrm{i}\, e \int_{x_-}^{x_+} \mathrm{d}\xi^\sigma A_\sigma(\xi) \right].$$
$$\tag{3.9.8}$$

Thus we obtain,

$$\partial_\mu P^{\mu 5}(x, \epsilon) = \mathrm{i}\, e P^{\mu 5}(x, \epsilon)\left( \epsilon^\nu F_{\mu\nu} + \mathcal{O}(\epsilon^3) \right). \tag{3.9.9}$$

For the matrix element of the divergence of the axial current between the vacuum states we may then write for $m = 0$, as a limit

$$\frac{\langle 0_+ | \partial_\mu j^{\mu 5}(x) | 0_- \rangle^{(e)}}{\langle 0_+ | 0_- \rangle^{(e)}} = -e\, \mathrm{Av}\left( \left[ \epsilon^\nu F_{\mu\nu} + \mathcal{O}(\epsilon^3) \right] \right.$$

$$\left. \times \mathrm{Tr}[S_+^A(x_-, x_+) \gamma^\mu \gamma^5] \exp\left[ \mathrm{i}\, e \int_{x_-}^{x_+} \mathrm{d}\xi^\mu A_\mu(\xi) \right] \right), \tag{3.9.10}$$

where Av stands for an average over $\epsilon^0 > 0$ and $\epsilon^0 < 0$, with $\epsilon$ in a space-like direction. As for the electromagnetic current in (3.6.25), it is expressed in terms of the propagator $S_+^A$ in (3.2.11).

We will show that only the *single* power of $\epsilon$ contributes in (3.9.10) in the limit $\epsilon \to 0$.

To the above end, we may infer from (3.2.12) in a power series in $eA_\mu$

$$\mathrm{Tr}[S_+^A(x_-, x_+) \gamma^\mu \gamma^5] = \mathrm{Tr}[S_+(x_- - x_+) \gamma^\mu \gamma^5] +$$

$$+ \sum_{N \geq 2} (e)^{N-1} \int (dx_2) \ldots (dx_N) A_{\mu_2}(x_2) \ldots A_{\mu_N}(x_N) \times$$

$$\times \mathrm{Tr}[S_+(x_- - x_N) \gamma^{\mu_N} S_+(x_N - x_{N-1}) \gamma^{\mu_{N-1}} \ldots \gamma^{\mu_2} S_+(x_2 - x_+) \gamma^\mu \gamma^5]. \tag{3.9.11}$$

Since $\mathrm{Tr}\,[\,S_+^A(x_-,x_+)\,\gamma^\mu\,\gamma^5\,] = \mathrm{Tr}\,[\,\gamma^\mu\,\gamma^5\,S_+^A(x_-,x_+)\,]$, we may equivalently rewrite (3.9.11) as

$$\mathrm{Tr}\,[\,\gamma^\mu\,\gamma^5\,S_+^A(x_-,x_+)\,] = \mathrm{Tr}\,[\,\gamma^\mu\gamma^5\,S_+(x_--x_+)\,] +$$

$$+ \sum_{N\geq 2}(e)^{N-1}\int (dx_2)\dots(dx_N)A_{\mu_2}(x_2)\dots A_{\mu_N}(x_N)\,\times$$

$$\times\,\mathrm{Tr}\,[\,\gamma^\mu\gamma^5 S_+(x_--x_2)\,\gamma^{\mu_2}S_+(x_2-x_3)\,\gamma^{\mu_3}\dots\gamma^{\mu_N}S_+(x_N-x_+)\,].$$

$$(3.9.12)$$

In (3.9.11), we use the fact that

$$\mathrm{Tr}\,[\,(.)\,] = \mathrm{Tr}\,[\,(.)^\top\,] = \mathrm{Tr}\,[\,\mathscr{C}^{-1}(.)^\top\mathscr{C}\,],$$

where $\mathscr{C}$ is the charge conjugation matrix (see (I.3)), and the key property $\gamma^\mu\gamma^5 = \mathscr{C}^{-1}(\gamma^\mu\gamma^5)^\top\mathscr{C}$, as well as the basic properties

$$\mathscr{C}^{-1}(\gamma^\mu)^\top\mathscr{C} = -\gamma^\mu, \qquad \mathscr{C}^{-1}(S_+(x-x'))^\top\mathscr{C} = S_+(x'-x), \qquad (3.9.13)$$

to rewrite (3.9.11) as the average of the resulting expression for it, just introduced, and of the one in (3.9.12):

$$\mathrm{Tr}\left[\,S_+^A(x_-,x_+)\gamma^\mu\gamma^5\,\right] = \frac{1}{2}\mathrm{Tr}\left[\,\gamma^\mu\gamma^5\,[S_+(x_--x_+) + S_+(x_+-x_-)]\,\right]$$

$$+ \sum_{N\geq 2}(e)^{N-1}\int (dx_2)\dots(dx_N)A_{\mu_2}(x_2)\dots A_{\mu_N}(x_N)\,\times$$

$$\times\,\frac{1}{2}\left[F^{\mu 5\mu_2\cdots\mu_N}(x_-,x_2,\dots,x_N,x_+) + (-1)^{N-1}F^{\mu 5\mu_2\cdots\mu_N}(x_+,x_2,\dots,x_N,x_-)\right],$$

$$(3.9.14)$$

$$F^{\mu 5\mu_2\cdots\mu_N}(x,x_2,\dots,x_N,y)$$

$$= \mathrm{Tr}\,[\,\gamma^\mu\gamma^5\,S_+(x-x_2)\,\gamma^{\mu_2}\,S_+(x_2-x_3)\,\cdots\,\gamma^{\mu_N}\,S_+(x_N-y)\,]. \qquad (3.9.15)$$

The importance of the $(-1)^{N-1}$ factor in (3.9.14) cannot be overemphasized.

Upon taking Fourier transforms, the following expression results for (3.9.14)

$$\mathrm{Tr}\,[\,S_+^A(x_-,x_+)\gamma^\mu\gamma^5\,] = \int\frac{(dp)}{(2\pi)^4}\frac{[e^{-ip\epsilon} + e^{ip\epsilon}]}{2}\mathrm{Tr}\,[\,\gamma^\mu\gamma^5 S_+(p)\,] +$$

$$+ \sum_{N\geq 2}e^{N-1}\int\frac{(dQ_2)}{(2\pi)^4}\cdots\frac{(dQ_N)}{(2\pi)^4}A_{\mu_2}(Q_2)\dots A_{\mu_N}(Q_N)\,e^{i(Q_2+\dots+Q_N)x}\,\times$$

$$\times \int \frac{(dp)}{(2\pi)^4} \frac{1}{2} \left( \left[ e^{-ip\epsilon} e^{i\sum' Q_i \epsilon/2} + (-1)^{N-1} e^{ip\epsilon} e^{-i\sum' Q_i \epsilon/2} \right] \times \right.$$

$$\left. \times \mathrm{Tr}\left[ \gamma^{\mu} \gamma^5 S_+(p + \frac{Q_2}{2}) \gamma^{\mu_2} S_+(p - \frac{Q_2}{2}) \dots \gamma^{\mu_N} S_+(p - \frac{Q_2}{2} - Q_3 - \dots - Q_N) \right] \right),$$

$$(3.9.16)$$

properly symmetrized. In reference to (3.9.10), the rest of the analysis rests on the following:

$$\epsilon^\nu \left[ e^{-ip\epsilon} e^{i\sum' Q_i \epsilon/2} + (-1)^{N-1} e^{ip\epsilon} e^{-i\sum' Q_i \epsilon/2} \right]$$

$$= i \frac{\partial}{\partial p_\nu} \left[ e^{-ip\epsilon} e^{i\sum' Q_i \epsilon/2} - (-1)^{N-1} e^{ip\epsilon} e^{-i\sum' Q_i \epsilon/2} \right], \qquad (3.9.17)$$

and for a function $f(p)$, as a product of Dirac propagators, with $p^0 \to ip^0$, such that $f(\lambda p) \to 1/\lambda^\kappa$, $\kappa \geq 4$, for $\lambda \to \infty$, $\int (dp) \, \partial f(p)/\partial p_\nu = 0$ by Gauss' Theorem, since the surface area grows like $\lambda^3$. The equality in (3.9.17) is used as follows. One integrates by parts over $p$ by transferring $-i\partial/\partial p_\nu$ to the last line in (3.9.16) involving the trace of products of Dirac propagators.

We multiply (3.9.16) by $\epsilon^\nu$, as indicated in (3.9.10), integrate over $p$ by parts by using, in the process, (3.9.17), to conclude from Gauss' Theorem that no contribution to (3.9.10) arises for $N \geq 4$, for $\epsilon \to 0$, since $S(\lambda p) = \mathcal{O}(\lambda^{-1})$. On the other hand, no contribution arises for $N = 3$, in the limit $\epsilon \to 0$. This is as a consequence of a net minus sign $(-(-1)^{N-1})$ that results in the middle within the square brackets, on the right-hand side of (3.9.17) in the process of integrating by parts. For two powers of $\epsilon$, one of the $\epsilon$'s introduces one additional power of $p$ in the denominator of $p$-integral. Hence the term with $N = 3$ does not contribute.

The very first term in (3.9.16) does not contribute for the simple reason that $\mathrm{Tr}[\gamma^\mu \gamma^5 S(p)] = 0$. For all those terms where Gauss' Theorem leads to no contribution to the anomaly, one does not have to set $m = 0$ in their initial expressions. Clearly only the $N = 2$ term contributes.

All told, in the limit $\epsilon \to 0$, the right-hand side of (3.9.10) becomes

$$- e^2 F_{\mu\nu}(x) \int \frac{(dQ)}{(2\pi)^4} e^{iQx} A_\sigma(Q) I^{\mu\nu\sigma}(Q), \qquad (3.9.18)$$

$$I^{\mu\nu\sigma}(Q) = -i \int \frac{(dp)}{(2\pi)^4} \frac{\partial}{\partial p_\nu} \mathrm{Tr}\left[ \gamma^\mu \gamma^5 S(p + \frac{Q}{2}) \gamma^\sigma S(p - \frac{Q}{2}) \right], \qquad (3.9.19)$$

in which $m$ is set equal to zero.

By using the identity $\mathrm{Tr}[\gamma^5 \gamma^\mu \gamma^\rho \gamma^\sigma \gamma^\lambda] = -4i\varepsilon^{\mu\rho\sigma\lambda}$, where the latter is totally anti-symmetric, $\varepsilon^{0123} = +1$, as well as the Feynman parameter representation given in (II.27) in Appendix II at the end of the book, we may write

$$\text{Tr}\left[\gamma^\mu \gamma^5 S(p + \frac{Q}{2})\gamma^\sigma S(p - \frac{Q}{2})\right] = \int_0^1 dx \frac{4i\,\varepsilon^{\mu\rho\lambda\sigma}Q_\lambda\,p_\rho}{[(p - \frac{Q}{2}(1-2x))^2 + Q^2x(1-x)]^2},$$
(3.9.20)

where we have used the anti-symmetry of $\varepsilon^{\mu\rho\sigma\lambda}$. If it weren't for the derivative $\partial/\partial p_\nu$ in (3.9.19), the $p$-integral will diverge. By the application of the derivative just mentioned, however, renders the integral finite and by shifting the integration of variable $p$ to $p + Q(1-2x)/2$ (see Appendix II, (II.14) at the end of the book), the $x$-integrand in (3.9.20) then reduces to

$$\frac{4i\,\varepsilon^{\mu\nu\lambda\sigma}Q_\lambda}{[p^2 + Q^2x(1-x)]^2} - \frac{16i\,\varepsilon^{\mu\rho\lambda\sigma}Q_\lambda\,p_\rho p^\nu}{[p^2 + Q^2x(1-x)]^3}.$$
(3.9.21)

With $p_\rho p^\nu$ effectively given by $p^2\delta_\rho{}^\nu/4$, (3.9.19) then simplifies to

$$I^{\mu\nu\sigma}(Q) = 4\int_0^1 dx \int \frac{(dp)}{(2\pi)^4} \frac{\varepsilon^{\mu\nu\lambda\sigma}Q_\lambda\,Q^2x(1-x)}{[p^2 + Q^2x(1-x)]^3} = \frac{i}{8\pi^2}\varepsilon^{\mu\nu\lambda\sigma}Q_\lambda,$$
(3.9.22)

where we have used the integral (II.8) in Appendix II.

From (3.9.18), (3.9.22), we finally obtain for (3.9.10),

$$\frac{\langle\,0_+ \,|\,\partial_\mu j^{\mu 5}(x)\,|\,0_-\rangle^{(e)}}{\langle\,0_+\,|\,0_-\rangle^{(e)}} = -\frac{e^2}{16\pi^2}\varepsilon^{\mu\nu\lambda\sigma}F_{\mu\nu}F_{\lambda\sigma},$$
(3.9.23)

which is (3.9.4) in the vector space of fermions and anti-fermions. The non-zero expression on the right-hand side of the above equation is referred to as an anomaly. It corresponds to the so-called abelian case where the gauge transformation of the Dirac field, as discussed below (3.6.22), is implemented by a phase transformation

$$\psi(x) \rightarrow U(x)\psi(x), \quad \text{with } U(x) = \exp[i\lambda(x)], \quad [U(x), U(x')] = 0,$$
(3.9.24)

i.e., with commuting group elements. The underlying symmetry group of such phase transformations is referred to as the U(1) group. In the next section, we encounter more generalized groups of gauge transformations, with non-commuting group elements, referred to as non-abelian.

The equation of anomaly in (3.9.23) appeared in an early classic paper of Schwinger [7, Eqs.(5.24), (3.30)]. Important work on anomalies was also carried out later in the late sixties in Adler [1], Bell and Jackiw [3], Jackiw and Johnson [5], and Adler and Bardeen [2].

The additional factor involving the line integral in defining gauge invariant currents, systematically used above, was introduced by Schwinger [8, 9], and, as mentioned earlier, is referred to as the Schwinger line-integral.

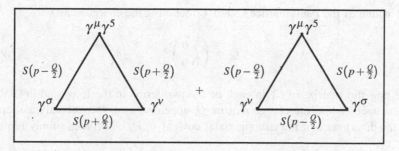

**Fig. 3.5** The integrand in (3.9.18)/(3.9.25) contributing to the anomaly may be represented by the above triangles with vertices as shown connected by massless fermion propagators

Before providing an experimental verification of the anomaly in the next subsection, we may gain further insight on the nature of the anomaly obtained above in the following manner. We carry out explicitly the partial derivative $\partial/\partial p_\nu$ in the integrand of the anomalous term in (3.9.19). This allows us to write

$$I^{\mu\nu\sigma}(Q) = i \int \frac{(dp)}{(2\pi)^4} \, \mathrm{Tr}\left[ \gamma^\mu \gamma^5 S(p + \frac{Q}{2}) \gamma^\nu S(p + \frac{Q}{2}) \gamma^\sigma S(p - \frac{Q}{2}) \right.$$
$$\left. + \gamma^\mu \gamma^5 S(p + \frac{Q}{2}) \gamma^\sigma S(p - \frac{Q}{2}) \gamma^\nu S(p - \frac{Q}{2}) \right],$$
(3.9.25)

where we have used the identity

$$\frac{\partial}{\partial p_\nu} S(p) = -S(p)\, \gamma^\nu\, S(p),$$
(3.9.26)

and the integrand in (3.9.25) may be represented diagrammatically as shown in Fig. 3.5, involving three vertex points identified by the matrices $(\gamma^\mu \gamma^5, \gamma^\nu, \gamma^\sigma)$, $(\gamma^\mu \gamma^5, \gamma^\sigma, \gamma^\nu)$, respectively, in the sum, and the anomaly in (3.9.23) is referred as an abelian triangle anomaly.

### 3.9.2 Experimental Verification of the Anomaly: $\pi^0 \to \gamma\gamma$ Decay

An experimental verification of the anomaly as given in (3.9.23) may be carried out by considering the decay of the neutral pion to two photons: $\pi^0 \to \gamma\gamma$.

To the above end, we present a simple description by replacing $\gamma^\mu \gamma^5$ in the axial-vector above by $\gamma^\mu \gamma^5 \sigma^a/2$, carrying SU(2) isospin, where $\sigma^1, \sigma^2, \sigma^3$ are the Pauli matrices, and replace $\gamma^\mu$ in the vector current by $\gamma^\mu Q$, that matches the

$2 \times 2$ nature of the Pauli matrices, with $Q$ denoting the charge matrix

$$Q = \begin{pmatrix} 1 & 0 \\ 0 & 0 \end{pmatrix}. \tag{3.9.27}$$

Since now the matrix, $\sigma^a/2$ in each of the two terms in the integrand in (3.9.25) appears once, while the charge matrix $Q$ appears twice, the anomaly associated with the divergence of the isotopic axial current $\partial_\mu j^{\mu 5 a}$ becomes simply replaced by

$$-\frac{e^2}{16\pi^2} \varepsilon^{\mu\nu\lambda\sigma} F_{\mu\nu} F_{\lambda\sigma} \, \mathrm{Tr}\left[ \frac{\sigma^a}{2} Q^2 \right] = -\delta^{a3} \frac{e^2}{32\pi^2} \varepsilon^{\mu\nu\lambda\sigma} F_{\mu\nu} F_{\lambda\sigma}. \tag{3.9.28}$$

Hence only the third component develops an anomaly, which is what is needed to investigate the $\pi^0 \to \gamma\gamma$ decay. It also multiplies the earlier anomalous factor by $1/2$. For $a = 3$, the anomaly may be rewritten as $-(e^2/8\pi^2)\varepsilon^{\mu\nu\lambda\sigma}\partial_\mu A_\nu \partial_\lambda A_\sigma$.

Consider a state $| k_1 e_1, k_2 e_2 \rangle$ with two photons of momenta $k_{1\mu}$, $k_{2\nu}$ and polarization vectors[12] $e_{1\lambda}, e_{2\sigma}$. Working to lowest order in the electromagnetic charge[13] for the process in question, we may write

$$\langle k e| A_\mu(x) = e_\mu^*(k) e^{-ikx} \langle \mathrm{vac}|, \tag{3.9.29}$$

and the matrix element of the anomaly $\langle k_1 e_1, k_2 e_2 | \partial_\mu j^{\mu 53}(x) | \mathrm{vac} \rangle$ becomes

$$\langle k_1 e_1, k_2 e_2 | \partial_\mu j^{\mu 53}(x) | \mathrm{vac} \rangle = \frac{e^2}{4\pi^2} k_{1\mu} e_\nu^* \, k_{2\lambda} e_\sigma^* \, \varepsilon^{\mu\nu\lambda\sigma} e^{-i(k_1+k_2)x}, \tag{3.9.30}$$

where the $8\pi^2$ factor has been replaced by $4\pi^2$ in the above coefficient because of the indistinguishability of the two photons thus introducing a multiplicative factor of 2.

Using the fact that the quantum field $\partial_\mu J^{\mu 53}$ is a pseudo-scalar field like the neutral pion field $\phi^3$ itself, the field $\partial_\mu j^{\mu 53}$ may be considered a source term to the pion field, i.e.,[14]

$$\langle k_1 e_1, k_2 e_2 | (-\Box + m_\pi^2) \phi^3(x) | \mathrm{vac} \rangle = \frac{1}{f_\pi} \langle k_1 e_1, k_2 e_2 | \partial_\mu j^{\mu 53}(x) | \mathrm{vac} \rangle, \tag{3.9.31}$$

---

[12]Polarization vectors for photons will be considered in Sect. 5.2. The underlying details are not needed here.

[13]This allows us to identify $F_{\mu\nu}, F_{\lambda\sigma}$ as free quantum fields in the anomaly in (3.9.28)

[14]Upon Fourier transform $(x \to p)$ of this equation, the latter reduces formally to what one refers to as the PCAC relation for $p^2 \approx 0$, where PCAC stands for partially conserved axial current.

where $1/f_\pi$ is a coupling parameter.[15] Hence

$$\langle k_1 e_1, k_2 e_2 \mid \phi^3(x) \mid \text{vac} \rangle = \frac{1}{f_\pi} \int (dx') \Delta_+(x - x') \langle k_1 e_1, k_2 e_2 \mid \partial_\mu j^{\mu 53}(x') \mid \text{vac} \rangle,$$

(3.9.32)

where

$$\Delta_+(x - x') = \int \frac{(dq)}{(2\pi)^4} \frac{e^{ip(x-x')}}{p^2 + m_\pi^2 - i\epsilon}, \qquad \epsilon \to +0.$$

(3.9.33)

With the necessary causal arrangement that the pion exists prior to its decay into two photons, i.e., by taking $x^0 < x'^0$ and applying (3.1.9)/(3.1.11), with $(-\gamma p + m)$ in them simply replaced by one, and the expression in (3.9.32) becomes

$$\langle k_1 e_1, k_2 e_2 \mid \phi^3(x) \mid \text{vac} \rangle = \frac{i}{f_\pi} \int (dx') \int \frac{d^3\mathbf{p}}{(2\pi)^3 2p^0}$$

$$\times e^{-ip(x-x')} \langle k_1 e_1, k_2 e_2 \mid \partial_\mu j^{\mu 53}(x') \mid \text{vac} \rangle.$$

(3.9.34)

Using the expression in (3.9.30), we may explicitly integrate over $x'$ to obtain

$$\langle k_1 e_1, k_2 e_2 \mid \phi^3(x) \mid \text{vac} \rangle = i \left( \frac{e^2}{4\pi^2 f_\pi} \right) \int \frac{d^3\mathbf{p}}{(2\pi)^3 2p^0} e^{-ipx}$$

$$\times (2\pi)^4 \delta^{(4)}(p - k_1 - k_2) \, k_{1\mu} e_\nu^* \, k_{2\lambda} e_\sigma^* \, \varepsilon^{\mu\nu\lambda\sigma}.$$

(3.9.35)

The neutral pion, prior to its decay into the two photons, was created by the pion field via the equation:

$$\phi^3(x) \mid \text{vac} \rangle = \int \mid p \rangle \frac{d^3\mathbf{p}}{(2\pi)^3 2p^0} e^{-ipx},$$

(3.9.36)

leading from (3.9.35) to

$$\int \langle k_1 e_1, k_2 e_2 \mid p \rangle \frac{d^3\mathbf{p}}{(2\pi)^3 2p^0} e^{-ipx}$$

$$= i \left( \frac{e^2}{4\pi^2 f_\pi} \right) \int \frac{d^3\mathbf{p}}{(2\pi)^3 2p^0} e^{-ipx} (2\pi)^4 \delta^{(4)}(p - k_1 - k_2) \, k_{1\mu} e_\nu^* \, k_{2\lambda} e_\sigma^* \, \varepsilon^{\mu\nu\lambda\sigma}.$$

(3.9.37)

---

[15]The reader should recognize, at this stage, that this is just the Klein-Gordon equation with a source term. See, e.g., Manoukian [6], p. 937.

On the left-hand side of this equation, we recognize $\langle k_1 e_1, k_2 e_2 \mid p \rangle$ as the amplitude $A(\pi^0 \to \gamma \gamma)$ of interest, and thus the following expression emerges for it

$$A(\pi^0 \to \gamma \gamma) = i \, (2\pi)^4 \delta^{(4)}(p - k_1 - k_2) \left( \frac{e^2}{4\pi^2 f_\pi} \right) k_{1\mu} e_\nu^* \, k_{2\lambda} e_\sigma^* \, \varepsilon^{\mu\nu\lambda\sigma}, \qquad (3.9.38)$$

which is, independently, a general expression for the process in question.

Experimentally,[16] $e^2/(4\pi^2 f_\pi) \approx 0.025 \, \text{GeV}^{-1}$, from which a non-vanishing value is obtained for $f_\pi \approx 93 \, \text{MeV}$. What is interesting here is that the decay constant $\sqrt{2} f_\pi \simeq 131 \text{MeV}$ may be measured, e.g, from the $\pi^+ \to \mu^+ \nu_\mu$ decay, and thus independently determines the anomaly factor $e^2/(4\pi^2 f_\pi)$ as obtained from applying the anomaly in deriving the decay rate $\pi^0 \to \gamma \gamma$.

It is quite interesting that a singular behavior, i.e., of divergence, of the axial vector current, leads to a finite non-zero result (an anomaly) which can be verified experimentally. This is an example of a good anomaly.[17]

## 3.10   Fermions and Anomalies in Field Theory: Non-Abelian Case

We generalize the equation of anomaly derived in the last section to the so-called non-abelian case to be defined shortly. Non-abelian gauge fields will be introduced and discussed in detail in Sect. 6.1 and in other sections following it. Here we introduce only the very basics to see how the equation of anomaly just mentioned is generalized to the non-abelian case.[18] As before, the result embodied in this section shows that a classical symmetry does not necessarily hold in the quantum world.

We consider several Dirac fields $\psi = (\psi_1(x), \ldots, \psi_n(x))$, satisfying Dirac equations,

$$\left[ \gamma^\mu \left( \frac{\partial_\mu}{i} - g A_\mu(x) \right) + m \right] \psi(x) = 0, \qquad (3.10.1)$$

written in matrix form, where $A_\mu(x)$ is an $n \times n$ matrix. The mass, $m$, may, in general, have a matrix structure as well. We consider a general transformation, by a

---

[16] The value of this constant may be inferred from the decay rate $\pi^0 \to \gamma \gamma$ [4].

[17] There are good anomalies and bad anomalies. An example of a bad anomaly is one which would destroy renormalizability. For example this would happen in non-abelian gauge theories if quarks, in general, come in other than three colors (see Sects. 6.4 *and* 6.15).

[18] Some readers may wish to read Sect. 6.1 simultaneously with the present one. In the present section we supply, however, enough details for the reader to be able to follow it.

unitary matrix $U(x)$,

$$\psi(x) \rightarrow U(x)\psi(x), \tag{3.10.2}$$

and infer from (3.10.1), that the latter remains invariant, if the so-called gauge field $A_\mu(x)$ transforms simultaneously as[19]

$$A_\mu(x) \rightarrow U(x)A_\mu(x)U^{-1}(x) + \frac{i}{g}U(x)\partial_\mu U^{-1}(x). \tag{3.10.3}$$

The gauge field $A_\mu(x)$ being an $n \times n$ matrix, may be expanded in terms of a basis set of matrices, assumed to have the following properties

$$A_\mu(x) = t_a A_{a\mu}(x), \quad [t_a, t_b] = i f_{abc} t_c, \tag{3.10.4}$$

where the $f_{abc}$ are called the structure constants of the group of transformations, and are totally anti-symmetric in their indices. If the structure constants are zero the group of transformations is referred to as abelian, otherwise it is called non-abelian. Non-abelian gauge groups will be studied in detail in Chap. 6. *In order not to confuse the indices just introduced with spinor indices we suppress the latter ones in this section.*

We may pose to consider a concrete case. For example, the group SU(3), consists of $3 \times 3$ unitary matrices of determinant one. Here U in SU(3) stands for unitary, $n = 3$ for the size of the matrices, and S for special, i.e., of determinant one. In quantum chromodynamics, $\psi$ corresponds to a quark field with three different colors specified by fields $\psi_1, \psi_2, \psi_3$. The Dirac field will also carry another index to specify the different flavors of quarks which we will also suppress here. The mass $m$ will carry a flavor index. The index $a$ introduced above in $A_{a\mu}(x)$ would take 8 values corresponding to eight gluons.

The matrices $U(x)$ will be expressed as

$$U(x) = \exp[i g t_a \vartheta_a(x)]. \tag{3.10.5}$$

Due to the nature of the SU(3) matrices, for example, the so-called generators $t_a$ of transformations, are not only Hermitian but also traceless since $\det U(x) = \exp[i g \vartheta_a(x) \operatorname{Tr} t_a]$.

The matrix elements of the massless free Dirac propagator, i.e., for $g = 0$, in the momentum representation, are simply given by

$$S(p) = \frac{-\gamma p}{p^2 - i\epsilon}. \tag{3.10.6}$$

---

[19]Note that $\partial_\mu(UU^{-1}) = 0$ implies that $U(\partial_\mu U^{-1}) = -(\partial_\mu U)U^{-1}$.

As in the abelian case, one naïvely obtains the conservation of the axial vector current $j^{\mu 5}(x) = [\overline{\psi}(x), (\gamma^\mu \gamma^5 \psi)(x)]/2$ from (3.10.1) for $m = 0$.

Consider the expression $(x_\pm = x \pm \epsilon/2)$

$$P^{\mu 5}(x, \epsilon) = \frac{1}{2}\left[\overline{\psi}(x_+), \left(1 + i g \int_{x_-}^{x_+} d\xi^\mu A_\mu(\xi)\right)(\gamma^\mu \gamma^5 \psi)(x_-)\right], \qquad (3.10.7)$$

The infinitesimal gauge transformations of the fields are given by

$$\psi(x) \rightarrow \psi(x) + i g \, \delta\vartheta_a(x) \, t_a \psi(x), \qquad (3.10.8)$$

$$\overline{\psi}(x) \rightarrow \overline{\psi}(x) - i g \, \overline{\psi}(x)\delta\vartheta_a(x) \, t_a, \qquad (3.10.9)$$

$$A_\mu(x) \rightarrow A_\mu(x) + \partial_\mu \, \delta\vartheta_a(x) \, t_a + i g \, \delta\vartheta_a(x) \, [t_a, A_\mu(x)]. \qquad (3.10.10)$$

The line integral is proportional to $\epsilon$, and is given by (see (3.9.8))

$$\int_{x_-}^{x_+} d\xi^\mu A_\mu(\xi) = \epsilon^\nu A_\nu(x) + \mathcal{O}(\epsilon^3). \qquad (3.10.11)$$

We will see that this expression, with the single power of $\epsilon$ will be sufficient to derive the full expression of the anomaly, and that higher powers of $\epsilon$ do not contribute. It is readily verified that $P^{\mu 5}(x, \epsilon)$ is gauge invariant for an infinitesimal gauge transformation implemented by an infinitesimal $\delta\vartheta^a(x)$ up to first order in $\epsilon^\nu$, which will eventually be taken to go to zero.

Upon taking the divergence of $P^{\mu 5}(x, \epsilon)$, we obtain[20]

$$\partial_\mu P^{\mu 5}(x, \epsilon) = i g \, [\overline{\psi}(x_+), K_\mu(x, \epsilon) \, (\gamma^\mu \gamma^5 \psi)(x_-)], \qquad (3.10.12)$$

$$K_\mu(x, \epsilon) = -\epsilon^\nu \partial_\nu A_\mu(x) - i g \, \epsilon^\nu \big(A_\mu(x)A_\nu(x) - A_\nu(x)A_\mu(x)\big) + \epsilon^\nu \, \partial_\mu A_\nu(x). \qquad (3.10.13)$$

Therefore upon introducing the matrix field

$$G_{\mu\nu}(x) = \partial_\mu A_\nu(x) - \partial_\nu A_\mu(x) - i g \, [A_\mu(x), A_\nu(x)], \qquad (3.10.14)$$

the following equation results

$$\partial_\mu P^{\mu 5}(x, \epsilon) = i g \, \epsilon^\nu \, [\overline{\psi}(x_+), G_{\mu\nu}(x) \, (\gamma^\mu \gamma^5 \psi)(x_-)]. \qquad (3.10.15)$$

From (3.10.4), $G_{\mu\nu}(x)$ may be also expanded in terms of the matrices $t^a$,

$$G_{\mu\nu}(x) = t_a \, G_{a\mu\nu}(x),$$

$$G_{a\mu\nu}(x) = \partial_\mu A_{a\nu}(x) - \partial_\nu A_{a\mu}(x) + g f_{abc} \, A_{b\mu}(x)A_{c\nu}(x). \qquad (3.10.16)$$

---

[20] See also (3.9.8), (3.9.9).

The matrix element between the vacuum states of the divergence of the axial vector current then takes, in a limiting sense, the form

$$\frac{\langle\, 0_+ \,|\, \partial_\mu j^{\mu 5}(x) \,|\, 0_-\rangle^{(g)}}{\langle\, 0_+ \,|\, 0_-\rangle^{(g)}} = - g\, G_{a\mu\nu}(x)\, \mathrm{Av}\left( \epsilon^\nu\, \mathrm{Tr}\,[\, S^A_+(x_-,x_+)\, t_a\, \gamma^{\,\mu}\, \gamma^{\,5}\,] \right),$$

(3.10.17)

expressed in terms of the Fourier transform of the propagator in (3.10.6), where Av stands for an average over $\epsilon^0 > 0$ and $\epsilon^0 < 0$, with $\epsilon$ in a space-like direction. In $\langle\, 0_+ \,|\, 0_-\rangle^{(g)}$, the products $\gamma^\mu A_\mu(x)$, in the abelian theory, being multiplied by the free Dirac propagators, are simply replaced by $\gamma^\mu t_a A_{a\mu}(x)$, in the non-abelian case, and the coupling parameter (charge) e by $g$.

Due to the simple diagonal structure of the free Dirac propagator in (3.10.6) with respect to the indices that label the matrix elements of the $t_a$ matrices, and that the $t_a$ matrices are of different nature than of the Dirac matrices, i.e., they commute with them, we can directly use the expression obtained for the abelian case in (3.9.16), (3.9.17), and factor out the trace of the products of the $t^a$ matrices as follows, to obtain, in a limiting sense for $\epsilon \to 0$,

$$\epsilon^\nu\, \mathrm{Tr}\,[\, S^A_+(x_-,x_+)\, t_a \gamma^\mu \gamma^{\,5}\,] = -\mathrm{i}\int \frac{(\mathrm{d}p)}{(2\pi)^4}\, \frac{[e^{-\mathrm{i}p\epsilon} - e^{\mathrm{i}p\epsilon}]}{2}\, \mathrm{Tr}(t_a)\, \frac{\partial}{\partial p_\nu} \mathrm{Tr}[\gamma^{\,\mu}\gamma^{\,5} S_+(p)]$$

$$-\mathrm{i}\sum_{N\ge 2} g^{N-1} \int \frac{(\mathrm{d}Q_2)}{(2\pi)^4} \cdots \frac{(\mathrm{d}Q_N)}{(2\pi)^4}\, A_{a_2\mu_2}(Q_2)\ldots A_{a_N\mu_N}(Q_N)\, e^{\mathrm{i}(Q_2+\ldots+Q_N)x}$$

$$\times \int \frac{(\mathrm{d}p)}{(2\pi)^4}\, \frac{1}{2}\bigg( \big[ \mathrm{Tr}(t_a t_{a_2}\ldots t_{a_N}) e^{-\mathrm{i}p\epsilon}\, e^{\mathrm{i}\sum' Q_i\epsilon/2} - \mathrm{Tr}(t_{a_N}\ldots t_{a_2}t_a)(-1)^{N-1} e^{\mathrm{i}p\epsilon}\, e^{-\mathrm{i}\sum' Q_i\epsilon/2} \big]$$

$$\times \frac{\partial}{\partial p_\nu} \mathrm{Tr}\Big[\gamma^\mu\gamma^{\,5} S_+\Big(p + \frac{Q_2}{2}\Big)\gamma^{\mu_2} S_+\Big(p - \frac{Q_2}{2}\Big) \ldots \gamma^{\mu_N} S_+\Big(p - \frac{Q_2}{2} - Q_3 - \ldots - Q_N\Big)\Big]\bigg),$$

(3.10.18)

symmetrized properly. Here we have already included the $\epsilon^\nu$ multiplicative factor, as shown on the left-hand side of the equation. Moreover, we have integrated over $p$ by parts to replace $\epsilon^\nu$ by $-\mathrm{i}\,\partial/\partial p_\nu$, with the latter operating to the right, and thus introduced an additional minus sign in the middle within the square brackets in the first and third lines involving $\exp(-\mathrm{i}p\epsilon)$ and $\exp(\mathrm{i}p\epsilon)$.

Clearly, we may invoke Gauss' Theorem to infer that all the terms with $N \ge 4$ are zero, because of the $\partial/\partial p_\nu$ factor and by simple power counting of powers of $p$.

The first term is obviously zero ($\mathrm{Tr}\,[t_a] = 0$).

The $N = 2$ term may be inferred directly from the abelian case in (3.9.16) and (3.9.17) to be

$$g\, \mathrm{Tr}(t^a t^{a_2})\, \frac{\varepsilon^{\mu\nu\lambda\sigma}}{8\pi^2}\, \partial_\lambda A_{a_2\sigma}(x) = g\, \mathrm{Tr}(t^a t^{a_2})\, \frac{\varepsilon^{\mu\nu\lambda\sigma}}{16\pi^2}\, (\partial_\lambda A_{a_2\sigma}(x) - \partial_\sigma A_{a_2\lambda}(x)).$$

(3.10.19)

For two powers of $\epsilon$, a net minus sign arises within the square brackets involving $\exp(-ip\epsilon)$ and $\exp(ip\epsilon)$, giving a null result. Gauss' Theorem guarantees that higher powers of $\epsilon$ do not contribute.

A new contribution, not present in the abelian case, now emerges from the $N = 3$ term. First note that for two and higher powers of $\epsilon$, one may invoke Gauss' Theorem to infer that they do not contribute. For the single power of $\epsilon$, as given in (3.10.18), the $N = 3$ term reads[21]

$$
-\mathrm{i}\,g^2 \int \frac{(\mathrm{d}Q_2)}{(2\pi)^4} \frac{(\mathrm{d}Q_3)}{(2\pi)^4} A_{a_2\mu_2}(Q_2) A_{a_3\mu_3}(Q_3)\, \mathrm{e}^{\mathrm{i}(Q_2+Q_3)x}
$$
$$
\times \operatorname{Tr}\left(t_a[t_{a_2}, t_{a_3}]\right) K^{\mu\nu\mu_2\mu_3}(Q_2, Q_3), \tag{3.10.20}
$$

$$
K^{\mu\nu\mu_2\mu_3}(Q_2, Q_3) = \int \frac{(\mathrm{d}p)}{(2\pi)^4} \frac{\partial}{\partial p_\nu} \operatorname{Tr}\left(\gamma^\mu \gamma^5 S_+\left(p + \frac{Q_2}{2}\right) \gamma^{\mu_2} S_+\left(p - \frac{Q_2}{2}\right)\right.
$$
$$
\left. \times \gamma^{\mu_3} S_+\left(p - \frac{Q_2}{2} - Q_3\right)\right). \tag{3.10.21}
$$

The product of the denominators of the Dirac propagators may be combined by using the Feynman parameter representation given in (II.35) of Appendix II, at the end of the book, for $m = 0$

$$
\frac{1}{(p + \frac{Q_2}{2})^2 (p - \frac{Q_2}{2})^2 (p - \frac{Q_2}{2} - Q_3)^2} = 2 \int_0^1 \mathrm{d}x_1 \int_0^{x_1} \mathrm{d}x_2 \frac{1}{[(p-k)^2 + M(Q_2, Q_3, x_1, x_2)]^3}, \tag{3.10.22}
$$

$$
k = \frac{1}{2}\left[Q_2(1 - 2x_2) + 2Q_3(1 - x_1)\right], \tag{3.10.23}
$$

$$
M(Q_2, Q_3, x_1, x_2) = Q_2^2 x_2(1 - x_2) + Q_3^2 x_1(1 - x_1) + 2Q_2 Q_3 x_2(1 - x_1). \tag{3.10.24}
$$

By carrying out the differentiation with respect to $p_\nu$, using, in the process, the integrals in (II.8), (II.9), (II.10), (II.11), and by now the familiar identity $\operatorname{Tr}[\gamma^5 \gamma^\mu \gamma^\nu \gamma^\sigma \gamma^\lambda] = -4\mathrm{i}\,\varepsilon^{\mu\nu\sigma\lambda}$, $K^{\mu\nu\mu_2\mu_3}(Q_2, Q_3)$ readily works out to be

$$
K^{\mu\nu\mu_2\mu_3}(Q_2, Q_3) = \frac{\varepsilon^{\mu\nu\mu_2\mu_3}}{16\pi^2}. \tag{3.10.25}
$$

---

[21] By carrying the differentiation with respect to $p_\nu$ in (3.10.21), we generate a diagram involving four connected fermion propagators referred to as a square diagram contributing to the anomaly as opposed to the triangle diagram contribution to the anomaly in (3.10.19) involving three fermion propagators. See also (3.10.25) which, *however* has exactly the same factor $\varepsilon^{\mu\nu\mu_2\mu_3}/(16\pi^2)$, with $\mu_2 = \lambda$, $\mu_3 = \sigma$, as the one in (3.10.19).

From (3.10.17), (3.10.19), (3.10.20), (3.10.25), and the expression for $G_{\mu\nu}$ in (3.10.14), we have

$$\frac{\langle\, 0_+ \,|\, \partial_\mu j^{\mu 5}(x)\,|\, 0_-\rangle^{(g)}}{\langle\, 0_+ \,|\, 0_-\rangle^{(g)}} = -\frac{g^2}{16\,\pi^2}\, \varepsilon^{\mu\nu\lambda\sigma}\,\mathrm{Tr}\,(G_{\mu\nu}G_{\lambda\sigma}), \qquad (3.10.26)$$

holding in the vector space of fermions and anti-fermions.

Note that this anomaly is the same as the abelian one in (3.9.23) with $F_{\mu\nu}F_{\lambda\sigma}$ replaced by $\mathrm{Tr}\,(G_{\mu\nu}\,G_{\lambda\sigma})$, and e by $g$.

The gauge transformation of $A_\mu(x)$ in (3.10.3), implies, in turn, the gauge transformation of $G_{\mu\nu}(x)$ in (3.10.14) to be (see Problem 3.14)

$$G_{\mu\nu}(x) \;\to\; U(x)\, G_{\mu\nu}(x)\, U^{-1}(x). \qquad (3.10.27)$$

Hence $\mathrm{Tr}\,(G^{\mu\nu}G_{\lambda\sigma})$ is gauge invariant, and so is the equation of anomaly in (3.10.26).

The above analysis of the anomaly will be crucial later on in Chap. 6 for the renormalizability of the standard model. The key point to realize about this at this point is that the standard model involves axial vector currents, and renormalizability of the model hinges on the assumption of the absence of such anomalies.

# Appendix A: Evaluation of $L^{\mu_1 \mu_2}$

The integral $L^{\mu_1 \mu_2}(Q,-Q)$ is defined in (3.6.29). Clearly, $L^{\mu_1 \mu_2}(Q,-Q)$ is zero for $Q = 0$. Also we will show that

$$Q_{\mu_1} L^{\mu_1 \mu_2}(Q,-Q) = 0, \qquad Q_{\mu_2} L^{\mu_1 \mu_2}(Q,-Q) = 0, \qquad (\text{A-3.1})$$

as required by gauge invariance under the transformation $A_{\mu_j}(Q_j) \to A_{\mu_j}(Q_j) + \lambda(Q_j)\, Q_{\mu_j}$ in (3.6.28). To this end, using the notation $S_+^{-1}(p) = \gamma\, p + m$, and the identity $\gamma\, Q = [S_+^{-1}(p - Q/2) - S_+^{-1}(p + Q/2)$, we note that when (3.6.29) is multiplied by $Q_{\mu_2}$ we obtain for the integrand $(\mathrm{Tr}\,\gamma^{\mu_1}[\,.\,])$, where

$$[\,.\,] = -[S_+\left(p + \frac{Q}{2}\right) - S_+\left(p - \frac{Q}{2}\right)] + Q\frac{\partial}{\partial p}S_+(p) + \frac{1}{24}\left(Q\frac{\partial}{\partial p}\right)^3 S_+(p)$$

$$= \mathscr{O}\left(\frac{2}{5!}\left(\frac{Q}{2}\frac{\partial}{\partial p}\right)^5\right)S_+(p), \qquad (\text{A-3.2})$$

and note that $(Q\partial/\partial p)S_+(p) = -S_+(p)\gamma Q S_+(p)$. We have used the useful identities

$$e^x - 1 = \int_0^1 d\lambda \, e^{\lambda x} x,$$

$$\vdots$$

$$e^x - 1 - x - \frac{x^2}{2!} - \frac{x^3}{3!} - \frac{x^4}{4!} = \int_0^1 d\lambda_5 \int_0^{\lambda_5} d\lambda_4 \int_0^{\lambda_4} d\lambda_3 \int_0^{\lambda_3} d\lambda_2 \int_0^{\lambda_2} d\lambda_1 \, e^{\lambda_1 x} x^5, \qquad \text{(A-3.3)}$$

to infer that

$$-\left(\exp(\frac{Q}{2}\frac{\partial}{\partial p}) - \exp(-\frac{Q}{2}\frac{\partial}{\partial p})\right) + \left(Q\frac{\partial}{\partial p}\right) + \frac{1}{24}\left(Q\frac{\partial}{\partial p}\right)^3$$

$$= -2 \int_0^1 d\lambda_5 \int_0^{\lambda_5} d\lambda_4 \int_0^{\lambda_4} d\lambda_3 \int_0^{\lambda_3} d\lambda_2 \int_0^{\lambda_2} d\lambda_1 \, \cosh\left(\lambda_1 \frac{Q}{2}\frac{\partial}{\partial p}\right)\left(\frac{Q}{2}\frac{\partial}{\partial p}\right)^5. \qquad \text{(A-3.4)}$$

We may thus invoke Gauss' Theorem to infer that the remainder in (A-3.2) integrates out to zero. A similar analysis applies for the case when the integral is multiplied by $Q_{\mu_1}$ thus establishing (A-3.1). This allows us to write

$$-i e^2 L^{\mu_1 \mu_2}(Q, -Q) = \left[ \eta^{\mu_1 \mu_2} Q^2 - Q^{\mu_1} Q^{\mu_2} \right] \Pi(Q^2). \qquad \text{(A-3.5)}$$

The i factor is chosen for convenience. Also $\left[ \eta^{\mu_1 \mu_2} Q^2 - Q^{\mu_1} Q^{\mu_2} \right] \Pi(Q^2)$ is the lowest order contribution to the so-called vacuum polarization tensor, with $L_{\mu\nu}$ given (3.6.29). To evaluate $\Pi(Q^2)$, we contract the indices $\mu_1$ and $\mu_2$ in (A-3.5). In particular, (3.6.29), leads us to evaluate the following integral

$$3(2\pi)^4 Q^2 \Pi(Q^2) = -i e^2 \int (dp)\left(\text{Tr}\left[\gamma^\mu S_+\left(p - \frac{Q}{2}\right)\gamma_\mu S_+\left(p + \frac{Q}{2}\right)\right]\right.$$

$$\left. + \frac{1}{24}\left(Q\frac{\partial}{\partial p}\right)^2 \frac{\partial}{\partial p_\mu}\text{Tr}\left[\gamma^\mu S_+(p)\right] - \text{Tr}\left[\gamma^\mu S_+(p)\gamma_\mu S_+(p)\right]\right). \qquad \text{(A-3.6)}$$

The following integral is easily evaluated from the integrals in (II.8)–(II.9) in Appendix II at the end of the book:

$$\int (dp) \frac{1}{24}\left(Q\frac{\partial}{\partial p}\right)^2 \frac{\partial}{\partial p_\mu}\text{Tr}\left[\gamma^\mu S_+(p)\right] =$$

$$-2 Q^2 \int (dp)\left[\frac{m^2}{(p^2 + m^2)^3} - \frac{m^2 p^2}{(p^2 + m^2)^4}\right] = -\frac{i\pi^2}{3} Q^2. \qquad \text{(A-3.7)}$$

For the *first* term in the integral in (A-3.6), we may use the Feynman parameter representation of the product of two denominators defined in (II.27) to write it, after carrying out the trace, up to the $-\mathrm{i}\,e^2$ factor, as

$$I(Q^2) = \int (\mathrm{d}p) \int_0^1 \mathrm{d}z \frac{-8p^2 + 2Q^2 - 16m^2}{[(p + \frac{Q}{2}(1-2z))^2 + Q^2 z(1-z) + m^2]^2}. \tag{A-3.8}$$

We may refer to the integral (II.23) and make a contraction over $\mu$ and $\nu$ there, to obtain, in reference to the term with $p^2$ in the numerator in the integrand in (A-3.8) the following integral

$$\int \frac{(\mathrm{d}p)\,p^2}{[(p-k)^2 + M^2(k^2)]^2} =$$

$$-\frac{3\,\mathrm{i}\,\pi^2}{2} k^2 + [k^2 - M^2(k^2)] \int \frac{(\mathrm{d}p)}{[p^2 + M^2(k^2)]^2} + \int \frac{(\mathrm{d}p)}{[p^2 + M^2(k^2)]}, \tag{A-3.9}$$

$$k = -\frac{Q}{2}(1-2z), \qquad M^2(k^2) = Q^2 z(1-z) + m^2. \tag{A-3.10}$$

We here define the degree of divergence of a 4D-integral as the power of $p$ in the numerator minus the power of $p$ in the denominator in the integrand plus four. Thus the degree of divergence of the integral in (A-3.8) restricted to the $(2Q^2 - 16m^2)$ term divided by the denominator is zero. This allows us (see Appendix II) to shift the integration variable $p$, restricted to this term, to $p - Q(1-2z)/2$. Accordingly, we may use (A-3.9), (A-3.10), to rewrite (A-3.8) as

$$I(Q^2) = 8\,\mathrm{i}\,\pi^2 \frac{3}{2} \frac{Q^2}{4} \int_0^1 \mathrm{d}z\,(1-2z)^2 +$$

$$\int (\mathrm{d}p) \int_0^1 \mathrm{d}z \frac{[2Q^2 - 8m^2 - 4Q^2(1-2z)^2 + 8Q^2 z(1-z)]}{[p^2 + Q^2 z(1-z) + m^2]^2}$$

$$-8 \int (\mathrm{d}p) \int_0^1 \mathrm{d}z \frac{1}{[p^2 + Q^2 z(1-z) + m^2]}. \tag{A-3.11}$$

Now we may integrate the last integral by parts over $z$

$$\int_0^1 \mathrm{d}z \frac{1}{[p^2 + Q^2 z(1-z) + m^2]} = \frac{z}{[p^2 + Q^2 z(1-z) + m^2]} \Big|_0^1$$

$$-\int_0^1 \mathrm{d}z \frac{-z Q^2(1-2z)}{[p^2 + Q^2 z(1-z) + m^2]^2}, \tag{A-3.12}$$

to combine its second part with the second integral in (A-3.11) to obtain

$$I(Q^2) = i\,\pi^2\,Q^2 + 8\,Q^2 \int (\mathrm{d}p) \int_0^1 \frac{z\,\mathrm{d}z}{[p^2 + Q^2 z(1-z) + m^2]^2}$$

$$- 8\,m^2 \int (\mathrm{d}p) \int_0^1 \frac{\mathrm{d}z}{[p^2 + Q^2 z(1-z) + m^2]^2} - 8 \int \frac{(\mathrm{d}p)}{[p^2 + m^2]}. \tag{A-3.13}$$

Integrating by parts one more time over $z$, using the integral (II.8), and writing $m^2 = (m^2 + Q^2 z(1-z) - Q^2 z(1-z))$, we obtain

$$I(Q^2) = i\,\pi^2 \frac{7}{3} Q^2 + \int (\mathrm{d}p) \frac{(-8p^2 - 16m^2)}{[p^2 + m^2]^2}$$

$$+ 4\,Q^2 \int \frac{(\mathrm{d}p)}{[p^2 + m^2]^2} + 4\,i\,\pi^2\,Q^2 \int_0^1 \mathrm{d}z \frac{(3z^2 - 2z^3)Q^2(1 - 2z)}{Q^2 z(1-z) + m^2}. \tag{A-3.14}$$

Here we recognize the $Q^2$-independent part as $I(0)$ corresponding to the third term in (A-3.6). The last integral in (A-3.14) may be integrated by parts over $z$ to finally obtain from (A-3.6), (A-3.7), (A-3.8),

$$\Pi(Q^2) = \frac{e^2}{12\,\pi^2} \left( \left[ \frac{1}{i\pi^2} \int \frac{(\mathrm{d}p)}{[p^2 + m^2]^2} + \frac{1}{2} \right] - 6 \int_0^1 \mathrm{d}z\, z(1-z) \ln\left[ 1 + \frac{Q^2}{m^2} z(1-z) \right] \right), \tag{A-3.15}$$

involving the logarithmically divergent integral $\int(\mathrm{d}p)/[p^2 + m^2]^2/(i\pi^2)$. We will have ample opportunities later on to interpret this constant (see Sect. 5.10.2). The additive factor $1/2$ to it depends on the rooting of the momentum $Q$ in the definition of the integral in (A-3.6).

## Appendix B: Infinitesimal Variation of the Exponential of a Matrix

For a matrix $M$, $\delta e^M \neq \delta M e^M$, in general, but $\mathrm{Tr}\,[\delta e^M] = \mathrm{Tr}\,[\delta M\,e^M]$. To see how the situation changes from one case to the other, and in passing derive an expression for $\delta e^M$, integrate the following, involving two matrices $N$, $M$, from $\lambda = 0$ to $\lambda = 1$:

$$\frac{\mathrm{d}}{\mathrm{d}\lambda} e^{\lambda N} e^{-\lambda M} = e^{\lambda N}(N - M) e^{-\lambda M}, \tag{B-3.1}$$

to obtain

$$e^N e^{-M} = 1 + \int_0^1 d\lambda \, e^{\lambda N} (N - M) \, e^{-\lambda M}, \tag{B-3.2}$$

or

$$e^N - e^M = \int_0^1 d\lambda \, e^{\lambda N} (N - M) \, e^{-\lambda M} \, e^M. \tag{B-3.3}$$

Let $N = M + \delta M$, then the latter equation reads

$$e^{M+\delta M} - e^M = \int_0^1 d\lambda \, e^{\lambda (M+\delta M)} (\delta M) \, e^{-\lambda M} \, e^M. \tag{B-3.4}$$

For an infinitesimal variation, we have the useful expression

$$\delta e^M = \int_0^1 d\lambda \, e^{\lambda M} (\delta M) \, e^{-\lambda M} \, e^M. \tag{B-3.5}$$

If we perform a trace of the above, we simply have

$$\mathrm{Tr}[\delta e^M] = \mathrm{Tr}\Big[\int_0^1 d\lambda \, \delta M \, e^M\Big],$$

$$= \mathrm{Tr}[\delta M \, e^M]. \tag{B 3.6}$$

# Problems

**3.1** Show that this remarkable positive function:

$$f(t) = \begin{cases} 0, & t < -a, \\ \frac{1}{2} \exp\big[-2\big(\frac{a}{t+a} - 1\big)\big], & -a \leq t < 0, \\ 1 - \frac{1}{2} \exp\big[-2\big(\frac{a}{-t+a} - 1\big)\big], & 0 \leq t < a, \\ 1, & a \leq t, \end{cases}$$

with $a > 0$, is not only *continuous* at $t = -a$, $t = 0$, $t = a$, it vanishes at $t = -a$, and is equal to 1 at $t = a$, but has a continuous derivative as well. This is a particular example of a function vanishing at a point and being also continuous. The keen reader will realize that this provides a continuous function representation for the (discontinuous) step function $\theta(t) = 1$ for $t > 0$ and $\theta(t) = 0$ for $t < 0$.

**3.2** Verify that the propagator $S_+(x' - x)$ also satisfies (3.1.13).

**3.3** Verify (3.2.15)/(3.2.16) for the vacuum expectation value of the Dirac field $\psi(x)$, in the presence of a weak external electromagnetic potential $A_\mu$, for $x^0$ corresponding to a time much later after the source has been switched off.

**3.4** Derive the expression for $\langle 0_+ \mid \psi(x) \mid 0_- \rangle / \langle 0_+ \mid 0_- \rangle$ in (3.3.16), where $x^0$ correspond to a point before the source $\eta(x')$ is switched on.

**3.5** Work out the details leading to the expression in (3.3.14)/(3.3.15) corresponding to the scattering of a positron $e^+(p) \to e^+(p')$ off a weak external electromagnetic potential.

**3.6** Show that the probability that a particle of spin 1/2 emitted by a source $\eta$ has a momentum $\mathbf{p}$ within a range $\Delta$ and spin $\sigma$ is given by

$$\frac{1}{N} \int_{\mathbf{p}\in\Delta} \frac{m}{p^0} \frac{\mathrm{d}^3\mathbf{p}}{(2\pi)^3} \mid \bar{u}(\mathbf{p},\sigma)\eta(p) \mid^2, \quad \text{where } N = \int \sum_\sigma \frac{m}{p^0} \frac{\mathrm{d}^3\mathbf{p}}{(2\pi)^3} \mid \bar{u}(\mathbf{p},\sigma)\eta(p) \mid^2.$$

**3.7** By averaging over the initial spins and summing over the final ones, derive (3.4.6) for $p^0 = p'^0$.

**3.8** Derive the anti-commutation relation in (3.5.12)/(3.5.13) for the Dirac quantum field valid for all times. Hint: Note that the Dirac field at any time may be expressed in terms of the time equal to zero, via the equation

$$\psi(x) = \int \sum_\sigma \frac{m}{p^0} \frac{\mathrm{d}^3\mathbf{p}}{(2\pi)^3} \int \mathrm{d}^3\mathbf{x}' \Big[ \mathrm{e}^{\mathrm{i}(x-x')p} u(\mathbf{p},\sigma) u^\dagger(\mathbf{p},\sigma)$$

$$+ \mathrm{e}^{-\mathrm{i}(x-x')p} v(\mathbf{p},\sigma) v^\dagger(\mathbf{p},\sigma) \Big] \psi(x'),$$

$x' = (0, \mathbf{x}')$. On the other hand, the orthogonal projections in (I.21), (I.22) allow one to write $\psi(x) = \int \mathrm{d}^3\mathbf{y}\, K(x,y)\, \psi(y)$, $y = (0, \mathbf{y})$,

$$K(x,y) = \int \frac{m}{p^0} \frac{\mathrm{d}^3\mathbf{p}}{(2\pi)^3} \Big[ \mathrm{e}^{\mathrm{i}(x-y)p}\, \mathbb{P}_+(p)\gamma^0 - \mathrm{e}^{-\mathrm{i}(x-y)p}\, \mathbb{P}_-(p)\gamma^0 \Big],$$

where $\mathbb{P}_+(p) = (-\gamma p + m)/(2m)$, $\mathbb{P}_-(p) = -(\gamma p + m)/(2m)$ as given, respectively, in (I.21), (I.22).

**3.9** Show that at equal times $x^0 = x'^0$, the anti-commutation relation obtained in the previous problem reduces to the one in (3.5.11).

**3.10** Consider the integral $I = \int[(\mathrm{d}Q)/(2\pi)^4]A_\mu(Q) \int_{x-(\epsilon/2)}^{x+(\epsilon/2)} \mathrm{d}\xi^\mu\, \mathrm{e}^{\mathrm{i}Q\xi}$. Setting $\xi^\mu = x^\mu + (\epsilon^\mu/2)\lambda$, with $\lambda$ as an integration variable, express the integral as a power series in $\epsilon$. This is a typical integral which occurs in Schwinger's point splitting method.[22]

**3.11** Show that $\frac{\mathrm{i}}{4}[\gamma^\mu,\gamma^\nu]F_{\mu\nu} = \begin{pmatrix} \sigma\cdot\mathbf{B} & -\mathrm{i}\,\sigma\cdot\mathbf{E} \\ -\mathrm{i}\,\sigma\cdot\mathbf{E} & \sigma\cdot\mathbf{B} \end{pmatrix}$, where $F_{\mu\nu} = \partial_\mu A_\nu - \partial_\nu A_\mu$.

**3.12** Derive the expression in (3.8.8) for the Schwinger effect by working in the complex $s$-plane, and elaborate on the importance of the term $(s|eE|)^2/3$ in (3.8.5).

**3.13** Prove the identity in (3.9.25) with the propagator $S(p) = 1/(\gamma p + m)$.

**3.14** Derive the transformation rule for $G_{\mu\nu}$ in (3.10.27) for non-abelian gauge theories.

**3.15** Show that

$$\exp[\mathrm{i}F(z)]\left[S_+^{-1}(z-x) + \delta^{(4)}(z-x)\gamma^\mu(\partial/\partial x^\mu)F(x)\right]$$
$$= S_+^{-1}(z-x)\exp[\mathrm{i}F(x)],$$

for a numerical function $F(z)$.[23]

# References

1. Adler, S. L. (1969). Axial-vector vertex in Spinor electrodynamics. *Physical Review, 177*, 2426–2438.
2. Adler, S. L., & Bardeen, W. (1969). Absence of higher-order corrections in the anomalous axial-vector divergence equation. *Nuovo Cimento A, 182*, 1517–1536.
3. Bell, J. S., & Jackiw, R. (1969). A PCAC Puzzle: $\pi^0 \to \gamma\gamma$ in the $\sigma$-Model. *Nuovo Cimento A, 60*, 47–61.
4. Beringer, J., et al. (2012). Particle data group. *Physical Review D, 86*, 010001.
5. Jackiw, R., & Johnson, K. (1969). Anomalies of the axial-vector current. *Physical Review, 182*, 1459–1469.
6. Manoukian, E. B. (2006). *Quantum theory: A wide spectrum*. Dordrecht: Springer.
7. Schwinger, J. (1951). On Gauge invariance and vacuum polarization. *Physical Review, 82*, 664–679.
8. Schwinger, J. (1959). Field theory commutators. *Physical Review Letters, 3*, 296–297.
9. Schwinger, J. (1962). Exterior algebra and the action principle I. *Proceedings of the National Academy of Sciences of the United States of America, 48*, 603–611.

---

[22]See Appendix IV at the end of the book.

[23]This result will be important in the analysis of Problem 5.20 in Chap. 5.

# Recommended Reading

1. Manoukian, E. B. (2006). *Quantum theory: A wide spectrum*. Dordrecht: Springer.
2. Schwinger, J. (1951). On gauge invariance and vacuum polarization. *Physical Review, 82*, 664–679.
3. Schwinger, J. (1970). *Particles, Sources, and Fields* (Vol. I). Reading: Addison-Wesley.

# Chapter 4
# Fundamental Aspects of Quantum Field Theory

At sufficiently high energies, relativity becomes the indispensable language to do physics and explain the exchange that takes place between energy and matter allowing the creation of an unlimited number of particles such that the number of particles need not be conserved. An appropriate description of such physical processes for which a variable number of particles may be created or destroyed is provided by the concept of a quantum field which we denote, generally, by $\chi(x)$ as a function of a spacetime variable $x = (x^0, \mathbf{x})$, for which the Dirac quantum field theory, as described in Chap. 3, is one example.

As the disturbance created by a measurement process cannot propagate faster than the speed of light, different regions of space, at any given fixed time, are dynamically independent in the sense that a measurement made in any region of space is incompatible with any other in a different region. Space, at any given time, defines a special case of a space-like (hyper-) surface with the latter defined in such a manner that any two points $x_1$, $x_2$ lying on it are space-like separated $(x_1 - x_2)^2 > 0$, and hence cannot be connected by any signal. The quantum fields as dynamical variables, providing degrees of freedom of a system, are thus labeled, by infinitesimal three dimensional regions of space, with the latter being independent in the sense just described, or in a local description they are labeled by the space coordinate points $\mathbf{x}$. At any given time $x^0 = t$, the space variable $\mathbf{x}$ together with any indices $a$ that independent fields may carry, such as spinor and/or vector indices, specify the infinite (uncountable) degrees of freedom as $\mathbf{x}$ is made to vary in space as well as the indices $a$ take on their specific values. Upon writing $\chi_a(x) = \chi_{\mathbf{x}a}(t)$, we note that $(\mathbf{x}a)$ replaces the index $i$ in the dynamical variables $q_i(t)$, in well known situations, dealing with finite number of degrees of freedom.

Independent dynamical variables $\chi_a(x)$ thus satisfy locality conditions, and for the elementary fields, for examples, consisting of a (non-Hermitian) scalar $\varphi(x)$, and the Dirac field $\psi(x)$, are specified by a commutation relation $[\varphi(x), \varphi^\dagger(x')] = 0$ for the scalar field (Sects. 4.5 and 4.7), and by anti-commutation relations

© Springer International Publishing Switzerland 2016

E.B. Manoukian, *Quantum Field Theory I*, Graduate Texts in Physics,
DOI 10.1007/978-3-319-30939-2_4

$\{\psi_a(x), \psi_b^\dagger(x')\} = 0$ for the Dirac field components (Sect. 3.5) for any two space-like separated points $x, x'$.

The significant physical role that a quantum field $\chi(x)$ acquires and the particle content of a theory emerges from the examination of the commutation relations of the field with the energy-momentum operator components $P^\mu$ (Sect. 4.1) corresponding to the underlying physical system.

The field concept, its significance and the meaning of wavefunction renormalizations are introduced in Sect. 4.1. The underlying algebra satisfied by the generators of the inhomogeneous Lorentz transformations (Poincaré algebgra) is worked out in Sect. 4.2. The action principle of quantum fields, together with its related consequences, is the subject matter of Sect. 4.3. The principle of stationary action encapsulates the dynamics of a system as the system develops between two space-like surfaces. The principle of stationary action, together with generators emerging from variations of these space-like boundary surfaces, summarize basic properties of an underlying theory. The mere fact that generators are to be introduced to generate field variations lead automatically to field equations – the so-called Euler-Lagrange equations. These field equations emerging from the theory, involve dynamical variables, as functions of time (and space), and describe the time evolution of theory. Section 4.4 deals with Lorentz invariance and the energy-momentum tensor, as well as the angular momentum density, providing, in particular, of a local description of the energy-momentum distribution of matter in space. The celebrated Spin and Statistics connection is studied in Sect. 4.5. The quantum dynamical principle (QDP) of quantum field theory is treated in Sect. 4.6 from which physical processes may be investigated.

A functional Fourier transform of expressions directly obtained from the QDP leads directly to the equivalent path integral formalism which is applied to gauge theories in subsequent chapters. The basic fields one encounters in field theory are covered in Sect. 4.7, with further applications of the QDP, involving some of these fields, in Sect. 4.8. Some intricacies with Lagrangians are spelled out in Sect. 4.9. Section 4.10 deals with the celebrated CPT Theorem, where, in particular, we will see that particle/antiparticle detectors and antiparticle/particle emitters are interchanged in a CPT "transformed" world.

One of the major difficulties in quantum field theory, is that not all components of the underlying theories are necessarily dynamically independent. This requires special attention to be given to them in carrying out a dynamical description of the theories. All of our present fundamental field theories are gauge theories, and involve such fields.

In remaining chapters we establish the connection between the fundamental theory developed here and the various interactions observed in nature. As emphasized in the introductory chapter to the book, a major theme of modern quantum field theory is to provide a unified description of nature. This calls for a look at the various approaches and contributions to the subject that appeared in the literature since the pioneering work of Dirac in the mid twenties, and several of them are treated in coming chapters and the accompanying book by the author.[1]

## 4.1 The Field Concept, Particle Aspect and Wavefunction Renormalization

The significant role that a quantum field $\chi(x)$ plays emerges from the examination of its commutation with the energy-momentum operator $P^\mu$, corresponding to the underlying physical system, and by considering, in particular, the vacuum state $|\text{vac}\rangle$ assumed to be an eigenstate of $P^\mu$ of zero energy and momentum $P^\mu |\text{vac}\rangle = 0$. As will be seen in the next section, the different components $P^\mu$ commute. The latter components consist of generators of spacetime translations, and they imply that

$$\chi(x) = e^{-ix_\mu P^\mu} \chi(0) e^{ix_\mu P^\mu}. \qquad (4.1.1)$$

The latter leads to the equation

$$-i\partial^\mu \chi(x) = [\chi(x), P^\mu]. \qquad (4.1.2)$$

Upon introducing the Fourier transform

$$\chi(x) = \int \frac{(dp)}{(2\pi)^4} e^{ixp} \chi(p), \qquad (4.1.3)$$

Eq. (4.1.2) gives

$$p^\mu \chi(p) = \chi(p) P^\mu - P^\mu \chi(p). \qquad (4.1.4)$$

Let $|p', \nu\rangle$ denote an eigenstate of the components $P^\mu$. That is, $P^\mu |p', \nu\rangle = p'^\mu | p', \nu\rangle$, where $\nu$ stands for other labels needed to specify a state $|p', \nu\rangle$. By working conveniently in terms of the bra $\langle p', \nu |$, and by rewriting (4.1.4) as $\chi(p) P^\mu = P^\mu \chi(p) + p^\mu \chi(p)$, we obtain

$$\langle p', \nu | \chi(p) P^\mu = (p'^\mu + p^\mu) \langle p', \nu | \chi(p). \qquad (4.1.5)$$

That is, for $p^0 \gtrless 0$, $\chi(p)$ injects/absorps a quantity of energy $|p^0|$. The idea that one learns in elementary treatments of (free) field theory that a *field may be expanded in terms of creation and annihilation operators* comes from this relation. The situation we are now in is far more complex, in general, and the physical

---

[1]Quantum Field Theory II: *Introductions to Quantum Gravity, Supersymmetry, and String Theory*, (2016), Springer.

interpretation of $\chi(x)$ in terms of creation and annihilation of just single isolated free particles (anti-particle) becomes incomplete.

In general, the field $\chi(x)$ satisfies coupled non-linear equations which in turn may lead to non-trivial reactions on the field leading to the creation of multi-particles of unlimited number altering the initial expected excitation that the field may provide by a mere examination of the free field theory counterpart. One may even encounter an extreme case that the particle expected to be created by the field is altogether absent. That is, the creation of the particle occurs with zero probability in the standard quantum mechanical language. All of this is best described by examining the effect of a field $\chi(x)$ on $\langle\,\mathrm{vac}\,|$ in the light of the spectrum of the energy-momentum operator $P^\mu$.

The spectral representation of the energy-momentum operator $P^\mu$, expressed through a unitary operator, is given by

$$\mathrm{e}^{\mathrm{i}xP} = \int (\mathrm{d}p)\,\mathrm{e}^{\mathrm{i}xp}\,\Lambda(p), \qquad \mathbb{1} = \int (\mathrm{d}p)\,\Lambda(p), \tag{4.1.6}$$

where $\Lambda(p)$ is a projection operator, which together with (4.1.1) and the condition $\langle\,\mathrm{vac}\,|\,P^\mu = 0$, allow one to write

$$\langle\,\mathrm{vac}\,|\,\chi(x) = \int (\mathrm{d}p)\,\mathrm{e}^{\mathrm{i}xp}\,\langle\,\mathrm{vac}\,|\,\chi(0)\Lambda(p). \tag{4.1.7}$$

We may decompose the spectral representation in terms of a single-particle plus multi-particle contributions to (4.1.7) and rewrite the latter as

$$\langle\,\mathrm{vac}\,|\,\chi(x) = \int \sum_\nu \frac{\mathrm{d}^3\underline{\mathbf{p}}}{2\underline{p}^{\,0}(2\pi)^3}\,\mathrm{e}^{\mathrm{i}x\underline{p}}\,\langle\,\mathrm{vac}\,|\,\chi(0)\,|\underline{\mathbf{p}},\nu\rangle\langle\underline{\mathbf{p}},\nu\,| + \int (\mathrm{d}p)\mathrm{e}^{\mathrm{i}xp}\,\langle\,\mathrm{vac}\,|\,\chi(0)\Lambda'(p), \tag{4.1.8}$$

where $\underline{p}^{\,0} = \sqrt{\underline{\mathbf{p}}^2 + m^2}$, $m$ is the mass of the particle associated with the field $\chi$,

$$\langle\underline{\mathbf{p}},\nu\,|\,\Lambda'(p') = 0, \qquad \langle\underline{\mathbf{p}},\nu\,|\,\underline{\mathbf{p}}',\nu'\rangle = (2\pi)^3\,2\underline{p}^{\,0}\,\delta_{\nu\nu'}\,\delta^3(\underline{\mathbf{p}} - \underline{\mathbf{p}}'). \tag{4.1.9}$$

Upon setting

$$\langle\,\mathrm{vac}\,|\,\chi(0)\,|\underline{\mathbf{p}},\nu\rangle = \sqrt{Z}\,U(\underline{\mathbf{p}},\nu), \tag{4.1.10}$$

up to a phase factor multiplying the real coefficient $\sqrt{Z}$, and where $U(\underline{\mathbf{p}},\nu)$ is the wavefunction of the particle in question, in particular, in the momentum description, we obtain

$$\langle\,\mathrm{vac}\,|\,\chi(x) = \int \sum_\nu \frac{\mathrm{d}^3\underline{\mathbf{p}}}{2\underline{p}^{\,0}(2\pi)^3}\,\mathrm{e}^{\mathrm{i}x\underline{p}}\,\sqrt{Z}\,U(\underline{\mathbf{p}},\nu)\langle\underline{\mathbf{p}},\nu\,| + \int (\mathrm{d}p)\,\mathrm{e}^{\mathrm{i}xp}\,\langle\,\mathrm{vac}\,|\,\chi(0)\Lambda'(p), \tag{4.1.11}$$

$$\langle \text{vac} | \, \chi(x) \, | \mathbf{p}, \nu \rangle = \sqrt{Z} \, U(\mathbf{p}, \nu) \, e^{i x \underline{p}}. \tag{4.1.12}$$

and where $U(\mathbf{p}, \nu) \, e^{ipx}$ denotes the wavefunction of the particle in question, and $\Lambda'(p)$ is a projection operator on the remaining spectrum of the energy-momentum of the system under study.

In the presence of multi-particle states, described by the second integral in (4.1.11), quantum mechanics implies the presence of a coefficient $\sqrt{Z}$ in the expansion (4.1.11), to describe a particle just described, with $Z$ as the probability of the creation of the particle in question by the field $\chi(x)$ out of the vacuum.

Situations may arise that a single isolated particle cannot be described, or that the underlying theory predicts only composite particles, in which cases the first term in (4.1.11) will be altogether absent. On the other hand, when dealing with a scalar particle $\phi(x)$ for which no symmetry arguments may be applied to dismiss its vacuum expectation value $\langle \text{vac} | \, \phi(0) \, | \, \text{vac} \rangle$, then an additional term of the form $\langle \text{vac} | \, \phi(0) \, | \, \text{vac} \rangle \langle \text{vac} |$ will appear on the right-hand of (4.1.11). Such a non-vanishing vacuum expectation value of a scalar field is what happens in describing the Higgs mechanism.

In studying a physical process and confronting theory with experiments, one may ask the question: *given* that a certain particle was observed, that is, this event happened with probability one, *then* what is the probability that it has emerged, say, with a specific momentum and a given spin projection?. In the language of probability theory, the underlying probability in such a question is referred to as a *conditional* probability:

> *given that something happened, i.e., with probability*
> *one, then what is the probability that the system*
> *under consideration has some given characteristics?*

In technical terms, this amounts to divide the field $\chi(x)$ by $\sqrt{Z}$, leading to the definition of the field

$$\chi_{\text{ren}}(x) = \chi(x)/\sqrt{Z}, \tag{4.1.13}$$

thus isolating the particle in question with probability one, i.e., with certainty, bringing the observed particle in evidence.

The process just described is called a wavefunction renormalization, which is done independently of any perturbation theory, and $Z$ is referred to as a *wavefunction renormalization constant*. The scaled field $\chi_{\text{ren}}(x)$ is the so-called renormalized field.

Let us go one step further to see how a wavefunction renormalization is carried out when one actually computes transition amplitudes. We will even generalize the above further, by bringing the treatment closer to an experimental set-up, where one would ask the question as to what is the amplitude that the particle is actually observed, i.e., detected?. The analysis, in turn, shows the role played by a field in computing transition amplitudes. As we are already familiar with the spin

1/2 propagator derived in Sect. 3.1, we consider for definiteness the field $\chi(x)$ to correspond to a Dirac field $\psi(x)$ in a general interacting field theory.

To the above end, let $\mathscr{O}(x_1,\ldots,x_n) \mid \text{vac})$ denote the product of $n$ arbitrary operators of spacetime arguments $x_1,\ldots,x_n$ acting on the vacuum state $\mid \text{vac}\rangle$, and consider the vacuum-expectation of the time-ordered product:

$$\langle \text{vac} \mid \big(\psi(x)\mathscr{O}(x_1,\ldots,x_n)\big)_+ \mid \text{vac}\rangle, \tag{4.1.14}$$

in which the operators are arranged from right to left in order of increasing time argument with a minus sign inserted for each permutation of Fermi-Dirac fields (Sect. 3.5). Here $\psi(x)$ is the quantum field in question under investigation. We emphasize that this matrix element is computed in a theory involving no external sources coupled to the fields such as $\eta$, $\bar{\eta}$ given in Chap. 3.

Now we multiply the above vacuum expectation value by a Fermi source $\bar{\eta}(x)$ (Sect. 3.1) from the left and integrate over $x$ to obtain

$$\int (\mathrm{d}x)\, \bar{\eta}(x)\langle \text{vac} \mid \big(\psi(x)\mathscr{O}(x_1,\ldots,x_n)\big)_+ \mid \text{vac}\rangle. \tag{4.1.15}$$

By doing this, we will concentrate on the particle in question associated with the field $\psi(x)$. A similar process may be carried out for the particles associated with the other fields in the time-ordered product by multiplying (4.1.14) by corresponding source functions. We recall that the field $\psi(x)$, in particular, is expected to satisfy a non-linear equation. Let $S_+(x-x')$ denote the free spin 1/2 propagator derived in Sect. 3.1. We use the property

$$\int (\mathrm{d}x')\, S_+(x-x')\, S_+^{-1}(x'-x'') = \delta^{(4)}(x-x''), \tag{4.1.16}$$

where $S_+^{-1}(x'-x'')$ is the inverse of the propagator. Consider the source $\bar{\eta}(x)$ to be switched on, i.e., operate, at large positive $x^0$ values, thus one may "follow" the particle for a sufficient long time to be able to measure its energy, momentum and other attributes. For all $x^0 > x'^0$, at which times the source $\bar{\eta}(x)$ operates, it is easily derived that

$$\int (\mathrm{d}x)\, \bar{\eta}(x) S_+(x-x') = \int \sum_\sigma \frac{\mathrm{d}^3 \mathbf{p}}{(2\pi)^3} \frac{m}{p^0}\, \mathrm{e}^{-\mathrm{i}x'p}\, [\,\mathrm{i}\,\bar{\eta}(p)u(\mathbf{p},\sigma)\,]\, \bar{u}(\mathbf{p},\sigma), \tag{4.1.17}$$

by using, in the process, the relations in (3.1.10) and (I.21). From (4.1.16), (4.1.17), and the identity

$$S_+^{-1}(x'-x'') = \Big(\frac{\gamma\partial'}{\mathrm{i}} + m\Big)\delta^{(4)}(x'-x''), \tag{4.1.18}$$

the following expression emerges for (4.1.15):

$$\int \sum_\sigma \frac{d^3\mathbf{p}}{(2\pi)^3} \frac{m}{p^0} [i\,\overline{\eta}(p)\,u(\mathbf{p},\sigma)]\sqrt{Z}$$

$$\times \left[ \int (dx')\,e^{-ix'p}\,\overline{u}(\mathbf{p},\sigma)\left(\frac{\gamma\partial'}{i}+m\right)\langle \text{vac}|\,(\psi_{\text{ren}}(x')\,\mathscr{O}(x_1,\ldots,x_n))_+|\text{vac}\rangle\right].$$

$$(4.1.19)$$

One may readily continue this process as applied to the other fields in the time ordered product in (4.1.14) by multiplying the latter by corresponding sources and noting that for sources switched on and operating at large positive times, one is dealing with particles outgoing from a process, while sources operating only at large negative times, one is dealing with ingoing particles in a process. The keen reader will realize that we have essentially obtained a formalism for computing transition amplitudes in field theory. This will be taken up later, dealing with the vacuum-to-vacuum transition amplitudes for various interactions, and specific applications will be carried out.

The expression in (4.1.19) brings us into contact with a formalism, referred to as the LSZ formalism,[2] except that we have now an *additional* piece of physics dealing with the detection (emission) of particles expressed by the first line in (4.1.19) depending on the source function $\overline{\eta}(p)$. From (3.2.8), we recognize the term $[i\,\overline{\eta}(p)\,u(\mathbf{p},\sigma)]$ as an amplitude for detection, i.e., of the actual observation of the particle in question. Having scaled the field in (4.1.13) to define the renormalized field $\psi_{\text{ren}}(x)$, we now have to define the renormalized source function $\overline{\eta}_{\text{ren}}(x) = \sqrt{Z}\,\overline{\eta}(x)$ as well for the physical interpretation of the latter corresponding to an amplitude of particle detection. When one adds source terms to a Lagrangian density such as $\overline{\eta}(x)\psi(x)$, corresponding, in particular, to the field $\psi(x)$, one has the invariance property $\overline{\eta}(x)\psi(x) = \overline{\eta}_{\text{ren}}(x)\psi_{\text{ren}}(x)$.

Introduction of sources coupled to the fields facilitates tremendously doing field theory. In practice, the transition amplitudes for particles (anti-particles) ingoing and outgoing in a process is obtained from the coefficient of the source terms in expressions like in (4.1.19) (see, e.g., Sects. 5.8 and 5.9).

## 4.2 Poincaré Algebra and Particle States

Poincaré (inhomogeneous Lorentz) transformations in Minkowski spacetime were given in (2.2.7), Sect. 2.2. Such a transformation being connected to the identity one, induces, according to Wigner's Theorem (Sect. 2.1), a unitary operator to act on particle states and quantum fields. To find the algebra satisfied by the

---

[2]See Lehmann et al. [2].

generators of such a transformation, we consider successive infinitesimal Poincaré transformations forming a closed path as follows

$$(\Lambda_2, b_2)^{-1}(\Lambda_1, b_1)^{-1}(\Lambda_2, b_2)(\Lambda_1, b_1) = (\Lambda, b), \tag{4.2.1}$$

represented pictorially by

emphasizing the reversal of the transformations in the third and the fourth segments of the path. The matrix elements of $\Lambda$ are explicitly given in (2.2.13), (2.2.14), (2.2.15), (2.2.16), (2.2.11), and for infinitesimal ones in (2.2.19), (2.2.20), (2.2.21) and (2.2.22).

The infinitesimal coordinate changes $\delta x^\mu$ associated with such transformations may be spelled out as

$$\delta x^\mu = x^\mu - x'^\mu = \delta b^\mu - \delta\omega^\mu{}_\nu x^\nu, \qquad \delta\omega^{\mu\nu} = -\delta\omega^{\nu\mu}, \tag{4.2.2}$$

where $\Lambda^{\mu\nu} = \eta^{\mu\nu} + \delta\omega^{\mu\nu}$.

From the group properties below (2.2.9) in Sect. 2.2, (4.2.1) readily leads to

$$\delta\omega^{\mu\nu} = \delta\omega_2{}^{\mu\rho}\,\delta\omega_{1\rho}{}^\nu - \delta\omega_2{}^{\nu\rho}\,\delta\omega_{1\rho}{}^\mu, \tag{4.2.3}$$

$$\delta b^\mu = \delta\omega_2{}^\mu{}_\rho\,\delta b_1{}^\rho - \delta\omega_1{}^\mu{}_\rho\,\delta b_2{}^\rho. \tag{4.2.4}$$

The unitary operator corresponding, for example, to $(\Lambda, b)$, for infinitesimal transformations, has the structure

$$U = 1 + \mathrm{i}\,G, \qquad G = \delta b_\mu P^\mu + \frac{1}{2}\delta\omega_{\mu\nu}J^{\mu\nu}, \tag{4.2.5}$$

with $P^\mu$, $J^{\mu\nu}$ denoting the energy-momentum and angular momentum operators, respectively, generating spacetime translations and homogeneous Lorentz transformations. The latter, corresponding to $J^{\mu\nu}$, consist of Lorentz boosts and spacial rotations.

Referring to the closed path in (4.2.1), one has

$$U_2^{-1}\,U_1^{-1}\,U_2\,U_1 = U, \tag{4.2.6}$$

which leads to

$$G = \frac{1}{\mathrm{i}}[G_1, G_2], \tag{4.2.7}$$

with the parameters of the generator $G$ given by $(\delta\omega^{\mu\nu}, \delta b^{\mu})$ spelled out in (4.2.3), (4.2.4), and those of $G_1$, $G_2$ are denoted by $(\delta\omega_1^{\mu\nu}, \delta b_1^{\mu})$, $(\delta\omega_2^{\mu\nu}, \delta b_2^{\mu})$.

Upon comparison of the coefficients of identical products of the above parameters on both sides of (4.2.7) gives the Poincaré algebra

$$[P^\mu, P^\nu] = 0 \tag{4.2.8}$$

$$[P^\mu, J^{\sigma\lambda}] = i\left(\eta^{\mu\lambda}P^\sigma - \eta^{\mu\sigma}P^\lambda\right), \tag{4.2.9}$$

$$[J^{\mu\nu}, J^{\sigma\lambda}] = i\left(\eta^{\mu\sigma}J^{\nu\lambda} - \eta^{\nu\sigma}J^{\mu\lambda} + \eta^{\nu\lambda}J^{\mu\sigma} - \eta^{\mu\lambda}J^{\nu\sigma}\right), \tag{4.2.10}$$

satisfied by the generators $P^\mu$, $J^{\nu\sigma}$.

The following operators $P^\mu P_\mu$ and $W^\mu W_\mu$, where $W^\mu$ is the (pseudo-) vector defined by

$$W^\mu = \frac{1}{2}\epsilon^{\mu\nu\sigma\lambda} P_\nu J_{\sigma\lambda}, \tag{4.2.11}$$

referred to as the Pauli-Lubanski (pseudo-) vector and whose physical significance will become clear shortly, are invariants of the corresponding group. That is, they commute with the Poincaré generators $P^\mu$, $J^{\mu\nu}$. Hence particle states may be labeled by their eigenvalues. Here $\epsilon^{\mu\nu\sigma\lambda}$ is totally anti-symmetric specified by $\epsilon^{0123} = +1$. Note that in (4.2.11): $\nu \neq \sigma$, $\nu \neq \lambda$, hence the operators $P_\nu$ and $J_{\sigma\lambda}$ in $W^\mu$ commute (see (4.2.9)).

$W^\mu$ satisfies the following commutation relations

$$[W^\mu, P^\sigma] = 0 \tag{4.2.12}$$

$$[W^\mu, W^\sigma] = i\epsilon^{\mu\sigma\rho\lambda} P_\rho W_\lambda, \tag{4.2.13}$$

$$[W^\mu, J^{\sigma\lambda}] = i(\eta^{\mu\lambda}W^\sigma - \eta^{\mu\sigma}W^\lambda), \tag{4.2.14}$$

and the conditions

$$W^\mu P_\mu = 0, \qquad W^\mu W_\mu \geq 0. \tag{4.2.15}$$

In detail, the components of $W^\mu$ are given by

$$W^0 = \mathbf{P}\cdot\mathbf{J}, \quad \mathbf{W} = P^0\mathbf{J} - \mathbf{P}\times\mathbf{N}, \quad N_k = -J_{0k}. \tag{4.2.16}$$

The representations of physical interest are those for which $P^\mu$

$$P^\mu P_\mu \equiv P^2 = -m^2 \leq 0, \tag{4.2.17}$$

with sgn $P^0 > 0$ corresponding to positive energy, for massive $(m \neq 0)$, i.e., $P^\mu$ time-like, and massless $(m = 0)$, i.e., $P^\mu$ light-like, particles. $P^\mu = 0$ corresponds to the vacuum state. Consider first massive particles.

▶ $m \neq 0$:

Since $W^\mu W_\mu$ is an invariant, it may be computed in any Lorentz frame. For a massive particle, we may consider its rest frame for which $P^\mu = (m, 0, 0, 0, 0)$. From (4.2.16), then

$$W^\mu W_\mu = m^2 \mathbf{J}^2. \tag{4.2.18}$$

In the rest frame of such a particle, (4.2.10) leads to the usual commutation relations of angular momentum components

$$[J^i, J^j] = i\epsilon^{ijk} J^k, \tag{4.2.19}$$

and

$$W^\mu W_\mu = m^2 j(j+1), \tag{4.2.20}$$

since $W^i$ becomes simply $mJ^i$ and $W^0$ reduces to zero.

Consider the eigenvalue problem

$$P^\mu \langle p | = p^\mu \langle p |, \tag{4.2.21}$$

where particle labels denoting other degrees of freedom are initially suppressed. In particular, in the rest frame of the particle, we have from (4.2.9)

$$[P^0, J^i]\langle p | = 0. \tag{4.2.22}$$

Accordingly, from (4.2.18), (4.2.19), (4.2.20), (4.2.21) and (4.2.22), we may label a massive particle state, in the rest frame of the particle, by its mass $m$, and the spin $j$, involving $(2j + 1)$ components. We will simply use the notation $\langle \mathbf{0}\sigma |$ for these states, with $\mathbf{0}$ corresponding to the momentum of the particle at rest, and $\sigma$ corresponding to the values $j$. All other labels necessary to specify such a state will be suppressed.

To find particle states for arbitrary momenta, we proceed as follows. Consider the state

$$\langle \mathbf{0}\sigma | \exp[-i\alpha J_{03}] \exp[-i\theta \mathbf{n} \cdot \mathbf{J}], \tag{4.2.23}$$

$$\mathbf{n} = (\sin\phi, -\cos\phi, 0), \quad \sinh\alpha = |\mathbf{p}|/m. \tag{4.2.24}$$

We will apply the operators $\mathbf{P}$ and $W^0$ to the state in (4.2.23). To this end, note that the vector operator

$$\mathbf{V}(\theta) = \exp[-i\,\theta\,\mathbf{n}\cdot\mathbf{J}]\,\mathbf{P}\,\exp[i\,\theta\,\mathbf{n}\cdot\mathbf{J}], \qquad (4.2.25)$$

satisfies the equation

$$\frac{d\mathbf{V}(\theta)}{d\theta} = -\mathbf{n}\times\mathbf{V}(\theta), \qquad \mathbf{V}(0) = \mathbf{P}. \qquad (4.2.26)$$

Hence from (2.2.10), (2.2.11) and (2.2.12),

$$V^i = R^{ij}P^j, \qquad (4.2.27)$$

with $[R^{ij}]$ denoting the rotation matrix spelled out in (2.2.11), and the unit vector $\mathbf{n}$ is defined in (4.2.24).

On the other hand, defining

$$Y^\mu = \exp[-i\,\alpha\,J_{03}]\,P^\mu\,\exp[i\,\alpha\,J_{03}], \qquad (4.2.28)$$

we obtain

$$\frac{dY^\mu}{d\alpha} = \eta^{\mu 3}\,Y^0 - \eta^{\mu 0}\,Y^3, \qquad \frac{d^2Y^\mu}{d\alpha^2} = \eta^{\mu 3}\,Y^3 - \eta^{\mu 0}\,Y^0. \qquad (4.2.29)$$

The solution of (4.2.29) is elementary and is given by

$$Y^3 = P^3\cosh\alpha + P^0\sinh\alpha, \qquad Y^0 = P^0\cosh\alpha + P^3\sinh\alpha, \qquad (4.2.30)$$

and

$$Y^a = P^a, \qquad \text{for} \quad a = 1,2. \qquad (4.2.31)$$

From (4.2.23), (4.2.24), (4.2.25), (4.2.26), (4.2.27), (4.2.28), (4.2.29), (4.2.30) and (4.2.31), the facts that $\langle 0\,\sigma\,|\,\mathbf{P} = 0$, and $\langle 0\,\sigma\,|\,P^0 = m\langle 0\,\sigma\,|$, we may infer that[3] (4.2.23) defines a state with momentum-energy $p^\mu$,

$$\langle\,0\,\sigma\,|\,\exp[-i\,\alpha\,J_{03}]\,\exp[-i\,\theta\,\mathbf{n}\cdot\mathbf{J}]\,P^\mu = p^\mu\,\langle\,0\,\sigma\,|\,\exp[-i\,\alpha\,J_{03}]\,\exp[-i\,\theta\,\mathbf{n}\cdot\mathbf{J}], \qquad (4.2.32)$$

with

$$p^0 = +\sqrt{\mathbf{p}^2 + m^2}, \qquad \mathbf{p} = |\mathbf{p}|(\sin\theta\,\cos\phi, \sin\theta\,\sin\phi, \cos\theta). \qquad (4.2.33)$$

---

[3] See Problem 4.1.

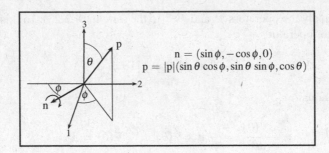

**Fig. 4.1** Rotation of a vector initially along the 3-axis c.w. by an angle $\theta$ about the unit vector **n** lying in the 1–2 plane

Here we note that a rotation of vector **p**, initially along the 3-axis, by an angle $\theta$, about the unit vector **n** in (4.2.24), gives the vector in (4.2.33) (Fig. 4.1).

A similar analysis[4] shows that $W^0$ applied to the state in (4.2.23) gives

$$\langle\, 0\sigma\,|\, \exp[-i\,\alpha\,J_{03}]\, \exp[-i\,\theta\,\hat{\mathbf{n}}\cdot\mathbf{J}]\, W^0$$
$$= m\sigma\,\sinh\alpha\,\langle\, 0\sigma\,|\, \exp[-i\,\alpha\,J_{03}]\, \exp[-i\,\theta\,\mathbf{n}\cdot\mathbf{J}]. \qquad (4.2.34)$$

Accordingly, up to a normalization factor, we denote the states in (4.2.23) by $\langle\,\mathbf{p}\sigma\,|$, providing an orthogonal basis of vectors of single-particle states. The orthogonality property follows from the facts that

$$0 = \langle\,\mathbf{p}'\sigma'\,|\,\mathbf{P} - \mathbf{P}\,|\,\mathbf{p}\sigma\,\rangle \quad = \quad (\mathbf{p}' - \mathbf{p})\langle\,\mathbf{p}'\sigma'\,|\,\mathbf{p}\sigma\,\rangle,$$
$$0 = \langle\,\mathbf{p}'\sigma'\,|\, W^0 - W^0\,|\,\mathbf{p}\sigma\,\rangle = m(\sigma'\,\sinh\alpha' - \sigma\,\sinh\alpha)\langle\,\mathbf{p}'\sigma'\,|\,\mathbf{p}\sigma\,\rangle.$$
$$(4.2.35)$$

From (4.2.24), $\sinh\alpha = |\mathbf{p}|/m$, hence the first equality in (4.2.35) shows that $\sinh\alpha = \sinh\alpha'$. The two equalities in (4.2.35) together then imply that $\langle\,\mathbf{p}'\sigma'\,|\,\mathbf{p}\sigma\,\rangle$ is proportional to $\delta^3(\mathbf{p} - \mathbf{p}')\,\delta_{\sigma\sigma'}$. Apart from the labels $m$ and $j$, i.e., of $\sigma$, due to the invariance of $P^\mu P_\mu$ and $W^\mu W_\mu$, and other labels, we will not dwell upon such additional labels here.

Using the Lorentz invariance of the measure[5] $d^3\mathbf{p}/(2\pi)^3 2p^0$, we may set

$$\langle\,\mathbf{p}'\sigma'\,|\,\mathbf{p}\sigma\,\rangle = (2\pi)^3\, 2p^0\, \delta^3(\mathbf{p} - \mathbf{p}')\,\delta_{\sigma\sigma'}. \qquad (4.2.36)$$

▶ $m = 0$:

Consider an eigenvalue problem as in (4.2.21). In the massless case, we may initially consider the three momentum of the particle to be along the 3-axis. In such a frame, we may write the energy-momentum of the particle as $k^\mu = (\omega, 0, 0, \omega)$.

---

[4]See Problem 4.2.
[5]See (3.1.18).

As a consequence of (4.2.9), (4.2.16), (4.2.12) the following relations are realized

$$[P^{\mu}, J^{12}]\langle\, \mathbf{k}\, | \, = 0, \quad W^{0}\langle\, \mathbf{k}\, | \, = \omega J^{3}\langle\, \mathbf{k}\, |, \quad W^{3}\langle\, \mathbf{k}\, | \, = \omega J^{3}\langle\, \mathbf{k}\, |, \qquad (4.2.37)$$

$$W^{1}\langle\, \mathbf{k}\, | \, = \omega T^{1}\langle\, \mathbf{k}\, |, \qquad\qquad W^{2}\langle\, \mathbf{k}\, | \, = \omega T^{2}\langle\, \mathbf{k}\, | \qquad (4.2.38)$$

where $(J^{12} = J^{3})$

$$T^{1} = J^{1} + N^{2}, \qquad T^{2} = J^{2} - N^{1}, \qquad N^{k} = J^{0k}, \qquad (4.2.39)$$

$$[T^{1}, T^{2}] = 0, \qquad [T^{1}, J^{3}] = -\mathrm{i}\, T^{2}, \qquad [T^{2}, J^{3}] = \mathrm{i}\, T^{1}. \qquad (4.2.40)$$

In the above, we have suppressed other labels in the states $\langle\, \mathbf{k}\, |$.

Upon comparison of the set of commutation relations in (4.2.40) with (1.2.8), (4.2.9), we learn that they correspond to translations in the 1–2 plane, via generators $T^{1}$, $T^{2}$, and rotations about the 3 - axis, via the generator $J^{3}$. $T^{1}$ and $T^{2}$ have continuous unbounded spectra. Incidentally, $(T^{1})^{2} + (T^{2})^{2} = (W^{\mu} W_{\mu})/\omega^{2}$, with $\omega > 0$, is an invariant of the underlying group, with seemingly a *continuous* non-negative spectrum. Unlike the energy and momentum of a particle, which possess continuous spectra, there are no physical states with such spectra of 'continuous' spins. This leaves room only for the cases that $T^{i}\langle\, \mathbf{k}\, | \, = 0$ for $i = 1, 2$, corresponding to $W^{\mu} W_{\mu} = 0$.

The interesting thing to note is that according to (4.2.15), we now have two orthogonal light-like vectors:

$$W^{\mu} P_{\mu} = 0, \quad W^{\mu} W_{\mu} = 0, \quad P^{\mu} P_{\mu} = 0, \quad p^{0} > 0, \qquad (4.2.41)$$

which means that $W^{\mu}$, $P^{\mu}$ are proportional to each other,[6] i.e.,

$$W^{\mu} = \lambda P^{\mu}. \qquad (4.2.42)$$

In particular, this equation applied to $W^{0}$ in (4.2.16) gives $\lambda P^{0} = \mathbf{P} \cdot \mathbf{J}$. That is, $\lambda$, referred to as the helicity, is determined from the component of spin along the momentum. Since $W^{\mu}$ is a (pseudo-) vector and $P^{\mu}$ is a vector, $\lambda$ is a (pseudo-) scalar. On account of (4.2.12), (4.2.14) and (4.2.42), $\lambda$ commutes with the generators $P^{\mu}$, $J^{\mu\nu}$ of the Poincaré group, hence it is an *invariant* and the physical states may be labeled by $\lambda$.

The invariance property of the helicity means that Nature would exhibit only a unique value for the helicity of a given massless particle. On the other hand, if parity is conserved for the interaction of a system under consideration, then this pseudo-scalar can also take the value $-\lambda$. Such a property is shared by photons

---

[6]See Problem 4.3.

$(\lambda = \pm 1)$ for the electromagnetic interaction, and by the graviton $(\lambda = \pm 2)$ for the gravitational interaction.

We may label the initial states considered above as $\langle \, \mathbf{k}\lambda \, |$. To find the states for arbitrary momenta is now straightforward since $\lambda$ is an invariant.

With $\omega = |\mathbf{p}|$, it is evident that, up to a normalization factor,

$$\langle \, \mathbf{p}\lambda \, | = \langle \, \mathbf{k}\lambda \, | \exp[-i\theta \mathbf{n} \cdot \mathbf{J}], \tag{4.2.43}$$

where $\mathbf{n}$ is defined in (4.2.24), and we verify, by using (4.2.25), (4.2.26) and (4.2.27), that

$$\langle \, \mathbf{p}\lambda \, | P^\mu = p^\mu \langle \, \mathbf{p}\lambda \, |, \tag{4.2.44}$$

with $p^0 = |\mathbf{p}|$, $\mathbf{p}$ given in (4.2.33). The orthogonality of the states $\langle \, \mathbf{p}\lambda \, |$ follows as before,

$$\langle \, \mathbf{p}'\lambda' \, | \, \mathbf{p}\lambda \, \rangle = (2\pi)^3 \, 2p^0 \, \delta^3(\mathbf{p} - \mathbf{p}') \, \delta_{\lambda\lambda'}, \tag{4.2.45}$$

written in an invariant manner.

Before closing this section, we note that due to the rotational invariance of the scalar product $\mathbf{P} \cdot \mathbf{J}$, i.e., $[\mathbf{P} \cdot \mathbf{J}, J^i,] = 0$, a rotation induced on a state in (4.2.43) about the momentum by an angle $\vartheta$, leads to

$$\langle \, \mathbf{p}\lambda \, | \exp[-i\vartheta \, \mathbf{P} \cdot \mathbf{J}/|\mathbf{p}|] = \exp[-i\vartheta\lambda] \langle \, \mathbf{p}\lambda \, |. \tag{4.2.46}$$

A rotation by an angle $\vartheta = 2\pi$ gives $+1$ or $-1$ for single or double valued representations corresponding to integer and half-odd integer values for $\lambda$.

## 4.3  Principle of Stationary Action of Quantum Field Theory: The Rationale Of

This section is involved in considering general variations, to be specified below, of transformation functions $\langle \sigma_2 | \sigma_1 \rangle$, associated with dynamical systems, between two space-like surfaces $\sigma_1$, $\sigma_2$. This brings us into the realm of a fundamental theory referred to as the "Principle of *Stationary Action*".

The Principle of Stationary Action provides the basic tool for deriving field equations. The latter field equations emerge from the variation of the action under the variations of the dynamical variables *within* specified boundaries. Variations of dynamical variables call for generators to implement the corresponding transformations. By introducing, in turn, such generators, the stationary aspect of the action under dynamical field variations *directly* follows giving automatically the field equations – the so-called Euler-Lagrange equations. Understanding this step should be comforting to the student who would, otherwise, blindly write down the Euler-Lagrange equations. Equally important, is that the concept of the action and of a Lagrangian density arise as a consequence of studying variations of transformation functions describing the evolution of a dynamical system in time, and hence are of physical interest. The principle also provides the way of expressing various generators in terms of the variables underlying the Lagrangian density and deals with conservation laws as well. Dynamics, beyond the field equations, such as providing a method for obtaining expressions for transition amplitudes, is the subject of Sect. 4.6 under the heading of the *Quantum Dynamical Principle*. The celebrated *Spin and Statistics Connection* is treated in Sect. 4.5.

A state defined on a space-like surface $\sigma$ will be denoted by $\langle\,\sigma\,|$, and labels needed to specify such a state will, in general, be suppressed. Since there is always a coordinate system for which such a space-like surface is the space itself, at a given time, i.e., represented by a surface $x^0 = t$ fixed, we may equally consider a state denoted by $\langle\,t\,|$. We are interested in changes that may occur in transformation functions $\langle\,t_2\,|t_1\rangle$ corresponding, in general, to *different* values of quantum numbers associated with an evolving system at the different times $t_1$ and $t_2$, under some basic transformations.

Under an infinitesimal transformation via a unitary operator $U = 1 + iG$, where $G$ defines the generator for such a transformation, an operator $\mathcal{O}$ changes to $\mathcal{O}' = U^\dagger \mathcal{O} U$ and with the change defined by

$$\delta\mathcal{O} = \mathcal{O} - \mathcal{O}',  \tag{4.3.1}$$

we have

$$\delta\mathcal{O} = \frac{1}{i}[\mathcal{O}, G].  \tag{4.3.2}$$

Hence for a wavefunction $\psi$, equivalence of the "primed" and "unprimed" depictions, implies that $\delta\langle\,\psi\,|\mathcal{O}|\,\psi\,\rangle = \langle\,\psi\,|\mathcal{O}|\,\psi\,\rangle - \langle\,\psi'\,|\mathcal{O}'|\,\psi'\,\rangle = 0$, and with $|\,\psi'\rangle = U^\dagger\,|\,\psi\rangle$ gives $|\,\psi\rangle - |\,\psi'\rangle = \delta|\psi\rangle = iG\,|\,\psi\rangle$. The invariance of the scalar product $\langle\phi'\,|\,\psi'\rangle = \langle\phi\,|\,UU^\dagger\psi\rangle = \langle\phi\,|\,\psi\rangle$, and the resolution of the identity $1 = \sum\,|\,a\rangle\langle a\,| = \sum U\,|\,a\rangle\langle a\,|\,U^\dagger$, for bases vectors $|\,a\rangle$, imply that $\delta\langle a\,| = \langle a\,| - \langle a\,|\,U^\dagger$ is given by

$$\delta\langle a\,| = i\langle a\,|G.  \tag{4.3.3}$$

We consider two types of changes of the transformation functions $\langle\,t_2\,|t_1\,\rangle$:

(i) One type is in response to, a priori, *imposed* variations of the underlying dynamical variables of the theory at all times $t_1 \le t \le t_2$, by keeping $t_1$ and $t_2$ *fixed*. In such cases, we will use the notation $\underline{\delta}\,\chi(x)$, as applied to the fields, for the variation that one may, a priori, impose on a field with a bar under $\delta$. Here the variations $\underline{\delta}\,\chi(x)$ are chosen at will and may be chosen to be numericals for unambiguous analyses, up to the *statistics* obeyed by the fields $\chi(x)$ and otherwise arbitrarily. Non-numerical changes of the fields will be discussed later.

**(ii)** A second type corresponds to variations of just the boundaries, i.e., of the surfaces $x^0 = t_1$ and $x^0 = t_2$ in *response* to a given transformation, taking into account of the *induced* variations that arise on the underlying dynamical variables themselves from such a transformation *on* the boundaries. For such *induced* variations on the fields, we will simply use the notation $\delta \chi(x)$ to distinguish them from ones one may, a priori impose on them.

The first type of variations and its consequence are considered in Sects. 4.3.1, 4.3.2, and 4.3.3, while the second type is considered in Sect. 4.3.4.

### 4.3.1  A Priori Imposed Variations of Dynamical Variables and Generators of Field Variations: Field Equations

We here consider variations of the fields $\chi_a(x)$, at *any* given time $t$ in $t_1 \leq t \leq t_2$, with the space-like surfaces $x^0 = t_1$ and $x^0 = t_2$, kept fixed. Let $\chi_{\mathbf{x}a}(t) \equiv \chi_a(x)$ denote the dynamical variables. This is an extension of a theory involving dynamical variables of a countable degrees of freedom $q_i(t)$, to a theory involving an infinite uncountable ones, with $(\mathbf{x}\,a)$ now replacing the index $i$, as discussed earlier in the introduction to this chapter.

An infinitesimal variation of a dynamical variable calls for a generator to carry out such a variation. We recall that in quantum mechanics, an infinitesimal coordinate translation $\delta x^i$ is carried out via a generator written as $P^i \delta x^i$, where $\delta x^i$ are numerical constants that commute with the operator $P^i$, and the latter is then identified with the momentum operator. By extending, in the process, the generator for the countable number of degrees of freedom case $\sum_i p^i(t)\delta q_i(t)$, to an infinite uncountable ones, the generator for infinitesimal variations, at a given time $t$, may be expressed as

$$G(t) = \int d^3\mathbf{x}\, \pi^{\,a}(x)\, \underline{\delta}\, \chi_a(x), \qquad x^0 = t. \tag{4.3.4}$$

In reference to the first type of variation with time fixed, we denote the changes of the dynamical variables by $\underline{\delta}\, \chi_a(x)$. The operators $\pi^{\,a}(x)$ are identified as the canonical conjugate momenta of the fields $\chi_a(x)$ and are to be *determined*. Following Schwinger, the variations $\underline{\delta}\, \chi_a(x)$ are taken as numericals. Non-numerical changes will be discussed later. The adjoints of the fields, as obtained from the fields themselves, will be suppressed for simplicity.

For the time derivative of (4.3.4), evaluated at any given time $t$: $t_1 \leq t \leq t_2$, we have

$$\frac{d}{dt}\, G(t) = \int d^3\mathbf{x}\Big[ \dot{\pi}^{\,a}(x)\, \underline{\delta}\, \chi_a(x) + \pi^{\,a}(x)\, \underline{\delta}\, \dot{\chi}_a(x) \Big]. \tag{4.3.5}$$

The integrand in (4.3.5) may be expressed in terms of the total infinitesimal variation $\underline{\delta}H$ of the Hamiltonian, at any given time, as follows. The latter, as a functional of $\chi_a(x)$ and $\pi^a(x)$, is given by

$$\underline{\delta}H = \int d^3x \left( \frac{1}{i} \underline{\delta}\pi^a(x) [\chi_a(x), H] - \frac{1}{i} [\pi^a(x), H] \underline{\delta}\chi_a(x) \right)$$

$$= \int d^3x \left( \underline{\delta}\pi^a(x) \dot{\chi}_a(x) - \dot{\pi}^a(x) \underline{\delta}\chi_a(x) \right). \tag{4.3.6}$$

Accordingly, upon solving for $\int d^3x \dot{\pi}^a(x)\underline{\delta}\chi_a(x)$ from the above equation, we may rewrite $dG(t)/dt$ as

$$\frac{d}{dt} G(t) = \int d^3x \underline{\delta}\left( \pi^a(x)\dot{\chi}_a(x) - \mathcal{H}(x) \right), \tag{4.3.7}$$

where $\mathcal{H}(x)$ is the Hamiltonian density

$$H(t) = \int d^3x \mathcal{H}(x), \qquad x^0 = t. \tag{4.3.8}$$

At this stage the concept of a Lagrangian density arises. Upon identifying the Lagrangian density $\mathcal{L}(x)$ as follows

$$\mathcal{L}(x) = \pi^a(x)\dot{\chi}_a(x) - \mathcal{H}(x), \tag{4.3.9}$$

we obtain

$$\frac{d}{dt} G(t) = \int d^3x \underline{\delta}\mathcal{L}(x). \tag{4.3.10}$$

Thus from (4.3.3), with $t_1$, $t_2$ fixed, we have

$$\underline{\delta}\langle t_2 | t_1 \rangle = i\langle t_2 |[ G(t_2) - G(t_1) ]| t_1 \rangle = i\langle t_2 | \int_{t_1}^{t_2} dt \frac{d}{dt} G(t)| t_1 \rangle, \tag{4.3.11}$$

and from (4.3.10), (4.3.11), we obtain in detail

$$\underline{\delta}\langle t_2 | t_1 \rangle = i\langle t_2 |[ G(t_2) - G(t_1) ]| t_1 \rangle = i\langle t_2 | \int_{x^0=t_1}^{x^0=t_2} (dx) \underline{\delta}\mathcal{L}(x)| t_1 \rangle, \tag{4.3.12}$$

$(dx) = dx^0 dx^1 dx^2 dx^3$, where we have used the notation $\underline{\delta}$ to remind us that we are considering variations of the dynamical variables only for fixed times $t_1$, $t_2$. $\underline{\delta}\mathcal{L}(x)$ in (4.3.12) means variations at any *given* $t$ in $t_1 \leq t \leq t_2$, that is including on the space-like surfaces $x^0 = t_1$, $x^0 = t_2$.

From the last equality in (4.3.12), we may infer that:

*Only the variations of the fields on the boundaries* $x^0 = t_1$, $x^0 = t_2$ *in the Lagrangian density may contribute to* $\underline{\delta}\langle t_2 | t_1 \rangle$, *as spelled out by the two surface terms in* $[G_1(t_2) - G_1(t_1)]$, *and variations of the fields in* $\underline{\delta}\mathscr{L}$ *within the boundaries do not contribute. The latter variations lead to the Euler-Lagrange equations.*

How do we isolate these two different parts in $\underline{\delta}\mathscr{L}$ ? To do this, suppose that the Lagrangian density $\mathscr{L}(x)$ depends on $\chi_a(x)$ and, on the first derivative $\partial_\mu \chi_a(x)$. Accordingly, quite generally, we may write

$$
\begin{aligned}
\underline{\delta}\mathscr{L}(x) &= F^a(x)\,\underline{\delta}\chi_a(x) + F^{a\mu}(x)\,\underline{\delta}\,(\partial_\mu \chi_a(x)) \\
&= [F^a(x) - \partial_\mu F^{a\mu}(x)]\,\underline{\delta}\chi_a(x) + \partial_\mu\big(F^{a\mu}(x)\,\underline{\delta}\chi_a(x)\big),
\end{aligned} \tag{4.3.13}
$$

thus separating the two contributions mentioned above, and the variations *within* the boundaries $x^0 = t_1$  $x^0 = t_2$, do not contribute, i.e., for arbitrary $\underline{\delta}\chi_a(x)$

$$
[F^a(x) - \partial_\mu F^{a\mu}(x)] = 0, \qquad t_1 < x^0 < t_2. \tag{4.3.14}
$$

These are the celebrated Euler-Lagrange Equations. On the other hand (4.3.13) also gives from (4.3.12)

$$
[G(t_2) - G(t_1)] = \int_{x^0 = t_1}^{x^0 = t_2} (dx)\, \partial_\mu\big(F^{a\mu}(x)\,\underline{\delta}\chi_a(x)\big). \tag{4.3.15}
$$

Upon introducing the hyper-surface elements

$$
d\sigma_\mu = (dx^1 dx^2 dx^3, \mathbf{0}), \tag{4.3.16}
$$

we may apply Gauss' Theorem to rewrite (4.3.15) as

$$
[G(t_2) - G(t_1)] = \int d^3\mathbf{x}\Big[F^{a0}(x)\,\underline{\delta}\chi_a(x)\Big]\Big|_{x^0 = t_2} - \int d^3\mathbf{x}\Big[F^{a0}(x)\,\underline{\delta}\chi_a(x)\Big]\Big|_{x^0 = t_1}. \tag{4.3.17}
$$

Hence upon comparison with (4.3.4), we obtain

$$
\pi^a(x) = F^{a0}(x). \tag{4.3.18}
$$

The above analysis, in turn, introduces the action integral

$$
A[\chi_a] = \int_{x^0 = t_1}^{x^0 = t_2} (dx)\,\mathscr{L}(x), \tag{4.3.19}
$$

a functional of the fields $\chi_a(x)$, as an integral over spacetime bounded by space-like surfaces $x^0 = t_1$, $x^0 = t_2$. Here the Lagrangian density $\mathscr{L}(x)$ is assumed to depend on $\chi_a(x)$ and on the derivatives $\partial_\mu \chi_a(x)$. It may be rewritten as

$$\mathcal{L}(x) = \mathcal{L}[\chi_a(x), \partial_\mu \chi_a(x)]. \tag{4.3.20}$$

The indices $a$ may stand for different types of fields and for any vector and/or spinor indices they may carry. The Lagrangian density is assumed to be written as the sum of products of the fields and their first derivatives.

To determine $F^a(x)$ and $F^{a\mu}(x)$, we proceed as follows. The variation of the action $\underline{\delta}A[\chi_a]$, with fixed space-like surfaces $x^0 = t_1$, $x^0 = t_2$, in response to the variation of the Lagrangian density $\underline{\delta}\mathcal{L}(x)$ is given by by

$$\underline{\delta}A[\chi_a] = \int_{x^0 = t_1}^{x^0 = t_2} (\mathrm{d}x)\, \underline{\delta}\mathcal{L}(x), \tag{4.3.21}$$

where

$$\underline{\delta}\mathcal{L}(x) = \mathcal{L}[\chi_a(x), \partial_\mu \chi_a(x)] - \mathcal{L}[\chi_a(x) - \underline{\delta}\chi_a(x), \partial_\mu(\chi_a(x) - \underline{\delta}\chi_a(x))], \tag{4.3.22}$$

arising solely from arbitrary variations $\underline{\delta}\chi_a(x)$ that one may a priori impose on the fields $\chi_a(x)$, where recall that the notation $\underline{\delta}\chi_a(x)$, with a bar under $\delta$, is used to *distinguish* these variations from ones that are *induced* on the fields by other given transformations such as by the rotation of a coordinate system or by a Lorentz transformation.

According to the Spin & Statistics Connection, investigated in Sect. 4.5, once the nature of a field has been established as being a Bose-Einstein or a Fermi-Dirac one, we consider infinitesimal numerical, now, c-number or Grassmann, variations $\underline{\delta}\chi_a(x)$, respectively, as the case may be, such that a variation commutes with all the fields and their derivatives, if the corresponding field $\chi_a$ is a Bose-Einstein one, or anti-commutes with all the Fermi-Dirac fields and their derivatives, and also commutes with all the Bose-Einstein fields and their derivatives, if the corresponding field $\chi_a$ is a Fermi-Dirac one.

## Field Equations

For infinitesimal $\underline{\delta}\chi_b(x)$, typical terms in (4.3.22) may have the structures

$$\chi_{a_1}(x)\, \partial_\sigma \chi_{a_2}(x) \ldots \underline{\delta}\chi_a(x) \ldots = (-\varepsilon)^{\Delta_n}\big(\chi_{a_1}(x)\, \partial_\sigma \chi_{a_2}(x) \ldots\big)\underline{\delta}\chi_a(x), \tag{4.3.23}$$

$$\partial_\sigma \chi_{b_1}(x)\, \chi_{b_2}(x) \ldots \partial_\mu \underline{\delta}\chi_b(x) \ldots = (-\varepsilon)^{\Delta_m}\partial_\sigma \chi_{b_1}(x)\, \chi_{b_2}(x) \ldots \big)\, \partial_\mu \underline{\delta}\chi_b(x), \tag{4.3.24}$$

up to product of numerical matrices. Here $\varepsilon = -1$ if the corresponding field $\chi_a(x)$, to $\underline{\delta}\chi_a(x)$ is a Bose-Einstein field, and $\varepsilon = +1$ if it is a Fermi-Dirac one.

$\Delta_n$, $\Delta_m$ denote the number of Fermi-Dirac fields to the *right* of $\underline{\delta}\,\chi_a(x)$, $\partial_\mu\underline{\delta}\,\chi_b(x)$, respectively. Each term contains an *even* number of fermion fields to satisfy the Bose character of a Lagrangian density. Therefore, if the number of fermion fields to the right of a *fermion* field is odd, then the number of fermion fields to its left must be even and so on.

It becomes clear that we obtain the *same* expression for $\underline{\delta}\mathscr{L}(x)$ in (4.3.13), (4.3.22), as the one that emerges from applying (4.3.23), (4.3.24) to all the terms in $\underline{\delta}\mathscr{L}(x)$, if we use instead the following alternative formal procedure.

We introduce partial derivatives $\partial/\partial\chi_a(x)$, $\partial/\partial(\partial_\mu\chi_a(x))$ to operate on the fields and their first derivatives, respectively, applied both from the *right* of the Lagrangian density in conformity with the identities in (4.3.23), (4.3.24), and both obeying the *same commutativity or anti-commutativity with the fields as* $\underline{\delta}\,\chi_a(x)$. A straightforward application of the chain rule gives

$$\underline{\delta}\mathscr{L}(x) = \left(\frac{\partial\mathscr{L}(x)}{\partial\chi_a(x)}\right)_r (\underline{\delta}\,\chi_a(x)) + \left(\frac{\partial\mathscr{L}(x)}{\partial(\partial_\mu\chi_a(x))}\right)_r (\partial_\mu\underline{\delta}\,\chi_a(x)). \qquad (4.3.25)$$

Here r stands for "from the right". The partial derivatives in this equation are defined by $\partial\chi_b(x)/\partial\chi_a(x) = \delta_b{}^a$, $\partial(\partial_\mu\chi_b(x))/\partial(\partial_\sigma\chi_a(x)) = \delta_\mu{}^\sigma\,\delta_b{}^a$, and so on, as opposed to functional derivatives which also involve Dirac delta functions (see below).

One may rewrite $\underline{\delta}\mathscr{L}(x)$ in (4.3.13)/(4.3.22) as

$$\underline{\delta}\mathscr{L}(x) = \left(\frac{\partial\mathscr{L}(x)}{\partial\chi_a(x)} - \partial_\mu\frac{\partial\mathscr{L}(x)}{\partial(\partial_\mu\chi_a(x))}\right)_r \underline{\delta}\,\chi_a(x) + \partial_\mu\left[\left(\frac{\partial\mathscr{L}(x)}{\partial(\partial_\mu\chi_a(x))}\right)_r \underline{\delta}\,\chi_a(x)\right].$$
$$(4.3.26)$$

Hence from (4.3.13), we have

$$F^a(x) = \left(\frac{\partial\mathscr{L}(x)}{\partial\chi_a(x)}\right)_r, \qquad F^{a\mu}(x) = \left(\frac{\partial\mathscr{L}(x)}{\partial(\partial_\mu\chi_a(x))}\right)_r, \qquad (4.3.27)$$

with all partial derivatives taken from the right.

Upon setting,

$$\left(\frac{\partial\mathscr{L}(x)}{\partial(\partial_\mu\chi_a(x))}\right)_r \equiv \pi^{a\mu}(x), \qquad (4.3.28)$$

the variation of the action $A[\chi_a]$ in (4.3.19) in response to the variation in (4.3.22) is then from (4.3.14), (4.3.15), (4.3.16), (4.3.17) and (4.3.18)

$$\delta A[\chi_a] = \int_{t_1}^{t_2}(dx)\,\partial_\mu\big[\pi^{a\mu}(x)\,\underline{\delta}\,\chi_a(x)\big], \qquad (4.3.29)$$

$$\left(\frac{\partial\mathscr{L}(x)}{\partial\chi_a(x)} - \partial_\mu\frac{\partial\mathscr{L}(x)}{\partial(\partial_\mu\chi_a(x))}\right)_r = 0. \qquad (4.3.30)$$

Using the Bose character of a Lagrangian density, we may replace the partial differentiations in (4.3.30) from the right by ones from the left since this means simply to multiply the whole equation, in the just mentioned equation, by $-1$ for a Fermi field. Accordingly the field equations (Euler-Lagrange Equations) may be rewritten as

$$\frac{\partial \mathscr{L}(x)}{\partial \chi_a(x)} - \partial_\mu \frac{\partial \mathscr{L}(x)}{\partial(\partial_\mu \chi_a(x))} = 0, \tag{4.3.31}$$

with the partial derivatives taken from the left, which is the conventional definition.

The Gauss's Theorem, of course, gives from (4.3.29)

$$\delta A[\chi_a] = \int d^3\mathbf{x} \left[\pi^{0a}(x)\underline{\delta}\,\chi_a(x)\right]\Big|_{x^0=t_2} - (t_2 \to t_1) = [G(t_2) - G(t_1)]. \tag{4.3.32}$$

A far simpler way of obtaining (4.3.31) is to take the functional derivative of the action $A[\chi_b]$, with respect to $\chi_a(x)$, with $A[\chi_a]$ as a functional of $\chi_b$, in the following manner.

To the above end, the functional derivative of $\chi_b(x')$ with respect to $\chi_a(x)$ is defined by

$$\frac{\delta\,\chi_b(x')}{\delta\,\chi_a(x)} = \delta_b{}^a\,\delta^{(4)}(x'-x). \tag{4.3.33}$$

For the functional derivative of a field $\partial'_\mu\,\chi_b(x')$ with respect to $\chi_a(x)$, we note that

$$\frac{\delta\,\partial'_\mu\,\chi_b(x')}{\delta\chi_a(x)} = -\int (dx'')\,\frac{\delta\,\chi_b(x'')}{\delta\,\chi_a(x)}\,\partial''_\mu\,\delta^{(4)}(x''-x')$$

$$= -\int (dx'')\,\delta_b{}^a\,\delta^{(4)}(x''-x)\,\partial''_\mu\,\delta^{(4)}(x''-x') = \delta_b{}^a\,\partial'_\mu\,\delta^{(4)}(x'-x), \tag{4.3.34}$$

The product rule then follows to be

$$\frac{\delta}{\delta\chi_a(x)}\,\chi_b(x')\,\partial'_\mu\,\chi_c(x') = \left(\delta_b{}^a\,(\partial'_\mu\,\chi_c(x')) - \varepsilon\,\chi_b(x')\,\delta_c{}^a\,\partial'_\mu\right)\delta^{(4)}(x'-x), \tag{4.3.35}$$

where $\varepsilon = -1$, if at least one the fields $\chi_a$, $\chi_b$ is a Bose-Einstein one, and $\varepsilon = +1$, if they are both Fermi-Dirac ones.

The Euler-Lagrange equations in (4.3.31) may be then simply replaced by the functional differentiation of the action $A[\chi_b]$ with respect to $\chi_a(x)$ expressed as follows

$$\frac{\delta}{\delta\chi_a(x)}A[\chi_b] = \frac{\delta}{\delta\chi_a(x)}\int (dx')\,\mathscr{L}[\chi_b(x'),\partial'_\mu\,\chi_b(x')] = 0, \tag{4.3.36}$$

using the simple rules in (4.3.33), (4.3.34) and (4.3.35) and the usual rule of commutativity or anti-commutativity with the fields as the case may be of the nature of the fields $\chi_a(x)$ in question applied. The integration in (4.3.36) is taken over all of spacetime in order not to introduce a surface integral since the latter is already accounted for by the integral in (4.3.29). Equation (4.3.36) treats the fields $\chi_b(x')$ and their derivatives $\partial'_\mu \chi_b(x')$ on an *equal* footing.

**Box 4.1**: Key points in the stationary action principle for *fixed* boundaries

$$\underline{\delta}\langle t_2 | t_1 \rangle = \langle t_2 | [G(t_2) - G(t_1)] | t_1 \rangle, \quad \text{with } t_1, t_2 \text{ fixed, where}$$

$$G(t) = \int d^3x \, \pi^a(x)\underline{\delta}\chi_a(x), \quad x^0 = t, \text{ for arbitrary infinitesimal}$$

variations of the fields $\chi_a(x)$, denoted by $\underline{\delta}\chi_a(x)$, for any *given*

$t$ in $t_1 \leq t \leq t_2$.

$$\pi^a(x) = \pi^{a0}(x), \quad \pi^{a\mu}(x) = \left( \frac{\partial \mathscr{L}(x)}{\partial(\partial_\mu \chi_a(x))} \right)_r, \text{ where the } \pi^a(x)$$

are the canonical conjugate momenta of the fields $\chi_a$.

$$A[\chi_b] = \int (dx) \mathscr{L}[\chi_b(x), \partial_\mu \chi_b(x)], \quad \text{denotes the action integral}$$

extended over all of spacetime, and

$$\frac{\delta}{\delta\chi_a(x)} A[\chi_b] = \frac{\delta}{\delta\chi_a(x)} \int (dx') \mathscr{L}[\chi_b(x'), \partial'_\mu \chi_b(x')] = 0$$

gives rise to the Euler-Lagrange equations upon functional

differentiations, using the simple rules in (4.3.33), (4.3.34) and (4.3.35).

In conclusion, the following should be also noted:

1. Not all the field equations following from (4.3.31) or (4.3.36) are necessarily equations of motion as such they may not involve the time derivative of a field in question and may lead to constraints for the latter field. For examples, working in .the Coulomb gauge in QED and QCD, the field equations obtained by varying the zeroth components $A^0$, $A_a{}^0$ of the vector potentials, respectively, do not involve their time derivatives and lead to constraints for them, and not all the remaining components are independent. The canonical conjugate momentum density of a dependent field, expressed in terms of independent fields, is zero, and its variation is induced by the generators of the independent fields just mentioned.
2. Later we will see how field equations are derived from Lagrangian densities for higher spin fields involving arbitrary orders in the space derivative $\partial$, due to constraints, but still of first order in the time derivative $\partial_0$, of the fields.

3. It is important to realize that all the results derived were obtained by a priori numerical variations of the fields obeying their respective statistics as established by the Spin & Statistics Connection in Sect. 4.5. Quantum variations applied *a priori* to a Lagrangian are certainly generalizations and as such they often involve additional assumptions and consequences not always of practical value, and further generalizations lead, for example, to generalized statistics such as para-statistics.

### 4.3.2   Commutation/Anti-commutation Relations

The generator for inducing infinitesimal changes $\delta \chi_a$, considering independent quantum fields $\chi_a(x)$, is given in (4.3.4), and from (4.3.2), this change is given by

$$i\underline{\delta} \chi_a(x) = [\chi_a(x), G(x^0)], \tag{4.3.37}$$

where [ , ] is the commutator. In detail,

$$i\underline{\delta} \chi_a(x) = \int d^3x' [\chi_a(x), \pi^b(x')\underline{\delta} \chi_b(x')] \qquad x'^0 = x^0. \tag{4.3.38}$$

We use the fact that

$$\underline{\delta} \chi_b(x') \chi_a(x) = -\varepsilon \chi_a(x) \underline{\delta} \chi_b(x'), \tag{4.3.39}$$

with $\varepsilon = -1$ if at least one of the fields $\chi_a(x)$, $\chi_b(x')$ is a Bose-Einstein one, and $\varepsilon = +1$ if both are Fermi-Dirac ones. We may thus rewrite (4.3.38) as

$$i\underline{\delta} \chi_a(x) = \int d^3x \left( \chi_a(x) \pi^b(x') + \varepsilon \pi^b(x') \chi_a(x) \right) \underline{\delta} \chi_b(x'), \quad x'^0 = x^0, \tag{4.3.40}$$

This equation implies the following equal-time anti-commutation/commutation rules:

$$\{\chi_a(x), \pi^b(x')\} = i\delta_a{}^b \delta^3(\mathbf{x} - \mathbf{x}'), \tag{4.3.41}$$

$$[\chi_a(x), \pi^b(x')] = i\delta_a{}^b \delta^3(\mathbf{x} - \mathbf{x}'), \tag{4.3.42}$$

for Fermi-Dirac and Bose-Einstein fields, respectively.

### 4.3.3  Generators for Quantum Responses to Field Variations: Internal Symmetry Groups

Consider the following commutation relation for $x^0 = x'^0$,

$$[\chi_c(x), \pi^a(x')\chi_b(x')] = \chi_c(x)\pi^a(x')\chi_b(x') - \pi^a(x')\chi_b(x')\chi_c(x)$$

$$= \left(\chi_c(x)\pi^a(x') + \varepsilon\pi^a(x')\chi_c(x)\right)\chi_b(x')$$

$$= \mathrm{i}\delta_c{}^a\delta^3(\mathbf{x} - \mathbf{x}')\chi_b(x'), \qquad (4.3.43)$$

where we have used (4.3.40), and, as before, $\varepsilon = \mp 1$ for Bose-Einstein/Fermi-Dirac fields, respectively. Let $M_a{}^b(x)$, $\Lambda_a(x)$ denote arbitrary numericals but infinitesimal depending, in general, on the spacetime coordinate. For $x'^0 = x^0$, we introduce the operator

$$\overline{G}(x^0) = \frac{1}{2}\int d^3\mathbf{x}'\left(M_a{}^b(x')\left(\pi^a(x')\chi_b(x') - \varepsilon\chi_b(x')\pi^a(x')\right)\right.$$

$$\left. + (1 - \varepsilon)\Lambda_a(x')\pi^a(x')\right), \qquad (4.3.44)$$

where note the minus sign multiplying $\varepsilon$, in this expression. Using the Fermi or Bose character of the fields, we may use (4.3.41) or (4.3.42) to infer from (4.3.44) that

$$[\chi_c(x), \overline{G}(x^0)] = \mathrm{i}\left(M_c{}^b(x)\chi_b(x) + \frac{1}{2}(1 - \varepsilon)\Lambda_c(x)\right), \qquad (4.3.45)$$

and hence the relation $\delta\chi_c(x) = (1/\mathrm{i})[\chi_c(x), \overline{G}(x^0)]$, induces a *quantum* variation to the field $\chi_c(x)$ given by

$$\delta\chi_c(x) = M_c{}^b(x)\chi_b(x) + \frac{1}{2}(1 - \varepsilon)\Lambda_c(x), \qquad (4.3.46)$$

as a linear combination of quantum field components $\chi_b(x)$ with, in general, space-time dependent coefficients plus, in general, of a spacetime dependent numerical shift for Bose-Einstein fields ($\varepsilon = -1$).

In particular, the infinitesimal gauge transformation of the non-abelian vector gauge field in (3.10.10) with

$$\delta A_\mu(x) = -\mathrm{i}g\,\delta\vartheta_a(x)\,[t_a, A_\mu(x)] - \partial_\mu\vartheta_a(x)\,t_a, \qquad (4.3.47)$$

may be expressed as

$$\delta A_{c\mu}(x) = A_{c\mu}(x) - A'_{c\mu}(x) = g\,\delta\vartheta_a(x)f_{abc}A_{b\mu}(x) - \partial_\mu\vartheta_c(x), \qquad (4.3.48)$$

where $A_\mu(x) = t_c A_{c\mu}(x)$, which is precisely of the form in (4.3.46), involving now a Lorentz index $\mu$.

### 4.3.4 Variations of Boundary Surfaces

In reference to the second type of variations corresponding only to change of the boundary surfaces, in response to a transformation $t \to t - \delta t$, the corresponding generator $G_B(t)$ is simply $-H(t)\,\delta t$, where $H(t)$ is the Hamiltonian. In terms of the Hamiltonian density $\mathscr{H}(x)$, in (4.3.8), this generator is explicitly given by

$$G_B(x^0) = -\int d^3\mathbf{x}\, \mathscr{H}(x)\, \delta x^0, \tag{4.3.49}$$

and

$$\delta\langle t| = i\langle t|G_B(t), \tag{4.3.50}$$

at a given time $t$, and for the transformation function $\langle t_2|t_1\rangle$,

$$\delta\langle t_2|t_1\rangle = i\langle t_2|[G_B(t_2) - G_B(t_1)]|t_1\rangle. \tag{4.3.51}$$

The latter may be equivalently rewritten as

$$\delta\langle t_2|t_1\rangle = i\langle t_2|\int_{t_1}^{t_2} dt\, \frac{d}{dt}\, G_B(t)|t_1\rangle, \tag{4.3.52}$$

or as

$$\delta_B\langle t_2|t_1\rangle = -i\langle t_2|\int_{t_1}^{t_2} dt\, \frac{d}{dt}\left(H(t)\,\delta t\right)|t_1\rangle, \tag{4.3.53}$$

where the subscript B is to remind us that only the boundary surfaces are being varied. In terms of the Hamiltonian density,

$$\delta_B\langle t_2|t_1\rangle = -i\langle t_2|\int_{x^0=t_1}^{x^0=t_2} (dx)\, \partial_0\left(\delta x^0\, \mathscr{H}(x)\right)|t_1\rangle. \tag{4.3.54}$$

We may re-express the above in terms of the Lagrangian density

$$\delta_B\langle t_2|t_1\rangle = -i\langle t_2|\int_{x^0=t_1}^{x^0=t_2} (dx)\, \partial_0\left(\delta x^0\, \pi^a(x)\partial_0\chi_a(x) - \delta x^0\mathscr{L}(x)\right)|t_1\rangle. \tag{4.3.55}$$

Under an infinitesimal time translation $x^0 \rightarrow x^0 - \delta x^0$ with $x' = (x^0 - \delta x^0, \mathbf{x})$ and hence $\chi'_a(x) = \chi_a(x) + \delta x^0 \, \partial_0 \, \chi_a(x)$, gives

$$\delta \chi_a(x) = \chi_a(x) - \chi'_a(x) = -\delta x^0 \, \partial_0 \, \chi_a(x). \tag{4.3.56}$$

We also recall that a canonical conjugate momentum $\pi^a(x)$ is the zeroth component of the field $\pi^{a\mu}(x)$ as we have seen in (4.3.18), (4.3.27), (4.3.28). Accordingly, (4.3.55) may be rewritten as

$$\delta_B \langle t_2 | t_1 \rangle = i \langle t_2 | \int_{x^0 = t_1}^{x^0 = t_2} (dx) \, \partial_0 \left( \delta x^0 \mathscr{L}(x) + \pi^{a0}(x) \, \delta \chi_a(x) \right) | t_1 \rangle. \tag{4.3.57}$$

We may use Gauss' Theorem, to rewrite the above expression as

$$\delta_B \langle t_2 | t_1 \rangle = i \langle t_2 | \int d^3\mathbf{x} \left( \delta x^0 \mathscr{L}(x) + \pi^{a0}(x) \, \delta \chi_a(x) \right) \Big|_{x^0 = t_2} - (t_2 \rightarrow t_1) \, | t_1 \rangle. \tag{4.3.58}$$

Accordingly, for a constant of motion given by

$$Q = \int d^3\mathbf{x} \left( \delta x^0 \mathscr{L}(x) + \pi^{a0}(x) \, \delta \chi_a(x) \right), \tag{4.3.59}$$

i.e., for $Q$, independent of time, $\delta_B \langle t_2 | t_1 \rangle = 0$.

With the Lagrangian density $\mathscr{L}(x)$ as a Lorentz scalar, we may use the invariance of a scalar product in spacetime, to readily generalize this result further for states associated with space-like surfaces $\sigma_1$, $\sigma_2$ corresponding to the change of the boundaries under a Lorentz transformation $x^\mu \rightarrow x^\mu - \delta x^\mu$ to

$$\delta_B \langle \sigma_2 | \sigma_1 \rangle = i \langle \sigma_2 | \int_{\sigma_1}^{\sigma_2} (dx) \, \partial_\mu \left( \delta x^\mu \mathscr{L}(x) + \pi^{a\mu} \delta \chi_a(x) \right) | \sigma_1 \rangle. \tag{4.3.60}$$

A local conservation law, corresponding to the conservation of the operator $Q$ in (4.3.59) may be also given, from the above integral, by considering an infinitesimal deviation of the hypersurface $\sigma_2$ from $\sigma_1$, about a point $x$, which leads to

$$\partial_\mu j^\mu(x) = 0, \quad j^\mu(x) = \delta x^\mu \mathscr{L}(x) + \pi^{a\mu} \delta \chi_a(x). \tag{4.3.61}$$

This is the content of Noether's Theorem . Of course the volume integral of $j^0(x)$ simply gives the expression for $Q$ in (4.3.59).

Finally, we note from (4.3.15), (4.3.28), (4.3.29), where $\underline{\delta} \chi_a(x)$ corresponds to, *a priori*, imposed arbitrary infinitesimal transformation, together with the expression for the generator in (4.3.60) that the generator which induces transformation of the fields for $x^\mu \rightarrow x^\mu - \delta x^\mu$ as well as other transformations, such as involved with internal symmetries, is given by the expression:

$$G(\sigma) = \int_\sigma (dx)\, \partial_\mu \left( \delta x^\mu \mathscr{L}(x) + \pi^{a\mu} \delta\chi_a(x) \right), \qquad (4.3.62)$$

where now $\delta\chi_a(x)$ is the infinitesimal response to the general transformations mentioned above.

## 4.4 Inhomogeneous Lorentz Transformations and Energy-Momentum Tensor

Under an infinitesimal inhomogeneous Lorentz transformation $x \to x - \delta x = x'$, as discussed in Sect. 2.2, where

$$\delta x_\nu = \delta b_\nu - \delta\omega_{\nu\lambda} x^\lambda, \qquad \delta\omega_{\nu\lambda} = -\delta\omega_{\lambda\nu}, \qquad (4.4.1)$$

a field $\chi_a(x)$ transforms as

$$\chi'_a(x') = \left( \delta_a{}^b + \frac{i}{2} \delta\omega^{\nu\lambda} S_{\nu\lambda a}{}^b \right) \chi_b(x), \qquad (4.4.2)$$

since the deviation from $\chi_a(x)$ must be proportional to $\delta\omega^{\nu\lambda}$. The matrix $[S_{\nu\lambda a}{}^b]$ defines the spin of the field. This gives

$$\delta\chi_a(x) = \chi_a(x) - \chi_a{}'(x') = -\left( \delta x^\nu \partial_\nu \chi_a(x) + \frac{i}{2} \delta\omega^{\nu\lambda} S_{\nu\lambda a}{}^b \chi_b(x) \right). \qquad (4.4.3)$$

Accordingly, from (4.3.60) we may write

$$\delta\langle \sigma_2 \mid \sigma_1 \rangle = \langle \sigma_2 \mid \int_{\sigma_1}^{\sigma_2} (dx)\partial_\mu \Big[ \delta x^\mu \mathscr{L}(x) - \pi^{a\mu} \delta x^\nu \partial_\nu \chi_a(x) $$
$$- \frac{i}{2} \delta\omega_{\nu\lambda} \pi^{a\mu} \big( S^{\nu\lambda} \chi(x) \big)_a \Big] \mid \sigma_1 \rangle. \qquad (4.4.4)$$

The antisymmetry property of $\delta\omega_{\nu\lambda}$ allows us to write

$$\delta\omega_{\nu\lambda} \pi^{a\mu} \big( S^{\nu\lambda} \chi(x) \big)_a = \delta\omega_{\nu\lambda} \Big( \pi^{a\mu} \big( S^{\nu\lambda} \chi(x) \big)_a + \pi^{a\lambda} \big( S^{\mu\nu} \chi(x) \big)_a + \pi^{a\nu} \big( S^{\mu\lambda} \chi(x) \big)_a \Big), \qquad (4.4.5)$$

which as a consequence of the symmetry in the last two terms in $(\nu, \lambda)$ is like adding zero to the left-hand side. This suggests to set

$$\Omega^{\mu\nu\lambda}(x) = \frac{i}{2} \Big( \pi^{a\mu} (S^{\nu\lambda} \chi(x))_a + \pi^{a\lambda} (S^{\mu\nu} \chi(x))_a + \pi^{a\nu} (S^{\mu\lambda} \chi(x))_a \Big), \qquad (4.4.6)$$

which is antisymmetric in the indices $(\mu, \lambda)$ since $S^{\mu\nu} = -S^{\nu\mu}$. On the other hand, one may write

$$\delta\omega_{\nu\lambda} = \frac{1}{2} [\partial_\nu \delta x_\lambda - \partial_\lambda \delta x_\nu], \tag{4.4.7}$$

and hence

$$\partial_\mu \left( \delta\omega_{\nu\lambda} \Omega^{\mu\nu\lambda} \right) = \frac{1}{2} \partial_\mu \left( [\partial_\nu \delta x_\lambda - \partial_\lambda \delta x_\nu] \Omega^{\mu\nu\lambda} \right). \tag{4.4.8}$$

Now

$$\partial_\mu \left( [\partial_\nu \delta x_\lambda] \Omega^{\mu\nu\lambda} \right) = \partial_\mu \left( -\omega_{\lambda\nu} \Omega^{\mu\nu\lambda} \right) = \partial_\mu \left( \omega_{\nu\lambda} \Omega^{\mu\nu\lambda} \right). \tag{4.4.9}$$

On the other hand

$$-\partial_\mu \left( [\partial_\lambda \delta x_\nu] \Omega^{\mu\nu\lambda} \right) = -\partial_\mu \left( \partial_\lambda [\delta x_\nu \Omega^{\mu\nu\lambda}] - \delta x_\nu \partial_\lambda \Omega^{\mu\nu\lambda} \right)$$

$$= -\partial_\mu \partial_\lambda \left( \delta x_\nu \Omega^{\mu\nu\lambda} \right) + \partial_\mu \left( \delta x_\nu \partial_\lambda \Omega^{\mu\nu\lambda} \right). \tag{4.4.10}$$

The first term on the right-hand of the last equality is zero since $\Omega^{\mu\nu\lambda}$ is antisymmetric in $(\mu, \lambda)$, while $\partial_\mu \partial_\lambda$ is symmetric. Hence from (4.4.9), (4.4.10), we may rewrite (4.4.8) as

$$\partial_\mu \left( \delta\omega_{\nu\lambda} \Omega^{\mu\nu\lambda} \right) = \partial_\mu \left( \frac{\delta\omega_{\nu\lambda}}{2} \Omega^{\mu\nu\lambda} + \frac{\delta b_\nu}{2} \partial_\lambda \Omega^{\mu\nu\lambda} - \frac{\delta\omega_{\nu\sigma}}{2} x^\sigma \partial_\lambda \Omega^{\mu\nu\lambda} \right). \tag{4.4.11}$$

The first term on the right-hand side is $1/2$ of the left-hand side, and hence using, in the process, the anti-symmetry property of $\delta\omega_{\nu\sigma}$, we get

$$\partial_\mu \left( \delta\omega_{\nu\lambda} \Omega^{\mu\nu\lambda} \right) = \partial_\mu \left( \delta b_\nu \partial_\lambda \Omega^{\mu\nu\lambda} + \frac{\delta\omega_{\nu\sigma}}{2} [x^\nu \partial_\lambda \Omega^{\mu\sigma\lambda} - x^\sigma \partial_\lambda \Omega^{\mu\nu\lambda}] \right). \tag{4.4.12}$$

Finally note that

$$\delta x^\mu \mathscr{L}(x) - \pi^{\mu a} \delta x^\nu \partial_\nu \chi_a(x) = \delta x_\nu \left( \eta^{\mu\nu} \mathscr{L}(x) - \pi^{\mu a} \partial^\nu \chi_a(x) \right). \tag{4.4.13}$$

Hence from (4.4.5)/(4.4.6), (4.4.12), (4.4.13) and (4.4.1), we may rewrite (4.4.4) in the form

$$\delta\langle \sigma_2 \mid \sigma_1 \rangle = \langle \sigma_2 \mid \int_{\sigma_1}^{\sigma_2} (dx) \, \partial_\mu j^\mu(x) \mid \sigma_1 \rangle, \tag{4.4.14}$$

$$j^{\mu}(x) = \delta b_{\nu}\, T^{\mu\nu}(x) + \frac{\delta\omega_{\sigma\nu}}{2}\left[x^{\sigma}T^{\mu\nu}(x) - x^{\nu}T^{\mu\sigma}(x)\right], \qquad (4.4.15)$$

where

$$T^{\mu\nu}(x) = \left(\eta^{\mu\nu}\mathscr{L}(x) - \pi^{a\mu}\partial^{\nu}\chi_a(x)\right) - \partial_{\lambda}\Omega^{\mu\nu\lambda}(x). \qquad (4.4.16)$$

The first term on the right-hand side between the round brackets is referred to as the canonical energy-momentum tensor.

We may set

$$M^{\mu\sigma\nu}(x) = x^{\sigma}\, T^{\mu\nu}(x) - x^{\nu}\, T^{\mu\sigma}(x), \qquad (4.4.17)$$

to re-express $j^{\mu}(x)$ as

$$j^{\mu}(x) = \delta b_{\nu}\, T^{\mu\nu}(x) + \frac{\delta\omega_{\sigma\nu}}{2}\, M^{\mu\sigma\nu}(x). \qquad (4.4.18)$$

Invoking invariance under inhomogeneous Lorentz transformations, and considering the hypersurface $\sigma_2$ deviating infinitesimally from the hypersurface $\sigma_1$, about a point $x$, gives, in a limiting sense, the local conservation law

$$\partial_{\mu}j^{\mu}(x) = 0. \qquad (4.4.19)$$

For arbitrary independent infinitesimal variations $\delta b_{\nu}$, $\delta\omega_{\sigma\mu}$, we may infer from (4.4.18), the conservations laws

$$\partial_{\mu}T^{\mu\nu}(x) = 0, \qquad (4.4.20)$$

$$\partial_{\mu}M^{\mu\sigma\nu}(x) = 0. \qquad (4.4.21)$$

On the other hand, applying $\partial_{\mu}$ directly to (4.4.17) and using (4.4.20) give

$$0 = \partial_{\mu}M^{\mu\sigma\nu}(x) = T^{\sigma\nu}(x) - T^{\nu\sigma}(x), \qquad (4.4.22)$$

establishing the symmetry of $T^{\nu\sigma}$. The latter is referred to as the symmetric energy-momentum tensor. This is unlike the canonical one which is not necessarily symmetric. This symmetry property is important as it is necessary for the conservation of the angular momentum density $M^{\mu\sigma\nu}$. Also general relativity requires that the energy-momentum tensor be symmetric.

The momentum and the angular momentum may be now written as

$$P^{\nu} = \int d\sigma_{\mu}\, T^{\mu\nu}(x), \qquad (4.4.23)$$

$$J^{\lambda\nu} = \int d\sigma_\mu \, M^{\mu\lambda\nu}(x). \tag{4.4.24}$$

They may be also rewritten as integrals over space

$$P^\nu = \int d^3x \, T^{0\nu}(x), \tag{4.4.25}$$

$$J^{\lambda\nu} = \int d^3x \, M^{0\lambda\nu}(x). \tag{4.4.26}$$

The generator for Lorentz transformations , may be also spelled out as

$$G = \delta b_\mu \, P^\mu + \frac{\delta\omega_{\mu\nu}}{2} \, J^{\mu\nu}, \tag{4.4.27}$$

expressed in terms of the variables of the Lagrangian density with infinitesimal variations $\delta b_\mu$, $\delta\omega_{\mu\nu}$.

For applications to the Maxwell field and the Dirac field, see Problems 4.6, 4.7 and 4.8. See also Problem 4.9.

## 4.5   Spin and Statistics Connection

The unique and important role that the Spin and Statistics Connection plays not only in physics but in all of the sciences, in general, and in the stability of matter and hence in our own existence, cannot be overemphasized. The Periodic Table of Elements in chemistry is based on the Pauli Exclusion Principle as a special case. Without it matter would be simply unstable and collapse.[7]

The Spin & Statistics Connection arises naturally and self consistently by the analysis of the action integral.

Suppose we are given a Lagrangian density having the structure

$$\mathcal{L}(x) = \frac{1}{4}\Big(\chi_a(x) \, Q^\mu_{ab} \, \partial_\mu \chi_b(x) - \partial_\mu \chi_a(x) \, Q^\mu_{ab} \, \chi_b(x)\Big) + \mathcal{F}[\chi(x)], \tag{4.5.1}$$

where the field $\chi(x) = (\chi_a(x))$ is real, $\mathcal{L}$ is of first order in the derivative of the field and of first degree, i.e., linear in $\partial_\mu \chi$ as shown, $Q^\mu$ is a numerical matrix, and $\mathcal{F}[\chi(x)]$ depends on $\chi(x)$. Hermiticity of $\mathcal{L}$, as the latter is related to the Hamiltonian, implies that

$$(Q^\mu)^\dagger = - \, Q^\mu. \tag{4.5.2}$$

---

[7]See also the introductory chapter to the book.

A complex field may be re-expressed as a real field with double the number of components consisting of its real and imaginary parts. For example, a four-component Dirac field may be re-expressed as a real one having eight components as we have seen in (3.5.3), Sect. 3.5. By appropriately choosing the fields in the theory to have even more components, such a first order Lagrangian density may, in general, be constructed. For the Dirac field, the structure given in (4.5.1) is obvious. The situation involved with other fields will be treated later. Further generalizations of (4.5.1) will be also discussed at the end of this section dealing with constraints.

Any square matrix $C$ may be written as the sum of a symmetric and an antisymmetric matrices $S$ and $A$, i.e., $C = S + A$, with

$$S_{ab} = (C_{ab} + C_{ba})/2, \qquad A_{ab} = (C_{ab} - C_{ba})/2, \qquad (4.5.3)$$

and $S^{\mathsf{T}} - S$, $A^{\mathsf{T}} - -A$. For a matrix $C$ such that $C^{\dagger} = -C$, we also have $S^{\dagger} = -S$, $A^{\dagger} = -A$.

Accordingly, we consider two systems for which

$$(Q^{\mu})^{\mathsf{T}} = Q^{\mu} \qquad \text{or} \qquad (Q^{\mu})^{\mathsf{T}} = -Q^{\mu}, \qquad (4.5.4)$$

together with the condition in (4.5.2).

As before we define the action integral, a functional of the field $\chi$,

$$A[\chi] = \int_{x^0 = t_1}^{x^0 = t_2} (dx)\, \mathscr{L}[\chi(x), \partial_\mu \chi(x)], \qquad (4.5.5)$$

as an integral over spacetime bounded by space-like surfaces $x^0 = t_1, x^0 = t_2$.

We investigate the nature of the variation of the action in response to the variation of the field defined by $\chi(x) \to \chi(x) - \underline{\delta}\,\chi(x)$. Following Schwinger, we consider variations of $A[\chi]$ arising from c-number variations $\underline{\delta}\,\chi(x)$. We will then encounter that such variations we were set to achieve with $\underline{\delta}\,\chi(x)$ commuting with the field itself is possible *only* for systems satisfying the second condition in (4.5.4), while $\underline{\delta}\,\chi(x)$ as a c-number Grassmann variable anti-commuting with the field is possible *only* for systems satisfying the first one.

We define the variation

$$\underline{\delta}\,\mathscr{L}[\chi(x)] = \mathscr{L}[\chi(x)] - \mathscr{L}[\chi(x) - \underline{\delta}\,\chi(x), \partial_\mu \chi(x) - \partial_\mu \underline{\delta}\,\chi(x)], \qquad (4.5.6)$$

paying special attention to the order in which a variation $\underline{\delta}\,\chi_a(x)$ appears relative to the fields $\chi_b(x)$. As in (4.3.26), one has

$$\underline{\delta} A[\chi] = \frac{1}{2}\int_{t_1}^{t_2}(dx)\,\partial_\mu\Big(\frac{1}{2}\,[\,\chi_a(x)\,Q^\mu_{ab}\,\underline{\delta}\,\chi_b(x) - \underline{\delta}\,\chi_a(x)\,Q^\mu_{ab}\,\chi_b(x)\,]\Big) +$$

$$\int_{t_1}^{t_2}(dx)\Big(\frac{1}{2}\,[\,\underline{\delta}\,\chi_a(x)\,Q^\mu_{ab}\,\partial_\mu\chi_b(x) - \partial_\mu\chi_a(x)\,Q^\mu_{ab}\,\underline{\delta}\,\chi_b(x)\,]\Big) + \int_{t_1}^{t_2}(dx)\,\underline{\delta}\,\mathscr{F}[\chi(x)], \qquad (4.5.7)$$

with the order in which the variations $\underline{\delta}\,\chi(x)$ appear kept *intact*. Note that the integrand of the first term is the integral of a total differential.

We set

$$Q^\mu_{ab} = \varepsilon\, Q^\mu_{ba}, \tag{4.5.8}$$

with $\varepsilon = \pm 1$ corresponding to the two cases in (4.5.4). By using a property of the Dirac delta function, the second integral on the right-hand side of (4.5.7) may, up to the addition of a term proportional to the first integral, be expressed as

$$-\frac{1}{2}\int_{t_1}^{t_2}(\mathrm{d}x)\int_{t_1}^{t_2}(\mathrm{d}x')\,Q^\mu_{ab}\left[\underline{\delta}\,\chi_a(x)\,\chi_b(x')-\varepsilon\,\chi_b(x')\,\underline{\delta}\,\chi_a(x)\right]\partial_\mu{}'\delta^{(4)}(x'-x). \tag{4.5.9}$$

From this we learn that a c-number variation $\underline{\delta}\,\chi_a(x)$ *commuting* with the fields $\chi_b(x)$ is possible *only* for $\varepsilon = -1$, i.e., for $(Q^\mu)^\mathsf{T} = -Q^\mu$, while a variation $\underline{\delta}\,\chi_a(x)$ as a c-number Grassmann variable *anti-commuting* with the fields $\chi_b(x)$ is possible *only* for $\varepsilon = +1$, i.e., for $(Q^\mu)^\mathsf{T} = Q^\mu$. Otherwise the expression within the square brackets in the integrand in (4.5.9) would be *zero* and no equations of motion for the fields involving their *derivatives* will result.

Therefore we may write,

$$[\underline{\delta}\,\chi_a(x),\chi_b(x')] = 0 \qquad \text{for} \qquad (Q^\mu)^\mathsf{T} = -Q^\mu,\ \epsilon = -1, \tag{4.5.10}$$

$$\{\underline{\delta}\,\chi_a(x),\chi_b(x')\} = 0 \qquad \text{for} \qquad (Q^\mu)^\mathsf{T} = +Q^\mu,\ \epsilon = +1. \tag{4.5.11}$$

That is, $\underline{\delta}\,\chi_a(x)\,\chi_b(x') = -\varepsilon\,\chi_b(x')\,\underline{\delta}\,\chi_a(x)$. In *both* cases, we have

$$\underline{\delta}A[\chi] = \int_{t_1}^{t_2}(\mathrm{d}x)\,\partial_\mu\left[\chi_a(x)\,Q^\mu_{ab}\frac{\underline{\delta}\,\chi_b(x)}{2}\right]$$

$$+\int_{t_1}^{t_2}(\mathrm{d}x)\left(\underline{\delta}\,\chi_a(x)\,Q^\mu_{ab}\,\partial_\mu\,\chi_b(x) + \underline{\delta}\,\mathscr{F}[\chi(x)]\right), \tag{4.5.12}$$

and $\underline{\delta}\,\mathscr{F}[\chi(x)] = \mathscr{F}[\chi(x)] - \mathscr{F}[\chi(x) - \underline{\delta}\,\chi(x)]$, by definition.

A typical term in $\underline{\delta}\,\mathscr{F}[\chi(x)]$ may be of the form

$$\chi_{a_1}\cdots\chi_{a_n}\underline{\delta}\,\chi_a\cdots = (-\varepsilon)^n\underline{\delta}\,\chi_a\,\chi_{a_1}\cdots\chi_{a_n}\cdots, \tag{4.5.13}$$

up to numerical matrices, and where on the left-hand side of this equation we have the product of $n$ field components appearing to the left $\underline{\delta}\,\chi_a$. Accordingly, we will get the same expression for $\underline{\delta}\,\mathscr{F}[\chi(x)]$ as prescribed above, if we set up the rule that

$$\underline{\delta}\,\mathscr{F}[\chi(x)] = \underline{\delta}\,\chi_a(x)\frac{\partial}{\partial\chi_a(x)}\,\mathscr{F}[\chi(x)], \tag{4.5.14}$$

with the partial derivative operating from the left and anti-commuting with the field components for the case $(Q^\mu)^\top = Q^\mu$, and commuting for $(Q^\mu)^\top = -Q^\mu$. The partial derivative is simply defined by $\partial\chi_b(x)/\partial\chi_a(x) = \delta_b{}^a$ as opposed to a functional derivative which also involves a Dirac delta function.

The principle of stationary action for arbitrary variations $\delta\chi_a(x)$ then gives the field equations

$$Q^\mu_{ab}\,\partial_\mu\,\chi_b(x) + \frac{\partial}{\partial\chi_a(x)}\,\mathcal{F}[\chi(x)] = 0. \tag{4.5.15}$$

An application of Gauss' Theorem to the first integral on the right-hand side of (4.5.12) gives

$$\int_{t_1}^{t_2} dx^0\,\partial_0\,G(x^0) = G(t_2) - G(t_1), \tag{4.5.16}$$

$$G(x^0) = \int d^3\mathbf{x}\,\chi_a(x^0,\mathbf{x})\,Q^0_{ab}\,\frac{\delta\chi_b(x^0,\mathbf{x})}{2}, \tag{4.5.17}$$

and $G(t_1)$, $G(t_2)$ correspond to the *variations* $\underline{\delta}\,\chi_c(x)$ of the field components on the space-like boundary surfaces $x^0 = t_1$, $x^0 = t_2$, respectively, i.e., for fixed times. As in Sect. 4.3.1, and as we will see at this section, $G(x^0)$ defines the generator for such a transformation, at a given time $x^0$, and accounts now for a change $\underline{\delta}\,\chi_c(x)/2$ via a unitary operator $U = 1 + iG$, and $U^\dagger\chi_c(x)U = \chi_c(x) - \underline{\delta}\,\chi_c(x)/2$ leads to

$$\frac{\underline{\delta}\,\chi_c(x)}{2} = \frac{1}{i}\,[\chi_c(x),G]. \tag{4.5.18}$$

From (4.5.10), (4.5.11) and (4.5.17), the above gives, at any given times $x^0 = x'^0$,

$$\frac{\underline{\delta}\,\chi_c(x)}{2} = \frac{1}{i}\,Q^0_{ab}\int d^3\mathbf{x}'\,\Big(\chi_c(x)\chi_a(x') + \varepsilon\,\chi_a(x')\chi_c(x)\Big)\frac{\underline{\delta}\,\chi_b(x')}{2}. \tag{4.5.19}$$

and the following commutator or anti-commutator, in turn, emerge

$$[\chi_c(x),\,\chi_a(x')]\,Q^0_{ab} = i\,\delta_{cb}\,\delta^3(\mathbf{x}-\mathbf{x}'), \qquad (Q^\mu)^\top = -Q^\mu, \tag{4.5.20}$$

$$\{\chi_c(x),\,\chi_a(x')\}\,Q^0_{ab} = i\,\delta_{cb}\,\delta^3(\mathbf{x}-\mathbf{x}'), \qquad (Q^\mu)^\top = +Q^\mu. \tag{4.5.21}$$

One may, for example, consider the case where $\mathcal{F}[\chi(x)]$ in (4.5.1) is of the form

$$\mathcal{F}[\chi(x)] = \frac{1}{2}\,\chi_a M_{ab}\chi_b, \tag{4.5.22}$$

where $M$ is a Hermitian numerical matrix.

According to (4.5.10), (4.5.11), we must have

$$(Q^\mu)^\top = -Q^\mu, \qquad M^\top = \phantom{-}M, \qquad (4.5.23)$$

$$(Q^\mu)^\top = \phantom{-}Q^\mu, \qquad M^\top = -M. \qquad (4.5.24)$$

### 4.5.1 Summary

Given a Lagrangian density as in (4.5.1), the statistics obeyed by the field in question depends on the anti-hermitian matrix $Q^\mu$ satisfying one of the conditions spelled out in (4.5.4). In particular, we learn that a variation $\underline{\delta}\,\chi_a(x)$ *commuting* with the fields $\chi_b(x)$ is possible *only* for $\varepsilon = -1$, i.e., for $(Q^\mu)^\top = -Q^\mu$, while a variation $\underline{\delta}\,\chi_a(x)$ as a Grassmann variable *anti-commuting* with the fields $\chi_b(x)$ is possible *only* for $\varepsilon = +1$, i.e., for $(Q^\mu)^\top = Q^\mu$. In the first case the theory is quantized via commutators derived in (4.5.20), while for the second case it is quantized via anti-commutators derived in (4.5.21). The field equations are given in (4.5.15), and the properties displayed in (4.5.23), (4.5.24) should be noted.

The method of investigation of the Spin and Statistics connection developed in this section is referred to as *Schwinger's Constructive Approach.*[8]

The above results have already been applied to the Dirac quantum field in Sect. 3.5, and special cases are discussed below. Applications to other fields are given in the next section. We first consider the Hamiltonian

### 4.5.2 The Hamiltonian

The Hamiltonian density $\mathscr{H}$ is directly obtained from (4.5.1) by subtracting off, in the process, the net term in it involving the time derivative. This gives

$$\mathscr{H}(x) = \frac{1}{4}\Big(\partial_k\chi_a(x)\,\mathbf{Q}_{ab}^k\chi_b(x) - \chi_a(x)\,\mathbf{Q}_{ab}^k\partial_k\chi_b(x)\Big) - \mathscr{F}[\chi(x)], \qquad (4.5.25)$$

and for the Hamiltonian

$$H = \int d^3\mathbf{x}\,\mathscr{H}(x). \qquad (4.5.26)$$

Under an infinitesimal variation of the fields $\underline{\delta}\,\chi_a(x)$, as before,

$$\underline{\delta}H = -\int d^3\mathbf{x}\,\underline{\delta}\,\chi_a(x)\Big(\mathbf{Q}_{ab}^k\partial_k\chi_b(x) + \frac{\partial}{\partial\chi_b(x)}\mathscr{F}[\chi(x)]\Big), \qquad (4.5.27)$$

---

[8]Schwinger [11–15], See also Manoukian [6], pp. 953–964.

and from the field equations (4.5.15), this simply becomes,

$$\underline{\delta} H = \int d^3\mathbf{x} \, \underline{\delta} \chi_a(x) \, Q^0_{ab} \, \partial_0 \, \chi_b(x). \tag{4.5.28}$$

The expression for the variation *induced* on the fields $\delta \chi_a(x)$, under a time displacement $x^0 \to x^0 - \delta x^0 = x'^0$, as opposed to the ones $\underline{\delta} \chi_a(x)$ that one may, *a priori*, impose on them as done above, follows from the consideration

$$\chi'(x') = \chi(x), \qquad x' = (x'^0, \mathbf{x}), \quad x = (x^0, \mathbf{x}), \tag{4.5.29}$$

$$\chi'(x^0, \mathbf{x}) = \chi(x^0 + \delta x^0, \mathbf{x}) = \chi(x) + \delta x^0 \partial_0 \chi(x), \tag{4.5.30}$$

$$\delta\chi(x) = \chi(x) - \chi'(x) = -\delta x^0 \partial_0 \chi(x). \tag{4.5.31}$$

With $H$ as the generator for time displacement via the unitary operator: $(1 - i\delta x^0 H)$, one has the induced variation $\delta\chi_a(x) = -i[H, \chi_a(x)]\delta x^0$, or

$$i\partial_0 \chi_a(x) = [\chi_a(x), H]. \tag{4.5.32}$$

Upon multiplying the latter by $Q^0_{ab} \underline{\delta} \chi_b(x)/2$, remembering the properties in (4.5.10), (4.5.11), using the Bose-Einstein character of the Hamiltonian, and finally integrating over $\mathbf{x}$ gives from (4.5.28)

$$i\frac{1}{2}\underline{\delta} H = [H, G], \tag{4.5.33}$$

where $G$ is given in (4.5.17). Since the integrand of $\underline{\delta} H$ in (4.5.28) is linear in $\underline{\delta}\chi_a(x)$, we see that this generator $G$ accounts for half $\underline{\delta}\chi_a(x)/2$ for the variation of the fields $\underline{\delta}\chi_a(x)$.

### 4.5.3 Constraints

Finally we note that for higher spins because of constraints, such as ones arising from gauge constraints, Lagrangian densities of the form in (4.5.1) may arise, where the matrix $Q^\mu$ may become dependent on the space derivative $\partial$ of arbitrary orders and so does $\mathscr{F}$, which would involve arbitrary orders of the space derivatives of the fields. That is, they become replaced by $Q^\mu(\partial)$, $\mathscr{F}[\chi(x), \partial\chi(x)]$ and are *even* in $\partial$: $Q^\mu(-\partial) = Q^\mu(\partial)$ and the same for $\mathscr{F}$. In particular, this means that $\partial$ operates equally to the left or right. In such cases the relations in (4.5.20), (4.5.21), with $x^0 = x'^0$, become simply replaced by

$$[\chi_c(x), Q^0_{ab}(\partial')\chi_a(x')] = i\delta_{cb}\delta^3(\mathbf{x} - \mathbf{x}'), \quad (Q^\mu)^\top(\partial) = -Q^\mu(\partial), \tag{4.5.34}$$

$$\{\chi_c(x), Q_{ab}^0(\partial')\chi_a(x')\} = i\delta_{cb}\delta^3(\mathbf{x}-\mathbf{x}'), \quad (Q^\mu)^\top(\partial) = +Q^\mu(\partial). \quad (4.5.35)$$

and the field equations are readily handled as will be seen later.

## 4.6  Quantum Dynamical Principle (QDP) of Field Theory

The Quantum Dynamical Principle (QDP) gives rise to a powerful and elegant formalism for studying dynamics in field theory and for carrying out all sorts of computations. It is particularly easy to apply and provides, in a compact way, a method for considering variations of so-called transformation functions, such as transition amplitudes, with respect to *parameters* that may appear in a theory as coupling parameters, as well as of variations with respect to external sources coupled to the underlying fields.

In the above sense, out of the QDP a functional *differential* formalism emerges for quantum field theory, where the dynamics is investigated and physical aspects of the theory are extracted simply by functional differentiations of some given generating functionals. In turn, it provides tremendous physical insight and an easy way for studying basic processes in a theory as particles/anti-particles move from emission to detection regions in a way similar to experimental situations.

By a functional Fourier transform (§ 2.6), the QDP also gives rise to its path integral counterpart, appropriately referred to as the path *integral* formalism, where continual *integrations* are to be carried out instead of *differentiations* of the differential QDP formalism, as will be seen in Chaps. 5 and 6.

The purpose of this section is to derive the QDP from first principles. Applications of this will be given in various stages, and some applications were already carried out in Sect. 3.3 and in Sect. 3.6. The reader will conveniently find a direct summary suitable for practical applications of the QDP at the end of this section.

Suppose one has a Hamiltonian $H(t,\lambda)$ depending on some parameters, denoted collectively by $\lambda$, such as coupling parameters and external sources that we wish to vary. The Hamiltonian may have an explicit time dependence assumed to come from such parameters. That is, the explicit time dependence is of numerical nature unaffected by transformations applied to operators in the underlying vector space in which $H(t,\lambda)$ operates. We may, in turn, introduce the Hamiltonian $H(t,0)$, and corresponding time development unitary operators $U^\dagger(t,\lambda)$, $U^\dagger(t,0)$, associated with the two Hamiltonians, satisfying

$$i\frac{d}{dt}U^\dagger(t,\lambda) = U^\dagger(t,\lambda)H(t,\lambda), \qquad i\frac{d}{dt}U^\dagger(t,0) = U^\dagger(t,0)H(t,0). \quad (4.6.1)$$

The Hamiltonian $H(t,0)$ may also depend on some other parameters that we do not wish to vary. $H(t,0)$, in general denotes the renormalized free Hamiltonian, that is the free Hamiltonian expressed in terms of physically measurable quantities.

Given a state $\langle a, \lambda = 0; t\,| \equiv \langle a, \lambda = 0\,|\ U^\dagger(t, 0)$, corresponding to a system governed by the Hamiltonian $H(t, 0)$, i.e., for $\lambda = 0$, we define a state

$$\langle a\,\tau, \lambda\,| = \langle a, \lambda = 0\,|\ U^\dagger(\tau, 0)\,U(\tau, \lambda), \tag{4.6.2}$$

developing in time via the unitary operator $U^\dagger(t, \lambda)$ i.e.,

$$\langle a\,\tau, \lambda; t\,| = \langle a\,\tau, \lambda\,|\ U^\dagger(t, \lambda). \tag{4.6.3}$$

Such a state, satisfies the Scrödinger equation

$$i\frac{\mathrm{d}}{\mathrm{d}t}\langle a\,\tau, \lambda; t\,| = \langle a\,\tau, \lambda; t\,|\ H(t, \lambda), \tag{4.6.4}$$

and has the remarkable property that for $t = \tau$, it simply becomes

$$\langle a\,\tau, \lambda; t\,|\ \Big|_{t=\tau} = \langle a, \lambda = 0\,|\ U^\dagger(\tau, 0)\,U(\tau, \lambda)\,U^\dagger(\tau, \lambda) = \langle a, \lambda = 0; \tau\,|\,. \tag{4.6.5}$$

In practice, one may choose $\tau \rightarrow \pm\infty$, with the states $\langle a \pm \infty, \lambda; t\,|$ describing scattering states. The reason why they are referred to as scattering states is that from (4.6.5),

$$\langle a \pm \infty, \lambda\,|\ U^\dagger(t, \lambda) = \langle a \pm \infty, \lambda; t\,| \rightarrow \langle a, \lambda = 0; t\,| = \langle a, \lambda = 0\,|\ U^\dagger(t, 0), \quad t \rightarrow \pm\infty.$$

The notation $\langle a\,\mathrm{out}\,|\ b\,\mathrm{in}\rangle$ for $\langle a + \infty, \lambda\,|\ b - \infty, \lambda\rangle$ in such limits is often used. Quite generally, however, we may introduce states:

$$\langle b\,\tau_2, \lambda; t\,|, \quad |a\,\tau_1, \lambda; t\rangle, \quad \text{and consider} \quad \tau_1 \leq t \leq \tau_2. \tag{4.6.6}$$

It is thus convenient to introduce the operator

$$V(t, \lambda) = U^\dagger(t, 0)\,U(t, \lambda), \tag{4.6.7}$$

and rewrite (4.6.2) as

$$\langle a\,\tau, \lambda\,| = \langle a, \lambda = 0\ |\ V(\tau, \lambda). \tag{4.6.8}$$

The operator $V(t, \lambda)$ satisfies the equation

$$i\frac{\mathrm{d}}{\mathrm{d}t}V^\dagger(t, \lambda) = U^\dagger(t, \lambda)[H(t, \lambda) - H(t, 0)]U(t, 0), \tag{4.6.9}$$

with $[H(t, \lambda) - H(t, 0)]$ denoting the part of the Hamiltonian depending on coupling parameters and external sources that we wish to vary, which, as mentioned above, are denoted collectively by $\lambda$. The Hamiltonian $H(t, 0)$ need not even to be specified and may depend on other parameters that we do not wish to vary.

We are interested in investigating the dynamics in a time interval $\tau_1 < t < \tau_2$. The QDP is particularly involved with the study of the variation of amplitudes $\langle b\, \tau_2, \lambda \mid a\, \tau_1, \lambda \rangle$ with respect to one or more of the components of $\lambda$, such as coupling parameters and of external sources. A derivative with respect to $\lambda$ standing for an external source will then denote a functional derivative (see, e.g., Sect. 3.6).

Using (4.6.9), the following identity is easily derived for $\tau \neq \tau_2$, $\tau \neq \tau_1$ and $\lambda \neq \lambda'$:

$$i\frac{d}{d\tau}\Big[ V(\tau_2, \lambda) V^\dagger(\tau, \lambda) V(\tau, \lambda') V^\dagger(\tau_1, \lambda') \Big]$$
$$= V(\tau_2, \lambda)\Big[ U^\dagger(\tau, \lambda)\big( H(\tau, \lambda) - H(\tau, \lambda') \big) U(\tau, \lambda') \Big] V^\dagger(\tau_1, \lambda'). \qquad (4.6.10)$$

We integrate the latter over $\tau$ from $\tau_1$ to $\tau_2$ to obtain

$$\Big[ V(\tau_2, \lambda) V^\dagger(\tau_1, \lambda) - V(\tau_2, \lambda') V^\dagger(\tau_1, \lambda') \Big]$$
$$= -i\, V(\tau_2, \lambda) \int_{\tau_1}^{\tau_2} d\tau \Big[ U^\dagger(\tau, \lambda)\big( H(\tau, \lambda) - H(\tau, \lambda') \big) U(\tau, \lambda') \Big] V^\dagger(\tau_1, \lambda').$$
$$(4.6.11)$$

For $\lambda'$ infinitesimally close to $\lambda$, the following variation arises for infinitesimal $\delta\lambda$,

$$\delta\Big[ V(\tau_2, \lambda) V^\dagger(\tau_1, \lambda) \Big] = -i\, V(\tau_2, \lambda)\Big[ \int_{\tau_1}^{\tau_2} d\tau\, U^\dagger(\tau, \lambda)\delta H(\tau, \lambda) U(\tau, \lambda) \Big] V^\dagger(\tau_1, \lambda).$$
$$(4.6.12)$$

We take the matrix element of the above between the $\lambda$- independent states $\langle b, \lambda = 0 \mid$, $\mid a, \lambda = 0 \rangle$, i.e., $\delta\langle b, \lambda = 0 \mid = 0$, $\delta\mid a, \lambda = 0 \rangle = 0$, and use the expression for the $\lambda$- dependent states in (4.6.2)/(4.6.8). This gives

$$\delta\langle b\, \tau_2, \lambda \mid a\, \tau_1, \lambda \rangle = -i\langle b\, \tau_2, \lambda \mid \int_{\tau_1}^{\tau_2} d\tau\, U^\dagger(\tau, \lambda)\, \delta H(\tau, \lambda)\, U(\tau, \lambda) \mid a\, \tau_1, \lambda \rangle.$$
$$(4.6.13)$$

The Hamiltonian $H(\tau, \lambda)$ is to be expressed in terms of the independent fields, denoted collectively by $\chi$, and their canonical conjugate momenta, denoted by $\pi$, as well as of the parameters $\lambda$. The independent fields are those for which their canonical conjugate momenta do not vanish. The Hamiltonian $H(t, \lambda)$ is then written as

$$H(t, \lambda) = H(\chi, \pi, \lambda, t). \qquad (4.6.14)$$

We may define the Heisenberg picture of $H(t, \lambda)$ by

$$\mathbb{H}(t, \lambda) = H(\chi(t), \pi(t), \lambda, t)) = U^\dagger(t, \lambda) H(\chi, \pi, \lambda, t) U(t, \lambda), \qquad (4.6.15)$$

relating the Schrödinger to the Heisenberg pictures, where now $\chi(t)$, $\pi(t)$, may depend on $\lambda$. The parameters $\lambda$ and the numerical $t$-dependence in $H(., ., \lambda, t)$ are not affected by the transformation. Accordingly, (4.6.13) may be rewritten as

$$\delta\langle b\,\tau_2, \lambda \mid a\,\tau_1, \lambda\rangle = -i\langle b\,\tau_2, \lambda \mid \int_{\tau_1}^{\tau_2} d\tau\ \delta\mathbb{H}(\tau, \lambda) \mid a, \tau_1, \lambda\rangle, \qquad (4.6.16)$$

*provided* it is understood that in the variation $\delta\mathbb{H}(\tau, \lambda)$, with respect to any of the components of $\lambda$, the fields $\chi(\tau)$, $\pi(\tau)$ *are kept fixed* as dictated by the expression in the integrand in (4.6.13) and by the steps going from (4.6.13), (4.6.14), (4.6.15) and (4.6.16).

To simplify the notation we will use the notation $\langle b\,t|$ for the $\lambda$-dependent states $\langle b\,t, \lambda\mid$ as no confusion may arise.

Equation (4.6.16) is the celebrated *Schwinger Dynamical Principle* also known as the *Quantum Dynamical Principle*, for parameters and external source variations, expressed in terms of the Hamiltonian and of the $\lambda$-*dependent* interacting states $\langle b\,\tau|, |a\,\tau\rangle$. The interesting thing to note is that although these states depend on $\lambda$, in the variation of $\langle b\,\tau_2 | a\,\tau_1\rangle$, with respect to $\lambda$, the entire variation on the right-hand side of (4.6.16) is applied directly to the Hamiltonian (with the fields $\chi(\tau)$, $\pi(\tau)$ kept fixed). This is thanks to the key identity derived in (4.6.10).

We also consider variations of the matrix elements of an arbitrary function of the variables indicated

$$B(\chi(t), \pi(t), \lambda, t) = \mathbb{B}(t, \lambda). \qquad (4.6.17)$$

To the above end, we subsequently use the identity

$$V(\tau_2, \lambda)\, \mathbb{B}(\tau, \lambda)\, V^\dagger(\tau_1, \lambda)$$

$$= V(\tau_2, \lambda)\, V^\dagger(\tau, \lambda)\, U^\dagger(\tau, 0)\, B(\chi, \pi, \lambda, \tau)\, U(\tau, 0)\, V(\tau, \lambda)\, V^\dagger(\tau_1, \lambda), \qquad (4.6.18)$$

where $B(\chi, \pi, \lambda, t)$ is in the Schrödinger picture.

By successive applications of (4.6.10), we have from (4.6.18), for $\tau_1 < \tau < \tau_2$

$$\delta[V(\tau_2, \lambda)\, \mathbb{B}(\tau, \lambda)\, V^\dagger(\tau_1, \lambda)] = -i\,V(\tau_2, \lambda)\int_{\tau}^{\tau_2} d\tau'\ \delta\mathbb{H}(\tau', \lambda)\, \mathbb{B}(\tau, \lambda)\, V^\dagger(\tau_1, \lambda)$$

$$+ V(t_2, \lambda)\, \delta\mathbb{B}(\tau, \lambda)\, V^\dagger(\tau_1, \lambda) - i\,V(t_2, \lambda)\int_{\tau_1}^{\tau} d\tau'\ \mathbb{B}(\tau, \lambda)\, \delta\mathbb{H}(\tau', \lambda)\, V^\dagger(\tau_1, \lambda).$$

$$(4.6.19)$$

where $\delta\mathbb{B}(\tau, \lambda)$ is defined the same way as $\delta\mathbb{H}(\tau, \lambda)$, i.e., by varying $\lambda$ and keeping the fields $\chi(t)$, $\pi(t)$ fixed.

Using the definition of the time-ordered product

$$\big(\delta\mathbb{H}(\tau', \lambda)\mathbb{B}(\tau, \lambda)\big)_+ = \delta\mathbb{H}(\tau', \lambda)\mathbb{B}(\tau, \lambda)\,\theta(\tau' - \tau) + \mathbb{B}(\tau, \lambda)\delta\mathbb{H}(\tau', \lambda)\,\theta(\tau - \tau'),$$
$$(4.6.20)$$

and taking the matrix element of (4.6.19) between the states $\langle a, \lambda = 0 |$, $| b, \lambda = 0 \rangle$, as before, the following equation emerges

$$\delta\langle b\,\tau_2 \mid \mathbb{B}(\tau, \lambda) \mid a\,\tau_1 \rangle$$

$$= -\mathrm{i}\,\langle b\,\tau_2 \mid \int_{\tau_1}^{\tau_2} \mathrm{d}\tau' \big(\delta\mathbb{H}(\tau', \lambda)\,\mathbb{B}(\tau, \lambda)\big)_+ \mid a\,\tau_1 \rangle + \langle b\,\tau_2 \mid \delta\mathbb{B}(\tau, \lambda) \mid a\,\tau_1 \rangle.$$
$$(4.6.21)$$

We may introduce an effective Lagrangian related to $H(\chi(t), \pi(t)\lambda, t)$ by

$$L_*(\chi(t), \dot{\chi}(t), \lambda, t) = \Big( \int \mathrm{d}^3\mathbf{x}\, \pi(x)\, \dot{\chi}(x)\Big) - H(\chi(t), \pi(t), \lambda, t), \qquad (4.6.22)$$

with a summation over the fields $\chi$ in the first term on the right-hand side understood. The canonical conjugate momenta $\pi(x)$ of the fields are defined through

$$L_*(\chi(t), \dot{\chi}(t), \lambda, t) - L_*(\chi(t), \dot{\chi}(t) - \delta\dot{\chi}(t), \lambda, t) = \int \mathrm{d}^3\mathbf{x}\, \pi(x)\delta\dot{\chi}(t). \qquad (4.6.23)$$

The variation of the expression in (4.6.22), with respect to $\lambda$, obtained by keeping $\chi(t)$, $\pi(t)$ fixed is then given by

$$\Big[ \int \mathrm{d}^3\mathbf{x}\, \pi(x)\, \delta\dot{\chi}(t) \Big] + \delta L_* \mid = \Big[ \int \mathrm{d}^3\mathbf{x}\, \pi(x)\, \delta\dot{\chi}(t) \Big] - \delta\mathbb{H} \mid, \qquad (4.6.24)$$

where the first term on the left-hand side comes from the variation of $L_*$, with respect to $\dot{\chi}(t)$, as given in (4.6.23), and hence

$$\delta L_* \mid = -\delta\mathbb{H} \mid. \qquad (4.6.25)$$

Here it is important to note that $\delta L_* \mid$ denotes variation with respect to $\lambda$ keeping $\chi(t)$, $\dot{\chi}(t)$ fixed, *while* $\delta\mathbb{H} \mid$ denotes variation with respect to $\lambda$ keeping $\chi(t)$, $\pi(t)$ fixed.

Consider the following two types of dependent fields:

1. Fields for which their canonical conjugate momenta vanish and for which the Euler-Lagrange equations give rise to *constraints* for them allowing one to eliminate them in favor of the independent fields $\chi(x)$ and the corresponding

canonical conjugate momenta $\pi(x)$ as well as the parameters $\lambda$, in general. When a Lagrangian does not involve the time derivative of a given field, its canonical conjugate momentum vanishes. The important thing to note here is that due to such constraints, these dependent fields may have an explicit dependence on the parameters $\lambda$. These dependent fields will be denoted collectively by $\varrho$, and the constraints are *derivable* ones. Examples of such fields, working in the Coulomb gauge, are the zeroth components $A^0(x)$ and $A_a^0(x)$ of the vector potentials in QED and QCD, respectively.

2. Consider dependent fields arising from constraints, one may, a priori, impose on functions of the fields, involving no time derivatives, and applied directly to the Lagrangian. That is, some fields or field components are written as functions of other fields or their components and may involve their space derivatives. In short, these dependent fields are expressed in terms of independent fields and perhaps their space derivatives. Since the variations encountered above, with respect to $\lambda$, are carried out with the independent fields kept fixed, such dependent fields are *automatically* kept fixed in these variations unless they are also chosen to have explicit $\lambda$- dependence. Such constraints are, a priori, *imposed* constraints. Examples of such constraints are the Coulomb gauges $\mathbf{V} \cdot \mathbf{A} = 0$, $\mathbf{V} \cdot \mathbf{A}_a = 0$ in QED and QCD, respectively. Here we see that one of the three components may be expressed in favor of the other two, and, needless to say, not all the components may be varied independently in deriving Euler-Lagrange equations.

After the dependent fields $\varrho$, as just discussed in point 1. above, are eliminated from the above constraints, the Lagrangian $L(t, \lambda) \equiv L(\chi(t), \dot\chi(t), \varrho(t), \lambda, t)$ of the system becomes replaced by the effective Lagrangian $L_*(\chi(t), \dot\chi(t), \lambda, t)$ considered earlier in (4.6.22), i.e.,

$$L(\chi(t), \dot\chi(t), \varrho(t), \lambda, t) = L_*(\chi(t), \dot\chi(t), \lambda, t). \tag{4.6.26}$$

The variation of the action $\int_\tau^{\tau'} dt\, L(t, \lambda)$ with respect to $\lambda$, with $\chi(t)$, $\dot\chi(t)$ kept fixed, comes from the explicit dependence of $L(t, \lambda)$ on $\lambda$, and may also come from the dependence of $\varrho$ on $\lambda$, as arising from the corresponding constraint equations. The fields $\varrho$ satisfy Euler-Lagrange equation obtained by carrying out functional derivatives of the action

$$E_\varrho(x) \equiv \frac{\delta}{\delta\varrho(x)} \int_{x^0=\tau}^{x'^0=\tau'} (dx')\, \mathscr{L}(x', \lambda) = 0, \tag{4.6.27}$$

for fixed $\tau$, $\tau'$, involving no surface terms, since, by definition, the canonical conjugate momenta $\pi^0 = \partial\mathscr{L}(x, \lambda)/\partial(\partial_0\varrho(x)) = 0$, due to the fact that $\mathscr{L}(x, \lambda)$ contains no time derivatives of the fields $\varrho(x)$. Here $\mathscr{L}(x, \lambda)$ is the Lagrangian density given through

$$L(t, \lambda) \equiv L(\chi(t), \dot\chi(t), \varrho(t), \lambda, t) = \int d^3\mathbf{x}\, \mathscr{L}(x, \lambda). \tag{4.6.28}$$

Accordingly the variation of the action in (4.6.27), with respect to $\lambda$, keeping $\chi(t)$, $\dot{\chi}(t)$ fixed, is from (4.6.26), (4.6.27)

$$\left[\int_{x^0=\tau}^{x^0=\tau'} (dx)\,\delta\varrho(x)E_\varrho(x)\right] + \int_{\tau}^{\tau'} dt\,\delta L(t,\lambda)| = \int_{\tau}^{\tau'} dt\,\delta L_*(t,\lambda)|. \qquad (4.6.29)$$

Hence,

$$\int_{\tau}^{\tau'} dt\,\delta L(t,\lambda)| = \int_{\tau}^{\tau'} dt\,\delta L_*(t,\lambda)|, \qquad (4.6.30)$$

*where* $\delta L(t,\lambda)|$ stands for variations with respect to $\lambda$, with $\chi(t)$, $\dot{\chi}(t)$ *as well as* the dependent fields $\varrho(t)$ all kept fixed.

In particular, upon taking the limits $\tau_1 \to -\infty, \tau_2 \to +\infty$, and using the notations $|a_-\rangle$ for $|a\,\tau_1\rangle$, $\langle b_+|$ for $\langle b\,\tau_2|$, in these limits, (4.6.16), (4.6.25), (4.6.28), (4.6.30), lead to

$$\delta\langle b_+ | a_-\rangle = i\langle b_+| \int (dx)\,\delta\mathscr{L}(x,\lambda)\,|a_-\rangle. \qquad (4.6.31)$$

The variations in $\delta\mathscr{L}(x,\lambda)$ in (4.6.31), with respect to $\lambda$, are from (4.6.28) carried out by keeping the independent fields $\chi(x)$ and the dependent fields $\varrho(x)$ as well as the derivatives $\partial_\mu\chi(x)$, $\nabla\varrho(x)$ all kept fixed.

Equation (4.6.31) is another version of the *Schwinger Dynamical Principle*, also known as the *Quantum Dynamical Principle* (QDP), for parameters and external source variations, expressed in terms of the Lagrangian density $\mathscr{L}(x,\lambda)$ and the $\lambda$-dependent interacting states $\langle b_+|$, $|a_-\rangle$.

The expression for the QDP in (4.6.31) takes on a particular simple and a very useful form by choosing $H(t,0) \equiv H_0$ to be the free time-independent Hamiltonian, and $|a,\lambda=0\rangle$, $|b,\lambda=0\rangle$ its ground (vacuum) state: $H_0\,|0\rangle = 0$. This gives for the variation of the vacuum-to-vacuum transition amplitude

$$\delta\langle 0_+|0_-\rangle = i\langle 0_+| \int (dx)\,\delta\mathscr{L}(x,\lambda)\,|0_-\rangle. \qquad (4.6.32)$$

One should note that other standard notations for the *vacuum-to-vacuum* transition amplitude $\langle 0_+ | 0_-\rangle$ one may encounter are $\langle 0\,\text{out} | 0\,\text{in}\rangle$, $\langle\text{vac(out)} | \text{vac(in)}\rangle$. As we have already seen in Sect. 3.6, $\langle 0_+ | 0_-\rangle$ is not necessarily just a phase factor.

The very compact equation in (4.6.32) provides a wealth of information in quantum field theory and is of central importance in applications as we have witnessed in Sect. 3.3 and in Sect. 3.6, and as will be seen again and again in later sections and subsequent chapters.

As an example, consider the Lagrangian density of a neutral scalar field[9] $\phi(x)$

$$\mathscr{L}(x,\lambda) = \mathscr{L}_0(x) + \lambda\mathscr{L}_I(x) + K(x)\phi(x), \qquad (4.6.33)$$

where $\mathscr{L}_0(x)$, $\mathscr{L}_I(x)$ may denote the free and interacting parts, $K(x)$ is an external (c-number) source, and $\lambda$ a coupling parameter. Hence

$$\frac{\partial}{\partial\lambda}\langle\, b_+\,|\,a_-\,\rangle = i\langle\, b_+\,|\int(dx)\,\mathscr{L}_I(x)\,|\,a_-\,\rangle, \qquad (4.6.34)$$

and, as a functional derivative,

$$(-i)\frac{\delta}{\delta K(x)}\langle\, b_+\,|\,a_-\,\rangle = \langle\, b_+\,|\int(dx')\frac{\delta}{\delta K(x)}\,\mathscr{L}(x',\lambda)\,|\,a_-\,\rangle = \langle\, b_+\,|\,\phi(x)\,|\,a_-\,\rangle, \qquad (4.6.35)$$

by using the functional derivative

$$\frac{\delta}{\delta K(x)}K(x') = \delta^{(4)}(x'-x). \qquad (4.6.36)$$

The QDP in terms of the vacuum-to-vacuum transition amplitude $\langle\, 0_+\,|\,0_-\,\rangle$ in (4.6.32) should be especially noted for applications. For earlier applications, see Sects. 3.3 and 3.6, and many others will follow in coming sections and chapters as mentioned above. A summary of the first steps needed for applications are spelled out below.

*** 

## 4.6.1 Summary

Now we provide a summary of how the QDP is applied to the vacuum-to-vacuum transition amplitude which is the starting point in many applications:

We are given a Lagrangian density $\mathscr{L}(x,\lambda)$ depending on parameters consisting of coupling constants and external sources, denoted collectively by $\lambda$. The variation of the vacuum-to-vacuum transition amplitude $\delta\langle\, 0_+\,|\,0_-\,\rangle$ with respect to any of the components of $\lambda$ is given by:

$$\delta\langle\, 0_+\,|\,0_-\,\rangle = i\langle\, 0_+\,|\int(dx)\,\delta\mathscr{L}(x,\lambda)\,|\,0_-\,\rangle, \qquad (4.6.37)$$

---

[9]The scalar field is just a field which remains invariant under Lorentz transformations: $\phi'(x') = \phi(x)$. Details on basic fields are given in the next section.

where the variation $\delta\mathscr{L}(x,\lambda)$ is carried out on the explicit dependence of $\mathscr{L}(x,\lambda)$ on $\lambda$, that is all the fields and their derivatives in it are kept fixed.

We may also be given a function $F(\chi(x),\pi(x),\varrho(x),\lambda)$ of the independent fields $\chi(x)$, their canonical conjugate momenta $\pi(x)$, and, as denoted, perhaps of dependent fields $\varrho(x)$ and of $\lambda$. Suppose through constraint equations as obtained, for example, from the Euler-Lagrange equations, the dependent fields may be expressed in terms of $\chi$, $\pi$, and at most of their spatial derivatives, and may be of (the components) of $\lambda$. Let the resulting expression for $F$ be denoted simply by $B(x,\lambda)$. Then the variation of $\langle 0_+ \mid B(x,\lambda) \mid 0_-\rangle$ with respect to any of the components of $\lambda$ is given by

$$\delta\langle 0_+|B(x,\lambda)|0_-\rangle = i\langle 0_+|\int(\mathrm{d}x')\,(\delta\mathscr{L}(x',\lambda)B(x,\lambda))_+|0_-\rangle + \langle\, 0_+|\delta B(x,\lambda)\,|0_-\rangle.$$

$$(4.6.38)$$

where the variation $\delta B(x,\lambda)$ is carried out with $\chi$, $\pi$, together with their spatial derivatives if any, kept fixed. Also $(\,.\,)_+$, as usual, denotes time-ordering. The second term will give a contribution only if $B(x,\lambda)$ depends explicitly on $\lambda$. In such cases the variation of the above matrix element will pick up a contribution coming from the second term on the right-hand side in addition to the first integral involving the time ordered products $(\delta\mathscr{L}(x',\lambda)B(x,\lambda))_+$. This causes some complications in the quantization of non-abelian gauge theories, and some abelian ones as well, but is easily handled by the application of (4.6.38), and is the origin of the so-called Faddeev-Popov factor, and its generalizations. These intricate details will be worked out later for such theories. An elementary application of this is given in Sect. 4.8 as well as other applications are given in coming chapters.[10] The path integral formalism will be also considered.

* **

## 4.7   A Panorama of Fields

This section deals with the basic quantum fields one often encounters in quantum field theory/high-energy physics. The fields in question are of spin 0, 1, 3/2 and 2, for massive and massless particles. The relevant technical *details* are relegated to their corresponding subsections that follow. The technical details concerning spin 1/2, the Dirac field, however, are spelled out in Chap. 3, while that of the massless spin 1, the photon, are the subject matter of Chap. 5. Needless to say, the corresponding subsections to follow of all these fields, form an integral part of this section and

---

[10]There is a long history with the quantum dynamical principle, see, Schwinger [10, 12, 14, 15], Lam [1], and Manoukian [3, 4].

should be read simultaneously with the text material. We now proceed to summarize some of the salient features of all these fields.

### 4.7.1   Summary of Salient Features of Some Basic Fields

Spin 0:

The Lagrangian density of a Hermitian scalar field $\varphi$ with corresponding field equation are

$$\mathscr{L}(x) = -\frac{1}{2}\,\partial_\mu\varphi(x)\partial^\mu\varphi(x) - \frac{m^2}{2}\,\varphi^2(x): \qquad (-\Box + m^2)\,\varphi(x) = 0, \qquad (4.7.1)$$

while the propagator is given by

$$\mathrm{i}\,\langle\mathrm{vac}\mid (\varphi(x)\varphi(x'))_+ \mid \mathrm{vac}\rangle = \Delta_+(x - x') = \int \frac{(\mathrm{d}p)}{(2\pi)^4}\frac{\mathrm{e}^{\mathrm{i}p\,(x-x')}}{p^2 + m^2 - \mathrm{i}\,\epsilon}, \qquad (4.7.2)$$

and satisfies

$$(-\Box + m^2)\,\Delta_+(x - x') = \delta^{(4)}(x - x'). \qquad (4.7.3)$$

The local commutativity relation reads

$$[\varphi(x), \varphi(x')] = \mathrm{i}\,\Delta(x - x'), \qquad \Delta(x - x') = \frac{1}{\mathrm{i}}\int\frac{(\mathrm{d}p)}{(2\pi)^3}\,\mathrm{e}^{\mathrm{i}p\,(x-x')}\,\delta(p^2 + m^2)\,\varepsilon(p^0),$$

$$(4.7.4)$$

where $\varepsilon(p^0) = +1$ for $p^0 > 0$, $= -1$ for $p^0 < 0$, and $\Delta(x - x')$ vanishes for space-like separations $(x - x')$.

In the presence of an external source $K$, described by the substitution $\mathscr{L}(x) \to \mathscr{L}(x) + K(x)\,\varphi(x)$, the vacuum-to-vacuum transition amplitude is given by

$$\langle\,0_+\mid 0_-\rangle = \exp\left[\frac{\mathrm{i}}{2}\int(\mathrm{d}x)(\mathrm{d}x')K(x)\Delta_+(x - x')K(x')\right]. \qquad (4.7.5)$$

For a non-Hermitian scalar field,

$$\mathscr{L}(x) = -\partial_\mu\varphi^\dagger(x)\partial^\mu\varphi(x) - m^2\varphi^\dagger(x)\varphi(x), \qquad (4.7.6)$$

$$(-\Box + m^2)\,\varphi(x) = 0, \quad (-\Box + m^2)\,\varphi^\dagger(x) = 0, \qquad (4.7.7)$$

$$[\varphi(x), \varphi(x')] = 0, \qquad [\varphi(x), \varphi^\dagger(x')] = \mathrm{i}\,\Delta(x - x'), \qquad (4.7.8)$$

$$i \langle \text{vac} \mid \left( \varphi(x) \, \varphi^\dagger(x') \right)_+ \mid \text{vac} \rangle = \Delta_+(x - x') = \int \frac{(\mathrm{d}p)}{(2\pi)^4} \frac{e^{ip\,(x-x')}}{p^2 + m^2 - i\epsilon}.$$

(4.7.9)

For $\mathscr{L}(x) \to \mathscr{L}(x) + K^\dagger(x) \, \varphi(x) + \varphi^\dagger(x) \, K(x)$,

$$\langle \, 0_+ \mid 0_- \rangle = \exp\!\left[ i \int (\mathrm{d}x)(\mathrm{d}x') K^\dagger(x) \Delta_+(x - x') K(x') \right].$$

(4.7.10)

Spin 1/2:

The Lagrangian density and the field equations are

$$\mathscr{L}(x) = -\frac{1}{2i}\left( \overline{\psi}(x) \gamma^\mu \partial_\mu \psi(x) - (\partial^\mu \overline{\psi})(x) \gamma^\mu \psi(x) \right) - m\,\overline{\psi}(x)\psi(x),$$

(4.7.11)

$$\left( \frac{\gamma \partial}{i} + m \right)\psi(x) = 0, \qquad \overline{\psi}(x)\left( -\frac{\gamma \overleftarrow{\partial}}{i} + m \right) = 0,$$

(4.7.12)

while the local anti-commutativity relation is

$$\{\psi_a(x), \overline{\psi}_b(x')\} = \left( -\frac{\gamma \partial}{i} + m \right)_{ab} \frac{1}{i}\, \Delta(x - x'),$$

(4.7.13)

where $\Delta(x - x')$ is defined in (4.7.4), and the propagator is given by

$$i \langle \text{vac} \mid \left( \psi(x)\overline{\psi}(x') \right)_+ \mid \text{vac} \rangle = S_+(x - x') = \int \frac{(\mathrm{d}p)}{(2\pi)^4} \frac{(-\gamma p + m)}{p^2 + m^2 - i\epsilon} e^{ip\,(x-x')},$$

(4.7.14)

satisfying

$$\left( \frac{\gamma \partial}{i} + m \right) S_+(x - x') = \delta^{(4)}(x - x'), \quad S_+(x' - x)\left( -\frac{\gamma \overleftarrow{\partial}}{i} + m \right) = \delta^{(4)}(x' - x).$$

(4.7.15)

In the presence of external sources $\eta, \overline{\eta}$: $\mathscr{L}(x) \to \mathscr{L}(x) + \overline{\eta}(x)\psi(x) + \overline{\psi}(x)\eta(x)$, the vacuum-to-vacuum transition amplitude is given by .

$$\langle \, 0_+ \mid 0_+ \rangle = \exp\!\left[ i \int (\mathrm{d}x)(\mathrm{d}x')\, \overline{\eta}(x)\, S_+(x - x')\, \eta(x') \right].$$

(4.7.16)

Spin 1:

. $m \neq 0$:

For a Hermitian massive vector field $V^\mu$ the Lagrangian density may be defined as

$$\mathscr{L}(x) = -\frac{1}{4} F_{\mu\nu}(x) F^{\mu\nu}(x) - \frac{m^2}{2} V_\mu(x) V^\mu(x), \qquad F_{\mu\nu}(x) = \partial_\mu V_\nu(x) - \partial_\nu V_\mu(x),$$

(4.7.17)

with corresponding field equations and a derived constraint

$$-\partial_\mu F^{\mu\nu} + m^2 V^\nu = 0, \quad \Rightarrow \quad \partial_\mu V^\mu = 0, \quad (-\Box + m^2) V^\mu = 0, \qquad (4.7.18)$$

and

$$[V^\mu(x), V^\nu(x')] = i\left(\eta^{\mu\nu} - \frac{\partial^\mu \partial^\nu}{m^2}\right) \Delta(x - x'), \qquad (4.7.19)$$

where $\Delta(x - x')$ is defined in (4.7.4). The propagator

$$\Delta_+^{\mu\nu}(x - x') = \int \frac{(dp)}{(2\pi)^4} \frac{e^{ip(x-x')}}{p^2 + m^2 - i\epsilon} \left(\eta^{\mu\nu} + \frac{p^\mu p^\nu}{m^2}\right), \qquad (4.7.20)$$

satisfies

$$\left[\eta_{\mu\nu}(-\Box + m^2) + \partial_\mu \partial_\nu\right] \Delta_+^{\nu\rho}(x - x') = \delta_\mu{}^\rho \delta^{(4)}(x - x'). \qquad (4.7.21)$$

For $\mathscr{L}(x) \to \mathscr{L}(x) + K^\mu(x) V_\mu(x)$,

$$\langle 0_+ | 0_- \rangle = \exp\left[\frac{i}{2} \int (dx)(dx') K_\mu(x) \Delta_+^{\mu\nu}(x - x') K_\nu(x')\right]. \qquad (4.7.22)$$

**. $m = 0$:**

For a massless (Hermitian) vector field in the presence of an external source $K^\mu$, the Lagrangian density in covariant gauges may be taken as

$$\mathscr{L} = -\frac{1}{4} F_{\mu\nu} F^{\mu\nu} - \chi \partial_\mu V^\mu + \frac{\lambda}{2} \chi^2 + K^\mu V_\mu, \quad F^{\nu\mu} = \partial^\nu V^\mu - \partial^\mu V^\nu, \qquad (4.7.23)$$

and $\chi$ is a field, referred to an auxiliary field, which may be eliminated and specifies the gauge. The field equations are:

$$\partial_\mu V^\mu = \lambda \chi, \qquad -\partial_\nu F^{\nu\mu} = K^\mu + (1 - \lambda)\partial^\mu \chi, \qquad -\Box \chi = \partial_\nu K^\nu, \qquad (4.7.24)$$

and the vacuum-to-vacuum transition amplitude takes the form

$$\langle 0_+ | 0_- \rangle = \exp\left[\frac{i}{2} \int (dx)(dx') K_\mu(x) D^{\mu\nu}(x - x') K_\nu(x')\right], \qquad (4.7.25)$$

where

$$D^{\mu\nu}(x - x') = \int \frac{(dp)}{(2\pi)^4} \left[\eta^{\mu\nu} - (1 - \lambda)\frac{p^\mu p^\nu}{p^2}\right] \frac{e^{ip(x-x')}}{p^2 - i\epsilon}, \qquad (4.7.26)$$

and is the covariant photon propagator for $K^\mu = 0$. Also the gauges defined by $\lambda = 1, \lambda = 0, \lambda = 3$, are, respectively, referred to as the Feynman gauge, the Landau gauge and the Yennie-Fried gauge.

In the so-called Coulomb gauge: $\partial_i V^i(x) = 0$ as applied to the lagrangian density

$$\mathcal{L} = -\frac{1}{4} F_{\mu\nu} F^{\mu\nu} + K^\mu V_\mu, \tag{4.7.27}$$

the field equations are:

$$- \partial_\mu F^{\mu\nu} = \left( \eta^{v\sigma} - \eta^{vj} \frac{1}{\partial^2} \partial^j \partial^\sigma \right) K_\sigma, \tag{4.7.28}$$

and

$$\langle 0_+ \mid 0_- \rangle = \exp\left[ \frac{i}{2} \int (dx)(dx') K_\mu(x) D_C^{\mu\nu}(x - x') K_\nu(x') \right], \tag{4.7.29}$$

where

$$D_C^{\mu\nu}(x - x') = \int \frac{(dQ)}{(2\pi)^4} \, e^{iQ(x-x')} D_C^{\mu\nu}(Q), \tag{4.7.30}$$

$$D_C^{00}(Q) = -\frac{1}{\mathbf{Q}^2}, \quad D_C^{ij}(Q) = \left( \delta^{ij} - \frac{Q^i Q^j}{\mathbf{Q}^2} \right) \frac{1}{Q^2 - i\epsilon}, \quad D_C^{0i}(Q) = 0, \tag{4.7.31}$$

defines the propagator in the Coulomb gauge for $K^\mu = 0$. For a conserved external current source $\partial_\mu K^\mu = 0$,

$$\langle 0_+ \mid 0_- \rangle = \exp\left[ \frac{i}{2} \int (dx)(dx') K_\mu(x) D_+(x - x') K^\mu(x') \right], \tag{4.7.32}$$

$$D_+(x - x') = \int \frac{(dp)}{(2\pi)^4} \frac{e^{ip(x-x')}}{p^2 - i\epsilon}. \tag{4.7.33}$$

Spin 3/2:
. $m \neq 0$:

The Lagrangian density of a massive spin 3/2 field, also referred to as the Rarita-Schwinger field,[11] may be taken as $(\gamma\partial \equiv \gamma^\mu \partial_\mu, \overset{\leftrightarrow}{\partial} = \overset{\rightarrow}{\partial} - \overset{\leftarrow}{\partial})$

$$\mathcal{L} = -\frac{1}{2i} \overline{\psi}_\mu \left( \eta^{\mu\nu} \gamma \overset{\leftrightarrow}{\partial} - (\gamma^\mu \overset{\leftrightarrow}{\partial}{}^\nu + \gamma^\nu \overset{\leftrightarrow}{\partial}{}^\mu) + \gamma^\mu (-\gamma \overset{\leftrightarrow}{\partial}) \gamma^\nu \right) \psi_\nu$$
$$- m \overline{\psi}_\mu \left[ \eta^{\mu\nu} + \gamma^\mu \gamma^\nu \right] \psi_\nu, \tag{4.7.34}$$

---

[11]Schwinger [18, p. 191] and Rarita and Schwinger [8].

with $\psi_{a\nu}$ carrying both a spinor index and a Lorentz one. The field equations are given by

$$\left[\eta^{\mu\nu}\left(\frac{\gamma\partial}{i}+m\right)-\left(\gamma^\mu\frac{\partial^\nu}{i}+\gamma^\nu\frac{\partial^\mu}{i}\right)+\gamma^\mu\left(-\frac{\gamma\partial}{i}+m\right)\gamma^\nu\right]\psi_\nu(x)=0,$$
(4.7.35)

or equivalently

$$\left(\frac{\gamma\partial}{i}+m\right)\psi^\mu(x)=0, \qquad \gamma_\mu\psi^\mu(x)=0, \quad \Rightarrow \quad \partial_\mu\psi^\mu(x)=0, \tag{4.7.36}$$

and with local anti-commutation relation

$$\{\psi_a^\mu(x),\overline{\psi}_b^\nu(x')\}=\frac{1}{i}\rho^{\mu\nu}\left(\frac{\partial}{i}\right)\Delta(x-x'), \tag{4.7.37}$$

$$\rho^{\mu\nu}\left(\frac{\partial}{i}\right)=\left(-\frac{\gamma\partial}{i}+m\right)\left(\eta^{\mu\nu}-\frac{\partial^\mu\partial^\nu}{m^2}\right)+\frac{1}{3}\left(\gamma^\mu+\frac{\partial^\mu}{im}\right)\left(\frac{\gamma\partial}{i}+m\right)\left(\gamma^\nu+\frac{\partial^\nu}{im}\right),$$
(4.7.38)

where $\Delta(x-x')$ is defined in (4.7.4). The propagator is given by

$$\Delta_{+\,ab}^{\;\mu\nu}(x-x')=\int\frac{(dp)}{(2\pi)^4}\frac{e^{ip(x-x')}}{p^2+m^2-i\epsilon}\rho_{ab}^{\mu\nu}(p), \tag{4.7.39}$$

$$\rho^{\mu\nu}(p)=(-\gamma p+m)\left[\eta^{\mu\nu}+\frac{1}{3}\left(\gamma^\mu\gamma^\nu+\gamma^\mu\frac{p^\nu}{m}-\gamma^\nu\frac{p^\mu}{m}+\frac{2}{m^2}p^\mu p^\nu\right)\right],$$
(4.7.40)

$$=(-\gamma p+m)\left(\eta^{\mu\nu}+\frac{p^\mu p^\nu}{m^2}\right)+\frac{1}{3}\left(\gamma^\mu+\frac{p^\mu}{m}\right)(\gamma p+m)\left(\gamma^\nu+\frac{p^\nu}{m}\right),$$
(4.7.41)

and satisfies

$$\left[\eta^{\mu\nu}\left(\frac{\gamma\partial}{i}+m\right)-\left(\gamma^\mu\frac{\partial^\nu}{i}+\gamma^\nu\frac{\partial^\mu}{i}\right)+\gamma^\mu\left(-\frac{\gamma\partial}{i}+m\right)\gamma^\nu\right]_{ab}\Delta_{+\,bc\,\nu\rho}(x-x')$$

$$=\delta_{ac}\delta^\mu_{\;\rho}\delta^{(4)}(x-x'). \tag{4.7.42}$$

In the presence of external sources $K^\mu$, $\overline{K}^\mu$, with $\mathscr{L}(x)\to\mathscr{L}(x)+\overline{K}^\mu(x)\psi_\mu(x)+\overline{\psi}^\mu(x)K_\mu(x)$,

$$\langle\,0_+|\,0_-\rangle=\exp\left[i\int(dx)(dx')\overline{K}_a^\mu(x)\,\Delta_{+\,ab\,\mu\nu}(x-x')\,K_b^\nu(x')\right]. \tag{4.7.43}$$

. $m = 0$:

The Lagrangian density for $m = 0$, in the presence of external sources, may be taken as $(\gamma\partial \equiv \gamma^\mu\partial_\mu, \; \boldsymbol{\gamma}\cdot\boldsymbol{\nabla} \equiv \gamma^i\partial_i, \square = \partial^\mu\partial_\mu)$[12]

$$\mathscr{L} = -\frac{1}{2i}\overline{\psi}_\mu\left(\eta^{\mu\nu}\gamma\overleftrightarrow{\partial} - (\gamma^\mu\overleftrightarrow{\partial}^\nu + \gamma^\nu\overleftrightarrow{\partial}^\mu) + \gamma^\mu(-\gamma\overleftrightarrow{\partial})\gamma^\nu\right)\psi_\nu + \overline{K}^\mu\psi_\mu + \overline{\psi}^\mu K_\mu.$$
(4.7.44)

We work in a Coulomb-like gauge $\partial_i\psi^i_a(x) = 0$. This leads to the field equations:

$$\gamma^i\psi^i = -i\frac{\boldsymbol{\gamma}\cdot\boldsymbol{\nabla}}{\nabla^2}\gamma^0 K_0, \tag{4.7.45}$$

$$\psi^0 = \frac{1}{2\nabla^2}\frac{\gamma\partial}{i}K^0 - \frac{1}{2\nabla^2}\left[\frac{\partial^i}{i} + \frac{\boldsymbol{\gamma}\cdot\boldsymbol{\nabla}\gamma^i}{i}\right]\gamma^0 K_i, \tag{4.7.46}$$

$$-\square\,\psi^i = -\frac{\gamma\partial}{i}\alpha^{ij}K^j - \frac{\square}{2\nabla^2}\left[\frac{\partial^i}{i} + \frac{\boldsymbol{\gamma}\cdot\boldsymbol{\nabla}\gamma^i}{i}\right]\gamma^0 K_0, \tag{4.7.47}$$

$$\alpha^{ij} = -\frac{1}{2}\left(\delta^{ik} - \frac{\partial^i\partial^k}{\nabla^2}\right)\gamma^\ell\gamma^k\left(\delta^{\ell j} - \frac{\partial^\ell\partial^j}{\nabla^2}\right). \tag{4.7.48}$$

The vacuum-to-vacuum transition amplitude becomes

$$\langle\, 0_+\,|\,0_-\rangle = \exp\Big[\,i\int(dx)(dx')\overline{K}_{a\mu}(x)D_{+\,ab}^{\ \ \mu\nu}(x-x')K_{b\nu}(x'), \tag{4.7.49}$$

where

$$D_{+\,ab}^{\ \ \mu\nu}(x-x') = \int\frac{(dp)}{(2\pi)^4}e^{ip(x-x')}D_{+\,ab}^{\ \ \mu\nu}(p), \tag{4.7.50}$$

$$D_+^{\ ij}(p) = \frac{(-\gamma p)}{p^2 - i\epsilon}\alpha^{ij}(p) = \frac{1}{p^2 - i\epsilon}\left(\delta^{ik} - \frac{p^ip^k}{\mathbf{p}^2}\right)\gamma^\ell\left(-\frac{\gamma p}{2}\right)\gamma^k\left(\delta^{\ell j} - \frac{p^\ell p^j}{\mathbf{p}^2}\right), \tag{4.7.51}$$

$$\alpha^{ij}(p) = -\frac{1}{2}\left(\delta^{ik} - \frac{p^ip^k}{\mathbf{p}^2}\right)\gamma^\ell\gamma^k\left(\delta^{\ell j} - \frac{p^\ell p^j}{\mathbf{p}^2}\right), \quad D_+^{\ 00}(p) = \frac{1}{2\mathbf{p}^2}(\gamma p), \tag{4.7.52}$$

$$D_+^{\ 0i}(p) = \frac{1}{2\mathbf{p}^2}(p^i + \boldsymbol{\gamma}\cdot\mathbf{p}\,\gamma^i)\gamma^0, \qquad D_+^{\ i0}(p) = -\frac{1}{2\mathbf{p}^2}(p^i + \boldsymbol{\gamma}\cdot\mathbf{p}\,\gamma^i)\gamma^0, \tag{4.7.53}$$

defines the propagator for $K^\mu = 0$, $\overline{K}^\mu = 0$.

---

[12]The massless Rarita-Schwinger field treatment is based on Manoukian [7].

Spin 2:

. $m \neq 0$:

The Lagrangian density of a massive Hermitian spin 2 field may be taken as

$$\mathscr{L} = -\frac{1}{2} \partial^\sigma U_{\mu\nu} \partial_\sigma U^{\mu\nu} + \partial_\mu U^{\mu\nu} \partial_\sigma U^\sigma{}_\nu - \partial_\sigma U^{\sigma\mu} \partial_\mu U$$

$$+ \frac{1}{2} \partial^\mu U \partial_\mu U - \frac{m^2}{2} \left( U^{\mu\nu} U_{\mu\nu} - U U \right), \tag{4.7.54}$$

where $U^{\mu\nu} = U^{\nu\mu}$, $U \equiv U^\mu{}_\mu$, and leads to the field equations

$$(-\Box + m^2) U_{\alpha\beta} + \partial_\alpha \partial_\sigma U^\sigma{}_\beta + \partial_\beta \partial_\sigma U^\sigma{}_\alpha - \partial_\alpha \partial_\beta U$$

$$- \eta_{\alpha\beta} \left[ (-\Box + m^2) U + \partial_\mu \partial_\nu U^{\mu\nu} \right] = 0. \tag{4.7.55}$$

The propagator is given by

$$\Gamma_+^{\alpha\beta;\mu\nu}(x - x') = \int \frac{(dp)}{(2\pi)^4} e^{ip(x-x')} \Gamma_+^{\alpha\beta;\mu\nu}(p), \tag{4.7.56}$$

$$\Gamma_+^{\alpha\beta;\mu\nu}(p) = \frac{1}{p^2 + m^2 - i\epsilon} \left( \frac{1}{2} \rho^{\alpha\mu}(p)\rho^{\beta\nu}(p) + \frac{1}{2} \rho^{\alpha\nu}(p)\rho^{\beta\mu}(p) \right.$$

$$\left. - \frac{1}{3} \rho^{\alpha\beta}(p)\rho^{\mu\nu}(p) \right), \qquad \rho^{\mu\nu}(p) = \left( \eta^{\mu\nu} + \frac{p^\mu p^\nu}{m^2} \right). \tag{4.7.57}$$

In the presence of an external source $T^{\mu\nu}$, $\mathscr{L}(x) \rightarrow \mathscr{L}(x) + T^{\mu\nu}(x) U_{\mu\nu}(x)$,

$$\langle 0_+ | 0_- \rangle = \exp \left[ \frac{i}{2} \int (dx)(dx') T^{\alpha\beta}(x) \Gamma_{+\,\alpha\beta;\mu\nu}(x - x') T^{\mu\nu}(x'). \tag{4.7.58}$$

. $m = 0$:

The Lagrangian density for the massless case in the presence of an external source $T^{\mu\nu}$, may be taken as

$$\mathscr{L} = -\frac{1}{2} \partial^\sigma U_{\mu\nu} \partial_\sigma U^{\mu\nu} + \partial_\mu U^{\mu\nu} \partial_\sigma U^\sigma{}_\nu - \partial_\sigma U^{\sigma\mu} \partial_\mu U$$

$$+ \frac{1}{2} \partial^\mu U \partial_\mu U + T^{\mu\nu} U_{\mu\nu}. \tag{4.7.59}$$

A covariant gauge treatment of massless spin 2 dealing with the graviton will be carried out in Sect. 2.3.1 of Volume II. On the other hand, in a Coulomb-like gauge $\partial_i U^{i\nu} = 0$, where $U^{\mu\nu} = U^{\nu\mu}$, $U \equiv U^\mu{}_\mu$, the propagator is given by

$$D_+^{\mu\nu;\sigma\rho}(x - x') = \int \frac{(dp)}{(2\pi)^4} e^{ip(x-x')} D_+^{\mu\nu;\sigma\rho}(p), \tag{4.7.60}$$

$$D_+^{ij;k\ell}(p) = \frac{\pi^{ij;k\ell}(\mathbf{p})}{p^2 - i\epsilon}, \qquad \pi^{ij}(\mathbf{p}) = \left(\delta^{ij} - \frac{p^i p^j}{\mathbf{p}^2}\right), \qquad (4.7.61)$$

$$\pi^{ij;k\ell}(\mathbf{p}) = \frac{1}{2}\left[\pi^{ik}(\mathbf{p})\pi^{j\ell}(\mathbf{p}) + \pi^{i\ell}(\mathbf{p})\pi^{jk}(\mathbf{p}) - \pi^{ij}(\mathbf{p})\pi^{k\ell}(\mathbf{p})\right], \qquad (4.7.62)$$

$$D_+^{ij;00}(p) = \frac{1}{2\mathbf{p}^2}\pi^{ij}(\mathbf{p}), \quad D_+^{00;00}(p) = \frac{1}{2\mathbf{p}^2}\frac{p^2}{\mathbf{p}^2}, \qquad (4.7.63)$$

$$D_+^{0i;0k}(p) = -\frac{1}{2\mathbf{p}^2}\pi^{ik}(\mathbf{p}), \quad D_+^{00;0i}(p) = 0, \quad D_+^{0i;jk}(p) = 0. \qquad (4.7.64)$$

and

$$\langle 0_+ | 0_- \rangle = \exp\left[\frac{i}{2}\int (dx)(dx')T_{\mu\nu}(x)D_+^{\mu\nu;\sigma\rho}(x-x')T_{\sigma\rho}(x')\right]. \qquad (4.7.65)$$

For a conserved source $\partial_\mu T^{\mu\nu} = 0$,

$$\langle 0_+ | 0_- \rangle = \exp\left[\frac{i}{2}\int (dx)(dx')\left(T^{\mu\nu}(x)T_{\mu\nu}(x') - \frac{1}{2}T^\mu{}_\mu(x)T^\nu{}_\nu(x')\right)D_+(x-x')\right], \qquad (4.7.66)$$

with $D_+(x-x')$ given in (4.7.33).

Now we carry out the details leading to the equations displayed above in this section.

## 4.7.2 Spin 0

A scalar field $\varphi(x)$ remains invariant under a Lorentz transformation, and hence, in particular, under an infinitesimal one

$$\varphi'(x') = \varphi(x), \qquad x'^\mu = x^\mu + \delta\omega^\mu{}_\nu x^\nu, \qquad \delta x^\mu = x^\mu - x'^\mu. \qquad (4.7.67)$$

The latter may be rewritten as

$$\varphi'(x) = \varphi(x + \delta x) = \varphi(x) + \delta x^\mu \partial_\mu \varphi(x), \qquad (4.7.68)$$

$$\delta\varphi(x) = \varphi(x) - \varphi(x') = -\delta\omega^{\mu\nu}x_\nu\partial_\mu\varphi(x). \qquad (4.7.69)$$

Or finally as

$$\delta\varphi(x) = \frac{i}{2}\delta\omega^{\mu\nu}\left(\frac{x_\mu\partial_\nu - x_\nu\partial_\mu}{i}\right), \qquad (4.7.70)$$

where $(x_\mu \partial_\nu - x_\nu \partial_\mu)/i$ corresponds to the orbital angular momentum, and a spin zero.

The field equation, in the presence of an external source $K(x)$, is defined by

$$(-\Box + m^2)\,\varphi(x) = K(x), \tag{4.7.71}$$

which may obtained from the Lagrangian density for a Hermitian scalar field $\varphi(x)$,

$$\mathcal{L}(x) = -\frac{1}{2}\,\partial_\mu\varphi(x)\,\partial^\mu\varphi(x) - \frac{1}{2}\,m^2\varphi(x)^2 + K(x)\varphi(x). \tag{4.7.72}$$

Upon taking the matrix elements $\langle\,0_+\,|\,.\,|\,0_-\rangle$ of (4.7.71), the following set of equations emerge by finally carrying out a functional integration over $K(x)$:

$$(-\Box + m^2)\langle\,0_+\,|\,\varphi(x)\,|\,0_-\rangle = K(x)\langle\,0_+\,|\,0_-\rangle, \tag{4.7.73}$$

$$\langle\,0_+\,|\,\varphi(x)\,|\,0_-\rangle = \int(dx')\Delta_+(x-x')K(x')\,\langle\,0_+\,|\,0_-\rangle, \tag{4.7.74}$$

$$(-i)\frac{\delta}{\delta K(x)}\langle\,0_+\,|\,0_-\rangle = \int(dx')\Delta_+(x-x')K(x')\langle\,0_+\,|\,0_-\rangle, \tag{4.7.75}$$

$$\langle\,0_+\,|\,0_-\rangle = \exp\left[\frac{i}{2}\int(dx)(dx')K(x)\Delta_+(x-x')K(x')\right], \tag{4.7.76}$$

where $\Delta_+(x-x')$ is the propagator

$$\Delta_+(x-x') = \int\frac{(dk)}{(2\pi)^4}\frac{e^{ik(x-x')}}{k^2+m^2-i\epsilon}, \quad (dk) \equiv dk^0 dk^1 dk^2 dk^3, \quad \epsilon\to 0. \tag{4.7.77}$$

The latter is defined as the vacuum expectation value of a time ordered product as follows

$$i\langle\,0_+\,|\,\big(\varphi(x)\varphi(x')\big)_+\,|\,0_-\rangle\Big|_{K=0,K^\dagger=0} = \Delta_+(x-x'). \tag{4.7.78}$$

The equal-time commutation relation emerges by introducing, in the process, $\pi^\mu = \partial\mathcal{L}/\partial(\partial_\mu\varphi) = -\partial^\mu\varphi$, $\pi = \partial_0\varphi = \dot{\varphi}$,

$$[\varphi(x),\dot{\varphi}(x')] = i\delta^3(\mathbf{x}-\mathbf{x}'), \quad x^0 = x'^{\,0}. \tag{4.7.79}$$

* * *

To discuss the spin and statistics connection, consider a first order formulation, with Lagrangian density

$$\mathcal{L} = -\frac{1}{4}\big(F^\mu\,\partial_\mu\varphi - \varphi\,\partial_\mu F^\mu\big) - \frac{1}{4}\big(\partial_\mu\varphi\,F^\mu - \partial_\mu F^\mu\,\varphi\big) + \frac{1}{2}F^\mu F_\mu - \frac{m^2}{2}\,\varphi^2, \tag{4.7.80}$$

which leads to

$$F^\mu = \partial^\mu \varphi, \qquad \partial_\mu F^\mu = m^2 \varphi. \tag{4.7.81}$$

Define $(\chi^a) = (F^\mu/\sqrt{m}, \sqrt{m}\,\varphi)$, having five components, and introduce the following matrices, referred to as Duffin-Kemmer-Petiau matrices,

$$\beta^0 = i \begin{pmatrix} 0 & 0 & 0 & 0 & 1 \\ 0 & 0 & 0 & 0 & 0 \\ 0 & 0 & 0 & 0 & 0 \\ 0 & 0 & 0 & 0 & 0 \\ -1 & 0 & 0 & 0 & 0 \end{pmatrix}, \quad \beta^1 = i \begin{pmatrix} 0 & 0 & 0 & 0 & 0 \\ 0 & 0 & 0 & 0 & -1 \\ 0 & 0 & 0 & 0 & 0 \\ 0 & 0 & 0 & 0 & 0 \\ 0 & -1 & 0 & 0 & 0 \end{pmatrix}, \tag{4.7.82}$$

$$\beta^2 = i \begin{pmatrix} 0 & 0 & 0 & 0 & 0 \\ 0 & 0 & 0 & 0 & 0 \\ 0 & 0 & 0 & 0 & -1 \\ 0 & 0 & 0 & 0 & 0 \\ 0 & 0 & -1 & 0 & 0 \end{pmatrix}, \quad \beta^3 = i \begin{pmatrix} 0 & 0 & 0 & 0 & 0 \\ 0 & 0 & 0 & 0 & 0 \\ 0 & 0 & 0 & 0 & 0 \\ 0 & 0 & 0 & 0 & -1 \\ 0 & 0 & 0 & -1 & 0 \end{pmatrix}, \tag{4.7.83}$$

and $\Lambda = \mathrm{diag}\,[-1, 1, 1, 1, -1]$, to rewrite (4.7.80) as

$$\mathscr{L} = \frac{1}{4} \left( \chi \, Q^\mu \, \partial_\mu \chi - \partial_\mu \chi \, Q^\mu \, \chi \right) + \frac{m}{2} \chi \, \Lambda \, \chi, \tag{4.7.84}$$

where

$$Q^\mu = \frac{\Lambda \beta^\mu}{i}, \quad (Q^0 \equiv i\beta^0), \tag{4.7.85}$$

satisfying

$$(Q^\mu)^\dagger = -Q^\mu, \qquad \Lambda^\dagger = \Lambda, \tag{4.7.86}$$

ensuring the Hermiticity of the Lagrangian density, and

$$(Q^\mu)^\mathsf{T} = -Q^\mu, \qquad \Lambda^\mathsf{T} = \Lambda, \tag{4.7.87}$$

establishing the Bose character of the field.[13] Here matrix multiplication is defined, e.g, as $\chi \Lambda \chi = \chi^a \Lambda_{ab} \chi^b$. The field equation and the equal-time commutation relation that follow from this formalism are (see Sect. 4.5)

$$\Lambda \left( \beta^\mu \frac{\partial_\mu}{i} + m \right) \chi = 0, \quad \delta(x^0 - x'^0)\,[\,\chi^a(x), \chi^b(x')\,]\,Q^0_{bc} = i\delta_{ac}\delta^{(4)}(x - x'), \tag{4.7.88}$$

---

[13] See (4.5.2), (4.5.20), (4.5.22), and (4.5.23).

and, in turn, lead to the field equation (for $K = 0$), and the equal-time commutation relation given in (4.7.71) and (4.7.79), respectively. In addition it gives rise to the local commutativity relation

$$\delta(x^0 - x'^0)[\varphi(x), \varphi(x')] = 0. \tag{4.7.89}$$

By an analysis very similar to the one carried out for the Dirac field leading to (3.5.12), we also have the general commutation relation

$$[\varphi(x), \varphi(x')] = i\,\Delta(x - x'), \tag{4.7.90}$$

where $\Delta(x - x')$ is defined in (3.5.13)) as

$$\Delta(x - x') = \int \frac{d^3\mathbf{p}}{(2\pi)^3}\, e^{i\mathbf{p}\cdot(\mathbf{x}-\mathbf{x}')}\, \frac{\sin p^0(x^0 - x'^0)}{p^0}, \quad p^0 = \sqrt{\mathbf{p}^2 + m^2}, \tag{4.7.91}$$

and is more conveniently rewritten as

$$\Delta(x - x') = \frac{1}{i} \int \frac{(dp)}{(2\pi)^3}\, e^{ip(x-x')}\, \delta(p^2 + m^2)\, \varepsilon(p^0), \tag{4.7.92}$$

where $\varepsilon(p^0) = +1$ for $p^0 > 0$, $= -1$ for $p^0 < 0$. The Lorentz invariance of (4.7.92) is evident since, in particular, a Lorentz transformation does not change the sign of $p^0$. That is, since from (4.7.91), $\Delta(x - x') = 0$ for $x^0 = x'^0$, and $\mathbf{x} \neq \mathbf{x}'$, the vector $(x - x')^\mu$ is space-like: $(\mathbf{x} - \mathbf{x}')^2 > 0$, we may infer from (4.7.92) that $\Delta(x - x') = 0$ vanishes for all space-like separations $(x - x') : (x - x')^2 > 0$. $\Delta(x - x') = 0$ satisfies the equation

$$(-\Box + m^2)\,\Delta(x - x') = 0, \tag{4.7.93}$$

as well as the key relation

$$\partial_0 \Delta(x - x')\Big|_{x^0 = x'^0} = \delta^3(\mathbf{x} - \mathbf{x}'). \tag{4.7.94}$$

* * *

For a non-Hermitian field, we may double the number of components consisting of real and imaginary parts of $F^\mu, \varphi$. The commutation relations become simply replaced by

$$\delta(x^0 - x'^0)[\varphi(x), \varphi^\dagger(x')] = 0. \tag{4.7.95}$$

$$\delta(x^0 - x'^0)[\varphi(x), \dot{\varphi}^\dagger(x')] = i\,\delta^{(4)}(x - x'), \quad \delta(x^0 - x'^0)[\varphi(x), \varphi(x')] = 0, \tag{4.7.96}$$

* * *

The Lagrangian density for a non-Hermitian field, in the presence of external sources $K$, $K^\dagger$ in a second order formalism, may be written as

$$\mathscr{L} = -\frac{1}{2}\left(\partial^\mu \varphi^\dagger\, \partial_\mu \varphi + \partial^\mu \varphi\, \partial_\mu \varphi^\dagger\right) - \frac{m^2}{2}\left(\varphi^\dagger \varphi + \varphi\, \varphi^\dagger\right) + K^\dagger \varphi + \varphi^\dagger K,$$

(4.7.97)

with

$$\langle\, 0_+ \mid 0_-\rangle = \exp\!\left[i \int (dx)(dx')K^\dagger(x)\Delta_+(x - x')K(x')\right],$$

(4.7.98)

and for $K = 0, K^\dagger = 0$,

$$i\langle\, 0_+ \mid \left(\varphi(x)\varphi^\dagger(x')\right)_+ \mid 0_-\rangle = \Delta_+(x - x'),$$

(4.7.99)

and the general commutation relations

$$[\varphi(x), \varphi(x')] = 0, \quad [\varphi^\dagger(x), \varphi^\dagger(x')] = 0, \quad [\varphi(x), \varphi^\dagger(x')] = i\,\Delta(x - x').$$

(4.7.100)

### 4.7.3  Spin 1

We introduce a vector field $V^\mu$ satisfying the equations

$$(-\Box + m^2)V^\mu = 0, \qquad \partial_\mu V^\mu = 0$$

(4.7.101)

We consider the case $m \neq 0$. The massless case, is the subject matter of Chap. 5, dealing with the photon. We will see that the above two equations describe a massive spin 1 particle.

The above equations, for a Hermitian vector field, may be obtained from the Lagrangian density

$$\mathscr{L} = -\frac{1}{4}F_{\mu\nu}F^{\mu\nu} - \frac{m^2}{2}V_\mu V^\mu, \qquad F_{\mu\nu} = \partial_\mu V_\nu - \partial_\nu V_\mu,$$

(4.7.102)

leading to

$$-\partial_\mu F^{\mu\nu} + m^2 V^\nu = 0.$$

(4.7.103)

Upon taking the partial drivative $\partial_\nu$ of the above equation, using the anti-symmetry of $F^{\mu\nu}$, and dividing, in the process, by $m^2$, lead to the two equations in (4.7.101).

In the presence of an external source $K^\mu$,

$$\mathcal{L} = -\frac{1}{4} F_{\mu\nu} F^{\mu\nu} - \frac{m^2}{2} V_\mu V^\mu + K_\mu V^\mu, \tag{4.7.104}$$

the field equations

$$\left[ \eta_{\mu\nu}(-\Box + m^2) + \partial_\mu \partial_\nu \right] V^\nu = K^\mu, \tag{4.7.105}$$

lead, in turn, to

$$\langle 0_+ | V^\mu(x) | 0_- \rangle = \int (dx') \Delta_+^{\mu\nu}(x - x') K_\nu(x') \langle 0_+ | 0_- \rangle \tag{4.7.106}$$

$$(-\mathrm{i}) \frac{\delta}{\delta K_\mu(x)} \langle 0_+ | 0_- \rangle = \int (dx') \Delta_+^{\mu\nu}(x - x') K_\nu(x') \langle 0_+ | 0_- \rangle. \tag{4.7.107}$$

$$\langle 0_+ | 0_- \rangle = \exp\left[ \frac{\mathrm{i}}{2} \int (dx)(dx') K_\mu(x) \Delta_+^{\mu\nu}(x - x') K_\nu(x) \right], \tag{4.7.108}$$

where the propagator $\Delta_+^{\mu\nu}(x - x')$ is given by

$$\Delta_+^{\mu\nu}(x - x') = \int \frac{(dk)}{(2\pi)^4} \frac{e^{\mathrm{i}k(x-x')}}{k^2 + m^2 - \mathrm{i}\epsilon} \left( \eta^{\mu\nu} + \frac{k^\mu k^\nu}{m^2} \right). \tag{4.7.109}$$

For $K^\mu = 0$, (4.7.103), gives, in particular, for $\nu = 0$, the constraint equation

$$V^0 - \frac{1}{m^2} \partial_k F^{k0}. \tag{4.7.110}$$

Also

$$\pi^{\nu\mu} = \frac{\partial \mathcal{L}}{\partial(\partial_\mu V_\nu)} = -F^{\mu\nu} = F^{\nu\mu}, \tag{4.7.111}$$

hence

$$V^0 = \frac{1}{m^2} \partial_k \pi^k, \tag{4.7.112}$$

with $\pi^k = F^{k0}$ denoting the canonical conjugate momenta of $V^k$, and

$$\delta(x^0 - x'^0) [V^j(x), F^{k0}(x')] = \mathrm{i} \delta^{jk} \delta^{(4)}(x - x'), \tag{4.7.113}$$

$$\delta(x^0 - x'^0) [V^j(x), V^0(x')] = \frac{1}{m^2} \mathrm{i} \partial'_j \delta^{(4)}(x - x'), \tag{4.7.114}$$

$$\delta(x^0 - x'^0) [V^j(x), V^k(x')] = 0, \tag{4.7.115}$$

$$\delta(x^0 - x'^0)\,[\,\pi^j(x), \pi^k(x')\,] = 0, \tag{4.7.116}$$

$$\delta(x^0 - x'^0)\,[\,V^0(x), V^0(x')\,] = 0. \tag{4.7.117}$$

Hence

$$\delta(x^0 - x^0)\,[\,V^\mu(x), F^{\sigma\lambda}(x')\,] = \mathrm{i}\left(\eta^{\mu\lambda}\eta^{\sigma 0} - \eta^{\mu\sigma}\eta^{\lambda 0}\right)\delta^{(4)}(x - x'), \tag{4.7.118}$$

$$[\,V^\mu(x), V^\nu(x')\,] = \mathrm{i}\left(\eta^{\mu\nu} - \frac{\partial^\mu \partial^\nu}{m^2}\right)\Delta(x - x'). \tag{4.7.119}$$

Under a Lorentz transformation $x \to x'$, $x'^\mu = x^\mu + \delta\omega^{\mu\nu} x_\nu$, a vector field has the transformation

$$V'^\mu(x') = \left(\delta^\mu{}_\nu + \delta\omega^\mu{}_\nu\right) V^\nu(x)$$

$$= \left(\delta^\mu{}_\nu + \frac{\mathrm{i}}{2}\delta\omega^{\alpha\beta}[S_{\alpha\beta}]^\mu{}_\nu\right) V^\nu(x), \tag{4.7.120}$$

where

$$[S_{\alpha\beta}]^\mu{}_\nu = \frac{1}{\mathrm{i}}\left(\delta_\alpha{}^\mu \eta_{\beta\nu} - \delta_\beta{}^\mu \eta_{\alpha\nu}\right). \tag{4.7.121}$$

From the submatrix

$$[S_{ij}]^a{}_b = \frac{1}{\mathrm{i}}\left(\delta_i{}^a \delta_{jb} - \delta_j{}^a \delta_{ib}\right), \quad i,j,a,b = 1,2,3, \tag{4.7.122}$$

as the coefficient of $\delta\omega^{ij}$, corresponding to pure 3D rotations (see Sect. 2.2), one may introduce the spin matrix $[S^k]$:

$$[S^k]^a{}_b = \frac{1}{2}\,\varepsilon^{ijk}\,[S_{ij}]^a{}_b, \tag{4.7.123}$$

where $\varepsilon^{ijk}$ is totally anti-symmetric with $\varepsilon^{123} = +1$, leading to

$$[\mathbf{S}^2]^a{}_c = [S_k]^a{}_b [S^k]^b{}_c = 2\,\delta^a{}_c, \tag{4.7.124}$$

establishing the spin $s = 1$ character of the vector field, with components satisfying the well known commutations relations,

$$[S^i, S^j] = \mathrm{i}\,\varepsilon^{ijk} S^k. \tag{4.7.125}$$

Matrix representations of the spin components are

$$S^1 = i \begin{pmatrix} 0 & 0 & 0 \\ 0 & 0 & -1 \\ 0 & 1 & 0 \end{pmatrix}, \quad S^2 = i \begin{pmatrix} 0 & 0 & 1 \\ 0 & 0 & 0 \\ -1 & 0 & 0 \end{pmatrix}, \quad S^3 = i \begin{pmatrix} 0 & -1 & 0 \\ 1 & 0 & 0 \\ 0 & 0 & 0 \end{pmatrix}. \tag{4.7.126}$$

It is interesting to note that the rotation matrix in Sect. 2.2 for the c.w. rotation of a 3-vector by an angle $\theta$ about a unit vector $\mathbf{n}$ may be rewritten as

$$R = \exp[i\,\theta\,\mathbf{n}\cdot\mathbf{S}]. \tag{4.7.127}$$

Polarization vectors may be defined for a massive vector particle $e_\lambda^\mu$, $\lambda = -1, 0, 1$, satisfying

$$e_\lambda^{\mu*}\, e_{\lambda'\mu} = \delta_{\lambda\lambda'}, \quad p_\mu e_\lambda^\mu = 0, \quad p^2 = -m^2, \tag{4.7.128}$$

as well as a completeness relation[14]

$$\eta^{\mu\nu} = -\frac{p^\mu p^\nu}{m^2} + \sum_{\lambda=-1,0,1} e_\lambda^\mu e_\lambda^{\nu*} = -\frac{p^\mu p^\nu}{m^2} + \sum_{\lambda=-1,0,1} e_\lambda^{\mu*} e_\lambda^\nu, \tag{4.7.129}$$

having expressions

$$e_+^\mu = (0, \mathbf{e}_+), \quad e_-^\mu = (0, \mathbf{e}_-), \quad e_0^\mu = \frac{1}{i\,m}\left(|\mathbf{p}|, p^0\frac{\mathbf{p}}{|\mathbf{p}|}\right), \tag{4.7.130}$$

$$e_\pm^{\mu*} = -e_\mp^\mu, \quad e_0^{\mu*} = -e_0^\mu. \tag{4.7.131}$$

⋆ ⋆⋆

A first order description of the vector field is given by the Lagrangian density

$$\mathscr{L} = -\frac{1}{4}F^{\mu\nu}\overleftrightarrow{\partial}_\mu V_\nu + \frac{1}{4}V_\nu\overleftrightarrow{\partial}_\mu F^{\mu\nu} + \frac{1}{4}F^{\mu\nu}F_{\mu\nu} - \frac{m^2}{2}V^\mu V_\mu. \tag{4.7.132}$$

$$(\chi^a) = \left(\sqrt{m}\,V^1, \sqrt{m}\,V^2, \sqrt{m}\,V^3, \frac{F^{01}}{\sqrt{m}}, \frac{F^{02}}{\sqrt{m}}, \frac{F^{03}}{\sqrt{m}}, \sqrt{m}\,V^0, \frac{F^{12}}{\sqrt{m}}, \frac{F^{13}}{\sqrt{m}}, \frac{F^{23}}{\sqrt{m}}\right). \tag{4.7.133}$$

In particular, the term involving the time derivative in (4.7.132) is given by

$$\frac{1}{4}\chi^a(Q^0)_{ab}\overleftrightarrow{\partial}_0\chi^b, \quad Q^0 = \left(\begin{array}{cc|c} 0 & I & 0 \\ -I & 0 & 0 \\ \hline 0 & 0 & 0 \end{array}\right), \tag{4.7.134}$$

---

[14]See also Problem 4.10.

where $Q^0$ is a $10 \times 10$ matrix, and $I$ is the $3 \times 3$ unit matrix. We note that $(Q^0)^\dagger = -Q^0$, for hermiticity, and $(Q^0)^\top = -Q^0$, in particular, establishing the Bose character of the vector field.

$$* * *$$

For a non-Hermitian vector field, the Lagrangian density may be written as

$$\mathscr{L} = -\frac{1}{2} F^{\mu\nu\dagger} F_{\mu\nu} - m^2 V^{\mu\dagger} V_\mu + K^{\mu\dagger} V_\mu + V^{\mu\dagger} K_\mu, \qquad (4.7.135)$$

and

$$\langle\, 0_+ \mid 0_-\rangle = \exp\Big[\, i \int (dx)(dx')K_\mu^\dagger(x)\triangle_+^{\mu\nu}(x - x\,')K_\nu(x')\Big], \qquad (4.7.136)$$

where the propagator $\triangle_+^{\mu\nu}(x - x')$ is given in (4.7.109).

### 4.7.4  Spin 3/2

We consider a field $\psi_a^{\,\mu}$ carrying a Lorentz index $\mu$ and a Dirac spinor index $a$. We will see, in particular, that the set of equations $(\gamma\partial \equiv \gamma^{\,\mu}\partial_\mu,\ \overset{\leftrightarrow}{\partial} = \overset{\rightarrow}{\partial} - \overset{\leftarrow}{\partial}\,)$

$$\Big(\frac{\gamma\partial}{i} + m\Big)\psi^\mu(x) = 0, \qquad \gamma_\mu\,\psi^\mu(x) = 0, \qquad (4.7.137)$$

suppressing the spinor index for simplicity of the notation, describes a spin $3/2$ field. The latter is referred to as the Rarita-Schwinger field.

Upon multiplying the above equation by $\gamma^{\,\sigma}$ and using the identity $\gamma^{\,\sigma}\gamma^{\,\mu} = -\gamma^{\,\mu}\gamma^{\,\sigma} - 2\eta^{\sigma\mu}$ leads to the condition

$$\partial_\mu\psi^\mu(x) = 0. \qquad (4.7.138)$$

The first equation in (4.7.137) together with the constraint on its right-hand side, may be obtained from the single equation[15]

$$\Big[\eta^{\mu\nu}\Big(\frac{\gamma\partial}{i} + m\Big) - \Big(\gamma^\mu\frac{\partial^\nu}{i} + \gamma^\nu\frac{\partial^\mu}{i}\Big) + \gamma^\mu\Big(-\frac{\gamma\partial}{i} + m\Big)\gamma^\nu\Big]\psi_\nu(x) = K^\mu(x),$$
$$(4.7.139)$$

---

[15]This is not the only equation which leads to the equations in (4.7.137). For example, the expressions within brackets in (4.7.139) replaced by $\eta^{\mu\nu}(\gamma\partial/i + m) - (1/3)\big(\gamma^{\,\mu}\partial^\nu/i +$ $\gamma^{\,\nu}\partial^\mu/i\big) + (1/3)\gamma^{\,\mu}\big(-\gamma\partial/i + m\big)\gamma^{\,\nu}$, also leads to the equations in question. This merely gives rise to adding a non-propagating term to the propagator in (4.7.144), i.e., a non-singular contribution on the mass shell $p^2 = -m^2$, see Problem 4.11.

for $K^\mu = 0$. For greater generality and for easily deriving the expression for the propagator, we have introduced a source term $K^\mu$ in the above equation.

Upon comparison of the two equation obtained by multiplying the above equation by $\gamma_\mu$ and $\partial_\mu$, respectively, one obtains for $m \neq 0$

$$\gamma_\mu \psi^\mu = -\frac{1}{3m}\left(\gamma^\sigma - 2\frac{\partial^\sigma}{im}\right)K_\sigma, \qquad (4.7.140)$$

$$\partial_\mu \psi^\mu = \frac{\gamma\partial}{3m}\left(\gamma^\sigma - 2\frac{\partial^\sigma}{im}\right)K_\sigma + \frac{\partial^\sigma K_\sigma}{m}. \qquad (4.7.141)$$

By substituting, in turn, these two equations in (4.7.139), gives

$$\left(\frac{\gamma\partial}{i} + m\right)\psi^\mu = K^\mu + \frac{1}{3}\left[\gamma^\mu\gamma^\sigma + \gamma^\mu\frac{\partial^\sigma}{im} - \gamma^\sigma\frac{\partial^\mu}{im} + \frac{2}{m^2}\frac{\partial^\mu}{i}\frac{\partial^\sigma}{i}\right]K_\sigma. \qquad (4.7.142)$$

Hence for $K_\sigma = 0$, the equations in question in (4.7.137), (4.7.138) are satisfied.

The expression for the propagator simply follows from (4.7.142) upon multiplying, in the process, by $(-\gamma\partial/i + m)$ leading to

$$\langle 0_+ | \psi_a^\mu(x) | 0_- \rangle = \int (dx')\Delta_{+\,ab}^{\mu\nu}(x - x')K_{b\nu}(x')\langle 0_+ | 0_-\rangle, \qquad (4.7.143)$$

$$\Delta_{+\,ab}^{\mu\nu}(x - x') = \int \frac{(dp)}{(2\pi)^4}\frac{e^{ip(x-x')}}{p^2 + m^2 - i\epsilon}\rho_{ab}^{\mu\nu}(p), \quad \epsilon \to 0, \qquad (4.7.144)$$

$$\rho^{\mu\nu}(p) = (-\gamma p + m)\left[\eta^{\mu\nu} + \frac{1}{3}\left(\gamma^\mu\gamma^\nu + \gamma^\mu\frac{p^\nu}{m} - \gamma^\nu\frac{p^\mu}{m} + \frac{2}{m^2}p^\mu p^\nu\right)\right]$$

$$= (-\gamma p + m)\left(\eta^{\mu\nu} + \frac{p^\mu p^\nu}{m^2}\right) + \frac{1}{3}\left(\gamma^\mu + \frac{p^\mu}{m}\right)(\gamma p + m)\left(\gamma^\nu + \frac{p^\nu}{m}\right). \qquad (4.7.145)$$

The field equations in (4.7.142) may be obtained from the Lagrangian density

$$\mathscr{L} = -\frac{1}{2i}\overline{\psi}_\mu\left(\eta^{\mu\nu}\gamma\overleftrightarrow{\partial} - (\gamma^\mu\overleftrightarrow{\partial}^\nu + \gamma^\nu\overleftrightarrow{\partial}^\mu) + \gamma^\mu(-\gamma\overleftrightarrow{\partial})\gamma^\nu\right)\psi_\nu$$

$$- m\overline{\psi}_\mu\left[\eta^{\mu\nu} + \gamma^\mu\gamma^\nu\right]\psi_\nu + \overline{K}^\mu\psi_\mu + \overline{\psi}^\mu K_\mu, \qquad (4.7.146)$$

where $\overline{\psi}_\mu = \psi^\dagger_\mu\gamma^0$ as usual.

From the QDP, $\langle 0_+ | \psi_a^\mu(x) | 0_-\rangle = -i\left(\delta/\delta\overline{K}_{a\mu}(x)\right)\langle 0_+ | 0_-\rangle$, we may integrate (4.7.143) to obtain

$$\langle 0_+ | 0_-\rangle = \exp\left[i\int(dx)(dx')\overline{K}_{a\mu}(x)\,\Delta_{+\,ab}^{\mu\nu}(x - x')\,K_{b\nu}(x')\right]. \qquad (4.7.147)$$

For $K^\mu = 0, \overline{K}^\mu = 0$, the field equation (4.7.142), together with the conditions in (4.7.140), (4.7.141) lead to a constraint on the components of $\psi^i$ as well (see Problems 4.12 and 4.13), given by

$$\left[\frac{\partial^i}{i} + \left(\frac{\gamma \cdot \nabla}{i} - m\right)\gamma^i\right]\psi^i(x) = 0, \tag{4.7.148}$$

involving no time derivatives. This constraint may be taken into account in setting up the equal-time anti-commutator of the spatial components $\psi^i(x)$ by defining

$$\beta^{ij} = \delta^{ij} + \frac{1}{3}\left(\gamma^i\gamma^j - \gamma^i\frac{\partial^j}{im} + \gamma^j\frac{\partial^i}{im} + \frac{2}{m^2}\frac{\partial^i}{i}\frac{\partial^j}{i}\right), \tag{4.7.149}$$

satisfying the constraint equation

$$\left[\frac{\partial^i}{i} + \left(\frac{\gamma \cdot \nabla}{i} - m\right)\gamma^i\right]\beta^{ij} = 0. \tag{4.7.150}$$

The term involving the time derivative in (4.7.146) is given by

$$\mathscr{L}\big|_{\text{time der.}} = -\frac{1}{2i}\,\psi^{i\dagger}\left(\delta^{ij} + \gamma^i\gamma^j\right)\overleftrightarrow{\partial_0}\,\psi^j. \tag{4.7.151}$$

But not all the components of $\psi^i$ are independent due to the constraint in (4.7.148). Thus we may introduce a field with components $\zeta^j(x)$, and set

$$\psi^i(x) = \beta^{ij}\zeta^j(x), \qquad \left[\frac{\partial^i}{i} + \left(\frac{\gamma \cdot \nabla}{i} - m\right)\gamma^i\right]\psi^i(x) = 0, \tag{4.7.152}$$

from which (4.7.151) becomes

$$\mathscr{L}\big|_{\text{time der.}} = -\frac{1}{2i}\,\psi^{i\dagger}\left(\delta^{ij} + \gamma^i\gamma^j\right)\beta^{jk}\overleftrightarrow{\partial_0}\zeta^k(x), \qquad \psi^{i\dagger} = (\beta^{ij}\zeta^j(x))^\dagger, \tag{4.7.153}$$

and we may vary the components of $\zeta^k(x)$ independently.

We now use the identity

$$\left(\delta^{ij} + \gamma^i\gamma^j\right)\left[\delta^{jk} + \frac{1}{3}\left(\gamma^j\gamma^k + \gamma^j\frac{\overleftarrow{\partial^k}}{im} - \gamma^k\frac{\overleftarrow{\partial^j}}{im} + \frac{2}{m^2}\frac{\overleftarrow{\partial^j}}{i}\frac{\overleftarrow{\partial^k}}{i}\right)\right]$$

$$= \delta^{ik} - \left[\frac{\overleftarrow{\partial^i}}{i} + \gamma^i\left(\frac{\gamma \cdot \overleftarrow{\nabla}}{i} - m\right)\right]\left(\frac{1}{3m}\gamma^k - \frac{2}{3m^2}\overleftarrow{\partial^k}\right), \tag{4.7.154}$$

together with the constraint (4.7.148) as applied to $\psi^{i\dagger}$

$$\psi^{i\dagger}\left[\frac{\overleftarrow{\partial^i}}{i} + \gamma^i\left(\frac{\gamma \cdot \overleftarrow{\nabla}}{i} - m\right)\right] = 0, \tag{4.7.155}$$

to infer from (4.7.153), that the canonical conjugate momentum of $\zeta_c^j(x)$ is simply $i\,\psi_c^{j\dagger}$. Thus the equal-time anti-commutation relations of $\zeta_c^k$ and of $i\,\psi_b^{j\dagger}$ is simply given by

$$\{\zeta_c^k(x), i\,\psi_b^{j\dagger}(x')\} = i\,\delta_{cb}\,\delta^{kj}\,\delta^3(\mathbf{x}-\mathbf{x}'),\tag{4.7.156}$$

and upon multiplying it by $(-i)\beta_{ac}^{ik}$, the following equal-time anti-commutation relations emerge

$$\{\psi_a^i(x), \psi_b^{j\dagger}(x')\} = \beta_{ab}^{ij}\,\delta^3(\mathbf{x}-\mathbf{x}'),\qquad x^0 = x'^0$$

$$\{\psi_a^i(x), \overline{\psi}_b^j(x')\} = (\beta^{ij}\gamma^0)_{ab}\,\delta^3(\mathbf{x}-\mathbf{x}'),\quad x^0 = x'^0,\tag{4.7.157}$$

generalizing that of the Dirac theory in (3.5.11) by taking into account of the constraint in (4.7.152). Since $\psi^0 = \gamma^0\gamma^i\psi^i$ this gives rise to the anti-commutations relations of all the components $\psi^\mu$.

Accordingly, as for the Dirac theory in (3.5.12), we have, from (4.7.154), (4.7.155), (4.7.157), (4.7.145), the following general anti-commutation relations[16]

$$\{\psi_a^\mu(x), \overline{\psi}_b^\nu(x')\} = \frac{1}{i}\,\rho_{ab}^{\mu\nu}\!\left(\frac{\partial}{i}\right)\Delta(x-x'),\tag{4.7.158}$$

$$\rho^{\mu\nu}\!\left(\frac{\partial}{i}\right) = \left(-\frac{\gamma\partial}{i}+m\right)\!\left(\eta^{\mu\nu}-\frac{\partial^\mu\partial^\nu}{m^2}\right)+\frac{1}{3}\!\left(\gamma^\mu+\frac{\partial^\mu}{im}\right)\!\left(\frac{\gamma\partial}{i}+m\right)\!\left(\gamma^\nu+\frac{\partial^\nu}{im}\right).\tag{4.7.159}$$

In the momentum description of the expression of the propagator in (4.7.145), let

$$P_+{}^{\mu\nu}(p) = \rho^{\mu\nu}(p),\quad P_-{}^{\mu\nu}(p) = \rho^{\mu\nu}(-p),\quad \text{for } p^0 = +\sqrt{\mathbf{p}^2+m^2}.\tag{4.7.160}$$

We reconsider the polarization vectors introduced for spin 1 in (4.7.128), (4.7.129), (4.7.130) and (4.7.131), and use the properties

$$\eta^{\mu\nu}+\frac{p^\mu p^\nu}{m^2} = \sum_{\lambda=-1,0,1} e_\lambda^\mu e_\lambda^{\nu*} = \sum_{\lambda=-1,0,1} e_\lambda^{\mu*} e_\lambda^\nu,\tag{4.7.161}$$

---

[16]Note that from (4.7.149)

$$\beta^{ij}\gamma^0 = \gamma^0\Big[\delta^{ij}+\frac{1}{3}\Big(\gamma^i\gamma^j+\gamma^i\frac{\partial^j}{im}-\gamma^j\frac{\partial^i}{im}+\frac{2}{m^2}\frac{\partial^i}{i}\frac{\partial^j}{i}\Big)\Big]$$

$$= \gamma^0\Big(\delta^{ij}-\frac{\partial^i\partial^j}{m^2}\Big)+\frac{1}{3}\Big(\gamma^i+\frac{\partial^i}{im}\Big)\big(-\gamma^0\big)\Big(\gamma^j+\frac{\partial^j}{im}\Big).$$

$$\gamma^\mu + \frac{p^\mu}{m} = \sum_{\lambda=-1,0,1} e^\mu_\lambda e^{\nu*}_\lambda \gamma_\nu + (-\gamma p + m)\frac{p^\mu}{m^2}, \tag{4.7.162}$$

$$(\gamma p + m) = \gamma^5(-\gamma p + m)\gamma^5, \tag{4.7.163}$$

$$\frac{(-\gamma p + m)}{2m} = \sum_\sigma u(\sigma, \mathbf{p})\, \bar{u}(\sigma, \mathbf{p}), \tag{4.7.164}$$

with the last relation given in (I.21), to write

$$P_+{}^{\mu\nu} = \left( \sum_\sigma \sum_\lambda e^\mu{}_\lambda u(\sigma)\bar{u}(\sigma)e^{\nu*}{}_\lambda - \frac{1}{3}\sum_\sigma \sum_{\lambda,\lambda'} e^\mu{}_\lambda A(\sigma,\lambda)\bar{A}(\sigma,\lambda')e^{\nu*}{}_\lambda \right), \tag{4.7.165}$$

suppressing the momentum variable for simplicity of writing, where

$$A(\sigma,\lambda) = e^{\mu*}_\lambda \gamma_\mu \gamma^5 u(\sigma), \qquad \bar{A}(\sigma,\lambda) = -\bar{u}(\sigma)\gamma^5\gamma_\mu e^\mu_\lambda. \tag{4.7.166}$$

The two terms in (4.7.165) may be combined to yield

$$P_+{}^{\mu\nu} = \sum_\xi U^\mu(\xi)\bar{U}^\nu(\xi), \tag{4.7.167}$$

where $\xi = -3/2, -1/2, 1/2, 3/2$,

$$U^\mu(3/2) = e^\mu_+ u(+1), \tag{4.7.168}$$

$$U^\mu(1/2) = \frac{1}{\sqrt{3}} e^\mu_+ u(-1) + \sqrt{\frac{2}{3}} e^\mu_0 u(+1), \tag{4.7.169}$$

$$U^\mu(-1/2) = \frac{1}{\sqrt{3}} e^\mu_- u(+1) - \sqrt{\frac{2}{3}} e^\mu_0 u(-1), \tag{4.7.170}$$

$$U^\mu(-3/2) = e^\mu_- u(-1), \tag{4.7.171}$$

establishing the spin 3/2 character of the field. A similar analysis may be carried out for $P_-{}^{\mu\nu}$, with the Dirac spinors replaced by the $v$ ones.

For a massless field $(m = 0)$, the Lagrangian density (4.7.146) becomes[17]

$$\mathcal{L} = -\frac{1}{2i}\,\bar{\psi}_\mu\left(\eta^{\mu\nu}\gamma\overleftrightarrow{\partial} - (\gamma^\mu\overleftrightarrow{\partial}^\nu + \gamma^\nu\overleftrightarrow{\partial}^\mu) + \gamma^\mu(-\gamma\overleftrightarrow{\partial})\gamma^\nu\right)\psi_\nu + \bar{K}^\mu\psi_\mu + \bar{\psi}^\mu K_\mu. \tag{4.7.172}$$

---

[17]The massless Rarita-Schwinger field treatment is based on Manoukian [7].

For $K = 0$, $\overline{K} = 0$, the Lagrangian density, up to a total derivative, is invariant under a gauge transformation

$$\psi^{\mu}_{a}(x) \rightarrow \psi^{\mu}_{a}(x) + \partial^{\mu}\Lambda_{a}(x). \tag{4.7.173}$$

We may work in a Coulomb-like gauge defined by

$$\partial_{i}\psi^{i}_{a}(x) = 0. \tag{4.7.174}$$

We may thus introduce fields $U^{i}_{b}$, $\rho_{b}$, and set $(\boldsymbol{\gamma} \cdot \boldsymbol{\nabla} \equiv \gamma^{j}\partial_{j})$

$$\psi^{i}_{a} = \alpha^{ij}_{ab} U^{j}_{b} - \frac{1}{2}\left(\gamma^{i} - \frac{\boldsymbol{\gamma} \cdot \boldsymbol{\nabla}\partial^{i}}{\nabla^{2}}\right)_{ab}\rho_{b}, \tag{4.7.175}$$

where

$$\alpha^{ij} = \left(\delta^{ij} - \frac{\partial^{i}\partial^{j}}{\nabla^{2}}\right) + \frac{1}{2}\left(\gamma^{i} - \frac{\boldsymbol{\gamma} \cdot \boldsymbol{\nabla}\partial^{i}}{\nabla^{2}}\right)\left(\gamma^{j} - \frac{\boldsymbol{\gamma} \cdot \boldsymbol{\nabla}\partial^{j}}{\nabla^{2}}\right), \tag{4.7.176}$$

$$= -\frac{1}{2}\left(\delta^{is} - \frac{\partial^{i}\partial^{s}}{\nabla^{2}}\right)\gamma^{\ell}\gamma^{s}\left(\delta^{\ell j} - \frac{\partial^{\ell}\partial^{j}}{\nabla^{2}}\right), \tag{4.7.177}$$

satisfying

$$\partial_{i}\alpha^{ij} = 0, \quad \partial_{j}\alpha^{ij} = 0, \quad \gamma^{i}\alpha^{ij} = 0, \quad \alpha^{ij}\gamma^{j} = 0, \quad \alpha^{ij}\alpha^{jk} = \alpha^{ik}. \tag{4.7.178}$$

Also $\psi^{i}$ defined in (4.7.175), satisfies the constraint in (4.7.174). One may now vary $U^{j}$, $\rho$, $\psi^{0}$, and as seen below, this will lead not only to field equations but to additional (derived) constraints.

To the above end, it is most convenient to express the field in the form

$$\psi^{\mu} = \delta^{\mu}_{\ 0}\psi^{0} + \delta^{\mu}_{\ i}\left[\alpha^{ij} U^{j} - \frac{1}{2}\left(\gamma^{i} - \frac{\boldsymbol{\gamma} \cdot \boldsymbol{\nabla}}{\nabla^{2}}\partial^{i}\right)\rho\right], \tag{4.7.179}$$

$$\overline{\psi}_{\mu} = \overline{\psi}^{0}\eta_{\mu 0} + \left[\overline{U}^{j}\overleftarrow{\alpha}^{j i} - \frac{1}{2}\overline{\rho}\left(\gamma^{i} - \frac{\boldsymbol{\gamma} \cdot \overleftarrow{\boldsymbol{\nabla}}\overleftarrow{\partial^{i}}}{\overleftarrow{\nabla^{2}}}\right)\right]\eta_{\mu i}. \tag{4.7.180}$$

The variation of the field $\overline{\psi}_{\mu}$ may be written in terms of the variation of $\overline{\psi}^{0}$, $\overline{U}^{j}$, $\overline{\rho}$, by using in the process (4.7.180). This leads from the Lagrangian density to $(\gamma\partial \equiv \gamma^{\mu}\partial_{\mu})$

$$\gamma^{i}\psi^{i} = -\mathrm{i}\frac{\boldsymbol{\gamma} \cdot \boldsymbol{\nabla}}{\nabla^{2}}\gamma^{0}K_{0}, \tag{4.7.181}$$

$$\psi^{0} = \frac{1}{2\nabla^{2}}\frac{\gamma\partial}{\mathrm{i}}K^{0} - \frac{1}{2\nabla^{2}}\left[\frac{\partial^{i}}{\mathrm{i}} + \frac{\boldsymbol{\gamma} \cdot \boldsymbol{\nabla}\gamma^{i}}{\mathrm{i}}\right]\gamma^{0}K_{i}, \tag{4.7.182}$$

$$-\Box \psi^i = -\frac{\gamma \partial}{i} \alpha^{ij} K^j - \frac{\Box}{2\nabla^2}\left[\frac{\partial^i}{i} + \frac{\gamma \cdot \nabla \gamma^i}{i}\right]\gamma^0 K_0. \tag{4.7.183}$$

Thus we derive, in particular, the constraints $\gamma^i \psi^i = 0$, $\psi^0 = 0$, for $K^\mu = 0$.

Upon taking the matrix elements $\langle 0_+ \mid . \mid 0_- \rangle$ of the expressions in (4.7.182), (4.7.183), and using the fact that

$$\langle 0_+ \mid \psi_a^\mu(x) \mid 0_- \rangle = (-i)\frac{\delta}{\delta \overline{K}_{a\mu}(x)} \langle 0_+ \mid 0_- \rangle,$$

we obtain by functional integration

$$\langle 0_+ \mid 0_- \rangle = \exp\left[i \int (dx)(dx')\overline{K}_{a\mu}(x)D_+{}^{\mu\nu}_{ab}(x-x')K_{b\nu}(x'), \tag{4.7.184}$$

where the propagator $D_+{}^{\mu\nu}_{ab}(x-x')$ is given by

$$D_+{}^{\mu\nu}_{ab}(x-x') = \int \frac{(dp)}{(2\pi)^4} e^{ip(x-x')}D_+{}^{\mu\nu}_{ab}(p), \tag{4.7.185}$$

with $(\gamma p = \gamma^\mu p_\mu, \ \gamma \cdot \mathbf{p} = \gamma^i p^i)$

$$D_+{}^{ij}(p) = \frac{(-\gamma p)}{p^2 - i\epsilon} \alpha^{ij}(p) = \frac{1}{p^2 - i\epsilon}\left(\delta^{ik} - \frac{p^i p^k}{\mathbf{p}^2}\right)\gamma^\ell\left(-\frac{\gamma p}{2}\right)\gamma^k\left(\delta^{\ell j} - \frac{p^\ell p^j}{\mathbf{p}^2}\right), \tag{4.7.186}$$

$$\alpha^{ij}(p) = -\frac{1}{2}\left(\delta^{ij} - \frac{p^i p^k}{\mathbf{p}^2}\right)\gamma^\ell \gamma^k\left(\delta^{\ell j} - \frac{p^\ell p^j}{\mathbf{p}^2}\right), \tag{4.7.187}$$

$$D_+{}^{00}(p) = \frac{1}{2\mathbf{p}^2}(\gamma p), \tag{4.7.188}$$

$$D_+{}^{0i}(p) = \frac{1}{2\mathbf{p}^2}(p^i + \gamma \cdot \mathbf{p}\gamma^i)\gamma^0, \tag{4.7.189}$$

$$D_+{}^{i0}(p) = -\frac{1}{2\mathbf{p}^2}(p^i + \gamma \cdot \mathbf{p}\gamma^i)\gamma^0. \tag{4.7.190}$$

$\epsilon \to 0$. Due to the presence of the denominator $(p^2 - i\epsilon)$ in (4.7.186), only the component $D_+{}^{ij}(p)$ has a propagation characteristic.

For $K^\mu = 0, \overline{K}^\mu = 0$, we have, in particular, two constraints on the spacial components $\psi^i$ given by $\gamma^i \psi^i = 0$, and the adopted Coulomb-like gauge $\partial^i \psi^i = 0$. These constraints are taken care by the projection operator $\alpha^{ij}$ as is evident in (4.7.175), (4.7.176), (4.7.177) and (4.7.178). The equal-time anti-commutation relations may be then expressed as

$$\{\psi^i(x), \psi^{j\dagger}(x')\} = -\frac{1}{2}\left(\delta^{ij} - \frac{\partial^i \partial^k}{\nabla^2}\right)\gamma^\ell \gamma^k \left(\delta^{\ell j} - \frac{\partial^\ell \partial^j}{\nabla^2}\right)\delta^3(\mathbf{x} - \mathbf{x}'), \quad x^0 = x'^0.$$
$$(4.7.191)$$

Consider the numerator of the propagator $D_+{}^{ij}(p)$ in (4.7.186), and set

$$\left(\delta^{ik} - \frac{p^i p^k}{\mathbf{p}^2}\right)\gamma^\ell\left(-\frac{\gamma p}{2}\right)\gamma^k \left(\delta^{\ell j} - \frac{p^\ell p^j}{\mathbf{p}^2}\right) = P_+{}^{ij}(p), \quad p^0 = |\mathbf{p}|. \quad (4.7.192)$$

By working in the chiral representation, the latter is explicitly given by (see Problem 4.14)

$$P_+{}^{ij}(p) = \sum_{\xi=\pm 3/2} U^i(\xi)\,\overline{U}^j(\xi), \qquad U^i(\pm 3/2) = e_\pm^i\, u(\pm). \quad (4.7.193)$$

### 4.7.5  Spin 2

We consider a symmetric Hermitian symmetric tensor field $U_{\mu\nu} = U_{\nu\mu}$, with $m \neq 0$, satisfying the equations

$$(-\Box + m^2)\, U_{\mu\nu}(x) = 0, \quad \partial^\mu U_{\mu\nu}(x) = 0, \quad U^\mu{}_\mu(x) \equiv U(x) = 0. \quad (4.7.194)$$

We will show that this system describes a massive spin 2 particle. The above equations may be derived from the single equation

$$(-\Box + m^2)\, U_{\alpha\beta} + \partial_\alpha \partial_\sigma U^\sigma{}_\beta + \partial_\beta \partial_\sigma U^\sigma{}_\alpha - \partial_\alpha \partial_\beta U$$
$$- \eta_{\alpha\beta}\left[(-\Box + m^2)\, U + \partial_\mu \partial_\nu U^{\mu\nu}\right] = T_{\alpha\beta}, \quad (4.7.195)$$

when the external source $T_{\alpha\beta} = T_{\beta\alpha}$ is set equal to zero.

By comparing the equations obtained from the above equation, one by taking its $\partial^\alpha$ derivative, and the one obtained from multiplying it by $\eta^{\alpha\beta}$, we obtain

$$\partial^\alpha U_{\alpha\beta} = \frac{1}{m^2}\left[\frac{1}{2}\left(\eta_{\mu\alpha}\eta_{\nu\beta} + \eta_{\nu\alpha}\eta_{\mu\beta}\right)\partial^\alpha - \frac{1}{3}\eta_{\mu\nu}\partial_\beta - \frac{2}{3}\frac{\partial_\mu \partial_\nu}{m^2}\partial_\beta\right]T^{\mu\nu},$$
$$(4.7.196)$$

$$U = -\frac{2}{3\,m^2}\left[\frac{1}{2}T^\alpha{}_\alpha + \frac{\partial_\alpha \partial_\beta}{m^2}T^{\alpha\beta}\right], \quad (4.7.197)$$

$$(-\Box + m^2)\, U_{\alpha\beta} = \left[\frac{1}{2}\rho_{\alpha\mu}\rho_{\beta\nu} + \frac{1}{2}\rho_{\alpha\nu}\rho_{\beta\mu} - \frac{1}{3}\rho_{\alpha\beta}\rho_{\mu\nu}\right]T^{\mu\nu}, \quad (4.7.198)$$

leading to the set of equations in (4.7.194) for $T^{\mu\nu} = 0$, where

$$\rho_{\mu\nu} = \left(\eta_{\mu\nu} - \frac{\partial_\mu \partial_\nu}{m^2}\right). \tag{4.7.199}$$

The field equations in (4.7.195) may be obtained from the Lagrangian density

$$\mathcal{L} = -\frac{1}{2} \partial^\sigma U_{\mu\nu} \, \partial_\sigma U^{\mu\nu} + \partial_\mu U^{\mu\nu} \, \partial_\sigma U^\sigma{}_\nu - \partial_\sigma U^{\sigma\mu} \, \partial_\mu U$$

$$+\frac{1}{2} \partial^\mu U \, \partial_\mu U - \frac{m^2}{2} \left(U^{\mu\nu} \, U_{\mu\nu} - U\,U\right) + T^{\mu\nu} \, U_{\mu\nu}. \tag{4.7.200}$$

Upon writing the vacuum expectation values of (4.7.198) in the form

$$\langle\, 0_+ |\, U^{\mu\nu}(x)\, |\, 0_-\rangle = \int (\mathrm{d}x')\, \Gamma_+^{\mu\nu;\sigma\lambda}(x - x')\, T_{\sigma\lambda}(x')\langle\, 0_+ |\, 0_-\rangle, \tag{4.7.201}$$

the propagator $\Gamma_+^{\mu\nu;\sigma\rho}(x - x')$ is readily extracted to be

$$\Gamma_+^{\mu\nu;\sigma\lambda}(x - x') = \int \frac{(\mathrm{d}p)}{(2\pi)^4}\, \mathrm{e}^{\mathrm{i}p\,(x-x')}\, \Gamma_+^{\mu\nu;\sigma\lambda}(p), \tag{4.7.202}$$

where

$$\Gamma_+^{\mu\nu;\sigma\lambda}(p) = \frac{\frac{1}{2}\rho^{\mu\sigma}(p)\rho^{\nu\lambda}(p) + \frac{1}{2}\rho^{\mu\lambda}(p)\rho^{\nu\sigma}(p) - \frac{1}{3}\rho^{\mu\nu}(p)\rho^{\sigma\lambda}(p)}{p^2 + m^2 - \mathrm{i}\epsilon}, \tag{4.7.203}$$

$$\rho^{\mu\nu}(p) = \left(\eta^{\mu\nu} + \frac{p^\mu p^\nu}{m^2}\right), \tag{4.7.204}$$

and the vacuum-to-vacuum transition amplitude takes the usual form

$$\langle\, 0_+ |\, 0_-\rangle = \exp\left[\frac{\mathrm{i}}{2} \int (\mathrm{d}x)(\mathrm{d}x')\, T^{\alpha\beta}(x)\, \Gamma_{+\alpha\beta;\mu\nu}(x - x')\, T^{\mu\nu}(x')\right]. \tag{4.7.205}$$

Consider the numerator of the propagator $\Gamma_{+\alpha\beta;\mu\nu}(p)$ in (4.7.203)/(4.7.205), and set

$$P^{\alpha\beta;\mu\nu}(p) = \frac{1}{2}\rho^{\alpha\mu}(p)\rho^{\beta\nu}(p) + \frac{1}{2}\rho^{\alpha\nu}(p)\rho^{\beta\mu}(p) - \frac{1}{3}\rho^{\alpha\beta}(p)\rho^{\mu\nu}(p), \tag{4.7.206}$$

for a particle on the mass shell $p^0 = \sqrt{\mathbf{p}^2 + m^2}$. Using the completeness relation in (4.7.129), the above expression may be rewritten as

$$P^{\alpha\beta;\mu\nu} = \sum_{\lambda,\lambda'}\left[\frac{1}{2}\left(\mathrm{e}_\lambda^\alpha \mathrm{e}_\lambda^{\mu*} \mathrm{e}_{\lambda'}^\beta \mathrm{e}_{\lambda'}^{\nu*} + \mathrm{e}_\lambda^\alpha \mathrm{e}_\lambda^{\nu*} \mathrm{e}_{\lambda'}^\beta \mathrm{e}_{\lambda'}^{\mu*}\right) - \frac{1}{3}\mathrm{e}_\lambda^\alpha \mathrm{e}_\lambda^{\beta*} \mathrm{e}_{\lambda'}^{\mu*} \mathrm{e}_{\lambda'}^\nu\right]. \tag{4.7.207}$$

Finally the conditions $e_{\pm}^{\mu*} = -e_{\mp}^{\mu}$, $e_0^{\mu*} = -e_0^{\mu}$ given in (4.7.131), give rise to the decomposition

$$P^{\alpha\beta;\mu\nu} = \sum_{\xi} u^{\alpha\beta}(\xi)\, u^{\mu\nu\dagger}(\xi), \qquad \xi = -2, -1, 0, 1, 2, \tag{4.7.208}$$

$$u^{\alpha\beta}(2) = e_+^{\alpha}\, e_+^{\beta}, \qquad u^{\alpha\beta}(-2) = e_-^{\alpha}\, e_-^{\beta}, \tag{4.7.209}$$

$$u^{\alpha\beta}(1) = \frac{1}{\sqrt{2}}\left(e_0^{\alpha}\, e_+^{\beta} + e_+^{\alpha}\, e_0^{\beta}\right), \qquad u^{\alpha\beta}(-1) = \frac{1}{\sqrt{2}}\left(e_0^{\alpha}\, e_-^{\beta} + e_-^{\alpha}\, e_0^{\beta}\right), \tag{4.7.210}$$

$$u^{\alpha\beta}(0) = \sqrt{\frac{2}{3}}\, e_0^{\alpha}\, e_0^{\beta} - \frac{1}{\sqrt{6}}\, e_+^{\alpha}\, e_-^{\beta} - \frac{1}{\sqrt{6}}\, e_-^{\alpha}\, e_+^{\beta}, \tag{4.7.211}$$

describing the 5 spin states of a massive spin 2 particle.

In the massless case, the Lagrangian density in (4.7.200) becomes

$$\mathscr{L} = -\frac{1}{2}\, \partial^{\sigma} U_{\mu\nu}\, \partial_{\sigma} U^{\mu\nu} + \partial_{\mu} U^{\mu\nu}\, \partial_{\sigma} U^{\sigma}{}_{\nu} - \partial_{\sigma} U^{\sigma\mu}\, \partial_{\mu} U + \frac{1}{2}\, \partial^{\mu} U\, \partial_{\mu} U + T^{\mu\nu}\, U_{\mu\nu}. \tag{4.7.212}$$

For $T^{\mu\nu} = 0$, the Lagrangian density is, up to a total derivative, invariant under the gauge transformation

$$U^{\mu\nu} \to U^{\mu\nu} + \partial^{\mu} \Lambda^{\nu} + \partial^{\nu} \Lambda^{\mu}, \tag{4.7.213}$$

for arbitrary $\Lambda^{\mu}$. A covariant gauge treatment of the massless spin 2 dealing with the graviton will be carried out in Sect. 2.3.1 of Volume II. On the other hand, in a Coulomb-like gauge we impose the constraint

$$\partial_i U^{i\nu} = 0, \qquad \nu = 0, 1, 2, 3. \tag{4.7.214}$$

The analysis becomes simplified by expressing the field in the following manner

$$U^{\mu\nu} = \eta^{\mu i}\eta^{\nu j}\, U^{ij} - \left(\eta^{\mu 0}\eta^{\nu i} + \eta^{\mu i}\eta^{\nu 0}\right) U^{0i} + \eta^{\mu 0}\eta^{\nu 0} U^{00}, \tag{4.7.215}$$

and set

$$U^{ij} = \frac{1}{2}\left(\pi^{ik}\pi^{j\ell} + \pi^{i\ell}\pi^{jk}\right) H^{k\ell}, \qquad U^{0i} = \pi^{ij}\, \varphi^j, \tag{4.7.216}$$

defined in terms of new unconstrained fields $H^{k\ell}$, $\varphi^j$, and

$$\pi^{ij} = \left(\delta^{ij} - \frac{\partial^i \partial^j}{\nabla^2}\right), \qquad \pi^{ij}\pi^{jk} = \pi^{ik}, \qquad \partial_i \pi^{ij} = 0. \tag{4.7.217}$$

The gauge condition (4.7.214) is automatically satisfied.

With the field components expressed as above, variations of the Lagrangian density in (4.7.212) with respect to $\varphi^i$, $U^{00}$, $H^{ij}$ lead upon defining

$$\pi^{ij\,k\ell} = \frac{1}{2}\left(\pi^{ik}\pi^{j\ell} + \pi^{i\ell}\pi^{jk} - \pi^{ij}\pi^{k\ell}\right), \tag{4.7.218}$$

to the following equations

$$U^{ii} = -\frac{1}{\nabla^2}\,T^{00}, \tag{4.7.219}$$

$$U^{00} = -\frac{1}{2\,\nabla^2}\left[\frac{\Box}{\nabla^2}\,T^{00} + \pi^{ij}T^j\right], \tag{4.7.220}$$

$$U^{0i} = -\frac{1}{\nabla^2}\,\pi^{ik}\,T^{0k}, \tag{4.7.221}$$

$$-\Box\,U^{ij} = \pi^{ijk\ell}\,T^{k\ell} + \frac{1}{2}\,\pi^{ij}\frac{\Box}{\nabla^2}\,T^{00}. \tag{4.7.222}$$

Thus we have derived additional constraints: $U^{ii} = 0$, $U^{00} = 0$, $U^{0i} = 0$, for $T^{\mu\nu} = 0$.

Upon writing the vacuum expectation values of (4.7.220), (4.7.221) and 4.7.222) in the form

$$\langle\,0_+\,|\,U^{\mu\nu}(x)\,|\,0_-\rangle = \int (dx')\,D_+^{\mu\nu\,;\sigma\rho}(x - x')\,T_{\sigma\rho}(x')\langle\,0_+\,|\,0_-\rangle, \tag{4.7.223}$$

the propagator $D_+^{\mu\nu\,;\sigma\rho}(x - x')$ is readily extracted to be

$$D_+^{\mu\nu\,;\sigma\rho}(x - x') = \int \frac{(dp)}{(2\pi)^4}\,e^{ip\,(x - x')}\,D_+^{\mu\nu\,;\sigma\rho}(p), \tag{4.7.224}$$

where $\Delta_+^{\mu\nu\,;\sigma\rho}(p)$ is symmetric in $(\mu, \nu)$, and $(\sigma, \rho)$, as well as in the interchange $(\mu, \nu) \leftrightarrow (\sigma, \rho)$, and

$$D_+^{ij;k\ell}(p) = \frac{\pi^{\,ij;k\ell}(\mathbf{p})}{p^2 - i\,\epsilon}, \qquad \epsilon \to 0, \qquad \pi^{ij}(\mathbf{p}) = \left(\delta^{ij} - \frac{p^i p^j}{\mathbf{p}^2}\right), \tag{4.7.225}$$

$$\pi^{\,ij;k\ell}(\mathbf{p}) = \frac{1}{2}\left[\pi^{ik}(\mathbf{p})\pi^{j\ell}(\mathbf{p}) + \pi^{i\ell}(\mathbf{p})\pi^{jk}(\mathbf{p}) - \pi^{ij}(\mathbf{p})\pi^{k\ell}(\mathbf{p})\right], \tag{4.7.226}$$

$$D_+^{ij;00}(p) = \frac{1}{2\,\mathbf{p}^2}\,\pi^{ij}(\mathbf{p}), \qquad D_+^{00;00}(p) = \frac{1}{2\,\mathbf{p}^2}\frac{p^2}{\mathbf{p}^2}, \qquad (4.7.227)$$

$$D_+^{0i;0k}(p) = -\frac{1}{2\,\mathbf{p}^2}\,\pi^{ik}(\mathbf{p}), \qquad D_+^{00;0i}(p) = 0, \qquad D_+^{0i;jk}(p) = 0. \quad (4.7.228)$$

The vacuum-to-vacuum transition amplitude has the usual form

$$\langle\,0_+\,|\,0_-\,\rangle = \exp\!\left[\frac{i}{2}\int (dx)(dx')T_{\mu\nu}(x)D_+^{\mu\nu;\sigma\rho}(x-x')T_{\sigma\rho}(x')\right], \qquad (4.7.229)$$

Consider the numerator $\pi^{ij;k\ell}(\mathbf{p})$ of the propagating part of the propagator in (4.7.225) for $p^0 = |\mathbf{p}|$. Using the polarization 3-vectors $\mathbf{e}_\pm$ in (4.7.130), with

$$\mathbf{e}_\lambda \cdot \mathbf{e}_{\lambda'}^* = \delta_{\lambda\lambda'}, \qquad \mathbf{p}\cdot\mathbf{e}_\pm = 0, \qquad \pi^{ij} = \sum_{\lambda=\pm} e_\lambda^i e_\lambda^{j*} = \sum_{\lambda=\pm} e_\lambda^{i*} e_\lambda^j, \qquad (4.7.230)$$

with the last relation defining a completeness relation in 3D, we may rewrite

$$\pi^{ij;k\ell}(\mathbf{p}) = \frac{1}{2}\sum_{\lambda,\lambda'}\left(e_\lambda^i\,e_\lambda^{k*}\,e_{\lambda'}^j e_{\lambda'}^{\ell*} + e_\lambda^i\,e_\lambda^{\ell*}\,e_{\lambda'}^j e_{\lambda'}^{k*} - e_\lambda^i\,e_\lambda^{j*}\,e_{\lambda'}^k e_{\lambda'}^{\ell*}\right). \qquad (4.7.231)$$

Using the following properties $\mathbf{e}_\pm^* = -\mathbf{e}_\mp$, given in the first relations in (4.7.131), the above expression simplifies to

$$\pi^{ij;k\ell}(\mathbf{p}) = \sum_{\xi=\pm 2} \varepsilon_\xi^{ij}\varepsilon_\xi^{k\ell*}, \qquad \varepsilon_{\pm 2}^{ij} = e_\pm^i e_\pm^j, \qquad (4.7.232)$$

describing the two helicity states of a massless spin 2 particle. The latter is identified with the graviton.

For a conserved source $\partial_\mu T^{\mu\nu}(x) = 0$, i.e., in the momentum description

$$p^0\,T^{0\nu}(p) = p^i\,T^{i\nu}(p), \qquad (4.7.233)$$

the expression for $\langle\,0_+\,|\,0_-\,\rangle$, simplifies to some extent and leads to a Lorentz invariant expression for it. To this end the integrand multiplying the $i$ factor, in the exponent in (4.7.172) including the 1/2 factor, may be rewritten as

$$\frac{1}{2}\left[\,T^{ij}D_+^{ij;k\ell}T^{k\ell} + 4\,T_{0i}D_+^{0i;0k}T_{0k} + T_{00}D_+^{00;00}T_{00} + 2\,T_{ij}D_+^{ij;00}T_{00}\,\right]. \qquad (4.7.234)$$

In the momentum description, we have

$$T^{ij}D_+^{ij;k\ell}T^{k\ell} = T^{ij}\frac{\pi^{ijk\ell}}{p^2 - i\epsilon}T^{k\ell} = \left(T^{\mu\nu}T_{\mu\nu} - \frac{1}{2}T^\mu{}_\mu T^\nu{}_\nu\right)\frac{1}{p^2 - i\epsilon}$$

$$+ \left(-T^{ii}T^{00} - T^{00}T^{00} + T^{00}\frac{p^2}{2\,\mathbf{p}^2}T^{00} + 2T^{0i}T^{0i}\right)\frac{1}{\mathbf{p}^2}, \qquad (4.7.235)$$

$$4\,T_{0i}\,D_+^{0i;0k}\,T_{0k} \;=\; -4\,T^{0i}\,\frac{\pi^{ik}}{2\,\mathbf{p}^2}\,T^{0k}$$

$$= \left(-\,2\,T^{0k}T^{0k} - 2\,T^{00}\frac{p^2}{\mathbf{p}^2}T^{00} + 2\,T^{00}T^{00}\right)\frac{1}{\mathbf{p}^2}, \tag{4.7.236}$$

$$T_{00}\,D_+^{00;00}\,T_{00} \;=\; T^{00}\,\frac{p^2}{2\,\mathbf{p}^4}\,T^{00}, \tag{4.7.237}$$

$$2\,T_{ij}\,D_+^{ij;00}\,T_{00} = 2\,T^{00}\,\frac{\pi^{ij}}{2\,\mathbf{p}^2}\,T^{ij} = \left(T^{i\,i}T^{00} + T^{00}\frac{p^2}{\mathbf{p}^2}T^{00} - T^{00}T^{00}\right)\frac{1}{\mathbf{p}^2}. \tag{4.7.238}$$

Hence from (4.7.234), (4.7.235), (4.7.236), (4.7.237), (4.7.238), and (4.7.229),

$$\langle\,0_+\,|\,0_-\rangle = \exp\!\left[\frac{\mathrm{i}}{2}\int(\mathrm{d}x)(\mathrm{d}x')\big[T^{\mu\nu}(x)T_{\mu\nu}(x') - \frac{1}{2}\,T^{\mu}{}_{\mu}(x)\,T^{\nu}{}_{\nu}(x')\big]D_+(x-x')\right], \tag{4.7.239}$$

$$D_+(x-x') = \int\frac{(\mathrm{d}p)}{(2\pi)^4}\frac{e^{ip\,(x-x')}}{p^2 - \mathrm{i}\,\epsilon}. \tag{4.7.240}$$

## 4.8  Further Illustrations and Applications of the QDP

Consider the simple Lagrangian density involving a neutral scalar field

$$\mathscr{L}(x,\lambda) \;=\; -\frac{1}{2}\,\partial_\mu\phi(x)\,\partial^\mu\phi(x) - \frac{m^2}{2}\,\phi^2(x) + \frac{\lambda}{4}\,\phi^4(x) + K(x)\,\phi(x), \tag{4.8.1}$$

where $K(x)$ is an external source, $\lambda$ is a coupling parameter, with the interaction Lagrangian defined by the term

$$\mathscr{L}_I(x,\lambda) \;=\; \frac{\lambda}{4}\,\phi^4(x). \tag{4.8.2}$$

Clearly, the canonical conjugate momentum $\pi(x)$ of the field $\phi(x)$ is given by $-\partial^0\phi(x)$, and the field equation is obtained by taking the functional derivative of the action with respect to $\phi(x)$, as given in (4.3.36), and using the simple rules (4.3.33), (4.3.34) and (4.3.35),

$$(-\Box + m^2)\,\phi(x) - \lambda\,\phi^3(x) \;=\; K(x). \tag{4.8.3}$$

The QDP in (4.6.37) for the derivative of the vacuum-to-vacuum transition amplitude $\langle\,0_+\,|\,0_-\rangle$, as given in the Summary at the end of Sect. 4.6., with respect to $\lambda$, gives

$$\frac{\partial}{\partial \lambda} \langle\, 0_+ \mid 0_- \rangle = i \langle\, 0_+ \mid \int (dx) \frac{1}{4} \phi^4(x) \mid 0_- \rangle. \tag{4.8.4}$$

Since $\phi(x)$ is an independent field, repeated applications of (4.6.38) gives

$$\langle\, 0_+ | \phi^4(x) \mid 0_- \rangle = \left(-i\frac{\delta}{\delta K(x)}\right)^4 \langle\, 0_+ \mid 0_- \rangle, \tag{4.8.5}$$

in a limiting sense of a time-ordered product whose nature will be discussed in the next section. Here we have used the fact that

$$\frac{\delta}{\delta K(x)} \int (dx') \mathcal{L}(x', \lambda) = \phi(x). \tag{4.8.6}$$

Upon setting

$$\widehat{\mathcal{L}}_I(x, \lambda) = \frac{\lambda}{4} \left(-i\frac{\delta}{\delta K(x)}\right)^4, \tag{4.8.7}$$

we may use (4.8.4) and (4.8.5) to integrate the former equation over $\lambda$ from 0 to a specific value of $\lambda$ to obtain

$$\langle\, 0_+ \mid 0_- \rangle_\lambda = \exp(i \int (dx) \widehat{\mathcal{L}}_I(x, \lambda)) \langle\, 0_+ \mid 0_- \rangle_{\lambda=0}, \tag{4.8.8}$$

where $\langle\, 0_+ \mid 0_- \rangle_{\lambda=0}$ corresponds to $\langle\, 0_+ \mid 0_- \rangle = \langle\, 0_+ \mid 0_- \rangle_\lambda$ with $\lambda$ set equal to zero. The expression for $\langle\, 0_+ \mid 0_- \rangle_{\lambda=0}$ for the scalar field is given in (4.7.5).

For the purpose of an illustration of a more involved case with a dependent field, consider the Lagrangian density

$$\mathcal{L}(x, \lambda) = \mathcal{L}_0(V^\nu(x)) + \mathcal{L}_0(\psi(x), \overline{\psi}(x)) - \frac{\lambda}{2} \overline{\psi}(x)\psi(x) V^\nu(x) V_\nu(x)$$
$$+ K_\nu(x) V^\nu(x) + \overline{\eta}(x)\psi(x) + \overline{\psi}(x)\eta(x), \tag{4.8.9}$$

where $\mathcal{L}_0(V^\nu(x))$ is the Lagrangian density of a neutral vector field $V^\nu(x)$ of mass $\mu > 0$

$$\mathcal{L}_0(V^\nu(x)) = -\frac{1}{4} \left(\partial_\mu V_\nu - \partial_\nu V_\mu\right)\left(\partial^\mu V^\nu - \partial^\nu V^\mu\right) - \frac{m^2}{2} V^\nu V_\nu, \tag{4.8.10}$$

(see (4.7.17)), and $\mathcal{L}_0(\psi(x), \overline{\psi}(x))$ is the Lagrangian density of the Dirac field in (4.7.11). Also $K_\nu(x)$, $\overline{\eta}(x)$, $\eta(x)$ are external sources coupled to the corresponding fields, with the latter two as Grassmann variables.

The interaction Lagrangian density of the coupling of the vector and the Dirac field is given by

$$\mathscr{L}_I(x, \lambda) = -\frac{\lambda}{2} \overline{\psi}(x) \psi(x) V^\nu(x) V_\nu(x), \tag{4.8.11}$$

and $\lambda > 0$ is a coupling parameter. To be more precise, we have to replace $\overline{\psi}_a(x) \psi_a(x)$, for example, by an antisymmetric average $\frac{1}{2}(\overline{\psi}_a(x)\psi_a(x) - \psi_a(x)\overline{\psi}_a(x))$ and such points will be elaborated upon in the next section.

We note that the Lagrangian density $\mathscr{L}_0(V^\nu(x))$ does not involves the time derivative $\partial_0 V^0$ of the field component $V^0$ of $V^\mu$, hence the canonical conjugate momentum of $V^0$ is zero, and $V^0$ is a *dependent* field.

The field equation for the vector field $V^\nu$ as obtained from (4.3.36), and from the rules (4.3.33)–(4.3.35), is

$$-\partial_\mu(\partial^\mu V^\nu(x) - \partial^\nu V^\mu(x)) + (m^2 + \lambda \overline{\psi}(x)\psi(x)) V^\nu(x) = K^\nu(x), \tag{4.8.12}$$

and the canonical conjugate momenta $\pi^k$ of the components $V^k$, $k = 1, 2, 3$, are given by

$$\pi^k(x) = \frac{\partial \mathscr{L}(x, \lambda)}{\partial(\partial_0 V^k)} = -(\partial^0 V^k - \partial^k V^0). \tag{4.8.13}$$

Upon taking the derivative $\partial_k$ of $\pi^k$, we note that

$$\left[ \nabla^2 V^0(x) - \partial^0(\nabla \cdot \mathbf{V}(x)) \right] = \nabla \cdot \boldsymbol{\pi}(x). \tag{4.8.14}$$

On the other hand, picking up the $\nu = 0$-component of (4.8.12) gives

$$-\left[ \nabla^2 V^0(x) - \partial^0(\nabla \cdot \mathbf{V}(x)) \right] + (m^2 + \lambda \overline{\psi}(x)\psi(x)) V^0(x) = K^0(x), \tag{4.8.15}$$

where we have used the definition $\Box = \nabla^2 - \partial^{02}$.

The last two equations then lead the following equation of constraint for $V^0$

$$(m^2 + \lambda \overline{\psi}(x)\psi(x)) V^0(x) = K^0(x) + \nabla \cdot \boldsymbol{\pi}(x). \tag{4.8.16}$$

The QDP in (4.6.37) gives

$$\frac{\partial}{\partial \lambda} \langle 0_+ | 0_- \rangle = i \langle 0_+ | \int (dx) \frac{\partial}{\partial \lambda} \mathscr{L}(x, \lambda) | 0_- \rangle$$

$$= -\frac{i}{2} \langle 0_+ | \int (dx) \left( \overline{\psi}(x)\psi(x) V^\mu(x) V_\mu(x) \right) | 0_- \rangle. \tag{4.8.17}$$

The field $V^0$ appearing in $\mathscr{L}_I(x, \lambda)$ in (4.8.11) is dependent. Since we will vary the components of $K^\mu(x)$ independently no constraints, such as $\partial_\mu K^\mu(x) = 0$, may be imposed on $K^\mu(x)$.

We use the properties

$$(-i)\frac{\delta}{\delta\overline{\eta}(x)}\int (dx')\,\mathscr{L}(x', \lambda) = -i\,\psi(x), \tag{4.8.18}$$

$$(+i)\frac{\delta}{\delta\eta(x)}\int (dx')\,\mathscr{L}(x', \lambda) = -i\,\overline{\psi}(x), \tag{4.8.19}$$

$$(-i)\frac{\delta}{\delta K^\nu(x)}\int (dx')\,\mathscr{L}(x', \lambda) = -i\,V_\nu(x), \tag{4.8.20}$$

where we recall that the Grassmann variables $\overline{\eta}$, $\eta$ anti-commute with the Dirac fields. Accordingly, by repeated application of (4.6.38), dealing with independent fields, gives

$$(i)\frac{\delta}{\delta\eta(x)}(-i)\frac{\delta}{\delta\overline{\eta}(x)}(-i)\frac{\delta}{\delta K_k(x)}(-i)\frac{\delta}{\delta K^k(x)}\langle\, 0_+ \mid 0_-\rangle$$
$$= \langle\, 0_+ \mid (\overline{\psi}(x)\psi(x)V^k(x)V_k(x)) \mid 0_-\rangle, \tag{4.8.21}$$

for $k = 1, 2, 3,$ as a limiting case of a time ordered product whose nature again will be elaborated upon in the next section.

On the other hand, an application of (4.6.38) for the dependent field $V^0(x)$ gives

$$(-i)\frac{\delta}{\delta K_0(x')}(-i)\frac{\delta}{\delta K^0(x)}\langle\, 0_+ \mid 0_-\rangle = (-i)\frac{\delta}{\delta K_0(x')}\langle\, 0_+ \mid V_0(x) \mid 0_-\rangle$$

$$= \langle\, 0_+ \mid (V^0(x')V_0(x))_+ \mid 0_-\rangle + (-i)\langle\, 0_+ \mid \frac{\delta}{\delta K_0(x')}V_0(x) \mid 0_-\rangle, \tag{4.8.22}$$

where $(\delta/\delta K_0(x'))V_0(x)$ is to be obtained from the constraint equation (4.8.16) by keeping the fields $\overline{\psi}(x)$, $\psi(x)$, and $\partial_k\pi^k(x)$ fixed. To this end, we introduce the following simplifying notations:

$$(-i)\frac{\delta}{\delta K^\mu(x)} \equiv \widehat{V}_\mu, \tag{4.8.23}$$

$$(i)\frac{\delta}{\delta\eta(x)}(-i)\frac{\delta}{\delta\overline{\eta}(x)} \equiv \widehat{S}(x). \tag{4.8.24}$$

From (4.8.16), we then have

$$\langle\, 0_+ \mid \frac{\delta}{\delta K_0(x')}V_0(x) \mid 0_-\rangle = \frac{1}{(m^2 + \lambda\,\widehat{S}(x))}\langle\, 0_+ \mid 0_-\rangle\,\delta^{(4)}(x - x'). \tag{4.8.25}$$

Upon carrying out the functional differentiations in $\widehat{S}(x)$, as applied to (4.8.22), and using the above equation, we obtain

$$\langle\, 0_+|\left(\overline{\psi}(x)\psi(x)V^0(x')V_0(x)\right)_+|\, 0_-\rangle$$

$$= \widehat{S}(x)\left(\widehat{V}^0(x')\widehat{V}_0(x) + i\,\frac{\delta^{(4)}(x-x')}{(m^2+\lambda\widehat{S}(x))}\right)\langle\, 0_+|\, 0_-\rangle, \qquad (4.8.26)$$

where note that $\widehat{S}(x)$, $\widehat{S}(x')$ commute, with each taken as the product defined in (4.8.24).

Finally we set

$$\widehat{\mathscr{L}}_I(x,\lambda) = -\frac{\lambda}{2}\,\widehat{S}(x)\widehat{V}^\mu(x)\widehat{V}_\mu(x), \qquad (4.8.27)$$

and use (4.8.21) and (4.8.26) to rewrite (4.8.17) as

$$\frac{\partial}{\partial\lambda}\langle\, 0_+|\, 0_-\rangle = i\langle\, 0_+|\int(\mathrm{d}x)\,\frac{\partial}{\partial\lambda}\mathscr{L}_I(x)|\, 0_-\rangle = i\left[\int(\mathrm{d}x)\,\frac{\partial}{\partial\lambda}\widehat{\mathscr{L}}_I(x,\lambda)\right.$$

$$\left. - \frac{i}{2}\int(\mathrm{d}x)\int(\mathrm{d}x')\,\delta^{(4)}(x'-x)\,\frac{\widehat{S}(x)}{[m^2+\lambda\widehat{S}(x)]}\,\delta^{(4)}(x-x')\right]\langle\, 0_+|\, 0_-\rangle.$$

$$(4.8.28)$$

By an integration over $\lambda$ from 0 to a specific value for $\lambda$, the following expression emerges for the vacuum-to-vacuum transition amplitude

$$\langle\, 0_+\,|\, 0_-\rangle_\lambda = \exp\left(i\int(\mathrm{d}x)\,\widehat{\mathscr{L}}_I(x,\lambda)\right)F\left(\widehat{S}(x)\right)\langle\, 0_+\,|\, 0_-\rangle_{\lambda=0}, \qquad (4.8.29)$$

where

$$F\left(\widehat{S}(x)\right) = \exp\left(\frac{1}{2}\int(\mathrm{d}x')\!\int(\mathrm{d}x)\,\delta^{(4)}(x'-x)\delta^{(4)}(x-x')\ln\!\left[1+\frac{\lambda}{m^2}\widehat{S}(x)\right]\right), \qquad (4.8.30)$$

$\widehat{S}(x)$ is defined in (4.8.24), and we have used the elementary integral

$$\int_0^\lambda\frac{a\,\mathrm{d}\lambda'}{[b+a\lambda']} = \ln\!\left[1+\lambda\,\frac{a}{b}\right]. \qquad (4.8.31)$$

The expression $\langle\, 0_+|\, 0_-\rangle_{\lambda=0}$ is the product of the corresponding one for the Dirac field as was already derived in (3.3.28), and given (4.7.16), and of the corresponding one for the vector field given in (4.7.22).

Upon comparison of the solution in (4.8.29) for the second example with the one in (4.8.8) for the first example, we learn the following. The integral $\int(\mathrm{d}x)\widehat{\mathscr{L}}_I(x,\lambda)$

for the second example is modified due to the $F$ factor in (4.8.30), and this is as a consequence of the presence of the dependent field $V^0(x)$. This means, in particular, that the action integral gets modified. Such a property is shared by non-abelian gauge theories in most gauges. We note that the double integral in (4.8.30) produces a $\delta^{(4)}(0)$ factor due to the product of delta functions at coincident spacetime points. Finally, we note that the presence of a dependent field is not necessary, in every case, for the presence of a modifying factor $F$ as encountered above beyond the term $\exp(\mathrm{i}\int(\mathrm{d}x)\,\widehat{\mathscr{L}}_I(x,\lambda))$. For example, if we replace the interaction Lagrangian density in (4.8.11), say, by $\lambda\,(\gamma^\mu)_{ab}[\overline{\psi}_a(x),\psi_b(x)]V_\mu/2$, which is linear in the vector field $V_\mu$, no such a modification arises (see Problem 4.16).

## 4.9  Time-Ordered Products, How to Write Down Lagrangians and Setting Up the Solution of Field Theory

This section deals with general aspects of the structure of Lagrangian densities in quantum field theory and on the way one may then proceed to set up the solution of an underlying theory. The way of actually constructing such Lagrangian densities to describe the fundamental interactions of nature will be carried out in remaining chapters. Before doing so, we recapitulate some aspects of time-ordered products. To this end, recall that one may add coupling terms, to a given Lagrangian density $\mathscr{L}(x)$, consisting of the interaction of the underlying fields in a theory, say, of a scalar field $\phi(x)$ and Dirac fields $\overline{\psi}(x)$, $\psi(x)$ with external sources

$$\mathscr{L}(x) \rightarrow \mathscr{L}(x) + K(x)\phi(x) + \overline{\eta}(x)\psi(x) + \overline{\psi}(x)\eta(x), \tag{4.9.1}$$

to generate matrix elements of the fields

$$(-\mathrm{i})\frac{\delta}{\delta\overline{\eta}(x)}\langle\,0_+\mid 0_-\rangle = \langle\,0_+|\psi(x)\mid 0_-\rangle, \tag{4.9.2}$$

$$(+\mathrm{i})\frac{\delta}{\delta\eta(x)}\langle\,0_+\mid 0_-\rangle = \langle\,0_+|\overline{\psi}(x)\mid 0_-\rangle, \tag{4.9.3}$$

$$(-\mathrm{i})\frac{\delta}{\delta K(x)}\langle\,0_+\mid 0_-\rangle = \langle\,0_+|\phi(x)\mid 0_-\rangle. \tag{4.9.4}$$

As a consequence of the Bose-Einstein character of a Lagrangian density, being related to the Hamiltonian density, the Dirac field sources $\overline{\eta}(x)$, $\eta(x')$ anti-commute as well as the functional derivatives with respect to them, i.e.,

$$\left\{\frac{\delta}{\delta\eta(x')},\frac{\delta}{\delta\overline{\eta}(x)}\right\} = 0. \tag{4.9.5}$$

Hence from

$$(+i)\frac{\delta}{\delta\eta_a(x')}(-i)\frac{\delta}{\delta\overline{\eta}_b(x)}\langle\,0_+\mid 0_-\rangle = \langle\,0_+\mid(\overline{\psi}_a(x')\psi_b(x))_+\mid 0_-\rangle, \qquad (4.9.6)$$

$$(-i)\frac{\delta}{\delta\overline{\eta}_b(x)}(+i)\frac{\delta}{\delta\eta_a(x')}\langle\,0_+\mid 0_-\rangle = \langle\,0_+\mid(\psi_b(x)\overline{\psi}_a(x'))_+\mid 0_-\rangle, \qquad (4.9.7)$$

we have the rule that *within* a time-ordered product, the Dirac quantum fields anti-commute, i.e.,

$$(\overline{\psi}_a(x')\psi_b(x))_+ = -(\psi_b(x)\overline{\psi}_a(x'))_+. \qquad (4.9.8)$$

Similarly, one has

$$(\phi(x')\phi(x))_+ = (\phi(x)\phi(x'))_+, \qquad (4.9.9)$$

$$(\phi(x')\psi_b(x))_+ = (\psi_b(x)\phi(x'))_+. \qquad (4.9.10)$$

The general rule is that *within* a time-ordered product, as far as their commutativity is concerned, two fields may be commuted if at least one of them is a Bose-Einstein field or may be anti-commuted if they both Fermi-Dirac fields.

In particular, the following time-ordered products take the form

$$(\psi_b(x)\overline{\psi}_a(x'))_+ = \psi_b(x)\overline{\psi}_a(x')\theta(x^0 - x'^0) - \overline{\psi}_a(x')\psi_b(x)\theta(x'^0 - x^0),$$
$$(4.9.11)$$

$$(\phi(x)\phi(x'))_+ = \phi(x)\phi(x')\theta(x^0 - x'^0) + \phi(x')\phi(x)\theta(x'^0 - x^0).$$
$$(4.9.12)$$

*How to Write Down Lagrangian Densities*:

1. *Lorentz Invariance and Other Symmetries*: In quantum field theory, a Lagrangian density $\mathscr{L}(x)$, not including couplings of the fields to external sources, is, *a priori*, chosen to be Lorentz invariant, that is, it behaves as a Lorentz scalar: $\mathscr{L}'(x') = \mathscr{L}(x)$ under a Lorentz transformation $x^\mu \to x'^\mu = \Lambda^\mu{}_\nu x^\nu$. This, however, does not mean that given such a Lagrangian density, which is formally a Lorentz scalar, one cannot subsequently use, for example, as may be done in QED or in QCD, non-covariant gauges such as the Coulomb one $\nabla \cdot \mathbf{A} = 0$, or $\nabla \cdot \mathbf{A}_a = 0$, respectively. The importance of using, for example, the Coulomb gauge in such gauge theories is that only the physical degrees of freedom associated with the vector fields are quantized ensuring the necessary positive definiteness of the underlying physical Hilbert space. One may then invoke the gauge *invariance* of transition probabilities associated with the underlying physical processes to infer that the results should be the same as the ones that may be obtained by using covariant gauges. By doing so,

one would finally recover a gauge and Lorentz invariant theoretical framework for the computation of physical results. Accordingly, it is misleading to state that a non-covariant gauge destroys the Lorentz invariance of a theory. In the same manner, when one chooses a specific gauge, such as a covariant gauge, the Lagrangian density becomes restricted to such a gauge and hence gauge invariance is destroyed *at* the Lagrangian level, but the theory dealing with physical processes is nevertheless gauge invariant. Symmetries relevant to the underlying theory may also be implemented, as we will see, for example, when supersymmetric field theories will be developed later. Symmetry breaking will be also discussed.

2. *Even in Fermi-Dirac Fields.* The Bose-Einstein character of the Lagrangian density implies that it must be even in the Fermi-Dirac fields constructed in such a manner so that this aspect is maintained.

3. *Limits of Time-Ordered Products.* We have seen in the previous section that the matrix element $\langle 0_+ | \mathscr{L}_I(x) | 0_- \rangle$ is generated by functional differentiations with external sources and leads an expression for $\mathscr{L}_I(x)$ always as a limit of time-ordered products. That is, the QDP self consistently requires that Lagrangian densities be defined as the limits of time-ordered products. We have also seen in (4.9.8), (4.9.9) and (4.9.10) that *within* a time-ordered product the fields obey commutativity or anti-commutativity properties as spelled out in the just mentioned equations. In reference to (4.9.11), for example, we consider the average of the limits $x \to x' + 0$, $x \to x' - 0$, with initial spacelike separation $(x - x')^2 > 0$, and similarly for the scalar field in (4.9.12). This, in particular, means that in writing down a Lagrangian density one is to take the anti-symmetric average of the product of two Fermi-Dirac fields and the symmetric average of the product of two Bose-Einstein fields. Hence, one makes the following replacement for the product of two Dirac fields

$$\overline{\psi}_a(x)\psi_b(x) \to \frac{1}{2}\left(\overline{\psi}_a(x)\psi_b(x) - \psi_b(x)\overline{\psi}_a(x)\right) = \frac{1}{2}[\overline{\psi}_a(x), \psi_b(x)],$$

$$(4.9.13)$$

and for the interaction Lagrangian density part, for example, in QED,

$$e_0 \overline{\psi}_a(x)(\gamma^\mu)_{ab}\psi_b(x) A_\mu(x) \to \frac{1}{2}\{j^\mu(x), A_\mu(x)\}, \qquad (4.9.14)$$

where the electromagnetic current, having a Bose-Einstein character, is given by

$$j^\mu(x) = e_0 \frac{1}{2} \gamma^\mu_{ab} [\overline{\psi}_a(x), \psi_b(x)], \qquad (4.9.15)$$

and $e_0$ is the (unrenormalized) charge. We have already seen the definition of the electromagnetic current defined as in (4.9.15) consistent with charge conjugation transformation in (3.6.9) in Sect. 3.6. It is important to note, as already emphasized in Sect. 3.6, that the functional derivative $\delta/\delta\overline{\psi}_c(x')$, for example, should anti-commute with $\psi_b(x)$ giving from (4.9.13)

$$\frac{\delta}{\delta\overline{\psi}_c(x')}\frac{1}{2}\left(\overline{\psi}_a(x)\psi_b(x) - \psi_b(x)\overline{\psi}_a(x)\right) = \delta_{ac}\,\psi_b(x)\,\delta^{(4)}(x-x'), \qquad (4.9.16)$$

otherwise one would get zero as a trivial result. The above analysis leads us to emphasize the following:

"Students often wonder *why one can commute (anti-commute) field compo-nents in the interaction Lagrangian density for a theory dealing with operators. With the interaction Lagrangian density defined as a limit of a time-ordered product this becomes evident*".

4. *Renormalizability*. The Lagrangian density is so chosen so the theory becomes renormalizable.

*Setting Up the Solution of Field Theory*:

The importance of deriving an expression for the vacuum-to-vacuum transition amplitude $\langle\,0_+\mid 0_-\rangle$ in field theory stems from the fact that one may carry out a unitarity expansion for the latter in terms of multi-particle states from which transition amplitudes for the underlying physical processes of a theory may be readily extracted.[18] At this stage the reader is urged to review the two examples provided in the last section on the application of the QDP before considering the general strategy for setting up the solution for $\langle\,0_+\mid 0_-\rangle$ given below.

Let $\chi(x)$ denote collectively the independent fields in a theory with canonical conjugate momenta denoted collectively by $\pi(x)$, and let $\varrho(x)$ denote collec-tively the dependent fields, i.e., with vanishing canonical conjugate momenta. The Lagrangian density will be expressed as

$$\mathscr{L}(x,\lambda) = \mathscr{L}_0(x) + \mathscr{L}_I(x,\lambda) + K(x)\chi(x) + S(x)\varrho(x), \qquad (4.9.17)$$

where

$$\mathscr{L}_I(x,0) = 0, \qquad (4.9.18)$$

and $K(x)$, $S(x)$ denote external sources coupled to these fields, respectively. The dependence of $\mathscr{L}(x,\lambda)$ on the adjoints of the fields is suppressed for simplicity of the notation.

A key equation for obtaining an expression for $\langle\,0_+\mid 0_-\rangle$ is (see (4.6.37))

$$\frac{\partial}{\partial\lambda}\langle\,0_+\mid 0_-\rangle = i\langle\,0_+\mid\int(\mathrm{d}x)\,\frac{\partial}{\partial\lambda}\mathscr{L}_I(x,\lambda)\mid 0_-\rangle. \qquad (4.9.19)$$

---

[18] See, e.g., (5.8.20), (5.8.21), (5.9.16), (5.9.17), (5.9.18), (5.9.19), (5.9.20) and (5.9.21).

Now the basic idea is to re-express $\langle 0_+ | (\partial/\partial\lambda)\mathscr{L}_I(x,\lambda) | 0_- \rangle$ as an operation involving the functional derivatives $(\delta/\delta K(x))$, $(\delta/\delta S(x))$ applied directly to $\langle 0_+ | 0_- \rangle$.
Suppose:

(i) The Euler-Lagrange equations allow one to eliminate the dependent fields $\varrho(x)$ in favor of $\chi(x)$, $\pi(x)$, $\lambda$, $K(x)$, $S(x)$, via equations of constraints, and, in turn, allow one to rewrite $\mathscr{L}_I(x,\lambda)$ as a sum of terms depending on the latter variables, with coefficients depending on $\lambda$, and

(ii) Repeated applications of (4.6.38), as described below it, and as also done in the second example of the last section, the above allows one, in turn, to rewrite

$$\langle 0_+ | \frac{\partial}{\partial\lambda} \mathscr{L}_I(x,\lambda) | 0_- \rangle = \left[ \frac{\partial}{\partial\lambda} \widehat{\mathscr{L}}_I(x,\lambda) + \widehat{\mathscr{F}}(x,\lambda) \right] \langle 0_+ | 0_- \rangle,$$

$$(4.9.20)$$

where $\widehat{\mathscr{L}}_I(x,\lambda)$ denotes $\mathscr{L}_I(x,\lambda)$, with the fields $(\chi(x), \varrho(x))$, simply replaced by $(\pm i)$ times $((\delta/\delta K(x)), (\delta/\delta S(x)))$, as the cases may be, (see, e.g., (4.9.2), (4.9.3) and (4.9.4)), and $\widehat{\mathscr{F}}(x,\lambda)$ is an operation expressed in terms of these functional derivatives and may depend on $\lambda$.

Upon integrating (4.9.20) over $\lambda$, the following expression for $\langle 0_+ | 0_- \rangle_\lambda = \langle 0_+ | 0_- \rangle$ emerges

$$\langle 0_+ | 0_- \rangle_\lambda = \exp\left( i \int (dx) \left[ \widehat{\mathscr{L}}_I(x,\lambda) + \int_0^\lambda d\lambda' \, \widehat{\mathscr{F}}(x,\lambda') \right] \right) \langle 0_+ | 0_- \rangle_{\lambda=0}.$$

$$(4.9.21)$$

Here we see that, in general, the interaction term $\widehat{\mathscr{L}}_I(x,\lambda)$, and hence also the action, get modified due to the presence of the second term in the exponent. The amplitude $\langle 0_+ | 0_- \rangle_{\lambda=0}$ has usually a simple structure. The full amplitude $\langle 0_+ | 0_- \rangle_\lambda$ is then obtained merely by functional differentiations of $\langle 0_+ | 0_- \rangle_{\lambda=0}$.

Applications of the above formula were given in the last section and other applications will be given in coming chapters.

## 4.10   CPT

We recall (Sect. 2.1) that charge conjugation transformation C consists in replacing every particle in a given process by its anti-particle, while parity P reverses the direction of the three-momentum of every particle, and time T reversal involves in reversing the direction in which a process evolves and as such, the particles initially going into it, become outgoing, and vice versa, while reversing the direction of their three-momenta and their spin projections. QFT predicts that the product of the three transformations CPT (in any order), taken together, is a symmetry of nature in the sense that any transition probability and its CPT counterpart are equal, and no experiment seems to contradict this. Clearly, the product CPT, as a net transformation, amounts to replacing every particle by its anti-particle, while reversing their spin projections, and interchanging the initial and final states. Recall also, due to the anti-unitary property of time reversal operation (Sect. 2.1), the CPT transformation is implemented by an anti-unitary operator.

It should be noted that establishing the CPT invariance of the action corresponding to a local Lagrangian density is only part of the analysis of the CPT Theorem, one also has to show how the interchange of the initial and the final states arises with the simultaneous replacements of particles by their antiparticles and vice versa. In this respect, we will see that in the present analysis, detectors of particles/antiparticles and emitters of antiparticles/particles are interchanged in a CPT "transformed" world. The latter property will automatically take care of the arbitrary the number of particles may be going in and those that may come out of a scattering process in a direct and simple manner.

Let us first consider how basic fields transform under the above transformation. To this end, note, in particular, that from (3.2.29), (3.2.31), (3.2.36), that for a Dirac field:

$$\mathrm{C}\,\psi(x)\,\mathrm{C}^{-1} = \mathscr{C}\overline{\psi}^{\mathsf{T}}(x), \; \mathrm{P}\,\psi(x)\,\mathrm{P}^{-1} = \gamma^0\psi(x'), \; \mathrm{T}\,\psi(x)\,\mathrm{T}^{-1} = \gamma^5\mathscr{C}\psi(x''),$$
$$(4.10.1)$$

with given fixed phases, where $x' = (x^0, -\mathbf{x})$, $x'' = (-x^0, \mathbf{x})$, $\mathscr{C} = \mathrm{i}\gamma^2\gamma^0$. With a rather conventional choices of phases, we may thus introduce $\Theta = $ CPT transformations, and quite generally, of the following basic fields, consisting of a Dirac field $\psi$, a generic scalar field $\varphi$, and a generic vector field $V^\mu$ with a direct generalization for the presence of several such fields for different species of particles:

$$\Theta\,\psi(x)\,\Theta^{-1} = \gamma^5\gamma^0\overline{\psi}^{\mathsf{T}}(-x), \qquad \Theta\,\overline{\psi}(x)\,\Theta^{-1} = -\psi^{\mathsf{T}}(-x)\gamma^0\gamma^5,$$
$$(4.10.2)$$

$$\Theta\,\varphi(x)\,\Theta^{-1} = \varphi^\dagger(-x), \qquad \Theta\,\varphi^\dagger(x)\,\Theta^{-1} = \varphi(-x), \qquad (4.10.3)$$

$$\Theta\,V^\mu(x)\,\Theta^{-1} = -V^{\mu\dagger}(-x), \qquad \Theta\,V^{\mu\dagger}(x)\,\Theta^{-1} = -V^\mu(-x). \qquad (4.10.4)$$

Opposite phases of the scalar and vector fields are taken since $\partial^\mu\varphi(x)$ should transform as a vector field.

Given a local Lagrangian density Lagrangian density $\mathscr{L}(x)$, it is not sufficient to establish the symmetry of the action integral $A = \int(\mathrm{d}x)\,\mathscr{L}(x)$ under a CPT transformation. One also has to describe how incoming states and outgoing states, for arbitrary processes, are interchanged and what happens to the spins of participating particles. The latter is best described by coupling the underlying fields to corresponding external sources to be able to identify *all* the particles that may be ingoing or outgoing in any process in a straightforward manner. As we will see, the latter avoids of introducing many-particle states for all possible processes that may take, and, in turn, leads to examine as to what effectively happens to the external source functions in a CPT "transformed" world.

To the above end, we consider a total action integral

$$A_{tot} = \int (dx) \Big( \mathscr{L}(x) + \mathscr{L}_S(x) \Big), \tag{4.10.5}$$

where $\mathscr{L}(x)$ a local Lagrangian density, and without loss of generality, restricting the analysis to the fields in (4.10.2), (4.10.3) and (4.10.4), with the labeling of different species suppressed,

$$\mathscr{L}_S(x) = \big[ \overline{\eta}(x)\psi(x) + \overline{\psi}(x)\eta(x) + K^\dagger(x)\phi(x)$$
$$+ \phi^\dagger(x)K(x) + K_\mu^\dagger(x)V^\mu(x) + V^{\mu\dagger}(x)K_\mu(x) \big], \tag{4.10.6}$$

where $K^\mu(x)$, $K^{\mu\dagger}(x)$, $K(x)$, $K^\dagger(x)$, $\eta(x)$, $\overline{\eta}(x)$ denote external sources with the latter two being Grassmann variables.

In the appendix to this chapter it is shown

$$\Theta \Big( \int (dx)\, \mathscr{L}(x) \Big) \Theta^{-1} = \int (dx)\, \mathscr{L}(x), \tag{4.10.7}$$

$$\Theta \left( \int (dx)\, [\overline{\psi}(x)\,\eta(x) + \overline{\eta}(x)\psi(x)] \right) \Theta^{-1}$$
$$= \int (dx)\, [\overline{\psi}(x)\,\gamma^5\eta(-x) - \overline{\eta}(-x)\gamma^5\,\psi(x)], \tag{4.10.8}$$

$$\Theta \left( \int (dx)\, [\varphi^\dagger(x)\,K(x) + K^\dagger(x)\varphi(x)] \right) \Theta^{-1}$$
$$= \int (dx)\, [\varphi^\dagger(x)\,K(-x) + K^\dagger(-x)\varphi(x)]. \tag{4.10.9}$$

$$\Theta \left( \int (dx)\, [V^{\mu\dagger}(x)\,K_\mu(x) + K_\mu^\dagger(x)V^\mu(x)] \right) \Theta^{-1}$$
$$= \int (dx)\, [-V^{\mu\dagger}(x)\,K_\mu(-x) - K_\mu^\dagger(-x)V^\mu(x)]. \tag{4.10.10}$$

Therefore, only the external sources change which, after all, are responsible of emitting and detecting all the particles in any given process. That is, CPT is a symmetry of the action of the local Lagrangian density $\mathscr{L}(x)$ in quantum field theory

$$A = \int (\mathrm{d}x)\, \mathscr{L}(x),$$ (4.10.11)

while the sources "change their roles" in the following manner:

$$\eta(x) \rightarrow \gamma^5 \eta(-x), \qquad\qquad \overline{\eta}(x) \rightarrow -\overline{\eta}(-x)\gamma^5,$$ (4.10.12)

$$K(x) \rightarrow K(-x), \qquad\qquad K^\dagger(x) \rightarrow K^\dagger(-x),$$ (4.10.13)

$$K_\mu(x) \rightarrow -K_\mu(-x), \qquad\qquad K_\mu^\dagger(x) \rightarrow -K_\mu^\dagger(-x).$$ (4.10.14)

signalling the fact of changes have occurred with the ingoing and outgoing particles/antiparticles. More precisely, the second substitution in (4.10.12), for example, implies that

$$\overline{\eta}(p)u(\mathbf{p}, \sigma) \rightarrow -\overline{\eta}(-p)\gamma^5 u(\mathbf{p}, \sigma) = -\overline{\eta}(-p)v(\mathbf{p}, -\sigma),$$ (4.10.15)

and so on, where we have used the definition of $v(\mathbf{p}, \sigma)$ in (I.17). Referring to (3.3.39), (3.3.42), we may infer that an outgoing particle of spin $\sigma$ has been replaced by an ingoing antiparticle of spin $-\sigma$. Upon introducing polarization vectors, as discussed in (4.7.130), (4.7.230), satisfying the properties in (4.7.131), for massive and massless vector bosons, we may then prepare Table 4.1 which shows the expected changes that occur with ingoing and outgoing states in the CPT "transformed" world.

The CPT-symmetric nature of the action of local field theory and the fate of the particles in a CPT "transformed" world in any given process, as described in Table 4.1, is the content of the celebrated CPT Theorem.

*Remarks*

1. It should be emphasized that the theoretical analysis of the nature of particles ingoing and outgoing in a physical process, as done above, under the CPT transformation, and not just of establishing the transformation property of the action of the local Lagrangian density in (4.10.11) in the absence of external

**Table 4.1** Source functions and corresponding particles/anti-particles in physical processes

| Type | | (IN) | CPT | | (OUT) |
|---|---|---|---|---|---|
| Dirac | Particle: | $\overline{u}(\mathbf{p}, \sigma)\eta(p)$ | $\rightleftarrows$ | Anti-particle: | $-\overline{v}(\mathbf{p}, -\sigma)\eta(-p)$ |
| | Anti-particle: | $-\overline{\eta}(-p)v(\mathbf{p}, -\sigma)$ | $\rightleftarrows$ | Particle: | $\overline{\eta}(p)u(\mathbf{p}, \sigma)$ |
| Spin-0 | Particle: | $K(p)$ | $\rightleftarrows$ | Anti-particle: | $K(-p)$ |
| | Anti-particle: | $K^\dagger(-p)$ | $\rightleftarrows$ | Particle: | $K^\dagger(p)$ |
| Vector | Particle: | $\mathrm{e}_\lambda^{\mu*}(p)K_\mu(p)$ | $\rightleftarrows$ | Anti-particle: | $\mathrm{e}_{-\lambda}^{\mu}(p)K_\mu(-p)$ |
| | Anti-particle: | $\mathrm{e}_{-\lambda}^{\mu*}(p)K_\mu^\dagger(-p)$ | $\rightleftarrows$ | Particle: | $\mathrm{e}_\lambda^{\mu}(p)K_\mu^\dagger(p)$ |

sources, is of great importance. Establishing the invariance property in (4.10.7) is only part of the story.

2. The introduction of the source functions simplifies the proof of the Theorem tremendously as they tell us what happens to all possible ingoing and outgoing particles/anti-particles in processes in a CPT "transformed" world. In particular, we saw how naturally particle/antiparticle detectors and antiparticle/particle emitters are interchanged in a CPT "transformed" world. They also provide much physical insight on the preparatory and final stages of physical processes in a theory.

3. From the way QFT has been developed over the years, the validity of the CPT Theorem is of no surprise, and its theoretical development historically was expected.

4. Although experiments do not contradict CPT invariance, they tell us, however, that one or two of C, P, T taken at a time are violated. The violation of CP has been particularly intriguing and, as pointed out by Sakharov [9] almost half a century ago, may give us a clue on the reason for the imbalance of matter and anti-matter with nature's preference of matter over anti-matter which, together with the Spin & Statistics Connection (Sect. 4.5), are contributing factors to our existence.

5. Note that the argument of the source functions in the second column with opposite sign of momentum to the right column, imply, through Fourier transform, that the corresponding particle/antiparticle, going into (loss) or coming out (gain) of the scattering process, as the case may be, carries the same energy-momentum before CPT transformation, but spin $\sigma$ and helicity state $\lambda$ change as seen in the Dirac spinors and polarization vectors, respectively.

## Appendix A: Basic Equalities Involving the CPT Operator

We have the following explicit equalities for the CPT operator $\Theta$ :

$$\Theta \left[ \overline{\eta}(x)\, \psi(x) + \overline{\psi}(x)\, \eta(x) \right] \Theta^{-1}$$

$$= \left[ -(\gamma^0)_{ab}\, (\gamma^5 \gamma^0)_{bc}\, \overline{\psi}_c(-x)\, \eta_a(x) + \eta_a^*(x)\, \psi_b(-x)(\gamma^0 \gamma^5)_{ba} \right]$$

$$= \left[ (\gamma^5)_{ac}\, \overline{\psi}_c(-x)\, \eta_a(x) - \eta_a^*(x)\, \psi_b(-x)(\gamma^5 \gamma^0)_{ba} \right]$$

$$= \left[ \overline{\psi}(-x)\, \gamma^5\, \eta(x) - \overline{\eta}(x)\, \gamma^5\, \psi(-x) \right]. \qquad (A\text{-}4.1)$$

Equation (4.10.8) now follows upon changing the variable of integration $x \to -x$. Equations (4.10.9), (4.10.10) are self evident.

Finally, to establish (4.10.7), we consider transformation properties of the product of fields and their derivatives that may appear in $\mathscr{L}(x)$. Consider first Dirac fields, and a general linear combination of, say, two Dirac fields

$$\frac{1}{2} M_{ab}^{\mu_1 \cdots \mu_k} [\overline{\psi}_a^1(x), \psi_b^2(x)], \tag{A-4.2}$$

where $M_{ab}^{\mu_1 \cdots \mu_k}$ is of the general form, or a linear combination of such tensors as, $(\gamma^{\mu_1} \ldots \gamma^{\mu_k})$, $(\gamma^{\mu_1} \ldots \gamma^{\mu_k} \gamma^5)$. Partial derivatives will be considered in studying the general structure of tensor components in (A-4.5) below. In view of the application of a CPT transformation, note that

$$[(\gamma^5 \gamma^0) \gamma^{\mu_1 *} \ldots \gamma^{\mu_k *} (\gamma^0 \gamma^5)]_{ab} = (-1)^k [\gamma^{\mu_1 \mathsf{T}} \ldots \gamma^{\mu_k \mathsf{T}}]_{ab} = (-1)^k [\gamma^{\mu_k} \ldots \gamma^{\mu_1}]_{ba}, \tag{A-4.3}$$

and similarly in the presence of an additional $\gamma^5$ matrix, where we have used the properties: $\gamma^5 \gamma^\mu = -\gamma^\mu \gamma^5$, $(\gamma^0)^2 = (\gamma^5)^2 = 1$, $\gamma^0 \gamma^{\mu *} = \gamma^{\mu \mathsf{T}} \gamma^0$, $\gamma^{0 \mathsf{T}} = \gamma^0$, $\gamma^{5 \mathsf{T}} = \gamma^5$. Hence from the transformation rule of a spinor given in (4.10.2), we obtain

$$\Theta \left( \frac{1}{2} M_{ab}^{\mu_1 \cdots \mu_k} [\overline{\psi}_a^1(x), \psi_b^2(x)] \right) \Theta^{-1} \Big|_{x \to -x} = (-1)^k \left( \frac{1}{2} M_{ab}^{\mu_1 \cdots \mu_k} [\overline{\psi}_a^1(x), \psi_b^2(x)] \right)^\dagger. \tag{A-4.4}$$

Note the importance of using the anti-symmetric average (see Sect. 4.9) in establishing this result.

On the other hand for any tensor component $\chi_{\sigma_1 \ldots \sigma_k}(x)$, such as constructed out of products of vector fields, of scalar fields and their derivatives, we obviously have under the CPT transformations (4.10.3) and (4.10.4),

$$\Theta \chi_{\sigma_1 \ldots \sigma_k}(x) \Theta^{-1} = (-1)^k \chi_{\sigma_1 \ldots \sigma_k}^\dagger(-x). \tag{A-4.5}$$

Upon defining

$$M_{ab}^{\mu_1 \cdots \mu_k} \frac{1}{2} [\overline{\psi}_a^1(x), \psi_b^2(x)] = \Omega^{\mu_1 \cdots \mu_k}(x), \tag{A-4.6}$$

we may infer from (A-4.4) and (A-4.5) that

$$\Theta \frac{1}{2} \{ \Omega^{\mu_1 \cdots \mu_k}(x), \chi_{\sigma_1 \ldots \sigma_k}(x) \} \Theta^{-1} \Big|_{x \to -x} = \left( \frac{1}{2} \{ \Omega^{\mu_1 \cdots \mu_k}(x), \chi_{\sigma_1 \ldots \sigma_k}(x) \} \right)^\dagger. \tag{A-4.7}$$

This establishes (4.10.7) since a Lagrangian density, must contain an even number of Lorentz indices and all products in it are symmetrized with respect to Bose-Einstein factors and anti-symmetrized with respect to Fermi-Dirac fields, and due to the Hermiticity of the Lagrangian density, and finally followed by a change of the variable of integration $x \to -x$.

# Problems

**4.1** Use (4.2.23), (4.2.24), (4.2.25), (4.2.26), (4.2.27), (4.2.28), (4.2.29), (4.2.30) and (4.2.31), with the definition of the unit vector, $\mathbf{n}$ as given in (4.2.24) and the explicit expression of the rotation matrix $[R^{ij}]$ in (2.2.11) with $\xi = \theta$ to derive (4.2.32).

**4.2** Show that

$$\exp[-i\alpha J_{03}]\exp[-i\theta\mathbf{n}\cdot\mathbf{J}]W^0$$

$$= (W^3\sinh\alpha + W^0\cosh\alpha)\exp[-i\alpha J_{03}]\exp[-i\theta\mathbf{n}\cdot\mathbf{J}],$$

thus verifying (4.2.34).

**4.3** We are given two orthogonal four vectors $P^\mu W_\mu = 0$, and $P^\mu P_\mu = 0$, $W^\mu W_\mu = 0$, $P^0 = |\mathbf{P}| > 0$. Show that these imply that, $W^\mu = \lambda P^\mu$ for some numerical $\lambda$.

**4.4** Derive the explicit expressions for $W^0, \mathbf{W}$ in (4.2.16).

**4.5** Derive the transformation law in (4.3.48) as it follows from (4.3.47).

**4.6** Under a Lorentz transformation: $x \to x - \delta x = x'$, $\delta x^\mu = -\delta\omega^{\mu\nu}x_\nu$, a vector field $A_\alpha(x)$ has the following transformation law[19]

$$\Lambda'_\alpha(x') = \left(\delta_\alpha{}^\beta + \frac{i}{2}\delta\omega^{\nu\lambda}(S_{\nu\lambda})_\alpha{}^\beta\right)A_\beta(x),$$

where note that the argument of $A'_\alpha$ on the left-hand-side is $x'$ and not $x$, and the spin structure is given by

$$(S_{\nu\lambda})_\alpha{}^\beta = \frac{1}{i}\left(\eta_{\nu\alpha}\eta_\lambda{}^\beta - \eta_{\lambda\alpha}\eta_\nu{}^\beta\right).$$

Given the Maxwell Lagrangian is: $\mathscr{L} = -(1/4)F^{\alpha\beta}F_{\alpha\beta}$ where $F^{\alpha\beta} = (\partial^\alpha A^\beta - \partial^\beta A^\alpha)$: (i) find the (symmetric) energy-momentum tensor, (ii) show it is gauge invariant, (iii) show directly that it is conserved. What is its trace ?. [Note that the canonical energy momentum tensor is not gauge invariant.]

**4.7** Use the formal expression of the Dirac Lagrangian density: $\mathscr{L} = -\overline{\psi}(\gamma\partial/i) + m)\psi$, to show that the (symmetric) energy-momentum tensor is given by: $T^{\mu\nu} = (1/4i)\overline{\psi}(\gamma^\mu\overleftrightarrow{\partial}{}^\nu + \gamma^\nu\overleftrightarrow{\partial}{}^\mu)\psi$, where $\overleftrightarrow{\partial}{}^\nu = \overrightarrow{\partial}{}^\nu - \overleftarrow{\partial}{}^\nu$.

---

[19]See also (4.4.2).

**4.8** Find the expressions of the components $T^{00}(x)$ $T^{0k}(x)$ of the energy-momentum tensor of the Dirac quantum field in Problem 4.7 involving no time derivatives.

**4.9** Derive the following fundamental equal-time commutation relation of the component $T^{00}$ of the energy-momentum tensor for the Dirac theory in Problems 4.7 and 4.8: $[T^{00}(x), T^{00}(x')] = -i[T^{0k}(x) + T^{0k}(x')]\partial_k \delta^3(\mathbf{x} - \mathbf{x}')$. This was derived by Schwinger [16, 17]. Commutations relations of all the components of the energy-momentum tensor in the full QED theory, taking into account of the gauge problem, are derived in Manoukian [5].

**4.10** Given three 3-vectors $\mathbf{e}_+$, $\mathbf{e}_-$, $\mathbf{p}/|\mathbf{p}|$, satisfying

$$\mathbf{e}_\pm \cdot \mathbf{e}_\mp^* = 0, \quad \mathbf{e}_\pm \cdot \mathbf{e}_\pm^* = 1, \quad \frac{\mathbf{p}}{|\mathbf{p}|} \cdot \mathbf{e}_\pm = 0, \quad \delta^{ij} = \frac{p^i}{|\mathbf{p}|}\frac{p^i}{|\mathbf{p}|} + e_+^{i*}e_+^j + e_-^{i*}e_-^j,$$

with the last relation providing a completeness relation in 3D space, verify explicitly the four dimensional completeness relation (4.7.129) in Minkowski spacetime.

**4.11** Show that the equation

$$\left[\eta^{\mu\nu}\left(\frac{\gamma\partial}{i} + m\right) - \frac{1}{3}\left(\gamma^\mu\frac{\partial^\nu}{i} + \gamma^\nu\frac{\partial^\mu}{i}\right) + \frac{1}{3}\gamma^\mu\left(-\frac{\gamma\partial}{i} + m\right)\gamma^\nu\right]\psi_\nu(x) = K^\mu(x),$$

also leads to the equations in (4.7.137), (4.7.138) for $K^\mu = 0$, and that it merely adds the non-propagating term $-(2/3 m^2)[p^\mu \gamma^\nu - p^\nu \gamma^\mu + (\gamma p - m)\gamma^\mu \gamma^\nu]$ to the propagator $\rho^{\mu\nu}(p)/[p^2 + m^2 - i\epsilon]$. This is the equation used, for example, by Takahashi [19].

**4.12** Derive the constraint on the spatial components $\psi^i(x)$ of the $m \neq 0$ Rarita-Schwinger field in (4.7.148).

**4.13** Derive the basic Eq. (4.7.150) needed for the $m \neq 0$ Rarita-Schwinger field components to satisfy the constraint in (4.7.148).

**4.14** Establish the spin 3/2 nature of the $m = 0$ Rarita-Schwinger field in (4.7.193).

**4.15** The propagator for a massive neutral vector field was derived in (4.7.106), (4.7.107), (4.7.108) and (4.7.109), and may be obtained by functional differentiation

$$\text{(i)} \ (-i)\frac{\delta}{\delta K_\mu(x)}(-i)\frac{\delta}{\delta K_\nu(x')}\langle 0_+ | 0_-\rangle\bigg|_{K^\sigma=0} = \Delta_+^{\mu\nu}(x - x'),$$

directly from the vacuum-to-vacuum transition amplitude $\langle 0_+ | 0_-\rangle$ given in (4.7.108). Use the identity in (4.6.38), following from the quantum dynamical principle, to show that the time ordered product of the vector fields $\langle 0_+ |$

$$\left(V^\mu(x)V^\nu(x\,')\right)_+ 0_-\Big)\Big|_{K^\sigma=0} \quad \text{satisfies}$$

$$\mathrm{i}\,\langle\, 0_+ \mid \left(V^\mu(x)V^\nu(x')\right)_+ 0_-\rangle\Big|_{K^\sigma=0} = \Delta_+{}^{\mu\nu}(x-x') - \frac{1}{m^2}\,\delta^\nu{}_0\eta^{0\,\mu}\,\delta^{(4)}(x-x').$$

That is, $(\mathrm{i} \times$ time-ordered-product) of the vector fields not only does not represent the propagator of the vector field, but is also not covariant.

**4.16** Show that if the interaction Lagrangian density in (4.8.11) is replaced, say, by $\mathscr{L}_I(x) = \lambda\,(\gamma^\mu)_{ab}\,[\overline{\psi}_a(x), \psi_a(x)]\,V_\mu(x)/2$, which is linear in $V_\mu(x)$, then there is no modifying factor as in (4.8.29) in the action integral.

# References

1. Lam, C. S. (1965). Feynman rules and Feynman integrals for systems with higher-spin fields. *Nuovo Cimento, 38*, 1755–1765.
2. Lehmann, H., Symanzik, K., & Zimmermann, W. (1955). The formulation of quantized field theories. *Nuovo Cimento, XI*, 205–222.
3. Manoukian, E. B. (1985). Quantum action principle and path integrals for long-range interactions. *Nuovo Cimento, 90Λ*, 295–307.
4. Manoukian, E. B. (1986). Action principle and quantization of Gauge fields. *Physical Review D, 34*, 3739–3749.
5. Manoukian, E. B. (1987). On the relativistic invariance of QED in the Coulomb Gauge and field transformations. *Journal of Physics G, 13*, 1013–1021.
6. Manoukian, E. B. (2006). *Quantum theory: A wide spectrum*. Dordrecht: Springer.
7. Manoukian, E. B. (2016). Rarita-Schwinger massless field in covariant and Coulomb-like gauges. *Modern Physics Letters A, 31*, 1650047, (1–8).
8. Rarita, W., & Schwinger, J. (1941). On a theory of particles with half-integral spin. *Physical Review, 60*, 61.
9. Sakharov, A. D. (1967). Violation of CP invariance, C asymmetry, and Baryon asymmetry of the Universe. *Soviet Physics JETP Letters, 5*, 24–27.
10. Schwinger, J. (1951a). On the Green's functions of quantized fields. I. *Proceedings of the National Academy of Sciences USA, 37*, 452–455.
11. Schwinger, J. (1951b). The theory of quantized fields. I. *Physical Review, 82*, 914–927.
12. Schwinger, J. (1953). The theory of quantized fields. II, III. *Physical Review, 91*, 713–728, 728–740.
13. Schwinger, J. (1958). Addendum to spin, statistics and the CPT theorem. *Proceedings of the National Academy of Sciences USA, 44*, 617–619.
14. Schwinger, J. (1960). Unitary transformations and the action principle. *Proceedings of the National Academy of Sciences USA, 46*, 883–897.
15. Schwinger, J. (1962). Exterior algebra and the action principle I. *Proceedings of the National Academy of Sciences USA, 48*, 603–611.
16. Schwinger, J. (1963a). Commutation relations and consevation laws. *Physical Review, 130*, 406–409.
17. Schwinger, J. (1963b). Energy-momentum density in field theory. *Physical Review, 130*, 800–805.
18. Schwinger, J. (1970). *Particles, sources, and fields* (Vol. I). Reading: Addison-Wesley.
19. Takahashi, Y. (1969). *An introduction to field quantization*. Oxford: Pergamon Press.

# Recommended Reading

1. Manoukian, E. B. (1986). Action principle and quantization of Gauge fields. *Physical Review D, 34*, 3739–3749.
2. Manoukian, E. B. (2006). *Quantum theory: A wide spectrum*. Dordrecht: Springer.
3. Manoukian, E. B. (2016). Rarita-Schwinger massless field in covariant and Coulomb-like gauges. *Modern Physics Letters A, 31*, 1650047, (1–8).
4. Schwinger, J. (1970). *Particles, sources, and fields* (Vol. I). Reading: Addison-Wesley.
5. Weinberg, S. (1995). *The quantum theory of fields. I: Foundations*. Cambridge: Cambridge University Press.

# Chapter 5
# Abelian Gauge Theories

Quantum electrodynamis (QED), describing the interactions of electrons (positrons) and photons, is, par excellence, an abelian gauge theory. It is one of the most successful theories we have in physics and a most cherished one. It stood the test of time, and provides the blue-print, as a first stage, for the development of modern quantum field theory interactions. A theory with a symmetry group in which the generators of the symmetry transformations commute is called an abelian theory. In QED, the generator which induces a phase change of a (non-Hermitian) charged field $\chi(x)$

$$\chi(x) \rightarrow e^{i\theta(x)} \chi(x), \tag{5.1}$$

is simply the identity and hence the underlying group of transformations is abelian denoted by U(1). The transformation rule in (5.1) of a charged field, considered as a complex entity with a real and imaginary part, is simply interpreted as a rotation by an angle $\theta(x)$, locally, in a two dimensional (2D) space, referred to as charge space.[1]

The covariant gauge description of QED as well as of the Coulomb one are both developed. Gauge transformations are worked out in the full theory not only between covariant gauges but also with the Coulomb one. Explicit expressions of generating functionals of QED are derived in the differential form, as follows from the quantum dynamical principle, as well as in the path integral form. A relatively simple demonstration of the renormalizability of QED is given, as well as of the renormalization group method is developed for investigating the effective charge. A renormalization group analysis is carried out for investigating the magnitude of the effective fine-structure at the energy corresponding to the mass of the neutral vector boson $Z^0$, based on all of the well known charged leptons and quarks of

---

[1] A geometrical description is set up for the development of abelian and non-abelian gauge theories in a unified manner in Sect. 6.1 and may be beneficial to the reader.

© Springer International Publishing Switzerland 2016

E.B. Manoukian, *Quantum Field Theory I*, Graduate Texts in Physics,
DOI 10.1007/978-3-319-30939-2_5

specific masses which would contribute to this end. This has become an important reference point for the electromagnetic coupling in present high-energy physics. The Lamb shift and the anomalous magnetic moment of the electron, which have much stimulated the development of quantum field theory in the early days, are both derived. We also include several applications to scattering processes as well as of the study of polarization correlations in scattering processes that have become quite interesting in recent years. The theory of spontaneous symmetry breaking is also worked out in a celebrated version of scalar boson electrodynamics and its remarkable consequences are spelled out. Several studies were already carried out in Chap. 3 which are certainly relevant to the present chapter, such as of the gauge invariant treatment of diagrams with closed fermion loops, fermion anomalies in field theory, as well as other applications.[2]

## 5.1  Spin One and the General Vector Field

Referring to Sect. 4.7.3, let us recapitulate, in a slightly different way, the spin 1 character of a vector field. Under an infinitesimal rotation c.c.w of a coordinate system, in 3D Euclidean space, by an infinitesimal angle $\delta\vartheta$ about a unit vector $\mathbf{N}$, a three-vector $\mathbf{x}$, now denoted by $\mathbf{x}'$ in the new coordinate system, is given by

$$\mathbf{x}' = \mathbf{x} - \delta\vartheta\,(\mathbf{N}\times\mathbf{x}) = \mathbf{x} + \delta\vartheta\,(\mathbf{x}\times\mathbf{N}), \tag{5.1.1}$$

$$x'^{i} = \left(\delta^{ij}+\delta\vartheta\,\varepsilon^{ijk}N^{k}\right)x^{j}, \qquad \delta\omega^{ij} = \delta\vartheta\,\varepsilon^{ijk}N^{k} = -\delta\omega^{ji}, \tag{5.1.2}$$

where $\varepsilon^{ijk}$ is totally anti-symmetric with $\varepsilon^{123} = +1$, from which the matrix elements of the rotation matrix for such an infinitesimal rotation are given by[3]

$$R^{ij} = \delta^{ij} + \delta\vartheta\,\varepsilon^{ijk}N^{k}. \tag{5.1.3}$$

Under such a coordinate transformation, a three-vector field $A^{i}(\mathbf{x})$, in the new coordinate system, is given by

$$A'^{i}(x') = \left(\delta^{ij} + \delta\omega^{ij}\right)A^{j}(x), \tag{5.1.4}$$

which may be rewritten as

$$A'^{i}(x') = \left(\delta^{ij} + \frac{i}{2}\,\delta\omega^{k\ell}\,[S^{k\ell}]^{ij}\right)A^{j}(x), \tag{5.1.5}$$

$$= \left(\delta^{ij} + \frac{i}{2}\,\delta\vartheta\,\varepsilon^{k\ell q}N^{q}\,[S^{k\ell}]^{ij}\right)A^{j}(x), \tag{5.1.6}$$

---

[2]It is worth knowing that the name "photon" was coined by Lewis [43].
[3]See also Eq. (2.2.11).

where

$$[S^{k\ell}]^{ij} = \frac{1}{i}\left(\delta^{ki}\eta^{\ell j} - \delta^{\ell i}\eta^{kj}\right), \qquad i,j,k,\ell = 1,2,3. \tag{5.1.7}$$

One may then introduce the spin matrices $[S^q]$, $q = 1,2,3$,

$$[S^q]^{ij} = \frac{1}{2}\varepsilon^{k\ell q}[S^{k\ell}]^{ij}, \tag{5.1.8}$$

and rewrite (5.1.6) as

$$A'^i(x') = \left(\delta^{ij} + i\,\delta\vartheta\,[\mathbf{S}]^{ij}\cdot\mathbf{N}\right)A^j(x). \tag{5.1.9}$$

It is readily verified that

$$[\mathbf{S}^2]^{ij} = \sum_{q=1}^{3}[S^q]^{ii'}[S^q]^{i'j} = 2\delta^{ij} = s(s+1)\delta^{ij}, \tag{5.1.10}$$

establishing the spin $s = 1$ character of the vector field, with the spin components satisfying the well known commutations relations

$$[S^q, S^{q'}] = i\varepsilon^{qq'k}S^k. \tag{5.1.11}$$

As a direct generalization of (5.1.5), (5.1.6) and (5.1.7), a vector field $A_\alpha(x)$ has the following transformation under a Lorentz transformation: $x \rightarrow x - \delta x = x'$, $\delta x^\mu = -\delta\omega^{\mu\nu}x_\nu$,

$$A'_\alpha(x') = \left(\delta_\alpha{}^\beta + \delta\omega_\alpha{}^\beta\right)A_\beta(x) \tag{5.1.12}$$

$$= \left(\delta_\alpha{}^\beta + \frac{i}{2}\delta\omega^{\nu\lambda}\left(S_{\nu\lambda}\right)_\alpha{}^\beta\right)A_\beta(x), \qquad \delta\omega^{\nu\lambda} = -\delta\omega^{\lambda\nu}, \tag{5.1.13}$$

for a covariant description,[4] where we note that the argument of $A'_\alpha$ on the left-hand side is again $x'$ and not $x$,

$$\left(S_{\nu\lambda}\right)_\alpha{}^\beta = \frac{1}{i}\left(\eta_{\nu\alpha}\eta_\lambda{}^\beta - \eta_{\lambda\alpha}\eta_\nu{}^\beta\right). \tag{5.1.14}$$

---

[4] See (4.7.120), (4.7.121), (2.2.17), (2.2.18), (2.2.19), (2.2.20), (2.2.21) and (2.2.22).

## 5.2   Polarization States of Photons

The polarization vectors $\mathbf{e}_1$, $\mathbf{e}_2$ of a photon are mutually orthogonal and are, in turn, orthogonal to its momentum vector $\mathbf{k}$. With the vector $\mathbf{k}$ chosen along the z-axis, we may then introduce three unit vectors

$$\mathbf{n} = \frac{\mathbf{k}}{|\mathbf{k}|} = (0, 0, 1), \quad \mathbf{e}_1 = (1, 0, 0), \quad \mathbf{e}_2 = (0, 1, 0), \tag{5.2.1}$$

with the latter two providing a real representation of the polarization vectors, satisfying

$$\mathbf{n} \cdot \mathbf{e}_\lambda = 0, \ \ \mathbf{e}_\lambda \cdot \mathbf{e}_{\lambda'} = \delta_{\lambda,\lambda'}, \ \ \delta^{ij} = n^i n^j + \sum_{\lambda=1,2} e^i_\lambda e^j_\lambda, \ \ \lambda, \lambda' = 1, 2, \ \ i, j = 1, 2, 3,$$
$$\tag{5.2.2}$$

where $\lambda$ specifies the two polarization vectors, and the index $i$ specifies the $i$th component of the vectors. The equality, involving $\delta^{ij}$, is a completeness relation in three dimensions for expanding a vector in terms of the three unit vectors $\mathbf{n}$, $\mathbf{e}_1$, $\mathbf{e}_2$.

One may also introduce a *complex* representation of the polarization vectors, such as, $(\mathbf{e}_\pm \equiv \mathbf{e}_{\pm 1})$

$$\mathbf{e}_+ = \frac{1}{\sqrt{2}} (-1, -i, 0), \quad \mathbf{e}_- = \frac{1}{\sqrt{2}} (1, -i, 0), \quad \mathbf{e}_\lambda \cdot \mathbf{e}^*_{\lambda'}, = \delta_{\lambda,\lambda'}, \ \lambda, \lambda' = \pm 1.$$
$$\tag{5.2.3}$$

The completeness relation now simply reads

$$\delta^{ij} = n^i n^j + \sum_{\lambda=\pm 1} e^i_\lambda e^{*j}_\lambda = n^i n^j + \sum_{\lambda=\pm 1} e^{*i}_\lambda e^j_\lambda, \tag{5.2.4}$$

as is easily checked by considering specific values for the indices $i, j$ specifying components of the vectors.

One would also like to have the general expressions of the polarization vectors, when the three momentum vector of a photon $\mathbf{k}$ has an arbitrary orientation

$$\mathbf{k} = |\mathbf{k}| (\cos\phi \, \sin\theta, \, \sin\phi \, \sin\theta, \, \cos\theta). \tag{5.2.5}$$

To achieve this, we rotate the initial coordinate system in which the vector $\mathbf{k}$ is initially along the z-axis, c.c.w. by an angle $\theta$ about the unit vector $\mathbf{N} = (\sin\phi, -\cos\phi, 0)$ as shown in (Fig. 5.1), by using the explicit structure of the

**Fig. 5.1** The initial frame is rotated c.c.w by an angle $\theta$ about the unit vector $\mathbf{N}$ so that $\mathbf{k}$ points in an arbitrary direction in the new frame

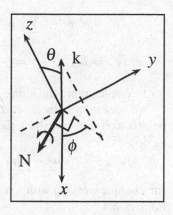

rotation matrix[5] with matrix elements

$$R^{ik} = \delta^{ik} - \varepsilon^{ijk} N^j \sin\theta + (\delta^{ik} - N^i N^k)(\cos\theta - 1), \quad i, j, k = 1, 2, 3, \quad (5.2.6)$$

where $\varepsilon^{ijk}$ is totally anti-symmetric with $\varepsilon^{123} = 1$.

The rotation matrix gives the following general expressions for the polarization vectors: (see Problem 5.1)

$$\mathbf{e}_1 = (\cos^2\phi \cos\theta + \sin^2\phi, \; \sin\phi \cos\phi (\cos\theta - 1), \; -\cos\phi \sin\theta),$$

$$(5.2.7)$$

$$\mathbf{e}_2 = (\sin\phi \cos\phi (\cos\theta - 1), \sin^2\phi \cos\theta + \cos^2\phi, \; -\sin\phi \sin\theta).$$

$$(5.2.8)$$

$$\mathbf{e}_\lambda \cdot \mathbf{e}_{\lambda'} = \delta_{\lambda\lambda'}, \quad \mathbf{k} \cdot \mathbf{e}_\lambda = 0, \quad \delta^{ij} = n^i n^j + \sum_{\lambda=1,2} e^i_\lambda e^j_\lambda, \quad \lambda, \lambda' = 1, 2. \quad (5.2.9)$$

for a real representation, and

$$\mathbf{e}_+ = \frac{1}{\sqrt{2}} (-\cos\phi \cos\theta + i \sin\phi, \; -\sin\phi \cos\theta - i \cos\phi, \; \sin\theta) e^{i\phi},$$

$$(5.2.10)$$

$$\mathbf{e}_- = \frac{1}{\sqrt{2}} (\cos\phi \cos\theta + i \sin\phi, \; \sin\phi \cos\theta - i \cos\phi, \; -\sin\theta) e^{-i\phi},$$

$$(5.2.11)$$

---

[5]A reader who is not familiar with this expression may find a derivation of it in Manoukian [56], p. 84. See also (2.2.11).

$$\delta^{ij} = n^i n^j + \sum_{\lambda=\pm 1} e^i_\lambda e^{*j}_\lambda = n^i n^j + \sum_{\lambda=\pm 1} e^{*i}_\lambda e^j_\lambda, \qquad (5.2.12)$$

$$\mathbf{e}_\lambda \cdot \mathbf{e}^*_{\lambda'} = \delta_{\lambda\lambda'}, \quad \mathbf{e}^*_\pm = -\mathbf{e}_\mp, \quad \mathbf{k} \cdot \mathbf{e}_\lambda = 0, \quad \lambda, \lambda' = \pm 1. \qquad (5.2.13)$$

for a complex representation.

We need to introduce a *covariant* description of polarization. Since we have two polarization states, we have the following orthogonality relations

$$\eta_{\mu\nu} e^\mu_\lambda e^{*\nu}_{\lambda'} = \delta_{\lambda\lambda'}, \quad \eta_{\mu\nu} k^\mu e^\nu_\lambda = 0, \quad \lambda, \lambda' = \pm 1, \quad k^2 = 0, \qquad (5.2.14)$$

for example working with a complex representation. The last orthogonality relation implies that

$$k^0 e^0_\lambda = \mathbf{k} \cdot \mathbf{e}_\lambda, \quad k^0 = |\mathbf{k}|, \qquad (5.2.15)$$

and with $\mathbf{k} \cdot \mathbf{e}_\lambda = 0$, we take $e^0_\lambda = 0$, and set

$$e^\mu_\lambda = (0, \mathbf{e}_\lambda). \qquad (5.2.16)$$

In order to write down a completeness relation in Minkowski spacetime, we may introduce two additional vectors[6] to $e^\mu_+$, $e^\mu_-$: $k^\mu = (k^0, -\mathbf{k})$, $\underline{k}^\mu = (k^0, \mathbf{k})$, where we note that $(k + \underline{k})$ is a time-like vector, while $(k - \underline{k})$ is a space-like one. Also

$$\eta_{\mu\nu} k^\mu e^\nu_\lambda = 0, \qquad \eta_{\mu\nu} \underline{k}^\mu e^\nu_\lambda = 0. \qquad (5.2.17)$$

The completeness relation simply reads as

$$\eta^{\mu\nu} = \frac{(k + \underline{k})^\mu (k + \underline{k})^\nu}{(k + \underline{k})^2} + \frac{(k - \underline{k})^\mu (k - \underline{k})^\nu}{(k - \underline{k})^2} + \sum_{\lambda=\pm 1} e^\mu_\lambda e^{*\nu}_\lambda, \qquad (5.2.18)$$

which on account of the facts that $k^2 = 0$, $\underline{k}^2 = 0$, this simplifies to

$$\eta^{\mu\nu} = \frac{k^\mu \underline{k}^\nu + \underline{k}^\mu k^\nu}{k\underline{k}} + \sum_{\lambda=\pm 1} e^\mu_\lambda e^{*\nu}_\lambda. \qquad (5.2.19)$$

## 5.3   Covariant Formulation of the Propagator

The gauge transformation of the Maxwell field $A^\mu(x)$ is defined by $A^\mu(x) \rightarrow A^\mu(x) + \partial^\mu \Lambda(x)$, and with arbitrary $\Lambda(x)$, it leaves the field stress tensor $F^{\mu\nu}(x) = \partial^\mu A^\nu(x) - \partial^\nu A^\mu(x)$ invariant. In particular, a covariant gauge choice for the

---

[6]We follow Schwinger's elegant construction [70].

electromagnetic field is $\partial_\mu A^\mu(x) = 0$. We will work with more general covariant gauges of the form

$$\partial_\mu A^\mu(x) = \lambda\,\chi(x), \tag{5.3.1}$$

where $\lambda$ is an arbitrary real parameter and $\chi(x)$ is a real scalar field. The gauge constraint in (5.3.1) may be *derived* from the following Lagrangian density

$$\mathscr{L} = -\frac{1}{4}F_{\mu\nu}F^{\mu\nu} + J^\mu A_\mu - \chi\partial_\mu A^\mu + \frac{\lambda}{2}\chi^2 \tag{5.3.2}$$

where $J^\mu(x)$ is an external, i.e., a classical, current. Variation with respect to $\chi$, gives (5.3.1), i.e., the gauge *constraint* is a *derived* one. While variation with respect to $A_\mu$ leads to[7]

$$-\partial_\nu F^{\nu\mu}(x) = J^\mu(x) + \partial^\mu\chi(x). \tag{5.3.3}$$

Using the expression $F^{\nu\mu} = \partial^\nu A^\mu - \partial^\mu A^\nu$, the above equation reads

$$-\Box A^\mu(x) = J^\mu(x) + (1 - \lambda)\,\partial^\mu\chi(x), \tag{5.3.4}$$

where we have used the derived gauge constraint in (5.3.1).

Upon taking the $\partial_\mu$ derivative of (5.3.3), we also obtain

$$-\Box\chi(x) = \partial_\nu J^\nu(x). \tag{5.3.5}$$

We consider the matrix element $\langle 0_+ |\,.\,| 0_- \rangle$ of (5.3.4), to obtain

$$\frac{\langle 0_+|A^\mu(x)|0_-\rangle}{\langle 0_+|0_-\rangle} = \int (dx')D_+(x-x')\Big(J^\mu(x') + (1-\lambda)\partial'^\mu\frac{\langle 0_+|\chi(x')|0_-\rangle}{\langle 0_+|0_-\rangle}\Big), \tag{5.3.6}$$

where $D_+(x - x')$ is the propagator $\big(-\Box D_+(x - x') = \delta^{(4)}(x - x')\big)$

$$D_+(x-x') = \int \frac{(dk)}{(2\pi)^4}\frac{e^{ik(x-x')}}{k^2 - i\epsilon}, \quad (dk) = dk^0 dk^1 dk^2 dk^3. \tag{5.3.7}$$

---

[7]Since the gauge constraint is now a derived one, one may vary all the components of the vector field independently.

Taking Fourier transforms of (5.3.6), and using (5.3.5), the following expression emerges

$$\frac{\langle\, 0_+ |A^\mu(x)\,|\, 0_-\rangle}{\langle\, 0_+\,|\,0_-\rangle} = \int (\mathrm{d}x')D^{\mu\nu}(x - x')J_\nu(x'), \qquad (5.3.8)$$

$$D^{\mu\nu}(x - x') = \int \frac{(\mathrm{d}k)}{(2\pi)^4}\left[\eta^{\mu\nu} - (1 - \lambda)\frac{k^\mu k^\nu}{k^2}\right]\frac{\mathrm{e}^{ik(x-x')}}{k^2 - i\epsilon}, \qquad (5.3.9)$$

defining covariant photon propagators[8] in gauges specified by the parameter $\lambda$. The gauge specified by the choice $\lambda = 1$ is referred to as the *Feynman gauge*, while the choices $\lambda = 0$ as the *Landau gauge*, and $\lambda = 3$ as the *Yennie-Fried gauge*.

Upon using

$$-\,\mathrm{i}\frac{\delta}{\delta J_\mu(x)}\,\langle\, 0_+\,|\,0_-\rangle = \langle\, 0_+ |A^\mu(x)\,|\, 0_-\rangle, \qquad (5.3.10)$$

we may integrate (5.3.8) to obtain

$$\langle\, 0_+\,|\,0_-\rangle = \exp\left[\frac{\mathrm{i}}{2}\int (\mathrm{d}x)(\mathrm{d}x')J_\mu(x)D^{\mu\nu}(x - x')J_\nu(x')\right] \qquad (5.3.11)$$

normalized to unity for $J_\mu(x) = 0$. The generating functional $\langle\, 0_+\,|\,0_-\rangle$ is determined in general covariant gauges specified by the values taken by the parameter $\lambda$ in (5.3.9)

The matrix element $\langle\, 0_+|F^{\mu\nu}(x)\,|\, 0_-\rangle$ of the field strength tensor $F^{\mu\nu}$, is given from (5.3.8) to be

$$\frac{\langle\, 0_+|F^{\mu\nu}(x)\,|\, 0_-\rangle}{\langle\, 0_+\,|\,0_-\rangle} = (\partial^\mu\,\eta^{\nu\sigma} - \partial^\nu\,\eta^{\mu\sigma})\int (\mathrm{d}x')D_+(x - x')\,J_\sigma(x'), \qquad (5.3.12)$$

and the gauge parameter $\lambda$ in (5.3.9) cancels out on the right-hand side of the equation.

It is important to note that for a conserved external current $\partial^\mu J_\mu(x) = 0$,

$$\langle\, 0_+\,|\,0_-\rangle\big|_{\partial^\mu J_\mu=0} = \exp\left[\frac{\mathrm{i}}{2}\int (\mathrm{d}x)(\mathrm{d}x')J_\mu(x)D_+(x - x')J^\mu(x')\right], \qquad (5.3.13)$$

is independent of the gauge parameter $\lambda$, and $D_+(x - x')$ is defined in (5.3.7).

---

[8]For $J^\mu = 0$.

**Fig. 5.2** The parallel plates
in question are placed
between two parallel plates
situated at large distances

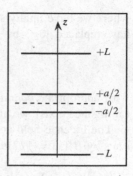

## 5.4  Casimir Effect

The Casimir effect, in its simplest theoretical description, is an electromagnetic
force of attraction between two parallel perfectly conducting neutral plates in
vacuum. It is purely quantum mechanical, i.e., it is attributed to the quantum
nature of the electromagnetic field, and to the nature of the underlying boundary
condition imposed on it by the presence of the plates. It is one of those mysterious
consequences of quantum theory, i.e., an $\hbar$-dependent result, that may be explained
by the response of the vacuum to external agents. By a careful treatment one may
introduce, in the process, a controlled environment, by placing the parallel plates
between two perfectly conducting plates placed, in turn, at very large macroscopic
distances (Fig. 5.2) from the two plates in question.[9] This analysis clearly shows
how a net finite attractive arises between the plates.

The electric field components, in particular, tangent to the plates satisfy the
boundary conditions

$$\mathbf{E}_\mathrm{T}(x^0, \mathbf{x}_\mathrm{T}, z)\Big|_{z = \pm a/2, \pm L} = 0. \tag{5.4.1}$$

Upon taking the functional derivative of (5.3.12), with respect to the external
current $J^\alpha$, we obtain, in the process, for $x'^0 > x^0$,

$$\mathrm{i}\,\langle \mathrm{vac}|F^{\alpha\beta}(x')\,F^{\mu\nu}(x)|\mathrm{vac}\rangle$$
$$= \left[\, \partial'^\alpha(\partial^\mu\eta^{\nu\beta} - \partial^\nu\eta^{\mu\beta}) - \partial'^\beta(\partial^\mu\eta^{\nu\alpha} - \partial^\nu\eta^{\mu\alpha})\,\right]D_+(x, x'), \tag{5.4.2}$$

---

[9]Schwinger [72] and Manoukian [50].

where we have finally set $J^\mu = 0$, and, in the absence of the external current, we have replaced $|0_\pm\rangle$ by $|\text{vac}\rangle$. Here $D_+(x,x')$ satisfies the equation

$$-\Box D_+(x,x') = \delta^{(4)}(x,x'),\tag{5.4.3}$$

with appropriate *boundary conditions*. Because translational invariance is broken (along the $z$-axis), we have replaced the arguments of $D_+$ and $\delta^{(4)}$ by $(x,x')$.

The Electric field components are given by $E^i = F^{0i}$, and the magnetic field ones by $B^i(x) = (1/2)\varepsilon^{ijk}F^{jk}$. Hence, in particular,[10]

$$\mathrm{i}\,\langle\text{vac}|\mathbf{E}_{\mathrm{T}}(x')\cdot\mathbf{E}_{\mathrm{T}}(x)|\text{vac}\rangle = \left(2\,\partial'^0\partial^0 - \nabla'_{\mathrm{T}}\cdot\nabla_{\mathrm{T}}\right)D_+^<(x,x'),\quad x^0 < x'^0.\tag{5.4.4}$$

In reference to this equation, corresponding to the tangential components of the electric fields, the boundary conditions in (5.4.1), implies a Fourier *sine* series for $\delta(z,z')$:

$$\delta(z,z') = \frac{2}{R}\sum_{n=1}^\infty \sin\frac{n\pi(z-d)}{R}\,\sin\frac{n\pi(z'-d)}{R},\tag{5.4.5}$$

in (5.4.3), where

$$\begin{aligned}
R &= L - a/2,\; d = a/2,\; \text{for}\; a/2 \le z,z' \le L\\
R &= a,\quad\;\; d = a/2,\; \text{for}\; -a/2 \le z,z' \le a/2\\
R &= L - a/2,\; d = -a/2,\; \text{for}\; -L \le z,z' \le -a/2
\end{aligned}\tag{5.4.6}$$

This leads to

$$D_+^<(x,x') = \mathrm{i}\frac{2}{R}\sum_{n=1}^\infty \int \frac{\mathrm{d}^2\mathbf{K}}{(2\pi)^2}\frac{\exp[\mathrm{i}\,\mathbf{K}\cdot(\mathbf{x}_{\mathrm{T}} - \mathbf{x}'_{\mathrm{T}})]}{2E_n(K,R)}$$

$$\times \sin\frac{n\pi(z-d)}{R}\,\sin\frac{n\pi(z'-d)}{R}\,\exp\left[-\mathrm{i}\,E_n(K,R)|x^0 - x'^0|\right],\tag{5.4.7}$$

$$E_n(K,R) = \sqrt{\mathbf{K}^2 + \frac{n^2\pi^2}{R^2}}\;.\tag{5.4.8}$$

---

[10] $D_+^<(x,x')$ stands for $D_+(x,x')$ for $x^0 < x'^0$.

Similarly, we may consider the case $x'^0 < x^0$, and upon taking the average of both cases, the following expression emerges

$$\langle \text{vac}| \frac{1}{2} \{ \mathbf{E}_T(x) \cdot, \mathbf{E}_T(x') \} | \text{vac} \rangle = \frac{2}{R} \sum_{n=1}^{\infty} \int \frac{d^2 K}{(2\pi)^2} \frac{\exp[i \mathbf{K} \cdot (\mathbf{x}_T - \mathbf{x}'_T)]}{2 E_n(K, R)} \left( \mathbf{K}^2 + 2 \frac{n^2 \pi^2}{R^2} \right)$$

$$\times \sin \frac{n\pi(z - d)}{R} \; \sin \frac{n\pi(z' - d)}{R} \; \exp[-i E_n(K, R)|x^0 - x'^0|]. \qquad (5.4.9)$$

For the component $E_3$, $\delta(z, z')$ in (5.4.3) is to be expressed in a Fourier cosine series expansion:

$$\delta(z, z') = \frac{1}{R} + \frac{2}{R} \sum_{n=1}^{\infty} \cos \frac{n\pi(z - d)}{R} \; \cos \frac{n\pi(z' - d)}{R}. \qquad (5.4.10)$$

This leads from (5.4.2) to

$$\langle \text{vac}| \frac{1}{2} \{ E_3(x), E_3(x') \} | \text{vac} \rangle = \frac{2}{R} \sum_{n=1}^{\infty} \int \frac{d^2 K}{(2\pi)^2} \frac{\exp[i \mathbf{K} \cdot (\mathbf{x}_T - \mathbf{x}'_T)]}{2 E_n(K, R)} \mathbf{K}^2$$

$$\times \cos \frac{n\pi(z - d)}{R} \; \cos \frac{n\pi(z' - d)}{R} \; \exp[-i E_n(K, R)|x^0 - x'^0|]. \qquad (5.4.11)$$

We may repeat a similar analysis for the magnetic field with corresponding boundary conditions: $B^3(x^0, \mathbf{x}_T, z)|_{z = \pm a/2, \pm L} = 0$. The total vacuum energy of the system may be then defined by ($x^0 - x'^0 \equiv T$, $L \to \infty$)

$$\mathcal{E} = \int_{-\infty}^{\infty} dT \delta(T) \int d^3 x \frac{1}{2} \langle \text{vac}| \frac{1}{2} \{ \mathbf{E}(x^0, \mathbf{x}) \cdot, \mathbf{E}(x^0, \mathbf{x}) \} + \frac{1}{2} \{ \mathbf{B}(x^0, \mathbf{x}) \cdot, \mathbf{B}(x^0, \mathbf{x}) \} | \text{vac} \rangle,$$
$$(5.4.12)$$

where the magnetic field contribution is identical to the electric one.

Upon using the elementary integrals

$$\frac{2}{R} \int_0^R dz \sin^2 \frac{n\pi z}{R} = 1 = \frac{2}{R} \int_0^R dz \cos^2 \frac{n\pi z}{R}, \qquad (5.4.13)$$

we obtain the simple expression

$$\mathcal{E} = A \sum_R \int_{-\infty}^{\infty} dT \delta(T) \int \frac{d^2 K}{(2\pi)^2} \sum_{n=1}^{\infty} \sqrt{\mathbf{K}^2 + \frac{n^2 \pi^2}{R^2}} \exp\left[ -i \sqrt{\mathbf{K}^2 + \frac{n^2 \pi^2}{R^2}} \, T \right],$$
$$(5.4.14)$$

where $A = \int d^2 x_T$, is the area of any of the plates, and the sum is over $R = L - a/2$ twice, and $R = a$ once.

Now we use the following basic identity (see Problem 5.3)

$$\frac{\partial}{\partial a}\left(\sqrt{\mathbf{K}^2 + \frac{n^2\pi^2}{R^2}}\ \exp\left[-i\sqrt{\mathbf{K}^2 + \frac{n^2\pi^2}{R^2}}\ T\right]\right)$$

$$= \frac{2n^2\pi^2}{R^3}\left(\frac{\partial R}{\partial a}\right)\frac{d^2}{dT^2}\left(\frac{\partial}{\partial \mathbf{K}^2}\frac{\exp\left[-i\sqrt{\mathbf{K}^2 + \frac{n^2\pi^2}{R^2}}\ T\right]}{\sqrt{\mathbf{K}^2 + \frac{n^2\pi^2}{R^2}}}\right), \qquad (5.4.15)$$

and the integral $(d^2\mathbf{K} = 2\pi|\mathbf{K}|d|\mathbf{K}| = \pi\,d\mathbf{K}^2)$

$$\int \frac{d^2\mathbf{K}}{(2\pi)^2}\frac{\partial}{\partial \mathbf{K}^2}\frac{\exp\left[-i\sqrt{\mathbf{K}^2 + \frac{n^2\pi^2}{R^2}}\ T\right]}{\sqrt{\mathbf{K}^2 + \frac{n^2\pi^2}{R^2}}} = -\frac{1}{4\pi}\frac{\exp[-in\pi T/R]}{n\pi/R}, \qquad (5.4.16)$$

to obtain

$$-\frac{1}{A}\frac{\partial \mathcal{E}}{\partial a} = \frac{1}{4\pi}\sum_R\int_{-\infty}^{\infty}dT\,\delta(T)\sum_{n=1}^{\infty}\frac{\partial R}{\partial a}\frac{d^2}{dT^2}\left(\frac{2n\pi}{R^2}\exp\left[\frac{-in\pi T}{R}\right]\right),$$

$$= \frac{i}{2\pi}\sum_R\int_{-\infty}^{\infty}dT\,\delta(T)\sum_{n=1}^{\infty}\frac{\partial R/\partial a}{R}\frac{d^3}{dT^3}\exp\left[\frac{-in\pi T}{R}\right], \qquad (5.4.17)$$

whose interpretation will soon follow. Upon carrying the elementary summations over $n$ and $R$, with the latter summation over $R$ as described below (5.4.14), the above equation becomes

$$-\frac{1}{A}\frac{\partial \mathcal{E}}{\partial a} = \frac{1}{4\pi a}\int_{-\infty}^{\infty}dT\,\delta(T)\frac{d^3}{dT^3}F(a,T,L), \qquad (5.4.18)$$

where

$$F(a,T,L) = \left[\frac{2i}{1 - e^{-i\pi T/a}} - \frac{2ia}{L - a/2}\frac{1}{1 - e^{-i\pi T/(L-a/2)}}\right]. \qquad (5.4.19)$$

The expansion

$$\frac{2i}{1 - e^{-ix}} = i + \left(\frac{2}{x} - \frac{x}{6} - \frac{x^3}{360} + \cdots\right), \qquad (5.4.20)$$

gives $(L \to \infty)$

$$F(a,T,L) = \left[\left(\frac{2a}{\pi T} - \frac{\pi T}{6a} - \frac{\pi^3 T^3}{360\,a^3} + \cdots\right) - \left(\frac{2a}{\pi T} - \frac{\pi aT}{6L^2} + \cdots\right)\right]. \qquad (5.4.21)$$

The interpretation of this expansion is now clear in view of its application in (5.4.18). The first term $2a/(\pi T)$ within each of the round brackets above gives each an infinite force per unit area on the plate *when taken each separately*, but these forces are equal and in opposite directions, and hence cancel out. The expression $(\mathrm{d}^3/\mathrm{d}T^3)F(a, T, L)$ will then lead to a *finite* attractive force between the plates in question, coming solely, for $L \to \infty$, from the third term $(-\pi^3 T^3/360 a^3)$ within the first round brackets in (5.4.21), leading from (5.4.18) to the final expression

$$-\frac{1}{A}\frac{\partial \mathscr{E}}{\partial a} = -\frac{\pi^2 \hbar c}{240\, a^4}, \tag{5.4.22}$$

where we have re-inserted the fundamental constants $\hbar$ and $c$ in the final expression.

The above beautiful result was first obtained by Casimir [16], and an early experiment by Sparnaay [73] was not in contradiction with Casimir's prediction. More recent experiments (Lamoreaux [41], and, e.g., Bressi et al. [13]), however, were more positive and showed agreement with theoretical predictions of the Casimir effect within a few percent. Casimir forces are not necessarily attractive and may be also repulsive depending on some factors such as on underlying geometrical situations.[11]

The Casimir effect may be also derived by the method of the Riemann zeta function regularization,[12] a method that we use, e.g., in string theory.[13] The above derivation, however, is physically more interesting, and clearly emphasizes the presence of arbitrary large forces, in opposite directions, within and out of the plates which precisely cancel out leading finally to a finite calculable result.

## 5.5  Emission and Detection of Photons

We consider the vacuum-to-vacuum transition amplitude $\langle 0_+ | 0_- \rangle\big|_{\partial^\mu J_\mu = 0}$ in (5.3.13) for the interaction of photons with a conserved external current $\partial^\mu J_\mu(x) = 0$. To simplify the notation only, we will simply write this amplitude as $\langle 0_+ | 0_- \rangle$. It may be rewritten as

$$\langle 0_+ | 0_- \rangle = \exp\left[\frac{i}{2}\int \frac{(\mathrm{d}k)}{(2\pi)^4} J_\mu^*(k)\frac{1}{k^2 - i\epsilon} J^\mu(k)\right], \tag{5.5.1}$$

where note that the reality of $J^\mu(x)$ implies that $J^{*\mu}(k) = J^\mu(-k)$.

---

[11] See, e.g. Kenneth et al. [36] and Milton et al. [64].

[12] See, e.g., Elizalde et al. [20] and Elizalde [19].

[13] See Vol. II: Quantum Field Theory II: Introductions to Quantum Gravity, Supersymmetry, and String Theory, (2016), Springer.

To compute the vacuum persistence probability $|\langle\, 0_+ \mid 0_-\rangle|^2$, we note that

$$\operatorname{Re} \frac{i}{k^2 - i\epsilon} = -\pi\delta(k^2) = -\pi \frac{[\delta(k^0 - |\mathbf{k}|) + \delta(k^0 + |\mathbf{k}|)]}{2\,|\mathbf{k}|}. \tag{5.5.2}$$

This gives

$$|\langle\, 0_+ \mid 0_-\rangle|^2 = \exp\left[-\int \sum_{\lambda=\pm} \frac{d^3\mathbf{k}}{(2\pi)^3 2k^0}\, (J_\mu^*(k)\, e_\lambda^\mu)\, (J_\nu(k)\, e_\lambda^{*\nu})\right] \le 1, \qquad k^0 = +|\mathbf{k}|, \tag{5.5.3}$$

where we have used the completeness relation, expressed in terms of the Minkowski metric, in (5.2.19): $\eta^{\mu\nu} = (k^\mu \underline{k}^\nu + \underline{k}^\mu k^\nu)/k\underline{k} + \sum_{\lambda=\pm 1} e_\lambda^\mu e_\lambda^{*\nu}$, with, e.g., complex representation of polarizations vectors, and used the conservation laws: $J_\mu^*(k)k^\mu = 0$, $k^\nu J_\nu(k) = 0$. This gives the consistent probabilistic result that the vacuum persistence probability does not exceed one.

We use the convenient notation for bookkeeping purposes[14]

$$i J_\mu(k)\, e_\lambda^{*\mu} \sqrt{\frac{d^3\mathbf{k}}{(2\pi)^3 2k^0}} = i J_{k\lambda}, \tag{5.5.4}$$

and don't let the notation scare you. We may then repeat the analysis given through Eqs. (3.3.28), (3.3.29), (3.3.30), (3.3.31), (3.3.32), (3.3.33), (3.3.34), (3.3.35) and (3.3.36), now applied to photons, and using the expressions in Eqs. (5.5.1), (5.5.3), (5.5.4), to infer from Eqs. (3.3.37) and (3.3.38), that the probability that an external source $J^\mu$ emits $N$ photons, $N_{k\lambda}$ of which have each momentum $\mathbf{k}$, and polasrization $e_\lambda^\mu$, and so on, is given by

$$\operatorname{Prob}[N] = \sum_{N=(N_{k_1\lambda_1}+N_{k_2\lambda_2}+\cdots)} \frac{\left(|J_{k_1\lambda_1}|^2\right)^{N_{k_1\lambda_1}}}{N_{k_1\lambda_1}!} \frac{\left(|J_{k_2\lambda_2}|^2\right)^{N_{k_2\lambda_2}}}{N_{k_2\lambda_2}!} \cdots |\langle\, 0_+ \mid 0_-\rangle|^2. \tag{5.5.5}$$

Now we use the multinomial expansion[15]

$$\sum_{N=(N_1+N_2+\cdots)} \frac{(|x_1|)^{N_1}(|x_2|)^{N_2}}{N_1!\ N_2!} \cdots = \frac{(|x_1| + |x_2| + \cdots)^N}{N!}, \tag{5.5.6}$$

---

[14] This was conveniently introduced by Schwinger [70].
[15] See, e.g., Manoukian [57].

and thanks to the convenient bookkeeping notation introduced above in (5.5.4), we also have

$$|J_{k_1\lambda_1}|^2 + |J_{k_2\lambda_2}|^2 + \cdots = \int \sum_\lambda \frac{d^3\mathbf{k}}{(2\pi)^3 2k^0} |J_\mu(k)\, e_\lambda^{*\mu}|^2. \tag{5.5.7}$$

This allows us to rewrite (5.5.5) as

$$\text{Prob}\,[N] = \frac{(\langle N\rangle)^N}{N!}\, e^{-\langle N\rangle}, \tag{5.5.8}$$

where

$$\langle N\rangle = \int \frac{d^3\mathbf{k}}{(2\pi)^3 2k^0}\, J_\mu^*(k) J^\mu(k), \tag{5.5.9}$$

and we have used (5.5.3) to write $|\langle\, 0_+ \mid 0_-\rangle|^2 = e^{-\langle N\rangle}$. We recognize $\text{Prob}\,[N]$, in (5.5.8), as defining the Poisson distribution[16] with $\langle N\rangle$ denoting the *average number* of photons emitted by the external source.

In evaluating $\langle N\rangle$, it is often more convenient to rewrite its expression involving integrals in spacetime. This may be obtained directly from (5.5.9) (see also (5.5.2) and (5.5.3)) to be

$$\langle N\rangle = \int (dx)(dx')\, J^\mu(x)\, [\,\text{Im}\, D_+(x-x')\,]\, J_\mu(x'). \tag{5.5.10}$$

where, as in (5.5.9), $\partial_\mu J^\mu = 0$, and $D_+(x-x')$ is defined in (5.3.7). Equation (5.5.10) may be equivalently rewritten as

$$\langle N\rangle = \int (dx)(dx')\, J^\mu(x)\, J_\mu(x') \int \frac{d^3\mathbf{k}}{(2\pi)^3 2k^0}\, e^{ik(x-x')}, \quad k^0 = |\mathbf{k}|, \tag{5.5.11}$$

using the reality condition of the current. For an application of the above expression in deriving the general classic radiation theory see Manoukian [58].

We may infer from Eqs. (3.3.39) and (3.3.41), that the amplitude of a current source, as a detector, to absorb a photon with momentum $\mathbf{k}$ and polarization $e_\lambda^\mu$, and the amplitude of a current source, as an emitter, to emit a photon with the same attributes which escapes this parent source, to be used in scattering theory, are given,

---

[16]See, e.g., *op. cit.*

respectively, by

$$\langle\, 0_+ |\mathbf{k}\lambda \rangle = \mathrm{i}\, J_\mu^*(k)\, e_\lambda^\mu \sqrt{\frac{\mathrm{d}^3\mathbf{k}}{(2\,\pi)^3\, 2k^0}}. \tag{5.5.12}$$

$$\langle\, \mathbf{k}\lambda \mid 0_- \rangle = \mathrm{i}\, J_\mu(k)\, e_\lambda^{*\mu} \sqrt{\frac{\mathrm{d}^3\mathbf{k}}{(2\,\pi)^3\, 2k^0}},\, \tag{5.5.13}$$

where we have omitted photons which are emitted and reabsorbed by the same source as they do not participate in a scattering process, by dividing by the corresponding vacuum-to-vacuum amplitudes $\langle\, 0_+ \mid 0_- \rangle$.[17]

## 5.6 Photons in a Medium

We consider a homogeneous and isotropic medium of permeability $\mu$, and permittivity $\varepsilon$. To describe photons in such a medium, one simply scales $F_{0\,i}F^{0\,i} \to \varepsilon F_{0\,i}F^{0\,i}$, and $F_{ij}F^{ij} \to F_{ij}F^{ij}/\mu$, in the Lagrangian density, where $F^{\mu\nu} = \partial^\mu A^\nu - \partial^\nu A^\mu$.[18] That is, the Lagrangian density becomes

$$\mathscr{L} = -\frac{1}{4\,\mu}\, F_{ij}\, F^{ij} - \frac{\varepsilon}{2}\, F_{0i}\, F^{0i} + J^\mu A_\mu. \tag{5.6.1}$$

We take advantage of our analysis in Sect. 5.5, and reduce the problem to the one carried out in that section. To this end, we consider the following scalings:

$$x^0 = \sqrt{\mu\varepsilon}\,\underline{x}^0, \quad \mathbf{x} = \underline{\mathbf{x}}, \qquad \partial_0 = \frac{1}{\sqrt{\mu\varepsilon}}\,\underline{\partial}_0, \quad \mathbf{\nabla} = \underline{\mathbf{\nabla}}, \tag{5.6.2}$$

$$A^0(x) = \frac{1}{\mu^{1/4}\,\varepsilon^{3/4}}\,\underline{A}^0(x), \qquad \mathbf{A}(x) = \frac{\mu^{1/4}}{\varepsilon^{1/4}}\,\underline{\mathbf{A}}(x), \tag{5.6.3}$$

$$J^0(x) = \frac{\varepsilon^{1/4}}{\mu^{1/4}}\,\underline{J}^0(x), \qquad \mathbf{J}(x) = \frac{1}{\mu^{3/4}\varepsilon^{1/4}}\,\underline{\mathbf{J}}(x). \tag{5.6.4}$$

---

[17] See discussion above Eq. (3.3.39).

[18] Note that the scaling factors are not $\varepsilon^2$, $1/\mu^2$, respectively, as one may naïvely expect. The reason is that functional differentiation of the action, with respect to the vector potential, involving the quadratic terms $\varepsilon F_{0\,i}F^{0\,i}$, $F_{ij}F^{ij}/\mu$, generate the linear terms corresponding to the electric and magnetic fields components which are just needed in deriving Maxwell's equations.

The action then simply becomes

$$W = \int (d\underline{x}) \left[ -\frac{1}{4} \left( \partial_\mu \underline{A}_\nu(x) - \partial_\nu \underline{A}_\mu(x) \right) \left( \partial^\mu \underline{A}^\nu(x) - \partial^\nu \underline{A}_\mu(x) \right) + \underline{J}^\mu(x) \underline{A}_\mu(x) \right].$$
(5.6.5)

up to a gauge fixing constraint. Note that the argument of $\underline{A}^\mu(x)$ is $x$ and not $\underline{x}$, also that

$$\partial_\mu \underline{J}^\mu(x) = \mu^{3/4} \varepsilon^{1/4} \partial_\mu J^\mu(x).$$
(5.6.6)

The field equation is given by

$$-\Box \frac{\langle 0_+ | \underline{A}^\nu(x) | 0_- \rangle}{\langle 0_+ | 0_- \rangle} = \underline{J}^\nu(x).$$
(5.6.7)

up to gauge fixing terms, proportional to $\partial^\nu$, which do not contribute when one finally imposes current conservation. Thus upon defining the propagator $D_+(\underline{x} - \underline{x}')$ satisfying

$$-\Box D_+(\underline{x} - \underline{x}') = \delta^{(4)}(\underline{x} - \underline{x}'),$$
(5.6.8)

we have

$$\frac{\langle 0_+ | \underline{A}^\nu(x) | 0_- \rangle}{\langle 0_+ | 0_- \rangle} = \int (d\underline{x}') D_+(x - \underline{x}') \underline{J}^\nu(x'),$$
(5.6.9)

where note that $\delta^{(4)}(\underline{x} - \underline{x}') (d\underline{x}') = \delta^{(4)}(x - x') (dx')$.

Hence from (5.5.10), we have for the average number of photons emitted by a current source in the medium

$$\langle N \rangle = \int (d\underline{x}) (d\underline{x}') \underline{J}^\mu(x) [\operatorname{Im} D_+(\underline{x} - \underline{x}')] \underline{J}_\mu(x'),$$
(5.6.10)

$$\langle 0_+ | 0_- \rangle = \exp\left[ \frac{i}{2} \int (d\underline{x}) (d\underline{x}') \underline{J}^\mu(x) D_+(\underline{x} - \underline{x}') \underline{J}_\mu(x') \right],$$
(5.6.11)

for a conserved current $\partial_\mu \underline{J}^\mu(x) = 0$ (see (5.6.6)). From (5.6.2) and (5.6.4), the above equation becomes

$$\langle 0_+ | 0_- \rangle = \exp\left[ \frac{i}{2} \frac{\mu^{1/2}}{\varepsilon^{1/2}} \int (dx) (dx') \left( \mathbf{J}(x) \cdot \mathbf{J}(x') - \frac{1}{\mu \varepsilon} J^0(x) J^0(x') \right) D_+(\underline{x} - \underline{x}') \right],$$
(5.6.12)

$$D_+(\underline{x} - \underline{x}') = \int \frac{(dk)}{(2\pi)^4} e^{i\mathbf{k} \cdot (\mathbf{x} - \mathbf{x}')} \frac{e^{-ik^0(x^0 - x'^0)/\sqrt{\mu \varepsilon}}}{k^2 - i\epsilon}.$$
(5.6.13)

Since the current $J^\mu(x)$ is real, we may rewrite (5.6.10) as

$$\langle N \rangle = \frac{\mu^{1/2}}{\varepsilon^{1/2}} \int (dx)\,(dx')\Big(\mathbf{J}(x)\cdot\mathbf{J}(x') - \frac{1}{\mu\varepsilon}J^0(x)J^0(x')\Big)$$

$$\times \int \frac{d^3\mathbf{k}}{(2\pi)^3\,2\,|\mathbf{k}|}\, e^{i\,\mathbf{k}\cdot(\mathbf{x}-\mathbf{x}')} e^{-i\,|\mathbf{k}|(x^0-x'^0)/\sqrt{\mu\varepsilon}}. \tag{5.6.14}$$

Upon inserting the identity

$$1 = \int_0^\infty d\omega\; \delta\Big(\omega - \frac{|\mathbf{k}|}{\sqrt{\mu\varepsilon}}\Big), \tag{5.6.15}$$

in (5.6.14), we get $\langle N \rangle = \int_0^\infty d\omega N(\omega)$, ( $\mathbf{k} = |\mathbf{k}|\mathbf{n}$ )

$$N(\omega) = \frac{\mu\sqrt{\mu\varepsilon}}{16\pi^3}\,\omega \int (dx)\,(dx')\Big(\mathbf{J}(x)\cdot\mathbf{J}(x') - \frac{1}{\mu\varepsilon}J^0(x)J^0(x')\Big)$$

$$\times e^{-i\omega(x^0-x'^0)} \int d\Omega\, e^{i\sqrt{\mu\varepsilon}\,\omega\,\mathbf{n}\cdot(\mathbf{x}-\mathbf{x}')}. \tag{5.6.16}$$

Consider a charged particle, of charge e, moving along the z-axis with constant speed v. Then

$$J^i(x) = e\,v\,\delta^{i3}\,\delta(x^1)\,\delta(x^2)\,\delta(x^3 - vx^0), \quad J^0(x) = e\,\delta(x^1)\,\delta(x^2)\,\delta(x^3 - vx^0). \tag{5.6.17}$$

Then $\mathbf{n}\cdot(\mathbf{x}-\mathbf{x}') = (x^3 - x'^3)\cos\theta$, $d\Omega = 2\pi\,d\cos\theta$. Upon integrating over $x^0$, $x'^0$, and then over $x'^3$, we obtain, per unit length,

$$N(\omega)\Big|_{\text{unit length}} = \frac{\mu e^2}{4\pi}\Big(1 - \frac{1}{\mu\varepsilon\,v^2}\Big)\int_{-1}^1 d\cos\theta\; \delta\Big(\cos\theta - \frac{1}{v\sqrt{\mu\varepsilon}}\Big). \tag{5.6.18}$$

where the unit length arises from the integral $\int dx^3$. The average number of photons emitted with angular frequency in the interval $(\omega, \omega + d\omega)$, as the charged particle traverses a unit length, is then

$$N(\omega)\Big|_{\text{unit length}} = \frac{\mu e^2}{4\pi}\Big(1 - \frac{1}{\mu\varepsilon\,v^2}\Big), \tag{5.6.19}$$

and, on account the property of a cosine function, the latter does not vanish only for $v > 1/\sqrt{\mu\varepsilon}$. For a large number of charged particles this number need not be small.

The expression in (5.6.19) is constant in $\omega$ and hence cannot be integrated over $\omega$ for arbitrary large $\omega$. A quantum correction treatment, however, provides a natural cut-off in $\omega$ emphasizing the importance of the inclusion of radiative corrections.[19]

This form of radiation is referred to as Čerenkov radiation. It is interesting that astronauts during Apollo missions have reported of "seeing" flashes of light even with their eyes closed. An explanation of this was attributed to high energy cosmic particles, encountered freely in outer space, that would pass through one's eyelids causing Čerenkov radiation to occur within one's eye itself.[20]

## 5.7 Quantum Electrodynamics, Covariant Gauges: Setting Up the Solution

We apply the functional differential formalism (Sects. 4.6 and 4.8), via the quantum dynamical principle (QDP), to derive an explicit expression of the full QED vacuum-to-vacuum transition amplitide $\langle\, 0_+ \mid 0_-\rangle$ in covariant gauges (Sect. 5.3). By carrying the relevant functional differentiations coupled to a functional Fourier transform (Sect. 2.6), the path integral form of $\langle\, 0_+ \mid 0_-\rangle$ is also derived. The corresponding expressions in the Coulomb gauge will be derived in Sect. 5.14. Gauge transformations of $\langle\, 0_+ \mid 0_-\rangle$ between covariant gauges as well as between covariant gauges and the Coulomb one will be derived in Sect. 5.15.

### 5.7.1 The Differential Formalism (QDP) and Solution of QED in Covariant Gauges

The purpose of this subsection is to find explicit expressions for $\langle\, 0_+ \mid 0_-\rangle$ of QED in covariant gauges. The vacuum-to-vacuum transition amplitude $\langle\, 0_+ \mid 0_-\rangle$ is a generating functional for all possible underlying physical processes and also for extracting various components of the theory, such a propagators and Green functions.

For the Lagrangian density of QED, we consider

$$\mathscr{L} = -\frac{1}{4} F_{\mu\nu}F^{\mu\nu} + \frac{1}{2}\left(\frac{\partial_\mu \overline{\psi}}{i}\gamma^\mu \psi - \overline{\psi}\gamma^\mu \frac{\partial_\mu \psi}{i}\right) - m_0 \overline{\psi}\psi$$

$$+ \frac{1}{2} e_0[\overline{\psi}, \gamma^\mu \psi]A_\mu + \overline{\eta}\,\psi + \overline{\psi}\,\eta + A_\mu J^\mu - \partial_\mu A^\mu \chi + \frac{\lambda}{2}\chi^2, \qquad (5.7.1)$$

---

[19] For these additional details, see, e.g., Manoukian and Charuchittapan [59].

[20] See, e.g., Fazio et al. [24], Pinsky et al. [66], and McNulty et al. [62].

where the field $\chi$ leads to a constraint on the vector potential $A^\mu$. That is, the underlying constraint is *derived* from the Lagrangian density. As we will also see $\chi$, in turn, may be eliminated in favor of $\partial_\mu A^\mu$. We have written the parameters $e_0$, $m_0$, with a subscript 0, to signal the facts that these are not the parameters directly measured, as discussed in the Introduction of the book. The electromagnetic current $j^\mu = e_0[\,\overline{\psi}, \gamma^\mu \psi\,]/2$, has been also written as a commutator, consistent with charge conjugation as discussed in Sect. 3.6 (see (3.6.9)).

The field equations together with the constraint are

$$-\partial_\mu F^{\mu\nu} = J^\nu + j^\nu + \partial^\nu \chi, \tag{5.7.2}$$

$$\left[\gamma^\mu\left(\frac{\partial_\mu}{i} - e_0 A_\mu\right) + m_0\right]\psi = \eta, \tag{5.7.3}$$

$$\overline{\psi}\left[-\gamma^\mu\left(\frac{\overleftarrow{\partial}_\mu}{i} + e_0 A_\mu\right) + m_0\right] = \overline{\eta}, \tag{5.7.4}$$

$$\partial_\mu A^\mu = \lambda \chi. \tag{5.7.5}$$

We take the expectation value of (5.7.2) between the vacuum states $\langle\, 0_+|\,.\,\,|\, 0_-\rangle$, take its $\partial_\nu$ derivative to derive the following equation

$$-\Box\, \langle\, 0_+|\chi(x)\,|\, 0_-\rangle = \partial_\nu\big[J^\nu(x)\langle\, 0_+\,|\, 0_-\rangle + \langle\, 0_+|j^\nu(x)\,|\, 0_-\rangle\big], \tag{5.7.6}$$

and use it in (5.7.5), to obtain the following equation of constraint

$$\partial_\mu \langle\, 0_+|A^\mu(x)\,|\, 0_-\rangle = \lambda \int (\mathrm{d}x')D_+(x-x')\partial'_\mu\big[J^\mu(x')\langle\, 0_+\,|\, 0_-\rangle + \langle\, 0_+|j^\mu(x')\,|\, 0_-\rangle\big]. \tag{5.7.7}$$

This, in turn, eliminates the $\chi$ field.

Upon taking the matrix element of (5.7.3) between the vacuum states, as indicated by the following notation $\langle\, 0_+|\,.\,\,|\, 0_-\rangle$, we obtain

$$\left[\gamma^\mu\left(\frac{\partial_\mu}{i} - e_0\,(-i)\frac{\delta}{\delta J^\mu}\right) + m_0\right]\langle\, 0_+|\psi\,\,|\, 0_-\rangle = \eta\langle\, 0_+\,|\, 0_-\rangle, \tag{5.7.8}$$

where we have used the quantum dynamical principle to write

$$\langle\, 0_+|(A_\mu(x')\psi(x))_+\,|\, 0_-\rangle = (-i)\frac{\delta}{\delta J^\mu(x')}\langle\, 0_+|\psi(x)\,|\, 0_-\rangle, \tag{5.7.9}$$

with the time ordered product in the limit $x' \leftrightarrow x$ understood as an average of the product of the two fields, as discussed in Sect. 4.9.

The vector potential $A_\mu$ has been eliminated in favor of the "classical field", represented by, $(-i)\delta/\delta J^\mu$ in (5.7.8). Thus introducing the "classical field"

$$(-i)\frac{\delta}{\delta J^\mu(x)} \equiv \widehat{A}_\mu(x), \tag{5.7.10}$$

and the Green function $S_+(x, x'; e_0\widehat{A})$ satisfying the equations

$$\left[\gamma^\mu\left(\frac{\partial_\mu}{i} - e_0\widehat{A}_\mu\right) + m_0\right]_{ab} S_{+bc}(x, x'; e_0\widehat{A}) = \delta^{(4)}(x, x')\delta_{ac}, \qquad (5.7.11)$$

$$\left[-\gamma^\mu\left(\frac{\partial_\mu}{i} + e_0\widehat{A}_\mu\right) + m_0\right]_{ba} S_{+cb}(x', x; e_0\widehat{A}) = \delta^{(4)}(x', x)\delta_{ac}, \qquad (5.7.12)$$

we may, from (5.7.8), conveniently write

$$\langle 0_+|\psi_a(x)|0_-\rangle = \int (dx')\, S_{+ab}(x, x'; e_0\widehat{A})\, \eta_b(x')\langle 0_+ | 0_-\rangle. \qquad (5.7.13)$$

Similarly we have

$$\langle 0_+|\overline{\psi}_a(x)|0_-\rangle = \int (dx')\, S_{+ca}(x', x; e_0\widehat{A})\, \overline{\eta}_c(x')\langle 0_+ | 0_-\rangle. \qquad (5.7.14)$$

As we have already seen in Sect. 3.2, (3.2.12), or directly from (5.7.11) and (5.7.12), the Green function $S_+(x, x'; e_0\widehat{A}_\mu)$ satisfies the integral equation

$$S_+(x, x'; e_0\widehat{A}) = S_+(x-x') + e_0 \int (dy)\, S_+(x-y)\, \gamma\widehat{A}(y)\, S_+(y, x'; e_0\widehat{A}), \qquad (5.7.15)$$

as is readily verified, where $S_+(x - x')$ is the free Dirac propagator satisfying,

$$\left(\frac{\gamma\partial}{i} + m_0\right)S_+(x-x') = \delta^{(4)}(x-x'), \quad S_+(x'-x)\left(-\frac{\gamma\partial}{i} + m_0\right) = \delta^{(4)}(x'-x). \qquad (5.7.16)$$

To obtain an expression for $\langle 0_+ | 0_-\rangle$ in QED, we use the quantum dynamical principle, and take its derivative with respect to $e_0$ to obtain[21] (see also (3.6.12), (3.6.13), (3.6.14) and (3.6.15))

$$(-i)\frac{\partial}{\partial e_0}\langle 0_+ | 0_-\rangle = \int (dx)\langle 0_+|\left(A^\mu(x)\overline{\psi}(x)\gamma^\mu\psi(x)\right)_+ | 0_-\rangle \qquad (5.7.17)$$

$$= \int (dx)(-i)\frac{\delta}{\delta J_\mu(x)}\langle 0_+|\left(\overline{\psi}(x)\gamma^\mu\psi(x)\right)_+ | 0_-\rangle \qquad (5.7.18)$$

$$= \int (dx)(-i)\frac{\delta}{\delta J_\mu(x)}(i)\frac{\delta}{\delta\eta(x)}\gamma_\mu(-i)\frac{\delta}{\delta\overline{\eta}(x)}\langle 0_+ | 0_-\rangle. \qquad (5.7.19)$$

---

[21]Recall from Sect. 3.9, that within a time ordered product the Dirac fields anti-commute.

Equation (5.7.19) may be readily integrated with respect to $e_0$, to obtain

$$\langle\, 0_+ \mid 0_-\rangle = \exp\!\left[\, e_0 \int (dx)\frac{\delta}{\delta J_\mu(x)}\frac{\delta}{\delta\eta(x)}\,\gamma_\mu\frac{\delta}{\delta\overline{\eta}(x)}\,\right]\!\langle\, 0_+ \mid 0_-\rangle_0, \qquad (5.7.20)$$

where from (3.3.28) and (5.3.11),

$$\langle\, 0_+ \mid 0_-\rangle_0 = \exp\!\left[\, i\int (dx)\,(dx')\,\overline{\eta}(x)S_+(x-x')\eta(x')\,\right]$$

$$\times \exp\!\left[\,\frac{i}{2}\int (dx)\,(dx')\,J_\mu(x)D^{\mu\nu}(x-x')J_\nu(x')\,\right], \qquad (5.7.21)$$

where $S_+(x-x')$ is given in (3.1.9), and $D^{\mu\nu}(x-x')$ is given in (5.3.9). Thus $\langle\, 0_+ \mid 0_-\rangle$ may be obtained by functional differentiations of the explicitly given expression of $\langle\, 0_+ \mid 0_-\rangle_0$.[22]

A far more interesting expression for $\langle\, 0_+ \mid 0_-\rangle$, and a more useful one for practical applications, is obtained by examining (5.7.18). To this end, upon taking the functional derivative of $\langle\, 0_+|\psi_a(x) \mid 0_-\rangle$, as given in (5.7.13), with respect to $i\delta/\delta\eta_c(x)$ and multiplying it by $\gamma^\mu_{ca}$, gives

$$\langle\, 0_+|j^\mu(x) \mid 0_-\rangle = i\,\mathrm{Tr}[\,\gamma^\mu S_+(x,x;e_0\widehat{A})\,]\langle\, 0_+ \mid 0_-\rangle$$

$$-\int (dx')\,S_{+\,ab}(x,x';e_0\widehat{A})\,\eta_b(x')\,\gamma^\mu_{ca}\,\langle\, 0_+|\overline{\psi}_c(x) \mid 0_-\rangle, \qquad (5.7.22)$$

using, in the process, the anti-commutativity of Grassmann sources, which from (5.7.14) leads to

$$\langle\, 0_+|(\overline{\psi}(x)\gamma^\mu\psi(x))_+ \mid 0_-\rangle = i\,\mathrm{Tr}[\,\gamma^\mu S_+(x,x;e_0\widehat{A})\,]\langle\, 0_+ \mid 0_-\rangle$$

$$+\int (dx')(dx'')\,\overline{\eta}(x'')\,S_+(x'',x;e_0\widehat{A})\,\gamma^\mu S_+(x,x';e_0\widehat{A})\,\eta(x')\,\langle\, 0_+ \mid 0_-\rangle. \tag{5.7.23}$$

Now we multiply this equation by $\widehat{A}_\mu(x)$, as defined in (5.7.10), and integrate over $x$, to obtain

$$\int (dx)\langle\, 0_+|(A_\mu(x)\overline{\psi}(x)\gamma^\mu\psi(x))_+ \mid 0_-\rangle = i\int (dx)\widehat{A}_\mu(x)\mathrm{Tr}[\gamma^\mu S_+(x,x;e_0\widehat{A})]\,\langle\, 0_+ \mid 0_-\rangle$$

$$+\int (dx)\,(dx')\,(dx'')\,\overline{\eta}(x'')\,S_+(x'',x;e_0\widehat{A})\,\gamma\widehat{A}(x)\,S_+(x,x';e_0\widehat{A})\,\eta(x')\langle\, 0_+ \mid 0_-\rangle. \tag{5.7.24}$$

---

[22]The path integral version of (5.7.20) is derived in the next subsection.

In Problem 5.5 it is shown that

$$\frac{\partial}{\partial e_0} S_+(x'', x'; e_0\widehat{A}) = \int (dx) S_+(x'', x; e_0\widehat{A}) \, \gamma\widehat{A}(x) S_+(x, x'; e_0\widehat{A}), \qquad (5.7.25)$$

which from (5.7.24) and (5.7.17) gives

$$(-i)\frac{\partial}{\partial e_0} \langle 0_+ \mid 0_- \rangle = i \int (dx) \widehat{A}_\mu(x) \, \mathrm{Tr}\left[\gamma^\mu S_+(x, x; e_0\widehat{A})\right] \langle 0_+ \mid 0_- \rangle$$

$$+ \left(\frac{\partial}{\partial e_0} \int (dx'') (dx') \, \overline{\eta}(x'') S_+(x'', x'; e_0\widehat{A})\eta(x')\right) \langle 0_+ \mid 0_- \rangle. \qquad (5.7.26)$$

We may now integrate over $e_0$ to obtain[23] $\left(\widehat{A}^\mu(x) \equiv (-i)\delta/\delta J_\mu(x)\right)$

$$\langle 0_+ \mid 0_- \rangle = \exp i \left[\int (dx)(dx') \, \overline{\eta}(x) S_+(x, x'; e_0\widehat{A}) \, \eta(x')\right]$$

$$\times \exp\left[-\int (dx) \int_0^{e_0} de' \, \mathrm{Tr}\left[\gamma\widehat{A}(x) S_+(x, x; e'\widehat{A})\right]\right]\langle 0_+ \mid 0_- \rangle_{0\gamma}, \qquad (5.7.27)$$

with the normalization condition

$$\langle 0_+ \mid 0_- \rangle\Big|_{J^\mu, \eta, \overline{\eta}\, = 0, \, e_0 \to 0} = 1, \qquad (5.7.28)$$

and where the Trace operation in (5.7.27) is over gamma matrix indices, and now $\langle 0_+ \mid 0_- \rangle_{0\gamma}$ is given by

$$\langle 0_+ \mid 0_- \rangle_{0\gamma} = \exp\left[\frac{i}{2}\int (dx)(dx') J_\mu(x) D^{\mu\nu}(x - x') J_\nu(x')\right], \qquad (5.7.29)$$

$$D^{\mu\nu}(x - x') = \int \frac{(dk)}{(2\pi)^4}\left[\eta^{\mu\nu} - (1 - \lambda)\frac{k^\mu k^\nu}{k^2}\right]\frac{e^{ik(x-x')}}{k^2 - i\epsilon}. \qquad (5.7.30)$$

We note that from (5.7.22), that $i\,\mathrm{Tr}\left[\gamma^\mu S_+(x, x; e_0\widehat{A})\right]$, with coincident points, which is to be applied to $\langle 0_+ \mid 0_- \rangle$, is nothing but the vacuum expectation value of the current $j^\mu(x)$, (for $\eta, \overline{\eta} = 0$), up to the factor $e_0$, and must, by itself, be gauge invariant under the replacement of $\widehat{A}^\mu(x)$ by $\widehat{A}^\mu(x) + \partial^\mu \Lambda(x)$. It is studied in detail in Sect. 3.6 and Appendix IV at the end of the book, and may be explicitly spelled out by simply making the obvious substitution $S_+(x, x'; e'A) \to S_+(x, x'; e'\widehat{A})$ in

---

[23]The expressions in (5.7.20) and (5.7.27) are appropriately referred to as generating functionals. As we have no occasion to deal with subtleties in defining larger vector spaces to accommodate covariant gauges, we will not go into such technicalities here.

there, for general $e'$. In Problem 5.6, it is shown that the constraint equation (5.7.7) is automatically satisfied, as expected and as it should.

Finally, we note that (5.7.27) may be also simply rewritten as

$$\langle\, 0_+ \mid 0_-\rangle = \exp\Big( \text{i} \Big[ \int (\text{d}x)(\text{d}x')\, \overline{\eta}(x)\, S_+(x, x'; e_0\widehat{A})\, \eta(x') \Big] \Big) \langle\, 0_+ \mid 0_-\rangle_\gamma,$$

(5.7.31)

*where* $\langle\, 0_+ \mid 0_-\rangle_\gamma$ represents the full QED theory with no external electron lines, and involves all the closed Fermion loops, with or without external photon lines, to all orders of the theory.

Before closing this section, we note that the analysis carried out through (3.6.7), (3.6.8) and (3.6.9), as applied to (5.7.3)/(5.7.4), shows that

$$\partial_\mu j^\mu(x) = \text{i}\, e_0\Big[\overline{\psi}(x)\eta(x) - \overline{\eta}(x)\psi(x)\Big], \qquad j^\mu(x) = e_0\frac{1}{2}\, [\overline{\psi}, \gamma^\mu\psi].$$

(5.7.32)

That is, in the absence of external Fermi sources, the current $j^\mu(x)$ is conserved.

In the next section, we examine the explicit expression for $\langle\, 0_+ \mid 0_-\rangle$ in (5.7.27), in some detail in view of applications in QED.

### 5.7.2  *From the Differential Formalism to the Path Integral Expression for* $\langle\, 0_+ \mid 0_-\rangle$

From (2.6.18), (2.6.19), (2.6.23), we explicitly have

$$\exp \text{i}\, \overline{\eta}S_+\eta = N_1^{-1} \int (\mathscr{D}\overline{\rho}\, \mathscr{D}\rho) \exp \text{i}\Big[ -\overline{\rho}\Big(\frac{\gamma\partial}{\text{i}} + m_0\Big)\rho + \overline{\rho}\eta + \overline{\eta}\rho\Big],$$

(5.7.33)

$$N_1 = \int (\mathscr{D}\overline{\rho}\, \mathscr{D}\rho) \exp \text{i}\Big[ -\overline{\rho}\Big(\frac{\gamma\partial}{\text{i}} + m_0\Big)\rho\Big],$$

(5.7.34)

with $S_+$ now defined in terms of $m_0$.

On the other hand, in the momentum description, the free photon propagator, in a covariant gauge specified by the parameter $\lambda$, is from (5.3.9) given by

$$D^{\mu\nu}(Q) = \Big(\eta^{\mu\nu} - (1-\lambda)\frac{Q^\mu Q^\nu}{Q^2}\Big) \frac{1}{Q^2 - \text{i}\epsilon},$$

(5.7.35)

from which its inverse follows to be

$$D_{\mu\nu}^{-1}(Q) = \eta_{\mu\nu}Q^2 - \Big(1 - \frac{1}{\lambda}\Big) Q_\mu Q_\nu, \qquad D_{\mu\rho}^{-1}(Q^2)\, D^\rho_{\ \nu}(Q^2) = \eta_{\mu\nu}.$$

(5.7.36)

We may now refer to (2.6.31) to write

$$\exp\left[\frac{i}{2}\int (dx)\,(dx')\,J_\mu(x)D^{\mu\nu}(x-x')J_\nu(x')\right]$$

$$= N_2^{-1}\int(\mathscr{D}a)\,\exp i\left[-\frac{1}{2}a^\mu\Big([-\eta_{\mu\nu}\,\Box+\partial_\mu\partial_\nu]-\frac{1}{\lambda}\partial_\mu\partial_\nu\Big)a^\nu+a^\mu J_\mu\right],$$

$$(5.7.37)$$

$$N_2 = \int(\mathscr{D}a)\,\exp i\left[-\frac{1}{2}a^\mu\Big([-\eta_{\mu\nu}\,\Box+\partial_\mu\partial_\nu]-\frac{1}{\lambda}\partial_\mu\partial_\nu\Big)a^\nu\right]. \quad (5.7.38)$$

where it is understood that $(\mathscr{D}a)$ involves a product over the indices of $a^\mu$ as well. By formally integrating by parts ($f^{\mu\nu}=\partial^\mu a^\nu-\partial^\nu a^\mu$), we obtain

$$\int(dx)\left[-\frac{1}{2}a^\mu\Big([-\eta_{\mu\nu}\,\Box+\partial_\mu\partial_\nu]-\frac{1}{\lambda}\partial_\mu\partial_\nu\Big)a^\nu\right]$$

$$=\int(dx)\left[-\frac{1}{4}f^{\mu\nu}(x)f_{\mu\nu}(x)-\frac{1}{2\lambda}\partial_\mu a^\mu(x)\,\partial_\nu a^\nu(x)\right]. \quad (5.7.39)$$

At this stage, we may introduce a scalar field $\varphi$ and, up to an overall unimportant multiplicative factor, write

$$\exp i\left[-\frac{1}{2\lambda}\partial_\mu a^\mu\,\partial_\nu\,a^\nu\right]=\int(\mathscr{D}\varphi)\,\delta(\partial_\nu\,a^\nu-\lambda\varphi)\exp i\left[-\varphi\,\partial_\mu\,a^\mu+\frac{\lambda}{2}\varphi^2\right].$$

$$(5.7.40)$$

Finally from (5.7.20) and (5.7.21) we thus obtain

$$\langle\,0_+\mid 0_-\rangle=\frac{1}{N_1N_2}\int(\mathscr{D}\overline{\rho}\,\mathscr{D}\rho)\,(\mathscr{D}a)\,(\mathscr{D}\varphi)\,\delta(\partial_\nu\,a^\nu-\lambda\,\varphi)\exp[\,i\int(dx)\mathscr{L}_c(x)],$$

$$(5.7.41)$$

up to an overall unimportant multiplicative constant. Here $\mathscr{L}_c$ is the Lagrangian density, including the external sources $J^\mu$, $\overline{\eta}$, $\eta$, obtained from the one in (5.7.1) upon carrying out the substitutions

$$\psi\to\rho,\quad\overline{\psi}\to\overline{\rho},\quad A^\mu\to a^\mu, \quad (5.7.42)$$

and with the auxiliary field $\chi\to\varphi$, where we recall that $\rho$, $\overline{\rho}$ are Grassmann variables, i.e., are anti-commuting.

This equation is quite interesting as it explicitly shows the gauge constraint via the delta functional $\delta(\partial_\nu\,a^\nu-\lambda\,\varphi)$. Needless to say, it is far easier to apply the differential formalism, given in (5.7.20), involving functional differentiations, than carrying out the functional integrations in (5.7.41), say, in a power series in $e_0$.

## 5.8   Low Order Contributions to $\ln\langle\, 0_+ \mid 0_-\rangle$

In the present section we obtain low order contributions to $\ln\langle\, 0_+ \mid 0_-\rangle$ in $e_0$ in QED. More precisely, upon writing

$$\frac{\langle\, 0_+ \mid 0_-\rangle}{\langle\, 0_+ \mid 0_-\rangle|_{J^\mu,\eta,\bar\eta=0}} \equiv e^{iW} = \exp i\big[\, a_0 + a_1 e_0 + a_2 e_0^2 + a_3 e_0^3 + \dots \,\big], \qquad (5.8.1)$$

we determine the coefficients $a_1$, $a_2$, $a_3$, containing a wealth of information on QED, consisting only of connected components of the theory, that is, having at least one propagator connecting any of its subparts. And from the normalization condition in (5.8.1), only so-called diagrams with *external* lines, connected to their respective sources, occur. These coefficients are analyzed and applied in the next couple of sections and their physical consequences are spelled out. The *power of the formalism*, is that all correct multiplicative factors in integrals, describing various components of the theory, such as physical processes, occur automatically and no guess work is required about such numerical factors, as powers of $\pi$'s, and proper normalization constants.

To the above end, we recall the basic equations needed for determining the above coefficients: (see (5.7.27)), $\widehat{A}_\mu(x) = (-\mathrm{i})\delta/\delta J^\mu(x)$)

$$\langle\, 0_+ \mid 0_-\rangle = \exp i\left[\int (\mathrm{d}x)(\mathrm{d}x')\,\bar\eta(x)\,S_+(x,x';e_0\widehat{A})\,\eta(x')\right]$$

$$\times \exp\left[-\int (\mathrm{d}x)\int_0^{e_0} \mathrm{d}e'\,\mathrm{Tr}[\gamma\widehat{A}(x)\,S_+(x,x;e'\widehat{A})]\right]\langle\, 0_+ \mid 0_-\rangle_{0\gamma}, \qquad (5.8.2)$$

$$\langle\, 0_+ \mid 0_-\rangle_{0\gamma} = \exp\frac{\mathrm{i}}{2}\left[\int (\mathrm{d}x)(\mathrm{d}x')J_\mu(x)D^{\mu\nu}(x-x')J_\nu(x')\right], \qquad (5.8.3)$$

$$S_+(x,x';e_0\widehat{A}) = S_+(x-x') + e_0\int(\mathrm{d}x'')\,S_+(x-x'')\gamma\widehat{A}(x'')S_+(x'',x';e_0\widehat{A}). \tag{5.8.4}$$

The latter may be expanded as follows

$$S_+(x,x';e_0\widehat{A}) = S_+(x-x') + e_0\int(\mathrm{d}x_1)\,S_+(x-x_1)\,\gamma\widehat{A}(x_1)\,S_+(x_1-x')$$

$$+ e_0^2\int(\mathrm{d}x_1)\,(\mathrm{d}x_2)\,S_+(x-x_1)\,\gamma\widehat{A}(x_1)\,S_+(x_1-x_2)\,\gamma\widehat{A}(x_2)\,S_+(x_2-x')$$

$$+ e_0^3\int(\mathrm{d}x_1)\,(\mathrm{d}x_2)\,(\mathrm{d}x_3)\,S_+(x-x_1)\,\gamma\widehat{A}(x_1)\,S_+(x_1-x_2)\,\gamma\widehat{A}(x_2)\,S_+(x_2-x_3)$$

$$\times \gamma\widehat{A}(x_3)\,S_+(x_3-x') + \cdots. \tag{5.8.5}$$

Also we have the expansion[24]

$$\int (\mathrm{d}x) \int_0^{e_0} \mathrm{d}e' \, \mathrm{Tr}\,[\,\gamma \widehat{A}(x) S_+(x,x\,;e'\widehat{A})\,]$$

$$= \frac{e_0^2}{2} \int \frac{(\mathrm{d}Q_1)}{(2\pi)^4} \frac{(\mathrm{d}Q_2)}{(2\pi)^4} (2\pi)^4 \, \delta^{(4)}(Q_1 + Q_2) \, \widehat{A}_{\mu_1}(Q_1) \widehat{A}_{\mu_2}(Q_2) \, L^{\mu_1\mu_2}(Q_1, Q_2) + \cdots$$

$$\equiv \frac{e_0^2}{2} \int (\mathrm{d}x_1)\,(\mathrm{d}x_2) \, K^{\mu_1\mu_2}(x_1, x_2)\, \widehat{A}_{\mu_1}(x_1)\, \widehat{A}_{\mu_2}(x_2) + \cdots, \tag{5.8.6}$$

as given through (3.6.26), (3.6.27), (3.6.28) and (3.6.29), and worked out in a gauge invariant manner, with (see Eqs. (3.6.30), (3.6.31) and (3.6.32))

$$L^{\mu_1\mu_2}(Q, \ Q) = [\,\eta^{\mu_1\mu_2} Q^2 - Q^{\mu_1} Q^{\mu_2}\,] \times$$

$$\frac{\mathrm{i}}{12\,\pi^2} \left( \left[ \frac{1}{\mathrm{i}\pi^2} \int \frac{(\mathrm{d}p)}{[p^2 + m^2]^2} + \frac{1}{2} \right] - 6 \int_0^1 \mathrm{d}z \; z(1-z) \ln\!\left[ 1 + \frac{Q^2}{m^2} z(1-z) \right] \right). \tag{5.8.7}$$

With expansions given above in (5.8.5) and (5.8.6), it is straightforward to obtain $iW$, up to third order in $e_0$, as indicated in (5.8.1) involving external sources.

The coefficients $a_1$, $a_2$, $a_3$ are worked out in Problem 5.7, and are given in detail through

$$a_0$$

$$= \int (\mathrm{d}x)\,(\mathrm{d}x')\, \overline{\eta}(x)\, S_+(x-x')\, \eta(x') + \frac{1}{2} \int (\mathrm{d}x)(\mathrm{d}x')\, J_\mu(x)\, D^{\mu\nu}(x-x') J_\nu(x'), \tag{5.8.8}$$

$$a_1$$

$$= \int (\mathrm{d}y)(\mathrm{d}y')(\mathrm{d}x_1)(\mathrm{d}z) \big[\, \overline{\eta}(y)\, S_+(y-x_1)\, \gamma_\mu\, S_+(x_1-y')\, \eta(y')\,\big] D^{\mu\nu}(x_1-z)\, J_\nu(z), \tag{5.8.9}$$

$$a_2$$

$$= \left\{ \int (\mathrm{d}x)(\mathrm{d}x')(\mathrm{d}x_1)(\mathrm{d}x_2) \big[\, \overline{\eta}(x)\, S_+(x-x_1)\, \gamma^{\mu_1}\, S_+(x_1-x_2)\, \gamma^{\mu_2}\, S_+(x_2-x')\, \eta(x')\,\big] \right.$$

---

[24]Note that the first order contribution involves the expression: $\mathrm{Tr}\,[\gamma^\mu\, S_+(0)] = -\mathrm{Tr}\,[\gamma^\mu \gamma^\sigma]\int ((\mathrm{d}p)/(2\pi)^4)p_\sigma/(p^2 + m_0^2) = 0$, as it has an odd integrand in $p$.

$$\times \left[ (-i)D_{\mu_1\mu_2}(x_1-x_2) + \int (dz_1)(dz_2)D_{\mu_1\nu_1}(x_1-z_1)J^{\nu_1}(z_1)D_{\mu_2\nu_2}(x_2-z_2)J^{\nu_2}(z_2) \right] \Bigg\}$$

$$+ \left\{ \frac{1}{2}\int (dy_1)(dy_2)(dx_1)(dx_2)(dy_1')(dy_2')\left[ \overline{\eta}(y_1)\,S_+(y_1-x_1)\,\gamma^{\mu_1}\,S_+(x_1-y_1')\,\eta(y_1') \right] \right.$$

$$\times D_{\mu_1\mu_2}(x_1-x_2)\left[ \overline{\eta}(y_2)\,S_+(y_2-x_2)\,\gamma^{\mu_2}S_+(x_2-y_2')\,\eta(y_2') \right] \Bigg\} + \left\{ \frac{i}{2}\int (dx_1)(dx_2) \right.$$

$$\times (dz_1)(dz_2)\,K^{\mu_1\mu_2}(x_1,x_2)\,D_{\mu_1\nu_1}(x_1-z_1)\,J^{\nu_1}(z_1)\,D_{\mu_2\nu_2}(x_2-z_2)\,J^{\nu_2}(z_2) \Bigg\},$$

$$(5.8.10)$$

$$a_3$$

$$= \left\{ \int (dy_1)(dy_1')(dx_1)(dx_2)(dx_3)\left[ \overline{\eta}(y_1)S_+(y_1-x_1)\,\gamma^{\mu_1}\,S_+(x_1-x_2)\,\gamma^{\mu_2}\,S_+(x_2-x_3) \right.\right.$$

$$\times \gamma^{\mu_3}\,S_+(x_3-y_1')\,\eta(y_1') \Big]\Big[(-i)\int (dz)\Big(D_{\mu_1\mu_2}(x_1-x_2)D_{\mu_3\nu_3}(x_3-z)J^{\nu_3}(z)$$

$$+ \ D_{\mu_1\mu_3}(x_1-x_3)D_{\mu_2\nu_2}(x_2-z)J^{\nu_2}(z) + D_{\mu_2\mu_3}(x_2-x_3)D_{\mu_1\nu_1}(x_1-z)J^{\nu_1}(z)\Big) +$$

$$\int (dz_1)(dz_2)(dz_3)D_{\mu_1\nu_1}(x_1-z_1)J^{\nu_1}(z_1)D_{\mu_2\nu_2}(x_2-z_2)J^{\nu_2}(z_2)D_{\mu_3\nu_3}(x_3-z_3)J^{\nu_3}(z_3) \Big]\Bigg\}$$

$$+ \ \left\{ \int (dy_1)(dy_1')(dy_2)(dy_2')(dx_1)(dx_2)(dx_3) \right.$$

$$\times \left[ \overline{\eta}(y_1)S_+(y_1-x_1)\gamma^{\mu_1}S_+(x_1-y_1')\eta(y_1') \right]$$

$$\times \left[ \overline{\eta}(y_2)S_+(y_2-x_2)\gamma^{\mu_2}S_+(x_2-x_3)\gamma^{\mu_3}S_+(x_3-y_2')\eta(y_2') \right]$$

$$\times \int (dz)\Big(D_{\mu_1\mu_2}(x_1-x_2)D_{\mu_3\nu_3}(x_3-z)J^{\nu_3}(z)+D_{\mu_1\mu_3}(x_1-x_3)D_{\mu_2\nu_2}(x_2-z)J^{\nu_2}(z)\Big)\Bigg\}$$

$$+ \left\{ \frac{i}{2}\int (dy)(dy')(dx_1)(dx_2)(dx_3)K^{\mu_1\mu_2}(x_1,x_2)\left[\overline{\eta}(y)S_+(y-x_3)\gamma^{\mu_3}S_+(x_3-y')\eta(y') \right]\right.$$

$$\times \int (dz)\left[ D_{\mu_1\mu_3}(x_1-x_3)D_{\mu_2\nu_2}(x_2-z)J^{\nu_2}(z) + D_{\mu_2\mu_3}(x_2-x_3)D_{\mu_1\nu_1}(x_1-z)J^{\nu_1}(z) \right]\Bigg\},$$

$$(5.8.11)$$

where $K^{\mu\nu}(x, y)$ is defined in (5.8.6), with

$$S_+(x-x') = \int \frac{(dp)}{(2\pi)^4} \frac{-\gamma p + m_0}{p^2 + m_0^2 - i\epsilon}\, e^{ip\,(x-x')}, \tag{5.8.12}$$

$$D^{\mu\nu}(x-x') = \int \frac{(dk)}{(2\pi)^4} \left[\, \eta^{\mu\nu} - (1-\lambda)\frac{k^\mu k^\nu}{k^2}\, \right] \frac{e^{ik(x-x')}}{k^2 - i\epsilon}. \tag{5.8.13}$$

Remember that there are only connected parts in Eqs.(5.8.10) and (5.8.11). For example, there is no $D_{\mu_1\mu_2}(x_1 - x_2)$ within the very last square brackets on the last line of (5.8.11) which will otherwise leave $[\,\overline{\eta}(y)\, S_+(y-x_3)\, \gamma^{\mu_3}\, S_+(x_3-y')\, \eta(y')\,]$ disconnected from the other part multiplying this factor.

To extract transition amplitudes for physical processes from the above expressions, we here recall the amplitudes of emission and detection by the external sources (acting as emitters or detectors) of the particles involved:

$$\langle\, \mathbf{p}\,\sigma, - \mid 0_-\rangle = [\,i\,\overline{u}(\mathbf{p}, \sigma)\, \eta(p)\,] \sqrt{\frac{m\,d^3\mathbf{p}}{p^0(2\pi)^3}}, \tag{5.8.14}$$

$$\langle\, \mathbf{p}\,\sigma, + \mid 0_-\rangle = [-i\,\overline{\eta}(-p)\, v(\mathbf{p}, \sigma)\,] \sqrt{\frac{m\,d^3\mathbf{p}}{p^0(2\pi)^3}}, \tag{5.8.15}$$

$$\langle\, 0_+ \mid \mathbf{p}\,\sigma, -\rangle = [\,i\,\overline{\eta}(p)\, u(\mathbf{p}, \sigma)\,] \sqrt{\frac{m\,d^3\mathbf{p}}{p^0(2\pi)^3}}, \tag{5.8.16}$$

$$\langle\, 0_+ \mid \mathbf{p}\,\sigma, +\rangle = [-i\,\overline{v}(\mathbf{p}, \sigma)\, \eta(-p)\,] \sqrt{\frac{m\,d^3\mathbf{p}}{p^0(2\pi)^3}}, \tag{5.8.17}$$

$$\langle\, \mathbf{k}\lambda \mid 0_-\rangle = i\, J_\mu(k)\, e_\lambda^{*\mu}(k) \sqrt{\frac{d^3\mathbf{k}}{(2\pi)^3\, 2k^0}}, \tag{5.8.18}$$

$$\langle\, 0_+ \mid \mathbf{k}\lambda\,\rangle = i\, e_\lambda^\mu(k)\, J_\mu^*(k) \sqrt{\frac{d^3\mathbf{k}}{(2\pi)^3\, 2k^0}}. \tag{5.8.19}$$

according to (3.3.39), (3.3.40), (3.3.41), (3.3.42), (5.5.12),(5.5.13).

In extracting transition amplitudes, the sources are eventually withdrawn, i.e., one finally sets $\eta = 0$, $\overline{\eta} = 0$, $J^\mu = 0$. Here we recall, for example, that $\langle\, \mathbf{p}\,\sigma, - \mid 0_-\rangle$ denotes the amplitude that the source $\eta$ emits an electron, thus acting as an emitter, and $\langle\, 0_+ \mid \mathbf{p}\,\sigma, -\rangle$ denotes the amplitude that a source $\overline{\eta}$ absorbs an electron, thus playing the role of a detector. Don't let the notations in these equations scare you. They provide a simple worry free formalism for getting correct numerical factors such as $(2\pi)^3$, and so on, as mentioned above, and facilitate further the analysis in obtaining transition amplitudes of physical processes.

The moral of introducing these amplitudes of emissions and detections, above is the following. Using the explicit expressions in (5.8.8), (5.8.9), (5.8.10) and (5.8.11), with the sources appropriately arranged causally to reflect the actual process where the particles in question are emitted and detected by these sources, the transition amplitudes for arbitrary processes are then simply given by the *coefficients* of these amplitudes of emissions followed by detection. This may be simply represented as follows:

$$\langle\, 0_+|\text{particles (out) detected by sources}\rangle$$
$$\times\langle\, \text{particles (out) on their way to detection}|\text{particles (in) emitted by sources}\rangle$$
$$\times\langle\, \text{particles (in) emitted by sources}|0_-\rangle,$$

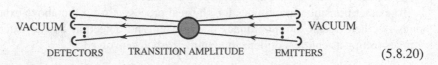

VACUUM   DETECTORS       TRANSITION AMPLITUDE       EMITTERS   VACUUM                (5.8.20)

as obtained from (5.8.8), (5.8.9), (5.8.10) and (5.8.11), corresponding to what happens experimentally, and where the transition amplitude is given by

$$\langle\, \text{particles (out)}|\text{particles (in)}\rangle$$

$$= \langle\, \text{particles (out) on their way to detection}|\text{particles (in) emitted by sources}\rangle.$$
$$(5.8.21)$$

This ingenious method which allows one to quickly extract transition amplitudes with correct multiplicative numerical factors is due to Schwinger, and no guessing is required as to what correct multiplicative factors should be in the formalism.

Before closing this section, we note that the interactions in processes are mediated by the propagators. In the momentum description, the photon propagator of momentum $Q$ develops a singularity at $Q^2 = 0$, i.e., for $Q$ as a light-like vector, corresponding to the masslessness-shell constraint of a real, i.e., detectable, photon. In general, we may have cases for which $Q^2 \neq 0$ corresponding to a space-like or a time-like $Q$. In the latter two cases, the four momentum of the photon is said to be off the masslessness-shell condition $Q^2 = 0$, and the photon thus cannot be detected by a detector. It is then referred to as a virtual particle since the photon does not have the appropriate relation between energy and momentum to be detectable. Let us investigate the meaning of this last inequality. For $Q$ space-like, i.e., for $Q^2 > 0$, for example, we have $\mathbf{Q}^2 > (Q^0)^2$, which means the photon lacks the appropriate energy, in comparison to its momentum, to be detectable. On the other hand, for $Q^2 < 0$ i.e., for $\mathbf{Q}^2 < (Q^0)^2$, the photon has a surplus of energy over its momentum to be detectable. To see how actually such a virtual photon arises, with a space-like or time-like momentum, in

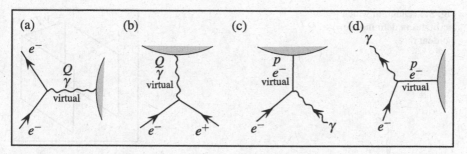

**Fig. 5.3** (a) Consider the diagram describing the scattering process of an electron, experiencing a change of its three-momentum, via the exchange of a (virtual) photon with the remaining part of the diagram (denoted by the *shaded area*). Conservation of momentum implies that the momentum of the virtual photon is space-like. (b) Consider an electron and a positron annihilating, e.g. in the c.m., into a (virtual) photon. The momentum of the virtual photon is time-like. (c) Consider the scattering of an electron and a photon to a (virtual) electron. The momentum $p$ of the virtual electron is off the mass-shell satisfying the relation $p^2 < -m^2$. (d) Consider an electron becoming virtual in the scattering of the electron with the emission of a photon. The momentum $p$ of the virtual electron is off the mass shell satisfying the relation $p^2 > -m^2$. The photon is denoted by a wavy line, while the electron (positron) by a solid one. A virtual particle which has too much energy or not enough energy to be on the mass shell, as the case may be, may, respectively, give off energy to another particle or absorb energy from another one and may eventually emerge as a real particle in an underlying *Feynman diagram description* of fundamental processes

the light of these observations, consider the diagrams depicting some processes in Fig. 5.3.[25]

Now let us proceed, move to the next section, and see how things work out, compute transition amplitudes by applying the above equations, and witness the wealth of information stored in them.

## 5.9 Basic Processes

Experimentally, scattering processes are quantified and their likelihood of occurrence are determined in terms of what one calls the cross section. This quantity arises in the following manner. Consider the scattering, say, of two particles of masses $m_1$ and $m_2$ leading finally to an arbitrary number of particles that may be allowed by the underlying theory. The transition probability per unit time $\text{Prob}|_{\text{unit time}}$ for the process to occur is first calculated. Consider a frame in which, say, the particle $m_2$, is (initially) at rest. Such a frame may be referred to as the laboratory or target (TF) frame. To determine the likelihood of the process to occur, and simultaneously obtain a measure of the interaction of the two particles, $\text{Prob}|_{\text{unit time}}$, in turn, is compared to, i.e., divided by, the probability per unit time

---

[25]These processes are assumed to involve no external sources or external potentials.

**Fig. 5.4** Diagram which
facilitates in defining the
incident flux

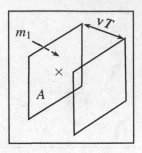

per unit *area*, denoted by $F$, that particle $m_1$ *crosses* at the position of particle of
mass $m_2$, represented by $\times$ in Fig. 5.4, *as if* the latter particle is *absent*, and hence
no interaction is involved.

The probability per unit area per unit time $F$, defined above, is called the *incident
flux*. Most importantly, the ratio defined above leads, upon integration, to a Lorentz
invariant expression for the cross section, and thus may be computed in any inertial
frame. By definition, the unit of cross section is that of an area and, intuitively, the
cross section provides an effective area for the process to occur. The differential
cross section $d\sigma$ is then simply defined by the above mentioned ratio, i.e., by
$\text{Prob}|_{\text{unit time}}$ per incident flux $F$: $d\sigma = \text{Prob}|_{\text{unit time}}/F$.

The probability of finding, the particle of mass $m_1$, of speed v, within a cube of
cross sectional area $A$ and width v$T$, during a time $T$, is given by $A\,\text{v}\,T/V$, where
$V$ denotes the volume of the 3D space in which the scattering process occurs. Thus
the probability that this particle crosses a unit area per unit time, at the position $\times$
of particle $m_2$, as if the latter is absent (see Fig. 5.4), is given by

$$\frac{\text{v}}{V} = \frac{1}{V}\frac{|\mathbf{p}_{1\,\text{TF}}|}{p_1^0} \equiv F. \tag{5.9.1}$$

For the initial particle of mass $m_2$ at rest, we note that the above expression
allows us to write

$$(p_1^0 p_2^0 V)\,F = m_2 |\mathbf{p}_{1\,\text{TF}}|. \tag{5.9.2}$$

On the other hand, for arbitrary sharp momenta $p_1$, $p_2$, of the initial particles, the
Lorentz invariant expression $\sqrt{(p_1 p_2)^2 - m_1^2 m_2^2}$, may be computed in any frame,
and, in particular, in the TF frame to be $(p_1 p_2 \equiv \eta_{\mu\nu} p_1^\mu p_2^\nu)$

$$\sqrt{(p_1 p_2)^2 - m_1^2 m_2^2} = m_2 |\mathbf{p}_{1\,\text{TF}}|, \tag{5.9.3}$$

The transition probability per unit time of a given process that emerges from the theory for the scattering, say, of two particles into an arbitrary number of particles, is given by an expression of the form[26]

$$\left| \langle p'_1 v'_1, p'_2 v'_2, \ldots | p_1 v_1, p_2 v_2 \rangle \right|^2_{\text{unit time}}$$
$$= (\Pi_j \, d\omega'_j) \, |\mathcal{M}|^2 (2\pi)^4 \delta^{(4)} \Big( \sum_j p'_j - p_1 - p_2 \Big) V \, d\omega_1 d\omega_2, \qquad (5.9.4)$$

$$d\omega_j = d^3 \mathbf{p}_j / [(2\pi)^3 \, 2p_j^0], \qquad (5.9.5)$$

where $|\mathcal{M}|^2$ is an invariant quantity, the symbol $v$ stands for any label needed to specify a particle, in addition to its momentum, and $V$ denotes the volume of the 3D space in which the scattering process occurs, as before. With the momenta of the initial particles prepared to have sharp values, box normalization implies to replace each of $d^3 \mathbf{p}_1 / (2\pi)^3$, $d^3 \mathbf{p}_2 / (2\pi)^3$, corresponding to the initial particles, by $1/V$. The differential cross section then takes the form

$$d\sigma = (\Pi_j \, d\omega'_j) \, |\mathcal{M}|^2 \frac{(2\pi)^4 \delta^{(4)} \Big( \sum_j p'_j - p_1 - p_2 \Big)}{4 \, p_1^0 p_2^0 \, VF}, \qquad (5.9.6)$$

which from (5.9.2) and (5.9.3), may be rewritten in the form

$$d\sigma = (\Pi_j \, d\omega'_j) \, |\mathcal{M}|^2 \frac{(2\pi)^4 \delta^{(4)} \Big( \sum_j p'_j - p_1 - p_2 \Big)}{4 \sqrt{(p_1 p_2)^2 - m_1^2 m_2^2}}, \qquad (5.9.7)$$

and is independent of $V$.

To obtain an expression of the differential cross section in the center-of-momentum (CM) frame, it is convenient to introduce the invariant variable $s = -(p_1 + p_2)^2$. Then some algebra gives

$$\sqrt{(p_1 p_2)^2 - m_1^2 m_2^2} = \frac{1}{2} \sqrt{s^2 - 2 (m_1^2 + m_2^2) \, s + (m_1^2 - m_2^2)^2}. \qquad (5.9.8)$$

In the center-of-momentum (CM) frame, $\mathbf{p}_1 + \mathbf{p}_2 = 0$ and $p_1{}^0 + p_2{}^0 = \sqrt{s}$, which by evaluating the expression on the left-hand side of (5.9.8), in the CM frame,

---

[26]The momentum conserving delta function occurs as follows. Invoking translational invariance, and the Hermiticity of the total momentum $P$ which may equally operate to right or left in $\langle p'_1 v'_1, p'_2 v'_2, \ldots | P | p_1 v_1, p_2 v_2 \rangle$, leading to the two equal expressions: $(p_1 + p_2) \langle p'_1 v'_1, p'_2 v'_2, \ldots | p_1 v_1, p_2 v_2 \rangle = (p'_1 + \ldots) \langle p'_1 v'_1, p'_2 v'_2, \ldots | p_1 v_1, p_2 v_2 \rangle$. Hence upon subtracting one expression from the other, we obtain $(p_1 + p_2 - (p'_1 + \ldots)) \langle p'_1 v'_1, p'_2 v'_2, \ldots | p_1 v_1, p_2 v_2 \rangle = 0$. On the other the relation $x f(x) = 0$ implies that $f(x)$ is proportional to $\delta(x)$.

allows us to write

$$|\mathbf{p}_{1\,CM}| = \frac{1}{2\sqrt{s}} \sqrt{s^2 - 2\,(m_1^2 + m_2^2)\,s + (m_1^2 - m_2^2)^2}. \tag{5.9.9}$$

Consider a process involving initially and finally two particles of masses $(m_1, m_2)$, $(m_1', m_2')$, and momenta $(p_1, p_2)$, $(p_1', p_2')$, respectively. We explicitly have $s = -(p_1 + p_2)^2 = -(p_1' + p_2')^2$, and in the CM frame we also have $\mathbf{p}_1' + \mathbf{p}_2' = 0$, $p_1^{\,0} + p_2^{\,0} = p_1'^{\,0} + p_2'^{\,0} = \sqrt{s}$. From the expression of $|\mathbf{p}_{1\,CM}|$ in (5.9.9), we similarly have

$$|\mathbf{p}'_{1\,CM}| = \frac{1}{2\sqrt{s}} \sqrt{s^2 - 2\,(m_1'^2 + m_2'^2)\,s + (m_1'^2 - m_2'^2)^2}. \tag{5.9.10}$$

In reference to CM frame, the following useful expression then emerges

$$\begin{aligned}
\int d\sigma &= \frac{1}{64\,\pi^2} \int \frac{|\mathcal{M}|^2}{|\mathbf{p}_{1\,CM}|\sqrt{s}} \frac{d^3\mathbf{p}_1'}{p_1'^{\,0}} \frac{d^3\mathbf{p}_2'}{p_2'^{\,0}} \delta^{(4)}(p_1' + p_2' - p_1 - p_2), \\
&= \frac{1}{64\,\pi^2} \int d\Omega_{CM} |\mathcal{M}|^2 \frac{|\mathbf{p}_1'|^2 \, d|\mathbf{p}_1'|}{p_1'^{\,0} p_2'^{\,0} |\mathbf{p}_{1\,CM}| \sqrt{s}} \delta(p_1'^{\,0} + p_2'^{\,0} - \sqrt{s}) \\
&= \frac{1}{64\,\pi^2} \int d\Omega_{CM} \frac{|\mathcal{M}|^2}{s} \frac{|\mathbf{p}'_{1\,CM}|}{|\mathbf{p}_{1\,CM}|}, \tag{5.9.11}
\end{aligned}$$

where in the first line, we have replaced $\sqrt{(p_1 p_2)^2 - m_1^2 m_2^2}$ in (5.9.11) by $|\mathbf{p}_{1\,CM}| \sqrt{s}$, referring to (5.9.8), (5.9.9) and (5.9.10). In the second line we have merely integrated over $\mathbf{p}_2'$. In the last line we have used the fact that

$$\delta(p_1'^{\,0} + p_2'^{\,0} - \sqrt{s}) = \frac{\delta(|\mathbf{p}_1'| - |\mathbf{p}'_{1\,CM}|)}{\left(|\mathbf{p}'_{1\,CM}| \sqrt{s}/(p_1'^{\,0} p_2'^{\,0})\right)}, \tag{5.9.12}$$

to integrate over $|\mathbf{p}_1'|$. Finally, note that $d\Omega_{CM} = \sin\theta d\theta \, d\phi$, where $\theta$ is the angle between $\mathbf{p}_{1\,CM}$ and $\mathbf{p}'_{1\,CM}$, and $|\mathbf{p}'_{1\,CM}|$, $|\mathbf{p}_{1\,CM}|$, may be replaced by the expressions on the right-hand sides of (5.9.9), and (5.9.10). Thus in the CM frame, we have the following expression for the differential cross section

$$\frac{d\sigma}{d\Omega_{CM}} = \frac{1}{64\,\pi^2} \frac{|\mathcal{M}|^2}{s} \frac{|\mathbf{p}'_{1\,CM}|}{|\mathbf{p}_{1\,CM}|}. \tag{5.9.13}$$

For future reference, we introduce the Mandelstam variables associated with a scattering process: $p_1 p_2 \to p'_1 p'_2$, including the invariant $s$ defined above:

$$s = -(p_1 + p_2)^2 = -(p'_1 + p'_2)^2, \quad t = -(p_1 - p'_1)^2 = -(p'_2 - p_2)^2,$$

$$u = -(p_1 - p'_2)^2 = -(p'_1 - p_2)^2, \quad s + t + u = m_1^2 + m_2^2 + m'^2_1 + m'^2_2. \tag{5.9.14}$$

In particular, in the CM frame, $dt = 2|\mathbf{p}'_{1\,CM}|\,|\mathbf{p}_{1\,CM}|\,d\cos\theta$, which from (5.9.13) leads to

$$\left.\frac{d\sigma}{dt}\right|_{CM} = \frac{1}{64\pi}\,\frac{1}{s\,|\mathbf{p}_{1\,CM}|^2}\,|\mathscr{M}|^2. \tag{5.9.15}$$

In this section, we consider several processes of interest to low orders, including polarizations correlations studies of outgoing particles in such processes which have been quite important in recent years. The derivations underlying the scattering processes are obtained directly from the explicit expression obtained for $\langle\,0_+\,|\,0_-\,\rangle$ in the previous section, to lowest orders. The formalism is powerful enough that all the numerical factors in transition amplitudes appear directly as well as various relations concerning the statistics of the underlying particles, and no guessing is necessary.

* **

For the convenience of the reader we here recall the transition made between box normalization and that of the infinite extension as arising from the Fourier transform, in the complex form, and that of the Fourier integral, used often in scattering theory:

$$-\infty < x < \infty, \qquad\qquad -L/2 \leq x \leq L/2,$$
$$-\infty < k < \infty, \qquad k = (2\pi n/L),\ n = 0, \pm 1, \pm 2, \ldots,$$
$$f(x) = \int_{-\infty}^{\infty} \frac{dk}{2\pi}\, e^{ikx}\widetilde{f}(k), \qquad f(x) = \frac{1}{L}\int_{-L/2}^{L/2} dk\, e^{ikx}\widetilde{f}(k),$$
$$\widetilde{f}(k) = \int_{-\infty}^{\infty} dx\, e^{-ikx}f(x), \qquad \widetilde{f}(k) = \int_{-L/2}^{L/2} dx\, e^{-ikx}f(x),$$
$$(2\pi)\,\delta(k) = \int_{-\infty}^{\infty} dx\, e^{ikx}, \qquad (L)\,\delta_{k,0} = \int_{-L/2}^{L/2} dx\, e^{ikx},$$
$$d^n\mathbf{k}/(2\pi)^n, \qquad 1/L_1 \times \ldots \times L_n,$$
$$(2\pi)^n\,\delta^{(n)}(\mathbf{k}). \qquad (L_1 \times \ldots \times L_n)\,\delta_{\mathbf{k},0}^{(n)}.$$

* **

## 5.9.1 $e^- e^- \to e^- e^-$, $e^+ e^- \to e^+ e^-$

We consider the processes $e^- e^- \to e^- e^-$, $e^+ e^- \to e^+ e^-$, referred to, respectively, as Møller scattering and Bhabha scattering, to second order. To this

**Fig. 5.5** Diagram
corresponding to Eq. (5.9.16)
including the presence of the
external sources denoted by
the *half circles*. Such a
diagram describes several
processes

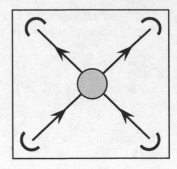

order, we may replace the parameters $e_0$, $m_0$ in the lagrangian density by their
zeroth order, i.e., by the corresponding physical values e, m. These processes are
described in terms of the expression within the curly brackets in the second term of
$a_2$ in (5.8.10)[27] multiplied by $i e^2$ as given in the exponent in (5.8.1) (see Fig. 5.5)

$$
\left\{ \frac{i e^2}{2} \int (dy_1)(dy_2)(dx_1)(dx_2)(dy'_1)(dy'_2) \left[ \overline{\eta}(y_1) \, S_+(y_1 - x_1) \, \gamma^{\mu_1} \, S_+(x_1 - y'_1) \, \eta(y'_1) \right] \right.
$$

$$
\left. \times D_{\mu_1 \mu_2}(x_1, x_2) \left[ \overline{\eta}(y_2) \, S_+(y_2 - x_2) \, \gamma^{\mu_2} \, S_+(x_2 - y'_2) \, \eta(y'_2) \right] \right\}.
$$

$$(5.9.16)$$

We may write the source as $\eta(x) = \eta_1(x) + \eta_2(x)$, where $\eta_1(x)$ is switched
on and then off in the remote past after the electrons are emitted, while $\eta_2(x)$
is switched on in the distant future and then switched off after the outgoing
electrons are detected (absorbed), with the interaction taking place later in time
than the emissions and earlier than detections. Thus for the first scattering process
in question, only the term involving the part $\eta_1(x)$ of $\eta(x)$, and the part $\overline{\eta}_2(z)$ of
$\overline{\eta}(z)$ will contribute.[28]

For all $y^0 > x^0$, we have (see (4.1.17))

$$
\int (dy) \, \overline{\eta}_2(y) S_+(y - x) = \int \sum_\sigma \frac{d^3\mathbf{k}}{(2\pi)^3} \frac{m}{k^0} \, e^{-i k x} \left[ i \, \overline{\eta}_2(p) \, u(\mathbf{k}, \sigma) \right] \overline{u}(\mathbf{k}, \sigma),
$$

$$(5.9.17)$$

---

[27]This expression involves also the process $e^+ e^+ \rightarrow e^+ e^+$.

[28]For the process $e^+ e^+ \rightarrow e^+ e^+$, only the term involving the pair $\eta_2(x)$ and $\overline{\eta}_1(x)$ will
contribute.

and similarly for all $x^0 > y^0$,

$$\int (dy) S_+(x-y) \eta_1(y) = \int \sum_\sigma \frac{d^3\mathbf{p}}{(2\pi)^3} \frac{m}{p^0} e^{ipx} u(\mathbf{p},\sigma) [i\,\overline{u}(\mathbf{p},\sigma) \eta_1(p)].$$

(5.9.18)

Accordingly, in reference to the first process in question, we obtain for (5.9.16)

$$\frac{i\,e^2}{2} \int \sum_{\sigma'_2, \sigma_2, \sigma'_1, \sigma_1} d\omega(\mathbf{k}'_1)\, d\omega(\mathbf{k}'_2)\, d\omega(\mathbf{k}_1)\, d\omega(\mathbf{k}_2)\, 16\,m^4\, (2\pi)^4\, \delta^4(k'_2 + k'_1 - k_1 - k_2)$$

$$\times (-)[i\,\overline{\eta}_2(k'_2) u(\mathbf{k}'_2, \sigma'_2)] [i\,\overline{\eta}_2(k'_1) u(\mathbf{k}'_1, \sigma'_1)] [i\,\overline{u}(\mathbf{k}_2, \sigma_2) \eta_1(k_2)] [i\,\overline{u}(\mathbf{k}_1, \sigma_1) \eta_1(k_1)]$$

$$\times \overline{u}(\mathbf{k}'_2, \sigma'_2) \gamma^{\mu_1} u(\mathbf{k}_2, \sigma_2) D_{\mu_1\mu_2}(k_2 - k'_2) \overline{u}(\mathbf{k}'_1, \sigma'_1) \gamma^{\mu_2} u(\mathbf{k}_1, \sigma_1),$$

(5.9.19)

where the overall $(-)$ sign arises because we have (anti-)commuted $\overline{\eta}_2(k'_1)$ and $\eta_1(k_2)$.

Now suppose that the momenta and spins of the initial electrons are prepared to be $(\mathbf{p}_1, \sigma_1)$, $(\mathbf{p}_2, \sigma_2)$. We encounter the two possibilities: $(\mathbf{k}_1 = \mathbf{p}_1,\ \mathbf{k}_2 = \mathbf{p}_2)$ or $(\mathbf{k}_1 = \mathbf{p}_2,\ \mathbf{k}_2 = \mathbf{p}_1)$ (and similarly for the corresponding spins). Clearly, since one has the integrations over $(\mathbf{k}'_2, \mathbf{k}'_1)$, and sums over $(\sigma'_1, \sigma'_2)$, we are free to carry out the relabelings $(\mathbf{p}'_1, \sigma'_1) \leftrightarrow (\mathbf{p}'_2, \sigma'_2)$, we have the complete symmetric relation[29] without change of the overall sign in (5.9.19) by simultaneously commuting $\eta_1(k_1)$, $\eta_1(k_2)$ and $\overline{\eta}(k'_1), \overline{\eta}(k'_2)$. This leads simply to multiplying the expression in (5.9.19) by a factor of 2.

We have to be also careful when considering the outgoing momenta.[30] If the momenta and spins of the outgoing (two identical particles) electrons are $(\mathbf{p}'_2, \sigma'_2)$, $(\mathbf{p}'_1, \sigma'_1)$ with $\mathbf{p}'_1 \neq \mathbf{p}'_2$, then the integral, say over $\mathbf{k}'_1$ will "meet" both values $(\mathbf{p}'_2, \mathbf{p}'_1)$.

Similarly in the case for the integral over $\mathbf{k}'_2$, the product $[i\,\overline{\eta}_2(p'_2) u(\mathbf{p}'_2, \sigma'_2)] \times [i\,\overline{\eta}_2(p'_1) u(\mathbf{p}'_1, \sigma'_1)]$ will occur twice with $(\mathbf{p}'_2, \sigma'_2) \leftrightarrow (\mathbf{p}'_1, \sigma'_1)$.

Accordingly, for the process, in question, with the momenta of the electrons taking values within infinitesimal intervals around the just specified non-coincident values, and for the values taken by their respective spins, we obtain

$$- i\,e^2\, d\omega(\mathbf{p}'_1)\, d\omega(\mathbf{p}'_2)\, d\omega(\mathbf{p}_1)\, d\omega(\mathbf{p}_2)\, 16\,m^4\, (2\pi)^4\, \delta^4(p'_2 + p'_1 - p_1 - p_2)$$

$$\times [i\,\overline{\eta}_2(p'_2) u(\mathbf{p}'_2, \sigma'_2)] [i\,\overline{\eta}_2(p'_1) u(\mathbf{p}'_1, \sigma'_1)] [i\,\overline{u}(\mathbf{p}_2, \sigma_2) \eta_1(p_2)] [i\,\overline{u}(\mathbf{p}_1, \sigma_1) \eta_1(p_1)]$$

$$\times \Big[ \overline{u}(\mathbf{p}'_2, \sigma'_2) \gamma^{\mu_1} u(\mathbf{p}_2, \sigma_2) D_{\mu_1\mu_2}(p_1 - p'_1) \overline{u}(\mathbf{p}'_1, \sigma'_1) \gamma^{\mu_2} u(\mathbf{p}_1, \sigma_1)$$

$$- \overline{u}(\mathbf{p}'_1, \sigma'_1) \gamma^{\mu_1} u(\mathbf{p}_2, \sigma_2) D_{\mu_1\mu_2}(p_1 - p'_2) \overline{u}(\mathbf{p}'_2, \sigma'_2) \gamma^{\mu_2} u(\mathbf{p}_1, \sigma_1) \Big].$$

(5.9.20)

---

[29] Also note that $D^{\mu\nu}(k_2 - k'_2) = D^{\mu\nu}(k_1 - k'_1)$, and the symmetry of $D^{\mu\nu}$ in $(\mu, \nu)$.

[30] On the other hand, if you fix the final momenta and spins first, then the argument is reversed.

Here we recall that we have two identical particles (electrons) in the final state, and that the *minus sign* between the two terms within the square brackets, in the last line, arises because we have (anti-)commuted $[i\,\overline{\eta}_2(p'_1)\,u(\mathbf{p}'_1,\sigma'_1)]$ and $[i\,\overline{\eta}_2(p'_2)\,u(\mathbf{p}'_2,\sigma'_2)]$, in the process of derivation, and thus it emerges as a consequence of the anti-commutativity of the Grassmann sources. The Fermi-Dirac statistics is automatically recovered (Fig. 5.6).

Upon comparing the above expression in (5.9.20) with the term in a unitarity sum (with connected terms in scattering) (see also (5.8.20)):

$$\langle 0_+|p'_2\sigma'_2, p'_1\sigma'_1\rangle\,\langle p'_1\sigma'_1, p'_2\sigma'_2|p_1\sigma_1, p_2\sigma_2\rangle\,\langle p_2\sigma_2, p_1\sigma_1\mid 0_-\rangle, \qquad (5.9.21)$$

we may infer from (5.9.20), that the amplitude of scattering in question is given by

$$\langle p'_1\sigma'_1, p'_2\sigma'_2|p_1\sigma_1, p_2\sigma_2\rangle = -i\,e^2\,\sqrt{2md\omega'_1}\,\sqrt{2md\omega'_2}\,\sqrt{2md\omega_1}\,\sqrt{2md\omega_1}$$

$$\times\,(2\pi)^4\,\delta^{(4)}(p'_2 + p'_1 - p_1 - p_2)\Big[\ \cdot\ \Big] \qquad (5.9.22)$$

where

$$\Big[\ \cdot\ \Big] = \Big[\overline{u}(\mathbf{p}'_2,\sigma'_2)\,\gamma^{\mu_2}\,u(\mathbf{p}_2,\sigma_2)\,D_{\mu_2\mu_1}(p_2 - p'_2)\,\overline{u}(\mathbf{p}'_1,\sigma'_1)\,\gamma^{\mu_1}u(\mathbf{p}_1,\sigma_1)$$

$$-\,\overline{u}(\mathbf{p}'_1,\sigma'_1)\,\gamma^{\mu_2}\,u(\mathbf{p}_2,\sigma_2)\,D_{\mu_2\mu_1}(p_2 - p'_1)\,\overline{u}(\mathbf{p}'_2,\sigma'_2)\,\gamma^{\mu_1}u(\mathbf{p}_1,\sigma_1)\Big],$$

$$(5.9.23)$$

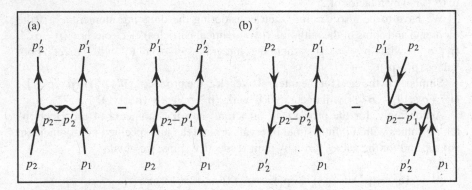

**Fig. 5.6** (a) Møller scattering: $e^-(p_1)\,e^-(p_2) \to e^-(p'_1)\,e^-(p'_2)$. The Fermi-Dirac statistics is automatically taken care of by the formalism. (b) Bhahba scattering: $e^-(p_1)\,e^+(q) \to e^-(p_1)\,e^+(q)$, obtained from the Møller one, in the process, by the substitutions $p'_2 \to -q$, $p_2 \to -q'$, $u(\mathbf{p}'_2,\sigma'_2) \to v(\mathbf{q},\sigma)$, $u(\mathbf{p}_2,\sigma_2) \to v(\mathbf{q}',\sigma')$, in reference to (5.9.41), with an overall change of sign of the amplitude

$$\langle p_2\sigma_2, p_1\sigma_1 \mid 0_-\rangle = [i\,\bar{u}(\mathbf{p}_2,\sigma_2)\,\eta_1(p_2)]\sqrt{2m\,d\omega_2}\,[i\,\bar{u}(\mathbf{p}_1,\sigma_1)\,\eta_1(p_1)]\sqrt{2md\omega_1},$$
$$(5.9.24)$$

$$\langle 0_+ \mid p_2'\sigma_2', p_1'\sigma_1'\rangle = [i\,\bar{\eta}_2(p_2')u(\mathbf{p}_2',\sigma_2')]\sqrt{2md\omega_2'}\,[i\,\bar{\eta}_2(p_1')\,u(\mathbf{p}_1',\sigma_1')]\sqrt{2m\,d\omega_1'},$$
$$(5.9.25)$$

and the latter two equations denote, respectively, the amplitudes of emissions and detection of these particles. (See (5.8.14) and (5.8.16)).

From the Dirac equations

$$\bar{u}(\mathbf{p},\sigma)(\gamma p + m) = 0, \qquad (\gamma p + m)u(\mathbf{p},\sigma_1) = 0, \qquad (5.9.26)$$

we obtain

$$(p_2 - p_2')_\mu\,\bar{u}(\mathbf{p}_2',\sigma_2')\,\gamma^\mu\,u(\mathbf{p}_2,\sigma_2) = 0, \quad (p_2 - p_1')_\mu\,\bar{u}(\mathbf{p}_1',\sigma_1')\,\gamma^\mu\,u(\mathbf{p}_2,\sigma_2) = 0,$$
$$(5.9.27)$$

and, in turn, conclude that only the part of the photon propagator $D_{\mu\nu}$ proportional to $\eta_{\mu\nu}$ will contribute to the amplitude in (5.9.22), thus establishing its gauge invariance.

Using the elementary property[31] $[(2\pi)^4\delta^{(4)}(p)]^2 = VT(2\pi)^4\delta^{(4)}(p)$, where $V$ is the normalization volume, $T$ is the total interaction time, we obtain for the transition probability per unit time for the process for unpolarized electrons

$$\text{Prob}\big|_{\text{unit time}} = e^4\,\frac{d^3\mathbf{p}_2'}{(2\pi)^3\,2p_2'^0}\,\frac{d^3\mathbf{p}_1'}{(2\pi)^3\,2p_1'^0}\,16\,m^4\Big(\frac{1}{4}\sum_{\text{spins}}\big|[\,\cdot\,]\big|^2\Big)$$

$$\times\,\frac{d^3\mathbf{p}_2}{(2\pi)^3\,2p_2^0}\,\frac{d^3\mathbf{p}_1}{(2\pi)^3\,2p_1^0}\,(2\pi)^4\,V\,\delta^{(4)}(p_2' + p_1' - p_1 - p_2). \qquad (5.9.28)$$

where the $1/4$ factor arises because one is averaging over the initial spins of the ingoing electrons.

For sharp initial momenta $\mathbf{p}_1, \mathbf{p}_2$, box normalization versus continuum normalization, via Fourier series (in complex form) versus Fourier transform: $(1/V)\sum_{\mathbf{p}}(.) \Leftrightarrow (1/(2\pi)^3)\int d^3\mathbf{p}(.)$, allows us to replace $(d^3\mathbf{p}_1/(2\pi)^3)(d^3\mathbf{p}_2/(2\pi)^3)$ by $1/V^2$. If $F$ denotes the incident flux, then according to (5.9.2) and (5.9.3),

$$p_1^0 p_2^0\,V\,F\Big|_{\text{TF}} = \sqrt{(p_1 p_2)^2 - m^4}. \qquad (5.9.29)$$

---

[31]This property is usually attributed to Fermi, and follows from the formal equality $\delta^{(4)}(p)|_{p=0} = \int(dx)/(2\pi)^4 = VT/(2\pi)^4$.

The cross section for the process for unpolarized electrons, is then given, from (5.9.6) and (5.9.7), by the Lorentz invariant expression

$$
\sigma = \frac{e^4}{4\pi^2} \int \frac{d^3\mathbf{p}_2'}{2p_2'^0} \frac{d^3\mathbf{p}_1'}{2p_1'^0} \, 16\,m^4 \Big( \frac{1}{4} \sum_{\text{spins}} \big|[ \; \cdot \; ]\big|^2 \Big) \frac{\delta^{(4)}(p_2' + p_1' - p_1 - p_2)}{4\sqrt{(p_1p_2)^2 - m^4}},
$$

(5.9.30)

The Lorentz invariance of this expression is evident since $\int d^3\mathbf{p}/2p^0[\cdots] = \int (dp)\,\theta(p^0) \times \delta(p^2 + m^2)[\cdots]$.

In reference to (5.9.7), $|\mathscr{M}|^2$, and its average $\overline{|\mathscr{M}|^2}$, may be written as

$$
|\mathscr{M}|^2 = e^4\,16\,m^4 \big|[ \; \cdot \; ]\big|^2, \qquad \overline{|\mathscr{M}|^2} = e^4 16\,m^4 \Big( \frac{1}{4} \sum_{\text{spins}} \big|[ \; \cdot \; ]\big|^2 \Big),
$$

(5.9.31)

After some algebra, the summation over spins leads to (see Problem 5.8) ($p_1p_2 = p_{1\mu}p_2{}^\mu$)

$$
\frac{1}{4} \sum_{\text{spins}} \big|[ \; \cdot \; ]\big|^2 = \frac{8}{(2m)^4} \left( \frac{[(p_1p_2)^2 + (p_1p_2')^2 + 2m^2(p_1p_1' + m^2)]}{(p_2 - p_2')^4} \right.
$$

$$
+ 2\,\frac{[m^4 + (p_1p_2)^2 + m^2(p_1p_2 + p_1p_1' + p_1p_2')]}{(p_2 - p_2')^2(p_2 - p_1')^2}
$$

$$
\left. + \frac{[(p_1p_2)^2 + (p_1p_1')^2 + 2m^2(p_1p_2' + m^2)]}{(p_2 - p_1')^4} \right).
$$

(5.9.32)

In the CM frame,

$$
\mathbf{p}_1 = -\mathbf{p}_2, \quad \mathbf{p}_1' = -\mathbf{p}_2', \quad p_1{}^0 = p_2{}^0 = p_1'{}^0 = p_1'{}^0,
$$

(5.9.33)

$$
\mathbf{p}_1 \cdot \mathbf{p}_1' = \gamma^2\beta^2\,m^2\cos\theta = (\gamma^2 - 1)\,m^2\cos\theta, \qquad \gamma^2 = \frac{1}{(1 - \beta^2)}.
$$

(5.9.34)

Using the general expression in (5.9.13), worked out in terms of CM variables, remembering that $|\mathbf{p}_1'|/|\mathbf{p}_1| = 1$, and using (5.9.32), the following expression for the differential cross section emerges

$$
\frac{d\sigma}{d\Omega_{\text{CM}}} = \frac{\alpha^2}{m^2} \frac{1}{\gamma^2(\gamma^2 - 1)^2} \Big[ (2\gamma^2 - 1)^2 \Big( \frac{1}{\sin^2\theta} - \frac{1}{4} \Big)^2 - \Big( \gamma^2 - \frac{3}{4} \Big) \Big( \frac{1}{\sin^2\theta} + \frac{1}{4} \Big) \Big],
$$

(5.9.35)

where $\alpha = e^2/(4\pi)$ is the fine-structure constant.

At high energies $\gamma^2 \gg 1$, the differential cross section takes the simple form

$$\frac{d\sigma}{d\Omega_{CM}} \simeq \frac{\alpha^2}{4m^2\gamma^2}\left(\frac{4}{\sin^2\theta} - 1\right)^2. \tag{5.9.36}$$

For the second process $e^+e^- \rightarrow e^+e^-$, clearly only the following terms contribute to (5.9.16)

$$2\left\{[\overline{\eta}_2 S_+ \gamma^\mu S_+ \eta_1]D_{\mu\nu}[\overline{\eta}_1 S_+ \gamma^\nu S_+ \eta_2] + [\overline{\eta}_1 \gamma^\mu S_+ \eta_1]D_{\mu\nu}[\overline{\eta}_2 S_+ \gamma^\nu S_+ \eta_2]\right\}, \tag{5.9.37}$$

where the factor 2 arises because of the symmetry that arises between the indices $1 \leftrightarrow 2$ of the sources that occurs in (5.9.16) upon writing $\eta = \eta_1 + \eta_2$. Here we note that for $x^0 < x'^0$, with $-x^0$ arbitrarily large[32]

$$\int (dx)\,\overline{\eta}_1(x)\,S_+(x-x') = -i\int \frac{m\,d^3\mathbf{p}}{p^0(2\pi)^3}\,e^{ipx'}\sum_\sigma (\overline{\eta}_1(-p)v(\mathbf{p},\sigma))\overline{v}(\mathbf{p},\sigma), \tag{5.9.38}$$

while for $y^0 < y'^0$, with $y'^0$ arbitrarily large,

$$\int (dy')\,S_+(y-y')\,\eta_2(y') = -i\int \frac{m\,d^3\mathbf{p}}{p^0(2\pi)^3}\,e^{-ipy}\sum_\sigma v(\mathbf{p},\sigma)(\overline{v}(\mathbf{p},\sigma)\eta_2(p)). \tag{5.9.39}$$

Using the Grassmann character of the sources, we obtain for the amplitude in question

$$A = ie^2\left\{[\overline{v}(\mathbf{q},\sigma)\,\gamma^\mu\,v(\mathbf{q}',\sigma')]\,D_{\mu\nu}(q-q')\,[\overline{u}(\mathbf{p}'_1,\sigma'_1)\,\gamma^\nu u(\mathbf{p}_1,\sigma_1)]\right.$$

$$\left. - [\overline{u}(\mathbf{p}'_1,\sigma'_1)\,\gamma^\mu\,v(\mathbf{q},\sigma')]\,D_{\mu\nu}(q+p_1)\,[\overline{v}(\mathbf{q},\sigma)\,\gamma^\nu u(\mathbf{p}_1,\sigma_1)]\right\}$$

$$\times \sqrt{2m\,d\omega(\mathbf{p}'_1)\,2m\,d\omega(\mathbf{q}')\,2m\,d\omega(\mathbf{p}_1)\,2m\,d\omega(\mathbf{q})}\,(2\pi)^4\delta^{(4)}(p'_1 + q' - p_1 - q). \tag{5.9.40}$$

Upon comparison of this amplitude of Bhabha scattering, with the Møller one in (5.9.22)/(5.9.23), we may infer that the latter amplitude is related to the former

---

[32]Note the sign change in the exponent of the following two equations.

one by the substitutions:

$$p_2' \to -q, \, p_2 \to -q', \, u(\mathbf{p}_2', \sigma_2') \to v(\mathbf{q}, \sigma), \, u(\mathbf{p}_2, \sigma_2) \to v(\mathbf{q}', \sigma'), \quad (5.9.41)$$

with an overall change of sign of the amplitude.

The above substitution rule leads, for unpolarized particles, to the following expression for the differential cross section, at high energies, in the CM frame

$$\frac{d\sigma}{d\Omega_{\text{CM}}} \simeq \frac{\alpha^2}{4m^2\gamma^2} \left( \frac{4}{\sin^2\theta} - 1 \right)^2 \cos^4\frac{\theta}{2}. \quad (5.9.42)$$

## 5.9.2  $e^-\gamma \to e^-\gamma$; $e^+e^- \to \gamma\gamma$ and Polarizations Correlations

We consider the process $e^-\gamma \to e^-\gamma$, well known as Compton scattering, and the related one of electron-positron annihilation into two photons: $e^+e^- \to \gamma\gamma$. These processes are described in terms of the expression

$$i e^2 \int (dx)\,(dx')\,(dx_1)\,(dx_2) \left[ \overline{\eta}(x) S_+(x-x_1)\gamma^{\mu_1} S_+(x_1-x_2)\gamma^{\mu_2} S_+(x_2-x')\eta(x') \right]$$

$$\times \left[ \int (dz_1)\,(dz_2)\,D_{\mu_1\nu_1}(x_1-z_1)\,J^{\nu_1}(z_1)\,D_{\mu_2\nu_2}(x_2-z_2)\,J^{\nu_2}(z_2) \right], \quad (5.9.43)$$

which is given by the second term within the curly brackets in the first term of $a_2$ in (5.8.10), multiplied by $i e^2$ as given in the exponent in (5.8.1).

To describe these processes, we write $\eta = \eta_1 + \eta_2$, $J^\mu = J_1^\mu + J_2^\mu$, where $\eta_1$, $J_1^\mu$ are switched on and then off in the remote past after the electron and the photon are emitted, while $\eta_2$, $J_2^\mu$, are switched on in the distant future and then switched off after the outgoing electron and photon are detected (absorbed), with the interaction taking place later in time than the emissions and earlier than detections. Clearly, the first process, is described by the explicit expression

$$i e^2 \int (dx)\,(dx')\,(dx_1)\,(dx_2)\,(dz_1)\,(dz_2)$$

$$\times \left[ \overline{\eta}_2(x) S_+(x-x_1)\,\gamma^{\mu_1} S_+(x_1-x_2)\,\gamma^{\mu_2} S_+(x_2-x')\,\eta_1(x') \right]$$

$$\times \left[ D_{\mu_1\nu_1}(x_1-z_1)\,J_2^{\nu_1}(z_1)\,D_{\mu_2\nu_2}(x_2-z_2)\,J_1^{\nu_2}(z_2) \right.$$

$$\left. + D_{\mu_1\nu_1}(x_1-z_1)\,J_1^{\nu_1}(z_1)\,D_{\mu_2\nu_2}(x_2-z_2)\,J_2^{\nu_2}(z_2) \right]. \quad (5.9.44)$$

Here $z_1{}^0 < x_1{}^0$, $z_2{}^0 > x_2{}^0$, with $-z_1{}^0$, $z_2{}^0$ arbitrarily large, hence invoking, in the process, gauge invariance with respect external photon lines, we may make the substitutions

$$\int (\mathrm{d}z_1)\, D_{\mu_1 \nu_1}(x_1 - z_1) J_1^{\nu_1}(z_1) \rightarrow \mathrm{i} \int \frac{\mathrm{d}^3 \mathbf{k}}{2k^0 (2\pi)^3}\, e^{\mathrm{i}kx_1} \sum_\lambda e_{\mu_1 \lambda}\big(e^*_{\nu\lambda} J_1^\nu(k)\big),$$

(5.9.45)

$$\int (\mathrm{d}z_2)\, D_{\mu_2 \nu_2}(x_2 - z_2) J_2^{\nu_2}(z_2) \rightarrow \mathrm{i} \int \frac{\mathrm{d}^3 \mathbf{k}}{2k^0 (2\pi)^3}\, e^{-\mathrm{i}kx_2} \sum_\lambda \big(J_2^{\nu*}(k)\, e_{\nu\lambda}\big) e^*_{\mu_2 \lambda}.$$

(5.9.46)

For the initial electron and photon with respective momenta $p$, $k$, and final momenta with $p'$, $k'$, with the latter two taking values within infinitesimal intervals around the just specified values, the above equation gives the amplitude (Fig. 5.7)

$$A = \mathrm{i}\,\mathscr{M}\, \sqrt{\frac{\mathrm{d}^3 \mathbf{p}'}{(2\pi)^3\, 2p'^0}}\, \sqrt{\frac{\mathrm{d}^3 \mathbf{k}'}{2k'^0 (2\pi)^3}}\, \frac{1}{V}\, \frac{1}{\sqrt{4 p^0 k^0}}\, (2\pi)^4 \delta^{(4)}(p' + k' - p - k),$$

(5.9.47)

$$\mathscr{M} = e^2\, 2m\, \overline{u}(\mathbf{p}', \sigma') \Big[ \gamma\, e'_{\lambda'}\, \frac{-\gamma(p+k)+m}{(p+k)^2 + m^2}\, \gamma\, e_\lambda + \gamma\, e_\lambda\, \frac{-\gamma(p-k')+m}{(p-k')^2 + m^2}\, \gamma\, e'_{\lambda'} \Big] u(\mathbf{p}, \sigma),$$

(5.9.48)

where we recall that $\mathrm{i}\, e_\lambda^\mu\, J_{1\mu}(k)\big(\mathrm{d}^3 \mathbf{k}/(2\pi)^3\, 2k^0\big)^{1/2}$ is the amplitude that the (emitter) source $J_{1\mu}$ emits a photon of momentum $k$ and polarization specified by the parameter $\lambda$, and similarly, for the amplitude of detection of a photon (see (5.5.4) and (5.5.5)) for a real polarization vector.

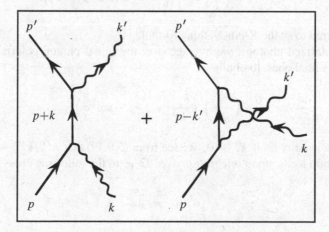

**Fig. 5.7** Compton scattering showing only the momenta of the particles involved. The photon is shown as a wavy line

We consider the initial electron at rest. Then upon squaring the momentum conserving equation $p' + k' = p - k$, using in the process $p'^0 + k'^0 = m + k^0$, and setting $\mathbf{k} \cdot \mathbf{k}' = k^0 k'^0 \cos\theta$, we obtain the well known expression relating the photon energy before and after the collision

$$\frac{k'^0}{k^0} = \frac{1}{1 + \frac{k^0}{m}(1 - \cos\theta)}. \tag{5.9.49}$$

We consider an experiment where the electron spins are not observed. Accordingly averaging over the initial spin of the electron and summing over the final spin, we obtain

$$\frac{1}{2}\sum_{\text{spins}} |\mathscr{M}|^2 = e^4\left(\frac{k'^0}{k^0} + \frac{k^0}{k'^0} + 2\left[2\,(e_\lambda e'_{\lambda'})^2 - 1\right]\right). \tag{5.9.50}$$

The $\mathbf{p}'$ integral, upon taking the absolute value squared of (5.9.47), is readily obtained by using in the rest frame of the electron

$$\int \frac{d^3\mathbf{p}'}{2p'^0}\,\delta^{(4)}(p' + k' - p - k) = \int (dp')\theta(p'^0)\delta(p'^2 + m^2)\delta^{(4)}(p' + k' - p - k)$$

$$= \theta(m + k^0 - k'^0)\delta\big((p + k - k')^2 + m^2\big) = \theta(m + k^0 - k'^0)\delta\big(-2kk' - 2m(k^0 - k'^0)\big), \tag{5.9.51}$$

and the $\mathbf{k}'$-integration may be now easily carried out. This leads for the differential cross section for polarized photons, in the rest frame of the electron (TF frame), the expression

$$\frac{d\sigma}{d\Omega_{\text{TF}}} = \frac{\alpha^2}{4\,m^2}\left(\frac{k'^0}{k^0}\right)^2\left(\frac{k'^0}{k^0} + \frac{k^0}{k'^0} + 2\left[2\,(e_\lambda e'_{\lambda'})^2 - 1\right]\right). \tag{5.9.52}$$

This is referred to as the Klein-Nishina formula.

For unpolarized photons, we average over the initial photon polarizations and sum over the final ones, to obtain

$$\frac{d\sigma}{d\Omega_{\text{TF}}} = \frac{\alpha^2}{2\,m^2}\left(\frac{k'^0}{k^0}\right)^2\left(\frac{k'^0}{k^0} + \frac{k^0}{k'^0} - \sin^2\theta\right). \tag{5.9.53}$$

In the low energy limit $k^0 \to 0$, we see from (5.9.49) that $k'^0/k^0 \to 1$, and the above equation leads, upon integration over $\Omega_{\text{TF}}$, to the following cross section

$$\sigma = \frac{8\pi}{3}\,r_0^2, \tag{5.9.54}$$

referred to as Thomson scattering cross section, and $r_0 = \alpha/m$ is the classical electron radius of the order $2.8 \times 10^{-13}$ cm.

Now we turn to the process of electron-positron annihilation into two photons. As an application, we will consider the polarizations correlations aspects of the emerging two photons. From (5.9.43), this process is extracted from the expression

$$i e^2 \int (dx)(dx')(dx_1)(dx_2)\left[\overline{\eta}_1(x)S_+(x-x_1)\gamma^{\mu_1}S_+(x_1-x_2)\gamma^{\mu_2}S_+(x_2-x')\eta_1(x')\right]$$

$$\times \left[\int (dz_1)(dz_2)D_{\mu_1\nu_1}(x_1-z_1)J_2^{\nu_1}(z_1)D_{\mu_2\nu_2}(x_2-z_2)J_2^{\nu_2}(z_2)\right]. \qquad (5.9.55)$$

Here $z_1^0 > x_1^0$, $z_2^0 > x_2^0$, with $z_1^0$, $z_2^0$ arbitrarily large, $x^0 < x_1^0$, $x'^0 < x_2^0$ with $-x^0$, $-x'^0$ arbitrarily large.

The amplitude for the process in question is readily extracted to be

$$A = i\mathscr{M}\sqrt{\frac{d\omega(\mathbf{k}_1)\,d\omega(\mathbf{k}_2)}{4p_1^0 p_2^0}}\,\frac{1}{V}\,(2\pi)^4\delta^{(4)}(k_1+k_2-p_1-p_2), \qquad (5.9.56)$$

$$\mathscr{M} = e^2 2m\overline{v}(\mathbf{p}_2,\sigma_2)\left[\gamma\,e_{\lambda_2}^{(2)}\,S_+(p_1-k_1)\gamma\,e_{\lambda_1}^{(1)} + \gamma\,e_{\lambda_1}^{(1)}\,S_+(p_1-k_2)\gamma\,e_{\lambda_2}^{(2)}\right]u(\mathbf{p}_1,\sigma_1). \qquad (5.9.57)$$

We work in the CM frame. To investigate the polarizations correlations of the outgoing photons, suppose that they have emerged in the experiment perpendicular to the direction of the electron positron pair (Fig. 5.8). Averaging over the initial

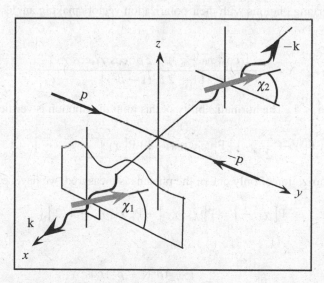

**Fig. 5.8** Photon polarization correlation experiment in $e^+e^- \to \gamma\gamma$ with unpolarized electron and positron, in which the outgoing photons have emerged and moved along the $x$-axis as shown. The polarization three-vectors lie in planes parallel to the $y-z$-plane. Only one of the possible outcomes of the experiment is shown here

spins of the electron and positron, we have in the CM frame corresponding to such a process, with

$$\mathbf{p}_1 \equiv \mathbf{p} = \gamma\, m\, \beta(0, 1, 0), \quad \mathbf{k}_1 = p^0(1, 0, 0), \tag{5.9.58}$$

$$\mathbf{e}_1^{(1)} = (0, 0, \cos\chi_1, \sin\chi_1), \quad \mathbf{e}_2^{(1)} = (0, 0, -\sin\chi_1, \cos\chi_1), \tag{5.9.59}$$

$$\mathbf{e}_1^{(2)} = (0, 0, \cos\chi_2, \sin\chi_2), \quad \mathbf{e}_2^{(2)} = (0, 0, -\sin\chi_2, \cos\chi_2), \tag{5.9.60}$$

where $\mathbf{e}_2^{(j)}$ is obtained from $\mathbf{e}_1^{(j)}$ by the substitutions

$$\chi_j \to \chi_j + \pi/2, \qquad j = 1, 2, \tag{5.9.61}$$

and for $\lambda_1 = 1$, $\lambda_2 = 1$, $(0 \le \beta^2 \le 1)$,

$$
m^2 \sum_{\text{spins}} \left| (\,.\,) \right|^2 \propto \left[ 1 - (\mathbf{e}_{\lambda_1}^{(1)} \mathbf{e}_{\lambda_2}^{(2)})^2 + 4(\mathbf{e}_{\lambda_1}^{(1)} \mathbf{e}_{\lambda_2}^{(2)}) \frac{(\mathbf{e}_{\lambda_1}^{(1)} p)(\mathbf{e}_{\lambda_2}^{(2)} p)}{(p^0)^2} - 4 \frac{(\mathbf{e}_{\lambda_1}^{(1)} p)^2 (\mathbf{e}_{\lambda_2}^{(2)} p)^2}{(p^0)^4} \right],
$$

$$= 1 - \left[ \cos(\chi_1 - \chi_2) - 2\beta^2 \cos\chi_1 \cos\chi_2 \right]^2,$$

$$\equiv 1 - \left[ (1 - \beta^2) \cos(\chi_1 - \chi_2) - \beta^2 \cos(\chi_1 + \chi_2) \right]^2, \tag{5.9.62}$$

up to an unimportant numerical proportionality factor at hand for the investigation in question. Therefore the properly normalized polarization correlations probability of the two emerging photons with their polarization vectors making angles $\chi_1$, $\chi_2$, as shown above, is

$$P[\chi_1, \chi_2] = \frac{1 - \left[ \cos(\chi_1 - \chi_2) - 2\beta^2 \cos\chi_1 \cos\chi_2 \right]^2}{2[1 + 2\beta^2(1 - \beta^2)]}, \tag{5.9.63}$$

for all $0 \le \beta \le 1$. The normalizability of this joint distribution is verified since

$$P[\chi_1, \chi_2] + P\left[\chi_1 + \frac{\pi}{2}, \chi_2\right] + P\left[\chi_1, \chi_2 + \frac{\pi}{2}\right] + P\left[\chi_1 + \frac{\pi}{2}, \chi_2 + \frac{\pi}{2}\right] = 1. \tag{5.9.64}$$

If the polarization of only one of the photons is measured, we have, e.g.,

$$P[\chi_1, -] = P[\chi_1, \chi_2] + P[\chi_1, \chi_2 + \pi/2],$$

and hence

$$P[\chi_1, -] = \frac{1 + 4\beta^2(1 - \beta^2)\cos^2\chi_1}{2[1 + 2\beta^2(1 - \beta^2)]}, \tag{5.9.65}$$

$$P[-, \chi_2] = \frac{1 + 4\beta^2(1 - \beta^2)\cos^2\chi_2}{2[1 + 2\beta^2(1 - \beta^2)]}. \tag{5.9.66}$$

The fact that the spin of one of the photons depends on the measurement of the spin of the other photon is expressed by the basic relation

$$P[\chi_1, \chi_2] \neq P[\chi_1, -]P[-, \chi_2]. \tag{5.9.67}$$

Re-iterating, this fundamental relation, in general, signifies the obvious correlation that exists between the measurement of one of the spins on the other.

The *speed* $\beta$ *dependence* of the probabilities cannot be overlooked. The moral of this is that polarization correlations probabilities, in general, depend on the underlying dynamics of the theory and, in particular, on the underlying energy (speed) and it is certainly misleading to attempt to derive them from just by combining spins.[33]

At low energies $\beta \to 0$, the probabilities in question become[34]

$$P[\chi_1, \chi_2] = \frac{1}{2} \sin^2(\chi_1 - \chi_2), \quad P[\chi_1, -] = P[-, \chi_2] - \frac{1}{2}. \tag{5.9.68}$$

### 5.9.3 $e^- \mu^- \to e^- \mu^-$

As another application we consider the scattering process $e^- \mu^- \to e^- \mu^-$, to the leading order (Fig. 5.9), by taking into account the fact that the mass of the muon $M$ is much larger than the mass of the electron. To this end, we add a QED Lagrangian for the muon to the QED Lagrangian of the electron.

Here the pertinent variables are given by

$$Q = k - k' - p' - p, \quad pQ = -\frac{Q^2}{2}, \quad kQ = \frac{Q^2}{2},$$

$$kk' = -m^2 - \frac{Q^2}{2} \approx -\frac{Q^2}{2}, \quad pp' = -M^2 - \frac{Q^2}{2}, \tag{5.9.69}$$

neglecting the mass of the electron, i.e., for $Q^2 \gg m^2$. We will also use the rather standard notations for the ingoing and outgoing electron energies:

$$k^0 \equiv E, \quad k'^0 \equiv E'. \tag{5.9.70}$$

---

[33]This subsection is based on Manoukian and Yongram [61]. For the significance of these polarizations correlations in their role in support of the monumental quantum theory against so-called hidden variables theories, see, e.g., the just mentioned paper, Manoukian [56], Bell [5], and Bell and Aspect [6]. For a detailed review, with applications to QED, the electroweak theory and strings, see Yongram and Manoukian [83]. We have observed the speed dependence of polarization correlations in actual *quantum field theory* calculations early (2003) (see [82])

[34]See also Problem 5.9.

**Fig. 5.9** The process
$e^-\mu^- \to e^-\mu^-$. The
momentum transfer carried
by the photon is denoted by
$Q$. The electromagnetic
charge e is shown for
convenience

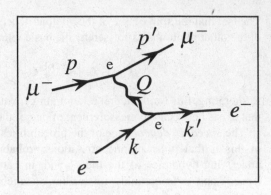

From our earlier treatments in the last two subsections, the expression of the differential cross section is, by now standard, and in the rest frame of the muon, is given by

$$d\sigma = \frac{d^3\mathbf{k}'}{(2\pi)^3 \, 2E'} \frac{d^3\mathbf{p}'}{(2\pi)^3 \, 2p'^0} \frac{(2\pi)^4 \delta^{(4)}(p + Q - p')}{4ME} \overline{|\mathcal{M}|^2}, \tag{5.9.71}$$

where $\overline{|\mathcal{M}|^2} = (e^4/Q^4)(1/2)(1/2)[\ . \ ]_1^{\mu\nu}[\ . \ ]_{2\,\mu\nu}$, and in an arbitrary frame

$$[\ . \ ]_1^{\mu\nu} = 4m^2 \sum_{\text{spins}} \text{Tr}\left[\bar{u}(\mathbf{k}',\sigma')\,\gamma^\mu u(\mathbf{k},\sigma)\,\bar{u}(\mathbf{k},\sigma)\,\gamma^\nu u(\mathbf{k}',\sigma')\right]$$

$$= 4\left[k'^\mu k^\nu + k'^\nu k^\mu - \eta^{\mu\nu} kk' - \eta^{\mu\nu} m^2\right], \tag{5.9.72}$$

$$[\ . \ ]_2^{\mu\nu} = 4M^2 \sum_{\text{spins}} \text{Tr}\left[\bar{u}(\mathbf{p}',\sigma')\,\gamma^\mu u(\mathbf{p},\sigma)\,\bar{u}(\mathbf{p},\sigma)\,\gamma^\nu u(\mathbf{p}',\sigma')\right]$$

$$= 4\left[p'^\mu p^\nu + p'^\nu p^\mu - \eta^{\mu\nu} pp' - \eta^{\mu\nu} M^2\right]. \tag{5.9.73}$$

In particular in the rest frame of muon $(p^0 = M)$, we may use (5.9.69), to obtain

$$[\ . \ ]_1^{\mu\nu}[\ . \ ]_{2\,\mu\nu} = 64M^2 EE'\left[\left(1 - \frac{Q^2}{4EE'}\right) + \frac{Q^2}{2M^2}\left(\frac{Q^2}{4EE'}\right)\right], \tag{5.9.74}$$

where we have neglected the mass $m$ of the electron in comparison to the mass $M$ of the muon inside the square brackets. We note that

$$Q^2 = -2pQ = 2M(E - E'), \quad Q^2 = 4EE' \sin^2\frac{\vartheta}{2}, \tag{5.9.75}$$

$$\sin^2\frac{\vartheta}{2} = \frac{Q^2}{4EE'}, \quad \cos^2\frac{\vartheta}{2} = 1 - \frac{Q^2}{4EE'}. \tag{5.9.76}$$

Now we use the integral

$$\int \frac{d^3\mathbf{p}'}{2p'^0} \delta^{(4)}(p + Q - p') = \int (dp')\theta(p'^0)\,\delta(p'^2 + M^2)\,\delta^{(4)}(p + Q - p')$$

$$= \delta((p + Q)^2 + M^2) = \delta(2pQ + Q^2), \qquad (5.9.77)$$

and in the rest frame of the muon, this gives

$$\int \frac{d^3\mathbf{p}'}{2p'^0} \delta^{(4)}(p + Q - p') = \delta(-2MQ^0 + Q^2)$$

$$= \frac{1}{2M} \delta\!\left(E' - \left(E - \frac{Q^2}{2M}\right)\right) = \frac{1}{2MD}\,\delta\!\left(E' - \frac{E}{D}\right) = \frac{E'}{2ME}\,\delta\!\left(E' - \frac{E}{D}\right),$$

$$(5.9.78)$$

where

$$D = \left(1 + 2\,\frac{E}{M}\,\sin^2 \frac{\vartheta}{2}\right), \qquad (5.9.79)$$

and we have used the relation $E' = (E - Q^2/2M)$, and the second equality in (5.9.75) to solve for $E'$.

Finally, we note that $d^3\mathbf{k}' = E'^2\,dE'\,d\Omega$, by neglecting the mass $m$ of the electron, to integrate over $E'$ by using, in the process, the constraint set by the delta function $\delta(E' - E/D)$ obtained above. All told, using, in particular (5.9.74) and (5.9.76), we obtain

$$\frac{d\sigma}{d\Omega} = \frac{\alpha^2}{4E^2 \sin^4 \frac{\vartheta}{2}}\,\frac{E'}{E} \left[\cos^2 \frac{\vartheta}{2} + \frac{Q^2}{2M^2}\,\sin^2 \frac{\vartheta}{2}\right]. \qquad (5.9.80)$$

## 5.10   Modified Propagators

To second order, consider the term in the exponential in (5.8.1) of $\langle\, 0_+ \mid 0_-\rangle$ involving only the bilinear form $\overline{\eta}\ldots\eta$, beyond the one in $a_0$, which naturally leads to a modification of the propagator from its "bare" counterpart $S_+$ due to the interaction. The corresponding expression is obtained from $i\,a_2$ in (5.8.10). Using a matrix multiplication notation in spacetime (as well as in Dirac indices), the total net contribution may be written as[35]

$$i\,\overline{\eta}\left[S_+ + e^2\,S_+\gamma^\mu S_+\gamma^\nu S_+(-i)\,D_{\mu\nu}\right]\eta. \qquad (5.10.1)$$

---

[35]For a simple structure we have $\int (dx)\,\overline{\eta}(x) S_+(x - x')\,\eta(x')\,(dx') = \overline{\eta}\,S_+\eta$, in matrix multiplication form.

In the momentum description this reads

$$i\int \frac{(dp)}{(2\pi)^4}\, \bar\eta(p)\, \widetilde{S}(p)\, \eta(p), \tag{5.10.2}$$

where $\widetilde{S}(p)$ is the modified fermion propagator due to the interaction, and is given by

$$\widetilde{S}(p) = S_+(p)\Big[1 - i\,e^2 \int \frac{(dk)}{(2\pi)^4}\, \gamma^\mu S_+(p-k)\gamma^\nu D_{\mu\nu}(k) S_+(p)\Big]. \tag{5.10.3}$$

It is more convenient to obtain the equation for the inverse $\widetilde{S}^{-1}$ of the modified propagator $(\widetilde{S}^{-1}\widetilde{S} = 1)$. To second order, it satisfies the equation

$$\widetilde{S}^{-1}(p) = S_+^{-1}(p) + i\,e^2 \int \frac{(dk)}{(2\pi)^4}\, \gamma^\mu S_+(p-k)\gamma^\nu D_{\mu\nu}(k). \tag{5.10.4}$$

We will investigate this equation and its physical interpretation in the first subsection below.

Now we also consider terms depending on the bilinear form $J^\mu \dots J^\nu$ which, again, naturally leads to the modified photon propagator, due to the interaction, from its "bare" counterpart $D_{\mu\nu}$. To this end, again using a matrix multiplication notation in spacetime, this is given from (5.8.8) and (5.8.10), to be

$$\frac{i}{2}J^{\nu_1}\big[D_{\nu_1\nu_2} + i\,e^2 D_{\nu_1\mu} K^{\mu\nu} D_{\nu\nu_2}\big]J^{\nu_2}, \tag{5.10.5}$$

where $K^{\mu\nu}$ is defined in (5.8.6) and (5.8.7). In the momentum description, the above equation reads

$$\frac{i}{2}\int \frac{(dk)}{(2\pi)^4}\, J^{*\mu}(k)\, \widetilde{D}_{\mu\nu}(k)\, J^\nu(k), \tag{5.10.6}$$

where $\widetilde{D}_{\mu\nu}$ is the modified photon propagator, due to the interaction, given by

$$\widetilde{D}_{\mu\nu}(k) = D_{\mu\mu_1}(k)\big[\delta^{\mu_1}{}_\nu + i\,e^2 L^{\mu_1\mu_2}(k,-k)D_{\mu_2\nu}(k)\big], \tag{5.10.7}$$

where $L_{\mu\nu}(k,-k)$ was explicitly computed in Appendix A of Chap. 3, and from (3.6.30), (3.6.31) and (3.6.32), it is given by

$$L^{\mu_1\mu_2}(k,-k) = [\eta^{\mu_1\mu_2}k^2 - k^{\mu_1}k^{\mu_2}]$$

$$\times \frac{i}{12\,\pi^2}\left(\Big[\frac{1}{i\,\pi^2}\int \frac{(dp)}{[p^2+m^2]^2} + \frac{1}{2}\Big] - 6\int_0^1 dz\, z(1-z)\ln\Big[1 + \frac{k^2}{m^2}z(1-z)\Big]\right). \tag{5.10.8}$$

Using the transversality condition of $L_{\mu\nu}(k, -k)$ in $k^\mu$, i.e., $k^\mu L_{\mu\nu}(k, -k) = 0$, $k^\nu L_{\mu\nu}(k, -k) = 0$, we have

$$\widetilde{D}_{\mu\nu}(k) = D_{\mu\nu}(k) + \frac{1}{k^2}\, \mathrm{i}\, e^2 L_{\mu\nu}(k, -k)\frac{1}{k^2}, \tag{5.10.9}$$

This equation will be analyzed in Sect. 5.10.2.

The quantum corrections to the propagators are referred to as radiative corrections. In general, a diagram depicting a process, involving no integrations over closed loops, i.e., no radiative corrections, is referred to as tree diagram.

## 5.10.1 Electron Self-Energy and Its Interpretation

To determine the modified electron propagator $\widetilde{S}(p)$, due to the interaction, and provide its physical interpretation, we evaluate the self energy-part $\Sigma(p)$, given by the integral (Fig. 5.10)

$$\Sigma(p) = -\mathrm{i}\, e^2 \int \frac{(\mathrm{d}k)}{(2\pi)^4}\, \gamma^\mu \frac{-\gamma(p-k) + m}{(p-k)^2 + m^2}\, \gamma^\nu D_{\mu\nu}(k), \tag{5.10.10}$$

where we have suppressed the $-\mathrm{i}\epsilon$ factor in the denominator for convenience of the notation. Also we work in arbitrary covariant gauge specified by the parameter $\lambda$ in the photon propagator, which we write as

$$D^{\mu\nu}(k) = \left(\eta^{\mu\nu} - (1-\lambda)\frac{k^\mu k^\nu}{k^2}\right)\frac{1}{k^2 + \mu^2}. \tag{5.10.11}$$

Here we have introduced a small fictitious mass $\mu \to 0$ for the photon to avoid divergences at low energies – the so called infrared divergence. Also note that due to the interaction, as we will see below, the actual mass $m$ of the electron is such that $m \neq m_0$, $m_0 = m[1 + \mathcal{O}(e^2)]$. Thus it is convenient to introduce the fee electron propagator $S(p)$, depending on the physical mass, and set in the sequel

$$S^{-1}(p) = \gamma p + m, \qquad S_+^{-1}(p) = \gamma p + m_0. \tag{5.10.12}$$

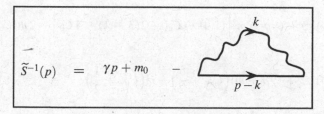

**Fig. 5.10** The inverse of the modified electron propagator, due to the interaction, to second order. The corresponding mathematical expression will be denoted by $\widetilde{S}^{-1}(p) = S_+^{-1}(p) - \Sigma(p)$. Note the minus sign between the latter two terms

Needless to say, we have used $S(p)$ in the integrand in (5.10.10) to the leading order.

The numerator $N$ of the integrand multiplying $(dk)/(2\pi)^4$ is given by

$$N = \left[ -(\lambda + 1)(\gamma p + m) - 2m \right] + 2\gamma k$$

$$- (\lambda - 1)\frac{\gamma k}{k^2}\left[ (p - k)^2 + m^2 \right] - (1 - \lambda)\frac{\gamma k}{k^2}[p^2 + m^2]. \tag{5.10.13}$$

The factor $\left[ (p - k)^2 + m^2 \right]$ in the third term cancels out with the denominator and leads to an odd integral in $k$, and hence the integral of the third term gives zero. We may now break the integral in question into three parts: $\Sigma(p) = \Sigma_1(p) + \Sigma_2(p) + \Sigma_3(p)$, where

$$\Sigma_1(p) = -i e^2 \left[ -(\lambda + 1)(\gamma p + m) - 2m \right] \int \frac{(dk)}{(2\pi)^4} \frac{1}{\left[ (p - k)^2 + m^2 \right](k^2 + \mu^2)}, \tag{5.10.14}$$

$$\Sigma_2(p) = -i e^2 \, (2) \int \frac{(dk)}{(2\pi)^4} \frac{\gamma k}{\left[ (p - k)^2 + m^2 \right](k^2 + \mu^2)}, \tag{5.10.15}$$

$$\Sigma_3(p) = i e^2 \, (1 - \lambda)[p^2 + m^2] \int \frac{(dk)}{(2\pi)^4} \frac{\gamma k}{k^2 \left[ (p - k)^2 + m^2 \right](k^2 + \mu^2)}. \tag{5.10.16}$$

The inverse electron propagator $\widetilde{S}^{-1}$ has the general structure $A + B(\gamma p + m) + \mathcal{O}((\gamma p + m)^2)$ where, $A$ and $B$ are constants. It diverges logarithmically at high energies, referred to as an ultraviolet divergence. Clearly from the dimensionality of $\widetilde{S}^{-1}$, only the constants $A$ and $B$ may have such a divergence. Any correction of the form $\mathcal{O}((\gamma p + m)^2)$ involves one additional power of $k$ in the integrand of $\Sigma(p)$ and hence necessarily is ultraviolet finite by simple power counting of powers of $k$ in its denominator and numerator. The main term $A + B(\gamma p + m)$ is the key term for examining the ultraviolet divergence of $\widetilde{S}^{-1}$.

Using the Feynman parameter representations to combine the denominators in these integrals as given in (II.27), (II.35), in Appendix II at the end of the book, and the momentum integrations given in (II.15), (II.16), (II.24), we obtain near the mass shell $p^2 \simeq -m^2$ (see Problem 5.13)

$$\frac{4\pi}{\alpha}\Sigma_1(p) \simeq -(\gamma p + m)\left[ (1 + \lambda)\, C_{\mathrm{uv}} + 2(3 + \lambda) - 4\, C_{\mathrm{ir}} \right] - 2\, m(C_{\mathrm{uv}} + 2), \tag{5.10.17}$$

$$\frac{4\pi}{\alpha}\Sigma_2(p) \simeq +(\gamma p + m)\left( C_{\mathrm{uv}} + \frac{5}{2} \right) - m\left( C_{\mathrm{uv}} + \frac{1}{2} \right), \tag{5.10.18}$$

$$\frac{4\pi}{\alpha}\Sigma_3(p) \simeq -(1 - \lambda)(\gamma p + m)\left( 1 - 2\, C_{\mathrm{ir}} \right), \tag{5.10.19}$$

up to ultraviolet finite terms $\mathscr{O}\big((\gamma p + m)^2\big)$, where we have used the identity

$$(p^2 + m^2) = 2m(\gamma p + m) - (\gamma p + m)^2, \tag{5.10.20}$$

$$C_{\text{uv}} = \frac{1}{i\pi^2} \int (dk) \frac{1}{(k^2 + m^2)^2}, \tag{5.10.21}$$

and for $\mu^2/m^2 \to 0$ (see Problem 5.12)

$$C_{\text{ir}} = \int_0^1 \frac{x\,dx}{\big[x^2 + \frac{\mu^2}{m^2}(1 - x)\big]} = -\frac{1}{2}\ln\Big(\frac{\mu^2}{m^2}\Big). \tag{5.10.22}$$

Here (uv) and (ir) stand for ultraviolet and infrared, respectively. The above Eqs. (5.10.17), (5.10.18) and (5.10.19) are given up to ultraviolet *finite* terms which approach zero near the mass shell. From (5.10.17), (5.10.18), (5.10.19), (5.10.20), (5.10.21) and (5.10.22), the inverse electron propagator, to second order, then has the following behavior near the mass shell

$$\widetilde{S}^{-1}(p) \simeq \frac{(\gamma p + m_0 + \delta m)}{Z_2} + \mathscr{O}\big((\gamma p + m)^2\big), \tag{5.10.23}$$

where

$$\delta m = \frac{3\alpha}{4\pi} m \Big(C_{\text{uv}} + \frac{3}{2}\Big), \tag{5.10.24}$$

$$\frac{1}{Z_2} = 1 + \frac{\alpha}{4\pi}\Big[\lambda\, C_{\text{uv}} - 2(3 - \lambda)\, C_{\text{ir}} + \lambda + \frac{9}{2}\Big]. \tag{5.10.25}$$

and $\delta m$ is called the self-mass of the electron. Thus the electron's mass is necessarily given by

$$m = m_0 + \delta m, \tag{5.10.26}$$

and is gauge independent, i.e., it is independent of the parameter $\lambda$. It is referred to as the renormalized mass, i.e., the experimentally observed electron mass, while $m_0$ is referred to as the unrenormalized mass or bare mass. We will elaborate further on this below.

The constant $C_{\text{uv}}$, may be defined in terms of an ultra-violet cut-off or by means of dimensional regularization.[36] We discuss both approaches. To this end, we may

---

[36]See Appendix III, at the end of the book, for dimensional regularization.

define $C_{uv}$ in terms of an ultraviolet cut-off $\Lambda$ as

$$C_{uv} = \frac{1}{i\pi^2} \int (dk) \frac{1}{(k^2 + m^2)^2} \frac{\Lambda^2}{(k^2 + \Lambda^2)} = \left[ \ln\left(\frac{\Lambda^2}{m^2}\right) - 1 \right], \qquad (5.10.27)$$

for $\Lambda^2 \to \infty$, where we have combined the denominators using (II.34), integrated over $k$, using (II.8), and finally carried out the elementary integral over the Feynman parameter $x$ (see Problem 5.11). One may also define $C_{uv}$ with dimensional regularization, by working in $(4 - \varepsilon)$ – dimsional spacetime, with $\varepsilon > 0$, by introducing, in the process, an arbitrary scale parameter, say, $\tau$, and using the definition of the dimensional-regularized integral (III.8) to obtain

$$C_{uv} = \frac{1}{\pi^{\varepsilon/2}} \left(\frac{\tau^2}{m^2}\right)^{\varepsilon/2} \Gamma\left(\frac{\varepsilon}{2}\right). \qquad (5.10.28)$$

Here the limit of the ultraviolet cut-off $\Lambda^2 \to \infty$, has been replaced by the limit of the parameter $\varepsilon \to +0$, where we note that

$$\Gamma\left(\frac{\varepsilon}{2}\right) = \frac{2}{\varepsilon} - \gamma_E + \mathcal{O}(\varepsilon). \qquad (5.10.29)$$

with $\gamma_E = 0.5772157\ldots$ denoting Euler's constant.

The renormalization constant $Z_2$ being the residue of the modified propagator, is gauge dependent and is, respectively, ultraviolet finite and infrared finite, in the following stated gauges

$$\frac{1}{Z_2} = 1 + \frac{\alpha}{4\pi}\left[\frac{9}{2} - 6\,C_{ir}\right], \qquad \lambda = 0, \qquad (5.10.30)$$

$$\frac{1}{Z_2} = 1 + \frac{\alpha}{4\pi}\left[3\,C_{uv} + \frac{15}{2}\right], \qquad \lambda = 3, \qquad (5.10.31)$$

referred to as the Landau gauge and the Yennie-Fried gauge, respectively. In the Feynman gauge $\lambda = 1$, it is ultraviolet cut-off and infrared cut-off dependent

$$\frac{1}{Z_2} = 1 + \frac{\alpha}{4\pi}\left[C_{uv} - 4\,C_{ir} + \frac{11}{2}\right], \qquad \lambda = 1. \qquad (5.10.32)$$

What the theory is telling us is that if $|\mathbf{p}, \sigma\rangle$ denotes an electron state with momentum $\mathbf{p}$, and spin $\sigma$, normalized as

$$\langle \mathbf{p}'\sigma' | \mathbf{p}\sigma \rangle = (2\pi)^3 \frac{p^0}{m} \delta^{(3)}(\mathbf{p}' - \mathbf{p}) \delta_{\sigma'\sigma}, \qquad (5.10.33)$$

then

$$\langle \mathbf{p}\sigma | \overline{\psi}(x) | \text{vac} \rangle = \sqrt{Z_2}\, \overline{u}(\mathbf{p}, \sigma)\, e^{-ipx}, \qquad (5.10.34)$$

defined in terms of the physical mass $m$. Moreover, with $x'^0 > x^0$,

$$
i \int \sum_\sigma \langle \text{vac}|\psi(x')|\mathbf{p}\,\sigma\rangle \frac{m\,d^3\mathbf{p}}{p^0\,(2\pi)^3} \langle \mathbf{p}\,\sigma|\overline{\psi}(x)|\text{vac}\rangle
$$

$$
= Z_2\, i \int \frac{d^3\mathbf{p}}{2p^0\,(2\pi)^3} (-\gamma p + m)\, e^{-ip\,(x-x')} = Z_2\, S(x'-x). \tag{5.10.35}
$$

In reference to (5.10.34), the particle aspect appropriate for describing scattering states dictates to scale $\psi(x) \to \psi(x)/(Z_2)^{1/2} = \psi_{\text{ren}}(x)$, thus introducing the concept of the renormalized electron field, and carrying a wavefunction renormalization.[37]

It is interesting to relate the wavefunction renormalization constant $Z_2$ in the arbitrary gauge, specified by the parameter $\lambda$, to the one in the Landau gauge in (5.10.30). This is obtained from (5.10.25) to be

$$
\frac{1}{Z_2}\Big|_\lambda = 1 + \frac{\alpha}{4\pi}\Big[\lambda \ln\Big(\frac{\Lambda^2}{m^2}\Big) + (3-\lambda)\ln\Big(\frac{\mu^2}{m^2}\Big) + \frac{9}{2}\Big]
$$

$$
= 1 + \frac{\alpha}{4\pi}\Big[\lambda \ln\Big(\frac{\Lambda^2}{\mu^2}\Big) + 3\ln\Big(\frac{\mu^2}{m^2}\Big) + \frac{9}{2}\Big] = \frac{1}{Z_2}\Big|_{\lambda=0} + \lambda\frac{\alpha}{4\pi}\ln\Big(\frac{\Lambda^2}{\mu^2}\Big). \tag{5.10.36}
$$

This will be compared with the full theory later on in Sect. 5.15 (see (5.15.30)).

Formally, $Z_2$ represents the probability that the (fermion) field creates a particle out of the vacuum. The electron field, however, is a gauge dependent object and one must be careful in such an interpretation here. The second point concerning $Z_2$ is, in general, its infrared cut-off dependence. This is due to the fact that an electron, being a charged particle, and the fact that the photon is massless, it cannot be rigorously defined, in isolation, without the presence of an arbitrary number of photons of arbitrarily small (soft) energies accompanying it, while $Z_2$ above is introduced in terms of an electron without such a cloud of soft photons. Finally, we note that this constant was introduced by going on the mass shell of the electron. This corresponds to defining scattering states of an electron as observed in emission and detection regions. This necessitates to scale the electron propagator by $1/Z_2$, and introduce, in turn, the renormalized electron propagator via the identification

$$
\widetilde{S}_{\text{ren}}(p) = \frac{1}{Z_2}\widetilde{S}(p), \tag{5.10.37}
$$

$$
\widetilde{S}_{\text{ren}}^{-1}(p) = (\gamma p + m) + \mathcal{O}((\gamma p + m)^2) \quad \text{near the mass shell.} \tag{5.10.38}
$$

---

[37] See also Sect. 4.1 for details on the concept of wavefunction renormalization.

In reference to (5.10.34), the above condition defines a boundary condition for the renormalized propagator near the mass shell, for a proper description of electron in scattering states, up to the presence of soft photons accompanying it. The ever presence of such soft photons will be discussed later in Sect. 5.12.

Now we come to the concept of mass renormalization. Using the ultraviolet cut-off dependence of $C_{uv}$ in (5.10.27), we may, from (5.10.24), write

$$\delta m = \frac{3\alpha}{4\pi} m \left[ \ln\left(\frac{\Lambda^2}{m^2}\right) + \frac{1}{2} \right]. \tag{5.10.39}$$

From (5.10.26), this suggests to define an effective mass which depends on an energy scale $\kappa$ by

$$m(\kappa^2) = m \left( 1 - \frac{3\alpha}{4\pi} \left[ \ln\left(\frac{\kappa^2}{m^2}\right) + \frac{1}{2} \right] \right), \tag{5.10.40}$$

with the so-called bare mass *given* by $m_0 = \lim_{\kappa^2 \to \Lambda^2} m(\kappa^2)$. The latter limit, for $\Lambda^2 \to \infty$, corresponds to going all the way to the very core of the electron for which, presumably, no justification can be given with our present physical theories. Through the process of renormalization, the parameter $m_0$ has been eliminated in favor of the physically observed mass $m$, as it should be done for a sensible description of the theory.

Equation (5.10.40) is still remarkable in the sense that even at high energies $\kappa$ but much smaller than $m \exp[(2\pi/3\alpha) - 1/4]$, so that $(3\alpha/4\pi)(\ln(\kappa^2/m^2) + 1/2) \ll 1$, $m(\kappa^2)$, in the just mentioned equation, is not drastically different from the renormalized (physically observed) mass $m$ measured at large distances in the lab.[38] At small distances, at the Planck length, for example, gravitation is expected to play an essential role in formulating quantum dynamical theories. Extrapolating to such distances, presumably, requires detailed generalizations in our physical theories.

## 5.10.2   Photon Self-Energy and Its Interpretation; Coulomb Potential

The modified photon propagator, from (5.10.8), (5.10.9) and (3.6.30), may be written as (Fig. 5.11)

$$\widetilde{D}_{\mu\nu}(k) = D_{\mu\nu}(k) - \frac{1}{k^2} \Pi_{\mu\nu}(k) \frac{1}{k^2}, \tag{5.10.41}$$

---

[38] This is due to the logarithmic nature on the cut-off dependence in QED. On the other hand, for the mass of the Higgs Boson, in the electroweak theory, one has a quadratic dependence on the cut-off $\Lambda$, and its unrenormalized and renormalized masses are drastically different.

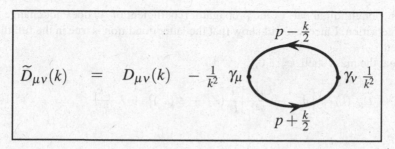

**Fig. 5.11** The modified photon propagator, due to the interaction, to second order, up to terms obtained from the expression of a gauge invariant current. The corresponding mathematical expression will be denoted by $\widetilde{D}_{\mu\nu}(k) = D_{\mu\nu}(k) - (1/k^2)\, \Pi_{\mu\nu}(k)(1/k^2)$

where $\Pi_{\mu\nu}(k)$ is given in (3.6.31) and (3.6.32) to be

$$\Pi_{\mu\nu}(k) = [\eta_{\mu\nu}\, k^2 - k_\mu k_\nu]\Pi(k^2), \tag{5.10.42}$$

and

$$\Pi(k^2) = \frac{e^2}{12\,\pi^2}\left(C_{uv} + \frac{1}{2} - 6\int_0^1 dz\, z(1-z)\ln\left[1 + \frac{k^2}{m^2}z(1-z)\right]\right), \tag{5.10.43}$$

where $C_{uv}$ is given in (5.10.27). $\Pi_{\mu\nu}(k)$ is referred to as the vacuum polarization tensor. Note the transversality of the latter: $k^\mu \Pi_{\mu\nu}(k) = 0$, $k^\nu \Pi_{\mu\nu}(k) = 0$ which, as mentioned in Sect. 3.6, is attributed to gauge invariance. Also note that

$$\Pi(0) = \frac{e^2}{12\,\pi^2}\left(C_{uv} + \frac{1}{2}\right) = \frac{\alpha}{3\pi}\left[\ln\left(\frac{\Lambda^2}{m^2}\right) - \frac{1}{2}\right], \tag{5.10.44}$$

$$\Pi(k^2) - \Pi(0) = -\frac{2\alpha}{\pi}\int_0^1 dz\, z(1-z)\ln\left[1 + \frac{k^2}{m^2}z(1-z)\right] \equiv \alpha\,\pi_c(k^2), \tag{5.10.45}$$

and the latter is a finite expression.

In an arbitrary covariant gauge, specified by the parameter $\lambda$,

$$D_{\mu\nu}(k) = \left[\eta_{\mu\nu} - \frac{k_\mu k_\nu}{k^2}\right]\frac{1}{k^2} + \lambda\,\frac{k_\mu k_\nu}{k^4}, \tag{5.10.46}$$

the modified photon propagator takes the form

$$\widetilde{D}_{\mu\nu}(k) = \left[\eta_{\mu\nu} - \frac{k_\mu k_\nu}{k^2}\right]\frac{1}{k^2}\left(1 - \Pi(k^2)\right) + \lambda\,\frac{k_\mu k_\nu}{k^4}, \tag{5.10.47}$$

and the longitudinal part of the propagator (coefficient of $\lambda$) does not change with the interaction. Later, we will show that the latter condition is true in the full theory as well.

Near the mass shell $k^2 \simeq 0$,

$$\widetilde{D}_{\mu\nu}(k) = \left[\eta_{\mu\nu} - \frac{k_\mu k_\nu}{k^2}\right]\frac{1}{k^2}\left(Z_3 + \mathscr{O}(k^2)\right) + \lambda\frac{k_\mu k_\nu}{k^4}, \tag{5.10.48}$$

where

$$Z_3 = 1 - \frac{\alpha}{3\pi}\left[C_{\mathrm{uv}} + \frac{1}{2}\right] = 1 - \frac{\alpha}{3\pi}\left[\ln\!\left(\frac{\Lambda^2}{m^2}\right) - \frac{1}{2}\right], \tag{5.10.49}$$

is gauge invariant, and is defined with an ultraviolet cut-off $\Lambda$, while with dimensional regularization,[39] it is given by

$$Z_3 = 1 - \frac{\alpha}{3\pi}\left[\frac{1}{\pi^{\varepsilon/2}}\left(\frac{\tau^2}{m^2}\right)^{\varepsilon/2}\Gamma\!\left(\frac{\varepsilon}{2}\right) + \frac{1}{2}\right]. \tag{5.10.50}$$

Again from our study of the field concept, $Z_3$ is interpreted as the probability that the electromagnetic field creates a photon out of the vacuum. A clear picture of the meaning of the renormalization constant $Z_3$ further arises from a so-called spectral decomposition of the photon propagator, which has also a counterpart in the full theory, and is obtained as follows.

To the above end, note that in reference to the integral in (5.10.45), we may generate the following chain of equalities

$$\int_0^1 dz\, z(1-z)\ln\!\left[1 + \frac{k^2}{m^2}z(1-z)\right] = -\int_0^1 dz\, z(1-z)\int_{m^2}^\infty d\mu^2 \frac{\partial}{\partial\mu^2}\ln\!\left[1 + \frac{k^2}{\mu^2}z(1-z)\right],$$

$$= k^2\int_0^1 dz\, z^2(1-z)^2\int_0^\infty \frac{d\mu^2}{\mu^2}\frac{\theta(\mu^2 - m^2)}{\mu^2 + k^2 z(1-z)}$$

$$= k^2\int_0^1 dz\, z^2(1-z)^2\int_{(2m)^2}^\infty dM^2\int_0^\infty \frac{d\mu^2}{\mu^2}\frac{\delta(M^2 z(1-z) - \mu^2)\,\theta(M^2 z(1-z) - m^2)}{k^2 + M^2},$$

$$\tag{5.10.51}$$

where the last equality is easily verified by integrating over $M^2$ and using, in the process, the delta function constraint. The lower limit $(2m)^2$ of the $M^2$-integral

---

[39] In the next chapter, the gluon propagator, the counterpart of the photon propagator in QCD, will be directly analyzed with dimensional regularization.

follows from the step function constraint and the fact that the roots of the equation $M^2 z(1-z) - m^2 = 0$ are

$$z_1 = \frac{1}{2}(1 - \sqrt{1 - 4m^2/M^2}), \quad z_2 = \frac{1}{2}(1 + \sqrt{1 - 4m^2/M^2}),$$

requiring that $M^2 \geq 4m^2$. Upon integrating over $\mu^2$, we obtain for the above integral

$$k^2 \int_{(2m)^2}^{\infty} \frac{dM^2}{M^2} \frac{1}{k^2 + M^2} \int_{z_1}^{z_2} dz \, z(1-z)$$

$$= \frac{k^2}{6} \int_{(2m)^2}^{\infty} \frac{dM^2}{M^2} \frac{1}{k^2 + M^2} \left(1 + \frac{2m^2}{M^2}\right) \sqrt{1 - \frac{4m^2}{M^2}}. \tag{5.10.52}$$

All told, we have from (5.10.45), (5.10.47), (5.10.49) the following expression for $\widetilde{D}_{\mu\nu}(k)$ emerges[40]

$$\widetilde{D}_{\mu\nu}(k) = \left(\eta_{\mu\nu} - \frac{k_\mu k_\nu}{k^2}\right) \widetilde{D}_+(k) + \lambda \frac{k_\mu k_\nu}{k^4}, \tag{5.10.53}$$

$$\widetilde{D}_+(k) = \int_0^{\infty} dM^2 \frac{\rho(M^2)}{k^2 + M^2 - i\epsilon}, \tag{5.10.54}$$

$$\rho(M^2) = Z_3 \, \delta(M^2) + \frac{\alpha}{3\pi} \theta(M^2 - (2m)^2) \frac{1}{M^2}\left(1 + \frac{2m^2}{M^2}\right) \sqrt{1 - \frac{4m^2}{M^2}}, \tag{5.10.55}$$

$\rho(M^2) \geq 0$, and is referred to as the spectral function of the photon propagator, to second order. In reference to (5.10.55), the first term in (5.10.55), corresponds to the photon, with residue $Z_3$, and the second to the electron-positron pair, having an invariant mass squared going from $(2m)^2$ up to arbitrary large values due to their arbitrary relative motion.

In the absence of radiative corrections, the counterpart of $\widetilde{D}_+(k)$ is simply $D_+(k) = 1/k^2$. The Coulomb interaction between two static charges,[41] each of charge $q$, may be extracted in the latter case as follows:

$$U(|\mathbf{x}|) = q^2 \int \frac{d^3\mathbf{k}}{(2\pi)^3} e^{i\mathbf{k}\cdot\mathbf{x}} D_+(k)|_{k^0=0} = \frac{q^2}{4\pi |\mathbf{x}|}. \tag{5.10.56}$$

---

[40]The presence of $-i\epsilon$ in $(k^2 + M^2 - i\epsilon)$ in the denominator is understood.
[41]A process involving no energy transfer.

In the presence of radiative corrections, with $D_+(k) \to \widetilde{D}_+(k)$, the Coulomb potential, in turn, using (5.10.54), is modified to

$$\widetilde{U}(|\mathbf{x}|) = q^2 \int \frac{d^3\mathbf{k}}{(2\pi)^3} \, e^{i\mathbf{k}\cdot\mathbf{x}} \, \widetilde{D}_+(k)|_{k^0=0} = \frac{q^2}{4\pi|\mathbf{x}|} \int_0^\infty dM^2 \, \rho(M^2) \, e^{-M|\mathbf{x}|},$$

$$= \frac{q^2}{4\pi|\mathbf{x}|} \left[ Z_3 + \frac{\alpha}{3\pi} \int_{(2m)^2}^\infty \frac{dM^2}{M^2} \left( 1 + \frac{2m^2}{M^2} \right) \sqrt{1 - \frac{4m^2}{M^2}} \, e^{-M|\mathbf{x}|} \right],$$

(5.10.57)

thus leading to the Coulomb potential, at large distances $|\mathbf{x}| \to \infty$, with modified renormalized charges,

$$\widetilde{U}(|\mathbf{x}|) \to \frac{q_{\mathrm{ren}}^2}{4\pi|\mathbf{x}|}, \quad q_{\mathrm{ren}} = q\sqrt{Z_3}. \qquad (5.10.58)$$

We will see later that the bare charge $e_0$ is renormalized in this way as well, and other renormalization constants such as $Z_2$, and a corresponding one for the vertex, do not contribute to charge renormalization. That is,

$$e_0 \sqrt{Z_3} = e, \quad \alpha_0 Z_3 = \alpha, \qquad (5.10.59)$$

where $e$ and $\alpha$, are the experimentally observed parameters measured at large distances from the electron.

As for the electron field renormalization, the Maxwell field is also renormalized as $A^\mu/\sqrt{Z_3} = A^\mu_{\mathrm{ren}}$, corresponding to scattering states, thus introducing the concept of a renormalized Maxqwell field. The renormalized photon propagator, then takes the form

$$\widetilde{D}^{\mu\nu}_{\mathrm{ren}} = \frac{1}{Z_3} \widetilde{D}^{\mu\nu}, \quad e^2 \widetilde{D}^{\mu\nu} = e_{\mathrm{ren}}^2 \widetilde{D}^{\mu\nu}_{\mathrm{ren}}, \qquad (5.10.60)$$

$$\widetilde{D}^{\mu\nu}_{\mathrm{ren}}(k) = D^{\mu\nu}(k) - \frac{1}{k^2} \left[ \eta^{\mu\nu} - \frac{k^\mu k^\nu}{k^2} \right] \Pi_{\mathrm{ren}}(k), \qquad (5.10.61)$$

$$\Pi_{\mathrm{ren}}(k) = \Pi(k) - \Pi(0) = -\frac{2\alpha}{\pi} \int_0^1 dz \, z(1-z) \ln\left[ 1 + \frac{k^2}{m^2} z(1-z) \right], \qquad (5.10.62)$$

$$\Pi_{\mathrm{ren}}(k) = -\frac{\alpha}{15\pi} \frac{k^2}{m^2} + \mathcal{O}\left( \left( \frac{k^2}{m^2} \right)^2 \right). \qquad (5.10.63)$$

The first relation defines the renormalized photon propagator whose transverse part behaves as the free propagator near the mass shell, corresponding to scattering states.

From (5.10.59), we may write

$$\alpha_0 = \alpha \left[ 1 + \frac{\alpha}{3\pi} \left( \ln\left(\frac{\kappa^2}{m^2}\right) - \frac{1}{2} \right) \right], \quad \kappa \to \infty. \tag{5.10.64}$$

This, in turn, suggests to define an effective fine-structure by the relation

$$\alpha(\kappa^2) = \alpha \left[ 1 + \frac{\alpha}{3\pi} \left( \ln\left(\frac{\kappa^2}{m^2}\right) - \frac{1}{2} \right) \right], \tag{5.10.65}$$

which increases as $\kappa^2$ increases, i.e., it increases as the distance between the charge decreases. This is physically expected, as from the diagram above, for example, one may infer that the creation of an electron-positron pair leads to screening of a charge when one is far away from the charge in question, and vice versa as one approaches the charge as discussed in the introductory chapter of the book. The bare fine-structure constant is then formally given by $\alpha_0 = \lim_{\kappa^2 \to \infty} \alpha(\kappa^2)$, which corresponds to moving all the way to the core of the electron through the cloud of the pairs of particles. The divergence encountered in this limit is of no surprise as our theories are not justified to describe nature at absolute zero distances. One, however, may re-express the theory in terms of physically observed quantities instead of limiting experimentally unattainable quantities. The good news about QED, and all renormalizable theories, is that when all unrenormalized quantities are eliminated in favor of renormalized ones, the theories become ultraviolet finite defined in terms of realistic physical quantities as the actual value of the fine-structure constant and the actual mass of the electron.

In (5.10.58)/(5.10.59) we have seen that the Coulomb potential is simply modified, at large distances by simply renormalizing the charges. On the other hand, for small distances $m|\mathbf{x}| \ll 1/2$, the Coulomb potential is also modified, and as given in Problem 5.14, to

$$\widetilde{U}(\mathbf{x}) \to \frac{q^2}{4\pi|\mathbf{x}|} \left( Z_3 + \frac{2\alpha}{3\pi} \left[ \ln\left(\frac{1}{m|\mathbf{x}|}\right) - \gamma_{\mathrm{E}} - \frac{5}{6} \right] \right), \tag{5.10.66}$$

where $\gamma_{\mathrm{E}} = 0.5772157\ldots$ denotes Euler's constant.

## 5.11  Vertex Part

We consider, up to third order, the combination of sources $\overline{\eta} \ldots \eta J^\mu$ within the exponential factor in $\langle 0_+ | 0_- \rangle$, in (5.8.1) which generate all diagrams with exactly two electron lines and one photon lines that are *attached* to these sources. The lowest order is given by $a_1$ in (5.8.9), and the corresponding generated diagram is shown in part (A) of Fig. 5.12. The other diagrams (B)–(E) are generated from the first and third expressions of $a_3$ in (5.8.11), within the curly brackets.

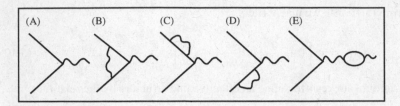

**Fig. 5.12** The set of all diagrams (A)–(E), up to third order, which may be connected to other diagrams by two electron lines and one photon line. This set of diagrams are all generated, with exact numerical factors, from the terms $a_1$, $a_3$ in (5.8.9), (5.8.11) involving the combination of sources $\bar{\eta} \ldots \eta J^\mu$

Diagram (C)–(E) are obtained from our investigation of the self energy parts of the electron and photon and are thus generated from the insertions of the latter in external electron and photon lines, respectively and will be treated in Sect. 5.11.2.

The diagram in (A) is generated from $a_1$, while the diagram in (B) is generated from the second term, within the first pair of curly brackets of $a_3$, leading, together, to the combination

$$i e_0 \int \frac{(dp')}{(2\pi)^4} \int \frac{(dp)}{(2\pi)^4} \int \frac{(dQ)}{(2\pi)^4} \left[ \bar{\eta}(p')S_+(p')\Gamma^\mu(p',p)S_+(p)\eta(p) \right]$$

$$\times D_{\mu\nu}(Q)J^\nu(Q)(2\pi)^4\delta^{(4)}(p'-p-Q), \tag{5.11.1}$$

where note that the overall coefficient multiplying the integral is the bare charge $e_0$,

$$\Gamma^\mu(p',p) = \gamma^\mu + \Lambda^\mu(p',p), \tag{5.11.2}$$

defining the vertex part,

$$\Lambda^\mu(p',p) = -i e^2 \int \frac{(dk)}{(2\pi)^4} D_{\sigma\rho}(k) \frac{\gamma^\sigma \left( -\gamma(p'-k) + m \right)}{(p'-k)^2 + m^2} \gamma^\mu \frac{\left( -\gamma(p-k) + m \right)\gamma^\rho}{(p-k)^2 + m^2}. \tag{5.11.3}$$

to second order, and here we have used the renormalized charge to this lowest order. The diagram in Fig. 5.13 shows the rooting of the momenta taken to run through $\Lambda^\mu(p',p)$, and the direction of the external momenta.

To investigate the ultraviolet divergence of this integral, it is sufficient to consider the external electron lines on the mass shell. We also consider initially the expression up to the linear term in $Q$. The resulting expression will be also useful in applications that will follow. The numerator coming from the $\eta_{\sigma\rho}$ part of the photon propagator (see (5.10.11)), with the understanding that it is to be sandwiched between $\bar{u}(\mathbf{p}',\sigma')$ and $u(\mathbf{p},\sigma)$, is

$$N^\mu = 2\gamma k \gamma^\mu \gamma k + 4 m k^\mu + 4(p^\mu + p'^\mu)\gamma k - 4(p'+p)k\gamma^\mu + 4pp'\gamma^\mu. \tag{5.11.4}$$

**Fig. 5.13** The rooting of the internal photon momentum taken in $\Lambda^\mu(p',p)$, where the vertex part is given by $\Gamma^\mu(p',p) = \gamma^\mu + \Lambda^\mu(p',p)$

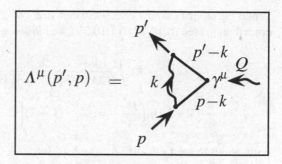

We combine the three propagators in (5.11.3) with $A = k^2$, $B = (p'-k)^2 + m^2$, $C = (p-k)^2 + m^2$, by using the Feynman parameter representation in (II.35) of Appendix II at the end of the book.

$$\frac{1}{ABC} = 2 \int_0^1 dx \int_0^x dz \frac{1}{[A(1-x) + Bz + C(x-z)]^3}. \tag{5.11.5}$$

Since by power counting the integral is only logarithmically divergent, we carry out, in turn, a change of variable: $k \to k + p'z + p(x-z)$, leading effectively to

$$A(1-x) + Bz + C(x-z) \to k^2 + m^2x^2 + \mu^2(1-x). \tag{5.11.6}$$

On the other hand for the resulting expression of the numerator $N^\mu$, we have

$$\int_0^x dz\, N^\mu = (k^2 + m^2x^2)x\,\gamma^\mu + mx^2(1-x)[\gamma^\mu, \gamma^\nu]Q_\nu + m^2x(x^2 + 4x - 4)\gamma^\mu. \tag{5.11.7}$$

Integrating over $k$ and $x$, give for the $\eta_{\sigma\rho}/(k^2 + \mu^2)$ part of the photon propagator contribution to (5.11.3)

$$\frac{\alpha}{4\pi}\left(\left[C_{uv} - 4C_{ir} + \frac{11}{2}\right]\gamma^\mu + \frac{1}{2m}[\gamma^\mu, \gamma^\nu]Q_\nu\right). \tag{5.11.8}$$

For the $-(1-\lambda)k^\sigma k^\rho/k^2(k^2+\mu^2)$ term (see (5.10.11)) in the photon propagator, note that the corresponding integrand in the vertex part (5.11.3) is simply

$$(\lambda - 1)\frac{k^\sigma k^\rho}{k^2(k^2 + \mu^2)}\gamma_\sigma S(p'-k)\gamma^\mu S(p-k)\gamma_\rho$$

$$= (\lambda - 1)\frac{1}{k^2(k^2 + \mu^2)}\left[1 - S^{-1}(p')S(p'-k)\right]\gamma^\mu\left[1 - S(p-k)S^{-1}(p)\right]. \tag{5.11.9}$$

When the latter is sandwiched between $\bar{u}(\mathbf{p}',\sigma')$, $u(\mathbf{p},\sigma)$, it leads to an additional contribution (see (5.10.22), (5.10.27), and Problem 5.12)

$$i\,e^2(1-\lambda)\int \frac{(dk)}{(2\pi)^4}\frac{1}{k^2(k^2+\mu^2)}\frac{\Lambda^2}{k^2+\Lambda^2}$$

$$=\frac{\alpha}{4\pi}(1-\lambda)\ln\!\left(\frac{\mu^2}{\Lambda^2}\right)=-\frac{\alpha}{4\pi}(1-\lambda)\Big[C_{\mathrm{uv}}+2\,C_{\mathrm{ir}}+1\Big], \tag{5.11.10}$$

with an ultraviolet cut-off $\Lambda^2$, and an infrared cut-off $\mu^2$. All told, the leading contribution in $Q$ to $\Gamma^\mu(p',p)$, on the mass shell, is given by

$$\Gamma^\mu(p',p)=\frac{1}{Z_1}\,\gamma^\mu+\frac{\alpha}{8\,\pi\,m}[\gamma^\mu,\gamma^\nu]\,Q_\nu, \qquad Q=p'-p, \tag{5.11.11}$$

$$\frac{1}{Z_1}=1+\frac{\alpha}{4\pi}\Big[\lambda\,C_{\mathrm{uv}}-2(3-\lambda)\,C_{\mathrm{ir}}+\lambda+\frac{9}{2}\Big]. \tag{5.11.12}$$

The renormalized vertex is then defined by

$$\Gamma^\mu_{\mathrm{ren}}(p',p)=Z_1\,\Gamma^\mu(p',p). \tag{5.11.13}$$

From (5.10.25), we note that the vertex function renormalization constant $Z_1$ is equal to $Z_2$ in all gauges. This is no accident, and we will see later that this is true to all orders.

An infrared cut-off may be also introduced by dimensional regularization by working in $4+\varepsilon$ dimensions of spacetime with $\varepsilon>0$. Still another approach, for example, is to simply restrict the momentum carried by a photon $|\mathbf{k}|>K$, with $K\ll m$. That is, in the latter approach, we may express the photon propagator part $D_+$ as

$$D_+(x-x')=\int_{|\mathbf{k}|>K}\frac{d^3\mathbf{k}}{(2\pi)^3}\int\frac{dk^0}{(2\pi)}\frac{1}{k^2-i\epsilon}\,e^{ik(x-x')}, \tag{5.11.14}$$

which may be rewritten as

$$D_+(x-x')=\mathrm{Lim}_{\mu\to0}\Bigg[\int\frac{(dk)}{(2\pi)^4}\frac{1}{k^2+\mu^2-i\epsilon}\,e^{ik(x-x')}$$

$$-\int_{|\mathbf{k}|<K}\frac{d^3\mathbf{k}}{(2\pi)^3}\int\frac{dk^0}{(2\pi)}\frac{1}{k^2+\mu^2-i\epsilon}\,e^{ik(x-x')}\Bigg]. \tag{5.11.15}$$

The first expression within the square brackets is what we have been using thus far. The second term, when substituted in the integral equation of the vertex part, in addition to the first one, allows one to eliminate the fictitious photon mass $\mu$ cut-off in favor of the infrared cut-off $K$, via the equation

$$C_{\text{ir}} = \int_0^1 \frac{x \, dx}{\left[x^2 + \frac{\mu^2}{m^2}(1-x)\right]} = -\frac{1}{2} \ln\left(\frac{\mu^2}{m^2}\right) = \ln\left(\frac{m}{2K}\right) + \frac{5}{6}. \qquad (5.11.16)$$

This transformation rule will be useful later on.

The consequences of the renormalization constants $Z_1$, $Z_2$, $Z_3$ is discussed next.

## 5.11.1 Charge renormalization and External lines

To second order, the inverse of the modified electron propagator is given by

$$\widetilde{S}^{-1}(p) = S_+^{-1}(p) - \Sigma(p), \qquad (5.11.17)$$

where $\Sigma(p)$ is defined in (5.10.10), and, as emphasized earlier, $S_+(p)$ depends on the bare mass $m_0$, while $S(p)$, given in (5.10.12), is defined in terms of the physical one $m$. Since $m_0 = m - \delta m$, we conveniently have

$$S_+^{-1}(p) = S^{-1}(p) - \delta m. \qquad (5.11.18)$$

Referring to Fig. 5.10, this allows us to re-express $\widetilde{S}^{-1}(p)$ as

$$\widetilde{S}^{-1}(p) = S^{-1}(p) - \big(\Sigma(p) + \delta m\big), \quad \widetilde{S}(p) = S(p) + S(p)\big(\Sigma(p) + \delta m\big)S(p), \qquad (5.11.19)$$

with the following boundary condition

$$\big(\Sigma(p) + \delta m\big)S(p)\big|_{\text{mass shell}} \to Z_2 - 1 = -(Z_2^{-1} - 1), \qquad (5.11.20)$$

and the latter equality holds true to second order. Accordingly, if the mass term $-m_0 \overline{\psi}\psi$ in the lagrangian density is simply rewritten as $-m\overline{\psi}\psi + \delta m \overline{\psi}\psi$, and we use the propagator $S(p)$, depending on the physical mass $m$, as we should, as the zeroth order and treat $\delta m$ in a perturbation expansion, thus generating another, though simple, vertex part, then we may simply replace $\Sigma(p)$ in our earlier expressions, wherever they appear, in $a_1$, $a_2$, $a_3$, in $\langle 0_+ \mid 0_- \rangle$, by $\Sigma(p) + \delta m$, and $S_+(p)$ by $S(p)$. Thus the totality of all diagrams in Fig. 5.12, with two external electron line and one photon line, up to the order considered, are generated by the corresponding terms in $a_1$, $a_3$, including now the perturbation term $\delta m \overline{\psi}\psi$, and is given in terms of momentum variables as

$$
i \int \frac{(\mathrm{d}p')}{(2\pi)^4} \int \frac{(\mathrm{d}p)}{(2\pi)^4} \int \frac{(\mathrm{d}Q)}{(2\pi)^4} (2\pi)^4 \, \delta(Q - p' + p)
$$

$$
\times \overline{\eta}(p') \, S(p') \, \Omega^\mu(p', p) \, S(p) \, \eta(p) \, D_{\mu\nu}(Q) \, J^\nu(Q), \tag{5.11.21}
$$

$$
\Omega^\mu(p', p) = e_0 \Big[ \gamma^\mu + \Big( \Sigma(p') + \delta m \Big) S(p') \gamma^\mu + \Lambda^\mu(p', p)
$$

$$
+ \gamma^\mu S(p) \Big( \Sigma(p) + \delta m \Big) - \gamma_\sigma \frac{1}{Q^2} \Pi^{\sigma\mu}(Q) \Big], \tag{5.11.22}
$$

where note that the overall charge multiplying the expression for $\Omega^\mu$ is $e_0$.

The coupling terms of the fields to their external sources in the Lagrangian density may be rewritten as

$$
\overline{\eta}\psi + \overline{\psi}\eta + A_\mu J^\mu = \overline{\eta}_{\mathrm{ren}}\psi_{\mathrm{ren}} + \overline{\psi}_{\mathrm{ren}}\eta_{\mathrm{ren}} + A_{\mathrm{ren}\mu}J_{\mathrm{ren}}^\mu, \tag{5.11.23}
$$

$$
\psi_{\mathrm{ren}} = \psi/\sqrt{Z_2}, \qquad \overline{\eta}_{\mathrm{ren}} = \sqrt{Z_2}\,\overline{\eta}, \tag{5.11.24}
$$

$$
\overline{\psi}_{\mathrm{ren}} = \overline{\psi}/\sqrt{Z_2}, \qquad \eta_{\mathrm{ren}} = \sqrt{Z_2}\,\eta, \tag{5.11.25}
$$

$$
A_{\mathrm{ren}\mu} = A_\mu/\sqrt{Z_3}, \qquad J_{\mathrm{ren}}^\mu = \sqrt{Z_3}\,J^\mu, \tag{5.11.26}
$$

and the external sources, now associated with *renormalized states*, i.e., states described by spinors $u$, $\overline{u}$, $v$, $\overline{v}$, expressed in terms of the physical mass and satisfy the earlier normalizations conditions, and with the photons described by appropriate polarization vectors as before.

The external sources now become expressed in terms of renormalized sources and hence $\Omega^\mu(p', p)$ becomes replaced by $(Z_2 Z_3 Z_2)^{-1/2} \Omega^\mu(p', p)$ $\equiv (Z_2 \sqrt{Z_3})^{-1} \times \Omega^\mu(p', p)$, which will be used below in (5.11.27).

The identity $\gamma^\mu k_\mu = (\gamma p' + m) - (\gamma p + m)$, implies the vanishing of the latter expression when sandwiched between the spinors $\overline{u}(p', \sigma')$ and $u(p, \sigma)$. When this is applied to the last term within the square brackets in (5.11.22), this term simply becomes

$$
\overline{u}(p', \sigma')\gamma^\mu u(p, \sigma)\Pi(Q^2),
$$

(see Problem 5.15). We introduce $\Omega_{\mathrm{ren}}^\mu(p', p)$ as the renormalized counterpart of $\Omega(p', p)$ with renormalized external lines. By using (5.11.19)/ (5.10.23), (5.11.11)/(5.11.2), (5.10.44)/(5.10.49), we obtain from (5.11.22) on the mass shell, the expression

$$\bar{u}(\mathbf{p}',\sigma')\,\Omega^\mu_{\text{ren}}(p',p)\,u(\mathbf{p},\sigma) = \Big[\frac{1}{(Z_2\sqrt{Z_3}}\Big]\bar{u}(\mathbf{p}',\sigma')\,\Omega^\mu(p',p)\,u(\mathbf{p},\sigma)$$

$$= e_0\Big[\frac{1}{Z_2\sqrt{Z_3}}\Big]\bar{u}(\mathbf{p}',\sigma')$$

$$\times\Big[\gamma^\mu\Big(1-2\big(\frac{1}{Z_2}-1\big)+\big(\frac{1}{Z_1}-1\big)-(1-Z_3)+\mathcal{O}(Q^2)\Big)+\mathcal{O}(Q^\mu)\Big]u(\mathbf{p},\sigma),$$

$$(5.11.27)$$

for $Q \simeq 0$.

*Now* we may take the limit $Q^\mu \to 0$, to infer that the right-hand side of the above equation becomes

$$\bar{u}(\mathbf{p},\sigma')\,\gamma^\mu\,u(\mathbf{p},\sigma)\,e_0\Big[\frac{1}{\sqrt{Z_3}}\Big]$$

$$\times\Big[1+\big(\frac{1}{Z_2}-1\big)\Big]\Big[1-2\big(\frac{1}{Z_2}-1\big)+\big(\frac{1}{Z_1}-1\big)-(1-Z_3)\Big]$$

$$= \sqrt{Z_3}\,e_0\,\bar{u}(\mathbf{p},\sigma')\,\gamma^\mu\,u(\mathbf{p},\sigma), \qquad (5.11.28)$$

where we have deliberately written $Z_2^{-1} = 1+(Z_2^{-1}-1)$ in the external electron lines renormalization constant, *multiplied* this factor with the expression within the square brackets following it, working to second order, thus simply adding the second order term $(Z_2^{-1} - 1)$ inside the square brackets, and finally used the fact that $Z_1 = Z_2$.

Thus $\Omega^\mu$, due to the external lines renormalization, is consequently renormalized as $\Omega^\mu_{\text{ren}} = (Z_2 Z_3 Z_2)^{-1/2}\Omega^\mu$, satisfying the boundary condition

$$\bar{u}(\mathbf{p}',\sigma')\,\Omega^\mu_{\text{ren}}(p',p)\,u(\mathbf{p},\sigma)\Big|_{Q=0} = e\,\bar{u}(\mathbf{p},\sigma')\,\gamma^\mu\,u(\mathbf{p},\sigma), \qquad (5.11.29)$$

defining the renormalized charge, i.e., the experimentally observed charge given by

$$e = \sqrt{Z_3}\,e_0, \qquad (5.11.30)$$

and the wavefunction renormalization constant $Z_2$ of the electron, and that of the vertex function $Z_1$ do not contribute to it, as a consequence of their equality - a result which also holds true to all orders, as we will see later.

The zero momentum transfer of the electron provides information only on the charge of the electron. To see any structure and obtain additional information on the electron, a non-zero momentum $Q$ transfer is necessary.

## 5.11.2 Anomalous Magnetic Moment of the Electron

We have seen earlier in (3.2.22), that the matrix element corresponding to the scattering amplitude of a Dirac particle off an external electromagnetic field $A^{\text{ext}}(x)$ has the following form

$$e\,[\bar{u}(\mathbf{p}',\sigma')\,\gamma^{\mu}\,u(\mathbf{p},\sigma)]\,A^{\text{ext}}_{\mu}(Q)$$

$$= e\,\bar{u}(\mathbf{p}',\sigma')\Big[\frac{(p^{\mu}+p'^{\mu})}{2m}+\frac{[\gamma^{\mu},\gamma^{\nu}]}{4m}\,Q_{\nu}\Big]\,u(\mathbf{p},\sigma)\,A^{\text{ext}}_{\mu}(Q), \qquad (5.11.31)$$

where $A^{\text{ext}}_{\mu}(Q)$ is the Fourier transform of $A^{\text{ext}}(x)$, $Q = (p'-p)$. On the other hand, radiative corrections, modify the vertex $e_0\,\gamma^{\mu}$, including the renormalization of charge, to the renormalized expression: $e(\gamma^{\mu} + \alpha\,[\gamma^{\mu},\gamma^{\nu}]\,Q_{\nu}/(8\,\pi\,m))$ at low momentum transfer (see (5.11.11)/(5.11.13)). Thus according to QED, the above amplitude becomes simply modified to

$$\bar{u}(\mathbf{p}',\sigma')\Big[e\,\frac{(p^{\mu}+p'^{\mu})}{2m}+\frac{[\gamma^{\mu},\gamma^{\nu}]}{4}\,Q_{\nu}\,\frac{e}{m}\Big(1+\frac{\alpha}{2\pi}\Big)\Big]\,u(\mathbf{p},\sigma)\,A^{\text{ext}}_{\mu}(Q),$$

$$(5.11.32)$$

providing a correction factor $(1 + \alpha/2\pi)$ to the electron magnetic moment,[42] a celebrated result first derived by Schwinger.[43] The additional factor $\alpha/2\pi$ is referred to as the anomaly. Needless to say higher order effects of this correction may be also carried out and have been successfully done over the years.[44]

## 5.12 Radiative Correction to Coulomb Scattering and Soft Photon Contribution

We consider radiative corrections to the Coulomb scattering for non-zero momentum transfer $Q = p' - p$, where $p, p'$ denote the initial and final momenta of the electron. To this end, for $Q^2 \neq 0$, we extend the analysis of the vertex part in Sect. 5.11 beyond the linear term in $Q$ to second order $Q^2$.

---

[42]See (3.2.22), (3.2.23), (3.2.24), (3.2.25) and (3.2.26).

[43]Schwinger [68].

[44]When the electron is treated non-relativistically, this correction has been also carefully computed (see Manoukian [56], p. 453 for a textbook treatment) to be: $[1+\kappa\,(\alpha/2\pi)]$, where $\kappa = (16/9) - 2\ln(3/2) \simeq 0.97$, and compares well with the QED result.

In this case, we note that the expression in (5.11.6) in the denominator becomes replaced by

$$A(1-x) + Bz + C(x-z) \rightarrow k^2 + m^2 x^2 + \mu^2(1-x) + Q^2 z(x-z), \quad (5.12.1)$$

where

$$2p'p = -Q^2 - 2m^2, \qquad Qp = -\frac{Q^2}{2}, \quad (5.12.2)$$

and, in particular, that

$$\frac{1}{i\pi^2} \int \frac{(dk)}{ABC} = C_{\text{ir}} + \frac{Q^2}{3\,m^2}\Big(C_{\text{ir}} - \frac{1}{2}\Big). \quad (5.12.3)$$

to second order in $Q$. The modification of $e_0\,\bar{u}(\mathbf{p}',\sigma')\,\Omega^\mu(p',p)\,u(\mathbf{p},\sigma)$, after mass and charge renormalization, is then readily obtained to second order in $Q$ to be

$$e\,\bar{u}(\mathbf{p}',\sigma')\,\Omega^\mu_{\text{ren}}(p',p)\,u(\mathbf{p},\sigma)$$

$$= e\,\bar{u}(\mathbf{p}',\sigma')\Big[\gamma^\mu\Big(1 - \frac{\alpha}{3\pi}\frac{Q^2}{m^2}\Big[-\frac{1}{2}\ln\Big(\frac{\mu^2}{m^2}\Big) - \frac{3}{8} - \frac{1}{5}\Big]\Big) + \frac{\alpha}{8\pi m}[\gamma^\mu, \gamma^\nu]\,Q_\nu\Big]u(\mathbf{p},\sigma),$$

$$(5.12.4)$$

where, in the process, we have, from (5.10.62) and (5.10.63), used the fact that $d\Pi_{\text{ren}}(Q)/dQ^2|_{Q^2=0} = -(\alpha/15\pi\,m^2)$ in (5.11.22) and (5.11.27), after sandwiching $\Omega^\mu_{\text{ren}}(p',p)$ between $\bar{u}(\mathbf{p}',\sigma')$, $u(\mathbf{p},\sigma)$.

Accordingly, for the scattering of an electron off a Coulomb potential

$$eA_\mu(x) = \Big(\frac{Ze^2}{4\pi|\mathbf{x}|}, \mathbf{0}\Big), \qquad eA_\mu(Q) = [-2\pi\delta(Q^0)\,\eta_{\mu 0}]\frac{Ze^2}{|\mathbf{Q}|^2}, \quad (5.12.5)$$

we have from (5.12.4) and this equation, that the transition probability per unit time, normalized in a unit volume, is given by $(Q = p' - p)$

$$\text{Prob}[p, \sigma \rightarrow p', \sigma']\big|_{\text{unit time}}$$

$$= \frac{m}{p'^0}\frac{d^3\mathbf{p}'}{(2\pi)^3}\frac{m}{p^0}\frac{2\pi\,\delta(Q^0)Z^2\,e^4}{|\mathbf{Q}|^4}\,\big|\bar{u}(\mathbf{p}',\sigma')\,\Omega^0_{\text{ren}}(p',p)\,u(\mathbf{p},\sigma)\big|^2, \quad (5.12.6)$$

for $|\mathbf{Q}|^2/m^2 \ll 1$. Upon averaging over the initial spin of the electron and summing over its final spin, it easily follows,[45] that the differential cross section for Coulomb

---

[45]See also (3.4.5), (3.4.6), (3.4.7), (3.4.8) and (3.4.9).

scattering, including radiative corrections, in the non-relativistic limit is given by

$$\frac{d\sigma}{d\Omega} = \frac{d\sigma}{d\Omega}\Big|_0\Big[1 - \frac{2\alpha}{3\pi}\frac{|\mathbf{Q}|^2}{m^2}\Big(-\frac{1}{2}\ln\Big(\frac{\mu^2}{m^2}\Big) - \frac{1}{5}\Big)\Big], \qquad \frac{|\mathbf{Q}|^2}{m^2} \ll 1, \qquad (5.12.7)$$

where $d\sigma/d\Omega|_0$ is the differential cross section for Coulomb scattering without radiative corrections (see (3.4.9), (3.4.10), (3.4.11), (3.4.12) and (3.4.13)).

At this point, it is wise to use the transformation rule in (5.11.16), to eliminate the fictitious photon mass in favor of the infrared cut-off and rewrite the differential cross section as

$$\frac{d\sigma}{d\Omega} = \frac{d\sigma}{d\Omega}\Big|_0\Big[1 - \frac{2\alpha}{3\pi}\frac{|\mathbf{Q}|^2}{m^2}\Big(\ln\Big(\frac{m}{2K}\Big) + \frac{5}{6} - \frac{1}{5}\Big)\Big], \qquad \frac{|\mathbf{Q}|^2}{m^2} \ll 1, \qquad (5.12.8)$$

involving a rigorously zero mass photon. This expression still depends on the unwanted infrared cut-off $K$. A charged particle, however, is accompanied by very soft photons, i.e., of arbitrary small energies and, due to the resolution involved in any experimental set up, such very soft photons cannot be experimentally detected. Thus to the lowest order considered above, we consider the emission of a very soft photon in the Coulomb scattering problem of Sect. 3.4. The experimental resolution will be defined by setting an upper bound for the energy of such a very soft by $\Delta$.

The diagrams in Fig. 5.14 show the emission of a photon of momentum $k$ from external electron lines in Coulomb scattering, involving no radiative corrections. The corresponding matrix element for the scattering process is given by

$$A_0(p - p' - k)\, e\, \bar{u}(\mathbf{p}', \sigma')\Big[\gamma^0\frac{-\gamma(p-k) + m}{(p-k)^2 + m^2}\gamma^\mu + \gamma^\mu\frac{-\gamma(p'+k) + m}{(p'+k)^2 + m^2}\gamma^0\Big]e_\mu u(\mathbf{p}, \sigma),$$
$$(5.12.9)$$

which for a very soft photon, we may set $k$ to be zero in the numerator, $p'^0 \simeq p^0$, and the matrix element above, upon using the Dirac equations for the spinors, reduces to

$$A_0(p - p')\, e\, \bar{u}(\mathbf{p}', \sigma')\gamma^0\Big[\frac{p'^\mu}{p'k} - \frac{p^\mu}{pk}\Big]e_\mu u(\mathbf{p}, \sigma). \qquad (5.12.10)$$

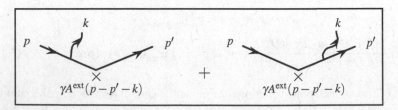

**Fig. 5.14** The emission of a photon of momentum $k$, respectively, by the ingoing and outgoing electron in Coulomb scattering not involving radiative corrections

For a non-relativistic electron $pk = p'k = mk^0$, and the differential cross section, with unpolarized electrons, corresponding to the emission of a very soft (unpolarized) photon, referred to as soft bremsstahlung (SB), is simply given by

$$\frac{d\sigma}{d\Omega}\Big|_{\text{SB}} = \frac{d\sigma}{d\Omega}\Big|_0 e^2 \int_{K<|\mathbf{k}|<\Delta} \frac{d^3\mathbf{k}}{(2\pi)^3\, 2\,|\mathbf{k}|} \frac{1}{m^2\,\mathbf{k}^2} Q_i Q_j \sum_\lambda e_\lambda^i\, e_\lambda^j, \qquad (5.12.11)$$

$Q^i = p^{\prime i} - p^i$. Using the completeness relation and the angular average given below,

$$\sum_\lambda e_\lambda^i\, e_\lambda^j = \eta^{ij} - \frac{k^i k^j}{\mathbf{k}^2}, \quad \left\langle \frac{k^i k^j}{\mathbf{k}^2} \right\rangle = \frac{4\pi}{3}\,\eta^{ij}, \qquad (5.12.12)$$

we obtain

$$\frac{d\sigma}{d\Omega}\Big|_{\text{SB}} = \frac{d\sigma}{d\Omega}\Big|_0 \frac{2\alpha}{3\pi} \frac{\mathbf{Q}^2}{m^2} \ln\!\left(\frac{\Delta}{K}\right), \quad \frac{|\mathbf{Q}|^2}{m^2} \ll 1. \qquad (5.12.13)$$

Therefore, for the inclusive process, for the scattering amplitude, with radiative corrections, and including the emission of a very soft photon, which would escape its detection, is, from (5.12.8) and (5.12.13), given by

$$\frac{d\sigma}{d\Omega}\Big|_{\text{incl}} = \frac{d\sigma}{d\Omega}\Big|_0 \left[ 1 - \frac{2\alpha}{3\pi} \frac{|\mathbf{Q}|^2}{m^2} \left( \ln\!\left(\frac{m}{2K}\right) + \frac{5}{6} - \frac{1}{5} - \ln\!\left(\frac{\Delta}{K}\right) \right) \right], \qquad (5.12.14)$$

or

$$\frac{d\sigma}{d\Omega}\Big|_{\text{incl}} = \frac{d\sigma}{d\Omega}\Big|_0 \left[ 1 - \frac{2\alpha}{3\pi} \frac{|\mathbf{Q}|^2}{m^2} \left( \ln\!\left(\frac{m}{2\Delta}\right) + \frac{5}{6} - \frac{1}{5} \right) \right], \quad \frac{|\mathbf{Q}|^2}{m^2} \ll 1, \qquad (5.12.15)$$

and the infrared cut-off $K$ cancels out eliminating the so-called infrared divergence. This cancelation occurs to all orders when emissions of soft photons are appropriately taken into consideration.[46]

## 5.13 Lamb Shift

Two experiments which have much stimulated the development of quantum field theory in the earlier days, were the observation of anomalous magnetic moment of the electron,[47] which was already derived in Sect. 5.11.2, and of the Lamb shift.[48]

---

[46]See, e.g., Yennie and Suura [81]. See also Manoukian [52].

[47]Kusch and Foley [39], cf., Foley and Kusch [25].

[48]See Lamb and Retherford [40] for the early experiment.

The Lamb shift in consideration here, is the splitting between the $2\,S_{1/2}$ and the $2\,P_{1/2}$ levels of the hydrogen atom and is due to the interaction of the atomic electron with radiation, that is, with the photon.[49] These levels are degenerate according to the Dirac theory.[50]

We provide a derivation of the Lamb shift which is physically easy to grasp, and parallels methods often used in formulating the eigenvalue problem in quantum mechanics courses and the reader will, hopefully, be able to easily follow. The basic idea is that with an infrared cut-off $m\alpha^2 \ll K \ll m$, as discussed in Sect. 5.11, we may use the expression of the vertex correction given in that section, by restricting the spatial photon momentum $\mathbf{k}$ as follows $|\mathbf{k}| > K$, and apply straightforward first order perturbation theory with the latter as a perturbation to the Dirac theory. On the other for $|\mathbf{k}| < K$, one is necessarily involved with a low energy non-relativistic regime, and a straightforward, though modified perturbative treatment in this region, may be developed and readily carried out. In this latter, case, special care needs to be taken as the analysis leads to a logarithmic expression in the fine-structure constant, and the Coulomb potential has to be treated exactly, while radiative corrections may be treated perturbatively. If we denote the contributions of the Lamb shift by $\delta E^{>}$ and $\delta E^{<}$, corresponding to both regions : $|\mathbf{k}| > K$, $|\mathbf{k}| < K$ respectively, then the total shift is given by $\delta E = \delta E^{>} + \delta E^{<}$, and the infrared cut-off $K$ cancels out between the two expressions.[51] This leads to a residual logarithmic dependence on the fine structure constant squared. Here we should remember that the infrared problem, in the radiative corrections to Coulomb scattering, arises from low energy photons, and in the atom, the energy of a photon is of the order of the difference between the energy levels of the Hydrogen atom, that is, of the order $m\alpha^2$.

• *Derivation and Evaluation of* $\delta E^{>}$:

Our starting point is the bound state problem of the Dirac equation in the presence of a Coulomb potential given by

$$\left(\gamma^{\mu}\frac{\partial_{\mu}}{\mathrm{i}} - \mathrm{e}\,\gamma^{0}A_0(\mathbf{x}) + m\right)\psi_n(x) = 0, \qquad (5.13.1)$$

with

$$\psi_n(x) = \psi_n(\mathbf{x})\,\mathrm{e}^{-\mathrm{i}E_n x^0}, \quad -\mathrm{e}A_0(\mathbf{x}) = -\frac{\alpha}{|\mathbf{x}|}, \qquad (5.13.2)$$

---

[49]Sometimes I have the feeling that there are as many derivations of the Lamb shift as there are field theory practitioners.

[50]A careful derivation of the Lamb shift was particularly carried out by Erickson and Yennie [22], see also Fox and Yennie [26]. There are also other careful derivations in the literature- too many to mention. A highly sophisticated and a very careful derivation, spelling out the finest of details, is also given by Schwinger [71].

[51]For a treatment and a pedagogical presentation of the Lamb shift with the electron treated non-relativistically, see Manoukian [56], pp. 391–403, which is in the spirit of Bethe's [8] classic treatment.

for an attractive (spherically symmetric) potential. The former equation may be rewritten as

$$\left(\gamma^0 \frac{\boldsymbol{\gamma} \cdot \boldsymbol{\nabla}}{i} - eA_0(\mathbf{x}) + \gamma^0 m\right)\psi_n(\mathbf{x}) = E_n \psi_n(\mathbf{x}). \tag{5.13.3}$$

Upon writing $\psi = (\phi \quad \varphi)^\top$, we have

$$\varphi = \frac{1}{2m}[1 + \mathscr{O}(\alpha^2)]\boldsymbol{\sigma} \cdot \mathbf{p}\,\phi, \tag{5.13.4}$$

in the hydrogen atom, where $eA_0 \sim m\alpha^2$, $E_n = m[1 + \mathscr{O}(\alpha^2)]$. The normalization condition $\int d^3x\,\psi_n^\dagger(\mathbf{x})\psi_n(\mathbf{x}) = 1$, dictates to set $\phi = (1 - \mathbf{p}^2/8\,m^2)\chi$ with $\int d^3x\,\chi^\dagger(\mathbf{x})\chi(\mathbf{x}) = 1$. That is, we may write

$$\psi_n = \begin{pmatrix} (1 - \frac{\mathbf{p}^2}{8m^2})\,\chi_n \\[2mm] \frac{\boldsymbol{\sigma} \cdot \mathbf{p}}{2m}\,\chi_n \end{pmatrix} \simeq \sqrt{\frac{p^0 + m}{2p^0}}\begin{pmatrix} \chi_n \\[2mm] \frac{\boldsymbol{\sigma} \cdot \mathbf{p}}{p^0 + m}\,\chi_n \end{pmatrix}, \tag{5.13.5}$$

with the latter expression properly normalized. We recall that the fine-structure of the hydrogen atom, as a perturbation, is of the order of $\alpha^4$ and non-relativistic wave-functions corresponding to $\chi_n$ above are sufficient to derive it. The Lamb shift is even of smaller magnitude, hence we may still consider non-relativistic wavefunctions.

The relativistic corrections arising from (5.13.3), necessitates combining spin and orbital momentum in an atom, thus introducing the total angular momentum by $\mathbf{J} = \mathbf{L} + \mathbf{S}$. For the Lamb shift in consideration, $j = 1/2$ and we need to consider the two values $\ell = 0, 1$.[52] This leads to the construction of wavefunctions for the treatment of relativistic corrections, which for the problem in consideration, with quantum number $n = 2$, we have,[53]

$$\chi_{2,\ell} = \frac{1}{N_\ell}\begin{pmatrix} \varphi_{2,\ell,+} \\ \varphi_{2,\ell,-} \end{pmatrix}, \quad N_0 = \sqrt{2}, \quad N_1 = \left(\sum_{m=0,-1}|a_m| + \sum_{m=0,+1}|b_m|\right)^{1/2}, \tag{5.13.6}$$

$$\varphi_{2,0,\pm}(r,\Omega) = R_{20}(r)\,Y_{0,0}(\Omega), \tag{5.13.7}$$

---

[52]Note that $m_j = m_\ell + m_s$, and for $m_s = +1/2$ : $m_j = +1/2$, $m_j = -1/2$, and we have, respectively, $m_\ell = 0, -1$. Similarly, for $m_s = -1/2$ : we have $m_\ell = +1, 0$.

[53]For details on constructions of wavefunctions corresponding to the addition of spin and orbital angular momentum see, e.g., Manoukian [56], pp. 383, 384, 407.

$$\varphi_{2,1,+}(r,\Omega) = R_{21}(r)\Big[\sum_{m=0,-1} a_m Y_{1,m}(\Omega)\Big], \qquad (5.13.8)$$

$$\varphi_{2,1,-}(r,\Omega) = R_{21}(r)\Big[\sum_{m=0,+1} b_m Y_{1,m}(\Omega)\Big], \qquad (5.13.9)$$

where a linear combination over spherical harmonics is involved for $\ell = 1$. The radial parts $R_{2\ell}(r)$ vanish exponentially $e^{-\alpha mr/2}$ for $r \to \infty$. The four component expressions may be then taken as (see (5.13.5) and (5.13.6))

$$\psi_{2,\ell}(\mathbf{p}) \simeq \sum_{\sigma=\pm} \sqrt{\frac{p^0+m}{2p^0}}\begin{pmatrix} \xi_\sigma \\ \frac{\sigma\cdot\mathbf{p}}{p^0+m}\xi_\sigma \end{pmatrix}\frac{\varphi_{2,\ell,\sigma}}{N_\ell} = \sum_{\sigma=\pm}\sqrt{\frac{m}{p^0}}u(\mathbf{p},\sigma)\frac{\varphi_{2,\ell,\sigma}(\mathbf{p})}{N_\ell},$$

$$(5.13.10)$$

conveniently denoted in a momentum description, where $u(\mathbf{p},\sigma)$ is the Dirac spinor, satisfying

$$(\gamma p+m)u(\mathbf{p},\sigma) = 0, \quad \text{normalized as} \quad u^\dagger(\mathbf{p},\sigma)u(\mathbf{p},\sigma') = \frac{p^0}{m}\delta_{\sigma\sigma'}, \qquad (5.13.11)$$

and the solutions $\varphi_{n,\ell,\sigma}(\mathbf{p})$ vanish rapidly for $|\mathbf{p}| \gg m\alpha$, which are both important conditions for the applicability of the explicit expression of the perturbation to be just discussed.

In the presence of a perturbation $\delta U(\mathbf{x})$,

$$-eA_0 \to -eA_0 + \delta U, \quad \psi_n \to \psi_n + \delta\psi_n, \quad E_n \to E_n + \delta E_n, \qquad (5.13.12)$$

leading from (5.13.3) to

$$\Big(\gamma^0\frac{\gamma\cdot\nabla}{i} - eA_0(\mathbf{x}) + \gamma^0 m\Big)\delta\psi_n(\mathbf{x}) + \delta U(\mathbf{x})\psi_n(\mathbf{x}) = E_n\delta\psi_n(\mathbf{x}) + \psi_n(\mathbf{x})\delta E_n.$$

$$(5.13.13)$$

Upon multiplying this equation by $\psi_n^\dagger(\mathbf{x})$, and integrating with respect to $\mathbf{x}$, we obtain the simple expression

$$\delta E_n = \int d^3\mathbf{x}\, \psi_n^\dagger(\mathbf{x})\, \delta U(\mathbf{x})\, \psi_n(\mathbf{x}), \qquad (5.13.14)$$

where we have, in the process, took the adjoint of (5.13.3), and used the fact that $(\gamma^i)^\dagger\gamma^0 = \gamma^0\gamma^i$, $\gamma^i\gamma^0 = -\gamma^0\gamma^i$ to cancel out the first term on the left-hand side of (5.3.13) with the first term on its right-hand side *after* multiplying the latter equation by $\psi_n^\dagger(\mathbf{x})$ and integrating with respect to $\mathbf{x}$ by parts.

In the presence of radiative corrections, the interaction of the electron with the external attractive Coulomb potential is provided by the replacement of $e\gamma^0 A_0(\mathbf{Q})$, in the momentum description, to the leading order, by the explicit expression (see (5.12.4))

$$\left\{ \gamma^0 \left[ 1 - \frac{\alpha}{3\pi} \frac{\mathbf{Q}^2}{m^2} \left( \ln\left( \frac{m}{2K} \right) + \frac{5}{6} - \frac{3}{8} - \frac{1}{5} \right) \right] + \frac{\alpha}{8\pi m} [\gamma^0, \gamma^i] Q_i \right\} eA_0(\mathbf{Q}),$$
(5.13.15)

as a modification of the coefficient of $\gamma^0$, where $K$ is an infrared cut off discussed in Sect. 5.11. In the atom, $|\mathbf{Q}|$, is of the order of the energy difference $m\alpha^2$ between its levels, hence, in particular, $\mathbf{Q}^2/m^2 \ll 1$, justifying the fact that only the order $\mathbf{Q}^2/m^2$ is retained in (5.13.15). Also note that $\psi_{n,\ell}(\mathbf{p})$ in (5.13.10) satisfies approximately the free, i.e. on mass shell, Dirac equation.

In the coordinate description, then $(\boldsymbol{\gamma} \cdot \mathbf{V} \equiv \gamma^i \partial_i)$

$$\delta U(\mathbf{x}) = \left[ \frac{\alpha}{3\pi} \left( \ln\left( \frac{m}{2K} \right) + \frac{5}{6} - \frac{3}{8} - \frac{1}{5} \right) \left( -\frac{\nabla^2}{m^2} \right) - \frac{\alpha}{4\pi m} \frac{\boldsymbol{\gamma} \cdot \mathbf{V}}{i} \right] \frac{\alpha}{|\mathbf{x}|}. \quad (5.13.16)$$

Using the relation $-\nabla^2 (1/|\mathbf{x}|) = 4\pi \delta^{(3)}(\mathbf{x})$, the first term contribution to $\delta E^>$ in (5.13.14), and using the notation discussed in the introductory paragraphs of this section, gives $\left( |\varphi_{n\ell}(0)|^2 - (m\alpha)^3 \delta_{\ell 0}/\pi n^3 \right)$

$$\delta^{(1)}E^> = \frac{4\alpha}{3} \left( \ln\left( \frac{m}{2K} \right) + \frac{5}{6} - \frac{3}{8} - \frac{1}{5} \right) \left( \frac{\alpha}{m^2} |\varphi_{2\ell}(0)|^2 \right)$$

$$= \frac{\alpha(\alpha)^4}{6\pi} m \left( \ln\left( \frac{m}{2K} \right) + \frac{5}{6} - \frac{3}{8} - \frac{1}{5} \right) \delta_{\ell 0}, \quad (5.13.17)$$

where we have used

$$|\varphi_{n\ell}(0)|^2 = (m\alpha)^3 \frac{\delta_{\ell 0}}{\pi n^3}.$$

For the second term, the relevant matrix element reduces to the evaluation

$$-\frac{\alpha^2}{4\pi m} \int d^3\mathbf{x} \left( \chi_{2,\ell}^\dagger \quad \left( \frac{\boldsymbol{\sigma} \cdot \mathbf{p}}{2m} \chi_{2,\ell} \right)^\dagger \right) \begin{pmatrix} 0 & \boldsymbol{\sigma} \cdot \mathbf{p} \\ -\boldsymbol{\sigma} \cdot \mathbf{p} & 0 \end{pmatrix} \frac{1}{|\mathbf{x}|} \begin{pmatrix} \chi_{2,\ell} \\ \frac{\boldsymbol{\sigma} \cdot \mathbf{p}}{2m} \chi_{2,\ell} \end{pmatrix}$$

$$= \frac{\alpha^2}{8\pi m^2} \int d^3\mathbf{x}\, \sigma_i \sigma_j \left\{ \chi_{2,\ell}^\dagger \left( -\partial_i \partial_j \frac{1}{|\mathbf{x}|} \right) \chi_{2,\ell} - \chi_{2,\ell}^\dagger \left( \partial_j \frac{1}{|\mathbf{x}|} \right) \partial_i \chi_{2,\ell} + \chi_{2,\ell}^\dagger \left( \partial_i \frac{1}{|\mathbf{x}|} \right) \partial_j \chi_{2,\ell} \right\}$$

$$= \frac{\alpha^2}{8\pi m^2} \int d^3\mathbf{x} \left\{ 4\pi\, \chi_{2,\ell}^\dagger\, \delta^{(3)}(\mathbf{x})\, \chi_{2,\ell} + 2\,i\, \varepsilon_{ijk}\, \chi_{2,\ell}^\dagger\, \sigma_k \left( \partial_i \frac{1}{|\mathbf{x}|} \right) \partial_j \chi_{2,\ell} \right\}, \quad (5.13.18)$$

where in the first equality we have integrated by parts, and in the second equality we have used the identity: $\sigma_i \sigma_j = \delta_{ij} + i \varepsilon_{ij\,k} \sigma_k$. As before, the first term contributes for $\ell = 0$ only. On the other hand, the second term is zero for $\ell = 0$.[54] For $\ell = 1$, the second term, within the curly brackets, may be written as $4 \chi^{\dagger}_{2,\ell} (\mathbf{S} \cdot \mathbf{L}/|\mathbf{x}|^3) \chi_{2\ell}$,[55] defining a spin-orbit coupling, whose expectation value is simply $-4 \langle 1/|\mathbf{x}|^3 \rangle = -4 (m\alpha)^3/24$, where we have used the fact that $\langle 1/|\mathbf{x}|^3 \rangle = (m\alpha)^3/3n^3$, with $n = 2$, for the hydrogen atom. All told, we have

$$\delta^{(2)} E^> = \frac{\alpha(\alpha)^4}{16\,\pi} m\, \delta_{\ell 0} - \frac{\alpha(\alpha)^4}{48\,\pi} m\, \delta_{\ell 1}. \tag{5.13.19}$$

Equations (5.13.17) and (5.13.19) then give

$$\delta E^> = \frac{\alpha(\alpha)^4}{2\,\pi} m \left[ \frac{1}{3}\left( \ln\left(\frac{m}{2K}\right) + \frac{5}{6} - \frac{3}{8} - \frac{1}{5} \right) + \frac{1}{8} \right] \delta_{\ell 0} - \frac{\alpha(\alpha)^4}{48\,\pi} m\, \delta_{\ell 1}. \tag{5.13.20}$$

- *Derivation and Evaluation of $\delta E^<$:*

  Our starting point for the derivation of the expression for $\delta E^<$, which is involved with very low energy much smaller than $m$, is the non-relativistic Hamiltonian for the interaction of the electron with radiation in the presence of the Coulomb potential. This is given by the expression

$$H = \frac{1}{2m} \left( \mathbf{p} - e\mathbf{A}_{\mathrm{RAD}} \right)^2 + H_{0,\mathrm{RAD}} - \frac{\alpha}{|\mathbf{x}|}, \tag{5.13.21}$$

where $H_{0,\mathrm{RAD}}$ is the free Hamiltonian of the radiation field whose detailed structure, as we will see below, is not needed. The Hamiltonian may be rewritten in a more convenient form as

$$H = H_{\mathrm{C}} + H_{0,\mathrm{RAD}} + H_{\mathrm{I}}, \tag{5.13.22}$$

where $H_{\mathrm{C}}$ is the Coulomb Hamiltonian

$$H_{\mathrm{C}} = \frac{\mathbf{p}^2}{2m} - \frac{\alpha}{|\mathbf{x}|}, \tag{5.13.23}$$

---

[54] A careful demonstration that this term is zero for $\ell = 0$, ($\mathbf{L}^2 = 0$), can be found on pp. 381–382 in Manoukian [56] in the analysis of the atomic fine-structure.

[55] Here one may use the identity $\mathbf{S} \cdot \mathbf{L} = (\mathbf{J}^2 - \mathbf{L}^2 - \mathbf{S}^2)/2$, and the fact that $(j(j+1) - \ell(\ell + 1) - 3/4)/2 = -1$ for $j = 1/2$, $\ell = 1$.

and $H_I$ is the interaction Hamiltonian of the electron-photon system

$$H_I = -\frac{e}{2m}\left(\mathbf{A}_{RAD}\cdot\mathbf{p} + \mathbf{p}\cdot\mathbf{A}_{RAD}\right) + \frac{e^2}{2m}\mathbf{A}_{RAD}^2. \tag{5.13.24}$$

We treat $H_I$ as a perturbation to the Coulomb Hamiltonian. As we will see below, the Coulomb potential has to be treated exactly.

To zeroth order, that is for no electron-photon interaction, let $|vac\rangle$, $|\mathbf{k},\lambda\rangle$ denote, respectively, the vacuum state and the single photon state of momentum $\mathbf{k}$ and (real) polarization $\mathbf{e}_\lambda(\mathbf{k})$, $\left(\mathbf{k}\cdot\mathbf{e}_\lambda(\mathbf{k}) = 0\right)$. Then, in particular,

$$\langle vac|\mathbf{A}_{RAD}(\mathbf{x})|vac\rangle = 0, \quad \langle vac|\mathbf{A}_{RAD}(\mathbf{x})|\mathbf{k}\lambda\rangle = e^{i\mathbf{k}\cdot\mathbf{x}}\,\mathbf{e}_\lambda(\mathbf{k}), \tag{5.13.25}$$

$$\langle vac|\mathbf{A}_{RAD}(\mathbf{x})|\mathbf{k}_1\lambda_1,\ldots,\mathbf{k}_a\lambda_a\rangle = 0, \quad \text{for} \quad a \geq 2, \tag{5.13.26}$$

$$\langle\mathbf{k}\lambda|\mathbf{A}_{RAD}(\mathbf{x}) = e^{-i\mathbf{k}\cdot\mathbf{x}}\,\mathbf{e}_\lambda(\mathbf{k})\,\langle vac|, \tag{5.13.27}$$

where

$$|\mathbf{k}_1\lambda_1,\ldots,\mathbf{k}_a\lambda_a\rangle = \Pi_i^a\,|\mathbf{k}_i\lambda_i\rangle.$$

The above expressions allow us to evaluate $\langle vac|\mathbf{A}_{RAD}^2(\mathbf{x})$ through the following chain of equalities

$$\langle vac|\mathbf{A}_{RAD}^2(\mathbf{x}) = \langle vac|\mathbf{A}_{RAD}(\mathbf{x})|vac\rangle \cdot \langle vac|\mathbf{A}_{RAD}(\mathbf{x})$$

$$+ \int\sum_{\lambda=1,2}\frac{d^3k}{(2\pi)^3\,2\,|\mathbf{k}|}\langle vac|\mathbf{A}_{RAD}(\mathbf{x})|\mathbf{k}\lambda\rangle \cdot \langle\mathbf{k}\lambda|\mathbf{A}_{RAD}(\mathbf{x})$$

$$= \int\sum_{\lambda=1,2}\frac{d^3k}{(2\pi)^3\,2\,|\mathbf{k}|}\,e^{i\mathbf{k}\cdot\mathbf{x}}\mathbf{e}_\lambda(\mathbf{k})\cdot\mathbf{e}_\lambda(\mathbf{k})\,e^{-i\mathbf{k}\cdot\mathbf{x}}\langle vac|$$

$$= \int\sum_{\lambda=1,2}\frac{d^3k}{(2\pi)^3\,2\,|\mathbf{k}|}\,\langle vac|, \tag{5.13.28}$$

where we have used the completeness relation for free photons consisting of projection on the vacuum state, plus one photon contribution, plus two photons contribution, and so on, given by

$$|vac\rangle\langle vac| + \left(\int\sum_{\lambda=1,2}\frac{d^3k}{(2\pi)^3\,2\,|\mathbf{k}|}|\mathbf{k}\,\lambda\rangle\langle\mathbf{k}\,\lambda|\right) + \cdots = 1. \tag{5.13.29}$$

Also note that $H_{0,RAD}|\mathbf{k}\,\lambda\rangle = |\mathbf{k}|\,|\mathbf{k}\,\lambda\rangle$.

To zeroth order, in the absence of the electron-photon interaction, the Hamiltonian is given by $H_C + H_{0,RAD}$, and the following equations, in particular, are satisfied

$$\left(H_C + H_{0,RAD}\right)|\varphi_n\rangle\, |\mathrm{vac}\rangle = \mathscr{E}_n |\varphi_n\rangle\, |\mathrm{vac}\rangle \qquad (5.13.30)$$

$$\left(H_C + H_{0,RAD}\right)|\varphi_n\rangle\, |\mathbf{k}\lambda\rangle = (\mathscr{E}_n + |\mathbf{k}|)|\varphi_n\rangle\, |\mathbf{k}\lambda\rangle, \qquad (5.13.31)$$

where the $\mathscr{E}_n$ denote the hydrogen atom energy levels with corresponding eigenvectors $|\varphi_n\rangle$.

In the presence of the perturbation, described by $H_I$, the bound-state eigenvalue equation reads

$$H|\phi_n\rangle = \widetilde{\mathscr{E}}_n |\phi_n\rangle. \qquad (5.13.32)$$

Upon multiplying the latter equation by $\langle\varphi_n|\langle\mathrm{vac}| \equiv \langle\varphi_n'|$, and using (5.13.30), give

$$\Delta\mathscr{E}_n = \frac{\langle\varphi_n'|H_I|\phi_n\rangle}{\langle\varphi_n'|\phi_n\rangle}, \qquad \Delta\mathscr{E}_n = \widetilde{\mathscr{E}}_n - \mathscr{E}_n. \qquad (5.13.33)$$

Let us first consider the contribution of $e^2\mathbf{A}_{RAD}^2/2m$ in $H_I$ to the latter equation. To this end, we have from (5.13.28)

$$\frac{\langle\varphi_n'|\mathbf{A}_{RAD}^2(\mathbf{x})|\phi_n\rangle}{\langle\varphi_n'|\phi_n\rangle} = \frac{\langle\varphi_n'|\phi_n\rangle}{\langle\varphi_n'|\phi_n\rangle} \int \sum_{\lambda=1,2} \frac{d^3\mathbf{k}}{(2\pi)^3\, 2\, |\mathbf{k}|}$$

$$= \int \sum_{\lambda=1,2} \frac{d^3\mathbf{k}}{(2\pi)^3\, 2\, |\mathbf{k}|}, \qquad (5.13.34)$$

which is independent of the quantum number $n$, $\ell$ and hence is common to all levels and does not contribute in computing to the energy differences.[56]

On the other hand for the first term in (5.13.24), we first multiply (5.13.32) by the single photon state $\langle\mathbf{k},\lambda|$, generating a state corresponding to $H_C$ only, given by

$$\langle\mathbf{k}\lambda|\phi_n\rangle = \frac{1}{(\widetilde{\mathscr{E}}_n - H_C - |\mathbf{k}|)} \langle\mathbf{k}\lambda|H_I|\phi_n\rangle. \qquad (5.13.35)$$

From (5.13.28), the matrix element of $\mathbf{A}_{RAD}$ between the vacuum states as well as between the vacuum states and multiphoton states are zero, and hence the contribution of the first term in (5.13.24), (5.13.25), (5.13.26), (5.13.27), (5.13.28),

---

[56] Although this common constant is infinite, it may be defined with a cut-off.

(5.13.29), (5.13.30), (5.13.31), (5.13.32) and (5.13.33) is, from (5.13.35), given by

$$
-\frac{e}{2m}\langle\varphi_n'|\mathbf{p}\cdot\mathbf{A}_{\mathrm{RAD}}(\mathbf{x})+\mathbf{A}_{\mathrm{RAD}}(\mathbf{x})\cdot\mathbf{p}\,|\phi_n\rangle=\frac{4\,\pi e^2}{m^2}\int\sum_\lambda\frac{d^3\mathbf{k}}{(2\,\pi)^3\,2\,|\mathbf{k}|}
$$

$$
\times\Big\langle\varphi_n\Big|\mathbf{p}\cdot\mathbf{e}_\lambda(\mathbf{k})\,e^{i\mathbf{k}\cdot\mathbf{x}}\frac{1}{(\mathscr{E}_n-H_{\mathrm{C}}-|\mathbf{k}|)}\,e^{-i\mathbf{k}\cdot\mathbf{x}}\,\mathbf{e}_\lambda(\mathbf{k})\cdot\mathbf{p}\Big|\varphi_n\Big\rangle,\qquad(5.13.36)
$$

where we have used the fact that $\mathbf{k}\cdot\mathbf{e}_\lambda(\mathbf{k})=0$, and replaced the denominator in (5.13.33) by one, and $\mathscr{E}_n'$ by $\mathscr{E}_n$, to the leading order, and obviously replaced $\langle\varphi_n'|$ by $\langle\varphi_n|$. Using the fact that $e^{i\mathbf{k}\cdot\mathbf{x}}$ is a translation operator of $\mathbf{p}$ to $\mathbf{p}-\mathbf{k}$ we have

$$
e^{i\mathbf{k}\cdot\mathbf{x}}\frac{1}{(\mathscr{E}_n^v-H_{\mathrm{C}}-|\mathbf{k}|)}\,e^{-i\mathbf{k}\cdot\mathbf{x}}=\frac{1}{(\mathscr{E}_n-\frac{(\mathbf{p}-\mathbf{k})^2}{2m}-U_{\mathrm{C}}-|\mathbf{k}|)}
$$

$$
=\frac{1}{(\mathscr{E}_n-\frac{(\mathbf{p}-\mathbf{k})^2}{2m}-H_{\mathrm{C}}+\frac{\mathbf{p}^2}{2m}-|\mathbf{k}|)},\qquad(5.13.37)
$$

where $U_{\mathrm{C}}$ is the Coulomb potential. Accordingly, we may rewrite (5.13.36) as

$$
-\frac{e}{2m}\langle\varphi_n'|\mathbf{p}\cdot\mathbf{A}_{\mathrm{RAD}}(\mathbf{x})+\mathbf{A}_{\mathrm{RAD}}(\mathbf{x})\cdot\mathbf{p}|\phi_n\rangle=\frac{4\,\pi e^2}{m^2}\int\sum_\lambda\frac{d^3\mathbf{k}}{(2\,\pi)^3\,2\,|\mathbf{k}|}
$$

$$
\times\Big\langle\varphi_n\Big|\mathbf{p}\cdot\mathbf{e}_\lambda(\mathbf{k})\frac{1}{(\mathscr{E}_n-\frac{(\mathbf{p}-\mathbf{k})^2}{2m}-H_{\mathrm{C}}+\frac{\mathbf{p}^2}{2m}-|\mathbf{k}|)}\,\mathbf{e}_\lambda(\mathbf{k})\cdot\mathbf{p}\Big|\varphi_n\Big\rangle.\qquad(5.13.38)
$$

The expression in (5.13.38) is not the final one for $\delta E^<$. The reason is that in the absence of the Coulomb interaction, such a term is not involved with the electron in the atom, i.e., the electron in the Coulomb potential. Accordingly, we have to subtract the corresponding conribution, with the substitutions $H_{\mathrm{C}}\to\mathbf{p}^2/2m$, $\mathscr{E}_n\to\mathbf{p}^2/2m$, in the denominator in (5.13.38), from its original expression as it appears in the latter equation. This leads to the replacement[57]

$$
\frac{1}{(\mathscr{E}_n-\frac{(\mathbf{p}-\mathbf{k})^2}{2m}-H_{\mathrm{C}}+\frac{\mathbf{p}^2}{2m}-|\mathbf{k}|)}\to
$$

$$
\frac{1}{(\mathscr{E}_n-\frac{(\mathbf{p}-\mathbf{k})^2}{2m}-H_{\mathrm{C}}+\frac{\mathbf{p}^2}{2m}-|\mathbf{k}|)}-\frac{1}{(-\frac{(\mathbf{p}-\mathbf{k})^2}{2m}+\frac{\mathbf{p}^2}{2m}-|\mathbf{k}|)}.\qquad(5.13.39)
$$

---

[57]This simply gives rise to a mass renormalization, see Manoukian [56], pp. 401–43. Also note that the matrix element $\langle\varphi_n|[\,.\,]|\varphi_n\rangle$, in (5.13.38) for $H_{\mathrm{C}}-\mathscr{E}_n\to0$, simply factors out as $\langle\varphi_n|\varphi_n\rangle\,[\,.\,]=[\,.\,]$, giving an expression independent of the Coulomb functions and its related quantum numbers.

This dictates, in turn, to modify the expression in (5.13.38) to the following one

$$
\frac{\alpha}{4\pi^2 m^2} \int \sum_\lambda \frac{d^3 k}{|\mathbf{k}|} \langle \varphi_n | \mathbf{p} \cdot \mathbf{e}_\lambda(\mathbf{k}) \frac{1}{\left[ \left(\frac{\mathbf{p}}{m}\right) \cdot \mathbf{k} - |\mathbf{k}| \left(1 + \frac{|\mathbf{k}|}{2m}\right) \right]}
$$

$$
\times (H_{\mathrm{C}} - \mathscr{E}_n) \frac{1}{\left[ \mathscr{E}_n - H_{\mathrm{C}} + \left(\frac{\mathbf{p}}{m}\right) \cdot \mathbf{k} - |\mathbf{k}| \left(1 + \frac{|\mathbf{k}|}{2m}\right) \right]} \mathbf{e}_\lambda(\mathbf{k}) \cdot \mathbf{p} \, | \varphi_n \rangle. \qquad (5.13.40)
$$

In the atom, $|\mathbf{p}|$ is of the order $m\alpha$, while $|\mathbf{k}|$ is of the order $m\alpha^2$, which is of the order of the difference between the energy levels of the hydrogen atom. Taking this into account and finally recalling that $\delta E^<$ is to be defined with an upper cut-off $|\mathbf{k}| < K$, we have

$$
\delta E^< = \frac{\alpha}{4\pi^2 m^2} \int_{|\mathbf{k}|<K} \sum_\lambda \frac{d^3 k}{|\mathbf{k}|^2} \left\langle \varphi_n \middle| \mathbf{p} \cdot \mathbf{e}_\lambda(\mathbf{k}) (H_{\mathrm{C}} - \mathscr{E}_n) \frac{1}{\left[ H_{\mathrm{C}} - \mathscr{E}_n + |\mathbf{k}| \right]} \mathbf{e}_\lambda(\mathbf{k}) \cdot \mathbf{p} \middle| \varphi_n \right\rangle.
$$

$$
(5.13.41)
$$

We note that $(H_{\mathrm{C}} - \mathscr{E}_n)$ is of an order comparable to $|\mathbf{k}|$ and should be kept in the denominator. As a matter of fact, the latter provides a lower bound cut-off in the infrared region $|\mathbf{k}| \to 0$.

Upon using the completeness relation and the following angular integration, respectively,

$$
\sum_\lambda \mathbf{e}_\lambda^i(\mathbf{k}) \, \mathbf{e}_\lambda^j(\mathbf{k}) = \delta^{ij} - \frac{k^i k^j}{k^2}, \qquad \int d\Omega \left( \delta^{ij} - \frac{k^i k^j}{k^2} \right) = \frac{8\pi}{3} \delta^{ij}, \qquad (5.13.42)
$$

we get

$$
\delta E^< = \frac{2\alpha}{3\pi m^2} \left\langle \varphi_n \middle| [\mathbf{p}, H_{\mathrm{C}}] \cdot \ln\left( \frac{K}{|H_{\mathrm{C}} - \mathscr{E}_n|} \right) \mathbf{p} \middle| \varphi_n \right\rangle, \qquad (5.13.43)
$$

where we have replaced $\mathbf{p}(H_{\mathrm{C}} - \mathscr{E}_n)$ by $[\mathbf{p}, H_{\mathrm{C}}]$ since $\left\langle \varphi_n \middle| (H_{\mathrm{C}} - \mathscr{E}_n) = 0 \right.$, and noted that $(H_{\mathrm{C}} - \mathscr{E}_n)$ is of the order $m\alpha^2$. The right-hand side of the above equation involves the Coulomb Hamiltonian $H_{\mathrm{C}}$, the hydrogenic atomic functions $\varphi_n$, and the eigenvalues $\mathscr{E}_n$, which are all *known*.

The expression in (5.13.43) may be rewritten, with $n = 2$, as

$$
\delta E^< = \frac{2\alpha}{3\pi m^2} \left[ \left\langle 2, \ell \middle| [\mathbf{p}, H_{\mathrm{C}}] \cdot \ln\left( \frac{\mathrm{Ry}}{|H_{\mathrm{C}} - \mathscr{E}_n|} \right) \mathbf{p} \middle| 2, \ell \right\rangle + \left\langle 2, \ell \middle| [\mathbf{p}, H_{\mathrm{C}}] \cdot \mathbf{p} \middle| 2, \ell \right\rangle \ln\left( \frac{K}{\mathrm{Ry}} \right) \right],
$$

$$
(5.13.44)
$$

where Ry stands for the Rydberg energy equal to $m\alpha^2/2$, and note that

$$\langle 2, \ell | [\mathbf{p}, H_{\mathrm{C}}] \cdot \mathbf{p} | 2, \ell \rangle = 2\pi\alpha |\varphi_2(0)|^2 = \frac{m^3}{4}\alpha^4 \delta_{\ell,0}, \tag{5.13.45}$$

as is readily shown.

The following expression, involving the above mentioned known quantities, has been evaluated numerically by several authors[58]

$$\frac{\langle 2, \ell | [\mathbf{p}, H_{\mathrm{C}}] \cdot \ln\left(\frac{\mathrm{Ry}}{|H_{\mathrm{C}} - \mathscr{E}_n|}\right) \mathbf{p} | 2, \ell \rangle}{\langle 2, 0 | [\mathbf{p}, H_{\mathrm{C}}] \cdot \mathbf{p} | 2, 0 \rangle} = -2.81\,\delta_{\ell,0} + 0.03\,\delta_{\ell,1}. \tag{5.13.46}$$

Hence

$$\delta E^< = \frac{\alpha\, m}{6\pi}(\alpha)^4\left[\left(-2.81 + \ln\left(\frac{2K}{m}\right) - \ln(\alpha^2)\right)\delta_{\ell,0} + 0.03\,\delta_{\ell,1}\right]. \tag{5.13.47}$$

- *The Total Shift $\delta E = \delta E^> + \delta E^<$* :
  From (5.13.20) and (5.13.47), we have

$$\delta E_{2,\ell} = \frac{\alpha\, m}{6\pi}(\alpha)^4\left[\frac{19}{30} - 2.81 - \ln(\alpha^2)\right]\delta_{\ell,0} + \frac{\alpha\, m}{6\pi}(\alpha)^4\left[-\frac{1}{8} + 0.03\right]\delta_{\ell,1}. \tag{5.13.48}$$

- *The Lamb Shift*:
  The Lamb shift is then given by

$$\frac{\delta E_{2,0} - \delta E_{2,1}}{h} = 1052.1\,\mathrm{MHz}, \tag{5.13.49}$$

and is in good agreement with the experimental value[59] which is about 1057.85 MHz. There are other small corrections which contribute to the Lamb shift, such as contributions coming, for example, from the proton recoil as well as from the finite structure of the proton, and one has to dwell on some phenomenological aspects (see also Problem 5.16).

---

[58] See, in particular, Bethe et al. [9] and Schwartz and Tiemann [67].

[59] See, e.g., Kinoshita and Yennie [37], p. 7.

## 5.14    Coulomb Gauge Formulation

In the present section, the vacuum-to-vacuum transition amplitude in the full QED is derived in the Coulomb gauge, in the differential form, by the application of the quantum dynamical principle. The path integral expression then also readily follows from the differential formalism. Gauge transformations of the vacuum-to-vacuum transition amplitude between the Coulomb and covariant gauges is given in the next section.

### 5.14.1    $\langle\, 0_+ \mid 0_- \rangle$ in the Coulomb Gauge in the Functional Differential Form

The Coulomb gauge is defined by

$$\nabla \cdot \mathbf{A}(x) = \partial_i A^i(x) = 0, \tag{5.14.1}$$

for which only physical transverse degrees of freedom of the photon are quantized. One may apply the Coulomb gauge directly to the Maxwell Lagrangian density, by solving one of the components of $\mathbf{A}$ in terms of the (derivatives) of the other components, and then varying the other two independent components. We may, for example, solve for $A^3$ as follows

$$A^3 = -\partial_3^{-1} \partial^a A^a, \tag{5.14.2}$$

with a sum over $a = 1, 2$. Accordingly, we may simply write

$$\delta A^i = \left[ \delta^{ia} - \delta^{i3} \partial_3^{-1} \partial^a \right] \delta A^a, \tag{5.14.3}$$

expressed solely in terms of variations of the components $A^1$, $A^2$, as one should.
    Now consider the Lagrangian density

$$\mathscr{L} = -\frac{1}{4} F_{\mu\nu} F^{\mu\nu} + A^\mu J_\mu. \tag{5.14.4}$$

Variations with respect to $A^a$, and $A^0$, respectively, give

$$\partial_\mu F^{\mu a} = \partial_3^{-1} \partial^a \big(\partial_\mu F^{\mu 3} + J^3\big) - J^a, \tag{5.14.5}$$

$$\partial_\mu F^{\mu 0} = -J^0, \quad \text{or} \quad \partial_i F^{i0} = -J^0. \tag{5.14.6}$$

We note that (5.14.5) also holds true if we replace $a$ by 3, since this simply gives the empty statement $\partial_\mu F^{\mu 3} = \partial_\mu F^{\mu 3}$. Accordingly, we may rewrite (5.14.5) as

$$\partial_\mu F^{\mu j} = \partial_3^{-1} \partial^j \big(\partial_\mu F^{\mu 3} + J^3\big) - J^j, \qquad (j = 1, 2, 3). \tag{5.14.7}$$

One may combine (5.14.6) and (5.14.7) as follows

$$\partial_\mu F^{\mu\nu} = \eta^{\nu j}\partial^j \partial_3^{-1}(\partial_\mu F^{\mu 3} + J^3) - J^\nu. \tag{5.14.8}$$

By applying $\partial_\nu$ to the above equation, and using $\partial_\mu \partial_\nu F^{\mu\nu} = 0$, we may, in turn, solve for $\partial_3^{-1}\partial^j(\partial_\mu F^{\mu 3} + J^3)$:

$$\partial_3^{-1}(\partial_\mu F^{\mu 3} + J^3) = \frac{1}{\nabla^2}\,\partial_\mu J^\mu. \tag{5.14.9}$$

Upon substituting this equality in (5.14.8) gives

$$\partial_\mu F^{\mu\nu} = -\left(\eta^{\nu\sigma} - \eta^{\nu j}\frac{\partial^j \partial^\sigma}{\nabla^2}\right)J_\sigma. \tag{5.14.10}$$

This leads to the basic equations[60]

$$-\nabla^2 A^0 = J^0, \tag{5.14.11}$$

$$-\Box A^i = \left(\delta^{ij} - \frac{\partial^i \partial^j}{\nabla^2}\right)J^j. \tag{5.14.12}$$

These two equations lead further to the equation for the matrix element of the vector potential

$$\frac{\langle\, 0_+ |A^\mu(x)\, |\, 0_-\rangle}{\langle\, 0_+ \,|\, 0_-\rangle} = \left[\eta^{\mu\nu} + \frac{\partial^\mu \partial^\nu}{\nabla^2} - \frac{\partial^\mu \partial^i}{\nabla^2}\eta^{\nu i} - \frac{\partial^\nu \partial^i}{\nabla^2}\eta^{\mu i}\right]\frac{1}{(-\Box - i\epsilon)}\, J_\nu(x), \tag{5.14.13}$$

giving

$$\frac{\langle\, 0_+ |A^\mu(x)\, |\, 0_-\rangle}{\langle\, 0_+ \,|\, 0_-\rangle} = \int (\mathrm{d}y) D_C^{\mu\nu}(x - y)J_\nu(y), \tag{5.14.14}$$

with $D_C^{\mu\nu}(x - y)$, specifying the photon propagator in the Coulomb gauge, given by

$$D_C^{\mu\nu}(x - x') = \int \frac{(\mathrm{d}Q)}{(2\pi)^4}\, e^{iQ(x-x')}\, D_C^{\mu\nu}(Q), \tag{5.14.15}$$

$$D_C^{00}(Q) = -\frac{1}{\mathbf{Q}^2}, \quad D_C^{ij}(Q) = \left(\delta^{ij} - \frac{Q^i Q^j}{\mathbf{Q}^2}\right)\frac{1}{Q^2 - i\epsilon}, \quad D_C^{0i}(Q) = 0. \tag{5.14.16}$$

---

[60]It is important to realize that the component $A^0$ is not a dynamical variable, i.e., its equation does not involve its time derivative, as is *derived* directly from the Lagrangian and need not, a priori assumed.

Only the components $D_C^{ij}(x-y) = D_C^{ji}(x-y)$ propagate. $D_C^{00}(x-y)$ is proportional to $\delta(x^0 - y^0)$.

Upon using the fact that

$$\langle \, 0_+ | A^\mu(x) \, | \, 0_- \rangle = (-i)\frac{\delta}{\delta J_\mu(x)}\langle \, 0_+ | \, 0_- \rangle,$$

we may functionally integrate (5.14.14) to obtain

$$\langle \, 0_+ \, | \, 0_- \rangle = \exp\left[\frac{i}{2}\int (dx)(dx')J_\mu(x)D_C^{\mu\nu}(x-x')J_\nu(x')\right], \qquad (5.14.17)$$

satisfying the B.C. $\langle \, 0_+ \, | \, 0_- \rangle = 1$ for $J^\mu = 0$.

In the presence of the interaction with the fermion field, the Lagrangian density is taken to be

$$\mathscr{L} = -\frac{1}{4}F_{\mu\nu}F^{\mu\nu} + \frac{1}{2}\left(\frac{\partial_\mu\overline{\psi}}{i}\gamma^\mu\psi - \overline{\psi}\gamma^\mu\frac{\partial_\mu\psi}{i}\right) - m_0\overline{\psi}\psi$$

$$+ \frac{1}{2}e_0[\overline{\psi},\gamma^\mu\psi]A_\mu + \overline{\eta}\psi + \overline{\psi}\eta + A_\mu J^\mu, \qquad (5.14.18)$$

from which the following field equations emerge

$$\partial_\mu F^{\mu\nu} = \eta^{\nu j}\frac{1}{\nabla^2}\,\partial^j\,\partial_\mu(J^\mu + j^\mu) - (J^\nu + j^\nu), \qquad (5.14.19)$$

$$\left[\gamma^\mu\left(\frac{\partial_\mu}{i} - e_0 A_\mu\right) + m_0\right]\psi = \eta, \qquad (5.14.20)$$

$$\overline{\psi}\left[-\gamma^\mu\left(\frac{\partial_\mu}{i} + e_0 A_\mu\right) + m_0\right] = \overline{\eta}, \qquad (5.14.21)$$

with $j_\nu = e_0[\overline{\psi},\gamma_\nu\psi]/2$. Also as in (5.7.32), we have the following useful equation

$$\partial_\mu j^\mu(x) = i\,e_0\left[\overline{\psi}(x)\eta(x) - \overline{\eta}(x)\psi(x)\right]. \qquad (5.14.22)$$

From (5.14.17), in particular, we then have, by following the procedure carried out for the corresponding covariant gauge formulation in Sect. 5.7 ($e_0 \neq 0$):

$$\langle \, 0_+ \, | \, 0_- \rangle = \exp\left[e_0\int (dx)\frac{\delta}{\delta\eta(x)}\gamma^\mu\frac{\delta}{\delta\overline{\eta}(x)}\frac{\delta}{\delta J^\mu(x)}\right]$$

$$\times \exp\left[\frac{i}{2}\int (dx)\,(dx')\,J_\mu(x)\,D_C^{\mu\nu}(x-x')\,J_\nu(x')\right]$$

$$\times \exp\left[i\int (dx)\,(dx')\,\overline{\eta}(x)\,S_+(x-x')\,\eta(x')\right], \qquad (5.14.23)$$

for the solution of QED in the celebrated Coulomb gauge, where $S_+(p) = 1/(\gamma p + m_0)$, in the momentum description, expressed in terms of the bare mass $m_0$. The corresponding path integral expression of the vacuum-to-vacuum amplitude is derived in the next subsection.

The canonical conjugate momenta $\partial \mathscr{L}/\partial(\partial_0 A^a)$ of the fields $A^1$, $A^2$, are given by[61]

$$\pi^a = F^{a0} - \partial_3^{-1} \partial^a F^{30}. \tag{5.14.24}$$

In particular, we note that this is also zero if we replace $a$ by 3: $\pi^3 = 0$, corresponding to the dependent field $A^3$. Also the canonical conjugate momentum of $A^0$ is zero, i.e., the latter is also a dependent field. Hence we may quite generally write

$$\pi^\mu = F^{\mu 0} - \eta^{\mu i} \partial_i \partial_3^{-1} F^{30}, \quad \mu = 0, 1, 2, 3. \tag{5.14.25}$$

Upon solving for $-\eta^{\mu i} \partial_i \partial_3^{-1} F^{30}$, and taking the $\partial_\mu$ derivative of the resulting equation leads to

$$-\nabla^2 \partial_3^{-1} F^{30} = \partial_\mu \pi^\mu - \partial_\mu F^{\mu 0} = \partial_\mu \pi^\mu + (J^0 + j^0), \tag{5.14.26}$$

where, with $\nu = 0$, we have used (5.14.19) in writing the last equality. This equation allows us to solve for $F^{\mu 0}$ from (5.14.25)

$$F^{\mu 0} = \pi^\mu - \eta^{\mu i} \frac{\partial_i}{\nabla^2} (J^0 + j^0 + \partial_a \pi^a), \tag{5.14.27}$$

expressed in terms of canonical conjugate momenta, their space derivatives, the current $j^0$ and the component of the *external* source $J^0$. Note that $F^{ij}$ may be expressed in terms of the independent fields $A^1$, $A^2$. For an application of this useful expression see Problem 5.17.

Equal-time commutation rules of the field components, may be derived as follows. We recall that

$$\pi^0 = 0, \quad \pi^3 = 0, \quad \pi^a = \partial_3^{-1} \partial^a \partial^0 A^3 - \partial^0 A^a, \qquad a = 1, 2,$$

and hence

$$[A^a(x^0, \mathbf{x}), \pi^b(x^0, \mathbf{x}')] = i \delta^{ab} \delta^{(3)}(\mathbf{x} - \mathbf{x}'). \tag{5.14.28}$$

---

[61]Note that the canonical conjugate momentum of $A^a$ is not $\partial_0 A^a$, for $a = 1, 2$.

This leads to the equal time commutator (see Problem 5.18)

$$[A^i(x^0, \mathbf{x}), \partial_0 A^j(x^0, \mathbf{x}')] = \mathrm{i}\left[\delta^{ij} - \frac{\partial^i \partial^j}{\mathbf{\nabla}^2}\right]\delta^{(3)}(\mathbf{x} - \mathbf{x}'). \tag{5.14.29}$$

In the next section, we consider the transformations rules that relate the full theory, via the vacuum-to-vacuum transition amplitudes, between the different covariant gauges as well as with the Coulomb one carried out above.[62] •

## 5.14.2  From the QDP to the Path Integral of $\langle 0_+ | 0_- \rangle$ in Coulomb Gauge

To derive the path integral expression of $\langle 0_+ | 0_- \rangle$ in the Coulomb gauge we use the elegant method of delta functionals developed in Sect. 2.7. To this end, we first establish that for the external current $J^\mu$,[63]

$$\Pi_\mu \delta\left(J_\mathrm{T}^\mu\right) = \delta\left(-\mathrm{i}\,\partial^k \frac{\delta}{\delta J^k}\right)\Pi_\sigma \delta(J^\sigma), \quad (\text{where } J_\mathrm{T}^\mu \equiv J^\mu - \eta^{\mu i}\frac{\partial^i \partial^\nu}{\mathbf{\nabla}^2}J_\nu) \tag{5.14.30}$$

up to a proportionality constant. Through a series of steps, the left-hand side of the above equation may be rewritten as

$$\Pi_\mu \delta\left(J^\mu - \eta^{\mu i}\frac{\partial^i \partial^\nu}{\mathbf{\nabla}^2}J_\nu\right)$$

$$= \delta(J^0) \int \Pi_i \mathscr{D}a^i \exp\left[\mathrm{i} \int (\mathrm{d}x)\left(a^i - \frac{\partial^i \partial^j}{\mathbf{\nabla}^2}a^j\right)J^i\right]$$

$$= \delta(J^0) \int \Pi_x \mathrm{d}\lambda(x)\, \Pi_i \mathscr{D}a^i\, \delta(\partial^j a^j - \lambda) \exp\left[\mathrm{i} \int (\mathrm{d}x)\left(a^i - \frac{\partial^i}{\mathbf{\nabla}^2}\lambda\right)J^i\right]$$

$$= \delta(J^0)\, \delta\left(-\mathrm{i}\,\partial^k \frac{\delta}{\delta J^k}\right) \int \Pi_x \mathrm{d}\lambda(x) \int \Pi_i \mathscr{D}a^i \exp\left[\mathrm{i} \int (\mathrm{d}x)\left(a^i - \frac{\partial^i}{\mathbf{\nabla}^2}\lambda\right)J^i\right]$$

$$= \delta(J^0)\, \delta\left(-\mathrm{i}\,\partial^k \frac{\delta}{\delta J^k}\right) \int \Pi_x \mathrm{d}\lambda(x) \int \Pi_i \mathscr{D}a^i \exp\left[\mathrm{i} \int (\mathrm{d}x)\, a^i J^i\right]$$

$$= \delta\left(-\mathrm{i}\partial^k \frac{\delta}{\delta J^k}\right)\Pi_\sigma \delta(J^\sigma), \tag{5.14.31}$$

---

[62]This section is based on Manoukian [48]. For the relativistic invariance of QED in the Coulomb gauge, see Manoukian [51]. For the gauge invariance of transition amplitudes, see Manoukian [53].
[63]Recall the notation $\Pi_\sigma \delta(J^\sigma) \equiv \Pi_\sigma \Pi_x \delta(J^\sigma(x))$.

up to the proportionality constant $\int \Pi_x \, \mathrm{d}\lambda(x)$,[64] and where in the step before the last one, we have made a change of variables $a^i \to a^i + \partial^i \lambda / \nabla^2$.

Set

$$\widehat{\mathscr{A}} = -\frac{1}{4} \int (\mathrm{d}x) \, \widehat{F}_{\mu\nu}(x) \, \widehat{F}^{\mu\nu}(x), \quad \text{where} \quad \widehat{A}_\mu(x) \equiv -\mathrm{i} \frac{\delta}{\delta J^\mu(x)}, \tag{5.14.32}$$

and note that

$$\left[ \frac{\delta}{\delta J^\mu(x)}, J^\nu(x') \right] = \delta_\mu{}^\nu \, \delta^{(4)}(x - x'), \quad \left[ \frac{\delta}{\delta J^\mu(x)}, \frac{\delta}{\delta J^\nu(x')} \right] = 0. \tag{5.14.33}$$

The following commutators then easily follow

$$[\mathrm{i}\,\widehat{\mathscr{A}}, J^\mu(x)] = \partial_\nu \widehat{F}^{\nu\mu}(x), \qquad \partial_\mu [\mathrm{i}\,\widehat{\mathscr{A}}, J^\mu(x)] = 0, \tag{5.14.34}$$

$$[\mathrm{i}\,\widehat{\mathscr{A}}, J_\mathrm{T}^\mu(x)] = \partial_\nu \widehat{F}^{\nu\mu}(x), \quad [\mathrm{i}\,\widehat{\mathscr{A}}, [\mathrm{i}\,\widehat{\mathscr{A}}, J_\mathrm{T}^\mu(x)]] = 0. \tag{5.14.35}$$

Now we use the convenient notation $\langle\, 0_+ \mid 0_-\rangle_{e_0 = 0, \gamma}$ for the right-hand side of (5.14.17). In reference to (5.14.10), we have

$$\partial_\nu \widehat{F}^{\nu\mu}(x) \langle\, 0_+ \mid 0_-\rangle_{e_0 = 0, \gamma} = -J_\mathrm{T}^\mu \langle\, 0_+ \mid 0_-\rangle_{e_0 = 0, \gamma}. \tag{5.14.36}$$

Hence the first commutator in (5.14.35) leads to

$$\Big( [\mathrm{i}\,\widehat{\mathscr{A}}, J_\mathrm{T}^\mu(x)] + J_\mathrm{T}^\mu(x) \Big) \langle\, 0_+ \mid 0_-\rangle_{e_0 = 0, \gamma} = 0. \tag{5.14.37}$$

We use the elementary relation that for two operators $A$, $B$, if $[A, [A, B]] = 0$, then $\mathrm{e}^A B \mathrm{e}^{-A} = [A, B] + B$. The second commutator in (5.14.35) then allows us to rewrite (5.14.37) as

$$\exp[\,\mathrm{i}\,\widehat{\mathscr{A}}\,] J_\mathrm{T}^\mu(x) \exp[-\mathrm{i}\,\widehat{\mathscr{A}}\,] \langle\, 0_+ \mid 0_-\rangle_{e_0 = 0, \gamma} = 0,$$

$$J_\mathrm{T}^\mu(x) \Big( \exp[-\mathrm{i}\,\widehat{\mathscr{A}}\,] \langle\, 0_+ \mid 0_-\rangle_{e_0 = 0, \gamma} \Big) = 0, \tag{5.14.38}$$

for *all* $x$. Using the property that from $x f(x) = 0$, one may infer that $f(x) = \text{Const.}\,\delta(x)$, (5.14.38) then leads to

$$\langle\, 0_+ \mid 0_-\rangle_{e_0 = 0, \gamma} = \exp[\,\mathrm{i}\,\widehat{\mathscr{A}}\,] \Pi_\mu \, \delta(J_\mathrm{T}^\mu), \tag{5.14.39}$$

---

[64]Don't let such an infinite proportionality constant scare you. We always divide by proportionality constants in $\langle\, 0_+ \mid 0_-\rangle$ when determining Green functions.

up to a multiplicative constant. From (5.14.30), we obtain the key equation

$$\langle\, 0_+ \mid 0_- \rangle_{e_0 = 0, \gamma} = \exp[\,i\,\widehat{\mathscr{A}}\,]\,\delta\!\left(-\,i\,\partial^k\frac{\delta}{\delta J^k}\right)\!\Pi_\sigma\delta(J^\sigma). \qquad (5.14.40)$$

The full QED $\langle\, 0_+ \mid 0_- \rangle$, in the Coulomb gauge, then follows from (5.14.23) to be

$$\langle\, 0_+ \mid 0_- \rangle = \int(\mathscr{D}\overline{\rho}\,\mathscr{D}\rho)\,(\mathscr{D}a)\,\delta(\partial_k\,a^k)\exp[\,i\int(dx)\mathscr{L}_c(x)], \qquad (5.14.41)$$

up to an unimportant multiplicative constant. Here $\mathscr{L}_c$ is the Lagrangian density, including the external sources $J^\mu$, $\overline{\eta}$, $\eta$, obtained from the one in (5.14.18) upon carrying out the substitutions

$$\psi \to \rho, \qquad \overline{\psi} \to \overline{\rho}, \qquad A^\mu \to a^\mu,$$

where we recall that $\rho$, $\overline{\rho}$ are Grassmann variables, i.e., are anti-commuting (see also (5.7.41) for the covariant gauges).

This equation is quite interesting as it explicitly shows the gauge constraint via the delta functional $\delta(\partial_k\,a^k)$.

Note that the expression for $\langle\, 0_+ \mid 0_- \rangle$ in (5.14.23), is a functional of the sources $J^\mu$, $\eta$, $\overline{\eta}$, while the coefficient of

$$\exp\Big[\,i\int(dx)\big(\overline{\eta}(x)\rho(x) + \overline{\rho}(x)\eta(x) + J^\mu(x)a_\mu(x)\big)\,\Big],$$

in the functional integrand in (5.14.41) is a functional of $(a_\mu, \overline{\rho}, \rho)$. Thus the path integral defines a functional Fourier transform of variables $(J^\mu, \eta, \overline{\eta})$ to $(a_\mu, \overline{\rho}, \rho)$.

## 5.15  Gauge Transformations of the Full Theory

Let us recall the expression for the vacuum-to-vacuum transition amplitude, or the generating functional, of QED in covariant gauges (see (5.7.20)/(5.7.21)), denoted here conveniently by $F[\eta, \overline{\eta}, J, \lambda]$ :

$$F[\eta, \overline{\eta}, J^\mu, \lambda] = \exp\Big[\,e_0\int(dx)\frac{\delta}{\delta\eta(x)}\,\gamma^\mu\,\frac{\delta}{\delta\overline{\eta}(x)}\,\frac{\delta}{\delta J^\mu(x)}\Big]$$

$$\times\exp\Big[\,\frac{i}{2}\int(dx)\,(dx')\,J_\mu(x)\,D^{\mu\nu}(x-x';\lambda)\,J_\nu(x')\Big] \qquad (5.15.1)$$

$$\times\exp\Big[\,i\int(dx)\,(dx')\,\overline{\eta}(x)\,S_+(x-x')\,\eta(x')\Big], \qquad (5.15.2)$$

$$D^{\mu\nu}(x-x';\lambda) = \int \frac{(dk)}{(2\pi)^4}\left[\eta^{\mu\nu} - (1-\lambda)\frac{k^\mu k^\nu}{k^2}\right]\frac{e^{ik(x-x')}}{k^2 - i\epsilon}, \tag{5.15.3}$$

where note that the exponential first factor involving the functional derivatives above, acts simply as a *translation* operator:

$$J^\mu \to J^\mu + e_0\left(\frac{\delta}{\delta\eta}\right)\gamma^\mu\left(\frac{\delta}{\delta\overline{\eta}}\right),$$

on any functional of $J^\mu$, via the functional derivative $\delta/\delta J^\mu$. This basic property will be used throughout this section.

Note that throughout $S_+$ is expressed in terms of the bare mass $m_0$, i.e. $S_+(p) = 1/(\gamma p + m_0)$ in the momentum description.

On the other hand, from the last section, the corresponding expression in the Coulomb gauge, and with corresponding sources $(\rho, \overline{\rho}, K^\mu)$, is given from (5.14.23) by

$$F_C[\rho,\overline{\rho},K^\mu] = \exp\left[e_0\int(dx)\frac{\delta}{\delta\rho(x)}\gamma^\mu\frac{\delta}{\delta\overline{\rho}(x)}\frac{\delta}{\delta K^\mu(x)}\right]$$

$$\times \exp\left[\frac{i}{2}\int(dx)(dx')\,K_\mu(x)\,D_C^{\mu\nu}(x-x')\,K_\nu(x')\right] \tag{5.15.4}$$

$$\times \exp\left[i\int(dx)(dx')\,\overline{\rho}(x)\,S_+(x-x')\,\rho(x')\right], \tag{5.15.5}$$

$$D_C^{\mu\nu}(x-x') = \int\frac{(dk)}{(2\pi)^4}D_C^{\mu\nu}(k)\,e^{ik(x-x')},$$

$$D_C^{ij}(k) = \left[\eta^{ij} - \frac{k^i k^j}{\mathbf{k}^2}\right]\frac{1}{k^2 - i\epsilon}, \quad D_C^{00}(k) = -\frac{1}{\mathbf{k}^2}, \quad D_C^{i0}(k) = 0 = D_C^{0i}(k). \tag{5.15.6}$$

Note the identity relating the free photon propagator in these gauges[65]

$$\left(\eta^{\mu\alpha} - \eta^{\mu i}\frac{\partial^i}{\nabla^2}\partial^\alpha\right)\left(\eta^{\nu\beta} - \eta^{\nu j}\frac{\partial^j}{\nabla^2}\partial^\beta\right)D_{\mu\nu}(x-x';\lambda) = D_C^{\alpha\beta}(x-x'). \tag{5.15.7}$$

Also note the following invariance property of the expression

$$\frac{\delta}{\delta\eta(x)}\gamma^\mu\frac{\delta}{\delta\overline{\eta}(x)}\frac{\delta}{\delta J^\mu(x)}, \tag{5.15.8}$$

---

[65]See Problem 5.19.

under the transformations

$$\eta(x) \to \eta(x)\, e^{i\Lambda(x)}, \quad \overline{\eta}(x) \to e^{-i\Lambda(x)}\, \overline{\eta}(x), \tag{5.15.9}$$

for an arbitrary numerical function $\Lambda(x)$.

We first derive the transformation rule relating the full theory in the Coulomb gauge to the full one in covariant gauges specified by the parameter $\lambda$. This reads:

$$F_C[\rho, \overline{\rho}, K^\mu] = \left[ e^{iW'} F[\eta, \overline{\eta}, J^\mu, \lambda] \right]\Big|_{\eta = 0, \overline{\eta} = 0, J^\mu = 0}, \tag{5.15.10}$$

$$
\begin{aligned}
W' &= \int (dx)\, \overline{\rho}(x) \exp\left[ -i\, e_0 c^\mu \frac{\delta}{i\delta J^\mu(x)} \right] \frac{\delta}{i\delta\overline{\eta}(x)} \\
&+ \int (dx) \frac{\delta}{(-i)\delta\eta(x)} \exp\left[ i\, e_0 c^\mu \frac{\delta}{i\delta J^\mu(x)} \right] \rho(x) \\
&+ \int (dx) \left[ \left( \eta^{\mu\nu} - c^\mu \partial^\nu \right) K_\nu(x) \right] \frac{\delta}{i\delta J^\mu(x)},
\end{aligned} \tag{5.15.11}
$$

$$c^\mu = \eta^{\mu j} \frac{\partial^j}{\nabla^2}. \tag{5.15.12}$$

To derive this, we begin with the right-hand side of (5.15.10). In a matrix notation, first note that

$$
e^{iW'} \exp\left[ i\, \overline{\eta}\, S_+ \eta \right] \exp\left[ \frac{i}{2} J_\mu\, D^{\mu\nu}(\lambda)\, J_\nu \right]
$$

$$
= \exp\left[ i\left( \overline{\eta} + \overline{\rho} \exp\left[ -i\, e_0 c^\mu \frac{\delta}{i\delta J^\mu} \right] \right) S_+ \left( \eta + \exp\left[ i\, e_0 c^\nu \frac{\delta}{i\delta J^\nu} \right] \rho \right) \right]
$$

$$
\times \exp\left[ \frac{i}{2} \left( J_\mu + \left( \eta_{\mu\sigma} - c_\mu \partial_\sigma \right) K^\sigma \right) D^{\mu\nu}(\lambda) \left( J_\nu + \left( \eta_{\nu\beta} - c_\nu \partial_\beta \right) K^\beta \right) \right]. \tag{5.15.13}
$$

On the other hand, since we eventually, and in particular, set $\overline{\eta} = 0$, $\eta = 0$, we may consider the transformations in (5.15.9), and use the fact that

$$
\exp\left[ e_0 \int (dx) \frac{\delta}{\delta\eta(x)} \gamma_\mu \frac{\delta}{\delta\overline{\eta}(x)} \frac{\delta}{\delta J_\mu(x)} \right]
$$

$$
= \exp\left[ e_0 \int (dx)\, e^{-i\Lambda(x)} \frac{\delta}{\delta\eta(x)} \gamma_\mu\, e^{i\Lambda(x)} \frac{\delta}{\delta\overline{\eta}(x)} \frac{\delta}{\delta J_\mu(x)} \right], \tag{5.15.14}
$$

for an arbitrary numerical $\Lambda(x)$. Hence for $\overline{\eta} = 0$, $\eta = 0$, in view of applying the analysis to the right-hand side of (5.15.10), we may, from (5.15.13), write the right-hand side of (5.15.10), before setting $J^\mu = 0$, as

$$\exp\Big[ e_0 \int (dx)\frac{\delta}{\delta\rho(x)}\, \gamma_\mu\, \frac{\delta}{\delta\overline{\rho}(x)}\, \frac{\delta}{\delta J_\mu(x)} \Big]$$

$$\times \exp\Big[ i\Big(\overline{\rho}\exp\Big[-i\,e_0 c^\mu \frac{\delta}{i\delta J^\mu}\Big]\Big) S_+ \Big(\exp\Big[ i\,e_0 c^\nu \frac{\delta}{i\delta J^\nu}\Big]\rho\Big)\Big]$$

$$\times \exp\Big[ \frac{i}{2}\Big(J_\mu + (\eta_{\mu\sigma} - c_\mu\partial_\sigma)K^\sigma\Big) D^{\mu\nu}(\lambda)\Big(J_\nu + (\eta_{\nu\beta} - c_\nu\partial_\beta)K^\beta\Big) \Big], \qquad (5.15.15)$$

rewritten in terms of the source variables $\overline{\rho}$, $\rho$. We eventually have to set $J^\mu = 0$ in this equation to compare the resulting expression with the one on the left-hand side of (5.15.10). To do this, we have to carry out the functional derivatives $\delta/\delta J^\mu$ in (5.15.15) explicitly first. To this end, we may use the identity

$$\exp\Big[ -\int (dx)\frac{\delta}{\delta\rho(x)}\, \gamma^\mu \frac{\delta}{\delta\overline{\rho}(x)}\, \partial_\mu\big(i\,\Lambda(x)\big)\Big]\exp\big[i\overline{\rho}\,S_+\,\rho\big]$$

$$= \exp\big[i\,\big(\overline{\rho}\,e^{-i\Lambda}\big)S_+\big(e^{i\Lambda}\rho\big)\big], \qquad (5.15.16)$$

for an arbitrary numerical $\Lambda(x)$, established in Problem 5.20, and making the identification

$$i\,\Lambda = i\,e_0 c^\sigma (\delta/i\,\delta J^\sigma),$$

to rewrite the expression in (5.15.15), for $J^\mu = 0$, as

$$\exp\Big[ e_0 \int (dx)\frac{\delta}{\delta\rho(x)}\, \gamma^\mu \frac{\delta}{\delta\overline{\rho}(x)}\, (\eta_{\mu\sigma} - c_\sigma\partial_\mu)\frac{\delta}{\delta J_\sigma(x)}\Big]\exp\big[i\overline{\rho}\,S_+\,\rho\big]$$

$$\times \exp\Big[ \frac{i}{2}\Big(J_\mu + (\eta_{\mu\sigma} - c_\mu\partial_\sigma)K^\sigma\Big) D^{\mu\nu}(\lambda)\Big(J_\nu + (\eta_{\nu\beta} - c_\nu\partial_\beta)K^\beta\Big)\Big]\Big|_{J^\mu=0}$$

$$= \exp\Big[ e_0 \int (dx)\frac{\delta}{\delta\rho(x)}\, \gamma^\mu \frac{\delta}{\delta\overline{\rho}(x)}\, \frac{\delta}{\delta K^\mu(x)}\Big]\exp\big[i\overline{\rho}\,S_+\,\rho\big]$$

$$\times \exp\Big[ \frac{i}{2}\Big((\eta_{\mu\sigma} - c_\mu\partial_\sigma)K^\sigma\Big) D^{\mu\nu}(\lambda)\Big((\eta_{\nu\beta} - c_\nu\partial_\beta)K^\beta\Big)\Big]. \qquad (5.15.17)$$

Now we may use (5.15.7) to conclude the statement made in (5.15.10).

To establish gauge transformations, say, from the Landau gauge, to covariant gauges specified by the parameter $\lambda$, we note that

$$\Big(\eta_{\mu\sigma} - \frac{\partial_\mu\partial_\sigma}{\Box}\Big) D^{\mu\nu}(\lambda)\Big(\eta_{\nu\beta} - \frac{\partial_\nu\partial_\beta}{\Box}\Big) = D_{\sigma\beta}(\lambda)\big|_{\lambda=0}, \qquad \Box \equiv \partial^\alpha\partial_\alpha, \qquad (5.15.18)$$

where

$$D^{\mu\nu}(\lambda)\big|_{\lambda=0} \equiv D_L^{\mu\nu},$$

is the celebrated free photon propagator in the Landau gauge:

$$D_L^{\mu\nu}(x-x') = \int \frac{(dk)}{(2\pi)^4}\, e^{ik(x-x')}\big[\eta^{\mu\nu} - \frac{k^\mu k^\nu}{k^2}\big]\frac{1}{k^2 - i\epsilon}. \tag{5.15.19}$$

An almost identical analysis as above, establishes the rule relating the gauge transformations of the full theory from covariant ones, specified by $\lambda$, to Landau's. This reads:

$$F[\eta,\overline{\eta},J^\mu,\lambda = 0] = \Big[e^{i\overline{W}'}F[\rho,\overline{\rho},K^\mu,\lambda]\Big]\Big|_{\rho=0,\overline{\rho}=0,K^\mu=0}, \tag{5.15.20}$$

where $\overline{W}'$ is the functional operator defined in detail by

$$\overline{W}' = \int (dx)\overline{\eta}(x)\exp\Big[-i\,e_0\widetilde{c}^{\,\mu}\frac{\delta}{i\delta K^\mu(x)}\Big]\frac{\delta}{i\delta\overline{\rho}(x)}$$

$$+ \int (dx)\frac{\delta}{(-i)\delta\rho(x)}\exp\Big[i\,e_0\widetilde{c}^{\,\mu}\frac{\delta}{i\delta K^\mu(x)}\Big]\eta(x)$$

$$+ \int (dx)\Big[(\eta^{\mu\nu} - \widetilde{c}^{\,\mu}\partial^\nu)J_\nu(x)\Big]\frac{\delta}{i\delta K^\mu(x)}, \tag{5.15.21}$$

$$\widetilde{c}^{\,\mu} = \frac{\partial^\mu}{\Box}. \tag{5.15.22}$$

By putting $\rho = 0, \overline{\rho} = 0, K^\mu = 0$ in (5.15.10), we immediately learn that in the absence of external sources, $F_C[0,0,0] = F[0,0,0,\lambda]$. Similarly, from (5.15.20) and (5.15.21), we learn that $F[0,0,0,0] = F[0,0,0,\lambda]$. Thus we have the following gauge invariance normalization conditions valid when the external sources are set equal to zero

$$F_C[0,0,0] = F[0,0,0,\lambda] = F[0,0,0,0]. \tag{5.15.23}$$

In the absence of external Fermi sources only, as shown in Problem 5.21, we also have

$$F[0,0,J^\mu,\lambda = 0] = \exp\Big[-\frac{i}{2}(\partial_\mu J^\mu)\,G\,(\partial_\nu J^\nu)\Big]F[0,0,J^\mu,\lambda], \tag{5.15.24}$$

$$G(x-x') = \lambda \int \frac{(dk)}{(2\pi)^4}\frac{e^{ik(x-x')}}{k^2(k^2+\mu^2)}\frac{\Lambda^2}{k^2+\Lambda^2}, \tag{5.15.25}$$

where we have defined $G(x - x')$ with infrared and ultraviolet cut-offs (see also (5.10.11) and (5.10.27)).[66]

The relation of the exact electron propagator in covariant gauges, specified by the parameter $\lambda$, to the exact one in the Landau gauge, in the presence of the external current $J^\mu$, is readily worked out and, as shown in Problem 5.22, is given by

$$\widetilde{S}(x, y; J^\mu)\big|_\lambda = \exp\left(i\, e_0^2\, [\, G(0) - G(x - y)\, ]\right)$$

$$\times \exp\left[\, i\, e_0 \int (dz) J_\mu(z)\, \partial_z^\mu [\, G(z - x) - G(z - y)\, ]\,\right]\left[\, \widetilde{S}(x, y; J^\mu)\big|_{\lambda=0}\,\right].$$

$$(5.15.26)$$

In particular, in the absence of the external current $J^\mu = 0$,

$$\widetilde{S}(x, y)\big|_\lambda = \exp\left(i\, e_0^2\, [\, G(0) - G(x - y)\, ]\right)\left[\, \widetilde{S}(x, y)\big|_{\lambda-0}\,\right]. \qquad (5.15.27)$$

For $|x - y| \to \infty$ in Euclidean region we have from (5.11.10) and (5.15.25),

$$i\, e_0^2\, [\, G(0) - G(x - y)\, ] \to \lambda\, \frac{\alpha_0}{4\pi} \ln\!\left(\frac{\mu^2}{\Lambda^2}\right), \qquad (5.15.28)$$

and near the mass shell of the electron propagators, we then have for the wavefunction renormalization constants in question

$$Z_2\big|_\lambda = \left(\frac{\mu^2}{\Lambda^2}\right)^{\lambda\, \alpha_0/4\pi} Z_2\big|_{\lambda=0}. \qquad (5.15.29)$$

Let us evaluate this to lowest order and compare it with our earlier result. To lowest order, this gives

$$Z_2\big|_\lambda = Z_2\big|_{\lambda=0} + \lambda\, \frac{\alpha}{4\pi} \ln\!\left(\frac{\mu^2}{\Lambda^2}\right), \qquad (5.15.30)$$

which coincides with our earlier low order result in (5.10.36) expressed there for $1/Z_2$.

This section on gauge transformations, is based on Manoukian and Siranan [60], which includes extensive references on the gauge problem. It also includes detailed treatments of the so-called axial gauge $n^\mu A_\mu = 0$, where $n^\mu$ is a fixed vector, as well as the Fock-Schwinger gauge $x^\mu A_\mu = 0$, in the full theory.[67]

---

[66]This is the Fourier transform of the longitudinal part of the free photon propagator $D^{\mu\nu}$, in the momentum description, with cut-offs, written as $\lambda\, k^\mu k^\nu \Lambda^2 / [k^2 (k^2 + \mu^2)(k^2 + \Lambda^2)]$.

[67]For the gauge invariance of transition amplitudes see Manoukian [53].

Before closing this section, we also provide the following useful and easily derived[68] identity of the generating functional in covariant gauges, specified by the parameter $\lambda$. It reads in matrix form

$$F[\varrho, \overline{\varrho}, \widetilde{J}^{\mu}, \lambda] = \exp\left[i\,\partial^{\mu}\varphi\,\widetilde{J}_{\mu}\right] \exp\left[-\frac{i}{2\lambda}\,\Box\varphi\,\Box\varphi\right] F[\eta, \overline{\eta}, J^{\mu}, \lambda], \quad (5.15.31)$$

$$\varrho = e^{i\,e_0\,\varphi}\eta, \qquad \overline{\varrho} = \overline{\eta}\,e^{-i\,e_0\,\varphi}, \qquad \widetilde{J}^{\mu} = J^{\mu} - \frac{\Box\,\partial^{\mu}}{\lambda}\varphi, \quad (5.15.32)$$

where $\varphi(x)$ is a real numerical dimensionless function. We will make use of this identity in the next section.

## 5.16  Vertex Function and Ward-Takahashi Identity; Full Propagators

In this section, going beyond perturbation theory, a classic identity, referred as the Ward-Takahashi identity, is derived which relates the derivative of the vertex function $\Gamma^{\mu}$ to the exact electron propagater, and establishes, in particular, the equality of the renormalization constants $Z_1 = Z_2$ to all orders. This is followed by deriving the equations satisfied by the electron propagator and vertex function without recourse to perturbation theory. The spectral decomposition of the photon propagator is then derived, generalizing the lowest order derivation given in Sect. 5.10.2, Eq. (5.10.55), with much emphasis put on charge renormalization. Finally an integral equation is derived for the vacuum polarization tensor.

### 5.16.1  Ward-Takahashi Identity

The generating functional of QED, in covariant gauges, specified by the parameter $\lambda$ in the presence of an arbitrary scalar function $\varphi(x)$ was given above in (5.15.31). The electron propagator follows from this identity, by functional differentiations defining the propagator, and is given by

$$\widetilde{S}(x, y) = \widetilde{S}(x, y)\big|_{\varphi=0} \exp\left[i\,e_0\big(\varphi(x) - \varphi(y)\big)\right], \quad (5.16.1)$$

where we have set $\eta = 0$, $\overline{\eta} = 0$. Also note that under a variation $\delta\varphi$ of $\varphi$

$$\delta\widetilde{S}(x, y) = i\,e_0 \int (dz)\big[\delta^{(4)}(z - x) - \delta^{(4)}(z - y)\big]\widetilde{S}(x, y)\,\delta\varphi(z). \quad (5.16.2)$$

---

[68]The derivation follows very closely to the ones in Problems 5.21 and 5.22.

Similarly[69] we have

$$\langle A^\mu(x) \rangle = \langle A^\mu(x) \rangle \big|_{\varphi=0} + \partial^\mu \varphi(x), \quad \delta \langle A^\mu(x) \rangle = \partial^\mu \delta\varphi(x). \tag{5.16.3}$$

The vertex function $\Gamma^\mu$ in the full theory is defined through

$$\frac{\delta}{\delta \langle A_\mu(z) \rangle} \widetilde{S}(x, y) = e_0 \int (dx')\,(dy')\,\widetilde{S}(x, x')\, \Gamma^\mu(x', y'; z)\, \widetilde{S}(y', y), \tag{5.16.4}$$

which will be justified in Sect. 5.16.2, and coincides with $\gamma^\mu$ to zeroth order. Hence from the chain rule, we obtain

$$\delta\widetilde{S}(x, y) = \int (dz) \frac{\delta}{\delta \langle A_\mu(z) \rangle} \widetilde{S}(x, y)\, \delta \langle A_\mu(z) \rangle$$

$$= e_0 \int (dx')\,(dy')\,(dz)\,\widetilde{S}(x, x')\, \Gamma^\mu(x', y'; z)\, \widetilde{S}(y', y)\, \partial_z^\mu \delta\varphi(z)$$

$$= - e_0 \int (dx')\,(dy')\,(dz)\,\widetilde{S}(x, x')\, \partial_z^\mu\, \Gamma^\mu(x', y'; z)\, \widetilde{S}(y', y)\, \delta\varphi(z). \tag{5.16.5}$$

Upon comparing this equation with that in (5.16.2), and multiplying, in the process, both equations from the right and the left by the inverse of the propagator $\widetilde{S}$, we obtain

$$\frac{\partial}{\partial z^\mu}\, \Gamma^\mu(x, y; z) = i\left[ \delta^{(4)}(z - x) - \delta^{(4)}(z - y) \right] \widetilde{S}^{-1}(x - y), \tag{5.16.6}$$

where all the external sources as well as $\varphi$ have been set equal to zero.

Introducing the Fourier transform

$$\Gamma^\mu(x, y; z) = \int \frac{(dp)}{(2\pi)^4} \frac{(dk)}{(2\pi)^4}\, e^{i(p+k)x}\, e^{-ipy}\, e^{-ikz}\, \Gamma^\mu(p + k, p), \tag{5.16.7}$$

we obtain the identity

$$k_\mu \Gamma^\mu(p + k, p) = \widetilde{S}^{-1}(p + k) - \widetilde{S}^{-1}(p), \tag{5.16.8}$$

referred to as the Ward-Takahashi Identity.[70] Upon taking the derivative of this equation with respect to $k^\mu$, and taking the limit $k^\mu \to 0$, we obtain the zero

---

[69]Recall that $\langle A^\mu(x) \rangle \equiv (-i)[\delta F/\delta J_\mu(x)]/F$.

[70]Takahashi [78]. The identity in differential form, that is for zero momentum transfer of the vertex function, is due to Ward [79]. Takahashi generalized it to non-zero momentum transfer.

momentum transfer version of the above identity given by

$$\Gamma^{\mu}(p,p) = \frac{\partial}{\partial p_{\mu}}\widetilde{S}^{-1}(p), \tag{5.16.9}$$

referred to as the Ward Identity. The wavefunction renormalization constant $Z_2$ of the electron propagator, and the vertex renormalization constant $Z_1$, may be defined through boundary conditions on the mass-shell $p^2 = -m^2$, where $m$ is the renormalized mass, as follow

$$\bar{u}(\mathbf{p},\sigma)\Big[\frac{\partial}{\partial p^{\mu}}\widetilde{S}^{-1}(p)\Big]u(\mathbf{p},\sigma) = \frac{1}{Z_2}\,\gamma_{\mu}, \quad \bar{u}(\mathbf{p},\sigma)\Gamma_{\mu}(p,p)u(\mathbf{p},\sigma) = \frac{1}{Z_1}\,\gamma_{\mu}. \tag{5.16.10}$$

These give the equality of the two renormalization constants

$$Z_1 = Z_2. \tag{5.16.11}$$

We have seen earlier in Sect. 5.11, that to lowest order, these two renormalization constants are equal. This is now seen to be true to all orders, and is a consequence of the Ward Identity. Due to this equality, charge renormalization occurs via the wavefunction renormalization constant $Z_3$ of the photon only. This universality property will be discussed in the next section.

In Sect. 5.16.2, the equations satisfied by the electron propagator and the vertex function are derived. Here the definition of the vertex function through (5.16.4) will be used. We have found it convenient to carry out a similar analysis for the vacuum polarization and the photon propagator, right in this section, in the following two subsections. In Sect. 5.16.3, the photon spectral decomposition of the photon propagator is carried out, generalizing the low order one in (5.10.55). The equation satisfied by the vacuum polarization is derived in Sect. 5.16.4.

### 5.16.2  Equations for the Electron Propagator and the Vertex Function

Upon taking the matrix element of the electron field between the vacuum states, we have

$$\Big[\gamma^{\mu}\Big(\frac{\partial_{\mu}}{i} - e_0\frac{1}{i}\frac{\delta}{\delta J^{\mu}(x)}\Big) + m_0\Big]\langle\,0_+|\psi(x)\,|\,0_-\rangle = \eta(x)\langle\,0_+\,|\,0_-\rangle. \tag{5.16.12}$$

The functional derivative $\delta/\delta\eta(y)$ of this equation gives

$$\left[\gamma^\mu\left(\frac{\partial_\mu}{i} - e_0\frac{1}{i}\frac{\delta}{\delta J^\mu(x)}\right) + m_0\right]i\langle\, 0_+|(\psi(x)\overline{\psi}(y))_+ \,|\, 0_-\rangle = \delta^{(4)}(x-y)\langle\, 0_+ \,|\, 0_-\rangle,$$
(5.16.13)

where now we have set the external Fermi sources equal to zero. Using the identity

$$\frac{1}{i}\frac{\delta}{\delta J^\mu(x)}\langle\, 0_+ \,|\, 0_-\rangle = \langle\, 0_+ \,|\, 0_-\rangle\left[\langle A_\mu(x)\rangle + \frac{1}{i}\frac{\delta}{\delta J^\mu(x)}\right],$$
(5.16.14)

we may rewrite the above equation as

$$\left[\gamma^\mu\left(\frac{\partial_\mu}{i} - e_0\langle A_\mu(x)\rangle\right) + m_0\right]\widetilde{S}(x,y) = \delta^{(4)}(x-y) + e_0\frac{1}{i}\frac{\delta}{\delta J^\mu(x)}\widetilde{S}(x,y).$$
(5.16.15)

Upon multiplying the latter from the right by $\widetilde{S}^{-1}(y,x')$, and integrating over $y$, gives

$$\widetilde{S}^{-1}(x,x') = \left[\gamma^\mu\left(\frac{\partial_\mu}{i} - e_0\langle A_\mu(x)\rangle\right) + m_0\right]\delta^{(4)}(x-x')$$

$$+ e_0\,\gamma^\mu\int (dy)\frac{1}{i}\widetilde{S}(x,y)\frac{\delta}{\delta J^\mu(x)}\widetilde{S}^{-1}(y,x'),$$
(5.16.16)

where the equality

$$\int (dy)\,(\delta\widetilde{S}(x,y))\,\widetilde{S}^{-1}(y,x') = -\int (dy)\,\widetilde{S}(x,y)\,\delta\widetilde{S}^{-1}(y,x'),$$

has been used. Using the definition of the exact photon propagator

$$\frac{\delta}{\delta J_\mu(x)}\langle A_\nu(z)\rangle = \widetilde{D}^\mu{}_\nu(x,z),$$
(5.16.17)

which will be treated in the next section, and the chain rule

$$\frac{\delta}{\delta J^\mu(x)}\widetilde{S}^{-1}(y,x') = \int (dz)\widetilde{D}_{\mu\nu}(x,z)\frac{\delta}{\delta\langle A_\nu(z)\rangle}\widetilde{S}^{-1}(y,x'),$$
(5.16.18)

we obtain from (5.16.16)

$$\widetilde{S}^{-1}(x,x') = \left[\gamma^\mu\left(\frac{\partial_\mu}{i} - e_0\langle A_\mu(x)\rangle\right) + m_0\right]\delta^{(4)}(x-x')$$

$$-i\,e_0\int (dy)(dz)\gamma^\mu\widetilde{S}(x,y)\frac{\delta}{\delta\langle A_\nu(z)\rangle}\widetilde{S}^{-1}(y,x')\widetilde{D}_{\mu\nu}(x,z).$$
(5.16.19)

This provides an iterative equation in terms of the exact electron and photon propagators, starting with

$$\widetilde{S}^{-1}(x,x') = \left[\gamma^\mu\left(\frac{\partial_\mu}{i} - e_0\langle A_\mu(x)\rangle\right) + m_0\right]\delta^{(4)}(x-x')$$

$$+ i e_0^2 \int (dy)(dz)\, \gamma^\mu\widetilde{S}(x,y)\, \gamma^\nu\delta^{(4)}(y-z)\, \delta^{(4)}(y-x')\widetilde{D}_{\mu\nu}(x,z) + \cdots .$$

$$(5.16.20)$$

It is easily verified that this coincides with the lowest order expression in (5.10.4), for

$$J^\mu \to 0, \quad \langle A_\mu(x)\rangle \to 0, \quad e_0 \to e,$$

and with the substitutions

$$\widetilde{S}(x,y) \to S_+(x-y), \quad \widetilde{D}_{\mu\nu} \to D_{\mu\nu},$$

made inside the integral.

Using the definition of the vertex function in (5.16.4), and the Fourier transform in (5.16.7), we obtain, in the absence of the external sources, the exact equation

$$\widetilde{S}^{-1}(p) = \gamma p + m_0 - \Sigma(p), \tag{5.16.21}$$

$$\Sigma(p) = -i e_0^2 \int \frac{(dk)}{(2\pi)^4}\, \gamma^\mu\, \widetilde{S}(p+k)\, \Gamma^\nu(p+k,p)\, \widetilde{D}_{\mu\nu}(k). \tag{5.16.22}$$

where from (5.16.4), we have used, in the process, the relation

$$\frac{\delta}{\delta\langle A_\mu(z)\rangle}\, \widetilde{S}^{-1}(x,y) = -e_0\, \Gamma^\mu(x,y;z). \tag{5.16.23}$$

In analogy to $\Sigma$ in (5.16.22), in the absence of external sources, we will still denote the second term in (5.16.19), in the presence of the external current $J^\mu$ by $-\Sigma$ as well :

$$\Sigma(x,x') = -i e_0^2 \int (dy)(dz)\gamma^\mu\widetilde{S}(x,y)\Gamma^\nu(y,x';z)\widetilde{D}_{\mu\nu}(x,z). \tag{5.16.24}$$

$\Sigma(x,x')$ is a functional of the exact propagators $\widetilde{S}$, $\widetilde{D}^{\mu\nu}$. By the chain rule

$$\frac{\delta}{\delta\langle A_\nu(z)\rangle}\Sigma(x,x') = \int (dy)(dy')\frac{\delta D^{\mu\nu}(y,y')}{\delta\langle A_\nu(z)\rangle}\frac{\delta}{\delta D^{\mu\nu}(y,y')}\Sigma(x,y)$$

$$+ \int (dy)(dy')\frac{\delta\widetilde{S}(y,y')}{\delta\langle A_\nu(z)\rangle}\frac{\delta}{\delta\widetilde{S}(y,y')}\Sigma(x,y). \tag{5.16.25}$$

From the definition of the exact photon propagator in (5.16.17), the first term necessarily includes an odd number (three) of external photon lines, for $J^\mu = 0$, $\langle A^\mu \rangle = 0$. Hence from Furry's Theorem,[71] the first term in the above equation is zero in the absence of external sources. At this stage, we introduce the $e^- e^+$ kernel $K$ defined by

$$\frac{\delta}{\delta \widetilde{S}(y,y')} \Sigma(x,x') = e_0^2 K(x,x';y',y). \qquad (5.16.26)$$

The factor $e_0^2$ is introduced to keep track of the order when carrying out a perturbation analysis. Upon taking the function derivative, via $(\delta/\delta\langle A_\sigma(z')\rangle)$, of (5.16.19), i.e., of

$$\widetilde{S}^{-1}(x,x') = \left[ \gamma^\mu \left( \frac{\partial_\mu}{i} - e_0 \langle A_\mu(x) \rangle \right) + m_0 \right] \delta^{(4)}(x-x') - \Sigma(x,x'), \qquad (5.16.27)$$

and using (5.16.23) and (5.16.25), we obtain, after dividing the resulting expression by $(-e_0)$

$$\Gamma^\sigma(x,x';z) = \gamma^\sigma \delta^{(4)}(z-x)\,\delta^{(4)}(x-x')$$

$$-i e_0^2 \int (dy)(dy')(dx'')(dy'') K(x,x';y',y)\, \widetilde{S}(y,x'')\, \Gamma^\sigma(x'',y'',z)\, \widetilde{S}(y'',y').$$

$$(5.16.28)$$

Upon defining the following Fourier transform, and using (5.16.7)

$$K(x,x';y',y) = \int \frac{(dp)}{(2\pi)^4} \frac{(dp')}{(2\pi)^4} \frac{(dQ)}{(2\pi)^4}\, K(p,p';Q)\, e^{ix(p+Q)} e^{iy(p'+Q)} e^{-ix'p} e^{-iy'p'},$$

$$(5.16.29)$$

we obtain from (5.16.28) the integral equation for the vertex function (Fig. 5.15):

$$\Gamma^\mu(p+Q,p) = \gamma^\mu - i e_0^2 \int \frac{(dp')}{(2\pi)^4} K(p,p';Q)\, \widetilde{S}(p'+Q)\, \Gamma^\mu(p'+Q,p')\, \widetilde{S}(p').$$

$$(5.16.30)$$

By definition, the $e^- e^+$-kernel $K$ is connected and the external electron lines are removed. Some graphs which are excluded in the kernel $K$ are given in Fig. 5.16.

The $e^- e^+$-kernel times $e_0^2 : e_0^2 K$ may be expanded in terms of the exact propagators and the exact vertex function as shown in Fig. 5.17. Such an expansion is quite useful in investigating aspects of the renormalized theory.

---

[71] See Appendix IV, below (IV.19), at the end of the book.

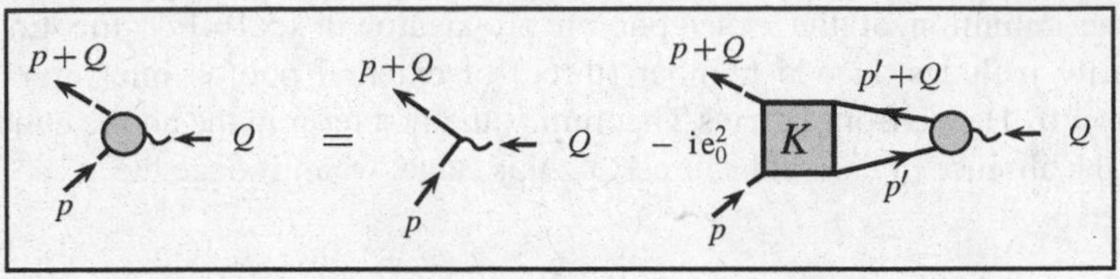

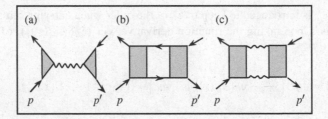

**Fig. 5.15** The integral equation for the vertex function. Note the momentum flow into the $e^-e^+$-kernel

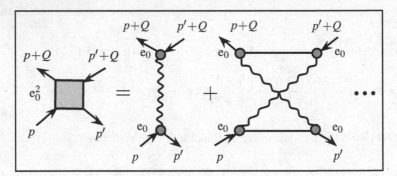

**Fig. 5.16** The kernel $K$ cannot be composed of two parts connected by a single photon line as shown in part (**a**). It cannot also be composed of two parts connected by one electron-positron pair as shown in part (**b**). The diagram in part (**c**), as well as those connected by an even number of photon lines, cannot contribute to the integral equation of the vertex function as they will give rise to subgraphs involving an odd number of external photon lines to the vertex in Fig. 5.15 which vanish by Furry's theorem

**Fig. 5.17** The $e^-e^+$- kernel times $e_0^2 : e_0^2 K$ expanded in terms of the exact propagators and the exact vertex function, referred to as a skeleton expansion. The *arrows* show the direction of momentum flow, and recall that the external electron-positron lines are removed in the definition of $K$

Finally we note from (5.16.24) and (5.16.6), we may also represent the integral equation for $\widetilde{S}^{-1}$ diagrammatically as in Fig. 5.18 (for $J^\mu \to 0$)[72]:

---

[72]Functional methods in deriving equations for propagators and vertex functions were particularly used over the years by Schwinger and by his former students, notably by K. Johnson (see, e.g., [34]), and by many others.

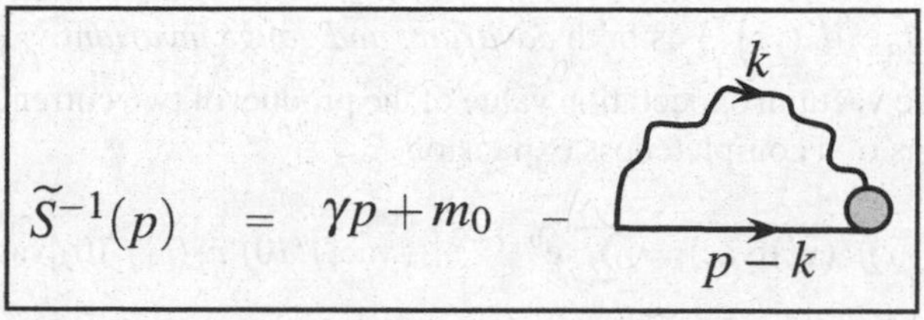

**Fig. 5.18** Integral equation of the inverse of the exact electron propagator $\widetilde{S}^{-1}(p) = \gamma p + m_0 - \sum(p)$, where $\sum(p)$ is defined in (5.16.22)

### 5.16.3 Spectral Representation of the Photon Propagator, Charge Renormalization, Coulomb Potential in Full QED, Unordered Products of Currents

In order to derive[73] a spectral representation[74] of the full photon propagator, as we did in (5.10.53), (5.10.54) and (5.10.55) to lowest order, we begin with the theory in the Coulomb gauge. The reason why we consider first the Coulomb gauge, is that in this gauge only physical degrees of freedom are quantized and one is dealing with an underlying vector space with a positive definite metric. From such an analysis, one may then infer the structure of the photon propagator in covariant gauges.

We first note that due to the simple fact that $[e^{i\Lambda(x)}(\delta/\delta\eta(x))]\,[e^{-i\Lambda(y)}(\delta/\delta\overline{\eta}(y))]$ is independent of $\Lambda$ for $x = y$, we learn from (5.15.10), (5.15.11), (5.15.12), (5.15.16), (5.15.17), and similarly, from (5.15.20), (5.15.21), (5.15.22), (5.15.18), as well as from (5.15.23), that

$$\left[\frac{1}{F_C}\left(\prod_{j=1}^{n}\frac{\delta}{\delta\eta(x_j)}\gamma^{\mu_j}\frac{\delta}{\delta\overline{\eta}(x_j)}\right)F_C\right]\Bigg|_{\eta=0,\overline{\eta}=0,J^\mu=0}$$

$$= \left[\frac{1}{F}\left(\prod_{j=1}^{n}\frac{\delta}{\delta\eta(x_j)}\gamma^{\mu_j}\frac{\delta}{\delta\overline{\eta}(x_j)}\right)F\right]\Bigg|_{\eta=0,\overline{\eta}=0,J^\mu=0,\lambda}. \tag{5.16.31}$$

In particular, this means that the current-current correlation function satisfies the relation[75]

$$\left\langle\left(j_\mu(x)j_\nu(y)\right)_+\right\rangle\Bigg|_{\text{Coulomb}} = \left\langle\left(j_\mu(x)j_\nu(y)\right)_+\right\rangle\Bigg|_{\text{Gauge}[\lambda]}. \tag{5.16.32}$$

---

[73] The details of the derivation of the spectral representation may be omitted in a first reading.

[74] The spectral representation of a propagator, in general, is usually attributed to Källen [35] and Lehmann [42], and is referred to as a Källen-Lehmann representation.

[75] $\langle . \rangle = \langle 0_+| . |0_-\rangle/\langle 0_+ | 0_-\rangle$, with now all the external sources set equal to zero.

That is, $\langle \left( j_\mu(x) j_\nu(y) \right)_+ \rangle$ is both *covariant and gauge invariant*.

Consider the vacuum expectation value of the product of two currents $j^\mu(x) j^\nu(x')$, written in terms of a completeness expansion

$$\langle \text{vac}| j^\mu(x) j^\nu(x')|\text{vac}\rangle = \sum_n e^{ip_n(x-x')} \langle \text{vac}| j^\mu(0)|n\rangle \langle n| j^\nu(0)|\text{vac}\rangle$$

$$= \int \frac{(dQ)}{(2\pi)^3} e^{iQ(x-x')} \theta(Q^0) \Sigma^{\mu\nu}(Q), \qquad (5.16.33)$$

$$\Sigma^{\mu\nu}(Q) = (2\pi)^3 \sum_n \delta^{(4)}(Q - p_n)\langle \text{vac}| j^\mu(0)|n\rangle \langle n| j^\nu(0)|\text{vac}\rangle$$

$$= \left[ Q^\mu Q^\nu - Q^2 \eta^{\mu\nu} \right] H(-Q^2), \qquad (5.16.34)$$

where $H(-Q^2)$ is a scalar and gauge invariant, $-Q^2 \geq 0$, $Q^0 \geq 0$. In writing the last equality above, we have used the fact that the current $j^\mu$ is conserved in the absence of external sources (see, in particular, (5.7.32) and (5.14.22)).

Now consider the matrix element $\langle \text{vac}|A^i(x)A^j(x')|\text{vac}\rangle_C$, in the Coulomb gauge, in the *absence* of external sources, to relate it to the vacuum expectation of the currents above. Using a completeness relation, we may also express the latter matrix element as

$$\langle \text{vac}|A^i(x)A^j(x')|\text{vac}\rangle_C = \sum_n e^{ip_n(x-x')} \langle \text{vac}|A^i(0)|n\rangle_C \langle n|A^j(0)|\text{vac}\rangle_C$$

$$= \int \frac{(dQ)}{(2\pi)^3} \left( \delta^{ij} - \frac{Q^i Q^j}{Q^2} \right) e^{iQ(x-x')} \theta(Q^0) \rho(Q), \qquad (5.16.35)$$

where we have used the transversality of the vector field $\mathbf{A}$: $\partial_i A^i(x) = 0$, and where

$$\rho(Q) = \frac{(2\pi)^3}{2} \sum_n \delta^{(4)}(Q - p_n)\langle \text{vac}|\mathbf{A}(0)|n\rangle_C \cdot \langle n|\mathbf{A}(0)|\text{vac}\rangle_C \geq 0. \qquad (5.16.36)$$

The important point to note here is that the matrix element $\langle \mathbf{A}(x) \cdot \mathbf{A}(x')\rangle_C$, is computed in a positive definite metric space.

To discuss the physical content of this equation, we write

$$1 = \sum_n |n\rangle\langle n| = \int \sum_\xi \frac{d^3 \mathbf{Q}}{2|\mathbf{Q}|(2\pi)^3} |\mathbf{Q}\,\xi\rangle\langle \mathbf{Q}\,\xi| + \sum_n {}'|n\rangle\langle n|, \qquad (5.16.37)$$

where the first term on the extreme right-hand side of the equation stands for a single photon contribution, and the parameter $\xi$ here, specifies the polarization of a photon,

$$\langle \mathbf{Q}\,\xi | \mathbf{Q}'\,\xi' \rangle = (2\pi)^3 2\,|\mathbf{Q}|\,\delta^{(3)}(\mathbf{Q} - \mathbf{Q}')\,\delta_{\xi\xi'}. \tag{5.16.38}$$

Also $\sum'_n |n\rangle\langle n|$ corresponds to all possible remaining states. In particular, we may write

$$\langle \mathbf{Q}\,\xi | \mathbf{A}(0) | \text{vac} \rangle = \sqrt{Z_3}\,\mathbf{e}_\xi^*(\mathbf{Q}). \tag{5.16.39}$$

with the $\mathbf{e}_\xi(\mathbf{Q})$ representing polarization vectors, satisfying

$$\mathbf{Q}\cdot\mathbf{e}_\xi(\mathbf{Q}) = 0, \quad \mathbf{e}_\xi(\mathbf{Q})\cdot\mathbf{e}_{\xi'}^*(\mathbf{Q}) = \delta_{\xi\xi'}, \quad \xi,\xi' = 1,2, \tag{5.16.40}$$

$$\sum_{\xi=1,2} e_\xi^i(\mathbf{Q})\,e_\xi^{*j}(\mathbf{Q}) = \left(\delta^{ij} - \frac{Q^i Q^j}{\mathbf{Q}^2}\right). \tag{5.16.41}$$

The single photon contribution to $\rho(Q)$ in (5.16.37), (5.16.38) and (5.16.39), follows from (5.16.36), to be simply $Z_3\,\delta(Q^2)$.

That is, we may write

$$\rho(Q) = Z_3\,\delta(Q^2) + \sigma(Q), \tag{5.16.42}$$

where $\sigma(Q) \geq 0$[76] denotes the remaining contribution to $\rho(Q)$.

Now we use the field equation in the absence of external sources in the Coulomb gauge

$$-\Box A^i(x) = \left(\delta^{ik} - \frac{\partial^i \partial^k}{\nabla^2}\right) j^k(x), \tag{5.16.43}$$

to relate $\sigma(Q)$ to $H(-Q^2)$ in (5.16.34). The above field equation leads from (5.16.35) to

$$(-\Box)(-\Box')\,\langle \text{vac}|\mathbf{A}(x)\cdot\mathbf{A}(x')|\text{vac}\rangle_{\text{C}}$$

$$= 2\int \frac{(dQ)}{(2\pi)^3}\,e^{iQ(x-x')}\,\theta(Q^0)(Q^2)^2 \rho(Q)$$

$$= \left(\delta^{i\ell} - \frac{\partial^i\partial^\ell}{\nabla^2}\right)\left(\delta^{i\ell'} - \frac{\partial'^i\partial'^{\ell'}}{\nabla'^2}\right)\langle \text{vac}|j^\ell(x)j^{\ell'}(x')|\text{vac}\rangle_{\text{C}}. \tag{5.16.44}$$

---

[76]See (5.16.36).

Hence upon using the identity

$$\left(\delta^{i\ell} - \frac{Q'^iQ'^\ell}{\mathbf{Q}^2}\right)\left(\delta^{i\ell'} - \frac{Q'^iQ'^{\ell'}}{\mathbf{Q}'^2}\right)\left(Q^\ell Q^{\ell'} - Q^2\,\delta^{\ell\ell'}\right) = -2\,Q^2, \tag{5.16.45}$$

Eqs. (5.16.33) and (5.16.34) lead to $(-Q^2 \geq 0,\ Q^0 \geq 0)$

$$0 \leq (Q^2)^2\,\rho(Q) = (Q^2)^2\,\sigma(Q) = (-Q^2)H(-Q^2), \tag{5.16.46}$$

$$\text{or} \qquad \sigma(Q) = H(-Q^2)/(-Q^2) \equiv \sigma(-Q^2), \tag{5.16.47}$$

where in the first equality in (5.16.46), we have used the fact that $(Q^2)^2\,\delta(Q^2) = 0$. Thus $\sigma(-Q^2) \geq 0$ is a scalar and gauge invariant.

We may rewrite (5.16.33) as

$$\langle\text{vac}|j^\mu(x)j^\nu(x')|\text{vac}\rangle_C = \int_0^\infty dM^2 H(M^2)\int\frac{(dQ)}{(2\pi)^3}$$

$$\times\ \delta(M^2 + Q^2)\left[Q^\mu Q^\nu - Q^2\eta^{\mu\nu}\right]e^{iQ(x-x')}\,\theta(Q^0). \tag{5.16.48}$$

Upon multiplying (5.16.48) by $\theta(x^0 - x'^0)$ and repeating the derivation of this equation for $\langle\text{vac}|j^\nu(x')j^\mu(x)|\text{vac}\rangle_C$, as well as multiplying it by $\theta(x'^0 - x^0)$, and finally adding the two contributions, we obtain

$$\langle\text{vac}|\left(j^\mu(x)j^\nu(x')\right)_+|\text{vac}\rangle_C$$

$$= \int_0^\infty dM^2 H(M^2)\int\frac{(dQ)}{(2\pi)^4}\left[Q^\mu Q^\nu - Q^2\,\eta^{\mu\nu}\right]\delta(Q^2 + M^2)$$

$$\times\ \theta(Q^0)\left[e^{iQ(x-x')}\theta(x^0 - x'^0) + e^{iQ(x'-x)}\theta(x'^0 - x^0)\right],$$

$$= \int_0^\infty dM^2 H(M^2)\int\frac{(dQ)}{(2\pi)^4}\left[Q^\mu Q^\nu - Q^2\,\eta^{\mu\nu}\right]\delta(Q^2 + M^2)$$

$$\times\ e^{iQ(x-x')}\left[\theta(Q^0)\theta(x^0 - x'^0) + \theta(-Q^0)\theta(x'^0 - x^0)\right], \tag{5.16.49}$$

or[77]

$$\langle\text{vac}|\left(j^\mu(x)j^\nu(x')\right)_+|\text{vac}\rangle_C$$

$$= -i\int_0^\infty dM^2 H(M^2)\int\frac{(dQ)}{(2\pi)^4}\left[Q^\mu Q^\nu - Q^2\eta^{\mu\nu}\right]\frac{e^{iQ(x-x')}}{Q^2 + M^2 - i\epsilon}. \tag{5.16.50}$$

---

[77]Note that $(Q^2 + M^2 - i\epsilon)$ in the denominator, means that $-Q^2$ and $M^2$ may be interchanged, in a function of $Q^2$ or $M^2$ multiplying $1/(Q^2 + M^2 - i\epsilon)$, when applying the residue theorem for $x^0 \lessgtr x'^0$. Also note that the integrands in (5.16.50), (5.16.51) are even function in the variable $Q^\mu$. See also Brown [14], pp. 461–463.

From (5.16.35) and (5.16.47), we may similarly write the corresponding expression for the field components $A^i$:

$$\langle\,\text{vac}|\left(A^i(x)A^j(x')\right)_+|\text{vac}\rangle_C = -i\int\frac{(\mathrm{d}Q)}{(2\pi)^4}\left(\eta^{ij}-\frac{Q^iQ^j}{\mathbf{Q}^2}\right)e^{iQ\,(x-x')}$$

$$\times\int_0^\infty \mathrm{d}M^2\frac{\rho(M^2)}{Q^2+M^2-i\epsilon}. \tag{5.16.51}$$

In particular for $x^0 > x'^0$,

$$\langle A^i(x)A^j(x')\rangle_C = \left(\delta^{ij}-\frac{\partial^i\partial^j}{\boldsymbol{\nabla}^2}\right)\int_0^\infty \mathrm{d}M^2\rho(M^2)\int\frac{\mathrm{d}^3\mathbf{Q}}{(2\pi)^3}\frac{e^{iQ\,(x-x')}}{2\sqrt{\mathbf{Q}^2+M^2}},\tag{5.16.52}$$

$Q^0 = \sqrt{\mathbf{Q}^2+M^2}$. Hence

$$\langle A^i(x)\dot{A}^j(x')\rangle_C = \frac{i}{2}\left(\delta^{ij}-\frac{\partial^i\partial^j}{\boldsymbol{\nabla}^2}\right)\int_0^\infty \mathrm{d}M^2\rho(M^2)\int\frac{\mathrm{d}^3\mathbf{Q}}{(2\pi)^3}e^{iQ\,(x-x')}.\tag{5.16.53}$$

Similarly, for $x'^0 > x^0$, we have

$$\langle\dot{A}^j(x')A^i(x)\rangle_C = \frac{-i}{2}\left(\delta^{ij}-\frac{\partial^i\partial^j}{\boldsymbol{\nabla}^2}\right)\int_0^\infty \mathrm{d}M^2\rho(M^2)\int\frac{\mathrm{d}^3\mathbf{Q}}{(2\pi)^3}e^{iQ\,(x'-x)}.\tag{5.16.54}$$

Upon taking the limit $x'^0 \to x^0 - 0$ in (5.16.53), and the limit $x'^0 \to x^0 + 0$ in (5.16.54), and subtracting the resulting latter equation from the former resulting one, we obtain

$$\langle\,[A^i(x^0,\mathbf{x}),\dot{A}^j(x^0,\mathbf{x}')]\rangle_C = i\left(\delta^{ij}-\frac{\partial^i\partial^j}{\boldsymbol{\nabla}^2}\right)\delta^{(3)}(\mathbf{x}-\mathbf{x}')\int_0^\infty \mathrm{d}M^2\,\rho(M^2).\tag{5.16.55}$$

Hence from the equal-time commutation relations (5.14.29), this gives

$$\int_0^\infty \mathrm{d}M^2\,\rho(M^2) = 1,\quad Z_3 + \int_0^\infty \mathrm{d}M^2\sigma(M^2) = 1,\quad 0 \le Z_3 \le 1,\tag{5.16.56}$$

with $Z_3$ denoting the probability that the field $\mathbf{A}$ creates a photon out of the vacuum.

At this stage, we note that the free photon propagator, in the Coulomb gauge, may be written as (see (5.14.13)) ($c^\mu = \eta^{\mu i}\partial^i/\nabla^2$)

$$D_C^{\mu\nu}(x-y) = \left[\eta^{\mu\nu} + \frac{\partial^\mu \partial^\nu}{\nabla^2} - \partial^\mu c^\nu - \partial^\nu c^\mu\right]\frac{1}{(-\Box - i\epsilon)}\,\delta^{(4)}(x-y), \quad (5.16.57)$$

$$\langle\, 0_+|A^\mu(x)\,|\,0_-\rangle_C = \left[\eta^{\mu\nu} + \frac{\partial^\mu \partial^\nu}{\nabla^2} - \frac{\partial^\mu \eta^{\nu i}\partial_i}{\nabla^2} - \frac{\partial^\nu \eta^{\mu i}\partial_i}{\nabla^2}\right]\frac{1}{-\Box - i\epsilon}$$
$$\times \left(J^\nu(x)\langle\, 0_+\,|\,0_-\rangle_C + \langle\, 0_+|j^\nu(x)\,|\,0_-\rangle_C\right). \quad (5.16.58)$$

The matrix element of the vector potential between the vacuum state, in the presence of external sources, which will eventually be set equal to zero, then follows to be

$$\langle\, 0_+|A^\mu(x)\,|\,0_-\rangle_C = \int (dy)\, D_C^{\mu\sigma}(x-y)\left[J_\sigma(y)\langle 0_+\,|\,0_-\rangle_C + \langle\, 0_+|j_\sigma(y)\,|\,0_-\rangle_C\right]. \quad (5.16.59)$$

We recall the equation giving the equation for $\partial_\nu j^\nu(x)$ in (5.14.22), expressed in terms of the expectation values between the vacuum state, in the presence of external Fermi sources:

$$\langle\, 0_+|\partial_\nu j^\nu(x)\,|\,0_-\rangle_C = i\,e_0\,\langle 0_+|\left(\overline{\psi}(x)\eta(x) - \overline{\eta}(x)\psi(x)\right)|\,0_-\rangle_C. \quad (5.16.60)$$

Hence

$$e_0(i)\frac{\delta}{\delta\eta(x')}\,\gamma^\mu(-i)\frac{\delta}{\delta\overline{\eta}(x')}\langle\, 0_+|\partial_\nu j^\nu(x)\,|\,0_-\rangle_C\Big|_{\eta=0,\overline{\eta}=0}$$
$$= i\,e_0^2\,\delta^{(4)}(x-x')\langle\, 0_+|\left[\left(\psi_b(x')\overline{\psi}_a(x)\right)_+ + \left(\overline{\psi}_a(x')\psi_b(x)\right)_+\right]|\,0_-\rangle_C\,\gamma_{ab}^\mu = 0. \quad (5.16.61)$$

Upon carrying out the functional differentiations of (5.16.59), with respect to

$$(i)\left(\frac{\delta}{\delta\eta(x')}\right)\gamma^\mu(-i)\left(\frac{\delta}{\delta\overline{\eta}(x')}\right),$$

and upon using, in the process, the expression for $D_C^{\mu\nu}$ in (5.16.57), the above equation leads for $J^\mu \to 0$ to ($c^\mu = \eta^{\mu i}\partial^i/\nabla^2$)

$$\langle\, 0_+|\left(A^\mu(x)j^\nu(x')\right)_+|\,0_-\rangle_C$$
$$= \left(\delta^\mu{}_\sigma - \partial^\mu c_\sigma\right)\frac{1}{(-\Box - i\epsilon)}\langle\, 0_+|\left(j^\sigma(x)j^\nu(x')\right)_+|\,0_-\rangle_C. \quad (5.16.62)$$

On the other hand, in the absence of external Fermi sources, we may use the conservation of the current $\partial_\sigma j^\sigma = 0$, to rewrite (5.16.59) as

$$\langle\, 0_+|A^\mu(x)\,|\,0_-\rangle_{\mathrm C} = \int (\mathrm{d}y)\, D_{\mathrm C}^{\mu\sigma}(x-y)J_\sigma(y)\langle\, 0_+\,|\,0_-\rangle_{\mathrm C}$$

$$+ \left(\delta^{\mu\sigma} - \partial^\mu c^\sigma\right)\frac{1}{(-\Box - i\epsilon)}\langle\, 0_+|j_\sigma(x)\,|\,0_-\rangle_{\mathrm C}. \qquad (5.16.63)$$

Therefore by taking the functional differentiation of this equation, with respect to, $\delta/\delta J_\nu(y)$, and setting $J^\mu = 0$, and then using (5.16.62), we finally obtain for the full photon propagator in the Coulomb gauge the expression

$$\widetilde{D}_{\mathrm C}^{\mu\nu}(x-x') = D_{\mathrm C}^{\mu\nu}(x-x')$$

$$+ \left[\delta^\mu{}_\alpha - \partial^\mu c_\alpha\right]\frac{1}{(-\Box - i\epsilon)}\left[\delta^\nu{}_\beta - \partial^\nu c_\beta\right]\frac{1}{(-\Box - i\epsilon)}\, i\,\langle\left(j^\alpha(x)j^\beta(x')\right)_+\rangle_{\mathrm C}.$$
$$(5.16.64)$$

Using the identity

$$\left[\delta^\mu{}_\alpha - \partial^\mu c_\alpha\right]\left[\delta^\nu{}_\beta - \partial^\nu c_\beta\right]\left[\partial^\alpha \partial^\beta - \Box\eta^{\alpha\beta}\right] = (-\Box)\left[\eta^{\mu\nu} + \frac{\partial^\mu \partial^\nu}{\nabla^2} - \partial^\mu c^\nu - \partial^\nu c^\mu\right],$$
$$(5.16.65)$$

and

$$\frac{i}{-\Box}\langle\mathrm{vac}|\left(j^\mu(x)j^\nu(x')\right)_+|\mathrm{vac}\rangle = \int \frac{(\mathrm{d}Q)}{(2\pi)^4}\left[Q^\mu Q^\nu - Q^2\eta^{\mu\nu}\right]e^{iQ(x-x')}$$

$$\times \int_0^\infty \mathrm{d}M^2\frac{M^2\sigma(M^2)}{Q^2(Q^+ M^2 - i\epsilon)}, \qquad (5.16.66)$$

we obtain

$$\widetilde{D}_{\mathrm C}^{\mu\nu}(x-x') = \left[\eta^{\mu\nu} + \frac{\partial^\mu \partial^\nu}{\nabla^2} - \partial^\mu c^\nu - \partial^\nu c^\mu\right]\frac{1}{(-\Box)}\int_0^\infty \mathrm{d}M^2$$

$$\times \left\{\left[Z_3\,\delta(M^2) + \sigma(M^2)\right]\delta^{(4)}(x-x') - \int \frac{(\mathrm{d}Q)}{(2\pi)^4}\, e^{iQ(x-x')}\frac{M^2\sigma(M^2)}{(Q^2 + M^2 - i\epsilon)}\right\},$$
$$(5.16.67)$$

where we have conveniently multiplied $\delta^{(4)}(x - x')$, within the curly brackets, by $1 = \int_0^\infty dM^2\left(Z_3\delta(M^2) + \sigma(M^2)\right)$ (see (5.16.56)), or $(c^\mu = \eta^{\mu i}\partial_i/\nabla^2)$

$$\widetilde{D}_C^{\mu\nu}(x - x') = \left[\eta^{\mu\nu} + \frac{\partial^\mu\partial^\nu}{\nabla^2} - \partial^\mu c^\nu - \partial^\nu c^\mu\right]h(x - x'),\qquad(5.16.68)$$

$$h(x - x') = \int \frac{(dQ)}{(2\pi)^4}\, e^{iQ(x-x')}\int_0^\infty dM^2\, \frac{\rho(M^2)}{(Q^2 + M^2 - i\epsilon)},\qquad(5.16.69)$$

where we have used the definition $\rho(M^2) = Z_3\delta(M^2) + \sigma(M^2)$ and $M^2\delta(M^2) = 0$.

The free photon propagator in a covariant gauge, specified by a parameter $\lambda$, may be simply written from (5.3.9), as

$$D^{\mu\nu}(x - y) = \left[\eta^{\mu\nu} - (1 - \lambda)\frac{\partial^\mu\partial^\nu}{\Box}\right]\frac{1}{(-\Box - i\epsilon)}\delta^{(4)}(x - y).\qquad(5.16.70)$$

The matrix elements of the vector potential between the vacuum states, in covariant gauges, then follows from (5.7.2)/(5.7.6), to be

$$\langle 0_+|A^\mu(x)|0_-\rangle = \int (dy)\, D^\mu{}_\sigma(x - y)\left[J^\sigma(y)\langle 0_+|0_-\rangle + \langle 0_+|j^\sigma(y)|0_-\rangle\right].\qquad(5.16.71)$$

Upon carrying out the functional differentiations of this equation with respect to (i)$(\delta/\delta\eta(x'))\, \gamma^\mu(-i)(\delta/\delta\overline{\eta}(x'))$, the above equation leads for $J^\mu \to 0$ to

$$\langle 0_+|\left(A^\mu(x)j^\nu(x')\right)_+|0_-\rangle = \frac{1}{(-\Box - i\epsilon)}\langle 0_+|\left(j^\mu(x)j^\nu(x')\right)_+|0_-\rangle.\qquad(5.16.72)$$

Hence by taking the functional differentiation of (5.16.71), with respect to, $\delta/\delta J_\nu(y)$, and setting $J^\mu = 0$, and then using (5.16.72), we finally obtain for the full photon propagator in covariant gauges

$$\widetilde{D}^{\mu\nu}(x - x') = D^{\mu\nu}(x - x') + \frac{1}{(-\Box - i\epsilon)}\frac{1}{(-\Box - i\epsilon)}\, i\langle\left(j^\mu(x)j^\nu(x')\right)_+\rangle.\qquad(5.16.73)$$

From (5.16.50), (5.16.47), (5.16.70), and by following a similar method as carried out in (5.16.66) and (5.16.67), the photon propagator in the different covariant gauges takes the form

$$\widetilde{D}^{\mu\nu}(x - x') = \left[\eta^{\mu\nu} - \frac{\partial^\mu\partial^\nu}{\Box}\right]h(x - x') + \lambda\frac{\partial^\mu\partial^\nu}{\Box}\frac{1}{(-\Box - i\epsilon)}\delta^{(4)}(x - y),\qquad(5.16.74)$$

where $h(x - x')$ is defined in (5.16.69). Note that the longitudinal part of the propagator does not change with the interaction.

In reference to (5.16.69), we may set

$$h(x - x') = \int \frac{(dk)}{(2\pi)^4} e^{iQ(x-x')} \frac{d(Q^2)}{Q^2}, \qquad \frac{d(Q^2)}{Q^2} = \int_0^\infty dM^2 \frac{\rho(M^2)}{(Q^2 + M^2 - i\epsilon)}.$$
(5.16.75)

Equation (5.16.56) then leads to the following formal boundary conditions for $Q^2 \to \infty$:

$$d(Q^2) \to 1, \qquad \alpha\, d_{\text{ren}}(Q^2) \equiv \frac{\alpha}{Z_3} d(Q^2) \to \alpha_0.$$
(5.16.76)

## Charge Renormalization and the Coulomb Potential

In the absence of radiative corrections, the Coulomb interaction between static charges,[78] each of charge $q$, may be extracted, in the Coulomb gauge, from the static part $D^0_{C0}(Q)$, of the photon propagator as follows:

$$U(|\mathbf{x} - \mathbf{x}'|) = q^2 \int \frac{d^3\mathbf{Q}}{(2\pi)^3} e^{iQ\cdot x} D^0_{C0}(Q)|_{Q^0=0} = \frac{q^2}{4\pi |\mathbf{x} - \mathbf{x}'|}.$$
(5.16.77)

In the presence of radiative corrections, in the Coulomb gauge, $D^0_{C0}(Q)$ is modified to $\widetilde{D}^0_{C0}(Q)$, in the full theory, and its Fourier transform is, from (5.16.68), (5.16.69), (5.16.56), given by

$$\widetilde{D}^0_{C0}(Q) = Z_3 \frac{1}{\mathbf{Q}^2} + \int_0^\infty dM^2 \sigma(M^2) \frac{Q^2}{Q^2(Q^2 + M^2)},$$
(5.16.78)

leading to a potential

$$\widetilde{U}(|\mathbf{x} - \mathbf{x}'|) = q^2 \int \frac{d^3\mathbf{Q}}{(2\pi)^3} e^{iQ\cdot(x-x')} \widetilde{D}^0{}_0(Q)|_{Q^0=0}$$

$$= \frac{q^2}{4\pi |\mathbf{x} - \mathbf{x}'|} \left[ Z_3 + \int_0^\infty dM^2 \sigma(M^2) e^{-M|\mathbf{x}-\mathbf{x}'|} \right],$$
(5.16.79)

---

[78] A process involving no energy transfer.

and, at large distance seaparation $|\mathbf{x} - \mathbf{x}'| \to \infty$, we recover the Coulomb potential from (5.16.56)

$$\widetilde{U}(|\mathbf{x} - \mathbf{x}'|) \to \frac{q_{\text{ren}}^2}{4\pi|\mathbf{x} - \mathbf{x}'|}, \quad q_{\text{ren}} = q\sqrt{Z_3}, \tag{5.16.80}$$

with modified renormalized charges. That the charge is renormalized via the photon renormalization constant $Z_3$ only and not by the renormalization constants $Z_2$, and $Z_1$, as well, is a consequence of the Ward identity, established in Sect. 5.16, implying that $Z_1 = Z_2$. This point will be further discussed in the next section.

An integral equation of the photon polarization tensor, to supplement those of the electron and vertex functions derived in the previous subsection, is derived in the following one.

### Fourier Transform of the Unordered Product of Two Currents

From (5.16.73) and (5.16.75), we may solve for the vacuum expectation value of the time-ordered product of two currents as follows

$$i\left\langle\left(j^\mu(x)j^\nu(0)\right)_+\right\rangle = \int \frac{(\mathrm{d}Q)}{(2\pi)^4} \, e^{iQx} \, (\eta^{\mu\nu}Q^2 - Q^\mu Q^\nu)Q^2 \int_0^\infty \mathrm{d}M^2 \, \frac{\rho(M^2)}{Q^2 + M^2 - i\epsilon}. \tag{5.16.81}$$

This means that

$$\theta(x^0)\langle j^\mu(x)j^\nu(0)\rangle = 2\pi \int_0^\infty \mathrm{d}M^2 \, \rho(M^2) \int \frac{(\mathrm{d}Q)}{(2\pi)^4} \delta(Q^2 + M^2)\theta(Q^0)$$

$$\times \, (\eta^{\mu\nu}Q^2 - Q^\mu Q^\nu)\, Q^2 \, e^{iQx}\, \theta(x^0). \tag{5.16.82}$$

and

$$\theta(-x^0)\langle j^\nu(0)j^\mu(x)\rangle = 2\pi \int_0^\infty \mathrm{d}M^2 \, \rho(M^2) \int \frac{(\mathrm{d}Q)}{(2\pi)^4} \delta(Q^2 + M^2)\theta(Q^0)$$

$$\times \, (\eta^{\mu\nu}Q^2 - Q^\mu Q^\nu)\, Q^2 \, e^{-iQx}\, \theta(-x^0), \tag{5.16.83}$$

where we have made a change of the variable of integration $Q \to -Q$ in the last integral.

Invoking hermiticity of the current operators and the normalization of the vacuum state, Eq. (5.16.83) above, upon taking its complex conjugate, becomes

$$\theta(-x^0)\langle j^\mu(x)j^\nu(0)\rangle = 2\pi \int_0^\infty dM^2 \rho(M^2) \int \frac{(dQ)}{(2\pi)^4} \delta(Q^2 + M^2)\theta(Q^0)$$
$$\times (\eta^{\mu\nu}Q^2 - Q^\mu Q^\nu)Q^2) e^{iQx} \theta(-x^0).$$

(5.16.84)

Finally adding (5.16.82) and (5.16.84), and using the relation $\theta(x^0) + \theta(-x^0) = 1$, gives for the vacuum expectation of the unordered product of two currents the relation

$$\langle j^\mu(x)j^\nu(0)\rangle = 2\pi \int_0^\infty dM^2 \rho(M^2) \int \frac{(dQ)}{(2\pi)^4} \delta(Q^2 + M^2)\theta(Q^0)$$
$$\times (\eta^{\mu\nu}Q^2 - Q^\mu Q^\nu)Q^2 e^{iQx}.$$

(5.16.85)

A Fourier transform, with $k^0 > 0$, then leads to the following useful expression

$$\int (dx) e^{-iky} \langle j^\mu(x)j^\nu(0)\rangle = 2\pi\sigma(-k^2)k^2(\eta^{\mu\nu}k^2 - k^\mu k^\nu),$$

(5.16.86)

where the part of the spectral function $\delta(M^2)$ of $\rho(M^2)$ does not contribute as it is multiplied by $M^2$ in (5.16.85), (5.16.86) with $M^2 = -Q^2$, $M^2 = -k^2$, respectively (see (5.16.42) and (5.16.56)).[79]

This equation will be used to study the fundamental process $e^+e^- \rightarrow$ hadrons in Sect. 6.4 in quantum chromodynamics.

### 5.16.4 Integral Equation for the Vacuum Polarization Tensor

To derive an integral equation for the photon polarization tensor, note that from (5.16.71), we may, by functional differentiation of $\langle A^\mu(x)\rangle$ with respect to $J^\nu(x')$ and using the chain rule, write

$$\widetilde{D}^{\mu\nu}(x, x') = D^{\mu\nu}(x, x') + \int (dy)(dz) D^\mu{}_\sigma(x, y)\frac{\delta}{\delta\langle A^\kappa(z)\rangle}\langle j^\sigma(y)\rangle \widetilde{D}^{\kappa\nu}(z, x').$$

(5.16.87)

---

[79]See also the QED lowest contribution to $\rho(M^2)$ in (5.10.55).

and we set the external sources equal to zero. This suggests to define the polarization tensor[80]

$$\Pi^\sigma{}_\kappa(y,z) = -\frac{\delta}{\delta\langle A^\kappa(z)\rangle}\langle j^\sigma(y)\rangle. \tag{5.16.88}$$

On the other hand, note that

$$\langle j^\sigma(y)\rangle = -e_0\,\gamma^\sigma_{ab}\,\langle\left(\psi_b(y)\overline{\psi}_a(y)\right)_+\rangle = i\,e_0\,\mathrm{Tr}\,[\gamma^\sigma\,\widetilde{S}(y,y)], \tag{5.16.89}$$

and from (5.16.4) and (5.16.88), we obtain, in the absence of external sources,

$$\Pi^{\mu\nu}(x-y) = -i\,e_0^2\int(dx')(dy')\,\mathrm{Tr}\,[\gamma^\mu\widetilde{S}(x-x')\,\Gamma^\nu(x',y';y)\,\widetilde{S}(y'-x)]. \tag{5.16.90}$$

Current conservation implies from (5.16.88) that $(\partial/\partial y^\sigma)\Pi^\sigma{}_\kappa(y,z) = 0$. On the hand, from (5.16.3) for $\delta\langle A^\kappa(z)\rangle = (\partial/\partial z^k)\delta\varphi(z)$ for an arbitrary $\delta\varphi(z)$, (5.16.1) gives $\delta\langle j^\sigma(y)\rangle = 0$,

$$0 = \delta\langle j^\sigma(y)\rangle = -\int(dz)\Pi^\sigma{}_\kappa(y,z)\delta\langle A^\kappa(z)\rangle = -\int(dz)\Pi^\sigma{}_\kappa(y,z)(\partial/\partial z_\kappa)\chi(z), \tag{5.16.91}$$

and hence we also have $(\partial/\partial z^\kappa)\,\Pi^{\sigma\kappa}(y,z) = 0$. That is, the Fourier transform $\Pi^{\mu\nu}(Q)$ has the general structure

$$\Pi^{\mu\nu}(Q) = (\eta^{\mu\nu}Q^2 - Q^\mu Q^\nu)\,\Pi(Q^2). \tag{5.16.92}$$

Compare with (5.10.42) to lowest order. Fourier transforming (5.16.90), and using (5.16.7) give

$$\Pi^{\mu\nu}(Q) = -i\,e_0^2\int\frac{(dp)}{(2\pi)^4}\mathrm{Tr}\,[\gamma^\mu\,\widetilde{S}(p+Q)\,\Gamma^\nu(p+Q,p)\,\widetilde{S}(p)]. \tag{5.16.93}$$

Upon multiplying this equation by $Q_\nu$ and using the Ward-Takahashi identity in (5.16.8), we obtain

$$\Pi^{\mu\nu}(Q)\,Q_\nu = -i\,e_0^2\int\frac{(dp)}{(2\pi)^4}\mathrm{Tr}\,[\gamma^\mu\,(\widetilde{S}(p) - \widetilde{S}(p+Q))]. \tag{5.16.94}$$

One is tempted to shift the variable of integration $p \to p-Q$, in the second term to obtain zero for the net contribution. This, however, is not justified as the integral is at least quadratically divergent. Since the transversality of the vacuum polarization

---

[80]The minus sign is introduced to be consistent with the notation of the lowest order.

tensor, giving rise to the structure in (5.16.92), *has* to be satisfied, on grounds of gauge invariance, one may introduce a strong ultraviolet cut-off in (5.16.94) so that the shift of the variable of integration is justified, thus eliminating the integral in question. Or one may carry out a careful definition of a gauge invariant current to define $\Pi^{\mu\nu}(Q)$, itself, as done in Sect. 3.9,[81] or one may equivalently extract the transverse part (5.16.93) directly from (5.16.93), as is often done in the literature.

Finally, we note from (5.16.87), (5.16.88), (5.16.92), that in the momentum description,

$$\widetilde{D}^{\mu\nu}(Q) = D^{\mu\nu}(Q) - \frac{1}{Q^2 - \mathrm{i}\epsilon}(\eta^{\mu}{}_{\sigma}Q^2 - Q^{\mu}Q_{\sigma})\Pi(Q^2)\widetilde{D}^{\sigma\nu}(Q), \qquad (5.16.95)$$

or

$$\widetilde{D}^{\mu\nu}(Q) = \left(\eta^{\mu\nu} - \frac{Q^{\mu}Q^{\nu}}{Q^2}\right)\frac{1}{(Q^2 + Q^2\Pi(Q^2))} + \lambda\frac{Q^{\mu}Q^{\nu}}{Q^4}. \qquad (5.16.96)$$

## 5.17   The Full Renormalized Theory

The renormalized basic components of the theory are defined by

$$\widetilde{S}_{\mathrm{ren}} = \widetilde{S}/Z_2, \ \ \widetilde{D}^{\mu\nu}_{\mathrm{ren}} = \widetilde{D}^{\mu\nu}/Z_3, \ \ \Gamma^{\mu}_{\mathrm{ren}} = Z_1\Gamma^{\mu}, \ \ \mathrm{e} = \sqrt{Z_3}\,\mathrm{e}_0, \ \ m = m_0 + \delta m. \tag{5.17.1}$$

*Renormalization is the process in which the experimentally unattainable parameters* $\mathrm{e}_0$ *and* $m_0$, *introduced earlier as limits at absolute zero distances going all the way to the core of the electron, are eliminated in favor of the experimentally observed physical quantities* $\mathrm{e}$ *and* $m$.

We, certainly, cannot claim that our present theories may be extended all the way to absolute zero distances. I find it rather surprising that there are still practitioners who are surprised when pathologies (divergences) are encountered when dealing with the physically unattainable unrenormalzed quantitites.

*On the other hand, the wavefunction renormalization constants emerge, irrespective of perturbation theory, in the following manner.*[82]

Consider the matrix elements of the fields, say, $\overline{\psi}$ and $A^{\mu}$, between single particle states and the vacuum as follows

$$\langle\,\mathbf{p}\,\sigma, -|\overline{\psi}(x)|\mathrm{vac}\rangle = \sqrt{Z_2}\,\mathrm{e}^{-\mathrm{i}px}\,\overline{u}(\mathbf{p},\sigma), \qquad (5.17.2)$$

$$\langle\,\mathbf{k}\,\lambda|A^{\mu}(x)|\mathrm{vac}\rangle = \sqrt{Z_3}\,\mathrm{e}^{-\mathrm{i}kx}\,e^{*\mu}_{\lambda}(\mathbf{k}). \qquad (5.17.3)$$

---

[81]One may also equivalently subtract a Taylor expansion in Q, of the integrand in (5.16.93), in the process, in finally evaluating $\Pi^{\mu\nu}(Q)$.

[82]See also Sect. 4.1.

This is what quantum theory says, irrespective of perturbation theory,[83] since the fields may create other particles as well out of the vacuum, which necessitates the presence of the normalization coefficients $\sqrt{Z_3}$, $\sqrt{Z_2}$ in a probabilistic context.

One may, in turn, introduce the renormalized fields

$$\psi_{\text{ren}} = \psi / \sqrt{Z_2}, \quad \overline{\psi}_{\text{ren}} = \overline{\psi} / \sqrt{Z_2}, \quad A_{\text{ren}}^{\mu} = A^{\mu} / \sqrt{Z_3}, \qquad (5.17.4)$$

which, in particular, now satisfy the conditions

$$\langle \mathbf{p}\,\sigma, - | \overline{\psi}_{\text{ren}}(x) | \text{vac} \rangle = e^{-ipx}\, \overline{u}(\mathbf{p}, \sigma), \quad \langle \mathbf{k}\lambda | A_{\text{ren}}^{\mu}(x) | \text{vac} \rangle = e^{-ikx}\, e_{\lambda}^{*\mu}(\mathbf{k})$$
$$(5.17.5)$$

creating, in addition to other particles, an electron and a photon, respectively, out of the vacuum with unit amplitudes, *emphasizing the particle aspect of the theory as "free" particles, described by properly normalized wavefunctions, when emerging into the detection regions by counters and when released by emitters.*

Thus carrying out wavefunction renormalizations results in the proper normalization of particles' wavefunctions in describing asymptotic states. It is a remarkable property of QED, as well as of a very special class of other cherished theories to be discussed later, that when the process of renormalization is carried out, in which unrenormalized expressions are eliminated in favor of the renormalized ones, the theory is finite.

An important step in studying the full renormalized QED theory, is to eliminate $\gamma^{\mu}$ in favor of the full vertex $\Gamma^{\mu}$, by using the integral equation for the latter given in (5.16.30). This shows that we may, in general, write

$$\gamma^{\mu} = \Gamma^{\mu}(p + Q, p) + i e_0^2 \int \frac{(dp')}{(2\pi)^4} K(p, p'; Q)\, \widetilde{S}(p' + Q)\, \Gamma^{\mu}(p' + Q, p')\, \widetilde{S}(p').$$
$$(5.17.6)$$

Let us see how the basic components of the theory respond to the scalings of the propagators and and vertex function defined in (5.17.1).

From Fig. 5.17 given earlier in Sect. 5.16.2, and Fig. 5.19, we note that for the $e^- e^+$- kernel $K$, the combination $e_0^2 K$ may be expressed in terms of renormalized quantitites multiplied by $(1/\sqrt{Z_2})^4 = (1/Z_2)^2$. That is,

$$e_0^2 K = \frac{1}{(Z_2)^2}\, e^2 K_{\text{ren}}, \qquad (5.17.7)$$

where $e^2 K_{\text{ren}}$ may be expressed in terms of the full renormalized electron and photon propagators, the renormalized vertex and the renormalized charge $e = \sqrt{Z_3}\, e_0$.

---

[83] See also the Introductory chapter of the book *and* Sect. 4.1.

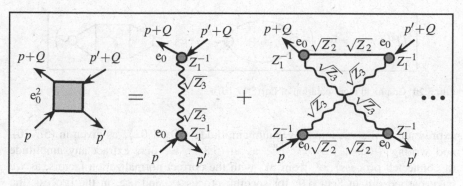

**Fig. 5.19** Upon writing $\widetilde{S} = Z_2 \widetilde{S}_{\text{ren}}$, $\widetilde{D}^{\mu\nu} = Z_3 \widetilde{D}_{\text{ren}}^{\mu\nu}$, $\Gamma^\mu = \Gamma_{\text{ren}}^\mu / Z_1$, and using the Ward identity relation $Z_1 = Z_2$, we note that we may rewrite $e_0^2 K$ as $e^2 K_{\text{ren}}/(Z_2)^2$, where $e^2 K_{\text{ren}}$ may be expressed in terms of the full renormalized electron and photon propagators, the renormalized vertex function and the renormalized charge $e = \sqrt{Z_3}\, e_0$

The integral equation for the renormalized vertex may be written as

$$\Gamma_{\text{ren}1}^\mu(p+Q,p)$$

$$= Z_1 \gamma^\mu - i\,e^2 \int \frac{(\mathrm{d}p')}{(2\pi)^4}\, K_{\text{ren}}(p,p';Q)\, \widetilde{S}_{\text{ren}}(p'+Q)\, \Gamma_{\text{ren}}^\mu(p'+Q,p')\, \widetilde{S}_{\text{ren}}(p').$$

$$(5.17.8)$$

as follows from (5.16.30), where we have used the Ward identity relation $Z_1 = Z_2$.

From (5.16.21)/(5.16.22), an integral equation for the inverse of the renormalized electron propagator may be also written. For studying its renormalizability, however, one may use the Ward-Takahashi identity in (5.16.8) to write

$$\widetilde{S}_{\text{ren}}^{-1}(p') = (p'-p)_\mu\, \Gamma_{\text{ren}}^\mu(p',p)\big|_{\gamma p \to -m}, \qquad (5.17.9)$$

since $\widetilde{S}^{-1}(p)\big|_{\gamma p \to -m} = 0$, where, in particular, $\big|_{\gamma p \to -m}$ attached to $\Gamma_{\text{ren}}^\mu(p',p)$ means that $p^2$ is set equal to $-m^2$ in it, and that $\gamma p$ on its extreme right-hand side is replaced by $-m$.

Finally we note that for the vacuum polarization tensor, we may, from (5.16.93) and (5.17.6), represent it symbolically as

$$\Pi^{\mu\nu} = -i\,e_0^2\Big(\text{Tr}\,[\widetilde{S}\,\Gamma^\mu\,\widetilde{S}\,\Gamma^\nu] + i\,e_0^2\,\text{Tr}\,[K\,\widetilde{S}\,\Gamma^\mu\,\widetilde{S}\widetilde{S}\,\Gamma^\nu\,\widetilde{S}]\Big), \qquad (5.17.10)$$

and may be represented diagrammatically as in Fig. 5.20, and its renormalized counterpart will be discussed in Sect. 5.18.2.

In Sects. 5.8 and 5.9, we have seen that the coupling of the external sources to external lines readily lead, in a straightforward manner, to the extraction of the scattering amplitude of any process desired. Specifically, upon using the explicit

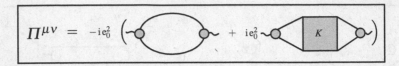

**Fig. 5.20**  Graphical representation of Eq. (5.17.10)

expression of the vacuum-to-vacuum amplitude $\langle\, 0_+ \mid 0_-\rangle$, as given in (5.7.20), and writing $\langle\, 0_+ \mid 0_-\rangle = e^{iW}$, as in (5.8.1), we may extract any amplitude (a connected process) $\mathscr{A}$ from $W$, with the correct normalization factors, as we have carried out in Sect. 5.9, for specific processes, and use, in the process, the expressions in (5.9.17), (5.9.18), (5.9.38), (5.9.39), (5.9.45), (5.9.46), which signify the role of the external sources in generating a scattering amplitude.

The external sources are coupled to the fields in the Lagrangian density in the following manner:

$$\bar{\eta}\psi + \bar{\psi}\eta + A_\mu J^\mu \equiv \bar{\eta}_{\mathrm{ren}}\psi_{\mathrm{ren}} + \bar{\psi}_{\mathrm{ren}}\eta_{\mathrm{ren}} + A_{\mathrm{ren}\mu}J^\mu_{\mathrm{ren}}, \qquad (5.17.11)$$

where $\eta = \eta_{\mathrm{ren}}/\sqrt{Z_2}$, $\bar{\eta} = \bar{\eta}_{\mathrm{ren}}/\sqrt{Z_2}$, $J^\mu = J^\mu_{\mathrm{ren}}/\sqrt{Z_3}$. The scalings of the propagators, the vertex function and the charge, as defined above, lead simply to the replacement of the sources by their renormalized scaled counterparts, emitting and detecting "free" particles, making the particle aspect of the theory evident.

Various vertex connections that may occur in processes are shown in the figure below, with the amplitudes expressed in terms of the full propagators and the full vertex function. By introducing the renormalization scaling factors, in defining the renormalized theory, one may, in turn, express amplitudes in terms of the full renormalized propagators, the full renormalized vertex function, and the renormalized charge $e = \sqrt{Z_3}\, e_0$.

Incidently, the above analysis, also shows that due to the Ward identity, the charge is renormalized by the photon wavefunction renormalzation only: $e = \sqrt{Z_3}\, e_0$.

## 5.18   Finiteness of the Renormalized Theory; Renormalized Vertex Function and Renormalized Propagators

The present section deals with the finiteness of renormalized theory order by order in perturbation theory by a direct graphical analysis. A different approach will be taken when dealing with non-abelian gauge fields.

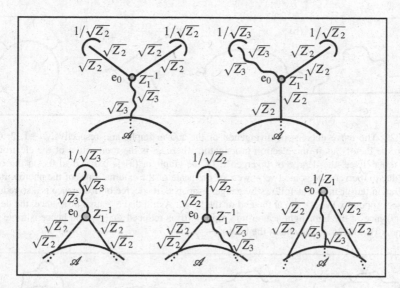

**Fig. 5.21** These are the types of vertex connections that may occur. Expressing the propagators, the vertex function and the charge by their renormalized counterparts, simply lead to the replacement of the sources by their renormalized scaled counterparts as defined below (5.17.11), and the particle aspect of the theory, discussed above, becomes evident. The sources are denoted by the *half circles* corresponding to electron/positron and photon emitters and detectors. By removing the renormalized sources and the renormalized external propagators attached to them, in the manner that was done in Sect. 5.9, as discussed above, the amplitudes $\mathscr{A}$ emerge expressed in terms of the full renormalized components of the theory

## 5.18.1   Finiteness of the Renormalized Theory

We establish the finiteness of the renormalized QED theory (Fig. 5.21) by a direct simple way. We then consider some applications of so-called renormalization group methods as applied to it in the next section.

To lowest order, we have encountered ultraviolet divergences in the theory when dealing with the electron and photon self energies, as well as the vertex function. These are shown in the figure below. With this in mind, we define the *degree of divergence* of a graph as:

the number of integration variables plus the highest power of momenta in the numerator minus the highest power of momenta in the denominator.

Due to the transversality property, in the external photon lines, of the light-light scattering graph, a result that follows by carefully invoking gauge invariance,[84] its naïve degree of divergence of 0 is reduced to $-4$, and hence is finite (Fig. 5.22).

---

[84]See Appendix IV at the end of the book.

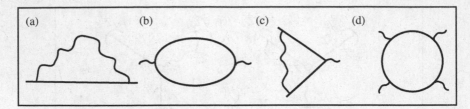

**Fig. 5.22** The naïve degrees of divergence of the above graphs are, respectively, $+1, 2, 0, 0$. Due to the factor $\gamma p$ in the electron propagator, where $p$ is the momentum of the electron on the external lines, the degree of divergence of the graph in (**a**) is reduced to 0. On the other hand, due to the very welcome two powers of momenta of the external lines of the photon, in the polarization tensor in (**b**), as follows by gauge invariance, its degree of divergence is also reduced to 0 (see Appendix IV, (IV.27) at the end of the book). Again due to gauge invariance, the degree of divergence of the light-light scattering graph in (**d**) is reduced to $-4$, and hence is finite (see Appendix IV, (IV.26) at the end of the book)

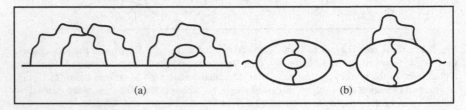

**Fig. 5.23** Some improper graphs

The coupling parameters in QED is dimensionless and a remarkable property of the theory is that the degree of divergence of a proper connected graph remains the same independently of the order of perturbation theory. To this end we define a graph to be proper if it cannot be broken into two disjoint graphs by cutting a single photon or a single electron line. Some improper graphs are shown in Fig. 5.23.

The degree of divergence $d(g)$ of a connected and proper graph in QED is easily checked to be given by

$$d(g) = \ell_{\text{integ}} - \ell_e^{\text{int}} - 2\,\ell_\gamma^{\text{int}}, \tag{5.18.1}$$

$$\ell_{\text{integ}} = \text{number of integration variables,}$$

$$\ell_e^{\text{int}} = \text{number of internal electron lines,} \tag{5.18.2}$$

$$\ell_\gamma^{\text{int}} = \text{number of internal photon lines.}$$

The remarkable property of this theory that the degree of divergence of a connected and proper graph with a fixed number of external lines does not change with the order of perturbation theory, that is with the increase of the number of internal lines, becomes evident when $d(g)$ is expressed in terms of the number of external lines

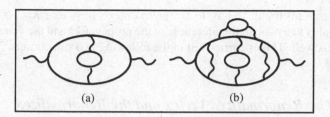

**Fig. 5.24** (a) Here $\ell_{\text{integ}} = 4 \times 3, \ell_e^{\text{int}} = 6, \ell_\gamma^{\text{int}} = 2$. (b) In this case, $\ell_{\text{integ}} = 4 \times 7, \ell_e^{\text{int}} = 14$, $\ell_\gamma^{\text{int}} = 6$. The degree of divergence of the graphs are $d(g) = 2$, and are *equal*, as expected, and this is effectively reduced to zero after subtracting two powers of momenta needed in the definition of the polarization tensor as discussed before

of a graph (e.g., Fig. 5.24). It is easily verified that this is simply given by

$$d(g) = 4 - \frac{3}{2}\,\ell_e^{\text{ext}} - \ell_\gamma^{\text{ext}}, \tag{5.18.3}$$

where $\ell_e^{\text{ext}}$, $\ell_\gamma^{\text{ext}}$ denote the number of external electron (positron) and photon lines, respectively.

To establish the finiteness of the renormalized theory to any order in the renormalized charge, we may proceed by induction.

To the above end, we have seen in Sects. 5.10 and 5.11, that to lowest order, the renormalized electron and photon propagators as well as the renormalized vertex functions are finite. Now we assume, as an in induction hypothesis, that in perturbation theory, the renormalized propagators and the renormalized vertex function are finite to all orders $\leq (n-2)$ in e. This will readily allow us to study the finiteness of $e^2 K_{\text{ren}}$, and any connected and proper graph $\mathscr{G}$, with four more external lines, to a given order $n$.

To the above end, we note, in reference to Figs. 5.19 and 5.21, $e^2 K_{\text{ren}}$, and a connected and proper graph $\mathscr{G}$, with at least four external lines, may be expanded in terms of the full renormalized propagators and the renormalized full vertex function, and have strictly negative degrees of divergence. The only divergence that may occur are from subgraphs within them which are self-energy insertions and vertex corrections. These, however, have been rendered finite, by the renormalization procedure, to any order $\leq (n-2)$, and has reduced their effective degrees of divergence to strictly negative values, according to the induction hypothesis. Also recall that a subgraph with only three external photon lines within them vanishes by Furry's theorem, that the effective degree of divergence of light-light scattering is equal to $-4$, and finally note that all other subgraphs with four or more external lines within them have strictly negative degrees of divergence. Hence $e^2 K_{\text{ren}}$ and $\mathscr{G}$, with at least four external lines, have all their subintegrations, as well as their overall integrations finite. That is, *to any given order*[85] $n$ in e, $e^2 K_{\mathrm{ren}}$, $\mathscr{G}$ *are finite*. Thus it remains to establish the finiteness of the propagators and the vertex function to order $n$ as well. This is carried out in the following two subsections.[86]

## 5.18.2   The Renormalized Vertex and the Renormalized Electron Propagator

From (5.17.8), the renormalized vertex has the structure

$$\Gamma_{\mathrm{ren}}^{\mu}(p',p) = Z_1 \gamma^{\mu} + e^2 \widetilde{\Lambda}^{\mu}(p',p). \qquad (5.18.4)$$

where $\widetilde{\Lambda}^{\mu}(p',p)$ is expressed in terms of renormalized components. Also

$$\Gamma_{\mathrm{ren}}^{\mu}(p,p)\Big|_{\gamma p \to -m} = \gamma^{\mu} = Z_1 \gamma^{\mu} + e^2 \widetilde{\Lambda}^{\mu}(p,p)\Big|_{\gamma p \to -m}, \qquad (5.18.5)$$

from which, by solving for $(Z_1 - 1)\gamma^{\mu}$, we may conveniently write

$$\Gamma_{\mathrm{ren}}^{\mu}(p+Q,p) = \gamma^{\mu} + e^2 \Big(\widetilde{\Lambda}^{\mu}(p+Q,p) - \widetilde{\Lambda}^{\mu}(p,p)\Big|_{\gamma p \to -m}\Big). \qquad (5.18.6)$$

The $n$th order contribution of the latter is given by

$$\Gamma_{\mathrm{ren}}^{\mu}(p+Q,p)\Big|^{(n)} = e^2 \Big(\widetilde{\Lambda}^{\mu}(p+Q,p) - \widetilde{\Lambda}^{\mu}(p,p)\Big|_{\gamma p \to -m}\Big)\Big|^{(n-2)}. \qquad (5.18.7)$$

In the induction hypothesis, we have assumed that the renormalized vertex and the renormalized electron and photon propagators are finite, and all subgraphs in them, are finite for all orders $\leq (n-2)$. In this subsection, we will see that this is true for the renormalized vertex and renormalized electron propagator to order $n$ as well. In the next subsection we also establish this for the renormalized photon propagator.

In Fig. 5.15 of Sect. 5.16.2, a graphical representation of the vertex function was given, and in Fig. 5.16, some graphs not contributing to the $e^-e^+$-kernel $K$, in the expression for the vertex function was also shown. A graphical representation of $\widetilde{\Lambda}^{\mu}(p',p)\Big|^{(n-2)}$ is then given by $K_{\mathrm{ren}}, \Gamma_{\mathrm{ren}}^{\mu}, \widetilde{S}_{\mathrm{ren}}$ as shown in Fig. 5.25, which may

---

[85]This is the content of a power counting theorem, due to Weinberg [80], which states that a graph is finite if all of its corresponding subintegrations, including the overall integration, are finite. For a pedagogical treatment of this technical problem see the author's book on renormalization [45].

[86]Finiteness is understood to refer to ultraviolet finiteness. The infrared problem is treated by the inclusion of soft photons in computing transition amplitudes. In the study of propagators and the vertex function, a non-zero photon mass is included in the analysis, as done earlier to the lowest order, to avoid infrared divergences in intermediate steps.

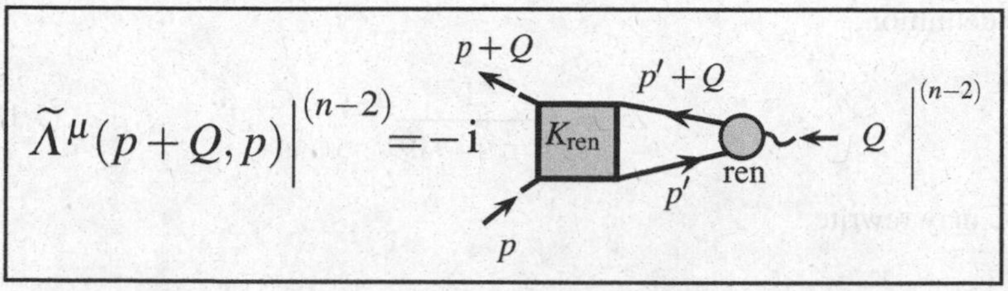

**Fig. 5.25** Graphical representation of $\widetilde{\Lambda}^\mu(p+Q,p)$. Only the overall integration diverges. The two electron propagators in the graph also denote renormalized ones. Note also the relation in (5.17.7)

contain subgraphs of orders $\leq (n-2)$ and the renormalization procedure has reduced their effective degrees of divergence to strictly negative values, and the only divergence[87] in $\widetilde{\Lambda}^\mu(p+Q,p)$, may come from the overall integration with a 0 degree of divergence, and the single subtraction in (5.18.7) renders this overall integration finite, giving rise to a renormalized vertex finite to order $n$ as well. The finiteness of the renormalized electron propagator, to order $n$, then also follows from (5.17.9), which we recall resulted from an elementary application of the Ward-Takahashi identity.

It remains to establish the finiteness of the renormalized photon propagator to order $n$. This is the subject of the next subsection.[88]

## 5.18.3    The Renormalized Photon Propagator

From (5.16.96), the full photon propagator, in covariant gauges, may be written as

$$\widetilde{D}^{\mu\nu}(Q) = \left(\eta^{\mu\nu} - \frac{Q^\mu Q^\nu}{Q^2}\right)\frac{d(Q^2)}{Q^2} + \lambda\,\frac{Q^\mu Q^\nu}{Q^4}. \tag{5.18.8}$$

We set

$$d(Q^2) = \frac{1}{1+\alpha_0\,\pi(Q^2)}, \qquad \Pi(Q^2) \equiv \alpha_0\,\pi(Q^2). \tag{5.18.9}$$

---

[87]Light-light scattering subgraphs, as noted before, are convergent.

[88]Finiteness of renormalized QED by induction was carefully treated in Bjorken and Drell [11], which was also applied to the renormalized photon propagator by a very elaborate method. We have preferred to treat the photon propagator by a method developed in a remarkable paper by Baker and Lee [4].

By definition,

$$Z_3 = \frac{1}{1 + \alpha_0 \, \pi(0)}, \tag{5.18.10}$$

and we may rewrite

$$d^{-1}(Q^2) = 1 + \alpha_0 \, \pi(0) + \alpha_0 \left(\pi(Q^2) - \pi(0)\right) = \frac{1}{Z_3}\left[1 + \alpha \, \pi_c(Q^2)\right], \tag{5.18.11}$$

where

$$\pi_c(Q^2) = \left[\pi(Q^2) - \pi(0)\right], \qquad \alpha \, \pi_c(0) = 0, \qquad \alpha = Z_3 \, \alpha_0, \tag{5.18.12}$$

and the latter relation for the fine-structure constant was defined earlier.

Using the renormalized version of the photon propagator

$$\widetilde{D}_{\mathrm{ren}}^{\mu\nu}(Q) = \left(\eta^{\mu\nu} - \frac{Q^\mu Q^\nu}{Q^2}\right)\frac{d_{\mathrm{ren}}(Q^2)}{Q^2} + \frac{1}{Z_3}\,\lambda\,\frac{Q^\mu Q^\nu}{Q^4}, \tag{5.18.13}$$

and (5.18.8), (5.18.11), we have

$$d_{\mathrm{ren}}^{-1}(Q^2) = \left[1 + \alpha \, \pi_c(Q^2)\right] = Z_3 \left[1 + \alpha_0 \, \pi(Q^2)\right]. \tag{5.18.14}$$

To establish the finiteness of the photon propagator to order $n$ as well, we first define the vacuum polarization tensor by factoring out the unrenormalized fine-structure constant $\alpha_0$, as done in (5.18.9), and write

$$\Pi^{\mu\nu}(Q) = \alpha_0 \, \pi^{\mu\nu}(Q) = \alpha_0 \, (\eta^{\mu\nu} Q^2 - Q^\mu Q^\nu) \, \pi(Q^2), \tag{5.18.15}$$

where from (5.17.10), (5.17.7), $\pi^{\mu\nu}(Q)$ has a skeleton expansion in terms of *renormalized* components of the theory given by

$$\pi^{\mu\nu} = -4\,\pi\,\mathrm{i}\Big(\mathrm{Tr}\left[\widetilde{S}_{\mathrm{ren}}\,\Gamma_{\mathrm{ren}}^\mu\,\widetilde{S}_{\mathrm{ren}}\,\Gamma_{\mathrm{ren}}^\nu\right] + \mathrm{i}\,\mathrm{e}^2\,\mathrm{Tr}\left[K_{\mathrm{ren}}\,\widetilde{S}_{\mathrm{ren}}\,\Gamma_{\mathrm{ren}}^\mu\,\widetilde{S}_{\mathrm{ren}}\,\widetilde{S}_{\mathrm{ren}}\,\Gamma_{\mathrm{ren}}^\nu\,\widetilde{S}_{\mathrm{ren}}\right]\Big), \tag{5.18.16}$$

and has the graphical representation in Fig. 5.26.

It is easily verified that $\pi(Q^2)$ is worked out to be given by

$$\pi(Q^2) = \frac{Q_\mu Q_\nu}{6\,Q^2}\,\pi^{\mu\nu}{}_{,\alpha}{}^\alpha(Q), \qquad \pi^{\mu\nu}{}_{,\alpha}{}^\alpha(Q) \equiv \frac{\partial}{\partial Q^\alpha}\frac{\partial}{\partial Q_\alpha}\,\pi^{\mu\nu}(Q). \tag{5.18.17}$$

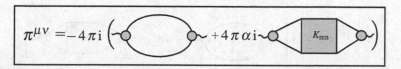

**Fig. 5.26** Graphical skeleton expansion representation of $\pi^{\mu\nu}(Q)$, where note that on account of the scaling factor as given in (5.17.1), and (5.17.7) for $e_0^2 K = e^2 K_{\text{ren}}/(Z_2)^2$, it is expressed in terms of the renormalized components of the theory. Accordingly, all the propagators, the vertex function, as well as the fine-structure constant, denote *renormalized* components

Upon scaling $Q$ by a parameter $\xi$, we then have from (5.18.12)

$$\pi_c(Q^2) = \pi(\xi^2 Q^2)\Big|_{\xi=0}^{\xi=1} = \frac{Q_\mu Q_\nu}{6 Q^2}\left(\pi^{\mu\nu},_\alpha{}^\alpha(Q) - \pi^{\mu\nu},_\alpha{}^\alpha(Q)\Big|_0\right). \tag{5.18.18}$$

Note that up to order $n$, $\alpha \pi_c^{\mu\nu}$ is expanded in terms of the renormalized propagators and the renormalized vertex, up to order $n-2$ in e.

To discuss the finiteness of the latter, we differentiate $\pi^{\mu\nu}$ in (5.18.16), with respect to $Q$, and thus differentiate all the (renormalized) components in it with respect to this variable $Q^\alpha$, wherever it appears in them. To this end, the renormalized vertex function may be written symbolically and conveniently as

$$\Gamma_{\text{ren}}^\nu = Z_1 \gamma^\nu - \mathrm{i}e^2 K_{\text{ren}} \widetilde{S}_{\text{ren}} \widetilde{S}_{\text{ren}} \Gamma_{\text{ren}}^\nu, \tag{5.18.19}$$

$$= Z_1 \gamma^\nu - \mathrm{i}e^2 \Gamma_{\text{ren}}^\nu \widetilde{S}_{\text{ren}} \widetilde{S}_{\text{ren}} K_{\text{ren}}. \tag{5.18.20}$$

Upon differentiating (5.18.19), we have,

$$\left(1 + \mathrm{i}e^2\left[K_{\text{ren}} \widetilde{S}_{\text{ren}} \widetilde{S}_{\text{ren}}\right]\right)\Gamma_{\text{ren}}^\nu,_\alpha = -\mathrm{i}e^2 \left(K_{\text{ren}} \widetilde{S}_{\text{ren}} \widetilde{S}_{\text{ren}}\right),_\alpha \Gamma_{\text{ren}}^\nu. \tag{5.18.21}$$

Multiplying this by $\Gamma_{\text{ren}}^\mu \widetilde{S}_{\text{ren}} \widetilde{S}_{\text{ren}}$ gives

$$\left(\Gamma_{\text{ren}}^\mu \widetilde{S}_{\text{ren}} \widetilde{S}_{\text{ren}} + \mathrm{i}e^2\left[\Gamma_{\text{ren}}^\mu \widetilde{S}_{\text{ren}} \widetilde{S}_{\text{ren}} K_{\text{ren}}\right]\widetilde{S}_{\text{ren}} \widetilde{S}_{\text{ren}}\right)\Gamma_{\text{ren}}^\nu,_\alpha$$

$$= -\mathrm{i}e^2 \Gamma_{\text{ren}}^\mu \widetilde{S}_{\text{ren}} \widetilde{S}_{\text{ren}} \left(K_{\text{ren}} \widetilde{S}_{\text{ren}} \widetilde{S}_{\text{ren}}\right),_\alpha \Gamma_{\text{ren}}^\nu. \tag{5.18.22}$$

From (5.18.20), we also have

$$\mathrm{i}e^2 \Gamma_{\text{ren}}^\mu \widetilde{S}_{\text{ren}} \widetilde{S}_{\text{ren}} K_{\text{ren}} = -(\Gamma_{\text{ren}}^\mu - Z_1 \gamma^\mu). \tag{5.18.23}$$

Thus the two terms within the round brackets on the left-hand side of (5.18.22) may be combined to obtain the useful identity

$$Z_1 \gamma^\mu \widetilde{S}_{\text{ren}} \widetilde{S}_{\text{ren}} \Gamma_{\text{ren}}^\nu,_\alpha = -\mathrm{i}e^2 \Gamma_{\text{ren}}^\mu \widetilde{S}_{\text{ren}} \widetilde{S}_{\text{ren}} \left(K_{\text{ren}} \widetilde{S}_{\text{ren}} \widetilde{S}_{\text{ren}}\right),_\alpha \Gamma_{\text{ren}}^\nu$$

$$= -\mathrm{i}e^2\left[\Gamma_{\text{ren}}^\mu \widetilde{S}_{\text{ren}} \widetilde{S}_{\text{ren}} K_{\text{ren}}\right]\left(\widetilde{S}_{\text{ren}} \widetilde{S}_{\text{ren}}\right),_\alpha \Gamma_{\text{ren}}^\nu - \mathrm{i}e^2 \Gamma_{\text{ren}}^\mu \widetilde{S}_{\text{ren}} \widetilde{S}_{\text{ren}} K_{\text{ren}},_\alpha \widetilde{S}_{\text{ren}} \widetilde{S}_{\text{ren}} \Gamma_{\text{ren}}^\nu. \tag{5.18.24}$$

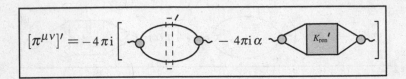

**Fig. 5.27** Graphical skeleton expansion representation of $\pi^{\mu\nu}(Q)_{,\alpha}$ expressed in terms of renormalized components. The prime $'$ denotes differentiation with respect to the momentum $Q^{\alpha}$ in $\pi^{\mu\nu}(Q)$. The prime $'$ in the first graph on the right-hand side stands for the differentiation of the two renormalized electron propagators $\left(\widetilde{S}_{\text{ren}}\widetilde{S}_{\text{ren}}\right)_{,\alpha}$

Using (5.18.23), the above equation may be equivalently rewritten as

$$Z_1 \gamma^{\mu}\widetilde{S}_{\text{ren}}\widetilde{S}_{\text{ren}} \Gamma^{\nu}_{\text{ren},\alpha} + Z_1 \gamma^{\mu}\left(\widetilde{S}_{\text{ren}}\widetilde{S}_{\text{ren}}\right)_{,\alpha} \Gamma^{\nu}_{\text{ren}}$$

$$= \Gamma^{\mu}_{\text{ren}}\left(\widetilde{S}_{\text{ren}}\widetilde{S}_{\text{ren}}\right)_{,\alpha} \Gamma^{\nu}_{\text{ren}} - \mathrm{i}\,e^2 \Gamma^{\mu}_{\text{ren}} \widetilde{S}_{\text{ren}}\widetilde{S}_{\text{ren}} K_{\text{ren},\alpha} \widetilde{S}_{\text{ren}}\widetilde{S}_{\text{ren}}\Gamma^{\nu}_{\text{ren}}. \qquad (5.18.25)$$

From (5.16.93) and (5.19.9), one may also express $\pi^{\mu\nu}$ symbolically in the convenient form

$$\pi^{\mu\nu} = -4\,\pi\mathrm{i}\,Z_1 \gamma^{\mu}\widetilde{S}_{\text{ren}}\widetilde{S}_{\text{ren}}\Gamma^{\nu}_{\text{ren}}, \qquad (5.18.26)$$

which upon using (5.18.25) gives

$$\pi^{\mu\nu}{}_{,\alpha} = -4\,\pi\,\mathrm{i}\left[\Gamma^{\mu}_{\text{ren}}\left(\widetilde{S}_{\text{ren}}\widetilde{S}_{\text{ren}}\right)_{,\alpha}\Gamma^{\nu}_{\text{ren}} - 4\pi\mathrm{i}\,\alpha\,\Gamma^{\mu}_{\text{ren}}\widetilde{S}_{\text{ren}}\widetilde{S}_{\text{ren}} K_{\text{ren},\alpha}\widetilde{S}_{\text{ren}}\widetilde{S}_{\text{ren}}\Gamma^{\nu}_{\text{ren}}\right].$$
$$(5.18.27)$$

This may be represented graphically as shown in Fig. 5.27.

By power counting, i.e., by simple dimensional analysis, we note that the function $\pi^{\mu\nu}{}_{,\alpha}{}^{\alpha}$ in (5.18.17), has only an overall logarithmic divergence contributing to its $n^{\text{th}}$ order, by invoking, in the process, of finiteness to lower orders by the induction hypothesis. This, in turn, means that $\pi(Q^2)$ has also only an overall logarithmic divergence. Clearly then, with one differentiation of $\pi^{\mu\nu}{}_{,\alpha}{}^{\alpha}$ with respect to $Q$ not only the sub-integrations of the latter are finite, by the induction hypothesis, but also the overall integration, as the single differentiation effectively decreases its degree of divergence to minus one from zero. Using the facts that

$$Q^2\frac{\partial}{\partial Q^2}\frac{Q^{\mu}Q^{\nu}}{Q^2} = 0, \qquad Q^2\frac{\partial}{\partial Q^2} = \frac{Q^{\beta}}{2}\frac{\partial}{\partial Q^{\beta}}, \qquad (5.18.28)$$

as obtained, in the process, from the chain rule, gives from (5.18.17)

$$Q^2\frac{\partial}{\partial Q^2}\pi(Q^2) = \frac{Q^{\mu}Q^{\nu}}{12Q^2}Q^{\beta}\frac{\partial}{\partial Q^{\beta}}\pi^{\mu\nu}{}_{,\alpha}{}^{\alpha}(Q) = \phi\left(\frac{Q^2}{m^2}\right) \equiv \text{a finite expression},$$
$$(5.18.29)$$

and we may write

$$\alpha \, \pi_c(Q^2) = \alpha\big(\pi(Q^2) - \pi(0)\big) = \alpha \int_0^{Q^2/m^2} \frac{d\xi}{\xi}\, \phi(\xi), \qquad (5.18.30)$$

which is finite up to order $n$ as well. This completes the demonstration of the renormalized theory to order $n$ as well.[89]

## 5.19   Effective Charge and the Renormalization Group

### 5.19.1   Renormalization Group Analysis

We introduce the concept of an effective charge which interpolates between the renormalized charge e and the unrenormalized charge $e_0$. To this end, we note from (5.18.9) that we may write

$$\alpha_0 \, d(Q^2) = \frac{\alpha_0}{1 + \alpha_0 \, \pi(Q^2)} = \frac{\alpha_0}{1 + \alpha_0 \pi(\xi^2) + \alpha_0\big[\pi(Q^2) - \pi(\xi^2)\big]}. \qquad (5.19.1)$$

In analogy to the fine-structure constant: $\alpha = \alpha_0/[1 + \alpha_0 \pi(0)]$, we may define an effective, energy dependent, fine-structure coupling $\alpha(\xi^2)$ by

$$\alpha(\xi^2) = \frac{\alpha_0}{1 + \alpha_0 \pi(\xi^2)}, \qquad (5.19.2)$$

with the conventional fine-structure constant $\alpha$ corresponding to $\xi = 0$, i.e. measured at large distance from an electron, given by

$$\alpha \equiv \alpha(0). \qquad (5.19.3)$$

From (5.19.1), (5.19.2) and (5.19.3), we obtain the basic equation

$$\alpha \, d_{\mathrm{ren}}(Q^2, \alpha) = \alpha(\xi^2) \, d_{\mathrm{ren}}\big(Q^2, \xi^2, \alpha(\xi^2)\big), \qquad (5.19.4)$$

where

$$d_{\mathrm{ren}}\big(Q^2, \xi^2, \alpha(\xi^2)\big) = \frac{1}{1 + \alpha(\xi^2)\,\big[\pi(Q^2) - \pi(\xi^2)\big]}, \qquad (5.19.5)$$

$$d_{\mathrm{ren}}(Q^2, 0, \alpha(0)) \equiv d_{\mathrm{ren}}(Q^2, \alpha). \qquad (5.19.6)$$

---

[89]Note from (5.18.14), that to order $n$, $d_{\mathrm{ren}}$ depends on $(\alpha\pi_c)$ to order $< n$ and on the $n$th order as well, and the latter has been also established to be finite.

Hence, in general, we have the following basic property encompassing the change under the redefinition of renormalized coupling parameter

$$\alpha(\xi_1^2)\, d_{\text{ren}}(Q^2, \xi_1^2, \alpha(\xi_1^2)) = \alpha(\xi_2^2)\, d_{\text{ren}}(Q^2, \xi_2^2, \alpha(\xi_2^2)). \tag{5.19.7}$$

Because of this basic property, such a procedure in renormalization theory has been referred to as "renormalization group" which is an unfortunate name as it doesn't really have to do with a group.[90]

The effective coupling $\alpha(\xi^2)$ may be obtained from (5.19.5) by choosing $Q^2 = \xi^2$ to give

$$\alpha(\xi^2) = \alpha\, d_{\text{ren}}(\xi^2, \alpha), \tag{5.19.8}$$

and from (5.16.76), satisfies the boundary conditions

$$\alpha(\xi^2) \rightarrow \alpha \ \text{ for } \ \xi^2 \rightarrow 0, \ \rightarrow \alpha_0 \ \text{ for } \ \xi^2 \rightarrow \infty, \tag{5.19.9}$$

where the latter limit may diverge.

For energies of the order ($\sim 91\,\text{GeV}$), corresponding to the mass of the vector boson $Z^0$, this effective coupling is experimentally of the order of $1/128$.[91] An estimate of the effective fine-structure at such an energy is carried out in Sect. 5.19.2 taking into account those contributing quarks and the charged leptons of the three generations by using the renormalization group.[92]

To study the high-energy asymptotic behavior of $\alpha\, d_{\text{ren}}(Q^2, \alpha)$, and hence of the effective coupling and thus investigate the small distance behavior of the theory, we first summarize some of the properties associated with the vacuum polarization tensor to lowest order in $\alpha$ (see (5.10.45), (5.10.49), also (5.18.9)). To this end,

$$\Pi(Q^2) = \alpha\, \pi(Q^2)$$

$$= \left( \left[ \frac{1}{Z_3} - 1 \right] - \frac{2\alpha}{\pi} \int_0^1 dz\, z(1-z) \ln\left[ 1 + \frac{Q^2}{m^2} z(1-z) \right] \right), \tag{5.19.10}$$

$$\frac{1}{Z_3} = 1 + \frac{\alpha}{3\pi} \left[ \ln\left( \frac{\Lambda^2}{m^2} \right) - \frac{1}{2} \right]. \tag{5.19.11}$$

---

[90]The renormalization group was introduced by Stueckelberg and Peterman [74], who also coined the name "renormalization group". This was followed by other important work by Gell-Mann and Low [27], Ovsyannikov [65], Bogoliubov and Shirkov [12] and by many others.

[91]See, e.g., Beringer et al. [7], Mele [63], and Erler [23].

[92]I remember when I was a graduate student, I used to wonder, together with my fellow students, as to why nature chooses the value $1/137$ rather than some other value. Now with the rapid development of quantum field theory, together with higher energy experiments, we understand this value to be a reflection of the energies "at which we were then". Effectively, this coupling changes as we move to higher energies attaining different numerical values as will be investigated in the next subsection. Although understanding why Nature chooses the low energy numerical value $1/137$ is still important, it does not seem to be as mysterious now as it was then.

From these two equations, we note, in particular, that to lowest order

$$\beta(\alpha) \equiv \frac{m}{Z_3} \frac{\mathrm{d}}{\mathrm{d}m} Z_3 = \frac{2\alpha}{3\pi}, \tag{5.19.12}$$

$$m\frac{\mathrm{d}}{\mathrm{d}m} \Pi(Q^2) = m\frac{\mathrm{d}}{\mathrm{d}m} \alpha \pi(Q^2) \sim -\frac{4\alpha}{\pi} \frac{m^2}{Q^2}, \qquad Q^2 \gg m^2, \tag{5.19.13}$$

$$\alpha \pi_c(Q^2) = \alpha \left[ \pi(Q^2) - \pi(0) \right] \sim -\frac{\alpha}{3\pi} \left[ \ln\left(\frac{Q^2}{m^2}\right) - \frac{5}{3} \right] \equiv \alpha \pi_c^{\mathrm{as}}(Q^2), \qquad Q^2 \gg m^2. \tag{5.19.14}$$

where "as" in $\pi_c^{\mathrm{as}}$ stands for asymptotic.

Now consider the full theory. We differentiate the equation

$$d_{\mathrm{ren}}^{-1}(Q^2) = \left[1 + \alpha \pi_c(Q^2)\right] = Z_3 \left[1 + \alpha_0 \pi(Q^2)\right]. \tag{5.19.15}$$

with respect to the mass $m$ of the electron by keeping the unrenormalized fine-structure constant $\alpha_0$ and the ultraviolet cut-off $\Lambda^2$ fixed with the latter arbitrarily large.[93] To this end we set[94]

$$\frac{1}{Z_3} m\frac{\mathrm{d}}{\mathrm{d}m} Z_3 = \beta(\alpha), \qquad m\frac{\mathrm{d}}{\mathrm{d}m} \pi(Q^2) = \Gamma_{\gamma\gamma S}. \tag{5.19.16}$$

With $\alpha_0$, $\Lambda^2$ fixed, the parameters that vary in $\pi(Q^2)$ on the extreme *right*-hand side of (5.19.15) with $m$ are then $\alpha$, and $m_0$. This leads to

$$m\frac{\mathrm{d}}{\mathrm{d}m} \left(Z_3[1 + \alpha_0 \pi(Q^2)]\right) = \beta(\alpha)\, d_{\mathrm{ren}}^{-1}(Q^2) + \alpha\, \Gamma_{\gamma\gamma S}. \tag{5.19.17}$$

On the other hand, the parameters that vary in the first expression in (5.19.15) are $\alpha$ and $m$ itself. That is,

$$m\frac{\mathrm{d}}{\mathrm{d}m} d_{\mathrm{ren}}^{-1}(Q^2) = \left[ m\frac{\partial}{\partial m} + \left(\frac{m}{\alpha} \frac{\mathrm{d}}{\mathrm{d}m} \alpha\right) \alpha \frac{\partial}{\partial \alpha} \right] d_{\mathrm{ren}}^{-1}(Q^2)$$

$$= \left[ m\frac{\partial}{\partial m} + \alpha\, \beta(\alpha) \frac{\partial}{\partial \alpha} \right] d_{\mathrm{ren}}^{-1}(Q^2). \tag{5.19.18}$$

---

[93]In carrying the differentiation of this equation with respect to $m$, we follow the elegant approach, and to some extent the notation, of Adler [1].

[94]Since $\pi(Q^2)$ depends on the mass $m$ through $m_0$, we may write: $m(\mathrm{d}/\mathrm{d}m) \pi(Q^2) = (m/m_0)(\mathrm{d}m_0/\mathrm{d}m) m_0 (\partial\pi(Q^2)/\partial m_0)$.

From (5.19.15), (5.19.17), (5.19.18), we obtain[95]

$$\left[ m\frac{\partial}{\partial m} + \beta(\alpha)\left( \alpha\frac{\partial}{\partial\alpha} - 1 \right) \right] d_{\text{ren}}^{-1}(Q^2) = \alpha\, \Gamma_{\gamma\gamma\, s}(Q^2). \qquad (5.19.19)$$

Now we let $Q^2/m^2$ become arbitrary large. Because of dimensional reasons, order by order in perturbation theory $\Gamma_{\gamma\gamma\, s}(Q^2)$ vanishes like $m^2/Q^2$ up to powers of $\ln(Q^2/m^2)$,[96] thus giving

$$\left[ m\frac{\partial}{\partial m} + \beta(\alpha)\left( \alpha\frac{\partial}{\partial\alpha} \right) \right] \alpha\, d_{\text{ren}}^{\text{as}}(Q^2/m^2, \alpha) = 0, \qquad (5.19.20)$$

*where* the function $d_{\text{ren}}^{\text{as}}(Q^2/m^2, \alpha)$ is obtained from $d_{\text{ren}}(Q^2/m^2, \alpha)$, by neglecting all the terms in the latter which vanish for $Q^2/m^2 \to \infty$. The function $d_{\text{ren}}^{\text{as}}(Q^2/m^2, \alpha)$, will then depend on powers of $\ln(Q^2/m^2)$ (see (5.19.33)). We note that due to the finiteness of $d_{\text{ren}}^{\text{as}}(Q^2/m^2, \alpha)$, this equation implies that $\beta(\alpha)$ is finite.

We will solve (5.19.20) for $d_{\text{ren}}^{\text{as}}(Q^2/m^2, \alpha)$ with the boundary condition

$$\alpha\, d_{\text{ren}}^{\text{as}}(1, \alpha) = q(\alpha). \qquad (5.19.21)$$

Up to second order in $\alpha$, $q(\alpha)$ may be directly read from (5.19.14), (5.19.15) to be

$$q(\alpha) = \alpha - \frac{5}{9\pi}\alpha^2 + \cdots . \qquad (5.19.22)$$

In reference to (5.19.20), we may use the chain rule, to replace the differentiation with respect to $\alpha$ by differentiation with respect to $q(\alpha)$, in the following manner

$$\frac{\partial}{\partial\alpha} = q'(\alpha)\frac{\partial}{\partial q(\alpha)}, \quad \text{or} \quad \frac{1}{2}\alpha\,\beta(\alpha)\frac{\partial}{\partial\alpha} = \frac{1}{2}\alpha\,\beta(\alpha)\,q'(\alpha)\frac{\partial}{\partial q(\alpha)}. \qquad (5.19.23)$$

where the factor $1/2$ is introduced for convenience. This suggests to introduce the function

$$\psi(q(\alpha)) = \frac{1}{2}\alpha\,\beta(\alpha)\,q'(\alpha), \qquad (5.19.24)$$

[95]Such a scaling equation is referred to as a Callan-Symanzik scaling equation [15, 75–77].

[96]This is fully justified by the application of a theorem due to Weinberg [80] when it is applied to the *renormalized* theory order by order [45], and essentially reduces to a dimensional analysis. The $m^2/Q^2$ factor is already seen in (5.19.13) to lowest order.

known as the Gell-Mann and low function. Equation (5.19.20) may be now rewritten as

$$\left[ \frac{\partial}{\partial \tau} - \psi(q(\alpha)) \frac{\partial}{\partial q(\alpha)} \right] \alpha\, d\,_{\text{ren}}^{\text{as}} = 0, \qquad \tau = \ln(Q^2/m^2), \qquad (5.19.25)$$

which, in particular, explains the choice of the factor $1/2$ in (5.19.23). Finally, we introduce the variable

$$y = \int_c^{q(\alpha)} \frac{dz}{\psi(z)}, \qquad \frac{\partial}{\partial y} = \psi(q(\alpha)) \frac{\partial}{\partial q(\alpha)}, \qquad (5.19.26)$$

where $c$ is an arbitrary constant, to rewrite (5.19.25) simply as

$$\left[ \frac{\partial}{\partial \tau} - \frac{\partial}{\partial y} \right] \alpha\, d\,_{\text{ren}}^{\text{as}} = 0. \qquad (5.19.27)$$

We will obtain two useful expressions from this equation. The solution of the differential equation is elementary and is given by

$$\alpha\, d\,_{\text{ren}}^{\text{as}}(Q^2/m^2, \alpha) = F[\tau + y] = F\left[ \ln \frac{Q^2}{m^2} + \int_c^{q(\alpha)} \frac{dz}{\psi(z)} \right], \qquad (5.19.28)$$

where $F$ is an arbitrary function, but with the normalization condition

$$\alpha\, d\,_{\text{ren}}^{\text{as}}(1, \alpha) = q(\alpha), \qquad (5.19.29)$$

it must satisfy the condition

$$F\left[ \int_c^{q(\alpha)} \frac{dz}{\psi(z)} \right] = q(\alpha). \qquad (5.19.30)$$

As a translation operation in the variable $y$, we may rewrite (5.19.28) as

$$\alpha\, d\,_{\text{ren}}^{\text{as}}(Q^2/m^2, \alpha) = \exp\left[ \tau \frac{\partial}{\partial y} \right] F[y]$$

$$= \exp\left[ \tau \psi(q(\alpha)) \frac{\partial}{\partial q(\alpha)} \right] q(\alpha), \qquad (5.19.31)$$

where we have used (5.19.30). Expanding the exponential factor in (5.19.31), gives

$$\alpha\, d\,_{\text{ren}}^{\text{as}}(Q^2/m^2, \alpha) = \sum_{n=0}^{\infty} \frac{(\tau)^n}{n!} \left( \psi(z) \frac{d}{dz} \right)^n z \bigg|_{z=q(\alpha)}, \qquad (5.19.32)$$

leading to the useful expression

$$\alpha \, d_{\text{ren}}^{\text{as}}(Q^2/m^2, \alpha) = q(\alpha) + \psi(q(\alpha)) \left[ \sum_{n=1}^{\infty} \frac{\left(\ln(Q^2/m^2)\right)^n}{n!} \frac{d}{dz} \left(\psi(z)\frac{d}{dz}\right)^{n-1} z \bigg|_{z=q(\alpha)} \right],$$

(5.19.33)

as an explicit expansion in powers of $\ln(Q^2/m^2)$.

Some expansions of $\beta(\alpha)$ and $\psi(z)$ which go beyond the lowest orders, are given, respectively,[97] by

$$\beta(\alpha) = \frac{2}{3} \frac{\alpha}{\pi} + \frac{1}{2} \frac{\alpha^2}{\pi^2} - \frac{121}{144} \frac{\alpha^3}{\pi^3} + \cdots, \tag{5.19.34}$$

$$\psi(z) = z \left[ \frac{1}{3} \frac{z}{\pi} + \frac{1}{4} \frac{z^2}{\pi^2} + \frac{1}{8} \left( \frac{8}{3} \zeta(3) - \frac{101}{36} \right) \frac{z^3}{\pi^3} + \cdots \right], \tag{5.19.35}$$

where $\zeta(3) \approx 1.202$, with $\zeta$ denoting the Riemann zeta function.

Another useful expression obtained from (5.19.25), in addition to the one in (5.19.33), for further analysis is obtained upon introducing the integral

$$I(\tau, \alpha) = \int_{q(\alpha)}^{\alpha \, d_{\text{ren}}^{\text{as}}(Q^2/m^2, \alpha)} \frac{dz}{\psi(z)}, \quad \text{with B.C.} \quad I(0, \alpha) = 0, \quad \tau = \ln\left(\frac{Q^2}{m^2}\right),$$

(5.19.36)

which from (5.19.25), it satisfies the equation

$$\left[ \frac{\partial}{\partial \tau} - \psi(q(\alpha)) \frac{\partial}{\partial q(\alpha)} \right] I(\tau, \alpha) = 1. \tag{5.19.37}$$

With $y$ defined in (5.19.26), the solution of the above equation is given by[98]

$$I(\tau, \alpha) = \tau + G[\tau + y], \quad \text{with} \quad G[y] = I(0, \alpha) = 0, \tag{5.19.38}$$

where $G[\,.\,]$ is arbitrary, but is to satisfy the just given B.C. Also $G[\tau + y] = \exp[\tau \partial/\partial y]G[y] = 0$, which leads from (5.19.36), (5.19.38) to

$$\ln\left(\frac{Q^2}{m^2}\right) = \int_{q(\alpha)}^{\alpha \, d_{\text{ren}}^{\text{as}}(Q^2/m^2, \alpha)} \frac{dz}{\psi(z)}, \tag{5.19.39}$$

satisfying the proper B.C.

---

[97] de Rafaël and Rosner [18] and Baker and Johnson [3]. See also Gorishnii et al. [30]. Note that we use Adler's $\beta(\alpha)$ function notation which is $\beta(\alpha)/\alpha$ of de Rafaël's and Rosner's.

[98] The solution may be also written as: $-y + \widetilde{G}[\tau + y] = \tau - (\tau + y) + \widetilde{G}[\tau + y] = \tau + G[\tau + y]$ where $\widetilde{G}[\tau + y] - (\tau + y) = G[\tau + y]$. Hence (5.19.38) is the general solution of (5.19.37).

Finally from the boundary condition on $d_{\text{ren}}^{-1}(Q^2) \to Z_3$ for $Q^2 \to \infty$, as inferred from (5.16.76), we also have from (5.19.20),

$$\left[ m \frac{\partial}{\partial m} + \beta(\alpha) \left( \alpha \frac{\partial}{\partial \alpha} - 1 \right) \right] Z_3 = 0. \tag{5.19.40}$$

A very attractive feature of the expansion in (5.19.33), which may provide a hint to determine the value of the fine-structure constant $\alpha$, if the latter is a zero of the beta function: $\beta(\alpha) = 0$,[99] i.e., formally from (5.19.24), $\psi(q(\alpha)) = 0$, then from the just mentioned expansion we may infer that $\alpha\, d_{\text{ren}}^{\text{as}}(Q^2/m^2, \alpha) = q(\alpha) = \alpha_0$. Such a solution would mean that $1/Z_3$ is also finite. This solution, in turn, implies that the interval of integration in (5.19.39) degenerates to a point at which $\psi(\alpha_0)$ vanishes. Not considering all possibilities, another situation that may arise is in which the integral in (5.19.39) would not diverge until $\alpha\, d_{\text{ren}}^{\text{as}}(Q^2) \to \infty$ as $Q^2 \to \infty$ and $1/Z_3$, as well as $\alpha_0$, would be infinite. The fine-structure constant would then, of course, be undetermined. Finally, one may also consider the situation, where $\alpha\, d_{\text{ren}}^{\text{as}}(Q^2) \to \alpha_0$, with the latter finite, and the Gell-Mann and Low function $\psi(z)$ develops a zero at such a point for consistency with (5.19.39). In this case $\alpha$ would be undetermined, and the series would sum up to a non-trivial function of $\ln(Q^2/m^2)$ which approaches $\alpha_0$ as $Q^2 \to \infty$. A key property of QED is that the slope of $\beta(\alpha)$ is positive near the origin, unlike non-abelian gauge theory ones, such as QCD, in which it is negative leading to a decrease of the effective charge with increase of energy, instead of increasing as in QED, a concept which is referred to as asymptotic freedom . This will be investigated in the next chapter.[100]

Unfortunately, the Gell-Mann and Low function of QED may not have a non-trivial zero,[101] and one would then expect that $e_0$, which is formally defined as the physically unattainable charge "measured" right at the core of an electron, to be infinite. The divergence encountered in this limit is of no surprise as our theories are not justified to describe nature at absolute zero distances. It is rather surprising that some practitioners, let alone the others, still find it surprising that one may encounter such an infinity when our theories are extended all the way with no limit. On the other hand, to see how robust QED is, note from (5.19.33), (5.19.22), (5.19.35), or directly from (5.19.14), (5.19.15), that to lowest order in $\alpha$, $(d_{\text{ren}}^{-1})^{\text{as}}(Q^2/m^2, \alpha) = 1 - (\alpha/3\pi)\ln(Q^2/m^2)$, leading to an unphysical pole at $Q^2 = m^2 \exp(3\pi/\alpha)$, referred as a Landau ghost. This seems to validate QED down to small distances corresponding to enormous energies $< m\,e^{645}$. Of course this is an estimate obtained by truncating an infinite series. But even with such a naïve estimate, this indicates, nevertheless, that QED may be valid even to corresponding experimentally unattainable small distances. At such distances, gravitation, in any case, may

---

[99] See Adler [1], and, e.g., Manoukian [44].

[100] One may similarly carry out an analysis of the wavefunction renormalization constant $Z_2$ and of the unrenormalized mass $m_0$ (see, e.g., [1]).

[101] See, e.g., Krasnikov [38].

play an important role, and extrapolating to such distances, presumably, requires detailed generalizations of our physical theories. Operationally, QED is a very successful theory and is one of our most cherished ones. By eliminating the physically unattainable parameters $(m_0, e_0)$ in favor of the physically measured ones $(m, e)$, as it should be done for a sensible description of the theory, and simultaneously carrying out wavefunction renormalizations for the corresponding particles, as a proper normalization of their wavefunctions, for their appropriate descriptions as "free particles" when impinging on detectors, the theory turns out to be finite.

## 5.19.2   The Fine-Structure Effective Coupling at High Energy Corresponding to the Mass of the Neutral $Z^0$ Vector Boson

At present available high energies, a standard of the fine-structure coupling is taken at an energy equal to the mass of the neutral $Z^0$ vector boson: $M_z = 91.188$ GeV. An estimate of its value is given here based on all the charged leptons and all those contributing quarks of the three generations. The effective fine-structure at an energy squared $Q^2$ may be written from (5.19.5), (5.19.6), (5.19.7) and (5.19.8) as

$$\alpha(Q^2) = \alpha d_{\text{ren}}(Q^2) = \frac{\alpha}{1 + \alpha \left[\pi(Q^2) - \pi(0)\right]}. \tag{5.19.41}$$

or

$$\frac{1}{\alpha(Q^2)} = \frac{1}{\alpha}\left(1 + \alpha \left[\pi(Q^2) - \pi(0)\right]\right). \tag{5.19.42}$$

As long as $\alpha \ln(Q^2/m^2) \ll 1$, for $m^2 \ll Q^2$, we may use (5.19.14) to write

$$\frac{1}{\alpha(Q^2)} = \frac{1}{\alpha}\left(1 - \frac{\alpha}{3\pi}\left[\ln\left(\frac{Q^2}{m^2}\right) - \frac{5}{3}\right]\right). \tag{5.19.43}$$

We may take the contributions of the quark loops, with quarks corresponding to the three colors, as well as of the charged lepton loops, of the three generations, with charges and approximate masses given in Table 5.1 which contribute for $Q^2 = M_Z^2$:

**Table 5.1** Scaled charges and approximate masses (in GeV) of the quarks of masses $\ll M_Z^2$ and the charged leptons

| Particle | $u$ | $d$ | $s$ | $c$ | $b$ | $e$ | $\mu$ | $\tau$ |
|----------|-----|-----|-----|-----|-----|-----|-------|--------|
| e | 2/3 | −1/3 | −1/3 | 2/3 | −1/3 | −1 | −1 | −1 |
| Mass | 0.0024 | 0.0049 | 0.095 | 1.275 | ∼4.2 | 0.00051 | 0.1057 | 1.7768 |

where e denotes the (scaled) charge, and note that the mass of the top quark $\gg$ $M_z = 91.188\,\text{GeV}$, and decouples from the analysis.[102] We may then write the contributions of the above quarks and leptons to the fine-structure constant at $Q^2 = M_Z^2$, with a factor 3 for the three quark colors, as

$$\frac{1}{\alpha(M_Z^2)} = \frac{1}{\alpha} - \frac{1}{3\pi}\left\{ \sum_{i,\text{quarks}} 3\,e_i^2\left[\ln\left(\frac{M_Z^2}{m_i^2}\right) - \frac{5}{3}\right] + \sum_{i,\text{leptons}} e_i^2\left[\ln\left(\frac{M_Z^2}{m_i^2}\right) - \frac{5}{3}\right] \right\},$$

(5.19.44)

leading to the approximate anticipated value $1/\alpha(M_Z^2) \simeq 128$.[103]

## 5.20 Scalar Boson Electrodynamics, Effective Action and Spontaneous Symmetry Breaking

When the symmetry of the action is not shared by the vacuum, the symmetry is said to be spontaneously broken. An interesting consequence of this is that for a spontaneously broken continuous symmetry, the theory involves massless spin zero boson(s), referred as Goldstone bosons. As we will see in a spontaneously broken symmetry, the vacuum is degenerate. Considering a continuous global transformation, we investigate in Sect. 5.20.2 to see how these Goldstone bosons arise. In gauge theories, abelian or non-abelian, with a priori massless vector bosons, we have invariance under local transformations, and the broken symmetry is local, with underlying gauge transformations existing, in particular, between charged scalar fields and the vector bosons. In these cases, the Goldstone bosons combine with the massless vector bosons resulting in a theory in which the vector bosons become massive borrowing the zeroth component of their helicities from the Goldstone bosons themselves, and the latter now become unobservable. This is worked out in a variation of scalar boson electrodynamics. The underlying mechanism for this transmutation is referred to as the Higgs mechanism. In the final theory, the number of degrees of freedom, associated with the fields, are the same as the original version. The importance of this result in the renormalization program, should be emphasized and the underlying idea is quite simple. We recall that the propagator of a massless vector boson is given by

$$D^{\mu\nu}(Q) = \left(\eta^{\mu\nu} - \frac{Q^\mu Q^\nu}{Q^2}\right)\frac{1}{Q^2 - i\epsilon},$$

(5.20.1)

---

[102]This is a consequence of the decoupling theorem and one carries out the analysis as if heavy masses in comparison to the energy of concern are not part of the dynamics. For intricacies, conditions and a proof of the decoupling theorem, see Manoukian [46, 47, 49] and other references therein.

[103]See below (6.16.22) for experimental value.

up to gauge terms, while for a massive one with mass $M$, we have from (4.7.20)

$$\Delta_+^{\mu\nu}(Q) = \left(\eta^{\mu\nu} + \frac{Q^\mu Q^\nu}{M^2}\right) \frac{1}{Q^2 + M^2 - i\epsilon}. \tag{5.20.2}$$

The massless propagator gives rise to damping $\sim 1/Q^2$ at high energies and is a very welcome property in establishing renormalizability of a theory. The massive one goes to a constant at high energies and renders the renormalizability of a theory obscure. On the other hand, if the theory is gauge invariant, one may hope to infer renormalizability of the theory from its massless vector boson counterpart. This has been a key result in the success of present non-abelian gauge theories with massive vector bosons.

In the following subsection, we consider a change of scalar field variables of integrations in a path integral of a generating functional, needed to establish the so-called Goldstone Theorem. Spontaneous symmetry breaking in the abelian case is treated in the following subsection with the generation of a massive vector boson. Spontaneous symmetry breaking in non-abelian gauge theories is considered in the next chapter.

## 5.20.1   Change of Real Field Variables of Integration in a Path Integral

Suppose we are given a Lagrangian density $\mathscr{L}$, which among other fields, depends on $n$ real scalar fields $\phi^1(x), \ldots, \phi^n(x)$, $n > 1$, and is invariant under the transformation

$$\phi^i(x) \rightarrow \left(\delta^{ij} + i\,\delta\varepsilon\,(t)^{ij}\right)\phi^j(x) = \widetilde{\phi}^j(x), \quad (t)^{ij} = -(t)^{ji}, \tag{5.20.3}$$

where $t$ is a Hermitian matrix and, due to the reality of the fields $\phi^1, \ldots, \phi^n$, it is *anti-symmetric* as stated above, and $\delta\varepsilon$ is infinitesimal.

We couple the fields $\phi^1, \ldots, \phi^n$ to external sources $J^1, \ldots, J^n$, and introduce the lagrangian density

$$\mathscr{L}_1(x) = \mathscr{L}(x) + J^i(x)\phi^i(x). \tag{5.20.4}$$

Under the transformation of the real fields in (5.20.3),

$$\mathscr{L}_1 \rightarrow \mathscr{L}_1 + i\,\delta\epsilon\, J^i\,(t)^{ij}\phi^j = \widetilde{\mathscr{L}}_1 \tag{5.20.5}$$

We investigate the nature of the path integral representation of the amplitude given below, associated with the Lagrangian $\mathscr{L}_1$, under a *change of the variables* of the field variables of integrations $\phi^1, \ldots, \phi^n$, up to first order in $\delta\varepsilon$, as defined in (5.20.3):

$$\langle\, 0_+ \mid 0_-\rangle = e^{iW[\mathbf{J}]}, \qquad \frac{\delta}{\delta J^j(x)} W = \frac{\langle\, 0_+|\phi^j(x) \mid 0_-\rangle}{\langle\, 0_+ \mid 0_-\rangle} \equiv \langle\phi^j(x)\rangle. \qquad (5.20.6)$$

where $\mathbf{J} = (J^1,\ldots,J^n)$. In particular, we note that due to the anti-symmetry of the matrix $t$, the Jacobian of the transformation will involve the determinant of a matrix with ones along the diagonal and with all the elements off the diagonal being of order $\delta\epsilon$. That is, in particular, $d\widetilde{\phi}^1(x)\, d\widetilde{\phi}^2(x) = (1 + \mathcal{O}(\delta\varepsilon)^2)d\phi^1(x)\, d\phi^2(x)$. Hence for part of the measure of the path integral, as a factor, involving the scalar fields above, we have, up to first order in $\delta\epsilon$,

$$\prod_i [\,\mathscr{D}\widetilde{\phi}^i\,] = \prod_i [\,\mathscr{D}\phi^i\,]. \qquad (5.20.7)$$

Also from (5.20.5),

$$e^{i\int(\mathrm{d}y)\widetilde{\mathscr{L}}_1(y)} = \left[1 - \int (\mathrm{d}x)\, \delta\varepsilon\, J^i(x)\, t^{ij}\, \phi^j(x)\right] e^{i\int(\mathrm{d}y)\widetilde{\mathscr{L}}_1(y)}. \qquad (5.20.8)$$

Hence, invoking invariance, to first order in $\delta\varepsilon$, we have

$$e^{iW[\mathbf{J}]} = \left[1 - \delta\epsilon \int (\mathrm{d}x)\, J^i(x)\, t^{ij}\, \langle\phi^j(x)\rangle\right] e^{iW[\mathbf{J}]}, \qquad (5.20.9)$$

leading to the equation

$$\int (\mathrm{d}x)\, J^i(x)\, t^{ij}\langle\phi^j(x)\rangle = 0. \qquad (5.20.10)$$

## 5.20.2   Goldstone Bosons and Spontaneous Symmetry Breaking

At this stage we carry out a so-called Legendre transform of $W[\mathbf{J}]$, as a functional of $\mathbf{J}$, to a functional $\Gamma[\langle\boldsymbol{\phi}\rangle]$ of $\langle\boldsymbol{\phi}\rangle$, which is defined by

$$\Gamma[\langle\boldsymbol{\phi}\rangle] = W[\mathbf{J}] - \int (\mathrm{d}x')\, J^j(x')\langle\phi^j(x')\rangle, \qquad (5.20.11)$$

referred to as an effective action, as it coincides with the classical action $\int(\mathrm{d}x)\,\mathscr{L}_c(x)$ written, in particular, in terms of $\langle\boldsymbol{\phi}\rangle$, when radiative corrections are neglected.

From the chain rule

$$\frac{\delta W[\mathbf{J}]}{\delta\langle\phi^i(x)\rangle} = \int (\mathrm{d}x')\, \frac{\delta W[\mathbf{J}]}{\delta J^j(x')} \frac{\delta J^j(x')}{\delta\langle\phi^i(x)\rangle}, \qquad (5.20.12)$$

from which we obtain

$$\frac{\delta \Gamma[\langle \phi \rangle]}{\delta \langle \phi^i(x) \rangle} = -J^i(x), \qquad (5.20.13)$$

and the basic Eq. (5.20.10) takes the form

$$\int (dx) \frac{\delta \Gamma[\langle \phi \rangle]}{\delta \langle \phi^i(x) \rangle} t^{ij} \langle \phi^j(x) \rangle = 0. \qquad (5.20.14)$$

We note that

$$\delta^{ik} \delta^{(4)}(x, x') = \frac{\delta J^i(x)}{\delta J^k(x')} = \int (dy) \frac{\delta J^i(x)}{\delta \langle \phi^j(y) \rangle} \frac{\delta \langle \phi^j(y) \rangle}{\delta J^k(x')} = \int (dy) \frac{\delta J^i(x)}{\delta \langle \phi^j(y) \rangle} \Delta_{jk}(y, x'), \qquad (5.20.15)$$

where the $\Delta_{jk}(y, x')$ denote propagators (correlations functions) of the fields $\phi^i$, and the above equation leads to

$$\frac{\delta J^i(x)}{\delta \langle \phi^j(y) \rangle} = \Delta_{ij}^{-1}(x, y). \qquad (5.20.16)$$

Hence (5.20.13) gives

$$\frac{\delta^2 \Gamma[\langle \phi \rangle]}{\delta \langle \phi^i(x) \rangle \delta \langle \phi^j(y) \rangle} = -\Delta_{ij}^{-1}(x, y). \qquad (5.20.17)$$

For a time independent sources $\mathbf{J}(\mathbf{x})$, $\langle 0_+ | 0_- \rangle$ is described through a time evolution from time $-\infty$ to $+\infty$. Accordingly, $w[\mathbf{J}]$ would then necessarily involve the multiplicative factor $T \to \infty$, corresponding to the total time just described, due to the unitarity of the time evolution operator. In the other extreme when $\mathbf{J}$ is independent of the space variable as well, then $w[\mathbf{J}]$ will involve the multiplicative factor $\Omega$, denoting the spacetime volume, and we recover a translational invariant theory. Finally, in the limit $\mathbf{J} \to \mathbf{0}$, we may write

$$\Gamma[\langle \phi \rangle] = -\Omega V_{\text{eff}}, \qquad (5.20.18)$$

where $V_{\text{eff}}$ is referred to as the effective potential, and (5.20.14) takes the form[104]

$$\frac{\partial V_{\text{eff}}}{\partial \phi_c^i} t^{ij} \phi_c^j = 0, \qquad (5.20.19)$$

---

[104]Note that functional derivatives have now become simply partial derivatives.

where due to translational invariance, the matrix elements $\langle \phi^i(x) \rangle$ become simply expressions independent of $x$, which we have conveniently denoted by $\phi^i_c$. Upon taking another derivative of (5.20.19), with respect to $\phi^k_c$, we obtain

$$\frac{\partial^2 V_{\text{eff}}}{\partial \phi^k_c \, \partial \phi^i_c} \, t^{ij} \, \phi^j_c \bigg|_{\phi_c = \overline{\phi}_c} = 0, \qquad (5.20.20)$$

where

$$\frac{\partial V_{\text{eff}}}{\partial \phi^i_c} \bigg|_{\phi_c = \overline{\phi}_c} = 0, \quad i = 1, \dots, n, \qquad (5.20.21)$$

and the $\overline{\phi}_c$ are at the minimum of the effective potential. On the other hand

$$\int (\mathrm{d}x)(\mathrm{d}y) \, \Delta^{-1}_{ij}(x - y) = \Omega \, \widetilde{\Delta}^{-1}_{ij}(0), \qquad (5.20.22)$$

where the $\widetilde{\Delta}^{-1}_{ij}(0)$ denote the inverse of the propagator functions in the *momentum* description at *zero* momentum.

From (5.20.17), (5.20.18), (5.20.22), we may then simply replace (5.20.20) by

$$\widetilde{\Delta}^{-1}_{ki}(0) \, t^{ij} \, \overline{\phi}^j_c = 0. \qquad (5.20.23)$$

Accordingly, for non-vanishing $t^{ij} \overline{\phi}^j_c$, for some $i$, corresponding to spontaneous symmetry breaking, the $\widetilde{\Delta}^{-1}_{ki}(0)$ vanish, indicating the presence of zero mass particles in the theory, referred to as Goldstone bosons. This is the content of the celebrated Goldstone Theorem.[105]

It is interesting to see how the symmetry of the Lagrangian density $\mathscr{L}$ is broken by the vacuum. To this end, from Wigner's symmetry transformations (see Sect. 2.1), the continuous transformation in (5.20.3) is implemented, in the vector space in which the field operators act, via a unitary operator, which for infinitesimal transformations would read as $U = I - i\delta\varepsilon \, G$, where $G$ is the generator of transformation - an operator. Then in matrix form we have the relation: $U^{-1}\phi \, U = (I + i\delta\varepsilon \, t) \, \phi$ which from (5.20.3) yields the relation

$$[G, \phi] = t \phi, \qquad (5.20.24)$$

where we recall that $t$ is a (numerical) matrix. Upon taking the vacuum expectation value of the above, we obtain

$$\langle \text{vac}| \, G\phi - \phi G \, |\text{vac}\rangle = t \langle \text{vac}|\phi|\text{vac}\rangle = t \overline{\phi}_c. \qquad (5.20.25)$$

---

[105]Goldstone [28] and Goldstone et al. [29].

Hence for a vector $t\overline{\phi}_c$ which is non-zero in (5.20.23), we cannot have $G|\text{vac}\rangle$ equal to zero for all of its matrix elements, showing that the symmetry is broken by the vacuum.

As an illustration of the above result, the example of interest here is the classic one with the following Lagrangian density involving a complex scalar field:

$$\mathscr{L} = -\partial^\mu \Phi^\dagger \partial_\mu \Phi + \mu^2 \, \Phi^\dagger \Phi - \lambda (\Phi^\dagger \Phi)^2, \tag{5.20.26}$$

$$V = -\mu^2 \, \Phi^\dagger \Phi + \lambda (\Phi^\dagger \Phi)^2. \tag{5.20.27}$$

First note that the coefficient of $\mu^2$ in the Lagrangian density in (5.20.26), is a "+" rather than a "−", and this is what we need for spontaneous symmetry breaking, i.e., for the field $\Phi$ to develop a non vanishing expectation value in the vacuum.[106] Also to avoid problems with unboundedness of the corresponding Hamiltonian from below, $\lambda$ is taken to be positive definite. Upon writing $\Phi = (\phi_1 + i\phi_2)/\sqrt{2}$, the above Lagrangian density becomes

$$\mathscr{L} = -\frac{1}{2} \partial^\mu \phi_1 \partial_\mu \phi_1 - \frac{1}{2} \partial^\mu \phi_2 \partial_\mu \phi_2 - V, \tag{5.20.28}$$

where now

$$V = -\frac{\mu^2}{2}(\phi_1^2 + \phi_2^2) + \frac{\lambda}{4}(\phi_1^2 + \phi_2^2)^2. \tag{5.20.29}$$

One may rotate the system to bring $\langle\phi_2\rangle \equiv \phi_{2c}$ to zero, and $\langle\phi_1\rangle \equiv \phi_{1c}$ to some value $v$, which is determined by minimizing the potential. By neglecting radiative corrections, for simplicity, we have[107]

$$\frac{\partial V}{\partial \phi_{1c}} = -\mu^2 \, \phi_{1c} + \lambda \phi_{1c} \left(\phi_{1c}^2 + \phi_{2c}^2\right) = 0, \quad \lambda v^2 = \mu^2. \tag{5.20.30}$$

Upon writing $\phi_1 = \varphi_1 + v$, $\phi_2 = \varphi_2$, with $\langle\varphi_i\rangle = 0$, the quadratic part of the Lagrangian density in (5.20.28) is easily worked out to be

$$\mathscr{L}_{\text{quad}} = -\frac{1}{2} \partial^\mu \varphi_1 \partial_\mu \varphi_1 - \frac{1}{2} \partial^\mu \varphi_2 \partial_\mu \varphi_2 - \frac{(2\lambda v^2)}{2}\varphi_1^2 + 0.\varphi_2^2, \tag{5.20.31}$$

with $\varphi_2$ representing the Goldstone boson, involving *no* mass.

---

[106]We will neglect radiative corrections in the subsequent analysis. Although we are not considering radiative corrections here, the $\lambda(\Phi^\dagger\Phi)^2$ term in the Lagrangian density is important for renormalizability of the theory.

[107]Note that if the coefficient of $\mu^2$ is of opposite sign to the one in (5.20.26), one obtains $v = 0$.

Now we consider scalar boson electrodynamics with the sign of the coefficient of the parameter $\mu^2$ is as given in (5.20.26). We now have a *local* broken symmetry, and, as we will see below, the long range force, associated with the gauge field, becomes screened in which the Goldstone boson combines with the gauge field to create a massive one. To this end, the Lagrangian density in question is given by[108]

$$\mathscr{L} = -\frac{1}{4} F^{\mu\nu} F_{\mu\nu} - \left(\partial^{\mu} + \mathrm{i}\,\mathrm{e}A^{\mu}\right)\Phi^{\dagger}\left(\partial^{\mu} - \mathrm{i}\,\mathrm{e}A^{\mu}\right)\Phi - V, \qquad (5.20.32)$$

where $V$ is given in (5.20.27), $F^{\mu\nu} = \partial^{\mu}A^{\nu} - \partial^{\nu}A^{\mu}$. Upon defining the field

$$V^{\mu} = A^{\mu} + \partial^{\mu}\phi_2/v\mathrm{e}, \qquad G^{\mu\nu} = \partial^{\mu}V^{\nu} - \partial^{\nu}V^{\mu}, \qquad (5.20.33)$$

the quadratic part of the above Lagrangian density is easily worked out to be

$$\mathscr{L}_{\text{quad}} = -\frac{1}{4} G^{\mu\nu} G_{\mu\nu} - \frac{\mathrm{e}^2 v^2}{2} V^{\mu} V_{\mu} - \frac{1}{2} \partial^{\mu}\varphi_1 \partial_{\mu}\varphi_1 - \frac{(2\lambda v^2)}{2} \varphi_1^2, \qquad (5.20.34)$$

and the Goldstone boson combined with the gauge field generating a massive vector boson. The resulting theory describes a massive neutral scalar particle and a massive vector boson. The scalar field $\phi_1(x)$ with non-vanishing vacuum expectation value $v$ is usually referred to as a Higgs field, and the resulting massive neutral spin zero particle as a Higgs boson. Finally the mechanism for the generation of the massive particles, as just described, is referred as the Higgs mechanism.[109] Also note that the spin zero massive neutral particle is described by the field $\varphi_1$, i.e., after the vacuum expectation value of $\phi_1$ has been subtracted out from the latter.

It is remarkable, that even if $\mu^2 = 0$, radiative corrections lead to spontaneous symmetry breaking generating again a theory with one massive neutral scalar particle and a massive vector boson.[110]

# Problems

**5.1** Derive the general expressions of the polarization vectors in (5.2.7), (5.2.8), (5.2.9), (5.2.10), (5.2.11), (5.2.12) and (5.2.13).

**5.2** Verify directly from (5.3.8) and (5.3.5) that $\partial_{\mu}\langle 0_{+}|A^{\mu}(x) \mid 0_{-}\rangle = \lambda\langle 0_{+}|\chi(x) \mid 0_{-}\rangle$, consistent with the gauge condition (5.3.1).

**5.3** Derive the key identity (5.4.15) in the Casimir Effect problem.

---

[108]Since we are not considering renormalizability aspects here, we use the notation for e.

[109]Englert and Brout [21] and Higgs [31–33]. See also Anderson [2].

[110]For the relevant details see Coleman and Weinberg [17].

**5.4** Consider two independent, but identical, assumed non-interacting, sources, emitting photons in uniformly arbitrary directions, situated at different positions $\mathbf{R}_1$, $\mathbf{R}_2$, and described by two current distributions giving rise to a resultant current distribution given by

$$J^\mu(x) = \sum_{j=1,2} \int \frac{(dk)}{(2\pi)^4} e^{i\mathbf{k}\cdot(\mathbf{x}-\mathbf{R}_j)} e^{-ik^0 x^0} J^\mu(k), \quad \text{where}$$

$$\mathbf{R}_1 = (0,0,a/2), \quad \mathbf{R}_2 = (0,0,-a/2), \quad \text{and} \quad J^\mu(k)\,e^*_{\mu\lambda} \equiv \frac{1}{\sqrt{2}} f(|\mathbf{k}|).$$

The angular correlation of the emerging photons is defined by[111]

$$\langle c \rangle = \int d\Omega_1 \int d\Omega_2 \, \mathbf{n}_1 \cdot \mathbf{n}_2 \, f_{ka}(\mathbf{n}_1, \mathbf{n}_2).$$

where $\mathbf{n}_i = \mathbf{k}_i/|\mathbf{k}_i|$ $i = 1,2$, $|\mathbf{k}_1| = |\mathbf{k}_2| \equiv k$, and $f_{ka}(\mathbf{n}_1, \mathbf{n}_2)$ is the corresponding probability density. Evaluate $\langle c \rangle$ as a function of $ka$. Argue that for small distance separation $a$, the two sources may not be considered as independent and non-interacting, and on other hand, for large separations, the correlation of photons is not a meaningful concept either. In the intermediate meaningful range, a large correlation, implies that the photons tend to travel in the same direction. (See [10, 54, 55].)

**5.5** Establish the identity in (5.7.25).

**5.6** Show that (5.7.7) may be rewritten as $\partial_\mu \langle\, 0_+ | A^\mu(x) \,|\, 0_- \rangle$ being equal to

$$\lambda \int (dx')D_+(x-x')\partial'_\mu J^\mu(x')\langle\, 0_+ \,|\, 0_- \rangle + i\lambda\, e_0 \int (dx')(dx'')\, D_+(x,x')$$

$$\times \left[ \overline{\eta}(x'')\, S_+(x'',x'; e_0\widehat{A})\, \eta(x') - \overline{\eta}(x')\, S_+(x',x''; e_0\widehat{A})\, \eta(x'') \right]\langle\, 0_+|0_- \rangle,$$

and verify that this equality also follows upon the application of the expression for $\langle\, 0_+ \,|\, 0_- \rangle$ in (5.7.27).

**5.7** Derive the explicit expressions for $a_0$, $a_1$, $a_2$, $a_3$ in the expansion of $\langle\, 0_+ \,|\, 0_- \rangle$ in terms of connected parts with external lines as given, respectively in (5.8.8), (5.8.9), (5.8.10), (5.8.11).

**5.8** Show that the summation over spins of the expression in (5.9.31) is given by (5.9.32).

---

[111]This was introduced by Bialinicki-Birula and Bialinicka-Birula [10].

**5.9** Consider the polarization correlations of the outgoing electrons in $e^- e^- \rightarrow e^- e^-$ scattering, in which the initial state has been prepared with one of the ingoing electrons moving with momentum $\mathbf{p}_1 = \mathbf{p}$ in the $y$-direction, with spin along the $z$-axis, and the other moving in the opposite direction with momentum $\mathbf{p}_2 = -\mathbf{p}$, and spin along the $-z$ direction. Suppose that the outgoing electrons have emerged with momenta in opposite direction of the $x$-axis. The spins are measured along unit vectors $\mathbf{n}_1$, $\mathbf{n}_2$, specified by angles $\chi_1$, $\chi_2$, as shown in the figure for some possible outcomes:

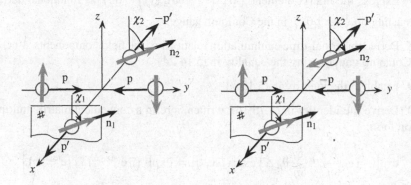

Show that the probability of simultaneous measurements of the spins along $\mathbf{n}_1$ $\mathbf{n}_2$ is given by $P[\chi_1, \chi_2] = \sin^2[(\chi_1 - \chi_2)/2]/2$. And if only one of the spins is measured, then $P[\chi_1, -] = P[-, \chi_2] = 1/2$. Interpret these results.

**5.10** Establish the identity $(p^2 + m^2) = 2m(\gamma p + m) - (\gamma p + m)^2$.

**5.11** The logarithmically ultraviolet divergent constant, that typically occurs in QED is given by the integral $C_{\mathrm{uv}}$ in (5.10.21). The latter may be rigorously defined with an ultraviolet cut-off as in (5.10.27). Show as indicated in that equation, with the cut-off, $C_{\mathrm{uv}} = [\ln(\Lambda^2/m^2) - 1]$ for $\Lambda^2 \rightarrow \infty$.

**5.12** Two basic infrared divergent constants that occur in QED, are given by the following two integrals, defined in terms of a fictitious mass $\mu$ of the photon,

$$C_{\mathrm{ir}} = \int_0^1 \frac{x \, dx}{[x^2 + \frac{\mu^2}{m^2}(1 - x)]}, \quad D_{\mathrm{ir}} = \int_0^1 \frac{x^3 \, dx}{[x^2 + \frac{\mu^2}{m^2}(1 - x)]^2}.$$

Show that for $(\mu^2/m^2) \rightarrow 0$, $C_{\mathrm{ir}} = -(1/2) \ln(\mu^2/m^2)$, $D_{\mathrm{ir}} = C_{\mathrm{ir}} - (1/2) = -(1/2)[\ln(\mu^2/m^2) + 1]$.

**5.13** Evaluate $\Sigma_1(p)$ in (5.10.14) to obtain the expression in (5.10.17).

**5.14** Use (5.10.57), to derive the expression of the modification of the Coulomb interaction, to second order, due to radiative corrections, at small distances, i.e., for $m |\mathbf{x}| \ll 1/2$, as given in (5.10.66).

**5.15** Show that $\bar{u}(\mathbf{p}',\sigma')\,\gamma^\mu\,\Pi_{\mu\nu}(k)\,(1/k^2)u(\mathbf{p},\sigma) = \bar{u}(\mathbf{p}',\sigma')\,\gamma_\nu\,u(\mathbf{p},\sigma)\,\Pi(k^2)$, which $\rightarrow \bar{u}(\mathbf{p}',\sigma')\gamma_\nu u(\mathbf{p},\sigma)\,(1-Z_3)$, for $k^2 \rightarrow 0$, where $\Pi_{\mu\nu}(k)$ is given in (5.10.42).

**5.16** Suppose that the charge distribution of the proton is given by $\rho(\mathbf{x}) = (1/8\pi\gamma^3)\,e^{-|\mathbf{x}|/\gamma}$, where $1/\gamma^2 = 0.81M_\mathrm{p}^2$, where $M_\mathrm{p}$ is the mass of the proton. Estimate the energy shift in a hydrogen $\ell = 0-$ energy level due to this distribution. Work in units with the Coulomb potential given by $-\alpha/|\mathbf{x}|$.

**5.17** Express the matrix element $\langle\,0_+|\big(F^{k\nu}(x)A_\nu(y)\big)_+|\,0_-\rangle$ as functional derivatives acting on $\langle\,0_+\mid 0_-\rangle$ in the Coulomb gauge.

**5.18** Derive the equal-time commutation relations of the field components $A^i(x)$ in the Coulomb gauge, using the equality in (5.14.24).

**5.19** Establish the identity in (5.15.7).

**5.20** Derive the identity in (5.15.16), written here in a convenient matrix multiplication form:

$$\exp\!\Big(\!-\mathrm{i}\,e_0\frac{\delta}{\delta\eta}\,\gamma^\mu\frac{\delta}{\delta\overline{\eta}}\,\partial_\mu\Lambda\Big)\exp[\mathrm{i}\,\overline{\eta}S_+\eta] = \exp\!\big[\mathrm{i}\,\big(\overline{\eta}\,\mathrm{e}^{-\mathrm{i}\,e_0\Lambda}\big)S_+\big(\mathrm{e}^{\mathrm{i}\,e_0\Lambda}\,\eta\big)\big].$$

**5.21** Show that in the absence of the external Fermi sources, the generating functionals in the Landau gauge, and in arbitrary covariant gauges, specified by the parameter $\lambda$, are simply related by (5.15.24).

**5.22** Show that the exact electron propagator, in an arbitrary covariant gauge, specified by the parameter $\lambda$, is related to the one in the Landau gauge, in the presence of a non-zero external current $J^\mu$, is as given in (5.15.26).

# References

1. Adler, S. L. (1972). Short-distance behavior of quantum electrodynamics and an eigenvalue condition for $\alpha$. *Physical Review D, 5*, 3021–3047.
2. Anderson, P. W. (1963). Plasmons, gauge invariance, and Mass. *Physical Review, 130*, 439–442.
3. Baker, M., & Johnson, K. (1969). Quantum electrodynamics at small distances. *Physical Review D, 183*, 1292–1299.
4. Baker, M., & Lee, C. (1977). Overlapping-divergence-free Skeleton expansion in non-Abelian gauge theories. *Physical Review D, 15*, 2201–2234.
5. Bell, J. S. (1989). *Speakable and unspeakable in quantum mechanics.* Cambridge: Cambridge University Press.
6. Bell, J. S., & Aspect, A. (2004). *Speakable and unspeakable in quantum mechanics: Collected papers on quantum philosophy* (2nd ed.). Cambridge: Cambridge University Press.
7. Beringer, J., et al. (2012). Particle data group. *Physical Review D, 86*, 010001.
8. Bethe, H. (1947). The electromagnetic shift of energy levels. *Physical Review, 72*, 339–341.

9. Bethe, H. A., Brown, L. M., & Stehn, J. R. (1950). Numerical value of the lamb shift. *Physical Review, 77*, 370–374.

10. Bialinicki-Birula, I., & Bialinicka-Birula, Z. (1990). Angular correlations of photons. *Physical Review, A42*, 2829–2838.

11. Bjorken, J. D., & Drell, S. D. (1965). *Relativistic quantum fields.* New York/San Francisco/London: McGraw-Hill.

12. Bogoliubov, N. N., & Shirkov, D. V. (1959). *Introduction to the theory of quantized fields.* New York: Interscience.

13. Bressi, G., Carugno, G., Onofrio, R., & Ruoso, G. (2002). Measurement of the Casimir force between parallel metallic surfaces. *Physical Review Letters, 88*, 041804:1–4.

14. Brown, L. S. (1995). *Quantum field theory.* Cambridge: Cambridge University Press.

15. Callan, C. G. (1970). Broken scale invariance in scalar field theory. *Physical Review, D2*, 1541–1547.

16. Casimir, H. G. (1948). On the attraction between two perfectly conducting plates. *Proceedings of the Koninklijke Nederlandse Akademie van Wetenschappen, B 51*, 793–795.

17. Coleman, S., & Weinberg, E. (1973). Radiative corrections as the orign of spontaneous symmetry breaking. *Physical Review, D7*, 1888–1910.

18. de Rafaël, E., & Rosner, J. L. (1974). Short-distance behavior of quantum electrodynanics and the Callan-Symanzik equation for the photon. *Annals of Physics (NY), 82*, 369–406.

19. Elizalde, E. (1995). *Ten physical applications of spectral zeta functions.* Berlin: Springer.

20. Elizalde, E., Odintsov, S. D., Romeo, A., Bytsenko, A. A., & Zerbini, S. (1994). *Zeta regularization techniques with applications.* Singapore: World Scientific.

21. Englert, F., & Brout, R. (1964). Broken symmetry and the mass of gauge vector bosons. *Physical Review Letters, 13*, 321–323.

22. Erickson, G. W., & Yennie, D. R. (1965). Radiative level shifts, I. Formulation and lowest order lamb shift. *Annals of Physics (NY), 35*, 271–313.

23. Erler, J. (1999). Calculation of the QED coupling $\hat{\alpha}(M_Z)$ in the modified minimal subtraction scheme. *Physical Review D, 59*, 054008, 1–7.

24. Fazio, G. G., Jelly, J. V., & Charman, W. N. (1970). Generation of Cherenkov light flashes by cosmic radiation within the eyes of the Apollo astronauts. *Nature, 228*, 260 261.

25. Foley, H. M., & Kusch, P. (1948). The magnetic moment of the electron. *Physical Review, 74*, 250–263.

26. Fox, J. A., & Yennie, D. R. (1973). Some formal aspects of the lamb shift problem. *Annals of Physics (NY), 81*, 438–480.

27. Gell-Mann, M., & Low, F. E. (1954). Quantum electrodynamics at small distances. *Physical Review, 95*, 1300–1312.

28. Goldstone, J. (1961). Field theories with ≪ superconductor ≫ solutions. *Il Nuovo Cimento, 19*, 154–164.

29. Goldstone, J., Salam A., & Weinberg, S. (1962). Broken symmetries. *Physical Review, 127*, 965–970.

30. Gorishnii, S. G., et al. (1991). The analytic four loop corrections to the QED beta function in the MS scheme and the QED Psi function: Total reevaluation. *Physics Letters, B256*, 81–86.

31. Higgs, P. W. (1964a). Broken symmetries, massles particles and gauge fields. *Physics Letters, 12*, 132–133.

32. Higgs, P. W. (1964b). Broken symmetries and the masses of gauge bosons. *Physical Review Letters, 13*, 508–509.

33. Higgs, P. W. (1966). Spontaneous symmetry breaking without massless particles. *Physical Review, 145*, 1156–1163.

34. Johnson, K. (1968). 9th Latin American Scool of Physics, Santiago de Chile. In K. Johnson & I. Saavedra (Eds.), *Solid state physics, nuclear physics, and particle physics.* New York: W. A. Benjamin

35. Källen, G. (1952). On the definition of renormalization constants in quantum electrodynamics. *Helvetica Physica Acta, 25*, 417–434.

36. Kenneth, O., Klich, I., Mann, A., & Revzen, M. (2002). Repulsive Casimir forces. *Physical Review Letters, 89*, 033001:1–4.

37. Kinoshita, T., & Yennie, D. R. (1990). High precision tests of quantum electrodynamics – an overview. In T. Kinoshita (Ed.), *Quantum electrodynamics: Advanced series on directions in high energy physics* (Vol. 7). Singapore: World Scientific.

38. Krasnikov, N. V. (1989). Is finite charge renormalization possible in quantum electrodynamics? *Physics Letters, B225*, 284–286.

39. Kusch, P., & Foley, H. M. (1947). Precision measurement of the ratio of the atomic 'g values' in the $^2P_{3/2}$ and $^2P_{1/2}$ states of Gallium. *Physical Review, 72*, 1256.

40. Lamb, W. E., Jr., & Retherford, R. C. (1947). Fine structure of the Hydrogen atom by a microwave method. *Zeitschrift für Physik, 72*, 241–243. Reprinted in Schwinger (1958).

41. Lamoreaux, S. K. (1997). Demonstration of the Casimir force in the 0.6 to 6 $\mu$m range. *Physical Review Letters, 78*, 5–8.

42. Lehmann, H. (1954). Properties of propagation functions and renormalization constants of quantized fields. *Il Nuovo Cimento, 11*, 342–357.

43. Lewis, G. N. (1926). The conservation of photons. *Nature, 118*, 874–875.

44. Manoukian, E. B. (1975). Fundamental identity for the infinite-order-zero nature in quantum electrodynamics. *Physical Review, D12*, 3365–3367.

45. Manoukian, E. B. (1983). *Renormalization*. New York/London/Paris: Academic.

46. Manoukian, E. B. (1984). Proof of the decoupling theorem of field theory in Minkowski space. *Journal of Mathematical Physics, 25*, 1519–1523.

47. Manoukian, E. B. (1985). Quantum action principle and path integrals for long-range interactions. *Nuovo Cimento, 90A*, 295–307.

48. Manoukian, E. B. (1986a). Action principle and quantization of gauge fields. *Physical Review, D34*, 3739–3749.

49. Manoukian, E. B. (1986b). Generalized conditions for the decoupling theorem of quantum field theory in Minkowski space with particles of vanishing small masses. *Journal of Mathematical Physics, 27*, 1879–1882.

50. Manoukian, E. B. (1987). Casimir effect, the gauge problem and seagull terms. *Journal of Physics A, 20*, 2827–2832.

51. Manoukian, E. B. (1987). On the relativistic invariance of QED in the Coulomb gauge and field transformations. *Journal of Physics, G13*, 1013–1021.

52. Manoukian, E. B. (1988a). Charged particle emission and detection sources in quantum field theory and infrared photons. *Fortschritte der Physik, 36*, 1–7.

53. Manoukian, E. B. (1988b). Gauge invariance properties of transition amplitudes in gauge theories. I. *International Journal of Theoretical Physics, 27*, 787–800.

54. Manoukian, E. B. (1992). Field-theoretical view of the angular correlation of photons. *Physical Review, A46*, 2962–2964.

55. Manoukian, E. B. (1994). Particle correlation in quantum field theory. *Fortschritte der Physik, 42*, 743–763.

56. Manoukian, E. B. (2006). *Quantum theory: A wide spectrum*. Dordrecht: Springer.

57. Manoukian, E. B. (2011). *Modern concepts and theorems of mathematical statistics*. New York: Springer. Paperback Edition.

58. Manoukian, E. B. (2015). Vacuum-to-vacuum transition amplitude and the classic radiation theory. *Radiation Physics and Chemistry, 106*, 268–270.

59. Manoukian, E. B., & Charuchittapan, D. (2000). Quantum electrodynamics of Čerenkov radiation at finite temperature. *International Journal of Theoretical Physics, 39*, 2197–2206.

60. Manoukian, E. B., & Siranan, S. (2005). Action principle and algebraic approach to gauge transformations in gauge theories. *International Journal of Theoretical Physics, 44*, 53–62.

61. Manoukian, E. B., & Yongram, N. (2004). Speed dependent polarization correlations in QED. *European Physical Journal, D 31*, 137–143.

62. McNulty, P. J., Pease, V. P., & Bond, V. P. (1976). Role of Cerenkov radiation in the eye-flashes observed by Apollo astronauts. *Life Sciences and Space Research, 14*, 205–217.

63. Mele, S. (2006). Measurements of the running of the electromagnetic coupling at LEP. In *XXVI Physics in Collision*, Búzios, Rio de Janeiro, 6–9 July.
64. Milton, K. A., et al. (2012). Repulsive Casimir and Casimir-Polder forces. *Journal of Physics, A 45*, 374006.
65. Ovsyannikov, L. V. (1956). General solution to renormalization group equations. *Doklady Akademii Nauk SSSR, 109*, 1112–1115.
66. Pinsky, L. S., et al. (1974). Light flashes observed by astronauts on Apollo 11 through Apollo 17. *Science, 183*, 957–959.
67. Schwartz, C., & Tiemann, J. J. (1959). New calculation of the numerical value of the lamb shift. *Annals of Physics (NY), 6*, 178–187.
68. Schwinger, J. (1948). On quantum-electrodynamics and the magnetic moment of the electron. *Physical Review, 73*, 416.
69. Schwinger, J. (Ed.) (1958). *Selected papers on quantum electrodynamics*. New York: Dover Publications.
70. Schwinger, J. (1969). *Particles and sources*. New York: Gordon and Breach.
71. Schwinger, J. (1973). *Particles, sources, and fields* (Vol. II). Reading: Addison-Wesley.
72. Schwinger, J. (1975). Casimir effect in source theory. *Letters in Mathematical Physics, 1*, 43–47.
73. Sparnaay, M. J. (1958). Measurement of attractive forces between flat plates. *Physica, 24*, 751–764.
74. Stueckelberg, E. C. G., & Peterman, A. (1953). La Normalisation des Constantes dans la Théorie des Quanta. *Helvetica Physica Acta, 26*, 499–520.
75. Symanzik, K. (1970). Small distance behaviour in field theory and power counting. *Communications in Mathematical Physics, 18*, 227–246.
76. Symanzik, K. (1971). Small distance behavior in field theory. In G. Höhler (Ed.), *Springer tracts in modern physics* (Vol. 57). New York: Springer.
77. Symanzik, K. (1971). Small distance behaviour analysis in field theory and Wilson expansions. *Communications in Mathematical Physics, 23*, 49–86.
78. Takahashi, Y. (1957). On the generalized ward identity. *Nuovo Cimento, 6*, 370–375.
79. Ward, J. C. (1950). An identity in quantum electrodynamics. *Physical Review, 78*, 182.
80. Weinberg, S. (1960). High-energy behavior in quantum field theory. *Physical Review, 118*, 838–849.
81. Yennie, D. R., & Suura, H. (1957). Higher order radiative corrections to electron scattering. *Physical Review, 105*, 1378–1382
82. Yongram, N., & Manoukian, E. B. (2003). Joint probabilities of photon polarization correlations in $e^+e^-$ annihilation. *International Journal of Theoretical Physics, 42*, 1755–1764.
83. Yongram, N., & Manoukian, E. B. (2013). Quantum field theory analysis of polarizations correlations, entanglement and Bell's inequality: Explicit processes. *Fortschritte der Physik, 61*, 668–684.

# Recommended Reading

1. DeWitt, B. (2014). *The global approach to quantum field theory*. Oxford: Oxford University Press.
2. Kinoshita, T. (Ed.) (1990). "Quantum electrodynamics": Advanced series on directions in high energy physics (Vol. 7). Singapore: World Scientific.
3. Manoukian, E. B. (1983). *Renormalization*. New York/London/Paris: Academic.
4. Manoukian, E. B. (1986). Action principle and quantization of gauge fields. *Physical Review, D34*, 3739–3749.

5. Manoukian, E. B. (2006). *Quantum theory: A wide spectrum*. Dordrecht: Springer.
6. Schwinger, J. (1973a). *Particles, sources, and fields* (Vol. II). Reading: Addison-Wesley.
7. Schwinger, J. (1973b). A report on quantum electrodynamics. In J. Mehra (Ed.), *The physicist's conception of nature*. Dordrecht/Holland: D. Reidel Publishing Company.
8. Weinberg, S. (1996). *The quantum theory of fields. II: Modern applications*. Cambridge: Cambridge University Press.
9. Yongram, N., Manoukian, E. B. (2013). Quantum field theory analysis of polarizations correlations, entanglement and Bell's inequality: Explicit processes. *Fortschritte der Physik, 61*, 668–684.

# Chapter 6
# Non-Abelian Gauge Theories

The present chapter[1] deals with the intricacies of non-abelian gauge field theories. We consider the extension of *local* gauge transformations of QED, with the gauge group U(1) of phase transformations, to SU($N$) groups. We recall that the group SU($N$) means the group of symmetry transformations consisting of $N \times N$ unitary matrices that are "special" in that they have determinant one. Unlike QED, SU($N$) involves ($N^2 - 1$) non-commuting generators, and hence field theories involving SU($N$) groups are referred to as non-abelian gauge field theories, with QED being an abelian one. Local gauge invariance imposed on a theory, under such an SU($N$) group, leads to the introduction of ($N^2 - 1$) interacting vector fields in a theory, referred to as Yang-Mills fields, which mediate interactions between particles carrying specific quantum numbers associated with the group. This happens in a similar way as in QED, where local gauge invariance leads to the introduction of ·the photon into a theory involving charged particles, and mediates the interaction between charged particles. A photon, however, is electrically neutral and does not have a direct self interaction. The vector particles of an SU($N$) group, however, do carry specific quantum numbers, such as charge, color, ... , and do respond directly to each other, i.e., they interact. We will see in Sect. 6.1, how such local symmetry groups of transformations are generated by insisting that one has the freedom in carrying transformations, depending explicitly on the spacetime point, at which a transformation is being carried out. These local gauge symmetry groups do not only assign specific quantum numbers to the particles in consideration but also give rise to the underlying dynamical theory of these particles, by being guided, in its development, by the QED theory.

The quantization of Yang-Mills field theories in the presence of matter is treated in Sects. 6.2 and 6.3. The first non-abelian gauge theory we consider is that of quantum chromodynamics as the modern theory of strong interaction and is

---

[1]It is a good idea to review the content of the introductory chapter of the book, especially of those aspects dealt with in this chapter.

© Springer International Publishing Switzerland 2016

E.B. Manoukian, *Quantum Field Theory I*, Graduate Texts in Physics,

DOI 10.1007/978-3-319-30939-2_6

developed in Sects. 6.4, 6.5, 6.6, 6.7, 6.8, 6.9, 6.10, 6.11, 6.12, and 6.13. It is based on the color group $SU(3)_C$, often denoted simply by $SU(3)$, involving quarks, coming in three colors, and 8 (massless) vector particles (the gluons). We will learn that there are many reasons why quarks should carry such a quantum number, and why three of them (Sect. 6.4). The renormalization group, which shows how the theory responds to scale transformations, in turn, leads to the description of the high-energy behavior of the theory simply in terms of an *effective coupling* which, thanks to the non-abelian gauge character of the theory, becomes weaker, with the increase of energy, and eventually becomes zero, a phenomenon referred to as asymptotic freedom. This is the subject of Sects. 6.6 and 6.7. We will see how the non-abelian gauge character of the theory with interacting gluons is responsible for this very welcome behavior of the theory at high energies. The smallness of the effective coupling constant at high energies allows one to carry out perturbation expansions in such a coupling at high energies such as in describing $e^- e^+$ annihilation (Sect. 6.8).

Asymptotic freedom is a key result in explaining several experiments such as the deep inelastic experiment (Sects. 6.9, 6.10, and 6.11) where a virtual photon imparts high energies to a proton, showing that a proton is considered to be consisting of weakly interacting quarks, quark-antiquark pairs and gluons, collectively called partons, within the extension of the proton. Some interesting aspects of quark and gluon splittings, as well as the formations of QCD jets, in $e^- e^+$ annihilation, are investigated as well (Sect. 6.10). Most commonly two jets are observed in opposite directions to a quark-antiquark produced in the CM frame of the annihilating $e^- e^+$ pair. Low energy (soft) gluons radiated by a parent quark (antiquark), give rise to an enhanced probability, due to an inherited infrared divergence as in QED (Sect. 5.12), when they are emitted in the same direction (collinear) as the parent quark (antiquark). Such a result, based on the observation of the 2 back-to-back jets events, not only gives a strong support of the existence of quarks but due to the orientations of the axis of the back-to-back jets imply the spin 1/2 character of quarks. As a consequence of asymptotic freedom, occasionally, i.e., with a smaller probability, due to the vanishing of the effective coupling at high energies, a high energy (hard) non-collinear gluon may be emitted as well leading to a three jets events and so on. As quarks are not observed, it is expected that the effective coupling between them becomes large (infrared slavery) when they separate beyond 1 fm, the approximate extension of a proton, and are thus confined within the proton. When a coupling becomes large, one has the tendency to carry out perturbation theory in the inverse of the coupling. Lattice gauge theory, where spacetime is considered to have a lattice structure, is quite suitable to do this and, in turn, allows one to investigate the nature of quark confinement. This underlying theory is developed in Sects. 6.12 and 6.13.

The electroweak theory, as a unification of the electromagnetic and weak interactions, based on the product group $SU(2) \times U(1)$ of leptons and the underlying vector bosons is treated in Sect. 6.14. Special attention is given to an experimental determination of $\sin^2 \theta_W$, where $\theta_W$ provides a measure of the mixing occurring between the electromagnetic and weak couplings. This section includes the study of the theory of spontaneous symmetry breaking, where masses are generated for some of the particles via the Higgs mechanism. Here by spontaneous symmetry breaking it is meant that a symmetry of the Hamiltonian is not obeyed by the vacuum. The problem of masses of neutrinos and the theory of "neutrino oscillations", very much relevant to neutrino masses and mass differences, are also presented. Quarks are incorporated into the theory in Sect. 6.15. Special attention is also given to the important aspect of the absence of anomalies in the standard model of elementary particles which is based on the product group $SU(3) \times SU(2) \times U(1)$, as investigated in Sect. 6.15, and is a key result for its renormalizability.

The standard model of elementary particles, based on the product of the above three groups, just mentioned, involves three different couplings, one for each group. The embedding of this product group into a larger group, such as $SU(5)$, leads to a unification of these couplings into a theory with a single coupling at some high energies, and the quantization of the electromagnetic fractional charges of quarks in units of $e/3$ emerges naturally. The underlying theory of this simplest grand unified theory is developed in Sect. 6.16 and its limitations are discussed.

## 6.1 Concept of Gauge Fields and Internal Degrees of Freedom: From Geometry to Dynamics

In a *geometrical* setting, gauge fields tell us something about systems without actually being considered as dynamical variables and, in turn, allow one to describe interactions between charged fields, or between fields carrying other internal quantum numbers, such as isospin and color, or may describe the interaction of all fields simply due to their energy-momentum content. In the present section, we consider aspects of gauge fields associated with internal quantum numbers. The gauge field corresponding to the last case will be treated in an accompanying volume by the author,[2] and is due to gravity felt by all particles.

We first consider the concept of a charged field, i.e., a non-Hermitian field $\chi(x)$, which, in turn, may be rewritten as $\chi = \chi^1 + i\chi^2$ *in* charge space (see Fig. 6.1). Upon a phase transformation by a phase factor $\vartheta(x)$ which may, in general, depend on spacetime coordinates, it is transformed into a field $\chi'(x)$, as shown in Fig. 6.1, and is given by

$$\chi(x) \rightarrow e^{i\vartheta(x)}\chi(x) = \chi'(x), \tag{6.1.1}$$

and is referred to as a gauge transformation. This may be equivalently achieved by the rotation of the coordinate system, of charge space, "set up" at spacetime point $x$, in opposite direction. Thus local gauge transformations of a charged field may

---

[2]Quantum Field Theory II: Introductions to Quantum Gravity, Supersymmetry, and String Theory, Springer (2016).

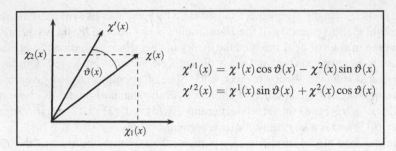

**Fig. 6.1** Phase transformation of a charged field in charge space with the field, at a spacetime point $x$, represented in terms of its real and imaginary parts

be described by "setting up", at every point of spacetime $x$, a coordinate system representing charge space.

The above is easily generalized to fields $\chi(x)$ with several internal degrees of freedom, represented, say, by $N$ components corresponding to various degrees of freedoms, which transform as

$$\chi^i(x) \;\rightarrow\; (e^{i\boldsymbol{\vartheta}(x)\cdot\mathbf{t}})^i{}_j\,\chi^j(x), \quad \boldsymbol{\vartheta}(x) = (\vartheta^1(x),\dots,\vartheta^M(x)), \quad i,j = 1,\dots,N. \tag{6.1.2}$$

where the indices $i,j$ specify the various components of a field in internal space, with $M$ matrices $t_a$: $\mathbf{t} = (t_1,\dots,t_M)$, represented by $N \times N$ matrices with basic properties to be spelled out below. By considering local transformations, where $\boldsymbol{\vartheta}(x)$, is real, and as indicated, may depend on the spacetime point $x$, and one, in particular, may describe both the spacetime position of a field and its internal (charge, isospin, color,...) state.

In order to describe dynamical theories, the derivative of a field should be also considered in the general case, in such a way that everything turns out to be invariant under the transformation rule in (6.1.2) with $\boldsymbol{\vartheta}(x)$ depending on the spacetime points. Consider a curve parametrized by a parameter $\lambda$, and coordinate labels $x^\mu(\lambda)$ in Minkowski spacetime. The field $\chi(x)$ will, in particular, depend on the parameter $\lambda$ through the dependence of $x^\mu$ on $\lambda$.

At every point $x$ of spacetime, we set up an $N$ dimensional Euclidean space, associated with internal space, with basis vector fields $\{e_i(x)\}$, and expand the field $\chi(x)$ at the point $x$ as follows

$$\chi(x) = \chi^i(x)\,e_i(x). \tag{6.1.3}$$

We need a structure to be able to *compare* the relative generalized "orientation" of a coordinate system, associated with internal space, at a spacetime point $x$ from one infinitesimally close to it as we move along such a curve (Fig. 6.2).

**Fig. 6.2** Relative coordinate rotations are considered and formally depicted in an abstract $N$ dimensional Euclidean space corresponding to two neighboring points on a curve in Minkowski spacetime. This leads to the concept of connection needed to relate these relative rotations

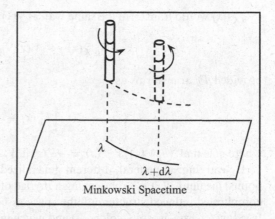

Minkowski Spacetime

Such a structure is called the connection and arises naturally in the following way. The change of the field $\chi(x(\lambda))$ as it moves along the curve may be represented from (6.1.3) by

$$d\chi(x(\lambda)) = \partial_\mu \chi^i(x)dx^\mu e_i(x) + \chi^i(x)\,de_i(x). \qquad (6.1.4)$$

Now $de_i(x)$ at point $x^\mu(\lambda)$, must vanish for $dx^\mu \to 0$, and it may be expanded in terms of the basis vectors $e_j$, i.e.,

$$de_i = -i\,\Gamma_{i\mu}{}^j(x)\,e_j(x)dx^\mu \qquad (6.1.5)$$

The totality of the expansion coefficients $\{\Gamma_{i\mu}{}^j(x)\}$ is referred to as the connection telling us how "fast" the coordinates get "rotated", locally, as we move from a spacetime point $x$ to an infinitesimally close one. The $-i$ factor in (6.1.5) is chosen for convenience.

From (6.1.4)), (6.1.5), we obtain

$$d\chi(x) = \left(\delta_i{}^j \partial_\mu - i\,\Gamma_{i\mu}{}^j(x)\right)\chi^i(x)e_j(x)dx^\mu. \qquad (6.1.6)$$

Hence $d\chi(x)$, as the field $\chi$ itself, may be expanded in terms of the basis vectors $e_i(x)$ in the local coordinate system of internal space. And we learn that the partial derivative $\partial_\mu$ has been replaced by a new structure $\nabla_\mu$ in the general case, referred to as the gauge-covariant derivative, which in matrix form, may be written as

$$(\nabla_\mu) = (\partial_\mu - i\,\Gamma_\mu). \qquad (6.1.7)$$

With the operator of transformation given in (6.1.2) written as

$$V(x) = \exp[i\boldsymbol{\vartheta}(x)\cdot\mathbf{t}], \qquad (6.1.8)$$

$(\nabla_\mu \chi)(x)$ would transform the same way as $\chi(x) \to (V\chi)(x)$, i.e.,

$$\nabla_\mu \, \chi(x) \; \to \; V(x) \, \nabla_\mu \, \chi(x), \qquad\qquad (6.1.9)$$

provided $\Gamma_\mu$ transforms as

$$\Gamma_\mu(x) \; \to \; V(x) \left[ \Gamma_\mu(x) - \frac{\partial_\mu}{i} \right] V^{-1}(x), \qquad\qquad (6.1.10)$$

where note that $\left( \partial_\mu V(x) \right) V^{-1}(x) = -V(x) \, \partial_\mu V^{-1}(x)$.

By carrying, in general, different generalized "rotations" at different spacetime points, the field $\chi(x)$ is experiencing a myriad of different twisting and providing a complicated induced structure of the space associated with the integral degrees of freedom (charge, isospin, color,...) and which manifest their existence. In turn, one may hold a structure "responsible" for such twisting of the field, in charge space, by introducing the concept of curvature, locally defined at each spacetime point, as a measure of the strength on such twisting. This gives rise to a geometrical way of introducing fields that interact with charges and other physical entities corresponding to other internal degrees of freedom.

The concept of curvature may be introduced in the following way. We first define the gauge-covariant derivative of $\chi(x)$ along a *Lorentz* vector $\mathscr{V}^\mu$

$$\mathscr{V}^\mu \nabla_\mu \, \chi(x) \equiv \mathscr{V} \nabla \, \chi(x). \qquad\qquad (6.1.11)$$

Consider now the transfer of $\chi(x)$ along a closed path in Minkowski spacetime defined by the parallelogram shown below along the Lorentz vectors $\delta\xi_1 v_1^\mu$, $\delta\xi_2 v_2^\mu$, $-\delta\xi_1 v_1^\mu$, $-\delta\xi_2 v_2^\mu$, with $v_1^\mu$, $v_2^\mu$ two different Lorentz vectors to form a parallelogram, with infinitesimal $\delta\xi_1$, $\delta\xi_2$ (Fig. 6.3).

At the point 2, we may write

$$\chi(x_2) = \chi^{\mathrm{I}}(x_1) + \delta\xi_1 v_1^\mu \nabla_{1\mu} \chi^{\mathrm{I}}(x_1) + \cdots$$

$$= (1 + \delta\xi_1 v_1 \nabla_1 + \cdots)\chi^{\mathrm{I}}(x_1), \qquad\qquad (6.1.12)$$

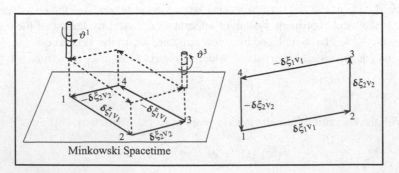

**Fig. 6.3** Diagrams corresponding to the transfer of $\chi(x)$ along a closed path in Minkowski spacetime defined by the parallelogram as described in the text

where $\chi^I(x_1)$ denotes the initial state, and $\nabla_1$ stands for $\nabla$ with $x = x_1$. Let $\chi^F(x_1)$ denote the final state, then

$$\chi^F(x_1) = (1 - \delta\xi_2 v_2 \nabla_4 + \ldots)(1 - \delta\xi_1 v_1 \nabla_3 + \ldots)$$
$$\times (1 + \delta\xi_2 v_2 \nabla_2 + \ldots)(1 + \delta\xi_1 v_1 \nabla_1 + \ldots)\chi^I(x_1). \qquad (6.1.13)$$

From which the following expression emerges for the change in twisting by going around the closed path

$$\chi^F(x_1) - \chi^I(x_1) = \delta\xi_1 \, \delta\xi_2 \, v_1^\mu \, v_2^\nu \, [\nabla_\mu, \nabla_\nu] \, \chi^I(x_1), \qquad (6.1.14)$$

where $[.,.]$ stands for the commutator. The curvature is defined by

$$R_{\mu\nu}(x) = i\,[\nabla_\mu, \nabla_\nu]$$
$$= \left(\partial_\mu \Gamma_\nu(x) - \partial_\nu \Gamma_\mu(x)\right) + \frac{1}{i}\left(\Gamma_\mu(x)\Gamma_\nu(x) - \Gamma_\nu(x)\Gamma_\mu(x)\right), \qquad (6.1.15)$$

responsible for the alteration of the state of $\chi^I(x)$ by going around a closed path. Note that the differential operators $\partial_\mu$, $\partial_\nu$ cancel out in the final expression for the commutator $[\nabla_\mu, \nabla_\nu]$. $R_{\mu\nu}(x)$ in (6.1.15)) is written in matrix form. Its matrix elements are given by

$$R^i{}_{\mu\nu j}(x) = \left(\partial_\mu \Gamma_{\nu j}{}^i(x) - \partial_\nu \Gamma_{\mu j}{}^i(x)\right) + \frac{1}{i}\left(\Gamma_{\mu k}{}^i(x)\Gamma_{\nu j}{}^k(x) - \Gamma_{\nu k}{}^i(x)\Gamma_{\mu j}{}^k(x)\right). \qquad (6.1.16)$$

Here $i, j, k = 1, \ldots, N$, where $N$ corresponds to the number of the components of the field $\chi(x)$, while $\mu, \nu = 0, 1, 2, 3$ correspond to spacetime indices.

From (6.1.10) and the definition in (6.1.15)), $R_{\mu\nu}$, as a matrix, has the following gauge transformation rule

$$R_{\mu\nu}(x) \to V(x)\, R_{\mu\nu}(x)\, V^{-1}(x), \quad \mathrm{Tr}\, R^{\mu\nu}(x)R_{\mu\nu}(x) \to \mathrm{Tr}\, R^{\mu\nu}(x)R_{\mu\nu}(x), \qquad (6.1.17)$$

with the second rule implying an important invariance property, where $\mathrm{Tr}(.) \equiv (.)^i{}_i$.

The operator in (6.1.8), corresponding to the transformation rule given in (6.1.2), gives rise to a representation of a continuous group whose elements are labeled by a set of smooth parameters referred to as a Lie group. The group multiplication is defined by

$$\exp(i\,\vartheta^1 \cdot \mathbf{t})\exp(i\,\vartheta^2 \cdot \mathbf{t}) = \exp(i\,\vartheta \cdot \mathbf{t}). \qquad (6.1.18)$$

In particular for any two components $t_a$, $t_b$ of $\mathbf{t}$, and a real number $\xi$, this group property implies that

$$\exp(-i\xi\, t_b)\exp(-i\xi\, t_a)\exp(i\xi\, t_b)\exp(i\xi\, t_a) = \exp(i\,\vartheta_c\, t_c), \qquad (6.1.19)$$

Clearly the coefficients $\vartheta_c$ depend on $t_a$, $t_b$ and $\to 0$ for $\xi \to 0$. We may use the classic Baker-Campbell-Hausdorff formula for two operators $A, B$

$$e^A e^B = e^{A+B+C}, \quad C = \frac{1}{2}[A, B] + \frac{1}{12}[A, [A, B]] + \frac{1}{12}[B, [B, A]] + \dots, \quad (6.1.20)$$

to infer from (6.1.19) that

$$\exp\left(-(i\xi)^2[t_a, t_b] + \mathcal{O}(\xi^3) + \dots\right) = \exp(i\,\vartheta_c\,t_c), \quad (6.1.21)$$

leading to $\vartheta_c = \xi^2 f_{abc} + \dots$ for a given pair $(a, b)$, with a choice of real numericals $f_{abc}$, and for $\xi \to 0$, the following commutation relation emerges

$$[t_a, t_b] = i f_{abc}\, t_c, \quad (6.1.22)$$

providing a representation of an algebra, referred to as a Lie algebra, with generators[3] $t_a$, $a = 1, \dots, N$. The numericals $f_{abc}$ are called the structure constants of the algebra, and due to the commutator on the left-hand side of (6.1.22), they are anti-symmetric in the indices $a, b$. The following identity, obtained by multiplying (6.1.22) by $t_d$ from the right, taking the trace, and using the property of a trace

$$i f_{abc} \text{Tr}(t_c\, t_d) = \text{Tr}([t_a, t_b]t_d) = \text{Tr}(t_d[t_a, t_b]) = \text{Tr}(t_a[t_b, t_d]) = \text{Tr}(t_b[t_d, t_a]), \quad (6.1.23)$$

is easily established. We will be exclusively involved with groups satisfying the normalization condition

$$\text{Tr}(t_c\, t_d) = \frac{1}{2}\delta_{cd}. \quad (6.1.24)$$

From the last two equalities in (6.1.23), we may then infer that the structure constants $f_{abd}$ are totally anti-symmetric in all of its three indices, as a result of the odd property that results in the interchange of any two generators, $t_a, t_b, t_d$ in their corresponding commutation relations.

Another representation of the algebra in (6.1.22) may be provided from the following consideration. For any three matrices, and hence, in particular, for matrices $t_a$, $t_b$, $t_c$, we may use the easily derived Jacobi identity

$$[t_a, [t_b, t_c]] + [t_b, [t_c, t_a]] + [t_c, [t_a, t_b]] = 0, \quad (6.1.25)$$

to conclude from (6.1.22) that

$$-f_{cbd}f_{dae} + f_{cad}f_{dbe} - f_{abd}f_{cde} = 0, \quad (6.1.26)$$

---

[3]For simplicity of the notation, we use the same symbol for a representation matrix, in general, as the abstract group element and no confusion should arise.

where in writing this equation, we have invoked the anti-symmetry nature of the structure constants $f_{abc}$. Accordingly, we may define matrices $T_a$ with matrix elements given below and satisfying the commutations relations in (6.1.22)

$$(T_b)_{ac} = \mathrm{i} f_{abc}, \qquad [T_a, T_b] = \mathrm{i} f_{abd} T_d. \tag{6.1.27}$$

These matrices provide a representation of the Lie algebra called the adjoint representation and is generated by the structure constants themselves.

An algebra for which all the generators commute is called an abelian algebra and otherwise it is called non-abelian.

A gauge vector field $A_\mu(x)$, referred to as a Yang-Mills field with corresponding field strength $G_{\mu\nu}(x)$, may be materialized and defined in terms of the connection in (6.1.7) and the curvature in (6.1.15), respectively, by introducing, in the process, a coupling parameter $g_0$

$$A_\mu(x) = \frac{1}{g_0} \Gamma_\mu(x), \quad \nabla_\mu = \partial_\mu - \mathrm{i} g_0 A_\mu(x), \tag{6.1.28}$$

$$G_{\mu\nu} = \frac{1}{g_0} R_{\mu\nu}(x) = \partial_\mu A_\nu(x) - \partial_\nu A_\mu(x) - \mathrm{i} g_0 [A_\mu, A_\nu]. \tag{6.1.29}$$

This, from a geometrical setting, provides a way of introducing vector fields, as dynamical variables, that interact with charge and physical entities corresponding to other internal degrees of freedom carried by a given field $\chi(x)$, referred to as a matter field, with $g_0$ defining an overall coupling parameter between the fields. The vector fields may themselves possess internal degrees of freedom themselves and the geometrical description with an associated curvature in charge (integral degrees of freedom) space produces naturally and in a simple way means to describe interactions between these curious entities we call gauge fields, and all with the same coupling parameter $g_0$.

For the subsequent treatment, we consider the parametrization $\vartheta(x) = g_0 \Lambda(x)$. These vector fields together with the given field $\chi(x)$ in question, have from (6.1.10) and (6.1.17) the following transformations

$$\chi(x) \to V(x)\chi(x), \quad A_\mu(x) \to V(x)\left[A_\mu(x) + \frac{\mathrm{i}}{g_0}\partial_\mu\right]V^{-1}(x), \tag{6.1.30}$$

$$G_{\mu\nu}(x) \to V(x)G_{\mu\nu}(x)V^{-1}(x), \quad \mathrm{Tr}\, G^{\mu\nu}(x)G_{\mu\nu}(x) \to \mathrm{Tr}\, G^{\mu\nu}(x)G_{\mu\nu}(x). \tag{6.1.31}$$

Of particular interest are infinitesimal transformations of the fields $\chi(x)$, $A_\mu(x)$, which are obtained from the above transformations to be

$$\chi(x) \to \chi(x) + \mathrm{i} g_0 \delta\Lambda_a(x) t_a \chi(x), \tag{6.1.32}$$

$$A_\mu(x) \to A_\mu(x) + \partial_\mu \delta\Lambda_a(x) t_a + \mathrm{i} g_0 \delta\Lambda_a(x) [t_a, A_\mu(x)]. \tag{6.1.33}$$

From the second term on the right-hand side of (6.1.33), we make use of the matrix nature of $A_\mu(x)$ to write

$$A_\mu(x) = A_{c\mu}(x)\, t_c, \quad A_{c\mu}(x) \to A_{c\mu}(x) + \partial_\mu \delta\Lambda_c(x) - g_0 f_{cab}\, \delta\Lambda_a(x) A_{b\mu}(x).$$
(6.1.34)

Here we see that in addition to a rather familiar change $\partial_\mu\, \delta\Lambda^c(x)$, internal degrees of freedom are associated with $A_{a\mu}(x)$ as well. It also transforms according to the adjoint representation.

The field strength $G_{\mu\nu}$ may be then rewritten as

$$G_{\mu\nu}(x) = G_{c\mu\nu}(x)\, t_c,$$
(6.1.35)

$$G_{c\mu\nu}(x) = \partial_\mu A_{c\nu}(x) - \partial_\nu A_{c\mu}(x) + g_0 f_{cab}\, A_{a\mu}(x) A_{b\nu}(x).$$
(6.1.36)

One may set up a so-called kinetic energy term for the set gauge vector fields, and, due to the invariance property on the right-hand side of (6.1.31), we are bound to consider an expression like

$$-\frac{1}{2}\,\mathrm{Tr}\, G^{\mu\nu}(x)G_{\mu\nu}(x) = -\frac{1}{2}\,\mathrm{Tr}(t_a\, t_b)\, G_{a\mu\nu}(x)G_{b\mu\nu}(x),$$
(6.1.37)

up to a normalization factor, involving a direct interaction between the gauge vector fields.

For the group U(1) corresponding to phase transformations, describing, in particular, electrodynamics for which the field strength $G_{\mu\nu}$ reduces to the field strength tensor $F_{\mu\nu}$, the coupling $g_0$ is identified with the electronic charge $e_0$, and one is dealing with an abelian gauge group, with properties

$$F_{\mu\nu} = \partial_\mu A_\nu(x) - \partial_\nu A_\nu(x), \quad A_\mu(x) \to A_\mu(x) + \partial_\mu \delta\Lambda(x),$$
(6.1.38)

$$\nabla_\mu = \partial_\mu - i\, e_0 A_\mu(x), \qquad \chi(x) \to \chi(x) + i\, e_0\, \delta\Lambda(x)\chi(x).$$
(6.1.39)

The group SU($N$), is defined in terms of $N \times N$ unitary and unimodular matrices, i.e., of determinant 1. This implies that for the transformations with group elements as given in (6.1.8), the corresponding matrices $t_a$ must be Hermitian to satisfy the unitary property of the group element, and must be of zero trace so that the group elements have determinant equal to one. The latter property is easily verified since referring to (6.1.8)

$$\det\left[\exp\left(i\,\vartheta_a\, t_a\right)\right] = \exp\left(i\vartheta_a\,\mathrm{Tr}\,[t_a]\right) = 1.$$

This involves of $M = N^2 - 1$ non-commuting generators $t_a$ represented by $N \times N$ Hermitian and traceless matrices. As shown at the end of this section, one has $(N-1)$ Hermitian diagonal matrices in this set. The number $(N-1)$ of diagonal matrices is referred to as the rank of the group. The construction of $a$ set of generators for SU($N$) is straightforward and is spelled out below in subsection 6.1.1. For the SU(3) color group of quantum chromodynamics, the quarks come in $N = 3$ different colors, and the theory involves $N^2 - 1 = 8$ gluons.

For the group SU(2), the lowest representation is given by a doublet with generators represented, as discussed at the end of this section by the three ($N^2 - 1 = 2^2 - 1 = 3$) Pauli matrices, as $2 \times 2$ Hermitian traceless matrices, divided by 2,

$$\chi(x) = \begin{pmatrix} \chi^1(x) \\ \chi^2(x) \end{pmatrix}, \quad \mathbf{t} = \frac{\sigma}{2}, \quad [t_a, t_b] = i\,\epsilon_{abc}\,t_c, \quad \sigma = (\sigma_1, \sigma_2, \sigma_3), \quad (6.1.40)$$

where $\epsilon_{abc}$ is totally anti-symmetric with $\epsilon_{123} = +1$. Here we have only one diagonal matrix, that is the group is of rank one, it is given by $t_3 = \text{diag}[1/2, -1/2]$, and a corresponding particle state may be labeled by its eigenvalues $\pm 1/2$. The adjoint representation is provided by the generators $T_a$, with matrix elements (see (6.1.27)) $(T_b)_{ac} - i\,\epsilon_{abc}$, i.e.,

$$T_1 = \begin{pmatrix} 0 & 0 & 0 \\ 0 & 0 & -i \\ 0 & i & 0 \end{pmatrix}, \quad T_2 = \begin{pmatrix} 0 & 0 & i \\ 0 & 0 & 0 \\ -i & 0 & 0 \end{pmatrix}, \quad T_3 = \begin{pmatrix} 0 & -i & 0 \\ i & 0 & 0 \\ 0 & 0 & 0 \end{pmatrix}, \quad (6.1.41)$$

which also satisfy the commutation relations

$$[T_a, T_b] = i\,\epsilon_{abc}\,T_c. \quad (6.1.42)$$

The next lowest representation of SU(2) is a triplet

$$\chi(x) = \begin{pmatrix} \chi^1(x) \\ \chi^2(x) \\ \chi^3(x) \end{pmatrix}. \quad (6.1.43)$$

A more convenient and equivalent representation of the above commutation relations by $3 \times 3$ Hermitian matrices in which $T_3'$ is diagonal and hence may explicitly display the eigenvalues of such a multiplet is obtained by a similarity transformation of these generators. That is, by multiplying (6.1.41) from the left by a matrix $S$ and from the right by its inverse $S^{-1}$, $(ST_aS^{-1} = T'_a)$ giving rise to the generators

$$T_1' = \frac{1}{\sqrt{2}} \begin{pmatrix} 0 & 1 & 0 \\ 1 & 0 & 1 \\ 0 & 1 & 0 \end{pmatrix}, \quad T_2' = \frac{i}{\sqrt{2}} \begin{pmatrix} 0 & -1 & 0 \\ 1 & 0 & -1 \\ 0 & 1 & 0 \end{pmatrix}, \quad T_3' = \begin{pmatrix} 1 & 0 & 0 \\ 0 & 0 & 0 \\ 0 & 0 & -1 \end{pmatrix}, \quad (6.1.44)$$

displaying the eigenvalues $(1, 0, -1)$ of a triplet, where

$$S = \begin{pmatrix} 1-i & 0 \\ 0 & 0 & -\sqrt{2} \\ -1-i & 0 \end{pmatrix}. \quad (6.1.45)$$

The lowest representation of SU(3) is defined by a triplet as in (6.1.43), referred to as the fundamental representation. SU(3) is a group of rank $2 : (3-1 = 2)$ that is, two of its generators $t_a$, $a = 1, \ldots, 8$, $(M = N^2-1 = 8)$, may be in diagonal form. Generators are obtained directly from the analysis to be carried out at the end of the section. They are $3 \times 3$ Hermitian traceless matrices and are given by $t_a = \lambda_a/2$, where $\lambda_a$ are called the Gell-Mann matrices

$$\lambda_1 = \begin{pmatrix} 0 & 1 & 0 \\ 1 & 0 & 0 \\ 0 & 0 & 0 \end{pmatrix}, \quad \lambda_2 = \begin{pmatrix} 0 & -i & 0 \\ i & 0 & 0 \\ 0 & 0 & 0 \end{pmatrix}, \quad \lambda_3 = \begin{pmatrix} 1 & 0 & 0 \\ 0 & -1 & 0 \\ 0 & 0 & 0 \end{pmatrix},$$

$$\lambda_4 = \begin{pmatrix} 0 & 0 & 1 \\ 0 & 0 & 0 \\ 1 & 0 & 0 \end{pmatrix}, \quad \lambda_5 = \begin{pmatrix} 0 & 0 & -i \\ 0 & 0 & 0 \\ i & 0 & 0 \end{pmatrix}, \quad \lambda_6 = \begin{pmatrix} 0 & 0 & 0 \\ 0 & 0 & 1 \\ 0 & 1 & 0 \end{pmatrix},$$

$$\lambda_7 = \begin{pmatrix} 0 & 0 & 0 \\ 0 & 0 & -i \\ 0 & i & 0 \end{pmatrix}, \quad \lambda_8 = \frac{1}{\sqrt{3}} \begin{pmatrix} 1 & 0 & 0 \\ 0 & 1 & 0 \\ 0 & 0 & -2 \end{pmatrix}. \tag{6.1.46}$$

The structure constants are easily evaluated and are given by

$$f_{123} = 1, \quad f_{147} = -f_{156} = f_{246} = f_{257} = f_{345} = -f_{367} = \frac{1}{2},$$

$$f_{458} = f_{678} = \frac{\sqrt{3}}{2}. \tag{6.1.47}$$

Due to the Hermitian property of the matrices $t_a = \lambda_a/2$, with the $\lambda_a/2$ defined in (6.1.46), the complex conjugates are given by $t_a^* = t_a^\mathsf{T}$, and the complex conjugate of the unitary matrix $(\exp(ig_0 \Lambda^a t_a))^*$ becomes simply $(\exp(ig_0 \Lambda^a (-t_a)^\mathsf{T}))$. Also as there is no matrix $S$ such that $S \lambda_a S^{-1} = -\lambda_a^\mathsf{T}$, $a = 1, \ldots, 8$, we have two inequivalent representations. This is easily seen by noting that $\lambda_8^\mathsf{T} = \lambda_8$, and the existence of such a matrix $S$ would imply that $\lambda_8$ and $-\lambda_8$ have the *same* eigenvalues thus leading to a contradiction. This new transformation is defined for the complex conjugate of the fields. One then introduces the complex conjugate transformation as

$$\chi^*(x) \rightarrow (\exp(ig_0 \Lambda^a (-t_a)^\mathsf{T})) \chi^*(x), \tag{6.1.48}$$

thus generating an inequivalent, that is new, representation. For SU(2), we may define the antisymmetric matrix $S = (\epsilon_{ab})$ with $\epsilon_{ab} = \epsilon^{ab}$, $\epsilon_{12} = +1$, with the result that for the Pauli matrices, $S\sigma S^{-1} = -\sigma^\mathsf{T}$, where we note that $\epsilon^{ab} \chi_b$ transforms the same way as $\chi^a$. That is, the complex conjugate provides an equivalent description for SU(2).

The above treatment will be very useful in setting up the formalism of gauge theories in later parts.

We close this section by considering some additional notions related to Lie algebras, as well as elaborate on the structure of the generators of $SU(N)$.

We recall from above, that two representations are equivalent, if the corresponding generators $t_a$, $t_a'$ are related by a similarity transformation: $S\, t_a\, S^{-1} = t_a'$.

A subalgebra is said to be invariant if the commutator of any one of its elements with any element of the full algebra yields an element of the subalgebra itself (or may be equal to zero).

An algebra which has no non-trivial invariant subalgebra is called simple. [The full algebra and the empty set are, by definition, trivial invariant subalgebras.]

A representation is said to be reducible if it is equivalent to one involving two (or more) sets of generators $\{t_a^1\}$, $\{t_{a'}^2\}$, and corresponding structure constants $f_{abc}^1$; $f_{a'b'c'}^2$, such that

$$[t_a^1, t_{b'}^2] = 0, \qquad [t_a^1, t_b^1] = i f_{abc}^1 t_c^1, \qquad [t_{a'}^2, t_{b'}^2] = i f_{a'b'c'}^2 t_{c'}^2. \qquad (6.1.49)$$

The latter equivalent representation is then said to be the direct sum of the two (or more) representations, with generators $\{t_a^1\}$, $\{t_{a'}^2\}$ which would act, in turn, in different vector spaces. Otherwise a representation is said to be irreducible.

## 6.1.1 Generators of SU(N)

We need to generate a complete set of $N^2 - 1$ Hermitian traceless $N \times N$ matrices. We may start by constructing elementary matrices which have the single element 1 in every possible way skipping the diagonal elements. Thus we generate $N^2 - N = N(N - 1)$ matrices, with necessarily zeros on their diagonal. From this we will construct Hermitian, properly normalized, matrices. We then proceed to generate $N - 1$ diagonal (and hence commuting) Hermitian matrices thus completing the set of generators.

Introduce the set of matrices:

$$E(a, b) = 1 \text{ at the } a\text{th row and } b\text{th column } for\ a \neq b,$$

$$\text{and zero otherwise,} \qquad a, b = 1, \ldots, N, \qquad (6.1.50)$$

thus generating $N^2 - N$ matrices each with a single element 1 and zeros everywhere else, including zeros for their diagonal elements. From these we will introduce below properly normalized Hermitian matrices. We now generate $N - 1$ diagonal Hermitian traceless matrices as follows.

We define these $N \times N$ matrices as

$$L_k = \frac{1}{\sqrt{2k(k+1)}} \operatorname{diag}[1, 1, \ldots, 1, -k, 0, .., 0], \quad k = 1, \ldots, (N-1). \quad (6.1.51)$$

where the number 1 appears $k$ number of times for each $k$. Clearly, for $k = N-1$ there are no zeros after the element $-k = -N+1$. We note that

$$L_j L_k \propto L_j \quad \text{for } j < k, \quad \operatorname{Tr}(L_i L_j) = \frac{1}{2} \delta_{ij} \quad i, j, k = 1, \ldots, (N-1). \quad (6.1.52)$$

From (6.1.50), we introduce the following two sets of $N(N-1)/2$ Hermitian traceless matrices

$$D_1(a, b) = \frac{1}{2} \left( E(a, b) + E(b, a) \right), \quad D_2(a, b) = \frac{1}{2i} \left( E(a, b) - E(b, a) \right),$$
$$(6.1.53)$$

$1 \le a < b \le N$, which we may denote by $L_N, \ldots, L_{(N^2-1)}$, thus completing the set of generators.

It is easy to verify that the generators $L_1, \ldots, L_{(N^2-1)}$ satisfy the normalization condition in (6.1.24). The number $(N-1)$ of diagonal matrices is called the rank of the group.

SU(2) is of rank 1, that is, it has one diagonal matrix and $L_1$ given above coincides with the Pauli matrix $\sigma_3/2$. Also $L_2, L_3$ coincide with $\sigma_1/2, \sigma_2/2$.

SU(3) is of rank 2, and $L_1, L_2$ coincide with $t_3 = \lambda_3/2$, $t_8 = \lambda_8/2$, where the Gell-Mann matrices $\lambda_a$ are defined in (6.1.46), and $L_3, \ldots, L_8$ coincide with the remaining Gell-Mann matrices divided by two.

## 6.2  Quantization of Non-Abelian Gauge Fields in the Coulomb Gauge

The gauge fields in non-abelian gauge theories $A_\mu^a$, unlike the gauge field in abelian gauge theories, the photon which has no charge, necessarily involves a direct self-interaction due to a "charge" it carries, and the simplest Lagrangian density for the gauge fields which is invariant under non-abelian gauge transformations reflects this fact.[4] A simple Lagrangian density for the gauge fields is given by

$$\mathscr{L}(x) = -\frac{1}{4} G_{a\mu\nu}(x) G_a^{\mu\nu}(x), \quad (6.2.1)$$

---

[4] A review of the content of Sect. 6.1 will be beneficial for the reader.

described in terms of the fields strengths $G_{a\mu\nu}(x)$ in (6.1.36), involving two derivatives of the fields, as a direct generalization of the corresponding one in electrodynamics, but also involves the interaction of the gauge field. The field strength is given in detail in terms of the gauge fields by the relation in (6.1.36)

$$G_{c\mu\nu} = \partial_\mu A_{c\nu} - \partial_\nu A_{c\mu} + g_0 f_{abc} A_{a\mu} A_{b\nu}. \tag{6.2.2}$$

where $f_{abc}$ are totally anti-symmetric structure constants of the underlying group of transformations, with generators $t_a$ satisfying (see (6.1.22), (6.1.24))

$$[t_a, t_b] = \mathrm{i} f_{abc} t_c, \quad \mathrm{Tr}(t_a t_b) = \frac{1}{2}\delta_{ab}. \tag{6.2.3}$$

The non-abelian symmetry gauge transformations are implemented by operators $V(x) = \exp[\mathrm{i}\, g_0\, \theta_a(x)\, t_a]$, and upon setting $A_{a\mu} t_a = A_\mu$, $G_{a\mu\nu} t_a = G_{\mu\nu}$, we have the following gauge transformation rules (see (6.1.30), (6.1.31)), for these fields

$$A_\mu(x) \to V(x)\Big[A_\mu(x) + \frac{\mathrm{i}}{g_0}\partial_\mu\Big]V^{-1}(x), \tag{6.2.4}$$

$$G_{\mu\nu}(x) \to V(x) G_{\mu\nu}(x) V^{-1}(x), \quad \mathrm{Tr}\,[\,G_{\mu\nu}(x) G^{\mu\nu}(x)] \to \mathrm{Tr}\,[\,G_{\mu\nu}(x) G^{\mu\nu}(x)]. \tag{6.2.5}$$

The last transformation rule for $\mathrm{Tr}\,[\,G_{\mu\nu} G^{\mu\nu}] \equiv G_a{}^{\mu\nu} G_{a\mu\nu}/2$, shows the gauge invariance of the Lagrangian density in (6.2.1) for pure gauge fields.

Upon defining

$$\nabla_{ab}^\mu = \big(\delta_{ab}\, \partial^\mu + g_0 f_{acb} A_c^\mu\big), \tag{6.2.6}$$

with matrix elements just defined here, we have the basic relations

$$[\nabla^\mu, \nabla^\nu]_{cb} = g_0 f_{cab}\, G_a^{\mu\nu}, \quad \nabla_{ab\mu} \nabla_{bc\nu}\, G_c^{\mu\nu} = 0, \tag{6.2.7}$$

as shown in Problems 6.2, 6.3, with the latter equality generalizing the equation $\partial_\mu \partial_\nu F^{\mu\nu} = 0$ in the abelian case, to the non-abelian one.

To develop the field theory dynamics, we couple the gauge fields $A_{a\mu}$ to external sources $K_a^\mu$ thus introducing the Lagrangian density

$$\mathscr{L} = -\frac{1}{4}\, G_{a\mu\nu}\, \tilde{G}_a^{\mu\nu} + K_a^\mu A_{a\mu}. \tag{6.2.8}$$

Now we must pick up a gauge to describe the dynamics. In the present section we work in the Coulomb gauge. Covariant gauges will be considered in the following section. The Coulomb gauge is defined by ($i = 1, 2, 3$)

$$\partial_i A_{ai}(x) = 0, \tag{6.2.9}$$

which allows us to solve, say, $A_a^3$ in terms of the other fields and their space derivatives

$$A_a^3 = -\partial_3^{-1}\partial^\kappa A_{a\kappa}, \quad \kappa = 1, 2, \tag{6.2.10}$$

and due to the dependence of $A_a^3$ on the other fields, the variation of the fields takes the form[5]

$$\delta A_{ai} = \left[\delta_{i\kappa} - \delta_{i3}\,\partial_3^{-1}\partial^\kappa\right]\delta A_{a\kappa}, \tag{6.2.11}$$

with a sum over $\kappa = 1, 2$, in the latter two equations understood.

The field equations then follow to be

$$\nabla_{abj}\,G_b^{j\,0} = \nabla_{ac\sigma}\,G_c^{\sigma\,0} = -K_a^0, \tag{6.2.12}$$

$$\nabla_{ab\mu}G_b^{\mu j} = \partial_3^{-1}\partial^j\big(\nabla_{ac\mu}G_c^{\mu 3} + K_a^3\big) - K_a^j, \tag{6.2.13}$$

where we note that for $j = 3$ the latter equation is trivially true.

The above two equations may be combined to

$$\nabla_{ab\mu}\,G_b^{\mu\nu} = \eta^{\nu j}\,\partial^j\,\partial_3^{-1}\big(\nabla_{ac\mu}G_c^{\mu 3} + K_a^3\big) - K_a^\nu. \tag{6.2.14}$$

By applying the operator $\nabla_{dav}$ to this equation, and using the second equation in (6.2.7), we obtain

$$\nabla_{daj}\partial^j\partial_3^{-1}\big(\nabla_{ac\mu}G_c^{\mu 3} + K_a^3\big) = \nabla_{dav}K_a^\nu, \quad \nabla_{daj}\partial^j \equiv \big(\delta_{da}\nabla^2 + g_0 f_{dca}A_{cj}\,\partial_j\big). \tag{6.2.15}$$

This suggests to introduce the Green function $D_{ab}$, to solve for $\partial_3^{-1}(\nabla_{ac\mu}G_c^{\mu 3} + K_a^3)$ in (6.2.14), satisfying the equation[6]

$$\big(\delta_{da}\nabla^2 + g_0 f_{dca}A_{cj}(x)\partial_j\big)D_{ab}(x, x') = \delta^{(4)}(x, x')\,\delta_{db}. \tag{6.2.16}$$

Thus from (6.2.15), (6.2.16), we may rewrite our field equation (6.2.14) finally as

$$\nabla_{ab\mu}\,G_b^{\mu\nu} = -\big(\delta_{ac}\,\delta^\nu{}_\sigma - \eta^{\nu j}\,\partial^j D_{ab}\nabla_{bc\sigma}\big)K_c^\sigma. \tag{6.2.17}$$

The canonical conjugate momenta of $A_a^\kappa$, $(\kappa = 1, 2)$, $A_a^0$ are readily worked to be

$$\pi_a^\kappa = G_a^{\kappa 0} - \partial_3^{-1}\partial^\kappa G_a^{30}, \quad \pi_a^0 = 0. \tag{6.2.18}$$

---

[5]Compare this with the Coulomb gauge problem in QED in Sect. 5.14.

[6]Note that since this equation involves no time derivative, $D_{ab}(x, x')$ involves a $\delta(x^0, x'^0)$ factor.

Therefore $A_a^0$ is a dependent field. Also the first expression $\pi_a^\kappa$ gives zero if $\kappa$ is replaced by 3 ($\pi_a^3 = 0$), as it should for a dependent field. Accordingly we may combine these components and write

$$\pi_a^\mu = G_a^{\mu 0} - \eta^{\mu j}\partial_3^{-1}\partial^j G_a^{30}, \qquad (6.2.19)$$

as is easily verified.

As we will see below, the expression for $G_a^{\mu 0}$ is necessary for finding the solution of the non-abelian gauge theory. To this end, we apply the operator $\nabla_{ba\mu}$ to the above equation to obtain

$$\nabla_{ba\mu}\pi_a^\mu = \nabla_{ba\mu}G_a^{\mu 0} - \nabla_{baj}\,\partial_3^{-1}\partial^j G_a^{30} = -K_b^{\ 0} - \nabla_{baj}\,\partial_3^{-1}\partial^j G_a^{30}, \qquad (6.2.20)$$

where we have used (6.2.12). Therefore

$$\nabla_{baj}\,\partial_3^{-1}\partial^j G_a^{30} = -\left(\nabla_{ba\mu}\pi_a^\mu + K_b^{\ 0}\right). \qquad (6.2.21)$$

From Eq. (6.2.16) of the Green function $D_{ab}$, we may then solve for $\eta^{\mu j}\partial_3^{-1}\partial^j G_a^{30}$ in (6.2.21), to finally obtain from (6.2.19)

$$G_a^{\mu 0} = \pi_a^\mu - \eta^{\mu j}\partial^j D_{ab}\left(\nabla_{bc\mu}\pi_c^\mu + K_b^{\ 0}\right), \qquad (6.2.22)$$

expressed in particular, in terms of canonical conjugate momenta and their space derivatives (no time derivatives). Since $A_a^0$ is a dependent field so is $G_a^{\kappa 0}$. The functional derivative of the latter with respect to the external source $K_c^\nu$ is then given by

$$\frac{\delta}{\delta K_c^\nu(x')}\,G_a^{\mu 0}(x) = -\delta^0_{\ \nu}\,\eta^{\mu j}\partial^j D_{ac}(x,x'), \qquad \frac{\delta}{\delta K_c^\nu(x')}\,G_a^{kl}(x) = 0. \qquad (6.2.23)$$

by keeping the independent fields $A^1$, $A^2$, and their canonical conjugate momenta, fixed, where we recall that $A^3$ is expressed in terms of $A^1$, $A^2$, and recall that $\pi^0 = 0$, $\pi^3 = 0$.

From the explicit expression of $G_{c\mu\nu}$ in (6.2.2), the partial derivative of the Lagrangian density in (6.2.8) with respect to the coupling parameter $g_0$, is given by

$$\frac{\partial}{\partial g_0}\mathscr{L} = -\frac{1}{2}f_{abc}\,G_a^{\mu\nu}A_{b\mu}A_{c\nu} = -f_{abc}\,G_a^{k0}A_{bk}A_{c0} - \frac{1}{2}f_{abc}\,G_a^{kl}A_{bk}A_{cl}. \qquad (6.2.24)$$

Now we use the basic functional differentiation rule, developed in Sect. 4.6, with a summary provided at the end of that section, that for an operator $\mathscr{O}(x)$

$$(-\mathrm{i})\frac{\delta}{\delta K_a^\mu(x')}\langle\, 0_+|\mathscr{O}(x)\, |\, 0_-\rangle = \langle\, 0_+|\big(A_{a\mu}(x')\mathscr{O}(x)\big)_+|\, 0_-\rangle$$

$$-\,\mathrm{i}\langle\, 0_+|\frac{\delta}{\delta K_a^\mu(x')}\mathscr{O}(x)|\, 0_-\rangle,\tag{6.2.25}$$

where $(\ldots)_+$, denotes the time-ordered product, and the functional derivative of $\mathscr{O}(x)$, in the second term on the right-hand side, is taken, as in (6.2.23), by keeping the (independent) fields and their canonical momenta fixed. Here *recall* that $A_a^3$ is expressed in terms of the independent fields and hence it is automatically kept fixed, and it involves no time derivatives.

Upon setting $-\mathrm{i}\delta/\delta K_a^\mu \equiv \widehat{A}_a^\mu$, an immediate application of (6.2.25), (6.2.23), leads to the following two key equations:

$$\langle\, 0_+|\big(A_{c0}A_{bk}\,G_a^{k0}\big)_+|\, 0_-\rangle = \widehat{A}_{c0}\langle\, 0_+|\big(A_{bk}\,G_a^{k0}\big)_+|\, 0_-\rangle$$

$$+\,\mathrm{i}\langle\, 0_+|\big(A_{bk}\frac{\delta}{\delta K_c^0}\,G_a^{k0}\big)_+|\, 0_-\rangle$$

$$=\widehat{A}_{c0}\widehat{A}_{bk}\widehat{G}_a^{k0}\langle\, 0_+\, |\, 0_-\rangle - \mathrm{i}\widehat{A}_{bk}\,\partial^k\widehat{D}_{ac}(x,x)\langle\, 0_+\, |\, 0_-\rangle,\tag{6.2.26}$$

$$\langle\, 0_+|G_a^{kl}A_{bk}A_{cl}\,|\, 0_-\rangle = \widehat{G}_a^{kl}\widehat{A}_{bk}\widehat{A}_{cl}\langle\, 0_+\, |\, 0_-\rangle.\tag{6.2.27}$$

Hence from (6.2.24)

$$\langle\, 0_+|\frac{\partial}{\partial g_0}\mathscr{L}(x)\, |\, 0_-\rangle = \Big[\frac{\partial}{\partial g_0}\widehat{\mathscr{L}}(x) - \mathrm{i}f_{abc}\widehat{A}_{bk}\,\partial^k\widehat{D}_{ca}(x,x)\Big]\langle\, 0_+\, |\, 0_-\rangle.\tag{6.2.28}$$

where we have used the anti-symmetry of $f_{abc}$ to rewrite the second expression within the square brackets. On the other hand, the quantum dynamical principle gives

$$\frac{\partial}{\partial g_0}\langle\, 0_+\, |\, 0_-\rangle = \mathrm{i}\int(\mathrm{d}x)\langle\, 0_+|\frac{\partial}{\partial g_0}\mathscr{L}(x)\, |\, 0_-\rangle,\tag{6.2.29}$$

and hence from (6.2.28)

$$\frac{\partial}{\partial g_0}\langle\, 0_+\, |\, 0_-\rangle = \int(\mathrm{d}x)\Big[\mathrm{i}\frac{\partial}{\partial g_0}\widehat{\mathscr{L}}(x) + f_{abc}\widehat{A}_{bk}\,\partial^k\widehat{D}_{ca}(x,x)\Big]\langle\, 0_+\, |\, 0_-\rangle.\tag{6.2.30}$$

We use the matrix notation as obtained from (6.2.16)

$$D_{ab}(x, x') = \left[ \left\langle x \left| \left( \frac{1}{\nabla^2 + g_0 (fA_k) \partial^k} \right) \right| x' \right\rangle \right]_{ab}, \qquad (6.2.31)$$

where $(f\widehat{A}_k)_{da} = f_{dca}\widehat{A}_{ck}$, the notation

$$\mathrm{Tr}\,[E] = \int (\mathrm{d}x) E_{aa}(x, x), \qquad (6.2.32)$$

to rewrite the second term within the square brackets in (6.2.30), integrated over $x$, as

$$\mathrm{Tr}\left[ (f\widehat{A}_k) \partial^k \frac{1}{[\nabla^2 + g_0 (f\widehat{A}_l) \partial^l]} \right]. \qquad (6.2.33)$$

An elementary integration of the latter over $g_0$ from 0 to $g_0$ gives the *multiplicative* factor

$$\det\left[ \delta_{ab} + g_0 f_{acb} \frac{1}{\nabla^2} \widehat{A}_{ck} \partial^k \right], \qquad (6.2.34)$$

to $\langle\, 0_+ \mid 0_- \rangle$ in (6.2.30). All told the following expression emerges for $\langle\, 0_+ \mid 0_- \rangle$:

$$\langle\, 0_+ \mid 0_- \rangle = \det\left[ \delta_{ab}\nabla^2 + g_0 f_{acb}\widehat{A}_{ck}\partial^k \right] \exp\left[ i \int (\mathrm{d}x) \widehat{\mathscr{L}}_{\mathrm{I}}(x) \right] \langle\, 0_+ \mid 0_- \rangle_0, \qquad (6.2.35)$$

up to an unimportant multiplicative constant factor, where in detail

$$\widehat{\mathscr{L}}_{\mathrm{I}} = -\frac{g_0}{2} f_{abc}\left( \partial_\mu \widehat{A}_{a\nu} - \partial_\nu \widehat{A}_{a\mu} \right) \widehat{A}_b^\mu \widehat{A}_c^\nu - \frac{g_0^2}{4} f_{abc}\widehat{A}_b^\mu \widehat{A}_c^\nu f_{ade}\widehat{A}_{d\mu}\widehat{A}_{e\nu}, \qquad (6.2.36)$$

and from (5.14.16), (5.15.17),

$$\langle\, 0_+ \mid 0_- \rangle_0 = \exp\left[ \frac{i}{2} \int (\mathrm{d}x)(\mathrm{d}x') K_a^\mu(x) \mathscr{D}_{ab\mu\nu}^{\mathrm{C}}(x - x') K_b^\nu(x') \right], \qquad (6.2.37)$$

$$\mathscr{D}_{ab\mu\nu}^{\mathrm{C}} = \delta_{ab}\,\mathscr{D}_{\mu\nu}^{\mathrm{C}},$$

$$\mathscr{D}_{00}^{\mathrm{C}} = -\frac{1}{\mathbf{Q}^2}, \quad \mathscr{D}_{ij}^{\mathrm{C}}(Q) = \left( \delta_{ij} - \frac{Q_i Q_j}{\mathbf{Q}^2} \right) \frac{1}{Q^2 - i\epsilon}, \quad \mathscr{D}_{0i}^{\mathrm{C}}(Q) = 0, \qquad (6.2.38)$$

in the momentum description. The determinant factor in (6.2.35) is referred to as a Faddeev-Popov determinant.[7]

The inclusion of matter field is straightforward, and with the $\psi$ field as a column vector,[8] with corresponding gauge transformation rules: $\psi \rightarrow V\psi$, $\overline{\psi} \rightarrow \overline{\psi}V^{-1}$, one may write the Lagrangian density as

$$\mathscr{L} = -\frac{1}{4}\, G_{a\mu\nu}G_a^{\mu\nu} + \frac{1}{2}\left(\frac{\partial_\mu \overline{\psi}}{i}\, \gamma^\mu\, \psi - \overline{\psi}\, \gamma^\mu\, \frac{\partial_\mu \psi}{i}\right) - \overline{\psi}M_0\psi$$

$$+ \frac{1}{2}\, g_0[\,\overline{\psi}, \gamma^\mu A_\mu \psi\,] + \overline{\eta}\, \psi + \overline{\psi}\, \eta + A_{a\mu}K_a^\mu, \qquad (6.2.39)$$

where $M_0$ is a mass matrix for the Fermion field. Such generalizations will be considered as we go along in coming sections.[9]

## 6.3   Functional Fourier Transform and Transition to Covariant Gauges; BRS Transformations and Renormalization of Gauge Theories

In the present section, dealing with the vacuum-to-vacuum transition amplitude, we develop a functional Fourier transform, working in the Coulomb gauge which then allows us to make a transition to covariant gauge. The final subsection deals with the renormalizability of the non-abelian gauge theories set up in the last section via an elegant approach carried out via so-called BRS transformations.[10]

### 6.3.1   Functional Fourier Transform and The Coulomb Gauge

The path integral expression for $\langle\, 0_+ \mid 0_-\rangle$, in the Coulomb gauge in (6.2.35) for non-abelian gauge fields, as a functional Fourier transform $K_a^\mu \rightarrow A_{a\mu}$, with the latter now a classical field, may be then inferred from (5.14.41), in the absence of

---

[7]Faddeev and Popov [37]. It is interesting to note that the Faddeev-Popov factor also appears in the functional differential equation satisfied by $\langle\, 0_+ \mid 0_-\rangle$, see Manoukian [86].

[8]See, for example, (6.1.43).

[9]This section is based on Manoukian [84].

[10]BRS stands for Becchi, Rouet and Stora: [15].

matter, and may be denoted by

$$\langle\, 0_+ \mid 0_-\rangle_C = \int \Pi_{ax\nu}\, \mathscr{D}A_{a\nu}(x)\delta\big(\partial_k A_a^k(x)\big)M_C[A]$$

$$\times \exp[\,i \int (\mathrm{d}x)W_{\mathrm{cl}}(x)\,]\, \exp[\,i \int (\mathrm{d}x)K_b^\mu(x)A_{b\mu}(x)\,], \qquad (6.3.1)$$

up to an unimportant multiplicative constant factor, with C in $\langle\, 0_+ \mid 0_-\rangle_C$ emphasizing that this expression corresponds to the Coulomb gauge, where[11]

$$M_C[A] = \det\Big[\partial_k\big(\delta_{ab}\partial^k + g_0 f_{acb}A_c{}^k\big)\Big], \qquad (6.3.2)$$

$$W_{\mathrm{cl}}(x) = -\frac{1}{4}\, G_a^{\mu\nu}(x)G_{a\mu\nu}(x), \qquad G_a^{\mu\nu} = \partial^\mu A_a^\nu - \partial^\nu A_a^\mu + g_0 f_{abc}\, A_b^\mu A_c^\nu. \qquad (6.3.3)$$

now defined in terms of classical fields. We here see both the Coulomb gauge constraint via the delta functional $\Pi_{ax}\delta(\partial_k A_a^k(x))$ explicitly, as well as a Faddeev-Popov determinant in the Coulomb gauge displayed in the functional integrand in (6.3.1). The expression for $\langle\, 0_+ \mid 0_-\rangle$ in (6.2.35), (6.2.36), (6.2.37) is a functional of $K_a^\mu$, and we here recognize the path-integral expression in (6.3.1) as a functional *Fourier integral* with the coefficient of $\exp[\,i \int (\mathrm{d}x)K_b^\mu(x)A_{b\mu}(x)]$ as a functional of $A_a^\mu$.

   To find the corresponding expression to the one in (6.3.1) in other gauges we proceed as follows. To this end, let us concentrate on the factor $\Pi_{ax}\delta\big(\partial_k A_a^k(x)\big)M_C[A]$, in the path integrand in (6.3.1), which originated from working in the Coulomb gauge. Consider all possible gauge transformations on the field $A_a^\mu(x) \rightarrow (A_a^\mu(x))^{(\theta)}$, via the group $g$ of transformations: $V(x) = \exp[\,i\, g_0\, \theta^a(x)t^a]$ for all $(\theta^a(x), a)$, with $\theta^a(x)$ real. In turn consider the sum of all these gauge transformations as applied to a given function $f(A_a^\mu(x))$, which we may express as

$$\sum_{\mathrm{all}\,(\theta^b(x),b)} f(A_a^\mu(x))^{(\theta)}. \qquad (6.3.4)$$

Taking a product over all spacetime points $x$, and all the values taken by the group indices $a$, we have

$$\Pi_{ax} \sum_{\mathrm{all}\,(\theta^b(x),b)} f(A_a^\mu(x))^{(\theta)} \rightarrow \int \Pi_x\, \mathrm{d}g(x)\Big(\Pi_{ax}f(A_a^\mu(x))^{(g)}\Big), \qquad (6.3.5)$$

where the expression on the right-hand side stands for the formal mathematical expression for the left-hand side.

---

[11]Note that in the determinant $\partial_k A_c^k \cdots = (\partial_k A_c^k) \cdots + A_c^k \partial_k \cdots$.

Let us use the notation

$$\int \Pi_x \, dg(x) \left( \Pi_{ax} f(A_a^\mu(x))^{(g)} \right) = B_f[A^\mu]. \tag{6.3.6}$$

Since the left-hand sides includes all possible transsformation defined by the group $g$ for all $(\theta^b(x), b)$, a further gauge transformation $(A_a^\mu(x))^{(g)} \to ((A_a^\mu(x))^{(g)})^{(g')}$ necessarily leaves the sum (integral) invariant as it is already included in the sum. That is, $B_f[A^\mu]$ is gauge invariant, which we record here as

$$B_f[(A^\mu)^{(g')}] = B_f[A^\mu]. \tag{6.3.7}$$

In view of application to the path integral in (6.3.1), consider the product

$$\Pi_{ax} \, \delta(\partial_k A_a^k(x)) \int \Pi_x \, dg(x) \left( \Pi_{ax} \, \delta \left( \partial_k A_a^k(x) \right)^{(g)} \right). \tag{6.3.8}$$

Since the expression on the left-hand side of the integral is the same as the one in the integrand corresponding to the identity transformation, i.e., for $\theta(x) \to 0$, it suffices to consider the integral for $\theta$ in this limit. From Problem 6.1, for infinitesimal $\theta$, we have

$$A_a^\mu \to A_a^\mu + \nabla_{ab}^\mu \theta_b, \qquad \nabla_{ab}^\mu = \delta_{ab} \, \partial^\mu + g_0 f_{acb} A_c^\mu, \tag{6.3.9}$$

hence for $\partial_k(A_a^k) = 0$, on account of the delta function $\delta(\partial_k A_a^k(x))$ on the left-hand side of the integral in (6.3.8) for all group indices $a$, and all spacetime points $x$, we may write

$$\partial_k(A_a^k)^{(\theta)} = \partial_k[(\delta_{ab} \, \partial^k + g_0 f_{acb} A_c^k)\theta_b]. \tag{6.3.10}$$

This allows us to rewrite (6.3.8) as

$$\Pi_{ax} \, \delta(\partial_k A_a^k(x)) \int \Pi_x \, dg(x) \left( \Pi_{ax} \, \delta \left( \partial_k A_a^k(x) \right)^{(g)} \right)$$

$$= \Pi_{ax} \, \delta(\partial_k A_a^k(x)) \int \Pi_{ex} \, d\theta_e(x) \left( \Pi_{ax} \, \delta \left( \partial_k[(\delta_{ab} \, \partial^k + g_0 f_{acb} A_c^k)\theta_b] \right) \right)$$

$$= \Pi_{ax} \, \delta(\partial_k A_a^k(x)) \frac{1}{M_C[A]} \int \Pi_{ex} \, d\theta_e(x) \delta(\theta_e(x))$$

$$= \Pi_{ax} \, \delta(\partial_k A_a^k(x)) \frac{1}{M_C[A]}, \tag{6.3.11}$$

where $M_C[A]$ is the determinant defined in (6.3.2).

On account of the multiplicative factor $\Pi_{ax}\,\delta(\partial_k A_a^k(x))$ in (6.3.8), we may write from (6.3.6) and (6.3.11),

$$\Pi_{ax}\,\delta(\partial_k A_a^k(x))\,\det\left[\partial_k(\delta_{ab}\partial^k + g_0 f_{acb}A_c^k)\right] = \Pi_{ax}\,\delta(\partial_k A_a^k(x))\frac{1}{B_f[A^\mu]}, \qquad (6.3.12)$$

$$f(A_a^\mu(x)) = \delta(\partial_k A_a^k(x)), \qquad (6.3.13)$$

and $B_f[A^\mu]$ is gauge invariant.

Now let us choose another function $f$ which we conveniently denote by

$$\tilde{f}(A_a^\mu(x)) = \delta\big(\partial_\mu A_a^\mu(x) - \lambda\varphi_a(x)\big), \qquad (6.3.14)$$

where the $\varphi_a(x)$ are some arbitrary functions, and $\lambda$ is an arbitrary parameter. Then almost an identical analysis leading to (6.3.12), gives

$$\Pi_{ax}\,\delta\big(\partial_\mu A_a^\mu(x) - \lambda\varphi_a(x)\big)\,\det\left[\partial_\mu(\delta_{ab}\partial^\mu + g_0 f_{acb}A_c^\mu)\right]$$

$$= \Pi_{ax}\,\delta\big(\partial_\mu A_a^\mu(x) - \lambda\varphi_a(x)\big)\frac{1}{B_{\tilde{f}}[A^\mu]}, \qquad (6.3.15)$$

where $B_{\tilde{f}}[A^\mu]$ is gauge invariant, defined by the general expression in (6.3.6) with $f$ now replaced by $\tilde{f}$. In particular we note the trivial identity

$$\frac{1}{B_{\tilde{f}}[A^\mu]}\int \Pi_x\,\mathrm{d}g(x)\left(\Pi_{ax}\,\delta\big(\partial_\mu(A_a^\mu(x))^{(g)} - \lambda\varphi_a(x)\big)\right) = 1. \qquad (6.3.16)$$

Upon multiplying the integrand in the path integral in (6.3.1) by the above expression, i.e., by *one*, the path integral, by using in the process (6.3.12), may be rewritten as

$$\langle\,0_+\mid 0_-\rangle_C = \int \Pi_{ax\nu}\,\mathscr{D}A_{a\nu}(x)\int \Pi_x\,\mathrm{d}g(x)\left(\Pi_{ax}\,\delta\big(\partial_\mu(A_a^\mu(x))^{(g)} - \lambda\varphi_a(x)\big)\right)$$

$$\times\,\delta\big(\partial_k A_a^k(x)\big)\frac{1}{B_{\tilde{f}}[A^\mu]\,B_f[A^\mu]}\exp\left[\mathrm{i}\int(\mathrm{d}x)W_{\mathrm{cl}}(x)\right]\exp\left[\mathrm{i}\int(\mathrm{d}x)K_b^\mu(x)A_{b\mu}(x)\right].$$

$$(6.3.17)$$

Using the invariance of the measure $\Pi_{ax}\mathscr{D}A_{ak}(x)$ under gauge transformations, we carry out an inverse gauge transformation of the field variables:

$A_a^\mu(x) \to (A_a^\mu(x))^{(g^{-1})}$ to obtain for the above path integral

$$\langle\, 0_+ \mid 0_-\rangle_C = \int \Pi_{axv}\,\mathscr{D}A_{av}(x)\,\delta\Big(\partial_\mu A_a^\mu(x) - \lambda\varphi_a(x)\Big)$$

$$\times \det\Big[\partial_\mu(\delta_{ab}\partial^\mu + g_0 f_{acb}A_c^\mu)\Big]\exp[\,\mathrm{i}\int (\mathrm{d}x)W_{\mathrm{cl}}(x)\,]\;\mathscr{H}[K,A], \qquad (6.3.18)$$

where

$$\mathscr{H}[K,A] = \frac{1}{B_f[A^\mu]}\int \Pi_x\,\mathrm{dg}(x)\left(\Pi_{ax}\,\delta\big(\partial_k(A_a^k(x))^{(g^{-1})}\big)\right)$$

$$\times \exp[\,\mathrm{i}\int (\mathrm{d}x)K_b^\mu(x)\,(A_{b\mu}(x))^{(g^{-1})}\,], \qquad (6.3.19)$$

and we have used the gauge invariance of $W_{\mathrm{cl}}(x)$, $B_f[A^\mu]$, $B_{\bar f}[A^\mu]$, and finally used (6.3.15).

In the *absence* of the external sources $K_b^\mu(x)$, $\mathscr{H}[0,A]$ is simply *one*. We may conveniently rewrite the expression in (6.3.18) simply as

$$\langle\, 0_+ \mid 0_-\rangle_C = \mathscr{H}\Big[K, -\mathrm{i}\frac{\delta}{\delta J}\Big]\int \Pi_{axv}\,\mathscr{D}A_{av}(x)\,\delta\Big(\partial_\mu A_a^\mu(x) - \lambda\varphi_a(x)\Big)$$

$$\times \det\Big[\partial_\mu(\delta_{ab}\partial^\mu + g_0 f_{acb}A_c^\mu)\Big]\exp[\,\mathrm{i}\int (\mathrm{d}x)W_{\mathrm{cl}}(x)\,]\exp[\,\mathrm{i}\int (\mathrm{d}x)J_b^\mu(x)\,A_{b\mu}(x)\,]\Big|_{J=0}.$$
$$(6.3.20)$$

Since $\langle\, 0_+ \mid 0_-\rangle_C$ on the left-hand of the above equation is independent of the functions $\varphi_a(x)$, we may multiply the above by any functional differential operator

$$\Pi_{ax}\exp\Big[-\frac{\mathrm{i}}{2\lambda}\,\mathscr{N}\,\Big(\frac{\delta}{\delta\varphi_a(x)}\Big)\Big], \qquad \mathscr{N}\,(w)\big|_{w=0} = 0. \qquad (6.3.21)$$

In particular for $\mathscr{N}[w_a(x)] = w_a^2(x)$, (6.3.20) becomes (see Problem 6.4)

$$\langle\, 0_+ \mid 0_-\rangle_C = \mathscr{H}\Big[K, -\mathrm{i}\frac{\delta}{\delta J}\Big]\int \Pi_{axv}\,\mathscr{D}A_{av}(x)$$

$$\times \det\Big[\partial_\mu(\delta_{ab}\partial^\mu + g_0 f_{acb}A_c^\mu)\Big]\exp[\,\mathrm{i}\int (\mathrm{d}x)\widetilde{W}_{\mathrm{cl}}(x)\,]\exp[\,\mathrm{i}\int (\mathrm{d}x)J_b^\mu(x)\,A_{b\mu}(x)\,]\Big|_{J=0},$$
$$(6.3.22)$$

up to an unimportant multiplicative constant, where

$$\int (\mathrm{d}x)\widetilde{W}_{\mathrm{cl}}(x) = \int (\mathrm{d}x)\big[-\frac{1}{4}\,G_a^{\mu v}(x)G_{a\mu v}(x) - \frac{1}{2\lambda}\,\partial_\mu A_a^\mu\,\partial_v A_a^v\big]$$
$$(6.3.23)$$

$$= -\frac{1}{2} \int (dx) A_a^\mu(x) \Big[ \eta_{\mu\nu}(-\Box) - \Big(\frac{1}{\lambda} - 1\Big)\partial_\mu \partial_\nu \Big] A_a^\nu(x) + \int (dx) \mathscr{L}_{\mathrm{Icl}}(x),$$

$$(6.3.24)$$

$$\mathscr{L}_{\mathrm{Icl}}(x) = -\frac{g_0}{2} f_{abc} (\partial_\mu A_{a\nu} - \partial_\nu A_{a\mu}) A_b^\mu A_c^\nu - \frac{g_0^2}{4} f_{abc} A_b^\mu A_c^\nu f_{ade} A_{d\mu} A_{e\nu}. \quad (6.3.25)$$

The generating functional $\langle 0_+ \mid 0_- \rangle_{\mathrm{C}}$, in the Coulomb gauge, may be denoted by $\mathrm{F_C}[K]$, and from (6.3.22), the generating functional $\mathrm{F}[J, \lambda]$, in *covariant gauges* specified by the parameter $\lambda$, by comparison, may be then explicitly given by

$$\mathrm{F}[J, \lambda] = \int \Pi_{ax\nu} \mathscr{D} A_{a\nu}(x) \, \det[\partial^\mu \nabla_{ab\mu}] \, \exp\Big[ \, \mathrm{i} \int (dx) \, \big( \widetilde{W}_{\mathrm{cl}}(x) + J_b^\mu(x) A_{b\mu}(x) \big) \Big],$$

$$(6.3.26)$$

and (6.3.22) relates the theory formulated in the Coulomb gauge with those in covariant gauges specified by the parameter $\lambda$, and is discussed further in the next subsection.

It is evident from the free part in the Lagrangian density in (6.3.24) that the gauge field free propagator in the momentum description, in the covariant gauges, is given by

$$D_{ab}^{\mu\nu}(Q) = \delta_{ab} \Big[ \eta^{\mu\nu} - (1 - \lambda)\frac{Q^\mu Q^\nu}{Q^2} \Big] \frac{1}{(Q^2 - \mathrm{i}\epsilon)}, \quad \big( Q_\mu Q_\nu D_{ab}^{\mu\nu}(Q) = \lambda \delta_{ab} \big).$$

$$(6.3.27)$$

Because of the presence of the Faddeev-Popov determinant $\det[\partial_\mu \nabla_{ab\mu}]$, in the quantization procedure, in covariant gauges, derived in (6.3.26), we may use the functional integral

$$\det M = \int \mathscr{D}\overline{\omega} \, \mathscr{D}\omega \, \exp[\mathrm{i} \int (dx) \, \overline{\omega}(x) M \omega(x)], \quad (6.3.28)$$

in (2.6.19), up to an unimportant multiplicative constant, corresponding to Grassmann variables, to write

$$\det[\partial_\mu \nabla_{ab}^\mu] = \int \Pi_{ax} \mathscr{D}\overline{\omega}_a(x) \, \mathscr{D}\omega_a(x) \, \exp[\mathrm{i} \int (dx) \, \overline{\omega}_a(x) \partial_\mu \nabla_{ab}^\mu \omega_b(x)], \quad (6.3.29)$$

generating a spin 0 field in the theory obeying Fermi-Dirac statistics. Because of this unusual property of a spin 0 field, it is referred to as a ghost. The mere fact that they emerge here is that they cancel out those contributions in the theory which would, otherwise, destroy gauge invariance. That is, in the present quantization procedure, because of these ghost fields a gauge invariant theory results.[12] As a matter of fact we will explicitly see in studying the gluon propagator in the next section how these ghost fields are responsible for restoring gauge invariance of the propagator. The ghost fields are not observable.

Now we are ready to write down the explicit effective Lagrangian density of the theory and work in covariant gauges, with the bare propagator of the gluon field given in (6.3.27):

$$\mathscr{L}_{\text{eff}}(x) = -\frac{1}{4} G_a^{\mu\nu}(x) G_{a\mu\nu}(x) - \frac{1}{2\lambda} \partial_\mu A_a^\mu \, \partial_\nu A_a^\nu + \overline{\omega}_a(x) \partial_\mu \nabla_{ab}^\mu \omega_b(x), \qquad (6.3.30)$$

where $\nabla_{ab}^\mu$ is defined in (6.3.9). The vacuum-to-vacuum transition amplitude, as the generating functional, depending on the external source $J_b^\mu$, in covariant gauges, specified by the parameter $\lambda$, is from (6.3.26), (6.3.29), (6.3.30), written in a convenient form, is given by

$$\langle \, 0_+ \mid 0_- \rangle = \int \Pi \, \mathscr{D}A \, \mathscr{D}\overline{\omega} \, \mathscr{D}\omega \, \exp\Big[\, \mathrm{i} \int (\mathrm{d}x)\big(\mathscr{L}_{\text{eff}}(x) \, + \, J_b^\mu(x) \, A_{b\mu}(x)\big)\Big].$$
$$(6.3.31)$$

It is worth recording how the ghost contribution $\overline{\omega}_a(x) \partial_\mu \nabla_{ab}^\mu \omega_b(x)$ arises in the Lagrangian density in (6.3.30). It is obtained in the following manner, by varying, in the process, the gauge constraint functional $F_a = \partial_\mu A_a^\mu - \lambda\varphi_a$, in (6.3.14), in response to a gauge transformation as given in (6.3.9), leading to

$$F_a \rightarrow \partial_\mu(A_a^\mu + \nabla_{ab}^\mu \theta_b - \lambda\varphi_a) \rightarrow \overline{\omega}_a\Big(\frac{\delta}{\delta\theta_b} F_a\Big)\omega_b \, = \, \overline{\omega}_a\Big(\frac{\delta}{\delta\theta_b} [\partial_\mu(\nabla_{ac}^\mu \theta_c)]\Big)\omega_b,$$
$$(6.3.32)$$

with the right-hand side coinciding with the ghost contribution. The transformation law from the Coulomb to covariant gauges of the theory is discussed further in the following subsection. The final subsection deals with the renormalizability of the theory.

---

[12]This also means that one may choose a specific gauge to make the ghost fields disappear. But there is no need to go into it here. Feynman, in his monumental work on the quantization problem of the gravitational field in 1963 [40], realized the presence of such ghost fields in his non-abelian gauge theory.

## 6.3.2 Trasformation Law from the Coulomb Gauge to Covariant Gauges in Non-abelian Gauge Field Theories

The transformation law from the Coulomb gauge to the covariant is expressed by the relation

$$F_C[K] = \mathscr{H}\left[K, -i\frac{\delta}{\delta J}\right] F[J, \lambda]\Big|_{J=0}. \tag{6.3.33}$$

where $F_C[K]$ and $F[J, \lambda]$ denote the generating functionals in the Coulomb and covariant gauges, specified by the parameter $\lambda$, respectively, in the presence of the external sources $K$ and $J$, and $\mathscr{H}\left[K, -i\frac{\delta}{\delta J}\right]$ is defined in (6.3.19). What the delta function $\delta\big(\partial_k(A_a^k)^{(g^{-1})}\big)$ in (6.3.19) is telling us is that we have to find $(A_a^\mu)^{(\theta)}$, as a function of $A_b^\nu$, for which $\partial_k(A_a^k)^{(\theta)} = 0$. This may be achieved by expanding $(\mathscr{A}_a^\mu)^{(\theta)}$ in powers of $A_b^\nu$ as shown in Problem 6.5. Using (6.3.6), i.e.,

$$\frac{1}{B_f[A^\mu]} \int \Pi_x \, dg(x) \left(\Pi_{ax}\, \delta\big(\partial_k(A_a^k(x))^{(g^{-1})}\big)\right) = 1, \tag{6.3.34}$$

we obtain from Problem 6.5

$$\mathscr{H}[K, A] = \exp i\int(dx) K_{a\mu}(x) \left[\left(\eta^{\mu\nu} - \frac{\partial^\mu \partial_k}{\nabla^2}\eta^{k\nu}\right) A_{a\nu}(x) + \mathscr{O}(A^2)\right], \tag{6.3.35}$$

$$\mathscr{H}\left[K, -i\frac{\delta}{\delta J}\right] = \exp i\int(dx) K_{a\mu}(x) \left[\left(\eta^{\mu\nu} - \frac{\partial^\mu \partial_k}{\nabla^2}\eta^{k\nu}\right)(-i)\frac{\delta}{\delta J_a^\nu(x)} + \mathscr{O}\left(\frac{\delta}{\delta J_a^\nu}\right)^2\right], \tag{6.3.36}$$

which should be compared with the corresponding expression in QED in (5.15.10), (5.15.11), (5.15.12), and is admittedly more complicated.

## 6.3.3 BRS Trasformations and Renormalization of Non-abelian Gauge Field Theories

Due to the particular covariant gauge chosen in the theory, described by the expression in (6.3.31) specified by a parameter $\lambda$, the effective Lagrangian density $\mathscr{L}_{eff}$ given in (6.3.30) is not gauge invariant. There exists, however, residual transformations, referred to as the BRS transformations,[13] for which $\mathscr{L}_{\text{eff}}$, as well as the functional integration measure in (6.3.31), are invariant, given by

$$\delta A_a^\mu(x) = \delta\xi \, \nabla_{ab}^\mu \, \omega_b(x), \tag{6.3.37}$$

$$\delta\omega_a(x) = -\delta\xi \, \frac{g_0}{2} f_{abc} \, \omega_b(x) \, \omega_c(x), \tag{6.3.38}$$

$$\delta\overline{\omega}_a(x) = \delta\xi \, \frac{1}{\lambda} \, \partial_\mu A_a^\mu(x), \tag{6.3.39}$$

as is readily verified, where $\delta\xi$ is an infinitesimal constant parameter which anti-commutes with $\omega_a$, $\overline{\omega}_a$, while it commutes with $A_c^\mu$. The ghost fields $\omega_a$ and $\overline{\omega}_a$ are treated independently. We couple the ghost fields $\omega_a$, $\overline{\omega}_a$ to external sources $\overline{\eta}_a$, and $\eta_a$, respectively. It is also convenient to introduce external sources $Q_{a\mu}$ and $L_a$ coupled to $\nabla_{ab}^\mu \, \omega_b$ and $-(g_0/2)f_{abc} \, \omega_a \, \omega_b$ appearing in (6.3.37), (6.3.38) as follows

$$Q_{a\mu}\nabla_{ab}^\mu\omega_b, \qquad -\frac{g_0}{2}L_a f_{abc} \, \omega_b \, \omega_c, \tag{6.3.40}$$

where it may be verified that the coefficients of the latter two external sources are invariant under the BRS transformations (6.3.37), (6.3.38), i.e.,

$$\delta(\nabla_{ab}^\mu \, \eta_b) = 0, \qquad \delta(f_{abc} \, \omega_b \, \omega_c) = 0. \tag{6.3.41}$$

The total Lagrangian density of the theory involving external sources may be now written as

$$\mathscr{L}_{\text{tot}} = \mathscr{L}_{\text{eff}} + J_{a\mu}A_a^\mu + \overline{\eta}_a\omega_a + \overline{\omega}_a\eta_a + Q_{a\mu}\nabla_{ab}^\mu\omega_b - \frac{g_0}{2}L_a f_{abc} \, \omega_b \, \omega_c, \tag{6.3.42}$$

where the sources $\overline{\eta}_a$, $\eta_a$, $Q_{a\mu}$ are Grassmann-like sources. Thus under the BSR transformations in (6.3.37), (6.3.38), (6.3.39), we have

$$\delta\mathscr{L}_{\text{tot}} = -\left[ J_{a\mu}\nabla_{ab}^\mu \, \omega_b(x) + \frac{g_0}{2} \, \overline{\eta}_a f_{abc} \, \omega_b(x) \, \omega_c(x) + \frac{1}{\lambda} \, \eta_a \, \partial_\mu A_a^\mu \right]\delta\xi. \tag{6.3.43}$$

Hence upon making infinitesimal changes of the functional integration variables in the generating functional

$$\langle \, 0_+ \mid 0_- \rangle = \int \Pi \, \mathscr{D}A \, \mathscr{D}\overline{\omega} \, \mathscr{D}\omega \, \exp\left[ i\int (dx)\mathscr{L}_{\text{tot}}(x) \right] \equiv \exp\left( i \, W[J, \overline{\eta}, \eta, Q, L; g_0, \lambda] \right), \tag{6.3.44}$$

---

[13]BRS stands for Becchi, Rouet and Stora: [15].

introduced in (6.3.37), (6.3.38), (6.3.39), we obtain[14]

$$\int (dx) \left[ -J_{a\mu} \frac{\delta}{\delta Q_{a\mu}} + \overline{\eta}_a \frac{\delta}{\delta L_a} - \frac{1}{\lambda} \eta_a \partial_\mu \frac{\delta}{\delta J_{a\mu}} \right] W[J, \overline{\eta}, \eta, Q, L; g_0, \lambda] = 0.$$

(6.3.45)

At this stage, we may establish the transversality of the polarization tensor associated with the vector particle in analogy to the polarization tensor of the photon in QED in (5.16.92), (5.16.96). To this end, we note that the field equation for $\omega_b$ is given by

$$\partial_\mu \nabla^\mu_{ab} \omega_b = -\eta_a,$$

(6.3.46)

which upon taking its vacuum expectation $\langle 0_+ | [ . ] | 0 \rangle$ value and dividing by $\langle 0_+ | 0_- \rangle$, gives the functional differential equation

$$\partial_\mu \frac{\delta}{\delta Q_{a\mu}} W = -\eta_a.$$

(6.3.47)

Hence upon taking the functional derivatives: $\delta/\delta\eta_a(x')\partial^y_\nu(\delta/\delta J_{b\nu}(y))$ of (6.3.45), and setting $\overline{\eta}_a = 0$, $J_{a\mu} = 0$ give

$$-\left[ \frac{\delta}{\delta\eta_a(x')} \partial^y_\nu \frac{\delta}{\delta Q_{b\nu}(y)} + \frac{1}{\lambda} \partial^{x'}_\mu \frac{\delta}{\delta J_{a\mu}(x')} \partial^y_\nu \frac{\delta}{\delta J_{b\nu}(y)} \right.$$

$$\left. - \int (dx) \frac{1}{\lambda} \eta_c(x) \frac{\delta}{\delta\eta_a(x')} \partial^x_\mu \frac{\delta}{\delta J_{c\mu}(x)} \partial^y_\nu \frac{\delta}{\delta J_{b\nu}(y)} \right] W \bigg|_{J=0} = 0.$$

(6.3.48)

Upon using (6.3.47), then setting $\eta_a = 0$, $Q_{a\mu} = 0$, and using the definition of the propagator of the vector field, here in a convenient notation, $D^{\mu\nu}_{ab}(x-y) = (\delta/\delta J_{a\mu}(x))(\delta/\delta J_{b\nu}(y)) W \big|_{J=0}$, we obtain

$$\partial^x_\mu \partial^y_\nu D^{\mu\nu}_{ab}(x-y) = \lambda \, \delta_{ab} \, \delta^{(4)}(x-y).$$

(6.3.49)

Upon Fourier transform this gives

$$k_\mu k_\nu D^{\mu\nu}_{ab}(k) = \lambda \, \delta_{ab},$$

(6.3.50)

from which the transversality of the corresponding polarization tensor, as in QED in (5.16.92), (5.16.96), easily follows.[15] Again, as in QED, elementary power

---

[14]This identity is referred to as a Slavnov-Taylor identity: Slavnov [110] and Taylor [117].

[15]This transversality property is explicitly established and worked out, e.g., to lowest order, with dimensional regularization, in (6.6.17), in the presence of matter in (6.6.18).

counting, in particular, shows that the ultraviolet divergences of the theory are only of logarithmic types.[16]

We introduce a Legendre transform as in (5.20.11), and recast the entire theory in terms of the effective action defined by

$$\Omega[A, \overline{\omega}, \omega, Q, L; g_0, \lambda] = W[J, \overline{\eta}, \eta, Q, L; g_0, \lambda] - \int (\mathrm{d}x)\,\left(J_{a\mu}A_a^\mu + \overline{\eta}_a \omega_a + \overline{\omega}_a \eta_a\right),$$

$$(6.3.51)$$

which takes into account all radiative corrections, generating full propagators and Green functions of the theory. We note that

$$\frac{\delta\Omega}{\delta A_a^\mu(x)} = -J_a^\mu(x), \quad \frac{\delta\Omega}{\delta\omega_a(x)} = \overline{\eta}_a(x), \quad \frac{\delta\Omega}{\delta\overline{\omega}_a(x)} = -\eta_a(x) = \partial_\mu \frac{\delta\Omega}{\delta Q_{a\mu}(x)},$$

$$(6.3.52)$$

where in writing the last equality we have used (6.3.47), which gives the classical field equation for $\omega_a(x)$ corresponding to (6.3.46). Accordingly, we may rewrite (6.3.45) as

$$\int (\mathrm{d}x)\left[\frac{\delta\Omega}{\delta A_{a\mu}(x)}\frac{\delta\Omega}{\delta Q_a^\mu(x)} + \frac{\delta\Omega}{\delta\omega_a(x)}\frac{\delta\Omega}{\delta L_a(x)} + \frac{1}{\lambda}\,\partial_\mu \frac{\delta\Omega}{\delta Q_{a\mu}(x)}\partial_\nu A_a^\nu(x)\right] = 0.$$

$$(6.3.53)$$

The first and the last terms within the square brackets may be combined, upon partial integration, giving rise to the integral corresponding to those two terms

$$\int (\mathrm{d}x)\,\frac{\delta\Omega}{\delta Q_{a\mu}(x)}\left[\frac{\delta\Omega}{\delta A_{a\mu}(x)} - \frac{1}{\lambda}\,\partial_\mu\partial_\nu A_a^\nu(x)\right].$$

$$(6.3.54)$$

This suggests to define an effective action $\Gamma$ by removing the gauge fixing term from $\Omega$ as follows

$$\Gamma[A, \overline{\omega}, \omega, Q, L; g_0, \lambda] = \Omega[A, \overline{\omega}, \omega, Q, L; g_0, \lambda] + \int (\mathrm{d}x)\,\frac{1}{2\lambda}\,\partial_\mu A_a^\mu \partial_\nu A_a^\nu.$$

$$(6.3.55)$$

In terms of $\Gamma$, Eq. (6.3.53) now simply becomes replaced by

$$\int (\mathrm{d}x)\left[\frac{\delta\Gamma}{\delta A_{a\mu}(x)}\frac{\delta\Gamma}{\delta Q_a^\mu(x)} + \frac{\delta\Gamma}{\delta\omega_a(x)}\frac{\delta\Gamma}{\delta L_a(x)}\right] = 0.$$

$$(6.3.56)$$

---

[16]See also below at the end of this subsection.

We define the following renormalized (scaled) fields and parameters through the equations

$$A_a^\mu = \sqrt{Z_3}\, A_{\text{ren}\,a}^\mu, \quad \omega_a = \sqrt{\tilde{Z}_3}\, \omega_{\text{ren}\,a}, \quad \overline{\omega}_a = \sqrt{\tilde{Z}_3}\, \overline{\omega}_{\text{ren}\,a},$$

$$L_a = \sqrt{Z_3}\, L_{\text{ren}\,a}, \quad Q_{a\mu} = \sqrt{\tilde{Z}_3}\, Q_{\text{ren}\,a\mu}, \quad g_0 = Z\, g_{\text{ren}}, \quad \lambda = Z_3\, \lambda_{\text{ren}}, \qquad (6.3.57)$$

We note, in particular, that for the gauge fixing term in (6.3.55),

$$\frac{1}{2\lambda}\, \partial_\mu A_a^\mu \partial_\nu A_a^\nu = \frac{1}{2\lambda_{\text{ren}}}\, \partial_\mu A_{\text{ren}\,a}^\mu \partial_\nu A_{\text{ren}\,a}^\nu. \qquad (6.3.58)$$

The above allows us to simply rewrite

$$\Gamma[A, \overline{\omega}, \omega, Q, L; g_0, \lambda] \equiv$$

$$\Gamma[\sqrt{Z_3}\, A_{\text{ren}}, \sqrt{\tilde{Z}_3}\, \overline{\omega}_{\text{ren}}, \sqrt{\tilde{Z}_3}, \omega_{\text{ren}}, \sqrt{\tilde{Z}_3}\, Q_{\text{ren}}, \sqrt{Z_3}\, L_{\text{ren}}; Z\, g_{\text{ren}}, Z_3\, \lambda_{\text{ren}}], \qquad (6.3.59)$$

as expressed in terms of renormalized components. Since

$$\sqrt{Z_3 \tilde{Z}_3}\, A_{\text{ren}\,a\mu}\, Q_{\text{ren}\,a}^\mu = A_{a\mu}\, Q_a^\mu, \qquad \sqrt{Z_3 \tilde{Z}_3}\, L_{\text{ren}\,a}\, \omega_{\text{ren}\,a} = L_a\, \omega_a, \qquad (6.3.60)$$

we may equivalently replace (6.3.56) by

$$\int (dx) \left[ \frac{\delta\Gamma}{\delta A_{\text{ren}\,a\mu}(x)} \frac{\delta\Gamma}{\delta Q_{\text{ren}\,a}^\mu(x)} + \frac{\delta\Gamma}{\delta\omega_{\text{ren}\,a}(x)} \frac{\delta\Gamma}{\delta L_{\text{ren}\,a}(x)} \right] = 0, \qquad (6.3.61)$$

From the last equation in (6.3.52) we may supplement this equation by

$$\partial_\mu \frac{\delta\Gamma}{\delta Q_{\text{ren}\,a\mu}(x)} - \frac{\delta\Gamma}{\delta\overline{\omega}_{\text{ren}\,a}(x)} = 0. \qquad (6.3.62)$$

By using the notation

$$\Gamma_1 * \Gamma_2 = \int (dx) \left[ \frac{\delta\Gamma_1}{\delta A_{\text{ren}\,a\mu}(x)} \frac{\delta\Gamma_2}{\delta Q_{\text{ren}\,a}^\mu(x)} + \frac{\delta\Gamma_1}{\delta\omega_{\text{ren}\,a}(x)} \frac{\delta\Gamma_2}{\delta L_{\text{ren}\,a}(x)} \right], \qquad (6.3.63)$$

the identity in (6.3.61) may be then denoted by

$$\Gamma * \Gamma = 0. \qquad (6.3.64)$$

Working with the renormalized (scaled) fields and parameters, we carry out a loop expansion of $\Gamma$ (in powers of $\hbar$) expressed as

$$\Gamma = \sum_{N \geq 0} \Gamma^{(N)}. \tag{6.3.65}$$

In particular, for $N = 0$, involving no loops, $\Gamma^{(0)}$ is by construction just the classical action written in terms of renormalized (scaled) fields and the renormalized parameter $g_{\text{ren}}$, not involving the gauge fixing term, and is explicitly given by

$$\Gamma^{(0)} = I[A_{\text{ren}\,a\mu}, \overline{\omega}_{\text{ren}\,a\mu}, \omega_{\text{ren}\,a\mu}, Q_{\text{ren}\,a\mu}, L_{\text{ren}\,a\mu}; g_{\text{ren}}]$$

$$= \int (\mathrm{d}x) \left[ -\frac{1}{4} G_{\text{ren}\,a\mu\nu} G_{\text{ren}\,a}^{\mu\nu} + (Q_{\text{ren}\,a\mu} + \overline{\omega}_{\text{ren}\,a} \partial_\mu) \nabla^\mu_{\text{ren}ab} \, \omega_{\text{ren}\,b} - \frac{g_{\text{ren}}}{2} L_{\text{ren}\,a} f_{abc} \, \omega_{\text{ren}\,b} \, \omega_{\text{ren}\,c} \right]. \tag{6.3.66}$$

Up to $N^{\text{th}}$ order, (6.3.64) reads

$$\sum_{n=0}^{N} \Gamma^{(n)} * \Gamma^{(N-n)} = 0. \tag{6.3.67}$$

To zeroth order, $\Gamma^{(0)}$ is given in (6.3.66) and is obviously finite since it involves no loop integrations. As an induction hypothesis, working with the renormalized fields and parameters, then suppose that all the infinities in (6.3.67) up to the loop order $(N-1)$ have been cancelled out by appropriate choices of renormalization constants up to that order, and (6.3.67) holds true with the $\Gamma^{(n)}$, for $n = 1, \ldots, (N-1)$, in it replaced by their finite parts. Hence the infinities in (6.3.67) can come only from the $n = N, n = 0$ terms. We denote by $\Gamma^{(N)}\big|_{\text{div}}$ the divergent part of $\Gamma^{(N)}$, i.e., $\Gamma^{(N)} - \Gamma^{(N)}\big|_{\text{div}}$ denotes its finite part, where we note, by the induction hypothesis, that divergences in subdiagrams of $\Gamma^{(N)}$ are removed up to $(N-1)$ loops. If

$$\Gamma^{(0)} * \Gamma^{(N)}\big|_{\text{div}} + \Gamma^{(N)}\big|_{\text{div}} * \Gamma^{(0)} = 0, \quad \partial_\mu \frac{\delta \Gamma^{(N)}\big|_{\text{div}}}{\delta Q_{\text{ren}\,a\mu}} - \frac{\delta \Gamma^{(N)}\big|_{\text{div}}}{\delta \overline{\omega}_{\text{ren}\,a}} = 0, \tag{6.3.68}$$

then (6.3.67) will be satisfied by the finite parts of all the $\Gamma^{(n)}$, for $n = 1, \ldots, N$, and all the infinities are cancelled out to the order of $N$ loops as well.

From (6.3.63), the first equation above may be conveniently rewritten as

$$\mathscr{O} \Gamma^{(N)}\big|_{\text{div}} = 0, \tag{6.3.69}$$

where[17]

$$
\mathcal{O} = \frac{\delta \Gamma^{(0)}}{\delta Q_{\mathrm{ren}\,a\mu}} \frac{\delta}{\delta A_{\mathrm{ren}\,a}^{\mu}} + \frac{\delta \Gamma^{(0)}}{\delta A_{\mathrm{ren}\,a}^{\mu}} \frac{\delta}{\delta Q_{\mathrm{ren}\,a\mu}} + \frac{\delta \Gamma^{(0)}}{\delta L_{\mathrm{ren}\,a}} \frac{\delta}{\delta \omega_{\mathrm{ren}\,a}} + \frac{\delta \Gamma^{(0)}}{\delta \omega_{\mathrm{ren}\,a}} \frac{\delta}{\delta L_{\mathrm{ren}\,a}},
$$
(6.3.70)

and $\Gamma^{(0)}$ is given in (6.3.66).

At this stage, we consider the $N$ loops contribution to the changes from the classical action $I[A_{\mathrm{ren}}, \overline{\omega}_{\mathrm{ren}}, \overline{\omega}_{\mathrm{ren}}, Q_{\mathrm{ren}}, L_{\mathrm{ren}}; g_{\mathrm{ren}}]$ with the latter written in terms of renormalized fields and renormalized coupling $g_{\mathrm{ren}}$, with changes of the renormalization constants as discussed below. To this end, if we let $(Z_3)^{1/2}\big|_N$ represents $(Z_3)^{1/2}$ determined up to $N$ loops, then we have the following expressions for the exact $N$ loops contributions:

$$
\delta \sqrt{Z_3} = \sqrt{Z_3}\big|_N - \sqrt{Z_3}\big|_{N-1}, \qquad \delta \sqrt{\tilde{Z}_3} - \sqrt{\tilde{Z}_3}\big|_N - \sqrt{\tilde{Z}_3}\big|_{N-1}, \qquad \delta Z = Z\big|_N - Z\big|_{N-1}.
$$
(6.3.71)

From the chain rule

$$
\delta \sqrt{Z_3} \frac{\delta}{\delta \sqrt{Z_3}} = \delta \sqrt{Z_3} \left[ \frac{\delta A_{\mathrm{ren}\,a\mu}}{\delta \sqrt{Z_3}} \frac{\delta}{\delta A_{\mathrm{ren}\,a\mu}} + \frac{\delta L_{\mathrm{ren}\,a}}{\delta \sqrt{Z_3}} \frac{\delta}{\delta L_{\mathrm{ren}\,a}} \right]
$$

$$
= -\frac{\delta \sqrt{Z_3}}{\sqrt{Z_3}} \left[ A_{\mathrm{ren}\,a\mu} \frac{\delta}{\delta A_{\mathrm{ren}\,a\mu}} + L_{\mathrm{ren}\,a} \frac{\delta}{\delta L_{\mathrm{ren}\,a}} \right],
$$
(6.3.72)

where we have used, in the process, that $A_{\mathrm{ren}\,a\mu} = A_{a\mu}/\sqrt{Z_3}$, $L_{\mathrm{ren}\,a} = L_a/\sqrt{Z_3}$. By a similar application to $\sqrt{\tilde{Z}_3}$ and $Z$, the following expression emerges for the $N$ loops contribution:

$$
\delta I[A_{\mathrm{ren}}, \overline{\omega}_{\mathrm{ren}}, \overline{\omega}_{\mathrm{ren}}, Q_{\mathrm{ren}}, L_{\mathrm{ren}}; g_{\mathrm{ren}}]_{(N)} = -\left\{ \delta \sqrt{Z_3} \left[ A_{\mathrm{ren}\,a\mu} \frac{\delta}{\delta A_{\mathrm{ren}\,a\mu}} + L_{\mathrm{ren}\,a} \frac{\delta}{\delta L_{\mathrm{ren}\,a}} \right] \right.
$$

$$
+ \delta \sqrt{\tilde{Z}_3} \left[ \overline{\omega}_{\mathrm{ren}\,a} \frac{\delta}{\delta \overline{\omega}_{\mathrm{ren}\,a}} + \omega_{\mathrm{ren}\,a} \frac{\delta}{\delta \omega_{\mathrm{ren}\,a}} + Q_{\mathrm{ren}\,a\mu} \frac{\delta}{\delta Q_{\mathrm{ren}\,a\mu}} \right] + \delta Z. g_{\mathrm{ren}} \frac{\partial}{\partial g_{\mathrm{ren}}} \right\}
$$

$$
\times I[A_{\mathrm{ren}}, \overline{\omega}_{\mathrm{ren}}, \omega_{\mathrm{ren}}, Q_{\mathrm{ren}}, L_{\mathrm{ren}}; g_{\mathrm{ren}}],
$$
(6.3.73)

---

[17]This operator was conveniently used, for example, in Lee [77] and Itzykson and Zuber [67], where it is understood that not only one sums over the repeated index $\mu$ in (6.3.70), but also one integrates over the coincident arguments $x$ of the functional differentiation variables in it.

where $\delta\sqrt{Z_3}$, $\delta\sqrt{\tilde{Z}_3}$, $\delta Z$ are given in (6.3.71)) involving each $N$ loops. Accordingly, we have replaced $\delta\sqrt{Z_3}/\sqrt{Z_3}$ by $\delta\sqrt{Z_3}$, and have done similarly for $\sqrt{\tilde{Z}_3}$ and $Z$, where recall that these renormalization constants are all equal to one for zero loops.

By commuting the operator $[\partial_\mu(\delta/\delta Q_{\mathrm{ren}\,a\mu}) - \delta/(\delta\overline{\omega}_{\mathrm{ren}\,a})]$ with the operator $\{\,.\,\}$ above in (6.3.73), and applying it directly to $I[A_{\mathrm{ren}}, \overline{\omega}_{\mathrm{ren}}, \omega_{\mathrm{ren}}, Q_{\mathrm{ren}}, L_{\mathrm{ren}}; g_{\mathrm{ren}}]$, we verify that the second equation in (6.3.68) is automatically satisfied on account of the combination of the structure $(Q_{\mathrm{ren}\,a\mu} + \overline{\omega}_{\mathrm{ren}\,a}\partial_\mu)$ appearing in $I$ as given in (6.3.66). On the other hand it is straightforward, though quite tedious, to verify that $\Gamma^{(N)}\big|_{\mathrm{div}}$ identified with the above expression, i.e.,

$$\Gamma^{(N)}\Big|_{\mathrm{div}} = \delta I[A_{\mathrm{ren}}, \overline{\omega}_{\mathrm{ren}}, \overline{\omega}_{\mathrm{ren}}, Q_{\mathrm{ren}}, L; g_{\mathrm{ren}}]_{(N)}, \qquad (6.3.74)$$

defined on the right-hand side (6.3.73), satisfies (6.3.69), by direct application of the functional differential operator $\mathcal{O}$ to the right-hand side of (6.3.73). That is, with (6.3.69) now satisfied, i.e., with $\Gamma^{(N)}\big|_{\mathrm{div}}$ determined and hence subtracted from $\Gamma^{(N)}$, we infer the finiteness of the theory up to $N$ loops as well.[18]

The transversality condition of the polarization tensor, as a consequence of (6.3.50), reduces the (naïve) degree of divergence of the graph with two external vector lines by 2, thus leading to a degree of divergence equal to zero. In particular, the derivative in one of the vertices involving an external ghost line can be transposed to $\overline{\omega}$, thus reducing an associated graph involving such a vertex by one.

We will see that when fermions are included, one may encounter anomalies of the type shown in Sect. 3.10. Such possible anomalies will be investigated in Sect. 6.15.2, and their absence, as will be clear, are important for the renormalizability of the standard model.

Graphs of some one loop contributions to the theory of Yang-Mills fields in the presence of matter (spinor) fields with utmost four external lines are given in Figs. 6.4 and 6.5. For explicit *one loop* computations, including that of renormalization constants, in the presence of matter (spinor) fields, with a regularization procedure to justify intermediate steps in the analysis, see Sects. 6.6 and 6.7.

---

[18]For additional details and other aspects, such as the inclusion of matter fields, see Lee [77], Itzykson and Zuber [67], Weinberg [121], and Joglekar and Lee [68].

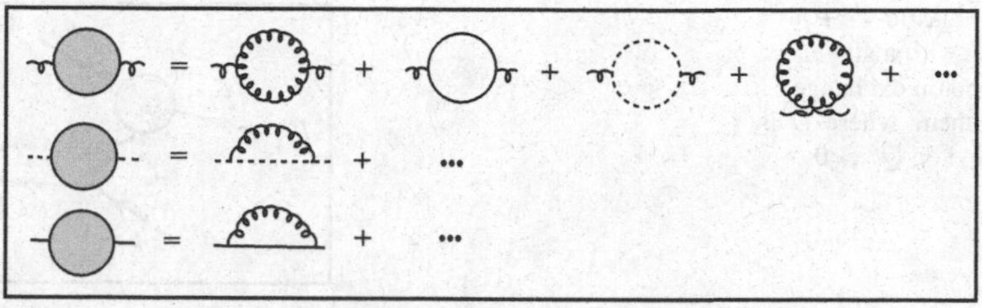

**Fig. 6.4** The Yang-Mills vectors, the ghosts and the spinors self energy parts. A curly line denotes a Yang-Mills field, a *dashed line* denotes a ghost and a *solid line* denotes a spinor

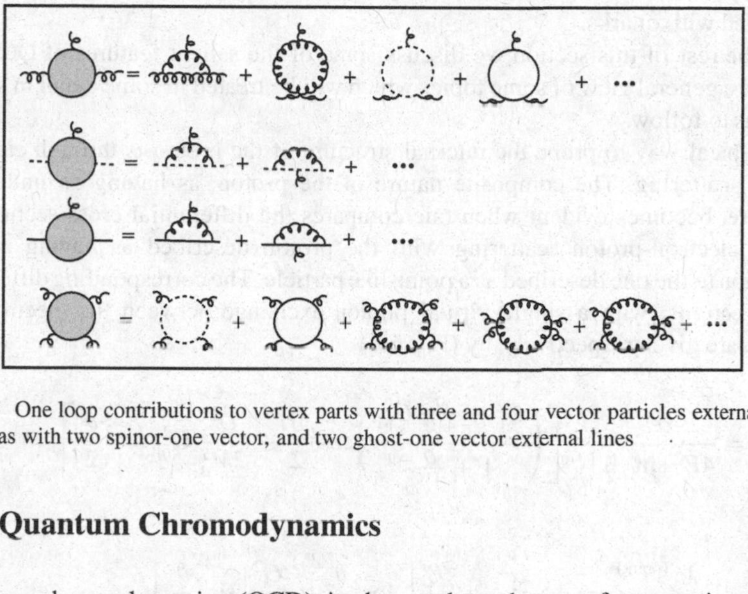

**Fig. 6.5** One loop contributions to vertex parts with three and four vector particles external lines, as well as with two spinor-one vector, and two ghost-one vector external lines

# 6.4   Quantum Chromodynamics

Quantum chromodynamics (QCD) is the modern theory of strong interaction with massless non-abelian gauge vector fields, referred to as Yang-Mills fields, interacting with quarks as well as with themselves. It is based on the symmetry color gauge group SU(3).[19] The renormalization group, which shows how the theory responds to scale transformations, in turn, leads to the description of the high-energy behavior of the theory simply in terms of an *effective coupling* which, thanks to the non-abelian gauge character of the theory, becomes weaker, with the increase of energy, and eventually becomes zero, a phenomenon referred to as asymptotic freedom. This has far reaching consequences as it allows one to develop perturbation theory at high energies, in the effective coupling, and carry out various applications which were not possible before the development of the theory, and

---

[19]Classic references for the quark description of hadrons are: Gell-Mann [52], Zweig [125], Gell-Mann [53], and for introducing color: Fritzsch and Gell-Mann [48], Fritzsch et al. [49], see also Han and Nambu [63] and Greenberg [58].

**Fig. 6.6** Elastic $e^- - p$ scattering with a single virtual photon exchange between them, where $Q$ is spacelike, i.e., $Q^2 > 0$

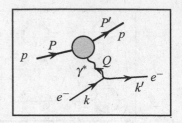

explain the results of several experiments such as the deep inelastic scattering (DIS) experiments[20] which confirm that a nucleon contains point-like constituents and are identified with quarks.

In the rest of this section we discuss some of the salient features of QCD and provide a general view of some topics which will be treated in some detail in several sections to follow.

A typical way to probe the internal structure of the proton is through electron-proton scattering. The composite nature of the proton, as having an underlying structure, becomes evident when one compares the differential cross sections for elastic electron-proton scattering with the proton described as having a finite extension to the one described as a point-like particle. The corresponding differential cross sections, with a single virtual photon exchange between the electron and proton, are given respectively, by (Fig. 6.6)

$$\frac{d\sigma}{d\Omega} = \frac{\alpha^2}{4E^2 \sin^4 \frac{\vartheta}{2}} \frac{E'}{E} \left[ \left( \frac{G_E^2 + \frac{Q^2}{4M_p^2} G_M^2}{1 + \frac{Q^2}{4M_p^2}} \right) \cos^2 \frac{\vartheta}{2} + \frac{Q^2}{2M_p^2} G_M^2 \sin^2 \frac{\vartheta}{2} \right], \qquad (6.4.1)$$

$$\frac{d\sigma^{(\text{point})}}{d\Omega} = \frac{\alpha^2}{4E^2 \sin^4 \frac{\vartheta}{2}} \frac{E'}{E} \left[ \cos^2 \frac{\vartheta}{2} + \frac{Q^2}{2M_p^2} \sin^2 \frac{\vartheta}{2} \right], \qquad (6.4.2)$$

where $E$, $E'$ denote the energies of the electron before and after scattering, respectively, and $\vartheta$ denotes the scattering angle: $\mathbf{k} \cdot \mathbf{k}' = |\mathbf{k}||\mathbf{k}'| \cos \vartheta$. The first expression above is referred to as the Rosenbluth formula and is derived in Problem 6.6 in the target frame. The second one is obtained directly from the one carried out for electron-muon elastic scattering in Sect. 5.9.3 (see (5.9.80)) by changing the mass of the muon to the proton mass $M_p$. In both expressions (6.4.1), (6.4.2), the mass of the electron is neglected in comparison to the mass of the proton. Here $G_E(Q^2)$, $G_M(Q^2)$ are the electric and magnetic form factors of the proton normalized as $G_E(0) = 1$, $G_M(0) = 1 + \kappa$, where $\kappa = 1.79$ is its anomalous magnetic moment. The obvious difference between the corresponding cross sections for large momentum transfer $Q^2$, for which the form factors $G_E(Q^2)$, $G_M(Q^2)$

---

[20]Panofski [95], Bloom et al. [23], and Friedman and Kendall [47].

vanish rapidly,[21] establishes the finite extension of the proton when compared with experiment.[22] A recent value for the root-mean-square charge radius of the proton is $\simeq .88 \times 10^{-13}$ cm.[23]

The proton is described as composite of three quarks: $u$, $u$, $d$. The three standard generations of quarks are given by

$u$ charge $+ 2/3|e|$, $\qquad$ $c$ charge $+ 2/3|e|$, $\qquad$ $t$ charge $+ 2/3|e|$,

$d$ charge $- 1/3|e|$, $\qquad$ $s$ charge $- 1/3|e|$, $\qquad$ $b$ charge $- 1/3|e|$,

making up six flavors all in all, with masses satisfying: $m_u < m_d < m_s < m_c < m_b < m_t$.[24]

The quarks also, necessarily, carry a quantum number called color.[25] There are several reasons for this. For example the baryon resonance $\Delta^{++}$, of spin 3/2 and charge $+2|e|$, is described by the triplet of quarks $(u\,u\,u)$, all with spin aligned in the ground state, and is thus given by a symmetric wavefunction in contradiction with the Spin & Statistics theorem. On the other hand, if the quarks carry an additional quantum number, one may form anti-symmetric linear combinations of the product wavefunctions of $(u^a u^b u^c)$ which certainly saves the day. Also the cross section of the process $e^+ e^- \to$ hadrons, with a single photon exchange between the $(e^+, e^-)$ - pair and the product, where the virtual photon creates quarks-antiquarks pairs, as charged particles, with the three possible colors of SU(3), relative to the cross section for the process $e^+ e^- \to \mu^+ \mu^-$, as we will see in Sect. 6.5, is simply derived to be (see also Problem 6.7)

$$R(s) = \frac{\sigma(e^+ e^- \to \text{ hadrons})}{\sigma(e^+ e^- \to \mu^+ \mu^-)} = 3 \sum_{i=1}^{n_f} (e_i^2/e^2),. \qquad (6.4.3)$$

with single photon exchange as shown in part (a) in Fig. 6.7, by neglecting radiative corrections as well as neglecting the mass of the electron and the masses of the quarks contributing to the sum. Here $s$ denotes the CM energy squared, and $n_f$ denotes the number of flavors to be retained in the sum with corresponding quarks having masses squared much less than $s$, for justifying the neglect of their masses, and the $e_i$ denote their charges. The critical factor 3 accounts for the three SU(3)

---

[21] See, e.g., Lepage and Brodsky [79, 80].

[22] For some recent experiments on these form factors see Puckett et al. [102], Punjabi et al. [103], and Andivahis et al. [9]. In earlier years a dipole fit was made for these form factors: $G_E(Q^2) = \mathcal{O}(1/Q^4)$, $G_M(Q^2) = \mathcal{O}(1/Q^4)$.

[23] Mohr et al. [93].

[24] See, for example, Beringer et al. [17] and the tabulated estimated values.

[25] Fritzsch and Gell-Mann [48] and Fritzsch et al. [49], see also Han and Nambu [63] and Greenberg [58].

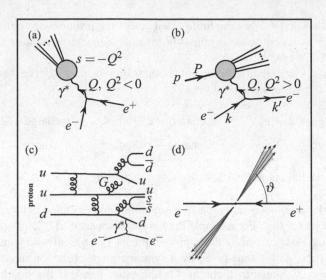

**Fig. 6.7** (**a**) The process $e^+e^- \to$ "anything". (**b**) The deep inelastic scattering process $e^- p \to$ $e^-$ "anything". (**c**) Quantum field theory at work: From valence quarks $(u, u, d)$ of the proton to partons consisting in addition all possible quark-antiquark pairs and gluons as constituents of a proton. (**d**) Production of two narrow jets of hadrons in $e^+e^-$- annihilation, in the CM frame, with jets axis angular distribution consistent with the distribution $(1 + \cos^2 \vartheta)$, corresponding to two outgoing spin 1/2 particles, confirming the spin 1/2 character of the initial quark and the antiquark (This will be investigated in Sect. 6.10.2)

colors. For example, in an earlier experiment[26] $R(s) = 3.55$ for $\sqrt{s} = 10.52$ GeV. For such an energy, we may include the 5 quarks $u, d, s, c, b$ since $m_b < 5.0$ GeV, in the sum above obtaining

$$R(s) = 3\left[\frac{4}{9} + \frac{1}{9} + \frac{1}{9} + \frac{4}{9} + \frac{1}{9}\right] = \frac{11}{3}, \qquad (6.4.4)$$

which compares well with the experimental result. Without the three factor, the result will be quite contradictory. We will also see in Sect. 6.15.2 that the assignment of three colors for quarks is essential for the internal consistency of the standard model, as it leads to the elimination of anomalies which is important for the renormalizability of the model.

The concept of color turns up to be of great importance. QCD is a theory invariant under transformations in color space, by the group generators of SU(3),[27] i.e., hadrons are necessarily colorless. Thus there is a built in constraint in the theory that single quark states, carrying a color index, are not invariant under such transformations and quarks are thus confined within the hadrons.

---

[26] Ammar et al. [8].

[27] See Sect. 6.1 for such transformations (rotations).

Upon the increase of the energy imparted to the proton in electron-proton scattering, beyond the resonance region,[28] experimentally the reaction changes "character" and the differential cross section for the underlying inelastic processes may be described (Sect. 6.9) by the expression (Sect. 6.9)

$$\frac{d^2\sigma}{dE'd\Omega} = \frac{\alpha^2 \cos^2 \vartheta/2}{4 E^2 \sin^4 \vartheta/2} [2 W_1(\nu, Q^2) \tan^2 \vartheta/2 + W_2(\nu, Q^2)], \qquad (6.4.5)$$

involving a single photon exchange, evaluated in the fixed target frame. Here $\mathbf{k} \cdot \mathbf{k}' = |\mathbf{k}||\mathbf{k}'| \cos \vartheta$, and $W_1$, $W_2$ are referred to as structure functions of the proton, have dimensionality mass$^{-1}$, and are functions of two invariant variables $(Q^2, \nu)$ which in the rest frame of the proton are given by: $Q^2 = (k - k')^2 = 4EE' \sin^2(\vartheta/2) > 0$, $\nu = -PQ/M_p = E - E' > 0$. Such processes are shown in part (b) of Fig. 6.7, giving rise to a general hadronic final state. Because of the rapid vanishing property of the electric and magnetic form factors with $Q^2$ as discussed below (6.4.2), it was expected that these structure functions will behave in a similar fashion at large momentum transfer squared $Q^2$. Early experiments,[29] mentioned before, have shown that for constant $x = Q^2/(2M_p\nu) < 1$, the dimensionless functions $M_p W_1(Q^2, \nu)$, and $\nu W_2(Q^2, \nu)$ remain approximately constant, independently of $Q^2$ instead.[30] This is reminiscent of scattering by point-like particles, and thus provided evidence of point-like constituents of the proton. This led to the development of the so-called parton model[31] in which these point-like particles within the proton are free and the virtual photon interacts with each of its charged constituents independently,[32] instead of interacting with the proton as a whole. That is, it assumes that the deep inelastic scattering cross section may be calculated from incoherent sums of elastic scattering processes of the electron with the constituents of the proton (Sect. 6.10).

Experimentally,[33] there are scaling violations of logarithmic type in $Q^2$, at present attainable high energies, and are accounted for through QCD radiative corrections. The parton model together with the scaling violation observed provides a picture of the proton as consisting of point-like particles which are nearly free inside the proton and have some interaction between them, at least at present high energies. Such an interaction between quarks supports the idea of the existence of other constituents within the proton which we call gluons mediating the interaction

---

[28]A typical resonance is the $\Delta^+$ particle of mass 1.232 GeV, consisting of a proton $p$ and a $\pi^0$ meson.

[29]Panofski [95], Bloom et al. [23], and Friedman and Kendall [47].

[30]This was predicted by Bjorken [18]. A strict $Q^2$- independence of $M_p W_1(Q^2, \nu)$, $\nu W_2(Q^2, \nu)$, for constant $x$ is referred to as Bjorken scaling.

[31]Feynman [41, 42] and Bjorken and Pachos [22].

[32]For an analogy to this, see part (b) of Fig. 1.7 in the introductory chapter to the book.

[33]See, e.g., Eidelman et al. [35] and Beringer et al. [17].

between the quarks in analogy to QED and bind the quarks through them to form hadrons. But unlike QED, the resulting theory would necessarily be a *non-abelian* gauge theory one to ensure its asymptotic freedom nature consistent with experiments. Such interactions, in turn, introduce the concept of partons as constituents of a proton in a very general way, as shown in part (c) in the Fig. 6.7, where in addition to the three quarks $u\,u\,d$, referred to as valence quarks, the proton may contain other quark-antiquark pairs and gluons. All these constituents are referred to as partons.[34] Later we will see that about half of the momentum of the proton is carried by the gluons.

The virtual photon impinging on the proton in the deep inelastic scattering process shown in part (b) of Fig. 6.7, is, by definition, not on the mass shell thus it acquires a longitudinal polarization in addition to the transverse ones. Thus one may define two cross sections $\sigma_T$, $\sigma_L$ associated with the transverse and longitudinal polarizations, respectively. A measurement of the ratio $R = \sigma_L/\sigma_T$ provides overwhelming evidence of the spin 1/2 structure of the quarks (Sect. 6.9) rather than that of spin 0.[35]

Before going into technical aspects of QCD and develop the underlying dynamics, we would like to comment on an intriguing observation related to quarks. In our presentation of the annihilation process $e^+e^- \rightarrow q\bar{q}$ in (6.4.3), in the CM, one may naively expect that the $q$, $\bar{q}$, produced, travel back to back and finally hit detectors and are observed. But this is not what happens. As soon as the products $q$ and $\bar{q}$ move apart by a separation distance of the order of 1 fm, color confinement prevents them to move further apart, and the increase[36] in potential energy with separation distance breaks the pair by creating two $q,\bar{q}$ pairs instead, with one pair moving essentially in the direction of the parent $q$ and the other essentially in the direction of the parent $\bar{q}$. Such a proliferation may continue generating more such pairs moving almost in the same directions as the parents $q$, and $\bar{q}$. As the transverse momenta are limited such break ups into pairs transversally are expected to be limited.[37] An emerging quark (antiquark) may, in turn, emit low energy (soft) gluons with an enhanced probability, due to an inherited infrared divergence as in QED (Sect. 5.12), to be emitted in the same direction as the parent quark (antiquark). Colored particles join together into color-singlets, and two narrow jets of such hadrons, are finally produced of well collimated hadrons, moving essentially in the same direction as, and as if sprayed by, the parents $q$, and $\bar{q}$, instead of

---

[34]Feynman [41, 42].

[35]Riordan et al. [105], Abe et al. [1], and Airapetian et al. [3].

[36]See Sect. 6.13.

[37]As a consequence of asymptotic freedom, a large momentum transfer to a constituent particle through the exchange of a high energy gluon is suppressed by the smallness of the effective coupling parameter. Also due the infrared-divergence problem, as in QED, the total rate for emitting a collinear gluon, due to its masslessness, by a quark is enhanced (see, e.g., the end of Sect. 6.10).

having a final state of hadrons distributed isotropically (see part (d) of Fig. 6.7).[38]
This has not only provided further overwhelming evidence of the existence of
quarks dynamically, but due to the fact that the jet axis angular distribution is
consistent with the distribution $(1 + \cos^2 \vartheta)$, corresponding to two outgoing spin
$1/2$ particles,[39] confirmed their spin $1/2$ character as well. This is investigated in
Sect. 6.10.2. As a consequence of asymptotic freedom, occasionally, i.e., with a
smaller probability, due to the vanishing of the effective coupling at high energies,
a high energy (hard) gluon may be emitted as well leading to a three jet event and
so on.

To develop the dynamics of QCD, we spell out the form of its Lagrangian density,
as directly obtained from (6.2.39), to be[40]

$$\mathscr{L} = -\frac{1}{4} G_{a\mu\nu} G_a^{\mu\nu} - \sum_{j=1}^{n_f} \overline{\psi}_j \left[ \gamma^\mu \left( \frac{\partial_\mu}{i} - g_0 A_\mu \right) + m_{0j} \right] \psi_j, \qquad (6.4.6)$$

where $n_f$ denotes the number of flavors, $A^\mu = A_a^\mu t_a$, and for the SU($N$) gauge
symmetry group, the generators of the group $\{t_a, a = 1, \ldots, N^2-1\}$ are represented
by $N \times N$ matrices (see Sect. 6.1 for details),

$$[t_a, t_b] = i f_{abc} t_c, \qquad \text{Tr}[t_a t_b] = \delta_{ab}/2, \qquad (6.4.7)$$

$$G_c^{\mu\nu} = \partial^\mu A_c^\nu - \partial^\nu A_c^\mu + g_0 f_{abc} A_a^\mu A_b^\nu. \qquad (6.4.8)$$

Now we are ready to write down the explicit effective Lagrangian density of the
theory and work in covariant gauges, specified by a gauge parameter $\lambda$, with the
bare propagator of the gluon field given in (6.3.27) with $\lambda = 1$, for example,
representing the Feynman gauge. This effective Lagrangian density which takes
into account the ghost fields as generated from the Faddeev-Popov determinant,
is from (6.4.6) and (6.3.30) given by

$$\mathscr{L}_{\text{eff}} = \mathscr{L} - \frac{1}{2\lambda} \partial_\mu A_a^\mu \partial_\nu A_a^\nu - \partial_\mu \overline{\omega}_a \nabla_{ab}^\mu \omega_b. \qquad (6.4.9)$$

with the last term obtained by partial integration, where $\nabla_{ab}^\mu$ is defined in (6.2.6).

---

[38]The formation of jets was suggested by Bjorken and Brodsky in [19]. For an early experiment
confirming this see Hanson et al. [64].

[39]See Problem 6.7.

[40]It is understood that $\partial_\mu$ is to be replaced by $(\overrightarrow{\partial}_\mu - \overleftarrow{\partial}_\mu)/2$ in (6.4.6), see (6.2.39). We have also
set the external sources equal to zero in (6.4.6).

From (6.4.6) and (6.4.9), the interaction part of the effective Lagrangian density becomes

$$\mathcal{L}_{\text{Ieff}} = -\frac{g_0}{2} f_{abc} \left( \partial_\mu A_{a\nu} - \partial_\nu A_{a\mu} \right) A_b^\mu A_c^\nu - \frac{g_0^2}{4} f_{abc} A_b^\mu A_c^\nu f_{ade} A_{d\mu} A_{e\nu}$$

$$+ \frac{1}{2} g_0 [\,\overline{\psi}_j, \gamma^\mu A_\mu \psi_j\,] - g_0 f_{acb} \, \partial_\mu \overline{\omega}_a A_c^\mu \, \omega_b. \qquad (6.4.10)$$

In high-energy QCD analysis, at an energy specified, say, by an energy scale parameter $\mu$, one may set all the masses $m_q$ of all those quarks, antiquarks $q, \bar{q}$, for which, $m_q \ll \mu$ equal to zero. On other hand, the propagator of a quark $q$ with sufficiently heavy mass $m_q$, behaves like $(-\gamma p + m_q)/(p^2 + m_q^2) = \mathcal{O}(1/m_q)$. The latter is an elementary example of the decoupling[41] of particles with heavy masses and if $m_q \gg \mu$ one carries out the analysis as if they are not part of the dynamics and neglect their contributions. As a working hypothesis, in high-energy QCD, one may then restrict the theory only to those quark flavors with masses $m_q \ll \mu$, and set these masses equal to zero in comparison to the energy scale in question, as a leading contribution. The number of relevant flavors $n_f$ is thus determined accordingly. For example in the application of $e^+ e^-$ annihilation in (6.4.4), we have neglected the contribution of the top quark whose mass is much larger than the specified energy, and $n_f = 5$.

In Sect. 6.5, $e^+ e^-$ annihilation is studied in view of applications of the theory with QCD radiative corrections in Sect. 6.8. Renormalized propagators and vertex functions are determined in Sect. 6.6, and the expression of the renormalized quark-gluon coupling is derived in Sect. 6.7.

The renormalization group is the subject matter of Sects. 6.8. DIS is considered in general in Sect. 6.9, with application to the parton model in Sect. 6.10, and with QCD corrections in Sect. 6.11. As the effective coupling is expected to become large at low energies, a perturbative treatment of it at low energies is not very promising in investigating the fundamental problem of quark confinement from this approach. An alternative approach to quark confinement is through the Wilson loop and lattice gauge theory, and is the subject matter of Sect. 6.12 and 6.13.

The effective interaction of the effective Lagrangian density defined in (6.4.10) generates the (bare) vertices in Fig. 6.8.

---

[41]For intricacies, conditions and a proof of the decoupling theorem, see Manoukian [82, 83, 85] and other references therein.

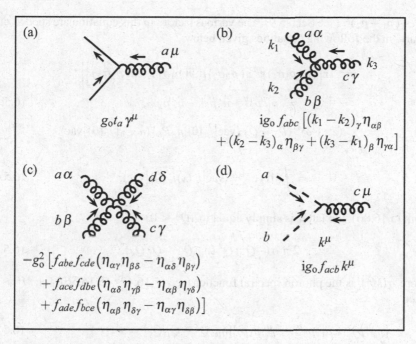

**Fig. 6.8** The bare QCD vertices: (**a**) fermion-gluon vertex, (**b**) three-gluon vertex, (**c**) four-gluon vertex, (**d**) gluon-ghost vertex. The *solid lines* denote fermions, the curly, spring-like ones, denote the gluons, while the *dashed ones* denote the ghosts

## 6.5  $e^+e^-$ Annihilation

A general expression is derived for $e^+e^-$ annihilation with a single photon exchange between the annihilated pair and the product particles as shown below in view of applications to the parton model in this section and in the investigation to the contribution of QCD radiative corrections in Sect. 6.8 after having introduced the renormalization group. At sufficiently high energies, we eventually neglect the mass of the electron (positron). To the above end, the cross section for the process in the CM frame, may be written as ($|p_1 p_2| = s/2$, see (5.9.8), also (5.9.7))

$$\sigma = e^2 \frac{1}{(Q^2)^2} \frac{1}{4} \sum_{\text{spins}} \frac{4m^2 \, \text{Tr} \left[ (\overline{v}(p_1\sigma_1)\gamma^\mu u(p_2\sigma_2))^* \, \overline{v}(p_1\sigma_1)\gamma^\nu u(p_2\sigma_2) \right]}{4(s/2)}$$

$$\times \sum_{n,P_n} (2\pi)^4 \delta^{(4)}(P_n - Q) \, \langle \text{vac}|j_\mu(0)|n, P_n\rangle \langle n, P_n|j_\nu(0)|\text{vac}\rangle,$$

$$(6.5.1)$$

$s = -(p_1+p_2)^2$ (see Sect. 5.9). The various factors in this equation are spelled out, in turn, in the following equalities given below

$$m^2 \operatorname{Tr}\left[\left(\overline{v}(p_1\sigma_1)\gamma^\mu u(p_2\sigma_2)\right)^* \overline{v}(p_1\sigma_1)\gamma^\nu u(p_2\sigma_2)\right]\Big|_{m\to 0}$$

$$= p_1{}^\mu p_2{}^\nu + p_1{}^\nu p_2{}^\mu - \eta^{\mu\nu}p_1 p_2, \tag{6.5.2}$$

$$\sum_{n,P_n}(2\pi)^4\delta^{(4)}(P_n - Q)\,\langle\mathrm{vac}|j_\mu(0)|n,P_n\rangle\langle n,P_n|j_\nu(0)|\mathrm{vac}\rangle$$

$$= \int(\mathrm{d}x)\,\mathrm{e}^{-\mathrm{i}Qx}\langle\mathrm{vac}|j_\mu(x)j_\nu(0)|\mathrm{vac}\rangle. \tag{6.5.3}$$

From (5.16.86), the latter is simply equal to $(Q^2 < 0)$

$$2\,\pi\,\sigma(-Q^2)\,Q^2\,(\eta_{\mu\nu}Q^2 - Q_\mu Q_\nu), \tag{6.5.4}$$

where $\sigma(M^2)$ is the photon spectral function (see (5.16.42), (5.16.56)) for $M^2 > 0$. Since

$$[p_1{}^\mu p_2{}^\nu + p_1{}^\nu p_2{}^\mu - \eta^{\mu\nu}p_1 p_2][\eta_{\mu\nu}Q^2 - Q_\mu Q_\nu] = -(Q^2)^2, \tag{6.5.5}$$

with $m$ set equal to zero, the following expression emerges for the cross section

$$\sigma(e^+e^- \to \text{hadrons}) = e^2\pi\sigma(s), \qquad s = -Q^2. \tag{6.5.6}$$

where $s$ is the CM-energy squared.

It is customary to compare the hadronic case with the process $e^+e^- \to \mu^+\mu^-$. To this end for $-Q^2 \gg m_\mu^2$, we may refer to the QED lowest contribution given in the second term on the right-hand side of (5.10.55), with $m \to m_\mu$, and neglect the mass of muon, giving for the spectral function $\sigma(s)_{\mu^+\mu^-} = \alpha/(3\pi s)$ for muons production. This leads to

$$\sigma(e^+e^- \to \mu^+\mu^-) = \frac{4\,\pi\alpha^2}{3s}, \tag{6.5.7}$$

and

$$R(s) = \frac{\sigma(e^+e^- \to \text{hadrons})}{\sigma(e^+e^- \to \mu^+\mu^-)} = 12\,\pi^2 s\,\frac{\sigma(s)}{e^2}. \tag{6.5.8}$$

In particular, if we neglect the QCD radiative corrections, only quark-antiquark $q,\bar{q}$ pairs, as charged particles, contribute to $\sigma(s)$, (Fig. 6.9). Gluon production necessarily involve the QCD coupling. Again at sufficiently high energies, we may neglect the masses of the quarks, and simply make the replacement of $\alpha$ in the expression $\alpha/3\pi s$, given above Eq. (6.5.7) for the muons, as follows:

**Fig. 6.9** (a) Expression for the process $e^+e^- \to q\bar{q}$, for $n_f$ flavors each with 3 colors. (b) The corresponding photon self-energy part for each quark $q_i$ with charge $e_i$

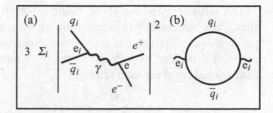

$\alpha \to 3 \times e_i^2/4\pi$, where $e_i$ is the charge of a given quark, with three different colors, and sum over flavors, $i = 1, \ldots, n_f$ to obtain by using, in the process, (6.5.6) with $\sigma(s) \to \sigma_{q\bar{q}}(s)$,

$$R(s)\big|_{q\bar{q}} = \frac{\sigma(e^+e^- \to q\bar{q})}{\sigma(e^+e^- \to \mu^+\mu^-)} = 3 \sum_{i=1}^{n_f} (e_i^2/e^2). \qquad (6.5.9)$$

This is the expression for the parton model for the hadronic process. The energy $s$ signifies how many flavors one may include in the sum. The higher the energies, the more flavors may be included. QCD corrections to $R(s)$ above may be determined after we investigate the nature of the effective coupling and the associated renormalization group as applied to the theory. This is done in Sect. 6.8. An assessment of the expression in (6.5.9) was done in (6.4.4).

## 6.6   Self-Energies and Vertex Functions in QCD

In this section we evaluate the self energies and vertex functions needed for determining the renormalized quark-gluon coupling to lowest order. The integrals given in Box 6.1 with dimensional regularization, the details of which are given Appendix III at the end of the book, are quite useful and straightforward to use in evaluating self energies.

**Box 6.1:** Integrals needed to evaluate the self-energies required for determining the renormalized quark-gluon coupling to the leading order. Details on the integrals are given in Appendix III

$$\int \frac{(dk)}{(2\pi)^4} \frac{1}{(Q-k)^2 \, k^2}\bigg|_{\mathrm{Reg}} = i\, F(\mu_D^2/Q^2, \varepsilon)$$

$$\int \frac{(dk)}{(2\pi)^4} \frac{k^\mu}{(Q-k)^2 \, k^2}\bigg|_{\mathrm{Reg}} = i\, \frac{Q^\mu}{2} F(\mu_D^2/Q^2, \varepsilon)$$

$$\int \frac{(dk)}{(2\pi)^4} \frac{k^\mu k^\nu}{(Q-k)^2 \, k^2}\bigg|_{\mathrm{Reg}} = \frac{i}{4(3-\varepsilon)} [-\eta^{\mu\nu} Q^2 + (4-\varepsilon)Q^\mu Q^\nu]\, F(\mu_D^2/Q^2, \varepsilon)$$

$$F(\mu_D^2/Q^2, \varepsilon) = \frac{1}{(4\pi)^{D/2}} \left(\frac{\mu_D^2}{Q^2}\right)^{\varepsilon/2} \frac{\Gamma^2(1-\frac{\varepsilon}{2})\Gamma(\frac{\varepsilon}{2})}{\Gamma(2-\varepsilon)}$$

The self energy of the ghost, being of order $g_0^2$ and having two external ghost lines, does not contribute in determining the lowest contribution the renormalized quark-gluon coupling, but only contributes to higher orders. Also the tadpole obtained from the four-gluon vertex by joining two of its lines is zero in the dimensional regularization scheme

$$: \int \frac{(dk)}{(2\pi)^4} \frac{1}{(Q-k)^2}\Bigg|_{\text{Reg}} = 0,$$

as obtained directly from the third integral from the table above by contracting over the $\mu$ and $\nu$ indices. We also need the following formulas involving the SU(N) generators and color factors[42]:

$$\text{Tr}\,[t_a t_b] = T_R\,\delta_{ab}, \quad [t_a t_a]_{ij} = C_F\,\delta_{ij}, \quad [t_b t_a t_b] = t_a\left[t_b t_b - \frac{1}{2} C_A\right] = -\frac{1}{2N} t_a,$$
$$(6.6.1)$$

$$f_{acd}\,f_{bcd} = C_A\delta_{ab}, \qquad i f_{abc}\,t_b t_c = -\frac{1}{2} C_A\,t_a, \tag{6.6.2}$$

$$T_R = 1/2, \qquad C_F = \frac{N^2-1}{2N}, \qquad C_A = N. \tag{6.6.3}$$

The self energies and vertex functions needed for determining the renormalized fermion-gluon coupling to lowest order are given below. We work in the Feynman gauge. All computations are carried out with massless quarks which will be justified below Eq. (6.7.15) in the next section by choosing a renormalization scale $\mu$ much larger than the neglected masses of these quarks in defining renormalization constants.

All the self energies below are obtained directly from the integrals in Box 6.1. Corresponding integrals for determining the vertex parts will be given later.

## 6.6.1   Fermion Inverse Propagator

The inverse of the fermion propagator is obtained from that of the electron in Sect. 5.10.1 by replacing e by g, for a massless particle, and multiplying the self energy part by $(t_a t_a)_{ij}$:

$$S_{ij}^{-1}(p) = \delta_{ij}\,\gamma p + i g^2 (t_a t_a)_{ij} \int \frac{(dk)}{(2\pi)^4} \frac{\gamma^\mu\big(-\gamma(p-k)\big)\gamma_\mu}{(p-k)^2\,k^2}\Bigg|_{\text{Reg}}, \tag{6.6.4}$$

---

[42] Here $F$ stands for fundamental, while $A$ stands for adjoint.

$$\gamma^{\mu}(-\gamma(p-k))\gamma_{\mu} = (2-\varepsilon)\,\gamma^{\mu}\,(k_{\mu}-p_{\mu}), \qquad (6.6.5)$$

which from Box. 6.1 gives[43]

$$S_{ij}^{-1} = \delta_{ij}\left[1 + g^2\left(1 - \frac{\varepsilon}{2}\right)C_F\,F(p^2,\varepsilon)\right]\gamma p, \quad C_F = (N^2-1)/2N. \qquad (6.6.6)$$

## 6.6.2   Inverse Gluon Propagator

This consists of three parts (see Fig. 6.8, Fig. 6.10)

$$D_{ab\mu\nu}^{-1}(Q) = Q^2\delta_{ab}\eta_{\mu\nu} + \Pi_{ab\mu\nu}^{(1)}(Q) + \Pi_{ab\mu\nu}^{(2)}(Q) + \Pi_{ab\mu\nu}^{(3)}(Q), \qquad (6.6.7)$$

$$\Pi_{ab\mu\nu}^{(1)}(Q) = ig^2\delta_{ii}\mathrm{Tr}\,[t_a t_b]\int\frac{(\mathrm{d}k)}{(2\pi)^4}\frac{\mathrm{Tr}\,[\gamma_{\mu}(-\gamma(Q-k))\gamma_{\nu}(-\gamma k)]}{(Q-k)^2\,k^2}\bigg|_{\mathrm{Reg}}, \qquad (6.6.8)$$

$$\mathrm{Tr}[\gamma^{\mu}\,\gamma(Q-k)\,\gamma^{\nu}\,\gamma k] = 4\,[-2\,k^{\mu}k^{\nu} + Q^{\mu}k^{\nu} + Q^{\nu}k^{\mu} + (k^2 - Qk)\,\eta^{\mu\nu}], \qquad (6.6.9)$$

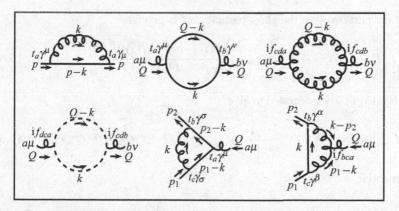

**Fig. 6.10** Self energies and vertex functions needed for determining the renormalized quark-gluon coupling to lowest order. The tadpole contribution to the gluon self-energy vanishes with dimensional regularization as shown above. These are expressed here in the Feynman gauge

---

[43]To simplify the notation we simply write $F(p^2,\varepsilon)$ for $F(\mu_D^2/p^2,\varepsilon)$ given in Box. 6.1.

which give

$$\Pi^{(1)}_{ab\mu\nu}(Q) = g^2 n_f \delta_{ab} \frac{(2-\varepsilon)}{(3-\varepsilon)}[\eta_{\mu\nu}Q^2 - Q_\mu Q_\nu]F(Q^2, \varepsilon). \tag{6.6.10}$$

For the second part, we have

$$\Pi^{(2)}_{ab\mu\nu}(Q) = ig^2\frac{1}{2}\left[f_{cda}f_{cdb}\right]\int\frac{(dk)}{(2\pi)^4}\frac{iV_{\alpha\beta\mu}(-Q+k, -k, Q)\,iV^{\alpha\beta}{}_\nu(Q-k, k, -Q)}{(Q-k)^2\,k^2}\Bigg|_{\text{Reg}},$$
$$\tag{6.6.11}$$

$$V_{\mu_1\mu_2\mu_3}(k_1, k_2, k_3) = (k_1-k_2)_{\mu_3}\eta_{\mu_1\mu_2} + (k_2-k_3)_{\mu_1}\eta_{\mu_2\mu_3} + (k_3-k_1)_{\mu_2}\eta_{\mu_3\mu_1}, \tag{6.6.12}$$

where the $1/2$ overall factor arises from having two identical bosons in a closed loop,

$$iV_{\alpha\beta\mu}(-Q+k, -k, Q)\,iV^{\alpha\beta}{}_\nu(Q-k, k, -Q)$$
$$= (2k^2 - 2kq + 5Q^2)\eta_{\mu\nu} - (2+\varepsilon)Q_\mu Q_\nu + (10-4\varepsilon)k_\mu k_\nu + (2\varepsilon-5)(Q_\mu k_\nu + Q_\nu k_\mu), \tag{6.6.13}$$

which give

$$\Pi^{(2)}_{ab\mu\nu}(Q) = -g^2\frac{1}{2}C_A\delta_{ab}\frac{1}{2(3-\varepsilon)}[(19-6\varepsilon)\eta_{\mu\nu}Q^2 - (22-7\varepsilon)Q_\mu Q_\nu]F(Q^2, \varepsilon). \tag{6.6.14}$$

The third part involves the ghost particle, and is given by

$$\Pi^{(3)}_{ab\mu\nu}(Q) = -ig^2 f_{dca}f_{cdb}\int\frac{(dk)}{(2\pi)^4}\frac{(-i(Q-k)_\mu)(ik_\nu)}{k^2(Q-k)^2}\Bigg|_{\text{Reg}}, \tag{6.6.15}$$

which is explicitly integrated to give

$$\Pi^{(3)}_{ab\mu\nu}(Q) = -g^2 C_A\delta_{ab}\frac{1}{4(3-\varepsilon)}[\eta_{\mu\nu}Q^2 + (2-\varepsilon)Q_\mu Q_\nu]F(Q^2, \varepsilon). \tag{6.6.16}$$

In particular, we note that

$$\Pi^{(2)}_{ab\mu\nu}(Q) + \Pi^{(3)}_{ab\mu\nu}(Q) = -g^2 C_A\delta_{ab}\frac{(10-3\varepsilon)}{2(3-\varepsilon)}[\eta_{\mu\nu}Q^2 - Q_\mu Q_\nu]F(Q^2, \varepsilon). \tag{6.6.17}$$

**Box 6.2**: Integrals needed to evaluate the vertex parts required for the determination of the renormalized quark-gluon coupling to the leading order. Details on the integrals are given in Problems 6.8, 6.9, with $\varepsilon > 0$ and $\delta > 0$, corresponding, respectively, to ultraviolet and infrared cutoffs, and appropriate mass scales $\mu_D$, $\mu_{D'}$

$$\int \frac{(dk)}{(2\pi)^4} \frac{1}{(p_1-k)^2(p_2-k)^2 k^2}\Big|_{\text{Reg}} = \frac{i}{(4\pi)^{D'/2}} \Big(\frac{1}{Q^2}\Big)\Big(\frac{\mu_{D'}^2}{Q^2}\Big)^{-\frac{\delta}{2}} \frac{\Gamma^2\big(\frac{\delta}{2}\big)\Gamma\big(1-\frac{\delta}{2}\big)}{\Gamma(1+\delta)}$$

$$\int \frac{(dk)}{(2\pi)^4} \frac{k^\mu}{(p_1-k)^2(p_2-k)^2 k^2}\Big|_{\text{Reg}} = \frac{i(p_1^\mu + p_2^\mu)}{(4\pi)^{D'/2}} \Big(\frac{1}{Q^2}\Big)\Big(\frac{\mu_{D'}^2}{Q^2}\Big)^{-\frac{\delta}{2}} \frac{\Gamma\big(\frac{\delta}{2}\big)\Gamma\big(1-\frac{\delta}{2}\big)\Gamma\big(1+\frac{\delta}{2}\big)}{\Gamma(2+\delta)}$$

$$\int \frac{(dk)}{(2\pi)^4} \frac{k^\mu k^\nu}{(p_1-k)^2(p_2-k)^2 k^2}\Big|_{\text{Reg}} = \frac{i\eta^{\mu\nu}}{(4\pi)^{D/2}}\Big(\frac{\mu_D^2}{Q^2}\Big)^{\frac{\varepsilon}{2}} \frac{\Gamma\big(\frac{\varepsilon}{2}\big)\Gamma^2\big(1-\frac{\varepsilon}{2}\big)}{2\,\Gamma(3-\varepsilon)} + \frac{i}{(4\pi)^{D'/2}}\Big(\frac{1}{Q^2}\Big)$$

$$\times \Big(\frac{\mu_{D'}^2}{Q^2}\Big)^{-\frac{\delta}{2}}\Gamma\Big(1-\frac{\delta}{2}\Big)\Big[(p_1^\mu p_1^\nu + p_2^\mu p_2^\nu)\frac{\Gamma\big(\frac{\delta}{2}\big)\Gamma\big(2+\frac{\delta}{2}\big)}{\Gamma(3+\delta)} + (p_1^\mu p_2^\nu + p_1^\nu p_2^\mu)\frac{\Gamma^2\big(1+\frac{\delta}{2}\big)}{\Gamma(3+\delta)}\Big]$$

Therefore for the inverse of the gluon propagator, we finally have

$$D_{ab\mu\nu}^{-1}(Q) = \delta_{ab}Q^2\Big[\eta_{\mu\nu} + g^2\Big(\eta_{\mu\nu} - \frac{Q_\mu Q_\nu}{Q^2}\Big)\frac{[2(2-\varepsilon)n_f - (10-3\varepsilon)C_A]}{2(3-\varepsilon)}F(Q^2,\varepsilon)\Big],$$

$$(6.6.18)$$

and most importantly, we note that the contribution of the ghost, coming from the third part, has restored the transversality of the gluon propagator, consistent with gauge invariance, when added to the second part.

In order to evaluate the vertex functions in Fig. 6.10, we need the integrals in Box 6.2, worked out in Problems 6.8, 6.9. Here $p_1^2 = 0 = p_2^2$.

Here we encounter infrared divergences, which are rigorously taken care of by continuing the dimensionality of spacetime from 4 to $4+\delta$ with $\delta > 0$ in integrals involving these divergences, and by introducing, in the process, a mass scale $\mu_{D'}$ to define the corresponding integrals, as done in ultraviolet regularization.[44]

### 6.6.3  Fermion-Gluon vertex

The fermion-gluon vertex consists of two parts:

$$\Gamma_{aij}^\mu(p_1,p_2) = \gamma^\mu [t_a]_{ij} + \Lambda_{aij}^{(1)\mu}(p_1,p_2) + \Lambda_{aij}^{(2)\mu}(p_1,p_2),\qquad (6.6.19)$$

---

[44]Dimensional regularization as a way for handling infrared divergences is not new, see, e.g., Field [44].

The first part is given by (see Fig. 6.8, Fig. 6.10)

$$\Lambda_{aij}^{(1)\mu}(p_1,p_2) = -ig^2[t_b t_a t_c]_{ij}\delta_{bc}\int \frac{(dk)}{(2\pi)^4}\frac{\gamma^\sigma(-\gamma(p_2-k))\gamma^\mu(-\gamma(p_1-k))\gamma_\sigma}{(p_1-k)^2(p_2-k)^2\,k^2}\bigg|_{\text{Reg}}.$$

(6.6.20)

Only the quadratic part in $k$ is relevant in the ultraviolet region. Using the identity $\gamma^\sigma\gamma k\gamma^\mu\gamma k\gamma_\sigma = \gamma^\sigma\gamma^\alpha\gamma^\mu\gamma^\beta\gamma_\sigma\,k_\alpha k_\beta$, gives

$$\Lambda_{aij}^{(1)\mu}(p_1,p_2) = [t_b t_a t_c]_{ij}\,\delta_{bc}\,\gamma^\mu\bigg[\frac{g^2}{(4\pi)^{D/2}}\Big(\frac{\mu_D^2}{Q^2}\Big)^{\varepsilon/2}(2-\varepsilon)^2\frac{\Gamma\big(\frac{\varepsilon}{2}\big)\Gamma^2\big(1-\frac{\varepsilon}{2}\big)}{2\,\Gamma(3-\varepsilon)}$$

$$+\frac{g^2}{(4\pi)^2}\,h_{\text{IR}}^{(1)}\big(\frac{\mu_{D'}^2}{Q^2},\delta\big)\bigg],$$

(6.6.21)

where $h_{\text{IR}}^{(1)}(\mu_{D'}^2/Q^2,\delta)$ is a function of the infrared regularization parameters, and is independent of the ultraviolet ones. Its infrared singular structure is spelled out in Problem 6.10. Its explicit expression will not be needed in the sequel. The second part is defined by

$$\Lambda_{aij}^{(2)\mu}(p_1,p_2) = -ig^2[f_{bca}t_b t_c]_{ij}\int \frac{(dk)}{(2\pi)^4}\frac{i V_{\alpha\beta}{}^\mu(k-p_2,p_1-k,Q)\gamma^\alpha(-\gamma k)\gamma^\beta}{(p_1-k)^2(p_2-k)^2\,k^2}\bigg|_{\text{Reg}}.$$

(6.6.22)

Again the ultraviolet relevant part is the quadratic part in $k$ in the numerator in the integrand, i.e.,

$$V_{\alpha\beta}{}^\mu(k-p_2,p_1-k,Q)\gamma^\alpha(-\gamma k)\gamma^\beta \to -2\,[\gamma^\mu\,k^2+(2-\varepsilon)\,k^\mu\gamma k],\qquad (6.6.23)$$

which leads for $\Lambda_{aij}^{(2)\mu}(p_1,p_2)$ the expression

$$-[if_{bca}t_b t_c]_{ij}\gamma^\mu\bigg[\frac{g^2}{(4\pi)^{D/2}}\Big(\frac{\mu_D^2}{Q^2}\Big)^{\varepsilon/2}\frac{\Gamma\big(\frac{\varepsilon}{2}\big)\Gamma^2\big(1-\frac{\varepsilon}{2}\big)}{\Gamma(3-\varepsilon)}\,2\,(3-\varepsilon)+\frac{g^2}{(4\pi)^2}h_{\text{IR}}^{(2)}\big(\frac{\mu_{D'}^2}{Q^2},\delta\big)\bigg],$$

(6.6.24)

where $h_{\text{IR}}^{(2)}(\mu_{D'}^2/Q^2,\delta)$ is a function of the infrared regularization parameters, and independent of the ultraviolet ones. Its infrared singular structure is spelled out in

Problem 6.10. Hence

$$\Lambda_{aij}^{(1)\mu}(p_1, p_2) + \Lambda_{aij}^{(2)\mu}(p_1, p_2) = \gamma^\mu \left[ \frac{g^2}{(4\pi)^{D/2}} \left( \frac{\mu_D^2}{Q^2} \right)^{\varepsilon/2} \right.$$

$$\times \left( [t_b t_a t_c]_{ij}\, \delta_{bc} \frac{(2-\varepsilon)^2}{2} - [i f_{bca} t_b t_c]_{ij}\, 2\,(3-\varepsilon) \right) \frac{\Gamma\left(\frac{\varepsilon}{2}\right)\Gamma^2\left(1-\frac{\varepsilon}{2}\right)}{\Gamma(3-\varepsilon)}$$

$$\left. + \frac{g^2}{(4\pi)^2}\, [t_a]_{ij}\, h_{IR}\left( \frac{\mu_D^2{}'}{Q^2}, \delta \right) \right], \tag{6.6.25}$$

where $h_{IR}(\mu_D^2{}'/Q^2, \delta)$ is a function of the infrared regularization parameters, and is independent of the ultraviolet ones. Its infrared singular structure is spelled out in Problem 6.10. The explicit expression of $h_{IR}(\mu_D^2{}'/Q^2, \delta)$, however, will not be needed again.

## 6.7 Renormalization Constants, Effective Coupling, Asymptotic Freedom, and What is Responsible for the Latter?

A renormalized quantity cannot depend on the ultraviolet regularization parameters $(\varepsilon, \mu_D^2)$. Referring to the expression for $S_{ij}^{-1}$, in (6.6.6), we note that

$$\left( 1 - \frac{\varepsilon}{2} \right) \frac{1}{(4\pi)^{D/2}} \left( \frac{\mu_D^2}{p^2} \right)^{\varepsilon/2} \Gamma\left( \frac{\varepsilon}{2} \right) \frac{\Gamma^2\left(1-\frac{\varepsilon}{2}\right)}{\Gamma(2-\varepsilon)}$$

$$= \frac{1}{(4\pi)^2} \left[ \frac{2}{\varepsilon} - \gamma_E - \ln\left( \frac{\mu^2}{4\pi\mu_D^2} \right) + \ln\left( \frac{\mu^2}{p^2} \right) + c_1 \right], \quad \text{for } \varepsilon \to 0, \tag{6.7.1}$$

where $c_1$ is some constant. Hence we may write

$$S_{ij}^{-1}(p) = \delta_{ij} \frac{1}{Z_2} \left[ 1 + \frac{g^2}{(4\pi)^2} C_F \left( \ln\left( \frac{\mu^2}{p^2} \right) + a_1 \right) \right] \gamma p, \tag{6.7.2}$$

$$\frac{1}{Z_2} = 1 + \frac{g^2}{(4\pi)^2} C_F \left[ \frac{2}{\varepsilon} - \ln\left( \frac{\mu^2}{\mu_D^2} \right) + a_0 \right], \tag{6.7.3}$$

where $a_0$, $a_1$ are constants, and $a_0 + a_1 = c_1 + \ln(4\pi)$. Hence a given choice for the value of $a_0$ determines the constant $a_1$. Note that the coefficient of $1/Z_2$ in (6.7.2), is independent of the ultraviolet regularization parameters. A given choice for the value of the constant $a_0$ defines a renormalization scheme.

The simplest choice is to choose $a_0 = 0$ which defines the so-called minimal subtraction scheme (MS). A more popular choice is to choose

$$a_0 = -\gamma_E + \ln(4\pi),$$

as suggested from the expression on the right-hand side of (6.7.1), and is referred to as the modified minimal subtraction scheme $(\overline{\text{MS}})$, which we adopt here.[45] The wavefunction renormalization constant $Z_2$ of a fermion may be then written in this renormalization scheme, to second order, as

$$\frac{1}{Z_2} = 1 + \frac{g^2}{(4\pi)^2} C_F \left[ \frac{2}{\varepsilon} - \gamma_E - \ln\left(\frac{\mu^2}{4\pi\mu_D^2}\right) \right], \qquad C_F = \frac{N^2 - 1}{2N}. \qquad (6.7.4)$$

One may refer to $\mu$ as a renormalization scale. An identical analysis gives directly from (6.6.18)

$$D_{ab\mu\nu}^{-1}(Q) = Q^2 \delta_{ab} \left[ \eta_{\mu\nu} + \left( \eta_{\mu\nu} - \frac{Q_\mu Q_\nu}{Q^2} \right) \left( \frac{1}{Z_3} [1 + A(\mu^2/Q^2)] - 1 \right) \right], \qquad (6.7.5)$$

$$A(\mu^2/Q^2) = \frac{g^2}{(4\pi)^2} \left( \frac{2}{3} n_f - \frac{5}{3} C_A \right) \left[ \ln\left(\frac{\mu^2}{Q^2}\right) + a_2 \right], \qquad (6.7.6)$$

$$D_{ab\mu\nu}(Q) = \delta_{ab} \eta_{\mu\nu} \frac{1}{Q^2} + \delta_{ab} \left( Z_3 \left[ 1 - A(\mu^2/Q^2) \right] - 1 \right) \left( \eta_{\mu\nu} - \frac{Q_\mu Q_\nu}{Q^2} \right) \frac{1}{Q^2}, \qquad (6.7.7)$$

where

$$\frac{1}{Z_3} = 1 + \frac{g^2}{(4\pi)^2} \left( \frac{2}{3} n_f - \frac{5}{3} C_A \right) \left[ \frac{2}{\varepsilon} - \gamma_E - \ln\left(\frac{\mu^2}{4\pi\mu_D^2}\right) \right], \qquad C_A = N. \qquad (6.7.8)$$

It also follows directly from (6.6.25) that

$$\Gamma_{aij}^\mu = [t_a]_{ij} \gamma^\mu \frac{1}{Z_1} \left[ 1 + \frac{g^2}{(4\pi)^2} [t_b t_b + C_A] \left( \ln\left(\frac{\mu^2}{Q^2}\right) + a_3 \right) + \frac{g^2}{(4\pi)^2} h_{\text{IR}}\left(\frac{\mu_{D'}^2}{Q^2}, \delta\right) \right], \qquad (6.7.9)$$

$$\frac{1}{Z_1} = 1 + \frac{g^2}{(4\pi)^2} [C_F + C_A] \left[ \frac{2}{\varepsilon} - \gamma_E - \ln\left(\frac{\mu^2}{4\pi\mu_D^2}\right) \right], \qquad (6.7.10)$$

---

[45]Higher order loops contributions depend on the subtraction scheme adopted.

$$t_b t_b = C_F I, \quad C_F = \frac{N^2 - 1}{2N}, \quad C_A = N, \tag{6.7.11}$$

where $h_{\mathrm{IR}}(\mu_D^2 / Q^2, \delta)$ is a function of the infrared regularization parameters, and independent of the ultraviolet ones. Its infrared singular structure is spelled out in Problem 6.10. That the vertex function is infrared divergent and requires an infrared cut-off should not be surprising from our earlier QED treatment. The explicit expression of $h_{\mathrm{IR}}(\mu_{D'}^2 / Q^2, \delta)$ will not be needed.

Therefore with $\psi = \sqrt{Z_2}\, \psi_{\mathrm{ren}}$, $\overline{\psi} = \sqrt{Z_2}\, \overline{\psi}_{\mathrm{ren}}$, $A^\mu = \sqrt{Z_3} A_{\mathrm{ren}}^\mu$, $\Gamma^\mu = \Gamma_{\mathrm{ren}}^\mu / Z_1$, the renormalized quark-gluon coupling is given by

$$g_s(\mu^2) = \frac{\sqrt{Z_3}\, Z_2}{Z_1} g_0, \quad \alpha_s(\mu^2) = g_s^2(\mu^2)/4\pi, \quad \alpha_{\mathrm{os}} = g_0^2/4\pi, \tag{6.7.12}$$

$$\alpha_s(\mu^2) = \alpha_{\mathrm{os}}\left(1 + \frac{\alpha_s(\mu^2)}{4\pi}\left[\frac{11\,N}{3} - \frac{2n_f}{3}\right]\left[\frac{2}{\varepsilon} - \gamma_E - \ln\left(\frac{\mu^2}{4\pi\mu_D^2}\right)\right]\right). \tag{6.7.13}$$

Upon setting

$$\beta_0 = \left(\frac{11N - 2n_f}{12\pi}\right), \quad \left[\beta_0 = \left(\frac{33 - 2n_f}{12\pi}\right) \text{ for SU(3)}\right], \tag{6.7.14}$$

we may isolate the $\mu^2$-independent part of (6.7.13) by rewriting it as

$$\left[\frac{2}{\varepsilon} - \gamma_E + \ln(4\pi\mu_D^2)\right] - \frac{1}{\beta_0\,\alpha_{\mathrm{os}}} - -\frac{1}{\beta_0\,\alpha_s(\mu^2)} + \ln(\mu^2). \tag{6.7.15}$$

Since the left-hand side of this equation is independent of $\mu^2$, this means that for all $\mu^2$ such that $n_f$ of the quark flavors have masses squared $\ll \mu^2$,[46] the combination on the right-hand side of the above equation, with each of the two terms depending on the $\mu^2$'s, is an invariant, which we denote by $\ln(\Lambda^2)$. In particular, the latter depends on $n_f$. This gives

$$\alpha_s(\mu^2) = \frac{1}{\beta_0 \ln(\mu^2/\Lambda^2)}. \tag{6.7.16}$$

In general $(33 - 2n_f) > 0$, and hence $\beta_0$ is strictly positive. As $\mu^2$ increases, the effective coupling $\alpha_s(\mu^2)$ becomes weaker and eventually vanishes. This property of non-abelian gauge theories is referred to as asymptotic freedom. This QCD mass scale $\Lambda$, is obtained experimentally as will be discussed below. The above equation

---

[46]This justifies the neglect of the masses of the $n_f$ flavors in comparison to $\mu^2$.

also gives

$$\mu^2 \frac{\mathrm{d}}{\mathrm{d}\mu^2} \alpha_{\mathrm{s}}(\mu^2) = -\beta_0 \alpha_{\mathrm{s}}^2(\mu^2), \qquad (6.7.17)$$

and the minus sign on the right-hand side of the equation is the origin of asymptotic freedom. This is unlike the situation in QED, where (5.10.65) gives, for the effective fine-structure constant, the relation $\kappa^2 (\mathrm{d}/\mathrm{d}\kappa^2) \alpha(\kappa^2) = +\alpha^2/3\pi + \cdots$. Higher order loops contribution to the right-hand side of (6.7.17), may be also evaluated. For example next to the leading term on the right-hand side of (6.7.17), due to two loops, is given by $-\beta_1 \alpha_s^3(\mu^2)$,[47] where $\beta_1 = (153 - 19n_f)/(24\pi^2)$, and depends on $n_f$.

When $\mu^2$ corresponds to an energy associated with a given process, $\alpha_{\mathrm{s}}(\mu^2)$ provides an effective coupling for the theory and denotes the strength of the strong interaction at that energy. This effective coupling must be small enough to carry out reliable perturbative computations and the choice of the value of $\Lambda$ is important. It is a function of the number of flavors $n_f$ retained in the theory. The effective coupling $\alpha_{\mathrm{s}}(\mu^2)$ for $\mu^2 = M_Z^2$, corresponding to the mass $M_Z \simeq 91$ GeV, of the $Z^0$ vector boson, is experimentally of the order[48] 0.12. The mass $M_Z$ provides a typical standard renormalization scale. The coupling parameter $\alpha_{\mathrm{s}}$ is determined at some momentum transfer from various experiments such as from DIS ones, where violation of Bjorken scaling is sensitive to the value of $\alpha_{\mathrm{s}}$ (as well as from other experiments such as $e^+e^-$-annihilation). Multi-loop analysis[49] of the renormalization group is then carried out to evolve such values of $\alpha_{\mathrm{s}}$ to the energy scale $M_Z$. Such analyses, in turn, require corresponding values for $\Lambda$ to be used consistently when solving the renormalization group equation involving the beta function.

If, $Q^2$ corresponds to a typical momentum squared associated with a given process, then from the invariance property established in (6.7.15), we may use (6.7.16) to express the effective coupling $\alpha_{\mathrm{s}}(Q^2)$, also referred to as running coupling, in terms of the renormalized coupling $\alpha_{\mathrm{s}}(\mu^2)$ as follows

$$\alpha_{\mathrm{s}}(Q^2) = \frac{\alpha_{\mathrm{s}}(\mu^2)}{1 + \beta_0 \alpha_{\mathrm{s}}(\mu^2) \ln(Q^2/\mu^2)} = \frac{1}{\beta_0 \ln(Q^2/\Lambda^2)}. \qquad (6.7.18)$$

---

[47] Jones [69] and Caswell [29].

[48] Typically, recommendations for the value of $\Lambda$ are: 205–221 MeV, 286–306 MeV, 329–349 MeV, for $n_f = 5, 4, 3$, respectively. For the underlying phenomenology, as well as the determination of $\alpha_{\mathrm{s}}(M_Z^2)$ from experiments, see Beringer et al. [17].

[49] When considering more than one loop contribution to the beta function, the renormalization scheme becomes important. The modified minimal subtraction method ($\overline{\mathrm{MS}}$) mentioned above, is the most widely used in this case.

The second equality gives rise to a transformation of the dependence of $\alpha_s(Q^2)$ on the dimensionless coupling $\alpha_s(\mu^2)$ to a dependence on an energy scale $\Lambda$ instead.[50]

### 6.7.1 What Part of the Dynamics is Responsible for Asymptotic Freedom?

Unlike QED, the vector particle, the gluon, has a direct self-interaction. Therefore it is natural to investigate the contribution of the gluon self-interaction to $\beta_0$ in (6.7.14) for the SU(3) symmetry group. To isolate this contribution, let us scale the three-gluon vertex $V_{\mu_1\mu_2\mu_3}$ contribution to it by a parameter, say $\kappa$, which we can later set equal to one. This vertex function appears in the gluon self-energy part $\Pi^{(2)}_{ab\mu\nu}$ in (6.6.11). We have seen in (6.6.17), that the self-energy part $\Pi^{(3)}_{ab\mu\nu}$, involving the ghost field in (6.6.15), must be added to it to ensure gauge invariance. Thus we must scale the expression for $(\Pi^{(2)}_{ab\mu\nu} + \Pi^{(3)}_{ab\mu\nu})$ in (6.6.17) by $\kappa$. The three-gluon vertex $V_{\mu_1\mu_2\mu_3}$ also appears in the fermion-gluon vertex part $\Lambda^{(2)\mu}_{aij}$ in (6.6.22). Thus we must also scale its expression given in (6.6.24) by $\kappa$ as well. An elementary algebra gives

$$\beta_0 = \frac{1}{12\pi}\left(14 C_A \kappa - 3 C_A - 2 n_f\right)\Big|_{\kappa=1}, \quad C_A = 3. \tag{6.7.19}$$

The gluon self-interaction is thus responsible for making $\beta_0$ strictly positive and responsible for asymptotic freedom. It is interesting to note that for $n_f = 6$, the gluon-self interaction contribution to $\beta_0$ is precisely twice as big as the remainder in magnitude.

## 6.8 Renormalization Group and QCD Corrections to $e^+e^-$ Annihilation

Consider a dimensionless[51] unrenormalized quantity $G(Q^2, \alpha_{os})$, and for simplicity of the presentation, suppose it is a function of a single momentum (squared) variable $Q^2$, and of the unrenormalized coupling $\alpha_{os}$, and is gauge independent, suppressing the cut-offs in the theory. The unrenormalized theory is independent of the renormalization scale $\mu$ we choose to define the renormalized theory. One chooses $\mu$ at will depending on the energy range one is interested in defining the

---

[50]This is, in general, referred to as dimensional transmutation.

[51]An expression which has a non-zero dimensionality is readily taken care of by a scaling argument.

renormalized theory. Hence we have the elementary relation

$$\mu^2 \frac{\mathrm{d}}{\mathrm{d}\mu^2} G(Q^2, \alpha_{\mathrm{os}}) = 0. \tag{6.8.1}$$

Suppose that this expression is multiplicatively renormalized via a renormalization constant $Z(\mu^2, \alpha_s)$, with $\alpha_s \equiv \alpha_s(\mu^2)$ denoting the renormalized coupling. That is,

$$G_{\mathrm{ren}}\left(\ln\left(\frac{Q^2}{\mu^2}\right), \alpha_s\right) = \frac{1}{Z(\mu^2, \alpha_s)} G(Q^2, \alpha_{\mathrm{os}}), \tag{6.8.2}$$

$$\mu^2 \frac{\mathrm{d}}{\mathrm{d}\mu^2} \alpha_s = \beta(\alpha_s), \tag{6.8.3}$$

$$\frac{1}{Z(\mu^2, \alpha_s)} \mu^2 \frac{\mathrm{d}}{\mathrm{d}\mu^2} Z(\mu^2, \alpha_s) = \gamma(\alpha_s). \tag{6.8.4}$$

Upon substituting (6.8.2), in (6.8.1), using (6.8.3), and the fact that

$$\mu^2 \frac{\mathrm{d}}{\mathrm{d}\mu^2} = \mu^2 \frac{\partial}{\partial\mu^2} + \beta(\alpha_s) \frac{\partial}{\partial\alpha_s}, \tag{6.8.5}$$

we obtain from (6.8.4), (6.8.2) and (6.8.1)

$$\left[\mu^2 \frac{\partial}{\partial\mu^2} + \beta(\alpha_s) \frac{\partial}{\partial\alpha_s} + \gamma(\alpha_s)\right] G_{\mathrm{ren}}\left(\ln\left(\frac{Q^2}{\mu^2}\right), \alpha_s\right) = 0. \tag{6.8.6}$$

This suggests to set

$$G_{\mathrm{ren}}\left(\ln\left(\frac{Q^2}{\mu^2}\right), \alpha_s\right) = \exp\left[-\int_c^{\alpha_s} \frac{\mathrm{d}z\, \gamma(z)}{\beta(z)}\right] \widetilde{G}_{\mathrm{ren}}\left(\ln\left(\frac{Q^2}{\mu^2}\right), \alpha_s\right), \tag{6.8.7}$$

where $c$ is some constant, to obtain

$$\left[\mu^2 \frac{\partial}{\partial\mu^2} + \beta(\alpha_s) \frac{\partial}{\partial\alpha_s}\right] \widetilde{G}_{\mathrm{ren}}\left(\ln\left(\frac{Q^2}{\mu^2}\right), \alpha_s\right) = 0. \tag{6.8.8}$$

Upon introducing the variable

$$y = \int_{c'}^{\alpha_s} \frac{\mathrm{d}z}{\beta(z)}, \tag{6.8.9}$$

where $c'$ is a constant, (6.8.8) may be rewritten as

$$\left[\frac{\partial}{\partial y} - \frac{\partial}{\partial \ln(Q^2/\mu^2)}\right] \widetilde{G}_{\mathrm{ren}}\left(\ln\left(\frac{Q^2}{\mu^2}\right), \alpha_s\right) = 0, \tag{6.8.10}$$

where we have used the chain rule $\partial/\partial\alpha_s = (\partial y/\partial\alpha_s)(\partial/\partial y)$. The solution of (6.8.10) is of the form

$$\widetilde{G}_{\text{ren}}\Big(\ln\Big(\frac{Q^2}{\mu^2}\Big),\alpha_s\Big) = \phi\Big[\int_{c'}^{\alpha_s}\frac{dz}{\beta(z)} + \ln\Big(\frac{Q^2}{\mu^2}\Big)\Big], \tag{6.8.11}$$

where we have used the definition of the variable $y$ in (6.8.9). From (6.8.8), this gives

$$G_{\text{ren}}\Big(\ln\Big(\frac{Q^2}{\mu^2}\Big),\alpha_s\Big) = \exp\Big[-\int_c^{\alpha_s}\frac{dz\,\gamma(z)}{\beta(z)}\Big]\phi\Big[\int_{c'}^{\alpha_s}\frac{dz}{\beta(z)} + \ln\Big(\frac{Q^2}{\mu^2}\Big)\Big]. \tag{6.8.12}$$

On the other hand, from (6.8.3), we may introduce the effective coupling measured at $Q^2$: $\alpha(Q^2)$, where recall that $\alpha_s \equiv \alpha_s(\mu^2)$, and write

$$\ln\Big(\frac{Q^2}{\mu^2}\Big) = \int_{\alpha_s}^{\alpha_s(Q^2)}\frac{dz}{\beta(z)}, \tag{6.8.13}$$

by simple integration, to re-express (6.8.12) as

$$G_{\text{ren}}\Big(\ln\Big(\frac{Q^2}{\mu^2}\Big),\alpha_s\Big) = \exp\Big[-\int_c^{\alpha_s}\frac{dz\,\gamma(z)}{\beta(z)}\Big]\phi\Big[\int_{c'}^{\alpha_s(Q^2)}\frac{dz}{\beta(z)} + 0\Big], \tag{6.8.14}$$

obtained by combining $\ln(Q^2/\mu^2)$ with the first integral within the square brackets in $\phi[\,.\,]$ in (6.8.12). Comparing the above equation with the one in (6.8.12), leads us to the inescapable conclusion that

$$\phi\Big[\int_{c'}^{\alpha_s(Q^2)}\frac{dz}{\beta(z)} + 0\Big] = \exp\Big[\int_c^{\alpha_s(Q^2)}\frac{dz\,\gamma(z)}{\beta(z)}\Big]G_{\text{ren}}\big(0,\alpha_s(Q^2)\big). \tag{6.8.15}$$

Upon substituting this back in (6.8.14), gives the following remarkable solution of the renormalization group equation in (6.8.6)

$$G_{\text{ren}}\Big(\ln\Big(\frac{Q^2}{\mu^2}\Big),\alpha_s\Big) = \exp\Big[\int_{\alpha_s}^{\alpha_s(Q^2)}\frac{dz\,\gamma(z)}{\beta(z)}\Big]G_{\text{ren}}\big(0,\alpha_s(Q^2)\big). \tag{6.8.16}$$

The fact that the effective coupling $\alpha_s(Q^2)$ becomes weaker at high energies allows one to use perturbation theory in this coupling. The moral in this is that the high energy behavior of the theory, described by $Q^2$ becoming large, has now shifted from the dependence of the theory on the momentum $Q$ to a dependence on the effective coupling instead which becomes small in the high energy limit. This, as we have seen in the previous section, is due to the non-abelian character of the gauge field and its inherited remarkable property one refers to as asymptotic freedom.

An elementary application of the renormalization group is to obtain the QCD corrections to $e^+e^-$ annihilation to that of the parton model. The underlying result has become quite important as the theory shows constancy, as one goes to higher energies, of $R(s)$,[52] defined by the ratio of the cross section $e^+e^- \to$ hadrons to that of $e^+e^- \to \mu^+\mu^-$, consistent with experiments.

To the above end, the transverse part of the photon propagator has the structure,

$$D_{\mu\nu}(Q) = \left(\eta_{\mu\nu} - \frac{Q_\mu Q_\nu}{Q^2}\right)\left(\frac{Z_3^{(\gamma)}}{Q^2 - i\epsilon} + \int_0^\infty dM^2 \frac{\sigma(M^2)}{Q^2 + M^2 - i\epsilon}\right), \qquad (6.8.17)$$

where the photon wavefunction renormalization constant $Z_3^{(\gamma)}$ is given by

$$Z_3^{(\gamma)} = 1 - \int_0^\infty dM^2 \sigma(M^2). \qquad (6.8.18)$$

Upon substituting this expression in (6.8.17) gives

$$D_{\mu\nu}(Q) = \left(\eta_{\mu\nu} - \frac{Q_\mu Q_\nu}{Q^2}\right)\frac{d(Q^2)}{Q^2 - i\epsilon}, \qquad (6.8.19)$$

$$d(Q^2) = 1 - \int_0^\infty dM^2 \frac{M^2 \sigma(M^2)}{Q^2 + M^2 - i\epsilon}. \qquad (6.8.20)$$

The renormalized counterpart of $d(Q^2)$, is given by the expression

$$d_{\text{ren}}(Q^2) = \frac{1}{Z_3^{(\gamma)}} d(Q^2). \qquad (6.8.21)$$

Sice $\sigma(M^2)$ has an overall factor $\alpha$ (the fine-structure constant), and is small, one may write

$$\frac{1}{Z_3^{(\gamma)}} = 1 + \int_0^\infty dM^2 \sigma(M^2), \qquad (6.8.22)$$

and we have

$$\begin{aligned} d_{\text{ren}}(Q^2) &= \left(1 + \int_0^\infty dM^2 \sigma(M^2)\right)\left(1 - \int_0^\infty dM^2 \frac{M^2 \sigma(M^2)}{Q^2 + M^2 - i\epsilon}\right) \\ &= 1 + \int_0^\infty dM^2 \sigma(M^2) - \int_0^\infty dM^2 \frac{M^2 \sigma(M^2)}{Q^2 + M^2 - i\epsilon}, \end{aligned} \qquad (6.8.23)$$

---

[52]See also Sect. 6.5.

leading to

$$d_{\text{ren}}(Q^2) = 1 + Q^2 \int_0^\infty dM^2 \frac{\sigma(M^2)}{Q^2 + M^2 - i\epsilon}, \qquad (6.8.24)$$

which is different from the expression of $d(Q^2)$ in (6.8.20). When we take their imaginary parts, however, they coincide, as seen from (6.8.24) and (6.8.20),

$$\text{Im}\, d_{\text{ren}}(Q^2) = \text{Im}\, d(Q^2) = \pi\, Q^2 \sigma(-Q^2), \qquad (6.8.25)$$

as a consequence of the assumption of working to the leading order in $\alpha$ with the corresponding approximation made in (6.8.22).

That is

$$\mu^2 \frac{d}{d\mu^2}[\,\text{Im}\, d_{\text{ren}}(Q^2)\,] = \mu^2 \frac{d}{d\mu^2}[\,\text{Im}\, d(Q^2)\,] = 0, \qquad (6.8.26)$$

where $\mu$ is a renormalization energy scale. Upon writing [53]

$$\text{Im}\, d_{\text{ren}}(Q^2) = \alpha\, G\big(\ln(s/\mu^2), \alpha_s\big), \quad s = -Q^2 > 0, \qquad (6.8.27)$$

using the relation

$$\mu^2 \frac{d}{d\mu^2} = \mu^2 \frac{\partial}{\partial \mu^2} + \mu^2 \frac{d\alpha}{d\mu^2} \frac{\partial}{\partial \alpha} + \mu^2 \frac{d\alpha_s}{d\mu^2} \frac{\partial}{\partial \alpha_s}, \qquad (6.8.28)$$

and the facts that

$$\mu^2 \frac{d\alpha}{d\mu^2} = \alpha \frac{\mu^2}{Z_3^{(\gamma)}} \frac{dZ_3^{(\gamma)}}{d\mu^2} = \mathcal{O}(\alpha^2), \qquad (6.8.29)$$

$$\mu^2 \frac{d\alpha_s}{d\mu^2} \frac{\partial}{\partial \alpha_s}\big[\, \alpha\, G\big(\ln(s/\mu^2), \alpha_s\big)\,\big] = \alpha\, \beta(\alpha_s) \frac{\partial}{\partial \alpha_s} G\big(\ln(s/\mu^2), \alpha_s\big) + \mathcal{O}(\alpha^2), \qquad (6.8.30)$$

lead to the simple equation

$$\Big[\mu^2 \frac{\partial}{\partial \mu^2} + \mu^2 \frac{d\alpha_s}{d\mu^2} \frac{\partial}{\partial \alpha_s}\Big] G\big(\ln(s/\mu^2), \alpha_s\big) = 0. \qquad (6.8.31)$$

---

[53]Note that since $Q$ is time-like, i.e., $Q^2 < 0$, means that $-Q^2 > 0$.

From (6.8.4), (6.8.8), (6.8.16), the solution is simply given by

$$G\big(\ln(s/\mu^2), \alpha_s\big) = G(0, \alpha_s(s)),  \tag{6.8.32}$$

from which one may carry out standard perturbation theory in $\alpha_s$ and then replace it by $\alpha_s(\mu^2)$.

From the relation in (6.8.25), (6.5.9), for single photon exchange, with $s$ denoting the CM energy squared, and with

$$\beta_0 = \frac{(11N - 2n_f)}{12\pi} = \frac{(33 - 2n_f)}{12\pi},$$

for the SU(3), color group (see (6.7.14)), the following expression then emerges, up to additional vanishing terms,

$$R(s) = \frac{\sigma(e^+e^- \to \text{hadrons})}{\sigma(e^+e^- \to \mu^+\mu^-)} = 3 \sum_{i=1}^{n_f} (e_i^2/e^2) \left[ 1 + c_1 \frac{1}{\beta_0 \ln(s/\Lambda^2)} + \cdots \right],  \tag{6.8.33}$$

and is properly normalized to the one in (6.5.9) in the absence of radiative corrections, where the perturbative expansion coefficients are numerical constants, the computations of which are quite tedious. For example, $c_1 = 1/\pi$,[54] and is obtained from the last three photon self-energy parts, of order $\alpha g^2$, shown in Fig. 6.11, together with the first part corresponding to the parton model.

The main thing is that the parton model is consistent with the above description at sufficiently high energies. The fact that $c_1 > 0$ means that the approach to the parton model is from above.

The simplicity of the parton model should, however, be emphasized as it provides a convenient starting point for straightforward computations in QCD. The next section is involved with the fundamental theory of deep inelastic scattering.

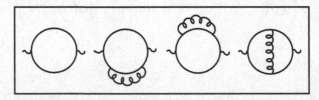

**Fig. 6.11** Photon self energy parts giving rise to the leading QCD contribution to $e^+e^-$ annihilation

---

[54]The value of this constant may be inferred from a computation carried out many years ago by Jost and Luttinger [70] in QED, see also Bjorken and Drell [21, p.344], Appelquist and Georgi [11], Zee [124] as well as Baikov et al. [13].

## 6.9  Deep Inelastic Scattering: Differential Cross Section and Structure Functions

The progress that has been made over the years in physics through scattering experiments has been quite significant. The emergence of the endless variety of particles by accelerators, the extraction of information on their interactions and the determination of the structure of matter are a few of the examples concerned with the analysis of scattering processes. Remarkable experimental and theoretical studies of resolving the structure of nucleons through the scattering of electrons and neutrinos off them have been, in particular, carried out. And, in the process, dynamical evidence for the existence of point-like[55] constituents was established. To probe and unravel the structure of a proton, for example, by $e^-p$ collision, via the exchange of a photon of spacelike momentum squared $Q^2$, imparted to the proton, the latter should be large enough.[56]

Here we are interested in the above inelastic $e^-p$ scattering beyond the resonance region,[57] in the neighborhood of a couple of GeVs, at which experimentally the reaction changes "character" and the scattering process may be appropriately represented by $e^-p \rightarrow e^- +$ "anything", where "anything" refers to any particles that may be created in the process with a final large invariant mass $M_n$, referred to as deep inelastic scattering, a process in which the proton loses its identity. The interest in such a process involving the proton (or a neutron) is to choose $Q^2(> 0)$ sufficiently large to unravel its underlying structure, as mentioned above, confirm its "quark description" with quarks as point-like constituents, and establish consistency of asymptotic freedom with experiments, with quarks inside the proton being nearly free at small distances. The process may be represented diagrammatically as shown in Fig. 6.12, where a one photon exchange is considered as a leading contribution to the process in question.

In the target frame, in which the ingoing proton is at rest, the momentum transferred squared $Q^2 > 0$ is given by $Q^2 = 2EE'(1 - \cos \vartheta) = 4EE' \sin^2 \vartheta/2$,

**Fig. 6.12**  The deep inelastic
$e^-p \rightarrow e^-$ "anything"
scattering process

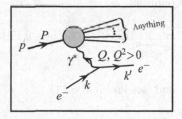

---

[55]Point-like at least at the energies in which such experiments were carried out.

[56] To obtain an order of magnitude of such a momentum squared needed, we note that since the size of the proton is of the order of 1 fm ($10^{-13}$ cm), we ought to have $h/\sqrt{Q^2} < 1$ fm, i.e., $2\pi/\sqrt{Q^2} < 1$ fm $\sim 5/$GeV or $1.6$ GeV$^2 < Q^2$ or even larger.

[57]A typical resonance is the $\Delta^+$ of mass 1.232 GeV, consisting of a proton $p$ and a $\pi^0$ meson.

where $\vartheta$ is the angle through which the electron is scattered: $\mathbf{k} \cdot \mathbf{k}' = |\mathbf{k}||\mathbf{k}'| \cos \vartheta$. If $P$, as indicated in the diagram, denotes the (four) momentum of the proton, the variable $v$ of the structure functions in (6.4.5), written in an invariant manner, is defined by

$$v = -\frac{PQ}{M}, \qquad Q = k - k', \tag{6.9.1}$$

where $M_p \equiv M$ is the mass of the proton. We note that if $P_n$ denotes the total momentum of all the particles created in the process corresponding to "anything" above, with an invariant mass squared $M_n^2 = -(P + Q)^2$, then $-PQ = (Q^2 + M_n^2 - M^2)/2$. In the rest frame of the proton, we then have

$$Q^0 = (Q^2 + M_n^2 - M^2)/2M > 0. \tag{6.9.2}$$

The expression for the differential cross section of the process is readily obtained. By averaging the initial spins of the electron and proton, and summing over the spin of the emerging electron, the differential cross section may be then simply written, in proton rest frame, as

$$d\sigma = \frac{(e)^4}{Q^4} \left(\frac{1}{4}\right) \sum_{\text{spins}} \sum_{(n,P_n)} (4m^2 2M) \frac{|\langle k'\sigma'; n, P_n|k\sigma, P\rangle|^2}{4ME} (2\pi)^4 \delta^4(P + Q - P_n) \frac{d^3\mathbf{k}'}{(2\pi)^3 2E'}, \tag{6.9.3}$$

where $E, E'$ are the initial and final energies of the electron, respectively, and $P_n$ is the total momentum of the particles in state $|n\rangle$,

$$\langle k'\sigma'; n, P_n|k\sigma, P\rangle = \bar{u}(\mathbf{k}', \sigma')\gamma^\mu u(\mathbf{k}, \sigma)\langle n, P_n|j_\mu(0)|P\rangle, \tag{6.9.4}$$

$j_\mu(0)$ is the electromagnetic current associated with the proton.

We use the facts that $\alpha = e^2/4\pi$, and for $m \to 0$, $d^3\mathbf{k}' = E'^2 dE' d\Omega'$, the relations

$$m^2 \sum_{\text{spins}} \text{Tr}[\bar{u}(\mathbf{k}', \sigma')\gamma^\mu u(\mathbf{k}, \sigma)\bar{u}(\mathbf{k}, \sigma)\gamma^\nu u(\mathbf{k}', \sigma')]\Big|_{m\to 0} = [k'^\mu k^\nu + k'^\nu k^\mu - \eta^{\mu\nu} kk'], \tag{6.9.5}$$

and (see Problem 6.11)

$$\frac{1}{2} \sum_\sigma \frac{1}{2\pi} \sum_{n,P_n} (2\pi)^4 \delta^4(P + Q - P_n)\langle P, \sigma|j_\mu(0)|n, P_n\rangle\langle n, P_n|j_\nu(0)|P, \sigma\rangle$$

$$= \frac{1}{2} \sum_\sigma \frac{1}{2\pi} \int (dy) e^{-iQy}\langle P, \sigma|j_\mu(y/2)j_\nu(-y/2)|P, \sigma\rangle \equiv W_{\mu\nu}(v, Q^2), \tag{6.9.6}$$

**Fig. 6.13** The process
$\gamma^* p \to$ "anything" for
determining general
expressions of the tensor $W_{\mu\nu}$

averaging over the spin of the proton, where $\int (dy)$ corresponds to an integration over spacetime, and the multiplicative factor $1/2\pi$ above is chosen for convenience.

To find the general expression for $W_{\mu\nu}$, we note that the corresponding relevant process for it is given in Fig. 6.13, involving a *virtual* photon ($Q^2 \neq 0$). Here we have at our disposal two vectors $Q$, $P$. Also current conservation $\partial_\mu j^\mu = 0$, implies that $Q^\mu W_{\mu\nu} = 0$, $Q^\nu W_{\mu\nu} = 0$. Hence $W_{\mu\nu}$ has the general structure

$$W_{\mu\nu} = \left(\eta_{\mu\nu} - \frac{Q_\mu Q_\nu}{Q^2}\right) W_1(\nu, Q^2) + \frac{1}{M^2}\left(P_\mu - \frac{PQ\, Q_\mu}{Q^2}\right)\left(P_\nu - \frac{PQ\, Q_\nu}{Q^2}\right) W_2(\nu, Q^2).$$
$$(6.9.7)$$

$W_1$, $W_2$ are referred to as structure functions of the proton, and have dimensionality mass$^{-1}$. They are functions of two invariant variables. It is more convenient to write the structure functions as functions of the *invariant* variable $Q^2$ and the dimensionless one

$$x = \frac{Q^2}{2M\nu}. \qquad (6.9.8)$$

Upon neglecting the mass of the electron in comparison to that of the proton, the differential cross section for the process in question, is, from (6.9.3), (6.9.4), (6.9.5), (6.9.6), (6.9.7), then given by

$$\frac{d^2\sigma}{dE' d\Omega'} = \frac{\alpha^2 \cos^2 \vartheta/2}{4\, E^2 \cdot \sin^4 \vartheta/2} \left[2\, W_1(x, Q^2) \tan^2 \vartheta/2 + W_2(x, Q^2)\right]. \qquad (6.9.9)$$

In particular, for elastic scattering, "anything" is replaced by an emerging proton, and the expression within the square brackets may be then expressed in terms of the electric and magnetic form factors of the proton as we have encountered in (6.4.1).

The invariant mass squared $M_n^2$ associated with final state may be conveniently written as

$$M_n^2 = M^2 + Q^2\left(\frac{1}{x} - 1\right). \qquad (6.9.10)$$

The above equality implies that $x \leq 1$. For elastic scattering $x = 1$. Since $\nu > 0$, we then have, in general,

$$0 \leq x \leq 1. \qquad (6.9.11)$$

Experimentally, as the available invariant mass $M_n$ of the final state is made to increase, by high momentum transfers squared $Q^2$, and large $v$, for fixed values of $x < 1$, defining the so-called Bjorken limit, the dimensionless functions $MW_1(x, Q^2)$, $vW_2(x, Q^2)$, become functions of the dimensionless ratio $x$ and depend on logarithms of $Q^2$, at present available high energies, depending on the value of $x$,[58] instead of rapidly vanishing like the elastic form factors, as one may naïvely expect. Note from (6.9.10), that as $Q^2 \to \infty$, $x < 1$, one is moving "deeply" into the inelastic region $M_n^2 \to \infty$, involved with "deep inelastic scattering".

One may directly derive positivity properties of the structure functions. Consider the target frame in which the proton is at rest: $P = (M, \mathbf{0})$. Suppose the virtual photon three-momentum is in the $z$ direction $Q^\mu = (v, 0, 0, \sqrt{Q^2 + v^2})$, where we have used the fact that $Q^2 = Q_3^2 - v^2$. The polarization vectors of the virtual photon $(Q^2 > 0)$ may be then taken as

$$\epsilon_1^\mu = (0, 1, 0, 0), \quad \epsilon_2^\mu = (0, 0, 1, 0), \quad \epsilon_0^\mu = (\sqrt{Q^2 + v^2}, 0, 0, v)/\sqrt{Q^2}.$$
(6.9.12)

A transversely polarized photon, leads to a cross section $\sigma_T \sim \epsilon_\lambda^\mu W_{\mu\rho} \epsilon_\lambda^\rho, = W_1$, for $\lambda = 1, 2$. Hence $W_1 \geq 0$. On the other hand, a longitudinal polarization,[59] leads to a cross section (see Problems 6.12)

$$\sigma_L \sim \epsilon_0^\mu W_{\mu\rho} \epsilon_0^\rho = [W_2 \frac{Q^2 + v^2}{Q^2} - W_1] \geq 0.$$
(6.9.13)

Accordingly, upon multiplying the above expression, by $2xM$, we are led to the basic inequalities

$$0 \leq 2xM W_1 \leq [1 + \frac{4M^2 x^2}{Q^2}] v W_2.$$
(6.9.14)

One often sets

$$M W_1(x, Q^2) \equiv F_1(x, Q^2), \quad v W_2(x, Q^2) \equiv F_2(x, Q^2),$$
(6.9.15)

expressed in terms of dimensionless functions. The basic inequalities in (6.9.14) then read

$$0 \leq 2x F_1(x, Q^2) \leq [1 + \frac{4M^2 x^2}{Q^2}] F_2(x, Q^2).$$
(6.9.16)

---

[58] See, e.g., Eidelman et al. [35].

[59] Longitudinal refers to the fact that the 3-vector of the polarization vector is parallel to $\mathbf{Q}$.

The necessity of having two structure functions is due to transverse as well as longitudinal polarization states available for virtual photons.[60]

The above also suggests to define transverse and longitudinal structure functions as follows:

$$F_T(x, Q^2) = 2 x F_1(x, Q^2), \tag{6.9.17}$$

$$F_L(x, Q^2) = \Big[ 1 + \frac{4 M^2 x^2}{Q^2} \Big] F_2(x, Q^2) - 2 x F_1(x, Q^2). \tag{6.9.18}$$

For future reference, we also define the ratio of these structure functions

$$R(x, Q^2) = \frac{F_L(x, Q^2)}{F_T(x, Q^2)}. \tag{6.9.19}$$

To find the proportionalities coefficients of the transverse and longitudinal parts of the cross section above, we define the flux factor (5.9.8), corresponding to the initial proton, with averaged spin, and virtual pair by $\sqrt{(PQ)^2 + Q^2 M^2} = M\nu \sqrt{1 + 2xM/\nu} \approx M\nu = Q^2/2x$. Hence in the deep inelastic region, we have from (6.9.6), (6.9.9), (5.9.7) for a given polarization

$$\sigma_\lambda = \frac{e^2}{4 (Q^2/2x)} (2M) \, \epsilon_\lambda^\mu (2\pi W_{\mu\rho}) \epsilon_\lambda^\rho, \tag{6.9.20}$$

where in reference to (5.9.7), $|\mathcal{M}|^2$ averaged over the spin of the proton $\overline{|\mathcal{M}|^2} = e^2 (2M) \, \epsilon_\lambda^\mu (2\pi W_{\mu\rho}) \, \epsilon_\lambda^\rho$. In particular, from (6.9.13), (6.9.15))

$$\frac{F_2(x, Q^2)}{x} = \frac{Q^2}{4\pi^2 \alpha \, x} (\sigma_T + \sigma_L), \qquad F_1(x, Q^2) = \frac{Q^2}{8\pi^2 \alpha \, x} \sigma_T, \tag{6.9.21}$$

for $Q^2 \to \infty$, $\nu \to \infty$, and $x \neq 0$ fixed.

We next consider the celebrated parton model to describe the above process in question, spell out the underlying assumptions involved in it, and derive the explicit expressions for the structure functions which show dependence only on the variable $x$, and are independent of $Q^2$. The latter condition is referred to as Bjorken *scaling*. As mentioned above, experiments also show logarithmic dependence on $Q^2$. As we will see later, a QCD treatment of the problem, leads to explicit calculable dependence on the logarithms of $Q^2$. The simpler description of deep inelastic scattering via the parton model is discussed next.

---

[60]The longitudinal polarization state arises as a consequence of the fact that for a virtual photon $Q^2 \neq 0$, and the photon ceases to be massless.

## 6.10   Deep Inelastic Scattering, The Parton Model, Parton Splitting; QCD Jets

### 6.10.1   The Parton Model

As the available invariant mass $M_n$ of the final state in $e^-p$ scattering is made to increase beyond the resonance region, by high momentum transfers squared $Q^2$, and large $\nu$, for fixed values of $x < 1$, experimentally the dimensionless functions $MW_1(x, Q^2)$, $\nu W_2(x, Q^2)$, become functions of the dimensionless ratio $x$, defined in (6.9.8), and depend on logarithms of $Q^2$, at present available high energies, for given values of $x$,[61] instead of rapidly vanishing like the elastic form factors, as one may naïvely expect. This non-vanishing property of the structure functions is reminiscent of scattering off point-particles. The parton model consists in assuming that the partons within the proton, as point-like, are free and the virtual photon interacts with each of its charged constituents independently, instead of interacting with the proton as a whole. That is, it assumes that the deep inelastic scattering cross section may be calculated from incoherent sums of elastic scattering processes of the electron with the charged constituents of the proton. If we represent interactions between partons by people holding hands, then if you pull one person you drag everybody else along. In the parton model the situation is illustrated by part (b) of the Fig. 1.7, in the introductory chapter, rather than by its part (a).

We will see below, that the parton model leads to the dependence of the structure functions on the dimensionless ratio $x$, but are independent of $Q^2$, a property referred to as Bjorken scaling, and the physical significance of this parameter naturally arises. QCD corrections studied in the next section accounts for the actual dependence of the structure functions on the logarithm of $Q^2$ as well.

The momentum carried by a quark, as a fraction of the momentum of the proton, is almost collinear with the momentum of the proton. The process for such a parton to acquire a significant transverse momentum through the exchange of a high energy gluon is suppressed by the smallness of the effective coupling at large momentum transfer. On the other hand, although the effective coupling becomes small at large momentum transfer, a quark may emit a collinear gluon, and due to the inherit infrared divergence associated with the masslessness of the gluons, as in QED, this compensates the smallness of the effective coupling and such a process becomes significant. This is discussed at the end of the section, and in detail in the next section in which QCD corrections, to the naïve parton model, are taken into account.

Let $\xi P$ $(0 \leq \xi \leq 1)$, denote the fraction of momentum of the proton carried by a parton. The differential cross section of elastic scattering of such a parton with the electron, with one photon exchange, as shown below, is simply obtained as

---

[61]See, e.g., Eidelman et al. [35].

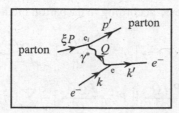

**Fig. 6.14** Elastic scattering of the electron, via a one photon exchange, with a parton of charge $e_i$ carrying a fraction $\xi P$ of the momentum of the proton. The gluon, of course, has no electromagnetic charge

follows. Let $e_i$ denote the charge of an $i^{\text{th}}$ parton (Fig. 6.14)).[62] The structure tensor $W_i^{\mu\nu}$ contribution due to the $i^{\text{th}}$ parton is then simply given by (see Problem 6.13)

$$
W_i^{\mu\nu} = \frac{e_i^2/e^2}{\xi M} \int \frac{d^3\mathbf{p}'}{2p'^0} \delta^{(4)}(\xi P + Q - p') \left[ 2\,\xi^2 P^\mu P^\nu + \xi(P^\mu Q^\nu + P^\nu Q^\mu) - \eta^{\mu\nu} \xi PQ \right].
$$
(6.10.1)

Now we use the integral

$$
\int \frac{d^3\mathbf{p}'}{2p'^0} \delta^{(4)}(\xi P + Q - p') = \frac{1}{2\,|PQ|}\,\delta(\xi - x),
$$
(6.10.2)

established in Problem 6.14, where $x$ is defined in (6.9.8). Using the facts that $QP/|QP| = -1$, $\xi = x$, according the delta function $\delta(\xi - x)$ constraint, the definition of the variable $x$, and conveniently grouping the various terms in (6.10.1), gives for the latter

$$
W_i^{\mu\nu}(\xi, Q^2) = \frac{e_i^2/e^2}{2M}\,\delta(\xi - x) \left[ \eta^{\mu\nu} - \frac{Q^\mu Q^\nu}{Q^2} \right]
$$

$$
+\ \frac{e_i^2/e^2}{\nu}\,x\,\delta(\xi - x)\frac{1}{M^2}\left[ P^\mu - \frac{PQ\,Q^\mu}{Q^2} \right]\left[ P^\nu - \frac{PQ\,Q^\nu}{Q^2} \right],
$$
(6.10.3)

where $\nu$ is defined in (6.9.1). We may infer that $x$ denotes the *fraction of momentum carried by a quark (antiquark)*. Let $f_i(\xi)$ denote the number densities of partons of type $i$ having momenta in the range $(\xi, \xi + d\xi)P$. Then the structure tensor $W^{\mu\nu}$

---

[62]Since gluons, as neutral partons, do not carry electromagnetic charge, they will have no role in this analysis. We will say more about this at the end of the section and elaborate on how they are taken into account in a proper QCD treatment which is the subject of the following section.

of the deep inelastic process, according to the parton model, is given by

$$\sum_i \int_0^1 d\xi f_i(\xi) W_i^{\mu\nu}(\xi, Q^2) = \left[ \eta^{\mu\nu} - \frac{Q^\mu Q^\nu}{Q^2} \right] \frac{1}{2M} \sum_i (e_i^2/e^2) f_i(x)$$

$$+ \frac{1}{M^2} \left[ P^\mu - \frac{PQ\,Q^\mu}{Q^2} \right]\left[ P^\nu - \frac{PQ\,Q^\nu}{Q^2} \right] \frac{1}{\nu} \sum_i (e_i^2/e^2)\, x f_i(x). \qquad (6.10.4)$$

The structure functions are then explicitly given by

$$2Mx\,W_1(x) = \sum_i (e_i^2/e^2)\, x f_i(x) \equiv 2x F_1(x), \qquad (6.10.5)$$

$$\nu\, W_2(x) = \sum_i (e_i^2/e^2)\, x f_i(x) \equiv F_2(x), \qquad (6.10.6)$$

where we have used the definitions of the dimensionless structure functions $F_1$, $F_2$ in (6.9.15). Thus Bjorken scaling holds in the parton model, with the structure constants depending only on $x$, and are independent of $Q^2$. It also gives the following equality involving structure functions[63]

$$2x F_1(x) = F_2(x). \qquad (6.10.7)$$

The above equality is quite significant as it provides ample dynamical evidence of the *spin 1/2 character of quarks*. To this end, using the definitions of the transverse and longitudinal structure functions in (6.9.17), (6.9.18), we obtain for the ratio $R$, defined in (6.9.19), the expression $R = 4M^2 x^2/Q^2 \sim 0$ for $Q^2 \to \infty$. On the other hand, if quarks are of spin 0, it is easily shown, as worked out in Problem 6.15, that $F_T = 0$, leading to exactly the *opposite* behavior $R \to \infty$. Although the strict equality in (6.10.7) does not hold in practice, the numerical data[64] for $2x F_1/F_2$ is concentrated around the value one, and hence from the definition of $R$, this implies consistency with a spin 1/2 character for quarks rather than of spin 0.[65]

It is worth noting that if one takes into account only the three basic quarks $(u, u, d)$, associated with the proton, into consideration,[66] and assumes that they occur with the same frequency, then we may write $\sum_i (e_i^2/e^2)\, x f_i(x)$ $= x f(x) \sum_i (e_i^2/e^2) = x f(x) = (1/3) \sum_i x f_i(x)$, where we have used the fact that $\sum_i (e_i^2/e^2) = (4/9) + (4/9) + (1/9) = 1$. It leads to the relation $\int_0^1 dx\, F_2(x) = (1/3) \int_0^1 dx \sum_i x f_i(x)$. This is consistent with rough estimates

---

[63]This equality is referred to as the Callan-Gross relation [28].

[64]See, e.g., Bodek et al. [24].

[65]The presence of such spin 0 bosons, if any, in addition to spin 1/2 ones, is possibly not of significance.

[66]These quarks are referred to as valence quarks and the remaining as sea quarks, and antiquarks.

$\int_0^1 dx \sum_i x f_i(x) \approx 0.54$, and $\int_0^1 dx\, F_2(x) \approx 0.18$ obtained by integrating over deep inelastic scattering experimental data.[67] They provide evidence of fractional charges of the quarks. Similar results hold for the deep inelastic scattering associated with the $e^- n$ system (see Problem 6.16).

In the development of the parton model above, gluons, as *neutral* partons, did not play any role as they have no electromagnetic charge. Hence it is expected that the sum of the fractions of the proton's momentum, carried by the quarks and antiquarks, do not add up to match the momentum of the proton. The experimental order of magnitude rough estimate $\int_0^1 dx \sum_i x f_i(x) \approx 0.54$ indicates that approximately the other half of the momentum of the proton is carried by neutral partons - the gluons. Hence such a result is hardly surprising. In the next section, dealing with QCD corrections, gluons are naturally taken into account as a result of their emissions by quarks and from the underlying QCD interactions.

Experimentally, Bjorken scaling is violated by logarithmic terms in the momentum transfer squared $Q^2$, for high values attainable in present experiments, and are accounted for in a QCD treatment of the problem as investigated in the next section.

### 6.10.2  Parton Splitting

In preparation for the analysis to be carried out in the next section, which takes into account QCD corrections to the naïve parton model, in deep inelastic scattering, at large $Q^2$, we investigate the nature of a quark emitting a gluon, as well as a gluon splitting, derive the probabilities of such events and see how these modify the corresponding parton distributions with a variation in $Q^2$.

To the above end, we recall the first expression

$$\frac{F_2(x, Q^2)}{x} = \frac{Q^2}{4\pi^2 \alpha x} \sigma, \tag{6.10.8}$$

in (6.9.21) in deep inelastic scattering, where $\sigma$ is the cross section for the scattering of a virtual photon off the proton: $\sigma(\gamma^* p \rightarrow$ "anything"). A quark with fractional momentum $\xi$ of that of the proton, upon the emission of a gluon, say, acquires a momentum with a fractional longitudinal momentum $z$ of the parent quark, thus having a final fractional longitudinal momentum $\xi z \equiv x$ of that of the proton. For the naïve parton model, we may write

$$f_i(x) = \int_0^1 d\xi\, f_i(\xi) \int_0^1 dz\, \delta(\xi z - x) \frac{Q^2}{4\pi^2 \alpha z} \left[ \frac{4\pi^2 \alpha z}{Q^2} \delta(z - 1) \right], \tag{6.10.9}$$

---

[67] Halzen and Martin [61, p.203]. See also Field and Feynman [45] for similar data.

easily verified upon integrations. For the emission of a gluon by a quark, the distribution function will change $f_i(x) \rightarrow f_i(x, Q^2)$, and from the above expressions given in (6.10.8), (6.10.9), we may write

$$f_i(x, Q^2) = \int_0^1 d\xi\, f_i(\xi) \int_0^1 dz\, \delta(\xi z - x)\, \frac{Q^2}{4\pi^2\alpha\, z}\left[\frac{4\pi^2\alpha\, z}{Q^2}\delta(z-1) + \frac{e^2}{e_i^2}\hat{\sigma}_i(z, Q^2)\right],$$

(6.10.10)

where $\hat{\sigma}_i(z, Q^2)$ is the cross section of the scattering: $\gamma^* q \rightarrow qG$, where $G$ denotes a gluon, and if the momentum of the initial quark is $p$, then the fractional longitudinal momentum of the final quark is $zp$. In the next section, we will see how to modify the above expression to take into account other relevant processes and also to reflect the dependence of these processes on $Q^2$ at high energies. To lowest order, there are two diagrams associated with the above process with the same final products of a quark and a gluon. They are given by

$$: \mathcal{M}_1 = e_i g\, \overline{u}(p', \sigma')\gamma^\mu \frac{-\gamma(p'-Q)}{(p'-Q)^2}\gamma^\nu t_c u(p, \sigma)\epsilon_\mu e_\nu,$$

$$: \mathcal{M}_2 = e_i g\, \overline{u}(p', \sigma')\gamma^\nu t_c \frac{-\gamma(p+Q)}{(p+Q)^2}\gamma^\mu u(p, \sigma)\epsilon_\mu e_\nu,$$

where the $e_\nu$ are polarization vectors of the gluon, and the $\epsilon_\mu$ are polarization vectors of the virtual photon.[68] We define Mandelstam variables (5.9.14) relevant to the parton scattering above in the following way:

$$\hat{s} = -(p+Q)^2, \quad \hat{t} = -(p-p')^2, \quad \hat{u} = -(Q-p')^2, \quad \hat{s}+\hat{t}+\hat{u} = -Q^2.$$

(6.10.11)

The masses of all the partons are set equal to zero. On the other hand, the photon being virtual means that it is not on the mass shell and $Q^2 \neq 0$, in general. Actually we are interested in the large $Q^2$ behavior. Upon using the identity $\mathrm{Tr}[t_c t_c] = 4$, for the SU(3) group, and averaging over spins, polarizations, and the three colors of the latter group, give for the average $\overline{|\mathcal{M}|^2}$ of $|\mathcal{M}|^2$, where $\mathcal{M} = 2m_i(\mathcal{M}_1 + \mathcal{M}_2)$,

---

[68]See (6.9.12).

the expression $(\alpha_s = g^2/4\pi)$

$$\overline{|\mathcal{M}|^2}\Big|_{m_i=0} = \frac{128\,\pi^2}{3}\,\alpha\,\alpha_s\left(\frac{e_i^2}{e^2}\right)\left[-\frac{\hat{t}}{\hat{s}} - \frac{\hat{s}}{\hat{t}} + 2\frac{\hat{u}\,Q^2}{\hat{s}\hat{t}}\right], \qquad (6.10.12)$$

where $m_i$ is the mass of the quark. The definition of the parameter $z$ as the fractional longitudinal momentum of the daughter quark relative to the parent quark momentum, and the definition of the invariant $\hat{s}$, given above, lead from the equalities $\hat{s} + Q^2 = -2pQ$, $z = -Q^2/2pQ$ (see (6.9.8)), to the relations

$$z = \frac{Q^2}{\hat{s} + Q^2}, \qquad \frac{\hat{s}}{\hat{s} + Q^2} = 1 - z, \qquad \hat{s} = Q^2\frac{(1-z)}{z}. \qquad (6.10.13)$$

Since $\hat{u} = -(\hat{s} + \hat{t} + Q^2)$, we may use the relations just given above, to rewrite the expression in the square brackets in (6.10.12) as

$$[\,.\,] = -\frac{Q^2}{z}\left[\frac{z^2}{Q^4(1-z)}\hat{t} + \left(\frac{1+z^2}{1-z}\right)\frac{1}{\hat{t}} + \frac{2z^2}{Q^2(1-z)}\right]. \qquad (6.10.14)$$

In the CM frame, (5.9.15) gives

$$\frac{d\hat{\sigma}_i}{d\hat{t}} = \frac{1}{64\pi\,\hat{s}\,\mathbf{p}_{\mathrm{CM}}^2}\overline{|\mathcal{M}|^2}, \qquad (6.10.15)$$

$$\hat{s}\,\mathbf{p}_{\mathrm{CM}}^2 = \frac{(\hat{s} + Q^2)^2}{4} = \frac{Q^4}{4z^2}, \qquad \hat{t} = \frac{Q^2}{2z}(\cos\theta - 1), \qquad (6.10.16)$$

and $0 \le |\hat{t}| \le Q^2/z$. For $Q^2$ large, (6.10.12), (6.10.13), (6.10.14), (6.10.15), (6.12.16) lead to

$$\hat{\sigma}_i(z, Q^2) = \frac{8\pi\alpha z\alpha_s}{3\,Q^2}\frac{e_i^2}{e^2}\left[\frac{1+z^2}{1-z}\right]\int_0^{Q^2/z}\frac{d|\hat{t}|}{|\hat{t}|}. \qquad (6.10.17)$$

The integral involves an infrared divergence coming from its lower end of integration, corresponding to emission of a collinear gluon, which in here may be rigorously defined by introducing an infrared cut-off. As we are interested in the variation of the distribution function in (6.10.10) with $Q^2$, we do not have to carry out this procedure here. Accordingly, we may write

$$\hat{\sigma}_i = \frac{4\pi^2\alpha z}{Q^2}\frac{e_i^2}{e^2}\frac{\alpha_s}{2\pi}\frac{4}{3}\left[\frac{1+z^2}{1-z}\right]\left[\ln(Q^2) + \cdots\right], \qquad (6.10.18)$$

where the remaining terms are independent of $Q^2$ for sufficiently large $Q^2$. From (6.10.10), the following expression emerges

$$\delta f_i(x, Q^2) = \int_0^1 d\xi\, f_i(\xi) \int_0^1 dz\, \delta(\xi z - x) \frac{\alpha_s}{2\pi} \frac{4}{3} \left[ \frac{1 + z^2}{1 - z} \right] \delta \ln(Q^2). \qquad (6.10.19)$$

The denominator $(1 - z)$ in $[(1 + z^2)/(1 - z)]$ is not integrable at $z = 1$. This, however, may be defined by a "+" operation, in the sense of distributions, which for a given smooth function $f(z)$, one sets

$$\int_0^1 dz \frac{f(z)}{(z - 1)_+} = \int_0^1 dz\, \frac{f(z) - f(1)}{(z - 1)}. \qquad (6.10.20)$$

Also to the same leading order above, we have to take into account the radiative (self-energy and vertex) corrections, where the quark reabsorbs the gluon and the quark remains a quark, it does not split into quark plus a gluon. This contribution may be represented by an expression $A\,\delta(z - 1)$ where $A$ is a constant, which modifies the expression within the square brackets in (6.10.19) to $\big( [(1 + z^2)/(1 - z)_+] + A\delta(z - 1) \big)$. Conservation of probability then leads from (6.10.10) to

$$\int_0^1 dz \left( \delta(z - 1) + \frac{\alpha_s}{2\pi} \frac{4}{3} \left[ \frac{1 + z^2}{(1 - z)_+} + A\delta(z - 1) \right] \ln Q^2 \right) = 1, \qquad (6.10.21)$$

giving $A = 3/2$, with $Q^2$ expressed relative to any scale. Thus we introduce the so-called splitting factor of a quark into a quark and a gluon, where the emerging quark has a fractional longitudinal momentum in the range $(z, z + dz)$ relative to the momentum of the parent quark:

$$\mathscr{P}_{q \to qG}(z) = \frac{4}{3} \left[ \frac{(1 + z^2)}{(1 - z)_+} + \frac{3}{2} \delta(z - 1) \right]. \qquad (6.10.22)$$

The same analysis gives $\mathscr{P}_{\bar{q} \to \bar{q}G}(z) = \mathscr{P}_{q \to qG}(z)$, and a similar analysis leads to the splitting factors $\mathscr{P}_{q \to Gq}(z) = \mathscr{P}_{\bar{q} \to G\bar{q}}(z)$, where now the gluon carries a fractional longitudinal momentum $z$ of the momentum of the quark (antiquark), and is recorded in (6.10.41).

Now we come to the interesting case of the splitting of a gluon. To the same order, the gluon may split to a quark antiquark pair, or split into two gluons, denoted by $\mathscr{P}_{G \to q\bar{q}}(z) = \mathscr{P}_{G \to \bar{q}q}(z)$, and $\mathscr{P}_{G \to GG}(z)$. We evidently have the constraint

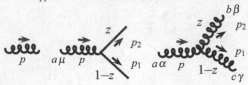

$$\int_0^1 dz\, z\, \delta(z-1) = 1, \qquad \int_0^1 dz\, z\left[ n_f \mathscr{P}_{G\to q\bar{q}}(z) + n_f \mathscr{P}_{G\to \bar{q}q}(z) + \mathscr{P}_{G\to GG}(z) \right] = 0,$$

$$(6.10.23)$$

corresponding to the splitting gluon with momentum $p$, where $n_f$ denotes the number of quark flavors considered, and note the directions of the momenta taken.

To obtain $\mathscr{P}_{G\to GG}$, we recall the expression for the three-gluon vertex in Fig. 6.8:

$$V_{\alpha\beta\gamma}^{abc} = ig f_{abc}\left[ (p+p_2)_\gamma \eta_{\alpha\beta} - (p_2-p_1)_\alpha \eta_{\beta\gamma} - (p_1+p)_\beta \eta_{\gamma\alpha} \right], \qquad (6.10.24)$$

As in (6.10.16), (6.10.17) a singularity arises in the zero mass limit for collinear momenta of the gluons with the parent one. The momenta are taken as

$$p = (k,0,0,k), \quad p_2 = \left(zk + \frac{\mathbf{k}_\perp^2}{2zk}, \mathbf{k}_\perp, zk\right), \quad p_1 = \left((1-z)k \; \frac{\mathbf{k}_\perp^2}{2zk}, -\mathbf{k}_\perp, (1-z)k\right),$$

$$(6.10.25)$$

where note that $p_2^0 = \sqrt{z^2k^2 + \mathbf{k}_\perp^2} \simeq zk + \mathbf{k}_\perp^2/2zk$, and $p_1^0$ follows by conservation of energy. Also $p_2^2 = \mathscr{O}(\mathbf{k}_\perp^4)$, that is the corresponding gluon is nearly on the mass shell, while

$$p_1^2 \simeq \frac{\mathbf{k}_\perp^2}{z}, \qquad (6.10.26)$$

and the corresponding gluon is to be treated as a virtual particle, the expression of which we record here for future reference.

Upon introducing polarization vectors $e^\alpha, e_2^\beta, e_1^\gamma$, as in (6.9.12), multiplying the latter vectors by the three-gluon vertex, taking its absolute value squared, and summing over the final polarizations states, and averaging over the initial ones, give

$$\overline{|V|^2} = C_A \frac{4g^2 \mathbf{k}_\perp^2}{z(1-z)}\left[ \frac{z}{1-z} + \frac{1-z}{z} + +z(1-z) \right], \qquad (6.10.27)$$

where $C_A$ is the group factor equal to 3 for the SU(3) group given in (6.6.2), (6.6.3). We consider, in turn, the cross section of a process: $AG \to BG$, via the exchange of a gluon, with the momentum of the incoming gluon given by $p$ in (6.10.25). The flux factor of the process as given in (5.9.7), (5.9.8) becomes

$$-\frac{1}{4pp_A} = \frac{1}{4kp_A^0(1 - p_A^z/p_A^0)} = \frac{(1-z)}{4(1-z)kp_A^0(1 - p_A^z/p_A^0)}$$

$$\simeq \frac{(1-z)}{4p_1^0 p_A^0(1 - p_A^z/p_A^0)} \simeq -\frac{(1-z)}{4p_1 p_A}, \qquad (6.10.28)$$

where note that $p_1 \simeq \big((1-z)k, 0, 0, (1-z)k\big) = (1-z)p$. By using

$$\frac{d^3\mathbf{p}_2}{(2\pi)^3 2p_2^0} \simeq \frac{\pi\, k\, dz\, d\mathbf{k}_\perp^2}{(2\pi)^3 2\, z\, k}, \qquad (6.10.29)$$

and the expression for the inverse of the squared of the gluon momentum squared $p_1^2$ as obtained from (6.10.26)

$$\frac{1}{p_1^4} = \frac{z^2}{\mathbf{k}_\perp^4}, \qquad (6.10.30)$$

the above mentioned cross section may be represented for $\mathbf{k}_\perp^2 \to 0$, by

$$\sigma\big(GA \to GB\big) = \int_0^1 dz \int \frac{\pi\, k\, d\mathbf{k}_\perp^2}{(2\pi)^3 2\, z\, k} \left( 12\frac{g^2\mathbf{k}_\perp^2}{z(1-z)}\Big[\frac{z}{1-z} + \frac{1-z}{z} + +z(1-z)\Big]\right)$$

$$\times \left(\frac{z^2}{\mathbf{k}_\perp^4}\right)[(1-z)\sigma\big(GA \to B\big)], \qquad (6.10.31)$$

where $\sigma\big(GA \to B\big)$ denotes the cross section with an incoming real transverse gluon having a longitudinal momentum $(1-z)k$, where the factor $(1-z)$ multiplies the cross section just mentioned occurring on the right-hand side of (6.10.31) as a consequence of the numerator in the last equality in (6.10.28). The above equation simplifies to

$$\sigma\big(GA \to GB\big) = \int_0^1 dz \int \frac{d\mathbf{k}_\perp^2}{\mathbf{k}_\perp^2}\, \frac{\alpha_s}{2\pi}\, \mathscr{P}_{G\to GG}(z)\, \sigma\big(GA \to B\big), \qquad (6.10.32)$$

displaying the mass zero singularity occurring in the integration over transverse momentum for a gluon with almost collinear momentum with the parent gluon, and

$(\alpha_s = g^2/4\pi)$

$$\mathscr{P}_{G \to GG}(z) = 6\left[\frac{z}{(1-z)_+} + \frac{1-z}{z} + z(1-z)\right], \tag{6.10.33}$$

where we have used the distributional definition of the term $1/(1-z)$ as given in (6.10.20). We have to add a term $C\delta(z-1)$ to the above expression corresponding to the emission of a soft gluon. The constant $C$ is determined from the constraint equation (6.10.23). Before determining the constant $C$, it is straightforward to work out, by an almost identical treatment as the one just given, and obtain the splitting factor of a gluon to a quark-antiquark pair:

$$\mathscr{P}_{G \to q\bar{q}}(z) = \frac{1}{2}[z^2 + (1-z)^2]. \tag{6.10.34}$$

In view of determining the constant $C$, note that

$$\int_0^1 dz\, z\,[z^2 + (1-z)^2] = \frac{1}{3}, \tag{6.10.35}$$

$$\int_0^1 dz\, z\left[\frac{z}{(1-z)_+} + \frac{1-z}{z} + z(1-z)\right] = -\frac{11}{12}, \tag{6.10.36}$$

from which (6.10.23) gives $C = (11/12 - n_f/18)$. Note the following equalities hold between the splitting factors:

$$\mathscr{P}_{q \to qG}(z) = \mathscr{P}_{\bar{q} \to \bar{q}G}(z), \tag{6.10.37}$$

$$\mathscr{P}_{q \to Gq}(z) = \mathscr{P}_{\bar{q} \to G\bar{q}}(z), \tag{6.10.38}$$

$$\mathscr{P}_{G \to q\bar{q}}(z) = \mathscr{P}_{G \to \bar{q}q}(z). \tag{6.10.39}$$

The last equality follows from that in (6.10.34) since the latter is invariant under the exchange $z \leftrightarrow (1-z)$. In $\mathscr{P}_{a \to bc}(z)$, parton $b$ has a fractional momentum $z$ relative to that of parton $a$. Accordingly, note that $\mathscr{P}_{q \to qG}(z) \neq \mathscr{P}_{q \to Gq}(z)$.

The splitting factors are then explicitly given by

$$\mathscr{P}_{q \to qG}(z) = \mathscr{P}_{\bar{q} \to \bar{q}G}(z) = \frac{4}{3}\left[\frac{1+z^2}{(1-z)_+} + \frac{3}{2}\delta(z-1)\right], \tag{6.10.40}$$

$$\mathscr{P}_{q \to Gq}(z) = \mathscr{P}_{\bar{q} \to G\bar{q}}(z) = \frac{4}{3}\left[\frac{1+(1-z)^2}{z}\right], \tag{6.10.41}$$

$$\mathscr{P}_{G \to q\bar{q}}(z) = \mathscr{P}_{G \to \bar{q}q}(z) = \frac{1}{2}[z^2 + (1-z)^2], \tag{6.10.42}$$

$$\mathscr{P}_{G \to GG}(z) = 6\left[\frac{z}{(1-z)_+} + \frac{1-z}{z} + z(1-z) + \left(\frac{11}{12} - \frac{n_f}{18}\right)\delta(z-1)\right].$$

(6.10.43)

As before, $n_f$ denotes the number of quark flavors, and $1/(1-z)_+$ is defined in (6.10.20).

A unified notation generalizing the one in (6.10.32) may be given as follows:

$$\sigma(aA \to bB) = \int_0^1 dz \int \frac{dk_\perp^2}{k_\perp^2} \frac{\alpha_s}{2\pi} \mathscr{P}_{a \to bc}(z) \sigma(cA \to B).$$

(6.10.44)

### 6.10.3 QCD Jets

One of the successful applications of QCD is in providing consistent descriptions of jet events observed in the process $e^-e^+ \to$ hadrons.[69] When a quark-antiquark pair is initially produced in such a process, these emerging particles develop bremsstrahlung cascades of narrowly collimated gluons, which in turn lead to the further productions of quark-antiquark pairs. These processes continue until all colored particles are combined into color singlet hadrons leading to the generation of jets of hadrons at large distances. The formation of hadrons from the $q, \bar{q}$ pair is called hadronization.[70] When a gluon, for example, is radiated from an initially emerging quark (antiquark) with a sufficiently large (hard) energy that one is able to detect it experimentally, and with a non-zero transverse momentum, such that its momentum is well separated (non-collinear) from the direction of the emerging quark (antiquark) momentum after it emits the gluon,[71] a system evolves into a system of 3 jets and so on for the generation of higher jets events. By the analysis of the angular distributions of the angular orientations of these jets, the spins of the initial particles in the annihilation process may, in principle, be predicted.

The purpose of this subsection is to investigate formally such jet structures as they follow from lowest order QCD. In the figure below the process of $e^-e^+$ annihilation, via a single photon exchange, is given, for the production of a $q\bar{q}$ pair, as well as the emission of a gluon $G$, and the emission and re-absorption of a gluon in the pair production process.

In the CM system of $e^-e^+$, $Q^2 = -(Q^0)^2$, and the longitudinal momentum of an emerging quark $q$ in the upper diagram in part (b) of Fig. 6.15, is a fraction $z$ of the intermediate (virtual) quark momentum, while the longitudinal momentum of an emerging gluon $G$ is of fraction $(1-z)$, and similarly for the lower diagram

---

[69] See, e.g., Hanson et al. [64] for experimental observations of two jet events, and Brandelik et al. [25], Berger et al. [16], and Bartel et al. [14] for three jet events.

[70] For specific illustrations of this, cf., Field and Feynman [46], Hoyer et al. [66], and Ali et al. [4].

[71] Such a gluon is referred to as a hard non-collinear gluon.

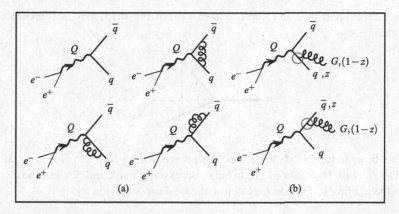

**Fig. 6.15** (a) Diagrams corresponding to the process $e^-e^+ \to q\bar{q}$, including radiative corrections, to lowest order, via the exchange of a photon. (b) Diagrams corresponding to the process $e^-e^+ \to q\bar{q}G$, to the leading order, via the exchange of a photon

involving the antiquark $\bar{q}$. From (6.10.44), and the expression for the splitting of a quark to a quark and a gluon in (6.10.40) for $z < 1$, we may write for $\mathbf{k}_\perp^2 \neq 0$,

$$\frac{d^2}{dz\, dk_\perp^2}\sigma(e^-e^+ \to q\bar{q}G) = 2\,\frac{\alpha_s}{2\pi}\frac{4}{3}\Big(\frac{1+z^2}{1-z}\Big)\frac{1}{k_\perp^2}\,\sigma(e^-e^+ \to q\bar{q}), \qquad (6.10.45)$$

where the overall factor 2 is due to the equivalence of the two diagrams in part (b) in Fig. 6.15 with $q \leftrightarrow \bar{q}$, and from (6.5.9)/6.5.7),

$$\sigma(e^-e^+ \to q\bar{q}) = \frac{4\pi\alpha^2}{E^2}\sum_f (e_f^2/e^2) \equiv \sigma_0, \quad E^2 = (Q^0)^2. \qquad (6.10.46)$$

In particular, note that the expression for the cross section in (6.10.45) will become arbitrarily large for $k_\perp^2 \simeq 0$, and all $0 \le (1-z) \le 1$, i.e., whether the gluon is soft or hard, with the gluon moving in the same direction (collinear gluon) as the parent quark (antiquark). On the other hand, as the value of $k_\perp^2$ deviates from a non-zero value, defined by the condition, say, $k_\perp^2/S \gtrsim \delta^2$, where $\delta$ is a small parameter: $0 < \delta \ll 1$, as a measure of departure from a collinear behavior of a radiated gluon, and $S$ is some scale, the cross section will take large values for a very soft gluon $(1-z) \simeq 0$. Such a soft gluon will not be observable if its energy is beyond the experimental resolution, i.e., say, for $(1-z) < \Delta$, for a small parameter: $0 < \Delta \ll 1$. In both cases where $k_\perp^2 \simeq 0, 0 \le (1-z) \le 1$, or $k_\perp^2/S \gtrsim \delta^2, (1-z) < \Delta$, one will be dealing with a 2-jet event where either a gluon is moving together with a quark (antiquark) and both in almost opposite direction to the antiquark (quark), or the quark and antiquark are moving in almost opposite directions, and in addition a non-collinear soft gluon is radiated of energy too low to be detected, respectively. In the CM of the $e^-e^+$ system, the momenta of the external particles, in reference to part (b) of Fig. 6.15, may be represented as follows:

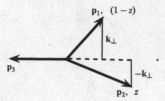

The above in turn, leads to define a 3-jet event, for $k_\perp^2/S \geq \delta^2$, $(1-z) \geq \Delta$, i.e., $z \leq 1 - \Delta$, with the radiated gluon being necessarily hard, and non-collinear.

To the leading order, one may define the total cross section by

$$\sigma(e^-e^+ \to \text{hadrons}) = \sigma_{2\text{jets}} + \sigma_{3\text{jets}}, \tag{6.10.47}$$

where

$$\sigma(e^-e^+ \to \text{hadrons}) = \sigma(e^-e^+ \to q\bar{q}) + \mathcal{O}(\alpha_s). \tag{6.10.48}$$

The expression for a 2-jet event may be the obtained by subtracting the 3-jet event contribution from the total cross section in (6.10.48). Since the integral of the expression in (6.10.45), over $z$ and $k_\perp^2$ is singular for $z \to 1$, $k_\perp^2 \to 0$, its leading behavior for the 3-jet event, will come from values of $z \lesssim 1-\Delta$, and of $k_\perp^2/S \gtrsim \delta^2$, corresponding to points very close to these singular points, and for which (6.10.45) is *valid*. Using the elementary integrals

$$\int_{\delta^2}^1 \frac{dk_\perp^2}{k_\perp^2} = -\ln \delta^2, \quad \int_0^{(1-\Delta)} dz \left(\frac{1+z^2}{1-z}\right) = -\left[\frac{3}{2} + 2\ln \Delta\right] + \mathcal{O}(\Delta), \tag{6.10.49}$$

to integrate the expression in (6.10.45), to obtain

$$\sigma_{3\text{jets}} = \sigma_0 \frac{4\alpha_s}{3\pi} \left[(3 + 4\ln \Delta)\ln \delta + \mathcal{O}(1)\right], \tag{6.10.50}$$

where $\mathcal{O}(1)$ is well defined for $\Delta \to 0$, $\delta \to 0$. From (6.10.47), (6.10.48), (6.10.46) and the just derived equation, we have to the leading order

$$\sigma_{2\text{jets}} = \sigma_0 \left\{1 - \frac{4\alpha_s}{3\pi}\left[(3 + 4\ln \Delta)\ln \delta + \mathcal{O}(1)\right]\right\}, \tag{6.10.51}$$

where $\sigma_0$ is given in (6.10.46).

Actually the expression within the curly brackets in the above equation is independent of the angle $\theta$ the jet axis makes with the one made by the annihilating $e^- e^+$ pair in the CM,[72] hence directly from Problem 6.7 we may write the expression for differential cross section corresponding to the cross section in (6.10.51) as follows

$$\frac{d\sigma_{2jets}}{d\Omega} = \alpha^2 \frac{(1 + \cos^2 \vartheta)}{4E^2} 3 \sum_f (e_f^2/e^2)\left\{1 - \frac{4\alpha_s}{3\pi}\left[(3 + 4\ln \Delta)\ln \delta + \mathcal{O}(1)\right]\right\}.$$

(6.10.52)

The angular distribution: $\propto (1 + \cos^2 \theta)$ of the initially emerging pair $q, \bar{q}$, is that of particles of spin 1/2. This jet axis angular distribution is consistent with experiment and is in good support of the spin 1/2 character of quarks.[73] The 3-jet event is consistent with a spin 1 for the gluon.

The advantage of deriving the expression for the 2-jet event from the 3-jet event one, to the leading order, is that there is a detailed cancelation of infrared divergence in part (a) of the diagrams in Fig. 6.15 of the same nature we have encountered in Sect. 5.12, while the 3-jet one involves no such infrared divergences.

## 6.11 Deep Inelastic Scattering: QCD Corrections

In deep inelastic scattering, one is, in particular, working at large momentum transfer squared $Q^2$ to the proton. At such a given momentum transfer squared, a quark may emit a gluon, as discussed at the end of the last section, and, in turn, a gluon may emit other gluons or a quark-antiquark pair. Accordingly a parton density distribution of finding a parton with fractional momentum in a range $(x, x \mid dx)$ will change, and will depend on $Q^2$, which we denote by $f_i(x, Q^2)$, with $i$ standing for quarks $(q)$, antiquarks $(\bar{q})$, or gluons $(G)$. Secondly, the coupling parameter will not be simply $\alpha_s$ but becomes $\alpha_s(Q^2)$, that is, it will also depend on $Q^2$. Examples of the splitting of partons into other partons are shown in Fig. 6.16.

Upon the increase of the momentum transfer $Q^2$ to $Q^2 + \delta Q^2$, a parton, say of type $a$, of longitudinal fractional momentum of the proton in $(\xi, \xi + d\xi)$ may emit partons $b$ and $c : a \rightarrow bc$, where parton $b$, say, may have a longitudinal fractional momentum of that of the proton in a range $(x, x + dx)$. We denote by $z$ the fractional longitudinal momentum carried by daughter parton $b$ relative to that of the parent

---

[72]See part (d) of Fig. 6.7 in Sect. 6.4.

[73]The explicit 2-jet structure in (6.10.51) was first obtained in Sterman and Weinberg [111]. See also Stevenson [112].

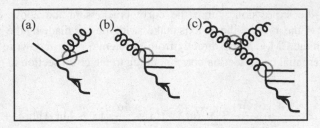

**Fig. 6.16** Splitting of partons into other partons. The corresponding measures of probabilities of splittings *at* each point of splitting, as sub-processes, are each of the order $\alpha_s$

parton $a$. That is[74]

$$z = -\frac{Q^2}{2\,\xi PQ} = \frac{x}{\xi}, \qquad z\xi = x. \tag{6.11.1}$$

The additional contribution to the structure function $\nu W_2$ in (6.10.6), to the parton model, due to a splitting of a parton $a$ into partons $b$ and $c$, may be extracted from the processes in Fig. 6.16, with the point of splitting denoting a vertex in the processes.

As we have done at the end of the last section, we introduce a splitting factor, denoted by $\mathscr{P}_{a\to bc}(z)$, corresponding to this "break up" of $a$ into $b$ and $c$ as follows:

"*Given* that a parton $a$, of fractional longitudinal momentum in $(\xi, \xi + d\xi)$ of that of the proton, is found in the proton, then $\mathscr{P}_{a\to bc}(z)$ corresponds to the process where parton $a$ splits into partons $b$ and $c: a \to bc$, with parton $b$ carrying a fractional longitudinal momentum $z$ of the parent parton $a$."

It modifies the parton distribution of type $b$ with the infinitesimal change $Q^2 \to Q^2 + \delta Q^2$, as discussed below. $\mathscr{P}_{a\to bc}(z)$, we recall is referred to as a splitting factor, and for various $a$, $b$, and $c$ partons, they were investigated in the last section and are spelled out in (6.10.40), (6.10.41), (6.10.42), (6.10.43).

The above allows us to write down a completeness probabilistic equation for the change in the parton distribution $f_b(x, Q^2)$, of parton of type $b$, with infinitesimal increase $\delta \ln Q^2$ as follows: (see also (6.10.19), (6.10.44))

$$\delta f_b(x, Q^2) = \frac{\alpha_s(Q^2)}{2\pi} \int_0^1 dz \int_0^1 d\xi \, \delta(x - \xi z) \sum_{a,c} f_a(\xi, Q^2)\, \mathscr{P}_{a\to bc}(z)\, \delta \ln(Q^2).$$

$$\tag{6.11.2}$$

---

[74]Here it is worth recalling that in the parton model, a parton in the proton has a reduced momentum $x$, compared to that of the proton, given by $x = -Q^2/(2\,PQ)$.

Read this equation as follows:

◁ "For all possible partons $a$ and $c$, the change of a parton $b$ distribution, upon infinitesimal change $\delta \ln Q^2$, is given by the splitting of a parton $a$, with fractional longitudinal momentum of the proton in $(\xi, \xi + d\xi)$ and with splitting factor $\mathscr{P}_{a \to bc}(z)$, into a parton $b$ and a parton $c$, where parton $b$ has fractional longitudinal momentum in $(z, z + dz)$ of the momentum of the parent $a$, for all $0 \le \xi \le 1$, $0 \le z \le 1$, such that $z\xi = x$". Here we have also used the effective coupling strength to be $\alpha_s(Q^2)$ at the momentum transferred squared $Q^2$ in question considered. ▷

For example, for a quark $q$, the following differential-integral equation emerges directly from (6.11.2)

$$Q^2 \frac{d}{dQ^2} f_q(x, Q^2) = \frac{\alpha_s(Q^2)}{2\pi} \int_0^1 d\xi \int_0^1 dz\, \delta(x - \xi z)$$

$$\times \left[ f_q(\xi, Q^2) \mathscr{P}_{q \to qG}(z) + f_G(\xi, Q^2) \mathscr{P}_{G \to q\bar{q}}(z) \right]. \tag{6.11.3}$$

Needless to say, a similar equation may be then written down for an antiquark $\bar{q}$.

Eqn. (6.11.3), may be also rewritten in a more familiar form as

$$Q^2 \frac{d}{dQ^2} f_q(x, Q^2) = \frac{\alpha_s(Q^2)}{2\pi} \int_x^1 \frac{d\xi}{\xi}$$

$$\times \left[ f_q(\xi, Q^2) \mathscr{P}_{q \to qG}\left(\frac{x}{\xi}\right) + f_G(\xi, Q^2) \mathscr{P}_{G \to q\bar{q}}\left(\frac{x}{\xi}\right) \right], \tag{6.11.4}$$

where, in the process of writing the above equation, we have used the fact that $(x/\xi) = z \le 1$ implies that $x \le \xi (\le 1)$.

It remains to adjoin to the equation of the quark distribution function in (6.11.3), an equation for the gluon distribution function $f_G(x, Q^2)$. It is simply given by a completeness probabilistic equation to be

$$Q^2 \frac{d}{dQ^2} f_G(x, Q^2) = \frac{\alpha_s(Q^2)}{2\pi} \int_0^1 d\xi \int_0^1 dz\, \delta(x - \xi z)$$

$$\times \left[ \sum_q f_q(\xi, Q^2) \mathscr{P}_{q \to Gq}(z) + \sum_{\bar{q}} f_{\bar{q}}(\xi, Q^2) \mathscr{P}_{\bar{q} \to G\bar{q}}(z) + f_G(\xi, Q^2) \mathscr{P}_{G \to GG}(z) \right]. \tag{6.11.5}$$

In $\mathscr{P}_{G \to GG}(z)$, one of the emerging gluons has a momentum reduced by a fraction $z$ relative to the parent gluon.[75]

----

[75]The evolution equations (6.11.3), (6.11.4), (6.11.5) in $Q^2$, are often referred to as Altarelli-Parisi equations [5].

The splitting factors $\mathscr{P}_{q\to qG}(z)$, $\mathscr{P}_{G\to q\bar{q}}(z)$, $\mathscr{P}_{q\to Gq}(z)$, $\mathscr{P}_{G\to GG}(z)$, are spelled at the end of the last section. They describe the breaking of partons into other partons. What we will need, however, are their moments $\int_0^1 dz\, z^{n-1} \mathscr{P}_{a\to bc}(z)$.

The evolution equations may be solved for the moments corresponding to an $i^{\text{th}}$ parton distribution functions defined by

$$M_i^n(Q^2) = \int_0^1 dx\, x^{n-1} f_i(x, Q^2). \tag{6.11.6}$$

by defining, in the process, the moments of the splitting factors, referred to as anomalous dimensions:

$$A_{bc}^n = \int_0^1 dz\, z^{n-1} \mathscr{P}_{a\to bc}(z). \tag{6.11.7}$$

Upon multiplying (6.11.3), (6.11.5) by $x^{n-1}$, and integrating over $x$ from 0 to 1, give the simultaneous equations:

$$Q^2 \frac{d}{dQ^2} M_q^n(Q^2) = \frac{\alpha_s(Q^2)}{2\pi} [M_q^n(Q^2) A_{qG}^n + M_G^n(Q^2) A_{q\bar{q}}^n], \tag{6.11.8}$$

$$Q^2 \frac{d}{dQ^2} M_{\bar{q}}^n(Q^2) = \frac{\alpha_s(Q^2)}{2\pi} [M_{\bar{q}}^n(Q^2) A_{\bar{q}G}^n + M_G^n(Q^2) A_{\bar{q}q}^n], \tag{6.11.9}$$

$$Q^2 \frac{d}{dQ^2} M_G^n(Q^2) = \frac{\alpha_s(Q^2)}{2\pi} \Big[ \Big[ \sum_q M_q^n(Q^2) + \sum_{\bar{q}} M_{\bar{q}}^n(Q^2) \Big] A_{Gq}^n + M_G^n(Q^2) A_{GG}^n \Big], \tag{6.11.10}$$

where we have simply noted that $\int_0^1 dx\, x^{n-1} \delta(x - \xi z) = \xi^{n-1} z^{n-1}$, and where we have used [76] $\mathscr{P}_{q\to qG}(z) = \mathscr{P}_{\bar{q}\to\bar{q}G}(z)$, implying that $A_{qG}^n = A_{\bar{q}G}^n$, also $\mathscr{P}_{q\to Gq}(z) = \mathscr{P}_{\bar{q}\to G\bar{q}}(z)$, implying that $A_{Gq}^n = A_{G\bar{q}}^n$, and finally $\mathscr{P}_{G\to q\bar{q}}(z) = \mathscr{P}_{G\to\bar{q}q}(z)$, implying that $A_{\bar{q}q}^n = A_{q\bar{q}}^n$.

The moments of the splitting factors in (6.11.7) are readily obtained by explicitly integrating the splitting factors in (6.10.40), (6.10.41), (6.10.42), (6.10.43), multiplied by $z^{n-1}$ as defined in (6.11.7), over $z$. They are given by

$$A_{qG}^n = A_{\bar{q}G}^n = \frac{4}{3} \Big[ \frac{1}{n(n+1)} - \frac{1}{2} - 2 \sum_{j=2}^n \frac{1}{j} \Big], \tag{6.11.11}$$

$$A_{Gq}^n = A_{G\bar{q}}^n = \frac{4}{3} \frac{n^2 + n + 2}{n(n^2 - 1)}, \tag{6.11.12}$$

---

[76]See (6.10.37), (6.10.38), (6.10.39), (6.10.40), (6.10.41), (6.10.42), and (6.10.43).

$$A_{q\bar{q}}^n = A_{qq}^n = \frac{1}{2}\frac{n^2 + n + 2}{n(n+1)(n+2)}, \tag{6.11.13}$$

$$A_{GG}^n = 3\left[\frac{2}{(n+1)(n+2)} + \frac{2}{n(n-1)} - \frac{1}{6} - \frac{n_f}{9} - 2\sum_{j=2}^n \frac{1}{j}\right], \tag{6.11.14}$$

where $n_f$ denotes the number of quark flavors.

The following values and rigorous bounds of these moments, established in Problem 6.17, are quite useful:

$$A_{qG}^2 = A_{\bar{q}G}^2 = -\frac{16}{9}, \ A_{Gq}^2 = A_{G\bar{q}}^2 = \frac{16}{9}, \ A_{q\bar{q}}^2 = A_{qq}^2 = \frac{1}{6}, \ A_{GG}^2 = -\frac{n_f}{3}, \tag{6.11.15}$$

and for $n = 3, 4, \ldots < \infty$, $\vartheta_4^n = \sum_{j=4}^n (1/j)$,

$$-\frac{26}{9} - \frac{8}{3}\vartheta_4^n < A_{qG}^n = A_{\bar{q}G}^n \le -\frac{25}{9} - \frac{8}{3}\vartheta_4^n < -\frac{25}{9}, \tag{6.11.16}$$

$$0 < A_{Gq}^n = A_{G\bar{q}}^n \le \frac{7}{9}, \qquad 0 < A_{q\bar{q}}^n = A_{qq}^n \le \frac{7}{60}, \tag{6.11.17}$$

$$-\frac{11}{2} - \frac{n_f}{3} - 6\vartheta_4^n < A_{GG}^n \le -\frac{21}{5} - \frac{n_f}{3} - 6\vartheta_4^n, \tag{6.11.18}$$

and where $\vartheta_4^n$ should be set equal to zero for $n = 3$.

To solve the evolution equations (6.11.8), (6.11.9), (6.11.10), we introduce the variable $\tau$ below, and re-express $Q^2 \, d/dQ^2$ in terms of a derivative with respect to it as follows (see Problem 6.18) (for the SU(3) color group)

$$\tau = \frac{1}{b_0}\ln\left[\frac{\alpha_s(Q_o^2)}{\alpha_s(Q^2)}\right] = \frac{1}{b_0}\ln\left[\frac{\ln(Q^2/\Lambda^2)}{\ln(Q_o^2/\Lambda^2)}\right], \quad b_0 = 2\pi\beta_0 = \frac{33 - 2n_f}{6}, \tag{6.11.19}$$

$$Q^2\frac{d}{dQ^2} = \frac{\alpha_s(Q^2)}{2\pi}\frac{d}{d\tau}, \quad \alpha_s(Q^2) = \frac{1}{\beta_0\ln(Q^2/\Lambda^2)}, \tag{6.11.20}$$

where $Q_o^2$ denotes some fixed reference point, $Q^2 > Q_o^2$, and where $\beta_0$ is defined in (6.7.14).

We may then rewrite the evolution equations (6.11.8), (6.11.9), (6.11.10) in the following convenients forms[77]

$$\frac{\mathrm{d}}{\mathrm{d}\tau}M_q^n(Q^2) = M_q^n(Q^2)A_{qG}^n + M_G^n(Q^2)A_{q\bar{q}}^n, \tag{6.11.21}$$

$$\frac{\mathrm{d}}{\mathrm{d}\tau}M_{\bar{q}}^n(Q^2) = M_{\bar{q}}^n(Q^2)A_{\bar{q}G}^n + M_G^n(Q^2)A_{\bar{q}q}^n, \tag{6.11.22}$$

$$\frac{\mathrm{d}}{\mathrm{d}\tau}M_G^n(Q^2) = \left[\sum_q M_q^n(Q^2) + \sum_{\bar{q}} M_{\bar{q}}^n(Q^2)\right]A_{Gq}^n + M_G^n(Q^2)A_{GG}^n. \tag{6.11.23}$$

Note the strict negativity of $A_{qG}^n = A_{\bar{q}G}^n$, and of $A_{GG}^n$, for all $n = 2, 3, \ldots < \infty$, as shown in (6.11.15), (6.11.16), (6.11.17), and (6.11.18).

A particular simple equation results for the difference $[M_q^n(Q^2) - M_{\bar{q}}^n(Q^2)]$, which follows from (6.11.21), (6.11.22) to be given by[78]

$$\frac{\mathrm{d}}{\mathrm{d}\tau}[M_q^n(Q^2) - M_{\bar{q}}^n(Q^2)] = [M_q^n(Q^2) - M_{\bar{q}}^n(Q^2)]A_{qG}^n, \tag{6.11.24}$$

whose solution is given by

$$[M_q^n(Q^2) - M_{\bar{q}}^n(Q^2)] = [M_q^n(Q_o^2) - M_{\bar{q}}^n(Q_o^2)]\left[\frac{\ln(Q^2/\Lambda^2)}{\ln(Q_o^2/\Lambda^2)}\right]^{\frac{1}{b_0}A_{qG}^n}. \tag{6.11.25}$$

The strict negativity condition of the anomalous dimensions $A_{qG}^n$, established above (see (6.11.15), (6.11.16)), implies that

$$[M_q^n(Q^2) - M_{\bar{q}}^n(Q^2)] \to 0, \qquad Q^2 \to \infty, \tag{6.11.26}$$

for $n \geq 2$. In particular, from (6.11.6), this means that for $n = 2$, the total fraction of momentum carried by the quarks and antiquarks, are asymptotically equal.

On the other hand, we may consider the combination[79]

$$\left[\sum_q M_q^n(Q^2) + \sum_{\bar{q}} M_{\bar{q}}^n(Q^2)\right] \equiv M^{n(S)}(Q^2). \tag{6.11.27}$$

---

[77]The equalities of some of the moments of the splitting factors in (6.11.11), (6.11.12), (6.11.13) should be noted.

[78]This equation is referred to as the non-singlet equation.

[79]$S$ here stands for singlet.

From (6.11.21), (6.11.22), (6.11.23), we then have the following simultaneous equations

$$\frac{d}{d\tau} M^{n(S)}(Q^2) = M^{n(S)}(Q^2) A^n_{qG} + 2 n_f M^n_G(Q^2) A^n_{q\bar{q}}, \qquad (6.11.28)$$

$$\frac{d}{d\tau} M^n_G(Q^2) = M^{n(S)}(Q^2) A^n_{Gq} + M^n_G(Q^2) A^n_{GG}, \qquad (6.11.29)$$

where the factor $n_f$ in the second term, on the right-hand side of (6.11.28) arises because we are summing over quarks in (6.11.27) corresponding to $n_f$ flavors.

These equations may be represented in an elegant matrix form as

$$\frac{d}{d\tau} \begin{pmatrix} M^{n(S)}(Q^2) \\ M^n_G(Q^2) \end{pmatrix} = \begin{pmatrix} A^n_{qG} & 2 n_f A^n_{q\bar{q}} \\ A^n_{Gq} & A^n_{GG} \end{pmatrix} \begin{pmatrix} M^{n(S)}(Q^2) \\ M^n_G(Q^2) \end{pmatrix}. \qquad (6.11.30)$$

To solve this equation, we need to diagonalize the $2 \times 2$ matrix on its right-hand side. To this end, we use the notations

$$A^n_{qG} \equiv A_n, \quad 2 n_f A^n_{q\bar{q}} \equiv B_n, \quad A^n_{Gq} \equiv C_n, \quad A^n_{GG} \equiv D_n. \qquad (6.11.31)$$

The matrix in question may be then diagonalized via the equation

$$M \begin{pmatrix} A_n & B_n \\ C_n & D_n \end{pmatrix} M^{-1} = \begin{pmatrix} \lambda^+_n & 0 \\ 0 & \lambda^-_n \end{pmatrix}, \qquad (6.11.32)$$

where

$$M = \frac{1}{\sqrt{(\lambda^+_n - \lambda^-_n) B_n}} \begin{pmatrix} -(\lambda^-_n - A_n) & B_n \\ (\lambda^+_n - A_n) & -B_n \end{pmatrix},$$

$$M^{-1} = \frac{1}{\sqrt{(\lambda^+_n - \lambda^-_n) B_n}} \begin{pmatrix} B_n & B_n \\ (\lambda^+_n - A_n) & (\lambda^-_n - A_n) \end{pmatrix}, \qquad (6.11.33)$$

and the eigenvalues $\lambda^\pm_n$, are given by

$$\lambda^\pm_n = \frac{1}{2} \left[ A_n + D_n \pm \sqrt{(A_n + D_n)^2 - 4(A_n D_n - B_n C_n)} \right]$$

$$= \frac{1}{2} \left[ A_n + D_n \pm \sqrt{(A_n - D_n)^2 + 4 B_n C_n} \right]. \qquad (6.11.34)$$

The eigenvalues are real for all $n$. In particular for $n = 2$, we have

$$\lambda_2^+ = 0, \qquad \lambda_2^- = -\frac{16}{9} - \frac{n_f}{3}, \qquad \text{i.e.,} \qquad \lambda_2^+ - \lambda_2^- > 0, \tag{6.11.35}$$

$$(\lambda_2^- - A_2) = -n_f/3, \qquad (\lambda_2^+ - A_2) = 16/9, \qquad B_2 = n_f/3, \tag{6.11.36}$$

and for $n = 3, 4, \ldots < \infty$, it is shown in Problem 6.19 that[80] $\lambda_n^- < \lambda_n^+ < 0$ and $(\lambda_n^\pm - A_n) \neq 0$ as well. Hence we may state

$$\lambda_n^+ - \lambda_n^- > 0, \quad (\lambda_n^\pm - A_n) \neq 0, \quad B_n > 0, \quad \text{for } n = 2, 3, \ldots < \infty, \tag{6.11.37}$$

where the positivity of $B_n = 2 n_f A_{q\bar{q}}^n$ follows from (6.11.17).

Upon multiplying (6.11.30) from left by the matrix $M$, we obtain the equation

$$\frac{d}{d\tau} \begin{pmatrix} -(\lambda_n^- - A_n)M^{n(S)} + B_n M_G^n \\ +(\lambda_n^+ - A_n)M^{n(S)} - B_n M_G^n \end{pmatrix} = \begin{pmatrix} \lambda_n^+ & 0 \\ 0 & \lambda_n^- \end{pmatrix} \begin{pmatrix} -(\lambda_n^- - A_n)M^{n(S)} + B_n M_G^n \\ +(\lambda_n^+ - A_n)M^{n(S)} - B_n M_G^n \end{pmatrix}. \tag{6.11.38}$$

whose solution is readily obtained for $n = 2$, on account that $\lambda_2^+ = 0$, and is given by

$$M^{2(S)}(Q^2) + M_G^2(Q^2) = M^{2(S)}(Q_o^2) + M_G^2(Q_o^2), \tag{6.11.39}$$

$$[\frac{16}{9} M^{2(S)}(Q^2) - \frac{n_f}{3} M_G^2(Q^2)] = [\frac{16}{9} M^{2(S)}(Q_o^2) - \frac{n_f}{3} M_G^2(Q_o^2)]$$

$$\times \left[ \frac{\ln(Q^2/\Lambda^2)}{\ln(Q_o^2/\Lambda^2)} \right]^{-\frac{1}{b_0}\left[\frac{16}{9} + \frac{n_f}{3}\right]}. \tag{6.11.40}$$

On the other hand, for $n \geq 3$, we obtain after elementary algebraic steps

$$M^{n(S)}(Q^2) = \frac{1}{(\lambda_n^+ - \lambda_n^-)} \left( [\cdot]_1 \left[ \frac{\ln(Q^2/\Lambda^2)}{\ln(Q_o^2/\Lambda^2)} \right]^{\frac{1}{b_0}\lambda_n^+} + [\cdot]_2 \left[ \frac{\ln(Q^2/\Lambda^2)}{\ln(Q_o^2/\Lambda^2)} \right]^{\frac{1}{b_0}\lambda_n^-} \right), \tag{6.11.41}$$

$$M_G^n(Q^2) = \frac{1}{(\lambda_n^+ - \lambda_n^-)} \left( [\cdot]_3 \left[ \frac{\ln(Q^2/\Lambda^2)}{\ln(Q_o^2/\Lambda^2)} \right]^{\frac{1}{b_0}\lambda_n^+} + [\cdot]_4 \left[ \frac{\ln(Q^2/\Lambda^2)}{\ln(Q_o^2/\Lambda^2)} \right]^{\frac{1}{b_0}\lambda_n^-} \right), \tag{6.11.42}$$

---

[80]A strictly negative upper bound for $\lambda_n^+$, and a lower bound for $\lambda_n^-$ are also given in Problem 6.19.

involving the very welcome strictly negative anomalous dimensions: $\lambda_n^- < \lambda_n^+ < 0$, and where

$$[.]_1 = -(\lambda_n^- - A_n)M^{n(S)}(Q_o^2) + B_n M_G^n(Q_o^2), \tag{6.11.43}$$

$$[.]_2 = +(\lambda_n^+ - A_n)M^{n(S)}(Q_o^2) - B_n M_G^n(Q_o^2), \tag{6.11.44}$$

$$[.]_3 = +\frac{(\lambda_n^+ - A_n)}{B_n}[.]_1, \tag{6.11.45}$$

$$[.]_4 = +\frac{(\lambda_n^- - A_n)}{B_n}[.]_2. \tag{6.11.46}$$

The above analysis provides us a lot of information. Let us elaborate on our findings.

We learn that for $n < \infty$, the *moments* of the parton density distributions become independent of $Q^2$, for $Q^2 \to \infty$, or more precisely that their limits exist, asymptotically for truly asymptotic $Q^2 \to \infty$.

For $3 \le n < \infty$, (6.11.27)/(6.11.41), (6.11.42), show that

$$\left[\sum_q M_q^n(Q^2) + \sum_{\bar{q}} M_{\bar{q}}^n(Q^2)\right] \to 0, \qquad Q^2 \to \infty, \tag{6.11.47}$$

$$M_G^n(Q^2) \to 0, \qquad Q^2 \to \infty. \tag{6.11.48}$$

This does not mean that the underlying densities: $\sum_q [f_q(x, Q^2) + f_{\bar{q}}(x, Q^2)], f_G(x, Q^2)$, vanish for large momentum transfer squared, but that they become mainly concentrated near $x = 0$, such that when multiplied by $x^{n-1}$, for $n \ge 3$, they give very small contributions to the corresponding moments in the sense of distributions. Such a behavior of these densities having large contributions near $x = 0$, at high momentum transfer squared $Q^2$, is well supported experimentally. It signals the fact that correspondingly a large number of partons may acquire only infinitesimally small fractions of the momentum of the proton to sum up to one, at high momentum transfer squared $Q^2$.

We note that (6.11.25) is also valid for $n = 1$, for which $A_{qG}^1 = A_{\bar{q}G}^1 = 0$ (see (6.11.11)). From the definition of the moments in (6.11.6), this then gives

$$\int_0^1 dx [f_q(x, Q^2) - f_{\bar{q}}(x, Q^2)] = \int_0^1 dx [f_q(x, Q_o^2) - f_{\bar{q}}(x, Q_o^2)]. \tag{6.11.49}$$

That is, the net number of quarks-antiquarks, of any given flavor, is conserved. In addition to the three basic quarks in a proton, there may be an arbitrary number of quark-antiquark of various flavors, within the proton, as the momentum transfer squared $Q^2$ is increased. We may also sum over all quarks and antiquarks to obtain from the above equation

$$\int_0^1 dx \left[ \sum_q f_q(x, Q^2) - \sum_{\bar{q}} f_{\bar{q}}(x, Q^2) \right] = \int_0^1 dx \left[ \sum_q f_q(x, Q_o^2) - \sum_{\bar{q}} f_{\bar{q}}(x, Q_o^2) \right].$$

$$(6.11.50)$$

For the contribution of all quarks and antiquarks in a proton, we then have a conservation law and a normaliation condition:

$$\delta_{Q^2} \int_0^1 dx \left[ \sum_q f_q(x, Q^2) - \sum_{\bar{q}} f_{\bar{q}}(x, Q^2) \right] = 0, \quad \int_0^1 dx \left[ \sum_q f_q(x, Q^2) - \sum_{\bar{q}} f_{\bar{q}}(x, Q^2) \right] = 3.$$

$$(6.11.51)$$

Equations (6.11.39)/(6.11.27), give the conservation of total momentum carried by the partons

$$\int_0^1 dx\, x \left[ \sum_q f_q(x, Q^2) + \sum_{\bar{q}} f_{\bar{q}}(x, Q^2) + f_G(x, Q^2) \right]$$

$$= \int_0^1 dx\, x \left[ \sum_q f_q(x, Q_o^2) + \sum_{\bar{q}} f_{\bar{q}}(x, Q_o^2) + f_G(x, Q_o^2) \right] = 1, \quad (6.11.52)$$

normalized to one. On the other hand, (6.11.40) gives

$$\frac{16}{9} M^{2(S)}(Q^2) - \frac{n_f}{3} M_G^2(Q^2) \to 0, \qquad Q^2 \to \infty, \quad (6.11.53)$$

which together (6.11.26), (6.11.27), lead finally from (6.11.52) to

$$\int_0^1 dx\, x \sum_q f_q(x, Q^2) \to \frac{3\, n_f}{32 + 6\, n_f}, \quad (6.11.54)$$

$$\int_0^1 dx\, x \sum_{\bar{q}} f_{\bar{q}}(x, Q^2) \to \frac{3\, n_f}{32 + 6\, n_f}, \quad (6.11.55)$$

$$\int_0^1 dx\, x f_G(x, Q^2) \to \frac{32}{32 + 6\, n_f}, \quad (6.11.56)$$

for $Q^2 \to \infty$, and the gluons carry a fraction $16/(16 + 3\, n_f)$ of the momentum of the proton.

## 6.12 From the Schwinger Line-Integral to the Wilson Loop and How the latter Emerges

We hold a quark at a space point $\mathbf{x}$, in the presence of a given classical gluon vector fields $A_a^\mu(x)$.[81] This means that for the quark propagator $S(x, x')$, with $\mathbf{x}$, $\mathbf{x}'$ denoting possible positions of such a quark, we must have $S(x, x') \propto \delta^{(3)}(\mathbf{x} - \mathbf{x}')$. That is, the effective Lagrangian density is not to involve a space derivative $\partial$. Gauge invariance, in turn, then says that it does not depend on the space component $\mathbf{A}$ as well. Hence, by inspection, the Lagrangian density must have the form

$$\mathscr{L} = \left( \frac{1}{2}\left[ \left(\frac{\partial_0}{i}\overline{\psi}\right)\gamma^0\psi - \overline{\psi}\gamma^0\frac{\partial_0}{i}\psi \right] - M\overline{\psi}\psi + g\overline{\psi}\gamma_0 A^0\psi \right), \qquad (6.12.1)$$

where $\psi$ is an appropriate spinor necessary to satisfy the requirement spelled out above, with $S(x, x') \propto \delta^{(3)}(\mathbf{x} - \mathbf{x}')$ for large $M$.

Technically, the above may indeed be achieved by a familiar Foldy-Wouthuysen-Tani[82] transformation in QCD, which simply allows one to consider the large mass behavior of the theory for the quarks. As we have seen above, this large $M$ behavior is easily written down by inspection.

To isolate field variables associated with a quark and its antiparticle the antiquark, we note that charge conjugation via $\mathscr{C} = i\gamma^2\gamma^0$ transforms a Dirac field, defined below, as follows[83]

$$\psi = \begin{pmatrix} q(x) \\ -i\sigma^2(q^\dagger(x))^\top \end{pmatrix} \;\rightarrow\; \psi^{\mathscr{C}}(x) = \mathscr{C}\overline{\psi}^\top(x) = \begin{pmatrix} \underline{q}(x) \\ -i\sigma^2(\underline{q}^\dagger(x))^\top \end{pmatrix}, \qquad (6.12.2)$$

simply exchanging particle with antiparticle $q(x) \leftrightarrow \underline{q}(x)$, where we have used the identity $-\sigma^2(\sigma^2)^\top = I$.

---

[81] The need of considering only a classical such field will become clear below as we will be working with functional derivatives with respect to external sources or equivalently in the path integral version.

[82] This was suggested in a remarkable paper by Brown and Weisberger [26]. For details of the Foldy-Wouthuysen-Tani transformation see Bjorken and Drell [20, pp. 46–52] and Manoukian [87, pp. 944–947].

[83] $\mathscr{C} = \mathscr{C}_D$, $\gamma^0$ are given, respectively, in (I.16), (I.8) in the Dirac representation in Appendix I at the end of the book.

Now by using the anti-commutativity of spinor fields, within a Lagrangian density, *and* integrating by parts with respect to $\partial_0$, (6.12.1) takes the simple form

$$\mathscr{L} = -q^\dagger \left( \frac{\partial_0}{i} - gA_0 + M \right) q - \underline{q}^\dagger \left( \frac{\partial_0}{i} - g\underline{A}_0 + M \right) \underline{q}, \tag{6.12.3}$$

$$A_0 = A_{c0}\, t_c, \qquad \underline{A}_0 = A_{c0}\, (-t_c^\top), \qquad t_c = \lambda_c/2, \tag{6.12.4}$$

with $\lambda_c$ denoting the Gell-Mann matrices of Sect. 6.1.

The equation of the propagator of a quark field is given by

$$\left( \frac{\partial_0}{i} - gA_0 + M \right) S(x, y) = \delta^{(4)}(x - y), \tag{6.12.5}$$

whose solution is readily verified to be

$$S(x, y) = i \left( \exp[+i g \int_{y^0}^{x^0} d\xi\, A_0(\xi, \mathbf{x})] \right)_+ \delta^{(3)}(\mathbf{x} - \mathbf{y})\, e^{-iM(x^0 - y^0)}\, \theta(x^0 - y^0), \tag{6.12.6}$$

where, with $x^0 > y^0$, the Schwinger line-integral, which is *a priori* introduced to generate gauge invariant expressions, is defined with a time-ordering operation

$$\left( \exp[+i g \int_{y^0}^{x^0} d\xi\, A_0(\xi, \mathbf{x})] \right)_+$$

$$= \sum_{n \geq 0} (+i g)^n \int_{y^0}^{x^0} d\xi_n \int_{y^0}^{\xi_n} d\xi_{n-1} \dots \int_{y^0}^{\xi_2} d\xi_1\, [A_0(\xi_n, \mathbf{x}) \dots A_0(\xi_1, \mathbf{x})]. \tag{6.12.7}$$

We recall that in a time-ordered product, one orders the product with the term at the latest time first, from the left. Also note that $x^0 \geq \xi_n \geq \xi_{n-1} \geq \dots \geq \xi_1 \geq y^0$. Note that although components of the fields $A_{c0}(\xi, \mathbf{x})$, as classical fields in the path integral expression or as functional derivatives, commute for different values of $\xi$, the time ordering "+" in (6.12.6), is necessary due to the non-commutativity of the group matrices $t_c$.

Differentiation with respect to $x^0$ gives (see Problem 6.20)

$$\partial_0 \left( \exp[+i g \int_{y^0}^{x^0} d\xi\, A_0(\xi, \mathbf{x})] \right)_+$$

$$= i g\, A_0(x^0, \mathbf{x}) \left( \exp[+i g \int_{y^0}^{x^0} d\xi\, A_0(\xi, \mathbf{x})] \right)_+, \qquad x^0 > y^0, \tag{6.12.8}$$

Therefore using the relations $\partial_0 \theta(x^0 - x'^0) = \delta(x^0 - x'^0)$, the solution of the propagator $S(x, x')$ in (6.12.6) is then easily verified, satisfying the basic relation $S(x, x') \propto \delta^{(3)}(\mathbf{x} - \mathbf{x}')$ sought.

The propagator $\underline{S}(x, y)$ for the antiquark satisfies the same equation as the one in (6.12.5) with $A_0(x)$ in it simply replaced by $\underline{A}_0$. That is,

$$\underline{S}(x, y) = i \left( \exp[+ig \int_{y^0}^{x^0} d\xi \, \underline{A}_0(\xi, \mathbf{x})] \right)_+ \delta^{(3)}(\mathbf{x} - \mathbf{y}) \, e^{-iM(x^0 - y^0)} \, \theta(x^0 - y^0),$$

(6.12.9)

and again the time-ordering is necessary.

The following two elementary identities emerge from the definition of time-ordered product in (6.12.7), the fact that the group matrices $t_c$ are Hermitian, and the definition of $\underline{A}_0$, for $x^0 > y^0$ :

$$\left( \exp[ig \int_{y^0}^{x^0} d\xi \underline{A}_0(\xi, \mathbf{x})] \right)_+ = \left[ \left( \exp[-ig \int_{y^0}^{x^0} d\xi A_0(\xi, \mathbf{x})] \right)_- \right]^{\mathsf{T}},$$

(6.12.10)

$$\left[ \left( \exp[ig \int_{y^0}^{x^0} d\xi A_0(\xi, \mathbf{x})] \right)_+ \right]^{\dagger} = \left( \exp[-ig \int_{y^0}^{x^0} d\xi A_0(\xi, \mathbf{x})] \right)_-,$$

(6.12.11)

where

$$\left( \exp[-ig \int_{y^0}^{x^0} d\xi \, A_0(\xi, \mathbf{x})] \right)_-$$

$$= \sum_{n \geq 0} (-ig)^n \int_{y^0}^{x^0} d\xi_n \int_{y^0}^{\xi_n} d\xi_{n-1} \ldots \int_{y^0}^{\xi_2} d\xi_1 \, [A_0(\xi_1, \mathbf{x}) \ldots A_0(\xi_n, \mathbf{x})].$$

(6.12.12)

is defined with a time anti-ordering operation. That is in time anti-ordering, one orders a product with the factor earliest first from the left, and note that $x^0 > \xi_n > \xi_{n-1} > \ldots > \xi_1 > y^0$.

Gauge transformations of Schwinger line-integrals, relevant to the subsequent work, are given in Box 6.3.

**Box 6.3**: We have introduced a unified notation for ordering and anti-ordering along a line $(0, R)$ in the same way as done for the time variable. Here
$$A_\mu(t, \mathbf{x}) \equiv A_\mu(t, a), V(t, a) = \exp[\,i\,\theta_c(t, a)\,t_c\,], \mathbf{x} = (a, 0, 0)$$

$$h_\pm(T, a; 0, a) \equiv \left(\exp[\pm i g \int_0^T d\xi\, A_0(\xi, a)]\right)_\pm, \qquad T > 0$$

$$h_\pm(t, R; t, 0) \equiv \left(\exp[\pm i g \int_0^R d\xi\, A_1(t, \xi)]\right)_\pm, \qquad R > 0$$

$$h_+(T, a; 0, a) \rightarrow V(T, a)\, h_+(T, a; 0, a)\, V^{-1}(0, a) \equiv h_+^V(T, a; 0, a)$$

$$h_-(T, a; 0, a) \rightarrow V(0, a)\, h_-(T, a; 0, a)\, V^{-1}(T, a) \equiv h_-^V(T, a; 0, a)$$

$$h_+(t, R; t, 0) \rightarrow V(t, R)\, h_+(t, R; t, 0)\, V^{-1}(t, 0) \equiv h_+^V(t, R; t, 0)$$

$$h_-(t, R; t, 0) \rightarrow V(t, 0)\, h_-(t, R; t, 0)\, V^{-1}(t, R) \equiv h_-^V(t, R; t, 0)$$

To prove the first transformation rule, define

$$Q_+(t) = V(t, a)\, h_+(t, a; 0, a), \qquad Q_+(0) = V(0, a), \quad t \geq 0. \qquad (6.12.13)$$

Using the facts that

$$\frac{d}{dt} h_+(t, a; 0, a) = igA_0(t, a) h_+(t, a; 0, a), \qquad (6.12.14)$$

$$\frac{dV(t, a)}{dt} V^{-1}(t, a) \equiv (\partial_0 V(t, a)) V^{-1}(t, a) = -V(t, a)\partial_0 V^{-1}(t, a), \qquad (6.12.15)$$

we have

$$\frac{d}{dt} Q_+(t) = \left[\frac{d}{dt} V(t, a) + V(t, a)\, ig\, A_0(t, a)\right] V^{-1}(t, a) V(t, a)\, h(t, a; 0, a)$$

$$= ig\, V(t, a)\left[A_0(t, a) - \frac{\partial_0}{ig}\right] V^{-1}(t, a)\, Q_+(t) \equiv igA_0^V(t, a)\, Q_+(t),$$
$$(6.12.16)$$

thus defining $A_0^V(t, a)$. Integrating from $0$ to $T$, gives

$$Q_+(T) = \left(\exp ig \int_0^T d\xi\, A_0^V(t, a)\right)_+ Q_+(0) \equiv h^V(t, a; 0, a)\, V(0, a), \qquad (6.12.17)$$

thus defining $h^V(t, a; 0, a)$. Upon multiplying the above equation from the right by $V^{-1}(0, a)$ and using the definition of $Q_+(T)$ in (6.12.13), the first transformation rule in Box 6.3 follows. The proof of the other three transformations are very similar (see Problem 6.21).

We are interested in deriving an expression of the static potential for the interaction between a quark and an antiquark held (i.e., for $M \to \infty$) at some distance $R$ apart. To this end, we place an antiquark at the origin $\mathbf{x} = (0,0,0)$, of a coordinate system, and a quark at a distance $R$ from it at $\mathbf{y} = (R,0,0)$. Accordingly, we introduce the following corresponding state for this configuration, and its corresponding time-developed one at time $T$:

$$|\phi_0\rangle = q^\dagger(0,R)\, C(0,R)\, \underline{q}^\dagger(0,0)|\text{vac}\rangle, \tag{6.12.18}$$

$$|\phi_T\rangle = q^\dagger(T,R)\, C(T,R)\, \underline{q}^\dagger(T,0)|\text{vac}\rangle, \tag{6.12.19}$$

where the kernel $C(0,R)$ will be specified below consistent with gauge invariance. Consider the amplitude

$$G(T\,0,T\,R;0\,R,0\,0) = \langle\phi_T|\phi_0\rangle \tag{6.12.20}$$

$$= \langle\text{vac}|\Big(\underline{q}(T,0)_D\, [C^\dagger(T,R)]_{DC}\, q_C(T,R)\, q_B^\dagger(0,R)\, [C(0,R)]_{BA}\, \underline{q}_A^\dagger(0,0)\Big)_+ |\text{vac}\rangle. \tag{6.12.21}$$

The basic idea is to ensure that this provides a gauge invariant expression.

From the explicit expressions of the amplitudes for $T > 0$ obtained, respectively, from the propagators[84] in (6.12.6) and (6.12.9)

$$\langle\text{vac}|q_C(T,R)\, q_B^\dagger(0,R)|\text{vac}\rangle_{\text{norm}} \to [h_+(T,R;0,R)]_{CB}, \tag{6.12.22}$$

$$\langle\text{vac}|\underline{q}_D(T,0)\, \underline{q}_A^\dagger(0,0)|\text{vac}\rangle_{\text{norm}} \to [h_-(T,0;0,0)]_{AD}, \tag{6.12.23}$$

properly normalized to one[85] for $T \to 0$, and from the transformation rules in Box 6.3, it is evident one needs to set

$$[C(0,R)]_{BA} = [h_+(0,R;00)]_{BA}, \quad [C^\dagger(T,R)]_{DC} = [h_-(T,R;T0)]_{DC}, \tag{6.12.24}$$

to obtain a gauge invariant expression

$$G(T\,0,T\,R;0\,0,0\,R)$$

$$\longrightarrow [h_-(T,0;00)]_{AD}\, [h_-(T,R;T0)]_{DC}\, [h_+(T,R;0,R)]_{CB}\, [h_+(0,R;00)]_{BA}$$

$$\equiv \text{Tr}\left(\exp\, i\,g \oint_C dx^\mu A_\mu(x)\right)_+, \tag{6.12.25}$$

---

[84]Recall that the propagator for a fermion field $\psi$ is defined by $S(x,x') = i\langle\text{vac}|(\psi(x)\overline{\psi}(x'))|\text{vac}\rangle$, for $x^0 > x'^0$, and for a properly normalized vacuum state, see, e.g., (3.6.21).

[85]These expressions will be applied in a discrete formulation of spacetime.

**Fig. 6.17** The Wilson loop, represented by the curve $C$. Here $C$ is along the sides of a rectangle of area $R \times T$

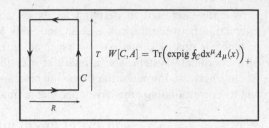

where

$$W[C,A] = \text{Tr}\left(\exp ig \oint_C dx^\mu A_\mu(x)\right)_+, \qquad (6.12.26)$$

is the celebrated Wilson loop and is automatically gauge invariant since the initial and end points of the path $C$ coincide (a closed loop) due to the nature of the trace operation. We see how the Wilson loop emerges from Schwinger line-integrals associated with the quark and antiquark propagators and by finally invoking gauge invariance, giving rise to four Schwinger line-integrals going around in a closed loop. The "+" operation, now, simply orders the group matrices $t^a$ around the loop. The closed loop, described by the curve $C$, is shown in Fig. 6.17.

We were in the course of defining the transformation function $\langle \phi_T | \phi_0 \rangle$, where the states $|\phi_0\rangle$, $|\phi_T\rangle$ are defined in (6.12.18), (6.12.19), corresponding to an antiquark set at $(0,0,0)$ and a quark set at $(R,0,0)$. We recall that fields may create various particles out of the vacuum in addition, in general, to creating the particle corresponding to it. Thus the state $\langle \phi_T |$, for example, in addition to *a quark and an antiquark, held at $R$ and $0$, respectively, at time $T$*, may also involve other complicated configurations. With this in mind, we may insert a completeness relation in the definition of the above Green function in (6.12.20), to write

$$\langle \phi_T | \phi_0 \rangle = \sum_n e^{-iE_n T} \langle \phi_0 | n \rangle \langle n | \phi_0 \rangle. \qquad (6.12.27)$$

To extract the total energy of the "heavy" quark-antiquark pair,[86] as just described above, we may make the replacement $T \to -iT$, and subsequently let $T \to \infty$, isolating the contribution of the lowest-energy intermediate state. In this limit, we obtain from (6.12.27)

$$\langle \phi_T | \phi_0 \rangle \to e^{-E(R)T}, \ T \to \infty, \qquad \langle \phi_0 | \phi_0 \rangle = 1, \qquad (6.12.28)$$

---

[86]Recall that the masses were taken to be large, and hence the kinetic energy $\mathbf{p}^2/2M$ of each particles goes to zero, as obtained from the limit of the relativistic energy, and only the mass $M$ survives in the latter.

where $E(R)$ provides the leading contribution to the sum over the intermediate states having the expected expression

$$E(R) = E_2 = 2M + V(R), \tag{6.12.29}$$

to be consistently verified, involving no kinetic energies, where the $2M$ are the contributions of the rest masses of the pair, and $V(R)$ denotes the static potential between the pair separated by a distance $R$. That $\langle \phi_0 | \phi_0 \rangle = 1$, follows from the fact that

$$\langle \phi_0 | \phi_0 \rangle = \langle \mathrm{vac} | [ \left( \exp -\mathrm{i}\, g \int_0^R \mathrm{d}\xi A_1(0, \xi) \right)_- ]_{AC}\ [ \left( \exp +\mathrm{i}\, g \int_0^R \mathrm{d}\xi A_1(0, \xi) \right)_+ ]_{CA} | \mathrm{vac} \rangle, \tag{6.12.30}$$

and from (6.12.11),

$$\left( \exp -\mathrm{i}\, g \int_0^R \mathrm{d}\xi A_1(0, \xi) \right)_- = \left( \left( \exp +\mathrm{i}\, g \int_0^R \mathrm{d}\xi A_1(0, \xi) \right)_+ \right)^\dagger, \tag{6.12.31}$$

and the result follows from the unitarity of the matrices. Since in the final analysis to follow, the problem is formulated in terms of pure gauge fields, the additive $2M$ term in (6.12.29), remnant of the Fermi fields, may be omitted subsequently, and one may introduce a corresponding normalization factor in the definition of the vacuum state.

To extract the potential in question, we are led to consider the expression

$$\frac{1}{\langle 0_+ | 0_- \rangle_J} \mathrm{Tr} \left( \exp \mathrm{i}\, g \oint_C \mathrm{d}x^\mu A'_\mu(x) \right)_+ \langle 0_+ | 0_- \rangle_J \Big|_{J=0}, \quad A'_\mu(.) \equiv t_c(-\mathrm{i}) \frac{\delta}{\delta J^\mu_c(.)}, \tag{6.12.32}$$

with $A_\mu$ represented by the functional derivative $A'_\mu$ given above. Here $\langle 0_+ | 0_- \rangle_J$ denotes the full vacuum-to-vacuum transition amplitude of the pure gauge field theory of gluons, including self energies and closed gluon loops as well as ghosts,[87] and depends on external sources $J^\mu_a(z)$ coupled to the gluons. The problem of quark confinement via the Wilson loop is the subject of the next section.

## 6.13 Lattices and Quark Confinement

In this section, we consider a variation of quantum field theory we have been considering so far, where a 4D Euclidean spacetime continuum is now replaced by a discrete one described by a lattice, with a finite small lattice spacing playing the

---

[87]In the limit $M \to \infty$ closed fermion loops vanish.

role of a cut-off. We will see that the lattice description gives a rise to a method for studying the strong coupling limit of QCD. The investigation of the strong coupling is relevant physically as it is expected that the effective coupling of QCD becomes strong at distances large enough for the confinement of quarks within hadrons. Thus this method provides a self-consistent method for the investigation of the problem of quark confinement. From the extrapolation between this strong coupling limit and the weak coupling one, emerging from the asymptotic free theory description, one may be optimistic that one is dealing with the same theory observed at different energy scales with a corresponding effective coupling.

To construct a lattice, with lattice spacing $a$, pick a point $x$ in 4D Euclidean space and draw a line of length $a$ in a given direction and another one in its opposite direction. Repeat this in the remaining mutually perpendicular directions. The end points of every such lines will then provide a starting point, as $x$ was, for repeating the same procedure in a self evident way. This is shown in parts (a) and (b) of Fig. 6.18. The points of intersections of the lines generated are referred to as lattice *sites*. A line connecting two lattice sites is called a *link*. A 2D dimensional square of area $a^2$ bounded by four links is referred to as a *plaquette* as shown in parts (b) and (c) of Fig. 6.18. No additional details on lattice description will be needed in the sequel.

We introduce an appropriate gauge invariant action integral, in Euclidean space, for the gauge field in lattice theory. To this end, we define four ordered Schwinger line-integrals $U(\ell_i)$ along the links $\ell_1$, $\ell_2$, $\ell_3$, $\ell_4$, in given *fixed* directions of vector components specified by the indices, $\mu$, and $\nu$ made evident by the definition of the links:

$$U(\ell_i) = \big(\exp[\,i\,a\,g\,A(\ell_i)]\big)_+, \tag{6.13.1}$$

where the $A(\ell_i)$ are integrals defined below. We suppress the other two variables along directions perpendicular to the directions *specified* by the indices $\mu$ and $\nu$.

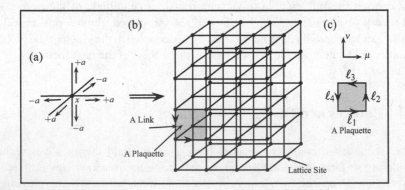

**Fig. 6.18** Construction of a lattice, with part (**b**) showing part of a 3 D version of a lattice

In particular for $a \to 0$, and fixed $\mu$ and $\nu$,

$$A(\ell_1) = \int_x^{x+a} dx^\mu A_\mu(x,y) \to a\, A_\mu(x,y) + \frac{a^2}{2} \frac{\partial A_\mu(x,y)}{\partial x^\mu}, \qquad (6.13.2)$$

$$A(\ell_2) = \int_y^{y+a} dy^\nu A_\nu(x+a,y) \to a\, A_\nu(x,y) + a^2 \frac{\partial A_\nu(x,y)}{\partial x^\mu} + \frac{a^2}{2} \frac{\partial A_\nu(x,y)}{\partial y^\nu}, \qquad (6.13.3)$$

$$A(\ell_3) = -\int_x^{x+a} dx^\mu A_\mu(x,y+a) \to -a\, A_\mu(x,y) - a^2 \frac{\partial A_\mu(x,y)}{\partial y^\nu} - \frac{a^2}{2} \frac{\partial A_\mu(x,y)}{\partial x^\mu}, \qquad (6.13.4)$$

$$A(\ell_4) = -\int_y^{y+a} dy^\nu A_\nu(x,y) \to -a\, A_\nu(x,y) - \frac{a^2}{2} \frac{\partial A_\nu(x,y)}{\partial y^\nu}, \qquad (6.13.5)$$

where *no* summation is implied over repeated indices here.

Upon using the Baker-Campbell-Hausdorff formula for two matrices $A$ and $B$: $\exp A \exp B = \exp[A+B+[A,B]/2+\ldots]$, (6.13.2), (6.13.3), (6.13.4), (6.13.5) give, for $a \to 0$,

$$\mathrm{Tr}\,[U(\ell_4)U(\ell_3)U(\ell_2)U(\ell_1)] \to \mathrm{Tr}\,\big[\exp[\mathrm{i}\,g\,a^2\,G_{\mu\nu}]\big], \qquad (6.13.6)$$

where recall that $G_{\mu\nu} = \partial_\mu A_\nu - \partial_\nu A_\mu - \mathrm{i}g[A_\mu, A_\nu]$.

Thus if we set

$$\mathrm{Tr}\,[U(\ell_4)U(\ell_3)U(\ell_2)U(\ell_1)] = W_p, \qquad (6.13.7)$$

and define

$$S_p = \beta\Big[1 - \frac{1}{N}\mathrm{Re}(W_p)\Big], \qquad S_{\mathrm{Euc}} = \sum_p S_p, \qquad (6.13.8)$$

where the latter sum is over lattice points, and where $\beta$ is a parameter to be specified below, we obtain for $a \to 0$

$$S_p \to \beta\Big(a^4 \frac{g^2}{2N} \mathrm{Tr}\,[G_{\mu\nu}^2]\Big), \qquad (6.13.9)$$

where we have used the fact that $\mathrm{Tr}[t_c] = 0$, $\mathrm{Tr}\,I = N$ for the SU($N$) group, and recall that $\mathrm{Tr}[t_c t_d] = \delta_{cd}/2$. Hence upon approximating the sum over all plaquettes

$a^4 \sum_p$ with a spacetime integral in Euclidean space, we see that for $a \to 0$,

$$S_{\text{Euc}} \to \beta \frac{g^2}{2N} \int \frac{1}{2} (\mathrm{d}x)_{\text{Euc}} \operatorname{Tr} [G_{\mu\nu}(x) G_{\mu\nu}(x)], \qquad (6.13.10)$$

where the $1/2$ factor in the integral arises because of the symmetry under the interchange of $\mu$ and $\nu$. Thus $S_{\text{Euc}}$ in (6.13.8) defines an appropriate expression for the action integral in lattice theory, with

$$\beta = 2N/g^2. \qquad (6.13.11)$$

With the strategy of obtaining a gauge invariant procedure for evaluating the expression in (6.12.32), involving the Wilson loop, integrations over the vector fields are replaced by integrations over Schwinger line-integral associated with all the links in the lattice, involving the vector field. That is, we introduce a measure of integration defined as a product involving all the links in the lattice normalized to unity

$$\mathrm{d}[U] = \Pi_{\ell_i} \mathrm{d}U(\ell_i), \qquad \int \mathrm{d}[U] = 1, \qquad (6.13.12)$$

and develop integration theory over such links by a gauge invariant procedure. This is done in a straightforward manner as follows. Consider the integral

$$\int \mathrm{d}[U]\, U(\ell_j). \qquad (6.13.13)$$

This is clearly not gauge invariant unless it is identically equal to zero. According to Box 6.3, $U(\ell_j)$ has a well defined gauge transformation, and hence gauge invariance alone implies that the above integral must be zero. On the hand, consider the integral

$$\int \mathrm{d}[U]\, [U(\ell_i)]_{AB}\, [U^\dagger(\ell_j)]_{CD}. \qquad (6.13.14)$$

According to the previous integral, this is zero unless $\ell_i$ and $\ell_j$ denote the same link. Thus the above integral must be proportional to $\delta_{\ell_i \ell_j}$. The above integral, in turn, is gauge invariant for $B = C$ or for $D = A$. For $B = C$, summed over, $[U(\ell_i)]_{AB}\, [U^\dagger(\ell_i)]_{BD} = \delta_{AD}$. Accordingly the above integral must be proportional to $\delta_{BC}\, \delta_{AD}\, \delta_{\ell_i \ell_j}$. Since the trace of the identity matrix is equal to $N$, for the SU($N$) group, the above integral is equal to $\delta_{BC}\, \delta_{AD}\, \delta_{\ell_i \ell_j}/N$. Summarizing then, we have the following elementary integration rules for integrations over the links in the lattice

$$\int \mathrm{d}[U] = 1, \qquad \int \mathrm{d}[U]\, U(\ell_i) = 0, \qquad \int \mathrm{d}[U]\, [U(\ell_i)]_{AB}\, [U^\dagger(\ell_j)]_{CD} = \frac{1}{N}\, \delta_{BC}\, \delta_{AD}\, \delta_{\ell_i \ell_j}.$$
$$(6.13.15)$$

These are the only three integrations rules we will need in the sequel. In particular, we learn:

(i) The integral over every link vanishes.
(ii) For every integral involving a link factor $U(\ell_j)$, we need a compensating factor $U^\dagger(\ell_j)$ to obtain a non-zero value for the integral.
(iii) The integral involving the pairing of two such factors, for each link, gives rise to a factor of $1/N$.

Thus we are led to consider the integral

$$\langle W[C,A] \rangle = \frac{1}{NZ} \int d[U]\, W[C,A] \exp[-S_{\text{Euc}}], \quad Z = \int d[U] \exp[-S_{\text{Euc}}], \tag{6.13.16}$$

where $W[C,A]$ is defined in Fig. 6.17, and the factor $1/N$ is a conventional factor, whose significance will emerge naturally below.

Clearly the constant factor $\exp[-\sum_p \beta]$ in $\exp[-S_{\text{Euc}}]$, with $S_{\text{Euc}}$ defined in (6.13.8), will cancel out in the numerator and the denominator in (6.13.16). Hence we may replace $\exp[-S_{\text{Euc}}]$ by $\Pi_p \exp[(\beta/2N)(W_p + W_p^{-1})]$ in the latter equation, where we have used the fact that $\text{Re}[W_p] = (W_p + W_p^{-1})/2$, and note that the plaquette is traversed in $W_p^{-1}$ in opposite direction to that of $W_p$, $\beta_N = 2N/g^2$.

In the strong coupling limit

$$\exp\left[\frac{1}{g^2}(W_p + W_p^{-1})\right] \to 1 + \left(\frac{1}{g^2}\right)(W_p + W_p^{-1}). \tag{6.13.17}$$

The area of each plaquette, being equal to $a^2$, the Wilson loop going around the sides of a rectangle of area $R \times T$, together with the elementary integration rules in (i) - (iii), spelled out above, dictate that we must have $L \times M$ plaquettes filling the area enclosed by the Wilson rectangle, where $L = R/a$, $M = T/a$. We may break the Wilson loop into the product of ordered Schwinger line-integrals going around c.c.w., and every link, specified by $U(\ell_i)$ in the Wilson loop, will then have exactly a partner $U(\ell_i)^\dagger$, coming from one of the sides of a plaquette touching the contour of the Wilson loop at the link $\ell_i$ in question, as shown in part (a) of Fig. 6.19. Hence to the *leading* order in the strong coupling, only the product of $L \times M$ plaquettes, arranged as shown in part (a) in Fig. 6.19, will contribution to (6.13.8).

Since we have precisely the product of $L \times M$ plaquettes, we may infer from (6.13.17), that $\langle W[C,A] \rangle$ in (6.13.16), involves a factor $(1/g^2)^{LM}$. Also at each link, due to the pairings $U(\ell_i)\, U^\dagger(\ell_i)$ for each link, as given by the rule of integration (ii) earlier, we get a factor $(1/N)^{(2LM+L+M)}$, where $(2LM + L + M)$ denotes the number of links in the two dimensional rectangle of $L \times M$ plaquettes within the Wilson loop (see Fig. 6.19). Due to the fact that the group generators are $N \times N$ matrices, and hence the trace of the identity matrix, obtained by contracting the Kronecker delta indices in rule (ii) of integration above, is equal to $N$, i.e.,

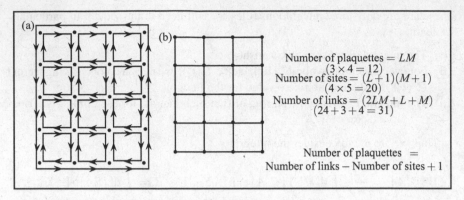

**Fig. 6.19** Part (a) shows the pairings $U(\ell_i)$, $U^\dagger(\ell_i)$ that occur for each link by an example. Part (b) facilitates counting the number of links and sites encountered in the problem by an example. The area of the Wilson rectangle, in this presentation, is actually equal to $R \times T$, with $R$ units horizontally and $T$ units vertically, $L = R/a$, $M = T/a$

equivalently there are $N$ "colors", per quark, associated with each lattice site, we also get a factor $N$ for each site, i.e., a factor $N^{(L+1)(M+1)}$. That is in the strong coupling limit, on account of $Z \to 1$, (6.13.16) becomes

$$\langle W[C,A]\rangle \to (1/N)(1/g^2)^{LM} (1/N)^{(2LM+L+M)} (N)^{(L+1)(M+1)}$$

$$= (1/Ng^2)^{LM} = (1/g^2N)^{RT/a^2} = \exp\left[-\left(\ln(g^2N)/a^2\right)RT\right], \qquad (6.13.18)$$

from which we conclude that $(N \geq 3)$

$$V(R) = KR, \qquad K = \frac{1}{a^2}\ln(Ng^2), \quad N = 3 \text{ for the SU(3) group}, \qquad (6.13.19)$$

giving a linearly rising potential in the separation distance $R$ between the pair. It takes an unlimited energy to separate the pair by an unlimited distance. $K$ is referred to as the string tension. This is in analogy to the situation where the two particles are joined by a string, with $K = V(R)/R$ denoting the fixed energy per unit length of the string, which is its tension. A rough estimate of $K$ may be obtained by taking $V$ to be of the order of the rest energy of a proton $\sim 1$ GeV, and $R$ to be its size $\sim 1$ fm $\sim 5/$GeV, giving $K \approx 0.2$ GeV$^2$.

With the inverse of the lattice spacing $1/a$ formally playing the role of an ultraviolet cut-off, one may use (6.7.17) for the beta function, to relate the coupling, in the weak coupling case, to the lattice spacing:

$$a\frac{dg^2(a)}{da} = \frac{\beta_0}{2\pi} g^4(a), \qquad (6.13.20)$$

where $\beta_0$ is given in (6.7.14), whose solution is

$$a = \frac{1}{\Lambda_L} \exp[-2\pi/\beta_0 g^2], \tag{6.13.21}$$

where $1/\Lambda_L$ is an integration constant characteristic of the lattice. With $a$ as the dimensional parameter in lattice theory, we may write $a^2 K = f(g)$. The string tension, being a physical quantity, should not depend on the cut-off $a$. That is, $dK/da = 0$. By the chain rule, this gives the following equation for $f(g)$

$$\left(a\frac{dg}{da}\right) \frac{d}{dg}f(g) = 2f(g), \quad \text{with solution} \quad f(g) = C\exp[-4\pi/\beta_0 g^2].$$
$$\tag{6.13.22}$$

Hence (6.13.21), and the relation $K = f(g)/a^2$ give

$$K = C\big(\exp[-2\pi/\beta_0 g^2]/a\big)^2 = C\Lambda_L^2, \tag{6.13.23}$$

where the coefficient $C$ is characteristic of the string tension.

It is not a *a priori* clear that the continuum limit as obtained from lattice theory will join smoothly its strong coupling limit, through intermediate coupling regimes, to the weak coupling case, obtained already via the phenomenon of asymptotic freedom, and establish that one is dealing with the same QCD theory all along. For example, the continuum limit theory from the strong coupling limit, and through intermediate coupling regimes, may finally lead to a constant non-vanishing value for its limiting coupling parameter, instead of approaching zero as given by the QCD asymptotic freedom. Analytical and numerical methods give strong support that the latter does not happen, and that the strong coupling limit and the weak coupling case are joined smoothly, through intermediate coupling regimes, indicating that one is dealing with one theory at different scales,[88] and QCD is not only asymptotically free but is confining as well. Although confinement may be established for the Abelian gauge theory through lattice theory in the strong coupling limit, as done above for the non-Abelian case, there are some investigations[89] indicating that the strong coupling limit and the weak coupling one may be separated by a phase transition and do not pertain to the same theory.

## 6.14   The Electroweak Theory I

In the present section, we develop the celebrated electroweak theory, starting from the early classic Fermi theory of weak interactions. The electroweak theory is based on the product symmetry groups $SU(2) \times U(1)$ which is spontaneously broken to

---

[88] See, e.g., Kogut [72], Kogut et al. [73], and Creutz [32].
[89] See, e.g., Guth [60] and Lautrup and Nauenberg [76].

the U(1) group. We then discuss how a key parameter "$\sin^2 \theta_W$" in the electroweak theory, where $\theta_W$ is referred to as the Weinberg angle or the weak-mixing angle, is experimentally determined. In the classic electroweak theory the neutrino masses are set equal to zero. Because of this, we elaborate on how the masses of neutrinos may be generated, and develop as well a straightforward theoretical description of "neutrino oscillations" experiments which have become quite popular in recent years providing a clear indication of neutrino mass differences.

## 6.14.1   Development of the Theory: From the Fermi Theory to the Electroweak Theory

The pioneering work on the weak interaction theory goes back to Fermi [38, 39] in an attempt to describe $\beta$ decay: $n \rightarrow p + e^- + \tilde{\nu}_e$, incorporating Pauli's postulated particle earlier in (1930), which Pauli[90] called neutron, while Fermi referred to it later as the neutrino. Now this particle is referred to as the anti-neutrino due to the fact that different lepton numbers are assigned to the neutrino and anti-neutrino, consistent with other reactions, and due the opposite helicities carried by them. The energy of the electron produced in the reaction, from a decaying neutron at rest, varied significantly experimentally up to the maximum energy $E_e = (M_n^2 - M_p^2 + m_e^2)/2M_n$, where the latter expression corresponds to the (hypothetical) reaction involving no (anti-) neutrino, and it was accordingly expected that the neutrino is extremely light. Thus Pauli considered it as small but actually non-zero.[91]

Fermi described weak interactions, by taking the interaction Hamiltonian density to involve the product of four fermions *at a point*[92] having the structure $\mathscr{H}_I \sim G_F[(\overline{\psi}_1 O_i \psi_2)(\overline{\psi}_3 O_i \psi_4) + h.c.]$ where the $O_i$ are, in general, expressed in terms of gamma matrices, and $G_F$ is the Fermi coupling consistently given by $G_F = 1.166 \times 10^{-5}/\text{GeV}^2$. The latter constant has dimensionality of $[\text{mass}]^{-2}$. Thus on dimensional grounds, a typical cross section for the scattering of four fermions will have the behavior $\sigma \sim (G_F)^2 E^2$, where $E$ is the energy associated with the process, while a formal partial wave analysis gives $\sigma \lesssim 1/E^2$. This unitarity bound would then be violated for $E \gtrsim (G_F)^{-1/2} \sim 300$ GeV. On the other hand consider some low orders radiative contributions to the scattering of four fermions in part (a) of Fig. 6.20.

By assigning a dimension $E^{-1}$ to every fermion propagator, and a dimension $E^4$ to every loop integration, in part (a) of Fig. 6.20, we see that even when there are no derivative couplings, the degree of divergences of the diagrams, including radiative

---

[90]See Pauli [96] for the historical account. The experimental discovery of the neutrino was made by Reines and Cowan [104].

[91]In this respect, see also Perrin [97] and Fermi [38, 39].

[92]A quick review of the part relevant to the weak interaction in the introductory chapter in the book is recommended.

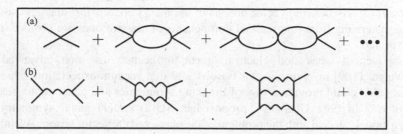

**Fig. 6.20** (**a**) The degrees of divergence increase with order of perturbation theory in this four fermion scattering with a four-fermi-interaction at a point without a bound. (**b**) Unless the longitudinal part of the propagator of the massive vector boson (represented by a wavy line) hitting a vertex vanishes, the same behavior occurs as in (**a**). The longitudinal part will not contribute if the vector boson is coupled to a conserved current

corrections, increase as $\Lambda^2$, $\Lambda^4$, ... with the orders of the interaction, as functions of an ultraviolet cut-off $\Lambda$, with no bound. The theory becomes out of control to give it any sense at all. To improve this divergence problem, an attempt was then made to describe weak interactions by introducing, in the process, intermediate massive vector bosons as mediators of interactions with two fermions and one vector boson meeting at a point, similar to what one encounters in QED with a (massless) vector boson being the photon, instead of four fermions meeting at a point of the original Fermi theory.

Massive vector bosons have one additional polarization state than that of a massless vector boson such as the photon. This extra polarization state, in turn, gives rise to a longitudinal component in the massive vector propagator which, as mentioned in the introductory chapter of the book, has, for example, the undesirable property that the propagator does not vanish at high energies,[93] unlike that of their massless counter parts which *vanish* rapidly. This makes the degree of divergences in such theories, involving non-vanishing longitudinal parts, in general, increase with the orders of perturbation theory without bound as seen above in part (b) of Fig. 6.20. These theories thus become again uncontrollable with no much hope to give them any sense and carry out practical computations as one does in QED through the renormalization programme. Details of this sort of such theories were particularly investigated by Salam and Ward [108]. The longitudinal polarization also behaves like $k^\mu/M$ at high energies also giving rise to problems with the high energy behavior of cross sections of processes involving external vector bosons. In QED, even if a longitudinal component arises, because of a specific choice of a gauge, this problem is eliminated in computing gauge invariant expressions because the photon is coupled to a conserved current.

A different approach had to be taken to deal with the weak interactions as discussed next. The emergence of a gauge invariant theory, such as QED or its

---

[93]It behaves like $k^\mu k^\nu / k^2 M^2$ for $k^2 \to \infty$.

generalization, where the vector bosons of the theory are *a priori* massless so that problems arising from longitudinal components of massive ones are avoided, is to be expected.

The idea of weak and electromagnetic unification was first advanced by Schwinger [109] involving vector bosons, and that vacuum expectation values of scalar fields could provide a way of breaking symmetries and generate masses for fermions.[94] In 1961, Glashow[95] proposed an $SU(2) \times U(1)$ gauge symmetry for such a model. To deal with the problem of the masses of the vector bosons, Weinberg [120] and independently Salam [107] proposed a renormalizable theory in which the vector bosons are taken to be *massless* in the action integral, and acquire masses, by spontaneous symmetry breaking, via the so-called Higgs mechanism, where the Higgs boson field has a non-vanishing vacuum expectation value. The associated photon field, however, remains massless, as it should, and the $SU(2) \times U(1)$ gauge symmetry is spontaneously broken to the $U(1)$ gauge symmetry of phase transformations of QED. This theory is popularly known as the Weinberg-Salam electroweak theory. The meaning of all this seemingly complicated "language", but hopefully and eventually comprehensible for the reader, will be spelled out, together with some of the extensions of this theory, in the remaining sections of this chapter.

With the discovery of violation of parity[96] and the observation in the 1950s that neutrinos are left-handed (and hence the anti-neutrinos are right-handed),[97] the leptonic current corresponding to the outgoing electron and the anti-neutrino in $\beta$ decay was taken as $J^\mu = \overline{e} \gamma^\mu (1 - \gamma^5) \nu$. To explain some processes which were observed not to happen,[98] such as $\mu^- \rightarrow e^- + \gamma$, and describe other processes, which happened, such as $\mu^- \rightarrow e^- + \nu + \tilde{\nu}$, it was inferred that there is a different[99] neutrino associated with $e^-$, $\mu^-$, and different quantum labelings (lepton numbers) (Table 6.1) were assigned to the pairs $(\nu_e, e^-)$, $(\nu_\mu, \mu^-)$, with *electronic* lepton-number: $L_e = +1$, and *muonic* lepton-number: $L_\mu = +1$, "$-1$" for their respective anti-particles. Separate conservation laws of electronic-type leptons and muonic-type leptons were then assumed. This, for example, explained why processes such as $\tilde{\nu}_\mu + p \rightarrow e^+ + n$ were not observed, while processes such as $\tilde{\nu}_\mu + p \rightarrow \mu^+ + n$ did happen. Landau [74], Lee and Yang [78], Salam [106], and Wu et al. [122] assumed that the neutrino is massless. The left-handedness of the neutrino was concluded in an experiment by Goldhaber et al. in 1958 [57]. By the late fifties, Feynman and Gell-Mann [43] and Sudarshan and Marshak [113] suggested that all the particles massive and massless are to be taken to be left-handed

---

[94]Martin and Glashow [88], p.16. That such a unification may lead to a renormalizable theory was expressed by Glashow in his Ph.D. thesis in 1958 as a graduate student of Schwinger.

[95]Glashow [55].

[96]For the experiment, see Wu et al. [122].

[97]The experiment was carried by Goldhaber et al. in 1958 [57].

[98]An upper bound for the branching ratio for the decay rates is $\Gamma(\mu^- \rightarrow e^- + \gamma)/\Gamma(\mu^- \rightarrow \text{all}) < 2.4 \times 10^{-12}$, at 90% confidence level (see [17]).

[99]The observation that there are different types of neutrinos was made by Danby et al. [33].

**Table 6.1** Lepton numbers
of the three generations of
leptons

| Lepton | $L_e$ | $L_\mu$ | $L_\tau$ |
|--------|-------|---------|----------|
| $\nu_e$ | 1 | 0 | 0 |
| $e$ | 1 | 0 | 0 |
| $\nu_\mu$ | 0 | 1 | 0 |
| $\mu$ | 0 | 1 | 0 |
| $\nu_\tau$ | 0 | 0 | 1 |
| $\tau$ | 0 | 0 | 1 |

in weak interactions theory generating what is called the $V-A$ theory involving vector minus axial vector (charge changing) currents.

The effective Fermi interaction Lagrangian density, based on four-fermions at a point, having the structure of a $V-A$ theory, incorporating parity violation, turns out to provide a good description for $\mu$ decay ($\mu^- \rightarrow e^- + \bar{\nu}_e + \nu_\mu$), and may be written as

$$\mathscr{L}_F = \frac{G_F}{\sqrt{2}} J_\alpha^\dagger J^\alpha, \quad J^\alpha = \left( \bar{e}\gamma^\rho(1-\gamma^5)\nu_e + \bar{\mu}\gamma^\rho(1-\gamma^5)\nu_\mu \right), \qquad (6.14.1)$$

from which the $\mu$ decay process may extracted, and the Fermi constant may be determined experimentally.[100]

After the discovery of parity violation, as discussed above, the charge exchanging interaction between leptons and hadrons, prior to the birth of the electroweak theory, was taken to have the structure

$$\frac{1}{\sqrt{2}} G_F \left( \bar{e}\gamma_\rho(1-\gamma^5)\nu_e + \bar{\mu}\gamma_\rho(1-\gamma^5)\nu_\mu \right) J_h^\rho + h.c., \qquad (6.14.2)$$

involving left-handed leptons (and right-handed for their anti-particles), where $J_h^\rho$ is a hadronic current to be specified later, with an inherit universality of the processes involving the pairs $(\nu_e, e^-)$ and $(\nu_\mu, \mu^-)$ with equal strength, as shown by the common coupling constant $G_F$. Upon including the pair $(\nu_\tau, \tau^-)$, using the anti-commutativity of $\gamma^5$ with the gamma matrices, and using the notation

$$\ell_1 = \begin{pmatrix} \nu_e \\ e^- \end{pmatrix}_L, \quad \ell_2 = \begin{pmatrix} \nu_\mu \\ \mu^- \end{pmatrix}_L, \quad \ell_3 = \begin{pmatrix} \nu_\tau \\ \tau^- \end{pmatrix}_L, \qquad (6.14.3)$$

---

[100]The decay rate of the muon is investigated in Problem 6.24.

where L stands for left-handed, we may write the charge leptonic-current as[101]

$$J^\rho = \sum_i \bar{\ell}_i \gamma^\rho T^- \ell_i, \quad T^\pm = T^1 \pm i T^2, \quad T^i = \frac{1}{2}\sigma^i, \qquad (6.14.4)$$

with the $\sigma^i$ denoting the Pauli matrices,[102] taking into account of the fact that the neutrinos have no charges. Invoking different lepton-number conservation, and due to the very structure of the charge current in (6.14.4), these left handed lepton pairs are considered, in a simplest possible way, as doublets of an SU(2) group, referred to as weak iso-doublets, with the latter group often denoted by $SU(2)_L$ to remind us that these lepton doublets are left-handed. In setting up the original electroweak theory, the neutrinos were assumed to be massless, as we do here,[103] thus parity violation allows only one helicity for them and the underlying theory would not involve right-handed neutrinos. Accordingly, the right-handed leptons $e_R^-$, $\mu_R^-$, and $\tau_R^-$, respectively, would not have right-handed neutrinos to be paired with and they are considered as singlets under the group. It is convenient to denote the charged right-handed leptons by $r_i$, $i = 1, 2, 3$. The transformation rules for a (left-handed) iso-doublet $\ell$, is given by

$$\ell \to e^{i\,\vartheta\cdot T}\,\ell, \qquad \mathbf{T} = \frac{1}{2}\,\sigma, \qquad (6.14.5)$$

and, of course, the right-handed ones do not respond to the transformations, i.e., $r \to r$. The weak isospin quantum number of the doublets is given by $I_L = 1/2$, with the third components of weak isospin for the upper and lower entries given by $I_3 = \pm 1/2$, respectively. On the other hand, $I_R = 0$. In terms of the electromagnetic charge $Q$ and the third components of the weak isospin $I_3$, we have the following equation for the left-handed (L) and right-handed (R) leptons:

$$L: Q = I_3 - \frac{1}{2}, \qquad R: Q = I_3 - 1 \; (I_3 = 0). \qquad (6.14.6)$$

This set of equations "cries out loud" to introduce a new quantum number, which we denote by $Y$, and rewrite the above equations simply as (Table 6.2)

$$Q = I_3 + \frac{1}{2}Y, \qquad L: Y = -1, \quad R: Y = -2. \qquad (6.14.7)$$

---

[101]In this section, we avoid using the standard notation $\tau = \sigma/2$ to avoid confusion with the notation $\tau$ used for the tau particle.

[102]In Sect. 6.1, we have seen that the Pauli matrices (divided by two) are generators of SU(2) transformations.

[103]The problem of the neutrino masses, together with the fact they are small, is addressed in Sect. 6.14.3 of the present section.

**Table 6.2** Electroweak quantum numbers of the leptons in the Standard Model. The anti-particles have the same $I$ but opposite $Q, I_3$ (for $Q \neq 0$, $I_3 \neq 0$), and $Y = 2(Q - I_3)$

| Particle | $I$ | $I_3$ | $Q$ | $Y$ |
|---|---|---|---|---|
| $(\nu_{\ell L}, \ell_L^-)$ | 1/2 | $(+1/2, -1/2)$ | $(0, -1)$ | $-1$ |
| $\ell_R^-$ | 0 | 0 | $-1$ | $-2$ |

This new quantum number is called weak hypercharge. In turn, this suggests to introduce the new symmetry group $U(1)_Y$ to account for this quantum number. The combined symmetry group generated thus far is then $SU(2)_L \times U(1)_Y$, with $Y$ corresponding to the weak hypercharge. Since the weak hypercharge of the right-handed leptons $r_i$ is twice as that of the left-handed ones $\ell_i$, we have the following transformation rules corresponding to the $U(1)_Y$ group as phase transformations

$$r \to e^{-i\beta} r, \qquad \ell \to e^{-i\beta/2} \ell. \tag{6.14.8}$$

Promoting the transformations in (6.14.5), (6.14.8) to local ones necessitates the introduction of four gauge vector fields: three vector bosons $W_1^\rho$, $W_2^\rho$, $W_3^\rho$ for $SU(2)_L$ group, and one vector boson $B^\rho$ for $U(1)_Y$, group to restore the symmetry in a lagrangian density involving the above mentioned leptons.

From what we have learnt from QCD, except now we are working with an $SU(2)$ group and we are dealing with three vector bosons instead of eight, and from what we have learnt from QED, except now the electromagnetic charge is replaced by weak hypercharge, we may immediately write down the Lagrangian density $\mathscr{L}_1$ for the leptons and the vector bosons, not considering, at this stage, the quarks, interactions with other fields and corresponding mass terms. The nature of such terms will be investigated and added later :

$$\mathscr{L}_1 = -\frac{1}{4} \sum_i G_{\mu\nu}^i G^{i\mu\nu} - \frac{1}{4} F_{\mu\nu} F^{\mu\nu} - \frac{1}{2} \sum_i \left( \bar{\ell}_i \gamma^\rho D_\rho \ell_i + \bar{r}_i \gamma^\rho D_\rho' r_i + h.c. \right),$$

$$\tag{6.14.9}$$

$$D_\rho = \frac{\partial_\rho}{i} + \frac{g'}{2} B_\rho - g\, \mathbf{T} \cdot \mathbf{W}_\rho, \quad D_\rho' = \frac{\partial^\rho}{i} + g' B_\rho, \quad F_{\mu\nu} = \partial_\mu B_\nu - \partial_\nu B_\mu,$$

$$\tag{6.14.10}$$

necessarily involving two coupling parameters $(g, g')$ for the product of two groups,[104]

$$\mathbf{G}_{\mu\nu} = \partial_\mu \mathbf{W}_\nu - \partial_\nu \mathbf{W}_\mu - g \mathbf{W}_\mu \times \mathbf{W}_\nu, \quad \mathbf{G}_{\mu\nu} = (G_{1\mu\nu}, G_{2\mu\nu}, G_{3\mu\nu}). \tag{6.14.11}$$

---

[104]Recall that the structure constants of $SU(2)$ are $\epsilon_{abc}$ .

To extract the physics from the Lagrangian density, we use, in the process, the identities[105]

$$\mathbf{T} \cdot \mathbf{W}_\rho = \frac{1}{\sqrt{2}} \left( T^- W_\rho + T^+ W_\rho^\dagger \right) + T_3 W_{3\rho}, \quad W_\rho = \frac{1}{\sqrt{2}} \left( W_{1\rho} + \mathrm{i}\, W_{2\rho} \right),$$

$$(6.14.12)$$

$$W_\rho^\dagger = \frac{1}{\sqrt{2}} \left( W_{1\rho} - \mathrm{i}\, W_{2\rho} \right), \quad \overline{\psi} \gamma^\mu \psi = \overline{\psi_L} \gamma^\mu \psi_L + \overline{\psi_R} \gamma^\mu \psi_R. \qquad (6.14.13)$$

By expressing the two fields $B_\rho$, $W_{3\rho}$, in terms of two other fields, together with the definitions that follow

$$B_\rho = A_\rho \cos \theta_W + Z_\rho \sin \theta_W, \qquad W_{3\rho} = -A_\rho \sin \theta_W + Z_\rho \cos \theta_W,$$

$$(6.14.14)$$

$$\tan \theta_W = g'/g, \quad e_1 = e^-, \quad e_2 = \mu^-, \quad e_3 = \tau^-, \quad \nu_1 = \nu_e, \quad \nu_2 = \nu_\mu, \quad \nu_3 = \nu_\tau,$$

$$(6.14.15)$$

where $T^\pm$ are defined in (6.14.4), the Lagrangian density (6.14.9) that describes the interaction of the leptons with the vector bosons may be written as

$$\mathscr{L}_1' = \frac{1}{2} \sum_i \left( \overline{\ell_i} \, \gamma^\rho \left[ g\, \mathbf{T} \cdot \mathbf{W}_\rho - \frac{g'}{2} B_\rho \right] \ell_i - g' \overline{r}_i \gamma^\rho r_i \, B_\rho + h.c. \right)$$

$$= \sum_i \left( - g \, \sin \theta_W \, \overline{e}_i \gamma^\rho e_i \, A_\rho + \right.$$

$$- g \sin \theta_W \left[ \overline{e}_i \, \gamma^\rho (\alpha_e - \beta_e \gamma^5) e_i + \alpha_\nu \overline{\nu}_i \, \gamma^\rho (1 - \gamma^5) \nu_i \right] Z_\rho$$

$$\left. + \frac{g}{\sqrt{2}} \left[ \overline{e_{iL}} \, \gamma^\rho \, \nu_{iL} \, W_\rho + \overline{\nu_{iL}} \, \gamma^\rho \, e_{iL} \, W_\rho^\dagger \right] \right). \qquad (6.14.16)$$

The first term in the summand on the right-hand side of the above equation, describes the interaction of a neutral vector field $A_\rho$ with the right-handed as well as the left-handed components of the electron, muon and the tau particles, and does not interact with the neutrinos. Accordingly, the latter field denotes the photon field, provided it does not acquire mass by symmetry breaking, which will be investigated next, and the combination $-g \sin \theta_W$ necessarily denotes the electron charge e. This together with the expressions of the parameters $\alpha_e$, $\beta_e$, $\alpha_\nu$ in (6.14.16) are given by

$$- e = g \sin \theta_W = g' \cos \theta_W, \quad \alpha_e = \frac{1}{4} (3 \tan \theta_W - \cot \theta_W), \qquad (6.14.17)$$

$$\beta_e = -\alpha_\nu, \quad \alpha_\nu = \frac{1}{4} (\tan \theta_W + \cot \theta_W) = \frac{1}{4 \sin \theta_W \cos \theta_W}. \qquad (6.14.18)$$

---

[105]Note that $\overline{\psi_{L/R}} = (\psi_{L/R})^\dagger \gamma^0$.

Since $g$ is given by $g = e/\sin\theta_W$, we see that everything so far may be determined in terms of the angle $\theta_W$. This angle is referred to as the weak-mixing angle or Weinberg angle. Also note that $Z$ couples to both left-handed and right-handed leptons $e_i$, that is, the neutral current is not a pure $V-A$ current for these leptons,[106] and there is a flavor universality of neutrinos $Z$ - couplings.

The experimental determination of $\sin\theta_W$ is discussed in the next subsection. There are aspects of the theory we have to investigate. First note that a lepton mass term of the general structure $\overline{\psi}\psi$, for a spinor $\psi$, cannot be added to the Lagrangian density since it would break the gauge symmetry. This is easily seen by noting that

$$\overline{\psi}\psi = \overline{\psi}_L\psi_R + \overline{\psi}_R\psi_L, \tag{6.14.19}$$

and since left-handed and right-handed components have different transformations, this expression is not invariant under the underlying symmetry transformation. It suggests, however, how one may use it to generate leptonic masses. To make the above expression invariant, we may multiply it by another doublet, referred to as a Higgs field, having the structure

$$\Phi = \begin{pmatrix} \phi^+ \\ \phi^0 \end{pmatrix}, \tag{6.14.20}$$

with necessarily a weak hypercharge $Y = 1$ for invariance,[107] and introduce, in turn, the gauge invariant simple Yukawa term in the Lagrangian density

$$\mathscr{L}_{\ell\Phi} = -\sum_i \left[ \lambda_i \overline{\begin{pmatrix} \nu_i \\ e_i \end{pmatrix}}_L \begin{pmatrix} \phi^+ \\ \phi^0 \end{pmatrix} e_{iR} + h.c. \right]. \tag{6.14.21}$$

From the doublet transformation rules, under $SU(2)_L \times U(1)_Y$, we must have

$$\begin{pmatrix} \phi^+ \\ \phi^0 \end{pmatrix} \to e^{i\varphi(x)\mathbf{n}\cdot\boldsymbol{\sigma}/2} \begin{pmatrix} \phi^+ \\ \phi^0 \end{pmatrix}, \tag{6.14.22}$$

$$\begin{pmatrix} \phi^+ \\ \phi^0 \end{pmatrix} \to e^{i\beta(x)/2} \begin{pmatrix} \phi^+ \\ \phi^0 \end{pmatrix}. \tag{6.14.23}$$

Now we introduce the potential for the scalar field in (6.14.20),

$$V(\overline{\Phi}\Phi) = -\frac{m^2}{2}\Phi^\dagger\Phi + \frac{1}{4}\lambda(\Phi^\dagger\Phi)^2, \qquad m^2 > 0, \ \lambda > 0. \tag{6.14.24}$$

---

[106]It is precisely because of this that the pure neutral leptonic current, for example, involving the electron, as discussed in Sect. 6.14.2, facilitates one to determine $\sin^2\theta_W$ experimentally.

[107]See (6.14.7), (6.14.8).

and subtract it from the Lagrangian density. The negative sign of the first term is important for spontaneous symmetry breaking and generation of masses of particles. Working in the tree approximation (i.e., with no radiative corrections), the stationary point of the potential is given by

$$\langle \text{vac}|\Phi^{\dagger}|\text{vac}\rangle\langle \text{vac}|\Phi|\text{vac}\rangle \equiv \frac{v^2}{2}, \quad \text{and} \quad \frac{v}{\sqrt{2}} = \sqrt{\frac{m^2}{\lambda}}. \tag{6.14.25}$$

As a matter of fact, we can always[108] find a real field $\varphi(x)$, a unit vector $\mathbf{n}$, and a real field $\beta(x)$, such that the transformations in (6.14.22), (6.14.23) reduce the doublet in (6.14.20) to the form

$$\begin{pmatrix} \phi^+ \\ \phi^0 \end{pmatrix} \to \begin{pmatrix} \tilde{\phi}^+ \\ \tilde{\phi}^0 \end{pmatrix}, \tag{6.14.26}$$

where $\tilde{\phi}^0$ is a Hermitian field, referred to as the Higgs boson field, and the vacuum expectation values of $\tilde{\phi}^0$, and $\tilde{\phi}^+$ are given by

$$\langle \text{vac}|\tilde{\phi}^0|\text{vac}\rangle = \frac{v}{\sqrt{2}}, \quad \text{and} \quad \langle \text{vac}|\tilde{\phi}^+|\text{vac}\rangle = 0, \tag{6.14.27}$$

corresponding to spontaneous symmetry breaking, with $v$ a non-negative number. This special gauge attained by the choice of the above transformation to achieve this structure is referred to as the *unitary* gauge.

From (6.14.21), (6.14.26), (6.14.27), we may infer that the leptons $e, \mu, \tau$ acquire masses given by[109]

$$m_i = \lambda_i \frac{v}{\sqrt{2}}. \tag{6.14.28}$$

Since *a priori* no right-handed neutrinos were included in the theory, the neutrinos do not acquire a mass.

The interaction of the vector bosons with the Higgs field arises through the standard expression

$$\mathscr{L}_{\text{vec.bosons},\Phi} = -\left|\left(\frac{\partial_\mu}{i} - g\mathbf{W}_\mu \cdot \mathbf{T} + \frac{g'}{2} B_\mu\right)\Phi\right|^2. \tag{6.14.29}$$

---

[108]See Problem 6.22.

[109]Here we have used the facts that $\overline{e_{R/L}}\, e_{R/L} = 0$, as a consequence of the facts that $\{\gamma^5, \gamma^0\} = 0$, and $(\gamma^5)^2 = I$.

From the latter term in the Lagrangian density (in the unitary gauge), the following expression emerges for the *mass* terms of the vector bosons

$$
-\left| \left( \frac{\partial_\mu}{i} - g \mathbf{W}_\mu \cdot \mathbf{T} + \frac{g'}{2} B_\mu \right) \begin{pmatrix} 0 \\ \frac{v}{\sqrt{2}} \end{pmatrix} \right|^2_{\text{mass}}
$$

$$
= -\frac{v^2}{2} \left( \frac{g^2}{2} W_\rho^\dagger W^\rho + \frac{1}{4} (g^2 + g'^2) Z^\rho Z_\rho \right). \tag{6.14.30}
$$

The acquired masses of the vector bosons are then given by $(g^2 + g'^2 = g^2 \sec^2 \theta_W)$

$$
M_W = \frac{v}{2} |g|, \qquad M_Z = \frac{v}{2} |g \sec \theta_W|, \qquad M_\gamma = 0, \tag{6.14.31}
$$

where we note that $W_\rho$ is a complex field, while $Z_\rho$ is a real one, which explains the presence of the overall $1/2$ factors in the above equations. As the photon field $A_\rho$ does not appear on the extreme right-hand side of (6.14.30), the photon remains massless.

The numerical value of the vacuum expectation value of the Higgs boson field $v/\sqrt{2}$ may be readily estimated as follows. By using the fact the propagator of the now massive vector boson $W$ propagator $\Delta_+^{\alpha\beta}(q^2)$ approaches[110]: $\eta^{\alpha\beta}/M_W^2$ at low energies $|q^2| \ll M_W^2$, we may infer from the last expression on the right-hand side of (6.14.16), that the exchange of $W$ vector boson at low energy between, $\overline{e}_L$ and $\mu_L$ type leptons, may be described by the effective Lagrangian density

$$
\mathscr{L}_{e\mu,\text{eff}} = \left( \frac{g}{\sqrt{2}} \right)^2 \frac{1}{M_W^2} \left( \left[ \overline{e} \gamma^\rho \frac{1 - \gamma^5}{2} v_e \right] \left[ \overline{v}_\mu \gamma_\rho \frac{1 - \gamma^5}{2} \mu \right] + h.c. \right). \tag{6.14.32}
$$

By extracting the equivalent expression from the effective Lagrangian density in (6.14.1) to the above one, and by making a direct comparison gives[111]

$$
\frac{g^2}{4M_W^2} = \sqrt{2}\, G_F. \tag{6.14.33}
$$

From this equation and the first equality in (6.14.31), we obtain the following key estimate in the electroweak theory

$$
v = \left( \frac{1}{\sqrt{2}\, G_F} \right)^{1/2} \simeq 246\, \text{GeV}. \tag{6.14.34}
$$

---

[110] See (4.7.20).

[111] Note that each $(1 - \gamma^5)$ in (6.14.32) is divided by a factor of 2 which accounts for the factor $1/4$ on the left-hand side of (6.14.33).

Eq. (6.14.31) and the first equality in (6.14.17), give the following masses for the vector bosons

$$M_W = \frac{v}{2}\left|\frac{e}{\sin\theta_W}\right|, \quad M_Z = v\left|\frac{e}{\sin 2\theta_W}\right|. \tag{6.14.35}$$

Simultaneous measurements of differential cross sections of processes involving electron-neutrinos scattering,[112] for example, as conveniently discussed in the next subsection, give the value

$$\sin^2\theta_W = 0.2324 \pm 0.0083. \tag{6.14.36}$$

This provides the following estimates for the masses of the vector bosons in (6.14.35): $M_W \simeq 77.4$ GeV, $M_Z \simeq 88.3$ GeV. On the other hand, taking the value of the effective fine-structure to be $\alpha \simeq 1/128$, as determined [113] at energies of the order 91 GeV, as discussed earlier below Eq. (5.19.9) and investigated theoretically in Sect. 5.19.2, gives for the masses estimates of the order 80.1 GeV, 91.4 GeV, respectively, which are comparable to experimental values.[114]

An expression for the Higgs boson mass may be obtained from the potential in (6.14.24) by working in the unitary gauge. Upon writing $\Phi \equiv (\rho + v)/\sqrt{2}$, with $\langle \text{vac}|\rho|\text{vac}\rangle = 0$, we then have

$$-V(\Phi^\dagger\Phi)\Big|_{\text{mass}} = -\frac{\lambda}{16}(\rho + v)^2\left(\rho^2 + 2\rho v - v^2\right)\Big|_{\text{mass}} = -\frac{1}{4}\lambda v^2\rho^2, \tag{6.14.37}$$

from which we may extract the mass of the Higgs boson to be

$$M_H = v\sqrt{\frac{\lambda}{2}}, \tag{6.14.38}$$

where $\lambda$ is undetermined.[115]

In setting up the electroweak theory, we recall that it was originally assumed that the neutrinos were massless, left-handed, and we have seen that the absence of right-handed components in the theory precludes the neutrinos from acquiring a mass. The problem of neutrino masses and their small values is investigated in Sect. 6.14.3. This is followed by Sect. 6.14.4 dealing with neutrino oscillations which is relevant to the masses of neutrinos. First, however, we consider an experimental determination of $\sin^2\theta_W$.

---

[112]Vilain et al. [119].

[113]See, e.g., Mele [89], Erler [36], and Beringer et al. [17].

[114]See, e.g., Eidelman et al. [35].

[115]If one is to rely on the validity of perturbation, and restricts $\lambda \in (0, 1)$, then formally for $\lambda \simeq 1/2$, for example, (6.14.34)/(6.14.38) give the value $M_H \simeq 123$ GeV. The coupling is, however, constrained by renormalization restrictions.

## 6.14.2   *Experimental Determination of* $\sin^2 \theta_W$

The interaction Lagrangian density for the description of purely leptonic *neutral* current, i.e., via the exchange of only the $Z$ vector boson, processes involving the following neutrinos-electron scattering: $\nu_\mu + e^- \rightarrow \nu_\mu + e^-$, $\tilde{\nu}_\mu + e^- \rightarrow \tilde{\nu}_\mu + e^-$, $\nu_e + e^- \rightarrow \nu_e + e^-$, $\tilde{\nu}_e + e^- \rightarrow \tilde{\nu}_e + e^-$, may be extracted from (6.14.16) to be

$$\mathscr{L}^{(Z)}_{\nu_e \nu_\mu e} = -g \sin \theta_W [\bar{e} \gamma^\rho (\alpha_e - \beta_e \gamma^5) e + \alpha_\nu \bar{\nu}_\mu \gamma^\rho (1 - \gamma^5) \nu_\mu + \alpha_\nu \bar{\nu}_e \gamma^\rho (1 - \gamma^5) \nu_e ] Z_\rho. \tag{6.14.39}$$

From (6.14.17), (6.14.18) this may be rewritten as

$$\mathscr{L}^{(Z)}_{\nu_e \nu_\mu e} = -\frac{g}{2 \cos \theta_W} \left[ e \gamma^\rho (g_V - g_A \gamma^5) e + \overline{\nu_{\mu L}} \gamma^\rho \nu_{\mu L} + \overline{\nu_{eL}} \gamma^\rho \nu_{eL} \right] Z_\rho, \tag{6.14.40}$$

$$g_V = 2 \sin^2 \theta_W - \frac{1}{2}, \quad g_A = -\frac{1}{2}. \tag{6.14.41}$$

On the other hand, (6.14.32), (6.14.33), (6.14.34) give

$$\frac{g}{2 \cos \theta_W} = M_Z (\sqrt{2} G_F)^{1/2}, \tag{6.14.42}$$

which leads to

$$\mathscr{L}^{(Z)}_{\nu_e \nu_\mu e} = -M_Z (\sqrt{2} G_F)^{1/2} \left[ \bar{e} \gamma^\rho (g_V - g_A \gamma^5) e + \overline{\nu_{\mu L}} \gamma^\rho \nu_{\mu L} + \overline{\nu_{eL}} \gamma^\rho \nu_{eL} \right] Z_\rho. \tag{6.14.43}$$

The scattering processes: $\nu_e + e^- \rightarrow \nu_e + e^-$, $\bar{\nu}_e + e^- \rightarrow \bar{\nu}_e + e^-$ from (6.14.16) also involve the exchange of a $W$ vector boson. From the just mentioned equation, the interaction Lagrangian density responsible for this exchange is then given by

$$\mathscr{L}^{(W)}_{\nu_e e} = \sqrt{2} M_W (\sqrt{2} G_F)^{1/2} \left[ \overline{e_L} \gamma^\rho \nu_{eL} W + \overline{\nu_{eL}} \gamma^\rho e_L W^\dagger \right]. \tag{6.14.44}$$

For low energy scattering of the above scattering processes in question, we may replace the $Z$ propagator by $\eta^{\alpha\beta} / M_Z^2$, and the $W$ propagator by $\eta^{\alpha\beta} / M_Z^2$. Thus the following effective interaction Lagrangian density emerges for corresponding to the exchange of the vector bosons $Z$ and $W$:

$$\mathscr{L}^{(Z)}_{\nu_e \nu_\mu e} \bigg|_{\text{eff}} = \sqrt{2} \, G_F [\bar{e} \gamma^\rho (g_V - g_A \gamma^5) e] [\overline{\nu_{\mu L}} \gamma_\rho \nu_{\mu L} + \overline{\nu_{eL}} \gamma_\rho \nu_{eL}], \tag{6.14.45}$$

$$\mathscr{L}^{(W)}_{\nu_e e} \bigg|_{\text{eff}} = 2 \sqrt{2} \, G_F [\overline{e_L} \gamma^\rho e_L] [\overline{\nu_{eL}} \gamma_\rho \nu_{eL}], \tag{6.14.46}$$

where in writing the last equation, we have used the classic Fierz identity (see, e.g., Appendix A of Chapter 8 in Vol. II), and the anti-commutativity of the fermion fields, to write (see Problem 6.23)

$$\left[\overline{e_L}\,\gamma^{\,\rho}\nu_{eL}\right]\left[\overline{\nu_{eL}}\,\gamma_\rho e_L\right] = \left[\overline{e_L}\,\gamma^{\,\rho}e_L\right]\left[\overline{\nu_{eL}}\,\gamma_\rho\nu_{eL}\right]. \tag{6.14.47}$$

Therefore the total effective interaction Lagrangian density for the processes mentioned in the beginning of this subsection is given by

$$\begin{aligned}\mathscr{L}^I_{\nu e\,\text{eff}} &= \sqrt{2}\,G_F\left[\overline{e}\,\gamma^{\,\rho}(g_V - g_A\gamma^{\,5})e\right]\left[\overline{\nu_{\mu L}}\,\gamma_\rho\nu_{\mu L}\right] \\ &\quad + \sqrt{2}\,G_F\left[\overline{e}\,\gamma^{\,\rho}\big([g_V+1]-[g_A+1]\gamma^{\,5}\big)e\right]\left[\overline{\nu_{eL}}\,\gamma_\rho\nu_{eL}\right],\end{aligned} \tag{6.14.48}$$

where $g_V$ and $g_A$ are defined in (6.14.41).

Precise simultaneous measurements[116] of the corresponding differential cross sections of $(\nu_\mu e)$, $(\tilde{\nu}_\mu e)$, $(\nu_e e)$, $(\tilde{\nu}_e e)$, and a comparison with the ones dictated[117] by the above expression in (6.14.48), allow the determination of the constants $g_V, g_A$,[118] and, in turn, give the value of $\sin^2\theta_W$ as recorded in (6.14.36).

### 6.14.3   Masses of the Neutrinos and the "Seesaw Mechanism"

There is an overwhelmingly large number of experiments[119] which are in favor of no-zero neutrino masses. In the original electroweak theory, no right-handed neutrinos where involved and thus the neutrinos remain massless under spontaneous symmetry breaking. By assuming their existence, the neutrinos may acquire masses through the Higgs mechanism as the charged leptons (and quarks Sect. 6.15). This method, however, fails to account for the expected small masses of the neutrinos. Perhaps, a far more interesting way to account for the small masses, and which is physically appealing, is through the so-called "seesaw mechanism"[120] which we now consider.

In the "seesaw mechanism", one assumes that lepton numbers are violated at some high-energy scales well beyond that of the electroweak energy scale. A simple way to describe this is to assign to each charged lepton, in the process, a

[116]Vilain et al. [119].

[117]These are obtained in the same way, e.g, as the differential cross section in Sect. 5.9.3. See 't Hooft [116].

[118]The value of $g_A$ is consistent with the theoretical value of $-1/2$.

[119]See, e.g., Davis et al. [34], Hirata et al. [65], Fukuda et al. [50], Ambrosio et al. [7], Athanassopoulos et al. [12], and Cleveland et al. [31]. The latter, in particular, is consistent with neutrino oscillations experiments which have been acknowledged by the 2015 Nobel Prize.

[120]See, e.g., Minkowski [90], Gell-Mann et al. [54], Yanagida [123], and Mohapatra and Senjanović [92].

right-handed neutrino. The underlying mechanism leads to two sets of massive Majorana particles, each with three particles, with one set corresponding to presently observed neutrinos, and another set consisting of neutrinos with very large masses beyond the presently available energies. We will see how these latter heavy neutrinos are to be handled in making explicit computations in field theory. We recall that a Majorana field is invariant under charge conjugation.

Recall that the charge conjugate of a spinor field $\chi$, is defined by $\chi^{\mathscr{C}} = \mathscr{C}\overline{\chi}^{\mathsf{T}}$ (see Sect. 2.5), where[121] $\mathscr{C}$ is the charge conjugation matrix. For a left-handed spinor $\psi_{\mathrm{L}}$, and a right-handed one $\psi_{\mathrm{R}}$, we have the following relations

$$(\psi_{\mathrm{L,R}})^{\mathscr{C}} = \mathscr{C}\,\overline{\psi_{\mathrm{L/R}}}^{\mathsf{T}}, \qquad \overline{\psi_{\mathrm{L/R}}} = -[(\psi_{\mathrm{L/R}})^{\mathscr{C}}]^{\mathsf{T}}\mathscr{C}^{-1}. \tag{6.14.49}$$

It is important to note that

$$\overline{\psi_{\mathrm{L/R}}} = \Big(\frac{1 \mp \gamma^5}{2}\psi\Big)^{\dagger}\gamma^0 = (\psi)^{\dagger}\frac{1 \mp \gamma^5}{2}\gamma^0 = \overline{\psi}\,\frac{1 \pm \gamma^5}{2} = \Big(\frac{1 \pm \gamma^5}{2}\overline{\psi}^{\mathsf{T}}\Big)^{\mathsf{T}},$$
$$\tag{6.14.50}$$

on account that $\{\gamma^5, \gamma^0\} = 0$, with $\gamma^5$, $\gamma^0$ as symmetric and Hermitian. That is, $\overline{\psi_{\mathrm{L}}}$, $(\psi_{\mathrm{L}})^{\mathscr{C}}$ are *right*-handed, while $\overline{\psi_{\mathrm{R}}}$, $(\psi_{\mathrm{R}})^{\mathscr{C}}$ are *left*-handed. These immediately imply that

$$\overline{\psi_{\mathrm{L}}}\,\psi_{\mathrm{L}} = 0, \qquad \overline{\psi_{\mathrm{R}}}\,\psi_{\mathrm{R}} = 0. \tag{6.14.51}$$

From the properties of the charge conjugation matrix, and the definition of charge conjugate spinor fields, we also have, in particular,

$$\overline{\psi_{1\mathrm{R}}}\,\psi_{2\mathrm{L}} = \overline{(\psi_{2\mathrm{L}})^{\mathscr{C}}}\,(\psi_{1\mathrm{R}})^{\mathscr{C}}. \tag{6.14.52}$$

For simplicity of the presentation, we consider a one generation neutrino with field $\nu_{\mathrm{L}}$, i.e., a one flavor neutrino. We introduce a right-handed neutrino, denoted by $N_{\mathrm{R}}$, and define, in turn, the following mass-Lagrangian density

$$\mathscr{L}_{\mathrm{mass}} = -m_D\big(\overline{N_{\mathrm{R}}}\,\nu_{\mathrm{L}} + \overline{\nu_{\mathrm{L}}}\,N_{\mathrm{R}}\big) - \frac{1}{2}M_{\mathrm{R}}\big(\overline{N_{\mathrm{R}}}\,(N_{\mathrm{R}})^{\mathscr{C}} + \overline{(N_{\mathrm{R}})^{\mathscr{C}}}\,N_{\mathrm{R}}\big), \tag{6.14.53}$$

where we will eventually take the mass parameters to satisfy the constraint $M_{\mathrm{R}} \gg m_D$. The reason for the $1/2$ factor in the second term is to avoid double counting, since we may, e.g., rewrite $\overline{N_{\mathrm{R}}^{\mathscr{C}}}\,N_{\mathrm{R}} = -(N_{\mathrm{R}})^{\mathsf{T}}\mathscr{C}^{-1}N_{\mathrm{R}}$ and the same field $N_{\mathrm{R}}$ appears twice which should be taken into account in the application of the Euler-Lagrange equation. The first term in (6.14.53) is referred to as a Dirac mass term, and is assumed to have been generated, via spontaneous symmetry breaking,

---

[121] Remember: $\mathscr{C}^{\dagger} = \mathscr{C}^{\mathsf{T}} = -\mathscr{C}$, $(\mathscr{C})^2 = -I$, $[\mathscr{C}, \gamma^5] = 0$.

through the Higgs mechanism, or simply as just an electroweak symmetry breaking term introduced into the theory. The second one is of different nature, referred to as a Majorana mass term, and with lepton number $+1$ for a neutrino, we note that this term does not conserve lepton number, and violates it by two units.

Using the basic properties in (6.14.49), (6.14.50), (6.14.51), (6.14.52), the mass-Lagrangian density in (6.14.53) may be conveniently rewritten as $\big(\mathscr{C}\psi^\top = -\overline{(\psi)^{\mathscr{C}}}\,\big)$

$$
\mathscr{L}_{\text{mass}} = \frac{1}{2}\Big\{ m_D\Big[ \mathscr{C}\big((N_R)^{\mathscr{C}}\big)^\top \nu_L + \mathscr{C}(\nu_L)^\top (N_R)^{\mathscr{C}} \Big] + M_R\Big[ \mathscr{C}\big((N_R)^{\mathscr{C}}\big)^\top (N_R)^{\mathscr{C}} \Big] \Big\} + h.c.,
\tag{6.14.54}
$$

where we have suitably grouped together terms in the first expression $\{.\}$. The above may be rewritten in simple matrix form as

$$
\mathscr{L}_{\text{mass}} = \frac{1}{2}\Big[ \mathscr{C}\begin{pmatrix} \nu_L \\ N_R^{\mathscr{C}} \end{pmatrix}^\top \Big] \begin{pmatrix} 0 & m_D \\ m_D & M_R \end{pmatrix} \begin{pmatrix} \nu_L \\ N_R^{\mathscr{C}} \end{pmatrix} + h.c.
\tag{6.14.55}
$$

It is easy to check that with

$$
V = \begin{pmatrix} i\cos\vartheta & -i\sin\vartheta \\ -\sin\vartheta & -\cos\vartheta \end{pmatrix}, \quad \sin\vartheta = \frac{\sqrt{m_1}}{\sqrt{m_1 + m_2}}, \quad \cos\vartheta = \frac{\sqrt{m_2}}{\sqrt{m_1 + m_2}},
\tag{6.14.56}
$$

$$
m_1 = \frac{1}{2}\big(\sqrt{4m_D^2 + M_R^2} - M_R\big), \quad m_2 = \frac{1}{2}\big(\sqrt{4m_D^2 + M_R^2} + M_R\big),
\tag{6.14.57}
$$

the following transformation emerges

$$
V^\top \begin{pmatrix} m_1 & 0 \\ 0 & m_2 \end{pmatrix} V = \begin{pmatrix} 0 & m_D \\ m_D & M_R \end{pmatrix}.
\tag{6.14.58}
$$

Hence upon setting

$$
V\begin{pmatrix} \nu_L \\ (N_R)^{\mathscr{C}} \end{pmatrix} = \begin{pmatrix} \nu_{1L} \\ \nu_{2L} \end{pmatrix},
\tag{6.14.59}
$$

we obtain $\big(\mathscr{C}\psi^\top = -\overline{(\psi)^{\mathscr{C}}}\,\big)$

$$
\mathscr{L}_{\text{mass}} = \frac{1}{2}\Big( \mathscr{C}\Big[ V\begin{pmatrix} \nu_L \\ (N_R)^{\mathscr{C}} \end{pmatrix} \Big]^\top \Big) \begin{pmatrix} m_1 & 0 \\ 0 & m_2 \end{pmatrix} V\begin{pmatrix} \nu_L \\ (N_R)^{\mathscr{C}} \end{pmatrix} + h.c.
$$

$$
= -\frac{m_1}{2}\Big[ \overline{(\nu_{1L})^{\mathscr{C}}}\,\nu_{1L} + \overline{\nu_{1L}}\,(\nu_{1L})^{\mathscr{C}} \Big] + \frac{m_2}{2}\Big[ \overline{(\nu_{2L})^{\mathscr{C}}}\,\nu_{2L} + \big(\overline{\nu_{2L}}\,(\nu_{2L})^{\mathscr{C}}\big) \Big].
\tag{6.14.60}
$$

Using the basic properties in (6.14.51), and defining the Majorana fields $\nu_i = \nu_{iL} + (\nu_{iL})^{\mathscr{C}}$, $i = 1, 2$, the following mass-Lagrangian density emerges

$$\mathscr{L}_{\text{mass}} = -\frac{1}{2}\left(m_1\,\overline{\nu_1}\,\nu_1 + m_2\,\overline{\nu_2}\,\nu_2\right). \qquad (6.14.61)$$

Thus, we have obtained two massive Majorana fields, and for $M_R \gg m_D$, we have from (6.14.57), one of mass $m_1 \simeq m_D^2/M_R$ which would be small for sufficiently large $M_R$, and one with corresponding large mass $m_2 \simeq M_R$.[122] Note that $\cos \vartheta \simeq 1, \sin \vartheta \simeq 0$. The $\nu_2$ does not participate in the dynamics, it is referred to as a "sterile" particle and decouples from the theory.

It is interesting that one may achieve small neutrino masses by the "seesaw mechanism". The analysis involving more than one generation of neutrino follows the same pattern as above.

### 6.14.4 Neutrino Oscillations: An Interlude

There are by now several experiments[123] which provide clear evidence of neutrino oscillations,[124] implying that transitions occur between different neutrino flavors, and that a given neutrino flavor may disappear, in a neutrino beam, and another neutrino flavor appears in an oscillatory manner depending on the distances traveled by the neutrinos. This is understood to be caused, in general, by nonzero neutrino masses and by so-called neutrino mixing. By neutrino mixing it is meant that the neutrino fields may be expressed as linear combinations of neutrinos, which we denote by $(\nu_1, \nu_2, \ldots)$, with specific masses $(m_1, m_2, \ldots)$.

The underlying aspect of the above physical phenomenon is that the transition probability between, say, two neutrino flavors $\nu^{(x)}$, $\nu^{(y)}$: $\text{Prob}\,[\nu^{(x)} \to \nu^{(y)}]$, is of an oscillatory nature, depending on the distance traveled by the $\nu^{(x)}$ neutrino, and a value of one taken by this probability, at a given distance, for example, specifies the "disappearance" of the $\nu^{(x)}$ neutrino.[125] Considering ultra-relativistic massive neutrinos, that is, with masses negligible in comparison to their momenta, we will see, by a simple derivation that for transitions involving two kinds of neutrinos, the amplitude of oscillations of the just mentioned probability is determined by the

---

[122]$M_R$ is often taken as large as $10^{15}$GeV according to a scale set up by GUT.

[123]See, e.g., Anselmann et al. [10], Cleveland et al. [31], Fukuda et al. [50, 51], Hampel et al. [62], and Altmann et al. [6]. It is rather interesting to point out that the theory of neutrino oscillations was written up in this book much earlier than the 2015 Nobel Prize in Physics was announced.

[124]Pontecorvo [99, 100] and Maki et al. [81]. See also Pontecorvo [101] and Gribov and Pontecorvo [59].

[125]That is, reduction in beam intensity of $\nu^{(x)}$ neutrinos, and hence a reduction of neutrinos, in general, will be observed if one is "looking" only for this flavor.

mixing of the neutrinos, while its oscillatory nature depends on the distance traveled by the $\nu^{(x)}$ neutrino and its energy.

In a theoretical description of a neutrino oscillating experiment, we have causally, an emission source of an ultra-relativistic neutrino flavor, followed by its propagation, described by the propagator taking into account of neutrino flavor mixing, and finally a detection source of a given flavor type : emitter × propagation × detection.

Before discussing the above process, let us quickly review the way one may describe the combined process of emission, propagation and detection of a single ultra-relativistic massive spin 1/2 particle of mass $m$. We know from § 3.3.1, that given two sources $\overline{\eta}^>(x)$, $\eta^<(x')$ with $x^0 > x'^0$ for all $x^0$, $x'^0$, for which, the respective sources do not vanish, this amplitude is given by

$$
\begin{aligned}
i\,W_{12} &= i \int (dx)(dx')\,\overline{\eta}^>(x) \left[ \int \frac{(dp)}{(2\pi)^4}\, e^{ip\,(x-x')}\, \frac{(-\gamma p + m)}{p^2 + m^2 - i\epsilon} \right] \eta^<(x') \\
&= \int (dx)(dx')\,(i)\,\overline{\eta}^>(x) \left[ \int \frac{d^3\mathbf{p}}{2p^0\,(2\pi)^3}\, e^{ip\,(x-x')}(-\gamma p + m) \right](i)\,\eta^<(x'),
\end{aligned}
$$

(6.14.62)

$p^0 = \sqrt{\mathbf{p}^2 + m^2}$. For an ultra-relativistic particle,

$$
p^0 = \sqrt{\mathbf{p}^2 + m^2} \simeq |\mathbf{p}| + \frac{m^2}{2|\mathbf{p}|},
$$

(6.14.63)

$$
\frac{-\gamma \cdot \mathbf{p} + \gamma^0 p^0 + m}{p^0} \simeq -\gamma \cdot \mathbf{N} + \gamma^0, \quad \mathbf{N} = \mathbf{p}/|\mathbf{p}|.
$$

(6.14.64)

We also note that

$$
-\gamma \cdot \mathbf{N} + \gamma^0 = \begin{pmatrix} 1 \\ \sigma \cdot \mathbf{N} \end{pmatrix} (1 \quad -\sigma \cdot \mathbf{N}).
$$

(6.14.65)

Accordingly by defining $\overline{u}(\mathbf{p}) = \sqrt{|\mathbf{p}|}\,(1 \quad -\sigma \cdot \mathbf{N})$,

$$
\rho(\mathbf{p}, x^0) = \int d^3\mathbf{x}\,\overline{u}(\mathbf{p})\,\eta(x)\, e^{-i(\mathbf{p}\cdot\mathbf{x} - |\mathbf{p}|x^0)}, \quad \rho^\dagger(\mathbf{p}, x^0) = \int d^3\mathbf{x}\,\overline{\eta}(x)u(\mathbf{p})\, e^{i(\mathbf{p}\cdot\mathbf{x} - |\mathbf{p}|x^0)},
$$

(6.14.66)

the amplitude $i\,W_{12}$ becomes simply

$$
i\,W_{12} = \int \frac{d^3\mathbf{p}}{2\,|\mathbf{p}|(2\pi)^3} [dx^0\,i\,\rho^{>\dagger}(\mathbf{p}, x^0)]e^{-i(m^2/2|\mathbf{p}|)(x^0 - x'^0)}\left[ dx'^0\,i\,\rho^<(\mathbf{p}, x'^0) \right].
$$

(6.14.67)

Here given that a particle was emitted by an emitter (source $\eta^<$) and detected by a detector (source $\eta^{>\dagger}$), then $e^{-i(m^2/2|\mathbf{p}|)(x^0 - x'^0)}$ denotes the amplitude that the particle has moved from the emitter to the detector with probability one, indicating the stability of the particle.

Now consider two spinor fields $\psi_1$, $\psi_2$ with corresponding masses $m_1$, $m_2$ respectively, with Lagrangian density

$$\mathscr{L} = \mathscr{L}_0 + \overline{\eta}_1 \psi_1 + \overline{\psi}_1 \eta_1 + \overline{\eta}_2 \psi_2 + \overline{\psi}_2 \eta_2, \tag{6.14.68}$$

where $\mathscr{L}_0$ are their free Lagrangian densities, and $\eta_1, \eta_2$ are external sources.

In experiments, involving neutrino oscillations, involving, for example, the transition $\nu_\mu \rightarrow \nu_e$ the latter are considered to correspond to states as linear combinations of states of definite masses $m_1$, $m_2$. Accordingly, if we let the sources of emission of $\nu_\mu$ be denoted by $\eta_{(\mu)}$, and that of the detection of $\nu_e$ by $\overline{\eta}_{(e)}$, then these sources may be defined as linear combinations of the ones in (6.14.68) corresponding to definite mass states, i.e., with normalized coefficients

$$\eta_{(e)} = \eta_1 \cos\theta + \eta_2 \sin\theta, \qquad \eta_{(\mu)} = -\eta_1 \sin\theta + \eta_2 \cos\theta,$$

or,

$$\eta_1 = \eta_{(e)} \cos\theta - \eta_{(\mu)} \sin\theta, \qquad \eta_2 = \eta_{(e)} \sin\theta + \eta_{(\mu)} \cos\theta. \tag{6.14.69}$$

Also by readily extracting the coefficient of $[\rho_{(e)}^{>\dagger} \rho_{(\mu)}^{<}]$ in

$$[\rho_1^{>\dagger} e^{-i(m_1^2/2|\mathbf{p}|)(x^0 - x'^0)} \rho_1^{<}] + [\rho_2^{>\dagger} e^{-i(m_2^2/2|\mathbf{p}|)(x^0 - x'^0)} \rho_2^{<}],) \tag{6.14.70}$$

where the sources denoted by $\rho$ are defined as in (6.14.66), we may immediately write down the amplitude of occurrence for the process: "emission of $\nu_\mu$ by the emitter, $\times$ transition amplitude $(\nu_\mu \rightarrow \nu_e) \times$ amplitude of detection of $\nu_e$" as follows:

$$\int \frac{d^3\mathbf{p}}{2|\mathbf{p}|(2\pi)^3} \, dx^0 dx'^0 \, i\rho_{(e)}^{>\dagger}(\mathbf{p}, x^0) \, A(\nu_\mu \rightarrow \nu_e) \, i\rho_{(\mu)}^{<}(\mathbf{p}, x'^0), \tag{6.14.71}$$

where $\rho_{(e)}^{>\dagger}$ is "switched on" after $\rho_{(\mu)}^{<}$ is "switched off",

$$A(\nu_\mu \rightarrow \nu_e) = \sin\theta \cos\theta \Big( \exp[-i\frac{m_2^2}{2|\mathbf{p}|}(x^0 - x'^0)] - \exp[-i\frac{m_1^2}{2|\mathbf{p}|}(x^0 - x'^0)] \Big). \tag{6.14.72}$$

Upon taking the absolute value squared of the above gives the probability

$$\text{Prob}[\nu_\mu \rightarrow \nu_e] = \sin^2(2\theta) \sin^2\Big(\frac{m_1^2 - m_2^2}{4} \frac{L}{E_\nu}\Big), \tag{6.14.73}$$

where we have taken $E_\nu \simeq |\mathbf{p}|$, and, due to the fact that the ultra-relativistic particle travels near with the speed of light $\Delta x/\Delta t \simeq 1$, identified $(x^0 - x'^0) \simeq L$, with $L$ denoting the distance traveled by the ultra-relativistic neutrino.[126] It is customary to define the oscillation length

$$L_\nu = \frac{(4\pi E_\nu)}{(m_2^2 - m_1^2)},$$  (6.14.74)

and rewrite the probability in question as

$$\text{Prob}[\nu_\mu \to \nu_e] = \sin^2(2\theta) \sin^2\left(\frac{\pi L}{L_\nu}\right).$$  (6.14.75)

We note that the oscillations of the probability are governed by the distance $L$ traveled by the ultra-relativistic muon neutrino and its energy $E_\nu$, while the amplitude $\sin^2(2\theta)$ of the oscillations is governed by the neutrino mixing. It is important to note that the probability of no transition: $\text{Prob}[\nu_\mu \to \nu_\mu]$, for example, referred to as the survival probability of the neutrino $\nu_\mu$, as obtained directly from the coefficient of $[\rho_{(\mu)}^{>\dagger} \rho_{(\mu)}^{<}]$ in (6.14.70), with emission and detection sources causally arranged, is given by

$$\text{Prob}[\nu_\mu \to \nu_\mu] = 1 - \sin^2(2\theta) \sin^2\left(\frac{\pi L}{L_\nu}\right),$$  (6.14.76)

and coincides with that obtained from a completeness relation

$$\text{Prob}[\nu_\mu \to \nu_\mu] = 1 - \text{Prob}[\nu_\mu \to \nu_e],$$  (6.14.77)

for consistency, as is easily checked.

Note that for $\pi L \ll L_\nu$, the neutrino $\nu_\mu$ has not traveled long enough to "disappear", and its survival probability $\text{Prob}[\nu_\mu \to \nu_\mu]$ will attain a value not far from one.

## 6.15  Electroweak Theory II: Incorporation of Quarks; Anomalies and Renormalizability

This section deals with the incorporation of quarks in the electroweak theory, generating what has been called the standard model based on the product symmetry groups $SU(3) \times SU(2) \times U(1)$. The electroweak theory as well as the standard model

---

[126]Needless to say, in a relativistic setting, one should not confuse distance traveled with length measurement, in general, with the latter, for example, involved in coincident time measurements of the end points, say, of a moving ruler.

involve axial vector currents and the problem of anomalies studied in Sects. 3.9 and 3.10 arises and their absence is critical for the renormalizability of the standard model. Based on the work in Sects. 3.9 and 3.10, we show that there are actually no anomalies in the standard model.

### 6.15.1  Quarks and the Electroweak Theory

In order to take into account strangeness changing interactions such as: $\Lambda^0 \to p + \pi^-$, and, e.g., suppress a decay (strangeness changing) process such as: $\Lambda^0 \to p + e^- + \tilde{\nu}_e$ in comparison to the classic beta decay process: $n \to p + e^- + \tilde{\nu}_e$, and take into account other processes involving flavor changing quarks, the three generations of quarks, in analogy to the leptons' generations, but now guided by the just mentioned processes, are taken of the form

$$q_{1L} = \begin{pmatrix} u \\ d' \end{pmatrix}_L, \quad q_{2L} = \begin{pmatrix} c \\ s' \end{pmatrix}_L, \quad q_{3L} = \begin{pmatrix} t \\ b' \end{pmatrix}_L,$$
$$u_R, d_R \qquad\qquad c_R, s_R \qquad\qquad t_R, b_R \tag{6.15.1}$$

with left-handed quarks placed in weak iso-doublets, while the right-handed ones in weak iso-singlets. Also $(d', s', b')$ are, on general grounds, linear combinations[127] (allowing necessarily cross generations transitions), of the form

$$\begin{pmatrix} d' \\ s' \\ b' \end{pmatrix} = \begin{pmatrix} V_{ud} & V_{us} & V_{ub} \\ V_{cd} & V_{cs} & V_{cb} \\ V_{td} & V_{ts} & V_{tb} \end{pmatrix} \begin{pmatrix} d \\ s \\ b \end{pmatrix}, \tag{6.15.2}$$

where the matrix $V_{CKM} = [V_{ab}]$ is referred to as the Cabibbo-Kobayashi-Maskawa matrix . $V_{ud}$ is a measure of the coupling of $u$ to $d$, and so on. The electroweak quantum numbers of the quarks are given in Table 6.3.

**Table 6.3** Electroweak quantum numbers of the quarks in the Standard Model: $q_> = u, c, t$; $q_< = d, s, b$. The antiquarks have the same $I$, but opposite $Q$, $I_3$ (for $I_3 \neq 0$), and $Y = 2(Q - I_3)$. Each of the quarks comes in three different colors

| Particle | $I$ | $I_3$ | $Q$ | $Y$ |
|---|---|---|---|---|
| $(q_{>L}, q_{<L})$ | 1/2 | $(+1/2, -1/2)$ | $(+2/3, -1/3)$ | 1/3 |
| $(q_{>R}, q_{<R})$ | 0 | $(0, 0)$ | $(+2/3, -1/3)$ | $(+4/3, -2/3)$ |

---

[127]Cabibbo [27], Glashow et al. [56], and Kobayashi and Maskawa [71].

The hadronic current may be then written as

$$J_h^\rho = \overline{\begin{pmatrix} u \\ c \\ t \end{pmatrix}} \gamma^\rho (1 - \gamma^5) \begin{pmatrix} d' \\ s' \\ b' \end{pmatrix},$$                    (6.15.3)

while an electromagnetic one is given by

$$J_{em}^\rho = \frac{2}{3} \sum_{q>} \overline{q_>} \gamma^\rho q_> - \frac{1}{3} \sum_{q<} \overline{q_<} \gamma^\rho q_<.$$                    (6.15.4)

A standard representation of the unitary Cabibbo-Kobayashi-Maskawa matrix is given by

$$V_{CKM} = \begin{pmatrix} c_{12}c_{13} & s_{12}c_{13} & s_{13}e^{-i\delta} \\ -s_{12}c_{23} - c_{12}s_{23}s_{13}e^{i\delta} & c_{12}c_{23} - s_{12}s_{23}s_{13}e^{i\delta} & s_{23}c_{13} \\ s_{12}s_{23} - c_{12}c_{23}s_{13}e^{i\delta} & -c_{12}s_{23} - s_{12}c_{23}s_{13}e^{i\delta} & c_{23}c_{13} \end{pmatrix},$$                    (6.15.5)

parametrized by three mixing angles $\theta_{12}$, $\theta_{13}$, $\theta_{23}$, and a phase $\delta$,[128] where $s_{ij} = \sin\theta_{ij}$, $c_{ij} = \cos\theta_{ij}$. Experimentally, approximate values of the magnitudes of the matrix elements are[129]

$$V_{CKM}\big|_{|V_{ab}|} = \begin{pmatrix} 0.974 & 0.225 & 0.004 \\ 0.225 & 0.973 & 0.041 \\ 0.009 & 0.040 & 0.999 \end{pmatrix},$$                    (6.15.6)

and the matrix deviates slightly from the identity.

In particular, in using the approximations $\theta_{13} = 0, \theta_{23} = 0, \delta = 0$, the CKM matrix reduces to

$$\begin{pmatrix} \cos\theta_{12} & \sin\theta_{12} & 0 \\ -\sin\theta_{12} & \cos\theta_{12} & 0 \\ 0 & 0 & 1 \end{pmatrix},$$                    (6.15.7)

making contact with earlier investigation of Cabibbo involving fermions mixing between the first two generations, with the single angle $\theta_{12} \equiv \theta_C$, $\sin\theta_C \simeq 0.225$, referred to as the Cabibbo angle. This, in particular, leads to $d' = d\cos\theta_C + s\sin\theta_C$, and it was originally proposed to suppress strangeness changing processes in comparison to strangeness non-changing ones by choosing $\sin\theta_C$ small enough. Examples of a strangeness changing-processes and a strangeness non-changing

---

[128]This phase is responsible for flavor changing $CP$-violating processes in the standard model.

[129]Beringer et al. [17].

one, involving now the $W$ boson, are $(s\,W^+,u)$ and $(d\,W^+,u)$, respectively. By intoducing, in the process, the Cabibbo angle, there was no need to introduce a new coupling to deal with strangeness changing processes and the Fermi coupling $G_F$ was taken as a universal weak interaction coupling.

To generate masses for the quarks, we may introduce three quark fields $Q_{i>}$, $i = 1,2,3$, of charges $2/3$, and three quark fields $Q_{i<}$, $i = 1,2,3$ of charges $-1/3$. We will then eventually obtain an expression of the fields associated with the quark fields $u,d,c,s,t,b$ in terms of these fields. In a unitary gauge, a Yukawa interaction may be defined by the interaction of quarks with the Higgs field as follows

$$
\mathcal{L}_{\mathrm{Yuk}} = -\sum_{i,j} \lambda_{ij} \overline{\begin{pmatrix} Q_{i>} \\ Q_{i<} \end{pmatrix}_L} \begin{pmatrix} (\rho + v)/\sqrt{2} \\ 0 \end{pmatrix} Q_{j>\mathrm{R}}
$$
$$
- \sum_{i,j} \eta_{ij} \overline{\begin{pmatrix} Q_{i>} \\ Q_{i<} \end{pmatrix}_L} \begin{pmatrix} 0 \\ (\rho + v)/\sqrt{2} \end{pmatrix} Q_{j<\mathrm{R}} + h.c., \tag{6.15.8}
$$

wit $i,j = 1,2,3$, leading to a mass-Lagrangian density having the structure

$$
\mathcal{L}^Q_{\mathrm{mass}} = -\overline{Q_{>\mathrm{L}}}\, M_1\, Q_{>\mathrm{R}} - \overline{Q_{<\mathrm{L}}}\, M_2\, Q_{<\mathrm{R}} + h.c., \tag{6.15.9}
$$

written in matrix form. We may now introduce unitary matrices $U_1, V_1, U_2, V_2$, which diagonalize the matrices $M_1, M_2$ as follows

$$
U_1 M_1 V_1^\dagger = m^{(1)}, \qquad U_2 M_2 V_2^\dagger = m^{(2)}, \tag{6.15.10}
$$

where $m^{(1)}$, $m^{(2)}$ are diagonal and real with non-negative entries. Accordingly, by setting $U_1 Q_{>\mathrm{L}} = Q'_{>\mathrm{L}}$, $V_1 Q_{>\mathrm{R}} = Q'_{>\mathrm{R}}$, $U_2 Q_{<\mathrm{L}} = Q'_{<\mathrm{L}}$, $V_2 Q_{<\mathrm{R}} = Q'_{<\mathrm{R}}$, we obtain

$$
\mathcal{L}^Q_{\mathrm{mass}} = -\left(\overline{Q'_{>\mathrm{L}}} + \overline{Q'_{>\mathrm{R}}}\right) m^{(1)} \left(Q'_{>\mathrm{L}} + Q'_{>\mathrm{R}}\right)
$$
$$
-\left(\overline{Q'_{<\mathrm{L}}} + \overline{Q'_{<\mathrm{R}}}\right) m^{(2)} \left(Q'_{<\mathrm{L}} + Q'_{<\mathrm{R}}\right), \tag{6.15.11}
$$

where due to the facts that $m^{(1)}$, $m^{(2)}$ are diagonal, we have used the properties $\overline{\psi}_{\mathrm{R/L}}\psi_{\mathrm{R/L}} = 0$, for a spinor $\psi$. With $(Q'_{>\mathrm{L}} + Q'_{>\mathrm{R}})$ interpreted as a column matrix with three elements, the latter may be identified with the quarks $u, c, t$. Similarly, $d, s, b$ are identified with $(Q'_{<\mathrm{L}} + Q'_{<\mathrm{R}})$.

## 6.15.2 Anomalies and Renormalizability

We first consider the standard model with massless particles. Gauge invariance is a key criterion for the renormalizability of the theory. In particular, gauge invariance requires that currents coupled to gauge fields must be conserved and hence implying the absence of anomalies of the type encountered in Sect. 3.10. The renormalizability of Yang-Mills field theories was considered in Sect. 6.3.3. Once we include fermions, anomalies associated with the divergence of currents may appear of the type investigated in Sect. 3.10. We will see that there are no such anomalies in the standard model.

To the above end, we first recall that the weak hypercharges assigned to the left- and right-handed fermions involved in the theory are as given Table 6.4 below as obtained from Tables 6.2 and 6.3.

We refer to the diagrams in Fig. 6.21 below, to investigate the nature of the anomalies that one may encounter in the theory, in general.

Here we recall that in the electroweak theory, the SU(2) gauge bosons couple only to the left-handed currents and the U(1) gauge boson couples to the left- and right-handed currents but with different U(1) charges. On the other hand in the color SU(3) theory, the gluons couple to the left- and right-handed fermions in the same way.

For further analysis, recall the following properties: $\{\gamma^\mu, \gamma^5\} = 0$, $S(p) = -\gamma p/p^2$, $\{S(p), \gamma^5\} = 0$, $(\gamma^5)^2 = 1$, $\text{Tr}[\gamma^5] = 0$, $\text{Tr}[1] = 4$. With the group structure $SU(3) \times SU(2)_L \times U(1)_Y$, with the standard model in mind, we define the following set of matrices:

$$\{\Gamma_a\} \equiv \left\{t_1, t_2, t_3, \tau_1, \tau_2, \tau_3, I\right\}, \tag{6.15.12}$$

where the identity $I$ corresponds to the generator $Y$, and the $a_i$, $b_i$ are some numerical constants. Recall that the generators of SU(3), and of SU(2) in Fig. 6.21 are given, respectively, by $t_a = \lambda_a/2$, with the $\lambda_a$ denoting the Gell-Mann matrices in (6.1.46), and $\tau_a = \sigma_a/2$. The $\Gamma^a$ matrices commute with the gamma ones. In reference to the above figure, diagram (b) is obtained from diagram (a) by carrying out the elementary substitutions

$$\text{Tr}\left[\gamma^{\mu_1}\gamma^5 1 \{\Gamma_b, \Gamma_c\}S\gamma^{\mu_2}S\gamma^{\mu_3}S\right] \Rightarrow$$

$$\text{Tr}\left[\gamma^{\mu_1}\gamma^5(a_1 + b_1\gamma^5)\Gamma_a\{\Gamma_b, \Gamma_c\}S\gamma^{\mu_2}(a_2 + b_2\gamma^5)S\gamma^{\mu_3}(a_3 + b_3\gamma^5)S\right]. \tag{6.15.13}$$

$$\text{Tr}\left[\{\Gamma_c, \Gamma_b\}\right] \Rightarrow 4\left[a_1(a_2 a_3 + b_2 b_3) + b_1(a_2 b_3 + a_3 b_2)\right]\text{Tr}\left[\Gamma_a\{\Gamma_c, \Gamma_b\}\right]. \tag{6.15.14}$$

where we have used, in the process, that $\gamma^{\mu_1}\gamma^5(a_1 + b_1\gamma^5) = (a_1 - b_1\gamma^5)\gamma^{\mu_1}\gamma^5$.

**Table 6.4** Weak hypercharge and charge classifications of the left-handed and right-handed Fermi particles in the first generation in the standard model. The corresponding quantum numbers of the other generations are identical. We use the definition $Y = 2(Q - I_3)$

| Particles | $\nu_L$ | $e_L$ | $u_L$ | $d_L$ | $e_R$ | $u_R$ | $d_R$ |
|-----------|---------|-------|-------|-------|-------|-------|-------|
| $Y$ | $-1$ | $-1$ | $1/3$ | $1/3$ | $-2$ | $4/3$ | $-2/3$ |
| $Q$ | $0$ | $-1$ | $2/3$ | $-1/3$ | $-1$ | $2/3$ | $-1/3$ |

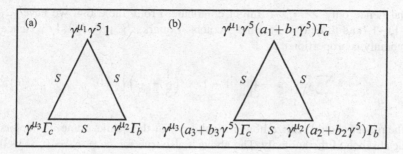

**Fig. 6.21** (a) Vertices in the triangle associated with the anomaly that may be extracted from the investigations in Sect. 3.10. (b) The vertices that may appear in the standard model, where the matrices $\Gamma_a$ are defined below.

$Y$, for the weak hypercharges, may be also written in matrix form. Using the basic properties, $\mathrm{Tr}[t_a] = 0$, $\mathrm{Tr}[\tau_a] = 0$, we have

$$\mathrm{Tr}\left[t_a\{\tau_b, \tau_c\}\right] = \mathrm{Tr}\left[t_a\right]\mathrm{Tr}\left[\{\tau_b, \tau_c\}\right] = 0,$$

$$\mathrm{Tr}\left[\tau_a\{t_b, t_c\}\right] = \mathrm{Tr}\left[\tau_a\right]\mathrm{Tr}\left[\{t_b, t_c\}\right] = 0,$$

$$\mathrm{Tr}\left[\tau_a, \{\tau_b, \tau_c\}\right] = \frac{1}{2}\delta_{bc}\mathrm{Tr}\left[\tau_a\right] = 0. \tag{6.15.15}$$

and we immediately infer the absence of anomalies for vertices corresponding to

- $SU(3) \times SU(2) \times SU(2)$,
- $SU(2) \times SU(3) \times SU(3)$,
- $SU(2) \times SU(2) \times SU(2)$,
- Two $U(1)$s.
    For
- $SU(3) \times SU(3) \times SU(3)$, $a_1 = 0, b_1 = 1, a_2 = a_3 = 1, b_2 = b_3 = 0$, imply from (6.15.14) the absence of an anomaly, where we have also used the fact that $(\gamma^5)^2 = I$. Refer to Fig. 6.21b.
    For
- One $U(1)$: For $b_1 = y_\varepsilon, a_1 = y_\varepsilon, \varepsilon = \mp 1$; $a_2 = a_3 = 1, b_2 = b_3 = -1$: where the coefficients $y_\varepsilon$ will be specified below by using Table 6.4, and $\varepsilon = \mp 1$ correspond to left-handed and right-handed particles, respectively. First note that

$$[a_1(a_2a_3 + b_2b_3) + b_1(a_2b_3 + a_3b_2)] = y_\varepsilon\left[\varepsilon(1+1) + (-1-1)\right] = 2y_\varepsilon(\varepsilon - 1),$$

$$\tag{6.15.16}$$

and hence only $\varepsilon = -1$ might contribute. From Table 6.4, we have $y_{-1} = -1, -1$, and three times for the 3 colors of quarks: $y_{-1} = 1/3, 1/3$. Hence the anomaly is proportional to

$$-4 \sum_{y-1} y_{-1} \equiv -4\left[-1 - 1 + 3\left(\frac{1}{3} + \frac{1}{3}\right)\right] = 0. \tag{6.15.17}$$

The reader should realize how critical it is that the quarks come in 3 colors for the absence of the anomaly. The above is also true for $b_2 = b_3 = 0$, as well for $b_2 = 1$, $b_3 = 0$, since $\sum_{y+1} y_{+1}$ is zero as well.
   For
•  $U(1) \times U(1) \times U(1)$. With the weak isospin written in a matrix form, the anomaly is proportional to

$$\text{Tr}\,[Y_L^3] - \text{Tr}\,[Y_R^3] = 2(-1)^3 + 3 \times 2\left(\frac{1}{3}\right)^3 - \left[1(-2)^3 + 3\left(\frac{4}{3}\right)^3 + 3\left(-\frac{2}{3}\right)^3\right] = 0, \tag{6.15.18}$$

and we note again that they vanish only if the quarks come in three colors. This provides further evidence of three color charges needed for quarks for consistency of the theory.

Therefore the standard model is anomaly-free, and lies on the fact that the theory involves several fermions whose separate contributions to the anomalies cancel out, and that quarks and leptons appear in Nature in equal numbers organize themselves into successive generations.

Gauge invariance of transition probabilities implies that one may work with the gauge fields in any gauge one pleases. To prove the renormalizability of spontaneously broken electroweak theory, where the vector particles, in particular, become massive, this gauge freedom may be invoked to choose a gauge where the propagator of a massive vector particle of mass $M$, may be expressed as

$$\Delta_+^{\mu\nu}(k, \xi; M) = \left[\eta^{\mu\nu} - \frac{k^\mu k^\nu}{k^2} \frac{k^2(1/\xi - 1)}{(k^2/\xi + M^2)}\right] \frac{1}{(k^2 + M^2)}, \tag{6.15.19}$$

as introduced by 't Hooft [114, 115]. In particular for finite parameter $\xi$, $\Delta_+^{\mu\nu}(k, \xi; M)$ has the very welcome high-energy behavior $\mathcal{O}(1/k^2)$ for $k^2 \to \infty$. Thus leading a class of gauges, denoted by $R_\xi$, for which renormalizability of the theory may be inferred as in QED. Thus renormalizability only results from the underlying gauge invariance. On the other hand for $\xi \to \infty$, $\Delta_+^{\mu\nu}(k, \xi; M)$ simply becomes the propagator of a massive vector particle, and one may formally infer, by invoking gauge invariance, the renormalizability of the spontaneously broken gauge theory with massive vector particles.

## 6.16   Grand Unification

A theory based on the product of the three symmetry groups $SU(3) \times SU(2) \times U(1)$ necessarily involves three different couplings, with $SU(3)$ corresponding to the color group of quarks, and $SU(2) \times U(1)$ corresponding to the electroweak theory. We may consider a larger group, such as $SU(5)$ which contains the product of the above three group just mentioned, and would necessarily involve one single coupling. The covariant derivative would then, in general, take the form

$$\nabla_\rho = \partial_\rho - i\, g_{\text{unif}} \sum_{a=1}^{24} V_\rho^a\, T_a, \qquad \text{Tr}\,[T_a T_b] = \frac{1}{2} \delta_{ab}. \qquad (6.16.1)$$

Here we recall that $SU(5)$ stands for all $5 \times 5$ unitary matrices of determinant one. Thus it involves $5^2 - 1 = 24$ generators and 24 gauge fields (see Sect. 6.1). We will explicitly obtain the generators corresponding to the fundamental representation and see, in the process, how the product group, given above, is contained in the $SU(5)$ group. The only quarks and leptons that arise in this unifying scheme are those of the electroweak theory.

Eleven of the twenty four generators denoted by $T_1, \ldots, T_{11}$ for the lowest dimensional representation of $SU(5)$ are given at the end of this section, corresponding simply to $SU(3)$, $SU(2)$ transformations, consisting of 8 plus 3 generators, respectively.

To construct the remaining generators, one may proceed as follows. It is easy to find the lowest five dimensional (the fundamental) representation represented by

$$\psi = \begin{pmatrix} a^1 \\ a^2 \\ a^3 \\ a^4 \\ a^5 \end{pmatrix}. \qquad (6.16.2)$$

Now consider, in turn, infinitesimal $SU(3)$, $SU(2)$ transformations via the operators

$$U = 1 + i \sum_{a=1}^{8} \delta \epsilon^a\, T_a + i \sum_{a=9,10,11} \delta \varepsilon^a T_a, \qquad (6.16.3)$$

applied to the above column vector. From the expression of the infinitesimal transformation, we note that, $(a_1, a_2, a_3)$ form an $SU(3)$ *triplet*, while it is unaffected by the $SU(2)$ transformation, and are thus *singlets* under the latter group. Referring to quarks from the first generation, we see from Table 6.3 in the last section that the right-handed quarks $u_R$, $d_R$ form the $SU(2)$ singlets. On the other hand, $(a^4, a^5)$, are unaffected by the $SU(3)$ are thus *singlets* under this group, but form a *doublet* under SU(2). From Table 6.2 in Sect. 6.14, one may be tempted to choose the SU(2) doublet to be $(\nu_{eL}, e_L)$. Since the generators for SU($N$), have to be traceless, the eigenvalues of the charge operator $Q$ add up to zero. This will the case if $(a^1, a^2, a^3)$ are chosen to be the right-handed $d$ quarks fields, of different colors, of charges $-1/3$, and, in turn, the SU(2) doublet is taken to be the following as charge conjugate of fields $((e_L)^{\mathscr{C}}, (\nu_{eL})^{\mathscr{C}})$, where we recall that the charge conjugate of a left-handed field is right-handed.[130] This incidently shows the consistency of the fractional charge $-1/3$ assigned to the (colored) $d$ quarks.[131] It is important to realize that a left handed state may be equivalently described in terms of its right-handed charge conjugate state. That is, we may write

$$
\psi = \begin{pmatrix} d_1 \\ d_2 \\ d_3 \\ (e_L)^{\mathscr{C}} \\ -(\nu_{eL})^{\mathscr{C}} \end{pmatrix}_R , \qquad
\begin{bmatrix} Q & Y/2 \\ \hline -1/3 & -1/3 \\ -1/3 & -1/3 \\ -1/3 & -1/3 \\ 1 & 1/2 \\ 0 & 1/2 \end{bmatrix} , \tag{6.16.4}
$$

denoted by [5], in a group theoretical notation. The minus sign multiplying $(\nu_{eL})^{\mathscr{C}}$ above arises in the following manner. In order to assign the correct weak isospin quantum numbers to $(e_L)^{\mathscr{C}}$ and $(\nu_e)^{\mathscr{C}}$, a rotation of $180°$ was made as follows:

$$
e^{-i\pi\sigma^2/2}\begin{pmatrix} (\nu_{eL})^{\mathscr{C}} \\ (e_L)^{\mathscr{C}} \end{pmatrix} = \begin{pmatrix} (e_L)^{\mathscr{C}} \\ -(\nu_{eL})^{\mathscr{C}} \end{pmatrix} .
$$

The 15 *left-handed* fermions, per generation, with the first generation consisting of the fermions:

$$
(\overline{d}_1, \overline{d}_2, \overline{d}_3, \nu_e, e^-, u_1, u_2, u_3, \overline{u}_1, \overline{u}_2, \overline{u}_3, d_1, d_2, d_3, e^+)_L,
$$

are fitted into the fundamental conjugate[132] representation as follows $(\overline{d}_1, \overline{d}_2, \overline{d}_3, \nu_e, e^-)_L$, denoted by $[\overline{5}]$, and the remaining 10 left-handed ones in a next higher dimensional representation of SU(5) denoted by [10]. The latter, consisting of 10 independent components, is represented by an anti-symmetric $5 \times 5$ matrix.

By referring to (6.16.4), the expression of this representation state allows us to introduce the weak hypercharge as a generator of SU(5), restricted to the above state $\psi$, given by $T_{12}| = \text{Const.} Y = \text{Const. diag}[-2/3, -2/3, -2/3, 1, 1]$.

---

[130]See below Eq. (6.14.50).

[131]Assignment of fractional charges to the quarks consistently as encountered here, lead eventually to such equalities as of the magnitudes of the charge of the electron and that of the proton.

[132]Conjugate representations are discussed in Sect. 6.1.

The normalization factor is obtained from $\mathrm{Tr}\left[(T_{12})^2\right] = 1/2$. This leads to the expression for $T_{12}$ matrix given at the end of the section.

Therefore, the remaining generators are of the form

$$T_{i+12} = \frac{1}{2}\left(\begin{array}{c|c} \mathbf{0} & (U^i)^{\dagger} \\ \hline (U^i) & \mathbf{0} \end{array}\right), \quad i = 1, \ldots, 12, \tag{6.16.5}$$

where the matrices $(U^i)_{3\times 2}$ may be readily defined with each having only one of their elements taking the value 1 or i, with the remaining elements equal to zero, as follows, going for example c.w. :

$$U^1 = \begin{pmatrix} 0 & 0 & 1 \\ 0 & 0 & 0 \end{pmatrix}, \ U^2 = \begin{pmatrix} 0 & 0 & i \\ 0 & 0 & 0 \end{pmatrix}, \ U^3 = \begin{pmatrix} 0 & 0 & 0 \\ 0 & 0 & 1 \end{pmatrix}, \ U^4 = \begin{pmatrix} 0 & 0 & 0 \\ 0 & 0 & i \end{pmatrix},$$

$$U^5 = \begin{pmatrix} 0 & 0 & 0 \\ 0 & 1 & 0 \end{pmatrix}, \ U^6 = \begin{pmatrix} 0 & 0 & 0 \\ 0 & i & 0 \end{pmatrix}, \ \ldots, \ U^{11} = \begin{pmatrix} 0 & 1 & 0 \\ 0 & 0 & 0 \end{pmatrix}, \ U^{12} = \begin{pmatrix} 0 & i & 0 \\ 0 & 0 & 0 \end{pmatrix},$$
$$\tag{6.16.6}$$

leading finally to the 24 generators as given at the end of this section.

One may now compare the SU(5) covariant derivative of $\psi$

$$\nabla_{\rho}\psi = \left(\partial_{\rho} - i\, g_{\mathrm{unif}} \sum_{a=1}^{24} V_{a\rho}\, T_a\right)\psi, \tag{6.16.7}$$

with the corresponding SU(3) $\times$ SU(2) $\times$ U(1) expression written conveniently as

$$\nabla_{\rho}\psi = \left(\partial_{\rho} - i\, g_s \sum_{a=1}^{8} A_{a\rho} T_a + i\, g'\sqrt{\frac{5}{3}}\, B_{\rho}\, T_{12} - i\, g \sum_{i=1,2,3} W_i T_{8+i}\right)\psi, \tag{6.16.8}$$

with $V_{a,\rho} = A_{a\rho}, a = 1, \ldots, 8$ denoting the gluons, and $V_{(8+i)} = W_i, i = 1, 2, 3$ from which the electroweak vector bosons, in (6.14.12), (6.14.14) are defined. From the comparison of the above two expressions, we may then infer that the equalities of the SU(3) $\times$ SU(2) $\times$ U(1) (couplings)$^2$ :

$$g_s^2(M^2) = \frac{5}{3}\, g'^2(M^2) = g^2(M^2), \tag{6.16.9}$$

at a unifying SU(5) energy scale $M$ which will be determined. This will be investigated below. Before doing this, we note that the SU(5) symmetry group introduces twelve additional vector bosons, which consist of 6 vector bosons, denoted by, $X_{\rho}^1, X_{\rho}^2, X_{\rho}^3\ Y_{\rho}^1, Y_{\rho}^2, Y_{\rho}^3$, and their 6 anti-particles, via the term $\sum_{a=13}^{24} V_{a\rho} T_a$ in the covariant derivative. This term may be written as

$$
\left(
\begin{array}{c|c}
\mathbf{0} & \begin{matrix} \bar{X}_1 \ \bar{Y}_1 \\ \bar{X}_2 \ \bar{Y}_2 \\ \bar{X}_3 \ \bar{Y}_3 \end{matrix} \\
\hline
\begin{matrix} X_1 \ X_2 \ X_3 \\ Y_1 \ Y_2 \ Y_3 \end{matrix} & \mathbf{0}
\end{array}
\right),
\tag{6.16.10}
$$

where the four-vector indices of the boson fields are suppressed.

Application of (6.14.15), (6.16.9): $\tan \theta_W = g'/g$, gives

$$
\sin^2 \theta_W \big|_{M^2} = 3/8,
\tag{6.16.11}
$$

at an energy scale $M$ of the order of, say, of the mass of the $X$ vector boson. For energies much less than $M_X$, specified by a parameter $\mu$, we may use the renormalization group equations, which relate the magnitudes of the couplings[2] $g_s^2$, $g^2$, $g'^2$ at different energy scales, to continue to lower energy scales available at present experiments. For the SU($N$) gauge group, we have from (6.7.17), (6.7.14),

$$
\mu^2 \frac{\mathrm{d}}{\mathrm{d}\mu^2} \frac{1}{\alpha_s(\mu^2)} = \beta_0 \equiv \frac{(11N - 4n_g)}{12\pi}, \qquad \alpha_s \equiv \frac{g_s^2}{4\pi},
\tag{6.16.12}
$$

expressed in terms of the number of generations of quarks $n_g$.

To obtain, the corresponding expression for U(1), Table 6.5, in this regard, spells out the weak hypercharges and charges of all the particles in the first generation and is useful. Accordingly for $n_g$ generations, with 15 fermions per generation, we have:

$$
\sum_i \left(\frac{Y_i}{2}\right)^2 = n_g \left[ 2\left(-\frac{1}{2}\right)^2 + (1)^2 + 3\left(2\left(\frac{1}{6}\right)^2 + \left(-\frac{2}{3}\right)^2 + \left(\frac{1}{3}\right)^2\right) \right] = \frac{10}{3} n_g.
\tag{6.16.13}
$$

We may infer from Sect. 5.19.1 in QED, which is simply based on the U(1) group, that

$$
\mu^2 \frac{\mathrm{d}}{\mathrm{d}\mu^2} \frac{1}{\alpha_{\mathrm{QED}}(\mu^2)} = -\frac{1}{3\pi} \sum_i Q_i^2,
\tag{6.16.14}
$$

**Table 6.5** Weak hypercharge and charge classifications of all the left-handed quarks and Leptons in the first generation. The corresponding quantum numbers of the other generations are identical

| Particles | $\nu$ | $e$ | $e^{\mathscr{C}}$ | $u$ | $u^{\mathscr{C}}$ | $d$ | $d^{\mathscr{C}}$ |
|-----------|-------|------|------|------|-------|------|------|
| $Y/2$ | $-1/2$ | $-1/2$ | $1$ | $1/6$ | $-2/3$ | $1/6$ | $1/3$ |
| $Q$ | $0$ | $-1$ | $1$ | $2/3$ | $-2/3$ | $-1/3$ | $1/3$ |

to lowest order in the charges, where the $Q_i$ (in units of $|e|$) are now the charges of quarks, taking into account color multiplicity, and of the charges of the charged leptons, and also the contributions of the underlying number of generations is understood to be taken. For the present $U(1)_Y$ group, we have to replace $\sum_i Q_i^2$ in (6.16.14) by the expression $(1/2) \sum_i (Y_i/2)^2$, with an overall $1/2$ factor because the expression in (6.16.13) takes into account the anti-particles as well.

The renormalization group equations then become[133] dealing with the incorporation of supersymmetry in the standard model.

$$\mu^2 \frac{d}{d\mu^2} \frac{1}{\alpha_s(\mu^2)} = +\frac{1}{12\pi}(33 - 4n_g), \qquad \left(\alpha_s = \frac{g_s^2}{4\pi} \text{ for } SU(3)\right), \qquad (6.16.15)$$

$$\mu^2 \frac{d}{d\mu^2} \frac{1}{\alpha(\mu^2)} = +\frac{1}{12\pi}(22 - 4n_g), \qquad \left(\alpha = \frac{g^2}{4\pi} \text{ for } SU(2)\right), \qquad (6.16.16)$$

$$\mu^2 \frac{d}{d\mu^2} \frac{1}{\alpha'(\mu^2)} = -\frac{1}{12\pi} \frac{(20 n_g)}{3}, \qquad \left(\alpha' = \frac{g'^2}{4\pi} \text{ for } U(1)\right), \qquad (6.16.17)$$

where in writing the right-hand sides of (6.16.15), (6.16.16), we have used the expression for $\beta_0$ in (6.16.14), with $N = 3$, $N = 2$, respectively.

In particular, these equations lead to

$$\mu^2 \frac{d}{d\mu^2} \left( \frac{1}{\alpha_s(\mu^2)} - \frac{1}{\alpha(\mu^2)} \right) = \frac{11}{12\pi}, \qquad (6.16.18)$$

$$\mu^2 \frac{d}{d\mu^2} \left( \frac{1}{\alpha(\mu^2)} - \frac{3}{5} \frac{1}{\alpha'(\mu^2)} \right) = \frac{22}{12\pi}. \qquad (6.16.19)$$

Upon subtracting Eq. (6.16.19) from twice of Eq. (6.16.18), integrating with respect to $\mu^2$ from $M_Z^2$, corresponding to the mass $M_Z$ of the $Z$ vector boson as a reference point, to $M$, the unifying energy specified by a mass parameter $M$, and finally using the unifying boundary conditions (6.16.9), we simply obtain

$$\frac{2}{\alpha_s(M_Z^2)} - \frac{3}{\alpha(M_Z^2)} + \frac{3}{5\alpha'(M_Z^2)} - 0. \qquad (6.16.20)$$

From the first two defining equations in (6.14.17) we may also write

$$\frac{1}{\alpha'} = \frac{\cos^2 \theta_W}{\alpha_{em}}, \quad \frac{1}{\alpha} = \frac{\sin^2 \theta_W}{\alpha_{em}}. \qquad (6.16.21)$$

---

[133]In the beta functions, on the right-hand sides of (6.16.16), (6.16.17), we have neglected tiny contributions due to the Higgs boson field. These will be spelled out in Sect. 3.13 of Volume II.

The latter two equations then lead to

$$\sin^2 \theta_W \bigg|_{M_Z^2} = \frac{1}{6} + \frac{5}{9} \frac{\alpha_{\mathrm{em}}(M_Z^2)}{\alpha_s(M_Z^2)}. \tag{6.16.22}$$

Experimentally,[134] $\alpha_s(M_Z^2) = 0.1184 \pm 0.0007$, $1/\alpha_{\mathrm{em}}(M_Z^2) = 127.916 \pm 0.015$. These give, $\sin^2 \theta_W \big|_{M_Z^2} \simeq 0.21$, which compares well with the experimental result in (6.14.36).

On the other hand, the second defining equation in (6.16.21), (6.16.18), and (6.16.22) give upon integration,

$$\ln\left(\frac{M^2}{M_Z^2}\right) = \frac{2\pi}{11\alpha_{\mathrm{em}}(M_Z^2)}\left(1 - \frac{8}{3}\frac{\alpha_{\mathrm{em}}(M_Z^2)}{\alpha_s(M_Z^2)}\right), \tag{6.16.23}$$

where we have used the boundary condition $\alpha_s(M^2) = \alpha(M^2)$. From the experimental values of $\alpha_{\mathrm{em}}(M_Z^2)$, $\alpha_s(M_Z^2)$ quoted above, we may determines the scale $M \simeq 1.1 \times 10^{15}\,\mathrm{GeV}$.

Finally (6.16.15) leads, upon integration with respect to $\mu^2$, the solution

$$\frac{1}{\alpha_s(M^2)} = \frac{1}{\alpha_s(M_Z^2)} + \frac{(33 - 4n_g)}{12\pi}\ln\left(\frac{M^2}{M_Z^2}\right), \tag{6.16.24}$$

which from (6.16.23) and the experimental input at $M_Z^2$, give $\alpha_s(M^2) \simeq 0.024$, and thus from (6.16.9) we also have $\alpha(M^2) \simeq 0.024$, $\alpha'(M^2) \simeq 0.0144$. Of particular interest is the value of $\alpha_{\mathrm{em}}(M^2)$ which is obtained from (6.16.11), (6.16.21) to be $\alpha_{\mathrm{em}}(M^2) \simeq (3/8) \times 0.024 = 0.009$.

To assess the departure from the strict equality of the couplings at a unifying scale, as given in (6.16.9), i.e., $\alpha_s(M^2) = (5/3)\alpha'(M^2) = \alpha(M^2)$, one may define the critical parameter[135]

$$\Delta = \frac{\alpha^{-1}(M_Z^2) - \alpha_s^{-1}(M_Z^2)}{(3/5)\alpha'^{-1}(M_Z^2) - \alpha^{-1}(M_Z^2)}, \tag{6.16.25}$$

which taking into account the experimental values quoted above, gives $\Delta_{\exp} \simeq 0.74$. On the other hand upon integrating (6.16.15), (6.16.16), (6.16.17) from

---

[134]Beringer et al. [17].
[135]Peskin [98].

$\mu^2 = M_Z^2$ to $\mu^2 = M^2$, denoting $\beta_s = (33 - 4n_g)/12\pi$, $\beta = (22 - 4n_g)/12\pi$, $(3/5)\beta' = -4n_g/12\pi$, and assuming the strict unifying equality stated above at $\mu^2 = M^2$, we have for for the right-hand side of $\Delta$ in (6.16.25),

$$\Delta_{\text{theor}} = \frac{\beta_s - \beta}{\beta - (3/5)\beta'} = \frac{11/12\pi}{22/12\pi} = 0.5, \qquad (6.16.26)$$

and the departure is significant. A supersymmetric treatment of this grand unifying theorem provides, in turn, excellent agreement for this critical parameter between theory and experiment as we will see in Sect. 3.13 of Volume II.

The new bosons $X$ and $Y$ are referred to as leptoquarks as they carry both color and lepton numbers. The mere fact that the SU(5) representations contain both quarks and leptons, implies that baryon number is not conserved, and the proton can decay. For example, $u + u \rightarrow X$, where, in turn, $X$ may convert to $e^+$ and $d^{\mathscr{C}}$, i.e., the charge of $X$ is $4/3$. Similarly, $u + d \rightarrow Y$, where, in turn, $Y$ may convert to $e^+$ and $u^{\mathscr{C}}$, i.e., the charge of $Y$ is $1/3$.

An order of magnitude of the lifetime of the proton is readily obtained from the expression of the muon lifetime as extracted from Problem 6.24 which is given by $\tau_\mu = 192\pi^3/G_F^2 M_\mu^5$, and from (6.14.33), $G_F^2 = g^4 M_W^{-4}/32$. By formally replacing $M_\mu^2$ by $M_p^2$, and $M_W^2$, $g^2$, in the expression for $G_F^2$, by $M^2 \sim M_X^2$, $g_{\text{unif}}^2$, respectively, one obtains the rough expression for the proton lifetime $\tau_p \sim M^4/g_{\text{unif}}^4 M_p^5$. A more precise calculation gives[136] $\tau_{p \rightarrow e^+ \pi^0} \sim 10^x$ years, $28.5 \leq x \leq 31.5$ which is too low in comparison to the experimental lower bound $> 2 \times 10^{33}$ years at 90% confidence limit.[137] Increasing the scale of unification[138] increases the theoretical estimate of the proton lifetime as it is proportional to $M^4$. Consistency with such a large experimental lower bound for the proton decay is certainly a basic problem in developing grand unified theories.

Although this unifying theory, predicts, for example, the key estimate for the value of $\sin^2\theta_W(M_Z^2)$ close enough to the experimental value, a discrepancy between theory and experiment that arises means that the strict equality of the fundamental constants as given in (6.16.9), does not hold rigorously. This situation is much improved in a supersymmetric setting.

---

[136]See Langacker [75].

[137]See Miura [91] and Olive et al. [94].

[138]In a supersymmetric extension $M \sim 10^{16}$ GeV, see § 2.13 in Volume II.

The fundamental energy scale that arises in the standard model from the vacuum expectation value of the Higgs boson sets the scale for the masses in the theory, such as for the masses of the vector bosons.[139] The enormous mass difference between the electroweak scale $\sim 10^2$ GeV and all the way up to the grand unified scale $\sim 10^{15}$ GeV, referred to as the hierarchy problem, should be of concern, as it is such an enormous shift in energy scale providing no hint as to what happens to the physics in between. The radiative corrections to the mass squared of the Higgs boson, as a *scalar* particle, are quadratically divergent,[140] and hence with a cut-off of the order of the grand unified scale, or of the Planck energy scale $\sim 10^{19}$ GeV, at which gravity is expected to be significant, it is necessary that the bare mass squared of the Higgs boson be correspondingly large to cancel such a quadratic dependence on the enormous cut-off to finally obtain a finite physical mass[141] for the Higgs boson of the order of the low electroweak energy scale, in comparison, and is highly unnatural.[142] This unnatural cancelation of enormously large numbers has been termed a facet of the hierarchy problem. Supersymmetry is of significance in dealing with the hierarchy problem, as in supersymmetric field theories, in general, cancelations of such large quadratic corrections, *a priori*, occur between graphs involving particles and their supersymmetric counterparts in a supersymmetric version of a non-supersymmetric field theory, up to possibly of divergences of logarithmic type which are tolerable, thus protecting a scalar particle from acquiring a large bare mass.

Finally, we note that in the above unifying theory, the quarks and leptons are those of the electroweak theory involving no right-handed neutrinos to generate masses to the presently known neutrinos by the "seesaw mechanism" as discussed in Sect. 6.14.3. The larger group SO(10), which has been also much studied over years as a grand unified theory, accommodates a right-handed neutrino and the "seesaw mechanism" may be then applied. Several other grand unified theories have been also proposed in recent years and we refer the reader to the vast literature on the subject.[143] The remarkable simplicity of the SU(5) grand unifying scheme should be noted, and that it provides the very first steps for the development of more complicated grand unifying schemes that may meet the challenge set up by experiments.

---

[139] We recall how accurate were the masses of the massive vector bosons obtained theoretically directly at the tree level via the vacuum expectation value of the Higgs field given through Eq. (6.14.34) which, in turn, is expressed simply in terms of the Fermi coupling.

[140] Veltman [118].

[141] Aad et al. [2] and Chatrchyan et al. [30].

[142] As mentioned above, the question, in turn, arises as to what amounts for such an enormous difference between the energy scale of grand unification and the energy scale that characterizes the standard model.

[143] See, e.g., Beringer et al. [17] and references therein.

In closing this section, we provide the fundamental representation of the SU(5) group generators:

$$T_{1,\ldots,8} = \frac{1}{2}\begin{pmatrix} \lambda_{1,\ldots,8} & \mathbf{0} \\ \mathbf{0} & \mathbf{0} \end{pmatrix}, \quad T_{9,10,11} = \frac{1}{2}\begin{pmatrix} \mathbf{0} & \mathbf{0} \\ \mathbf{0} & \sigma^{1,2,3} \end{pmatrix},$$

$$T_{12} = \frac{1}{2}\sqrt{\frac{3}{5}}\left(\begin{array}{ccc|c} -2/3 & 0 & 0 & \\ 0 & -2/3 & 0 & \mathbf{0} \\ 0 & 0 & -2/3 & \\ \hline & \mathbf{0} & & \begin{matrix}1 & 0\\0 & 1\end{matrix} \end{array}\right),$$

$$T_{13} = \frac{1}{2}\left(\begin{array}{c|c} \mathbf{0} & \begin{matrix}0 & 0\\0 & 0\\1 & 0\end{matrix} \\ \hline \begin{matrix}0 & 0 & 1\\0 & 0 & 0\end{matrix} & \mathbf{0} \end{array}\right), \quad T_{14} = \frac{1}{2}\left(\begin{array}{c|c} \mathbf{0} & \begin{matrix}0 & 0\\0 & 0\\-i & 0\end{matrix} \\ \hline \begin{matrix}0 & 0 & i\\0 & 0 & 0\end{matrix} & \mathbf{0} \end{array}\right),$$

$$T_{15} = \frac{1}{2}\left(\begin{array}{c|c} \mathbf{0} & \begin{matrix}0 & 0\\0 & 0\\0 & 1\end{matrix} \\ \hline \begin{matrix}0 & 0 & 0\\0 & 0 & 1\end{matrix} & \mathbf{0} \end{array}\right), \quad T_{16} = \frac{1}{2}\left(\begin{array}{c|c} \mathbf{0} & \begin{matrix}0 & 0\\0 & 0\\0 & -i\end{matrix} \\ \hline \begin{matrix}0 & 0 & 0\\0 & 0 & i\end{matrix} & \mathbf{0} \end{array}\right),$$

$$T_{17} = \frac{1}{2}\left(\begin{array}{c|c} \mathbf{0} & \begin{matrix}0 & 0\\0 & 1\\0 & 0\end{matrix} \\ \hline \begin{matrix}0 & 0 & 0\\0 & 1 & 0\end{matrix} & \mathbf{0} \end{array}\right), \quad T_{18} = \frac{1}{2}\left(\begin{array}{c|c} \mathbf{0} & \begin{matrix}0 & 0\\0 & -i\\0 & 0\end{matrix} \\ \hline \begin{matrix}0 & 0 & 0\\0 & i & 0\end{matrix} & \mathbf{0} \end{array}\right),$$

$$T_{19} = \frac{1}{2}\left(\begin{array}{c|c} \mathbf{0} & \begin{matrix}0 & 1\\0 & 0\\0 & 0\end{matrix} \\ \hline \begin{matrix}0 & 0 & 0\\1 & 0 & 0\end{matrix} & \mathbf{0} \end{array}\right), \quad T_{20} = \frac{1}{2}\left(\begin{array}{c|c} \mathbf{0} & \begin{matrix}0 & -i\\0 & 0\\0 & 0\end{matrix} \\ \hline \begin{matrix}0 & 0 & 0\\i & 0 & 0\end{matrix} & \mathbf{0} \end{array}\right),$$

$$T_{21} = \frac{1}{2}\left(\begin{array}{c|c} \mathbf{0} & \begin{matrix}1 & 0\\0 & 0\\0 & 0\end{matrix} \\ \hline \begin{matrix}1 & 0 & 0\\0 & 0 & 0\end{matrix} & \mathbf{0} \end{array}\right), \quad T_{22} = \frac{1}{2}\left(\begin{array}{c|c} \mathbf{0} & \begin{matrix}-i & 0\\0 & 0\\0 & 0\end{matrix} \\ \hline \begin{matrix}i & 0 & 0\\0 & 0 & 0\end{matrix} & \mathbf{0} \end{array}\right),$$

$$
T_{23} = \frac{1}{2}\left(\begin{array}{c|cc} \mathbf{0} & \begin{array}{cc} 0 & 0 \\ 1 & 0 \\ 0 & 0 \end{array} \\ \hline \begin{array}{ccc} 0 & 1 & 0 \\ 0 & 0 & 0 \end{array} & \mathbf{0} \end{array}\right), \quad T_{24} = \frac{1}{2}\left(\begin{array}{c|cc} \mathbf{0} & \begin{array}{cc} 0 & 0 \\ -\mathrm{i} & 0 \\ 0 & 0 \end{array} \\ \hline \begin{array}{ccc} 0 & \mathrm{i} & 0 \\ 0 & 0 & 0 \end{array} & \mathbf{0} \end{array}\right),
$$

where the Gell-Mann matrices $\lambda_1, \ldots, \lambda_8$ are defined in Sect. 6.1.

## Problems

**6.1** Find the infinitesimal gauge transformation of the non-abelian gauge fields $A_a^\mu$ via the unitary operators $V(x) = \exp[\mathrm{i}\, g_0\, \theta_c(x) t_c]$.

**6.2** Show that $[\nabla^\mu, \nabla^\nu]_{cb} = g_0 f_{cab} G_a^{\mu\nu}$, where $(\nabla^\mu)_{ac} = \delta_{ac}\, \partial^\mu + g_0 f_{abc} A_b^\mu$.

**6.3** Show that $\nabla_{ab\mu} \nabla_{bc\nu}\, G_c^{\mu\nu} = 0$.

**6.4** Show that

$$
\exp - \frac{\mathrm{i}}{2\lambda}\left[\int (\mathrm{d}x) \frac{\delta}{\delta\varphi_c(x)} \frac{\delta}{\delta\varphi_c(x)}\right] \Pi_{ax} \delta\big(\partial_\nu \mathscr{A}_a^\nu(x) - \lambda\varphi_a(x)\big)\bigg|_{\varphi=0}
$$

$$
= \exp\left[-\frac{\mathrm{i}}{2\lambda}\int (\mathrm{d}x)\partial_\mu \mathscr{A}_a^\mu(x)\, \partial_\nu \mathscr{A}_a^\nu(x)\right],
$$

up to an unimportant multiplicative constant.

**6.5** Obtain the leading contribution of the gauge transformed non-abelian gauge field as a function of the original field if the former satisfies the Coulomb gauge.

**6.6** Derive Rosenbluth formula in (6.4.1), for elastic electron-proton scattering, for a finite extended description of the proton, via a single photon exchange, by neglecting the electron mass.

**6.7** Derive the differential cross section for $e^- e^+ \rightarrow q\bar{q}$, via a single photon exchange, in the CM, by neglecting the mass of the electron (positron) as well as of the quark (antiquark). Also neglect radiative corrections.

**6.8** Evaluate the following infrared regularized integrals, where the dimensionality of spacetime is replaced by

$$
D' = 4 + \delta, \quad \text{with} \quad \delta > 0, \quad p_1^2 = 0 = p_2^2, \quad -2p_1 p_2 \equiv Q^2,
$$

(i) $\displaystyle \int \frac{\mathrm{d}^{D'}k}{(2\pi)^{D'}} \frac{1}{(p_1-k)^2 (p_2-k)^2\, k^2} = \frac{\mathrm{i}(\mu_{D'}^2)^{-\delta/2}}{(4\pi)^{D'/2}}\left(\frac{1}{Q^2}\right)^{1-\delta/2} \frac{\Gamma^2\!\left(\frac{\delta}{2}\right)\Gamma\!\left(1-\frac{\delta}{2}\right)}{\Gamma(1+\delta)}.$

(ii) $\int \dfrac{d^{D'}k}{(2\pi)^{D'}} \dfrac{k^\mu}{(p_1-k)^2(p_2-k)^2\,k^2} = \dfrac{i(\mu_{D'}^2)^{-\delta/2}}{(4\pi)^{D'/2}}\left(\dfrac{1}{Q^2}\right)^{1-\delta/2}$

$\qquad\qquad \times\; \dfrac{\Gamma\!\left(\frac{\delta}{2}\right)\Gamma\!\left(1-\frac{\delta}{2}\right)\Gamma\!\left(1+\frac{\delta}{2}\right)}{\Gamma(2+\delta)}\left(p_1^\mu + p_2^\mu\right).$

**6.9** Show by dimensional regularization of the following integral in its ultraviolet and infrared divergent parts, that it may written as:

$$\int \dfrac{(dk)}{(2\pi)^4}\dfrac{k^\mu k^\nu}{(p_1-k)^2(p_2-k)^2 k^2}\bigg|_{\text{Reg}} = \dfrac{i\,\eta^{\mu\nu}}{(4\pi)^{D/2}}\left(\dfrac{\mu_D^2}{Q^2}\right)^{\frac{\varepsilon}{2}}\dfrac{\Gamma\!\left(\frac{\varepsilon}{2}\right)\Gamma^2\!\left(1-\frac{\varepsilon}{2}\right)}{2\,\Gamma(3-\varepsilon)}$$

$$+\dfrac{i}{(4\pi)^{D'/2}}\left(\dfrac{1}{Q^2}\right)\left(\dfrac{\mu_{D'}^2}{Q^2}\right)^{-\frac{\delta}{2}}\Gamma\!\left(1-\dfrac{\delta}{2}\right)\left[(p_1^\mu p_1^\nu + p_2^\mu p_2^\nu)\dfrac{\Gamma\!\left(\frac{\delta}{2}\right)\Gamma\!\left(2+\frac{\delta}{2}\right)}{\Gamma(3+\delta)}\right.$$

$$+\left.(p_1^\mu p_2^\nu + p_1^\nu p_2^\mu)\dfrac{\Gamma^2\!\left(1+\frac{\delta}{2}\right)}{\Gamma(3+\delta)}\right],\; \text{where } D = 4-\varepsilon,\, D' = 4+\delta,\, \varepsilon > 0,\, \delta > 0.$$

**6.10** Derive the explicit general structure of the infrared regularized function $h_{\text{IR}}(\mu_D^2/Q^2,\delta)$ in (6.6.25), $h_{\text{IR}}^{(1)}(\mu_D^2/Q^2,\delta)$ in (6.6.21), $h_{\text{IR}}^{(2)}(\mu_D^2/Q^2,\delta)$ in (6.6.24).

**6.11** Derive the equality in (6.9.6).

**6.12** Prove the expressions in (6.9.13), (6.9.14), (6.9.16).

**6.13** Show that the structure tensor $W_i^{\mu\nu}$ contribution due to the $i^{\text{th}}$ parton may be written as in (6.10.1).

**6.14** Show that $\int \dfrac{d^3 p'}{2p'^0}\delta^{(4)}(\xi P + Q - p') = \dfrac{1}{2|PQ|}\delta\!\left(\xi + \dfrac{Q^2}{2PQ}\right)$, where $p'^2 = p^2 = -\xi^2 M^2$.

**6.15** Show that if quarks (anti-quarks) are of spin 0, then $W_1 = 0$. This leads to the results that $F_T = 0$, for the transverse structure function in (6.9.17)/(6.10.5), and $R \to \infty$, defined in (6.9.19).

**6.16** Show that if one takes into account only the three basic quarks $(u, d, d)$, associated with the neutron, into consideration, and assumes that they occur with the same frequency, then this leads to the relation $\int_0^1 dx\, F_2^{en}(x) = (2/9)\int_0^1 dx\, \sum_i x f_i(x)$, consistent with approximate experimental values

$$\int_0^1 dx\, F_2^{en}(x) \approx 0.12, \qquad \int_0^1 dx\, \sum_i x f_i(x) \approx 0.54.$$

This, again, provides evidence of fractional charges of the quarks.

**6.17** Derive the rigorous bounds in (6.11.16), (6.11.17), (6.11.18) for the moments of the splitting functions in (6.11.11), (6.11.12), (6.11.13), (6.11.14), for $n = 3, 4, \ldots < \infty$.

**6.18** Establish the relation: $Q^2 \, d/(dQ^2) = [\alpha_s(Q^2)/(2\pi)] \, d/(d\tau)$ in (6.11.20).

**6.19** (i) Verify equation (6.11.32) which diagonalizes the $2 \times 2$ matrix on the left-hand side of (6.11.32). For all $n = 3, 4, \ldots < \infty$,
(ii) Show that $\lambda_n^+ < \lambda_n^+ < 0$, for the eigenvalues.
(iii) Derive the inequality:

$$\sqrt{2}\left[-\frac{11}{2} - \frac{n_f}{3} - 6\vartheta_4^n\right] < \lambda_n^- < \lambda^+ < \frac{1}{2}\left[-\frac{50}{9} + \sqrt{(98/135)\, n_f}\,\right].$$

(iv) Show that $\lambda_n^+ - A_n > 0$, $\lambda_n^- - A_n < 0$. That is, $\lambda_n^\pm - A_n \neq 0$.

**6.20** Derive the relation in (6.12.8).

**6.21** Establish the gauge transformation rules for the remaining three line-integrals in Box 6.3 in Sect. 6.12.

**6.22** Show by a combined $SU(2)_L \times U(1)_Y$ transformation, as in (6.14.22), (6.14.23), one may always find one to reduce the Higgs field doublet to the form in (6.14.26) satisfying (6.14.27).

**6.23** Show that $[\overline{e_L}\gamma^\rho \nu_L][\overline{\nu_L}\gamma^\rho e_L] = [\overline{e_L}\gamma_\rho e_L][\overline{\nu_L}\gamma^\rho \nu_L]$.

**6.24** According to (6.14.1), the amplitude for muon decay: $\mu^- \to e^- + \tilde{\nu}_e + \nu_\mu$ may be taken in a tree approximation as

$$\mathscr{A} = \frac{G_F}{\sqrt{2}} \, [\bar{u}_{\nu_\mu}(k_2)\,\gamma^\rho(1-\gamma^5)u_\mu(p)]\,[\bar{u}_e(k)\gamma_\rho(1-\gamma^5)v_{\nu_e}(k_1)].$$

By neglecting the masses of the electron and neutrinos in comparison to the mass $M_\mu$ of the muon, and working in the rest frame of the muon, show that its decay rate is given by $\Gamma = G_F^2 M_\mu^5/192\pi^3$.

# References

1. Abe, K. et al. (1999). Measurement of $R = \sigma_L/\sigma_T$ for $.03 < x < 0.1$ and fit to world data. *Physics Letters, B452*, 194–200.
2. Aad, G. et al. (2012). Observation of a new particle in the search for the Standard Model Higgs Boson with the ATLAS detector at the LHC. *Physics Letters, B716*, 1–29.
3. Airapetian, A. et al. (2002). Measurement of $R = \sigma_L/\sigma_T$ in deep ineastic scattering. arXiv:hep-ex/0210068.
4. Ali, A. et al. (1980). A QCD analysis of the high energy $e^+e^-$ data from PETRA. *Physics Letters, 93B*, 155–160.
5. Altarelli, G., & Parisi, G. (1977). Asymptotic freedom in parton language. *Nuclear Physics, B126*, 298–318.
6. Altmann, M. et al. (2005). Complete results for five years of GNO solar neutrino observations. *Physics Letters, B616*, 174–190.

7. Ambrosio, M. et al. (1998). Measurement of the atmospheric neutrino induced upgoing muon flux using MACRO. *Physics Letters, B434*, 451–457.

8. Ammar, R. et al. (1998). Measurement of the total cross section for $e^+e^- \rightarrow$ Hadrons at $\sqrt{s} = 10.52$ GeV. *Physical Review, D57*, 1350–1358.

9. Andivahis, L. et al. (1994). Measurements of the electric and magnetic form factors of the proton from $Q^2 = 1.75$ to 8.83 $(GeV/c)^2$. *Physical Review, D50*, 5491–5517.

10. Anselmann, P. et al. (1992). Solar neutrinos observed by GALLEX at Gram Sasso. *Physics Letters, B285*, 376–389.

11. Appelquist, T., & Georgi, H. (1973). $e^+e^-$ annihilation in gauge theories of strong interactions. *Physical Review, D8*, 4000–4002.

12. Athanassopoulos, C. et al. (1998). Results on $\nu_\mu \rightarrow \nu_e$ neutrino oscillations from the LSND experiment. *Physical Review Letters, 81*, 1774–1777.

13. Baikov, P. A., Chetyrkin, K. G., & Khun, J. H. (2008). Order $\alpha_s^4$ QCD corrections to $Z$ and $\tau$ decays. *Physical Review Letters, 101*, 012002.

14. Bartel, W. et al. (1980). Observations of planar three-jet events in $e^-e^+$ annihilation and evidence for gluon bremsstrahlung. *Physics Letters, 91B*, 142–147.

15. Becchi, C., Rouet, A., & Stora, R. (1975). Renormalization of the Abelian Higgs Kibble model. *Communications in Mathematical Physics, 42*, 127–162.

16. Berger, C. et al. (1979). Evidence for gluon bremsstrahlung in $e^+e^-$ annihilation at high energies. *Physics Letters, 86B*, 418–425.

17. Beringer, J. et al. (2012). Particle data group. *Physical Review D, 86*, 010001.

18. Bjorken, J. D. (1969). Asymptotic sum rules at infinite momentum. *Physical Review, 179*, 1547–1553.

19. Bjorken, J. D., & Bodsky, S. J. (1970). Statistical model for electron-positron annihilation into hadrons. *Physical Review, D1*, 1416–1420.

20. Bjorken, J. D., & Drell, S. D. (1964). *Relativistic quantum mechanics*. New York/San Francisco/London: McGraw-Hill.

21. Bjorken, J. D., & Drell, S. D. (1965). *Relativistic quantum fields*. New York/San Francisco/London: McGraw-Hill.

22. Bjorken, J. D., & Pachos, E. A. (1969). Inelastic electron-proton and y proton scattering and the structure of the nucleon. *Physical Review, 185*, 1975–1982.

23. Bloom, E. D. et al. (1969). High- Energy Inelastic e-p Scattering at 6° and 10°. *Physical Review Letters, 23*, 93–934.

24. Bodek, A. et al. (1979). Experimental studies of the neutron and proton electromagnetic structure functions. *Physical Review, D20*, 1471–1552.

25. Brandelik, R. et al. (1979). Evidence for planar events in $e^+e^-$ annihilation at high energies. *Physics Letters, 86B*, 243–249.

26. Brown, L. S., & Weisberger, W. I. (1979). Remarks on the static potential in quantum chromodynamics. *Physical Review, D20*, 3239–3245.

27. Cabibbo, N. (1963). Unitary symmetry and leptonic decays. *Physical Review Letters, 10*, 531–533.

28. Callan, C. G., & Gross, D. J. (1969). High-energy electroproduction and the constitution of the electric current. *Physical Review Letters, 22*, 156–159.

29. Caswell, W. E. (1974). Asymptotic behavior of Non-Abelian gauge theories to two-loop order. *Physical Review Letters, 33*, 244–246.

30. Chatrchyan, S. et al. (2012). Observation of a new boson at mass 125 GeV with the CMS Experiment at LHC. *Physics Letters, B716*, 30–61.

31. Cleveland, B. T. et al. (1998). Measurement of the solar electron neutrino flux with the homestake chlorine detector. *Astrophysics Journal, 496*, 505–526.

32. Creutz, M. (1983). *Quarks, gluons and lattices*. Cambridge: Cambridge University Press.

33. Danby, G. et al. (1962). Observation of high-energy neutrino reactions and the existence of two kinds of neutrinos. *Physical Review Letters, 9*, 36–44.

34. Davis, R. et al. (1968). Search for neutrinos from the sun. *Physical Review Letters, 20*, 1205–1209.

35. Eidelman, S. et al. (2004). Particle data group. *Physics Letters, B592*, 1.
36. Erler, J. (1999). Calculation of the QED coupling $\hat{\alpha}(M_Z)$ in the modified minimal subtraction scheme. *Physical Review D, 59*, 054008, 1–7.
37. Faddeev, L. D., & Popov, V. N. (1967). Feynman diagrams for the Yang-Mills field. *Physics Letters, B25*, 29–30.
38. Fermi, E. (1934a). Tentativo di una teoria dei raggi $\beta$. *Nuovo Cimento, 11*, 1–19.
39. Fermi, E. (1934b). Versuch einer Theorie der $\beta$ - Strahlen. *Zeitschrift fur Physik, 88*, 161–171.
40. Feynman, R. P. (1963). Quantum theory of gravitation. *Acta Physica Polonica, 24*, 697–722.
41. Feynman, R. P. (1969a). The behavior of hadron collisions at extreme energies. In *Proceedings of the 3rd Topical Conference on High Energy Collisions*, Stony Brook. New York: Gordon & Breach.
42. Feynman, R. P. (1969b). Very high-energy collisions of hadrons. *Physical Review Letters, 23*, 1415–1417.
43. Feynman, R. P., & Gell-Mann, M. (1958). Theory of fermi interaction. *Physical Review, 109*, 193–198.
44. Field, R. D. (1989). *Applications of perturbative QCD*. Redwood City: Addison-Welry.
45. Field, R. D., & Feynman, R. P. (1977). Quark elastic scattering as a source of high-transverse-momentum mesons. *Physical Review, D15*, 2590–2616.
46. Field, R. D., & Feynman, R. P. (1978). A parametrization of the properties of quark jets. *Nuclear Physics, B136*, 1–76.
47. Friedman, J. I., & Kendall, H. W. (1972). Deep inelastic electron scattering. *Annual Review of Nuclear and Particle Science, 22*, 203–254.
48. Fritzsch, H., & Gell-Mann, M. (1972). Quatks and what else? In J. D. Jackson & A. Roberts (Eds.), *Proceedings of the XVI International Conference on High Energy Physics* (Vol. 2). Chicago: Chicago University Press.
49. Fritzsch, H., Gell-Mann, M., & Leutwyler, H. (1973). Advantages of the color octet gluon. *Physics Letters, B47*, 365–368.
50. Fukuda, Y. et al. (1998). Evidence for oscillations of atmospheric neutrinos. *Physical Review Letters, 81*, 1562–1567.
51. Fukuda, Y. et al. (2002). Determination of solar neutrino oscillation parameters using 1496 days of Super-Kamiokande-I data. *Physics Letters, B539*, 179–187.
52. Gell-Mann, M. (1964). A schematic model of baryons and mesons. *Physics Letters, 8*, 214–215.
53. Gell-Mann, M. (1972). Quarks. *Acta Physica Austriaca Supplement IX, 9*, 733–761.
54. Gell-Mann, M., Raymond, P., & Slansky, R. (1979). Complex spinors and unified theories. In P. van Nieuwenhuizen & D. Z. Friedman (Eds.), *Supergravity*. Amsterdam: North-Holland.
55. Glashow, S. L. (1961). Partial symmetries of weak interactions. *Nuclear Physics, 22*, 579–588.
56. Glashow, S. L., Iliopoulos, J., & Maiani, L. (1970). Weak interactions with Lepton-Hadron symmetry. *Physical Review, D2*, 1285–1292.
57. Goldhaber, M., Grodzins, L., & Sunyar, A. W. (1958). Helicity of the neutrinos. *Physical Review, 109*, 1015–1017.
58. Greenberg, O. W. (1964). Spin and unitary spin independence in a paraquark model of baryons and mesons. *Physical Review Letters, 13*, 598–602.
59. Gribov, V. N., & Pontecorvo, B. (1969). Neutrino astronomy and lepton charge. *Physics Letters, B616*, 174–190.
60. Guth, A. H. (1980). Existence proof of a nonconfining phase in four-dimensional U(1) lattice theory. *Physical Review, D21*, 2291–2307.
61. Halzen, F., & Martin, A. D. (1984). *Quarks and leptons: An introductory course in modern particle physics*. New York: Wiley.
62. Hampel, W. et al. (1999). GALLEX solar neutrino observations: Results for GALLEX IV. *Physics Letters, B447*, 127–133.
63. Han, M. Y., & Nambu, Y. (1965). Three-triplet model with double SU(3) symmetry. *Physical Review, 139*, B1006–B1010.

64. Hanson, G. et al. (1975). Evidence for jet structures in hadron production by $e^+e^-$ annihilation. *Physical Review Letters, 35*, 1609–1612.
65. Hirata, K. S. et al. (1996). Solar neutrino data covering solar cycle 22. *Physical Review Letters, 77*, 1683–1686.
66. Hoyer, P. et al. (1979). Quantum chromodynamics and jets in $e^+e^-$. *Nuclear Physics, B161*, 349–372.
67. Itzykson, C., & Zuber, J.-B. (1980). *Quantum field theory*. New York/Toronto: McGraw-Hill.
68. Joglekar, S. D., & Lee, B. W. (1976). General theory of renormalization of gauge invariant operators. *Annals of Physics, 97*, 160–215.
69. Jones, D. R. T. (1974). Two-loop diagrams in Yang-Mills theory. *Nuclear Physics, B75*, 531–538.
70. Jost, R., & Luttinger, J. M. (1950). Vacuum polarization and $e^4$ charge renormalization for electrons. *Helvetica Physica Acta, 23*, 201.
71. Kobayashi, M., & Maskawa, K. (1973). CP violation in the renormalizable theory of weak interaction. *Progress of Theoretical Physics, 49*, 652–657.
72. Kogut, J. B. (1980). Progress in lattice theory. *Physics Reports, 67*, 67–102.
73. Kogut, J. B., Pearson, R. P., & Shigemitsu, J. (1981). The string tension, confinement and roughening in SU(3) Hamiltonian lattice gauge theory. *Physics Letters, 98B*, 63–68.
74. Landau, L. D. (1957). On the Conservation Laws for Weak Interactions. *Nuclear Physics, 3*, 127–131.
75. Langacker, P. (1981). Grand unified theories and proton decay. *Physics Reports, 72*, 185–385.
76. Lautrup, B., & Nauenberg, M. (1980). Phase transition in four-dimensional compact QED. *Physics Letters, 95B*, 63–66.
77. Lee, B. (1976). In R. Balian & J. Zinn-Justin (Eds.), *Methods in field theory*. Amsterdam: North Holland.
78. Lee, T. D., & Yang, C. N. (1956). Question of parity conservation in weak interactions. *Physical Review, 104*, 254–258. See also *ibid., 106*, 1671 (1957).
79. Lepage, G. P., & Brodsky, S. J. (1979). Exclusive processes in quantum chromodynamics: The form factors of baryons at large momentum transfer. *Physical Review Letters, 43*, 545–549.
80. Lepage, G. P., & Brodsky, S. J. (1980). Exclusive processes in perturbative quantum chromodynamics. *Physical Review, D22*, 2157–2198.
81. Maki, Z., Nakagawa, M., & Sakata, S. (1962). Remarks on the unified model of elementary particles. *Progress of Theoretical Physics, 28*, 870–880.
82. Manoukian, E. B. (1984a). Proof of the decoupling theorem of field theory in Minkowski space. *Journal of Mathematics and Physics, 25*, 1519–1523.
83. Manoukian, E. B. (1985). Quantum action principle and path integrals for long-range interactions. *Nuovo Cimento, 90A*, 295–307.
84. Manoukian, E. B. (1986a). Action principle and quantization of gauge fields. *Physical Review, D34*, 3739–3749.
85. Manoukian, E. B. (1986b). Generalized conditions for the decoupling theorem of quantum field theory in Minkowski space with particles of vanishing small masses. *Journal of Mathematics and Physics, 27*, 1879–1882.
86. Manoukian, E. B. (1987). Functional differential equations for gauge theories. *Physical Review, D35*, 2047–2048.
87. Manoukian, E. B. (2006). *Quantum theory: A wide spectrum*. Dordrecht: Springer.
88. Martin, P. C., & Glashow, S. L. (2008). Julian Schwinger 1918-1994: A biographical memoir. *National Academy of Sciences*, Washington, DC, Copyright 2008.
89. Mele, S. (2006). Measurements of the running of the electromagnetic coupling at LEP. In *XXVI Physics in Collision*, 6–9 July 2006, Búzios, Rio de Janeiro.
90. Minkowski, P. (1977). $\mu \rightarrow ey$ at a rate of one out of $10^9$ muon decays? *Physics Letters, B67*, 421–428.
91. Miura, M. (2010). Search for nucleon decays in Super-Kamiokande. *ICHEP, Paris, Session, 10*, 408–412.

92. Mohapatra, R. N., & Senjanović, G. (1980). Neutrino mass and spontaneous parity noncon-servation. *Physical Review Letters, 44*, 912–915.

93. Mohr, P. J. et al. (2008). CODATA recommended values of the fundamental physical constants. *Reviews of Modern Physics, 80*, 633–730.

94. Olive, K. A. et al. (2014). Particle data group. *Chinese Physics C, 38*, 090001.

95. Panofski, W. (1968). In *Proceedings of the 14th International Conference on High Energy Physics* (pp. 23–39). CERN Scientific Information, Vienna.

96. Pauli, W. (1957). On the earlier and more recent history of the neutrino. In: Winter, K. (Ed.), *Neutrino physics*. Cambridge: Cambridge University Press (1991).

97. Perrin, F. (1933). Possibilité d'Emission de Particules Neutres de Masse Nulle dans les Radioactivités $\beta$. *Comptes Rendus Academie des Sciences Paris, 197*, 1625.

98. Peskin, M. (1997). Beyond the Standard Model. In N. Ellis & M. Neubert (Eds.), *European School of High-Energy Physics 1996*, CERN-97-03, Genève.

99. Pontecorvo, B. (1957). Mesonium and Anti-Mesonium. *Soviet Physics JETP, 33*, 549–551. Original Russian Version: *Zhurnal Experimental'noi i Teoreticheskoi Fiziki, 6*, 429–431 (1957).

100. Pontecorvo, B. (1958). Inverse beta processes and nonconservation of lepton charge. *Soviet Physics JETP, 34*, 247–248. Original Russian Version: *Zhurnal Experimental'noi i Teoretich-eskoi Fiziki, 7*, 172–173 (1957).

101. Pontecorvo, B. (1968). Neutrino experiments and the problem of conservation of Leptonic charge. *Soviet Physics JETP, 53*, 1717–1725. Original Russian Version: *Zhurnal Experimen-tal'noi i Teoreticheskoi Fiziki, 26*, 984–988 (1967).

102. Puckett, A. J. R. et al. (2012). Final analysis of proton form factor ratio at $Q^2 = 4.0$, 4.8 and 5.6 GeV$^2$. *Physical Review, C85*, 045203.

103. Punjabi, V. et al. (2005). Proton elastic From factor ratios to $Q^2 = 3, 5 \, \text{GeV}^2$ by polarization transfer. *Physical Review, C71*, 055202.

104. Reines, F., & Cowan, C. L. (1956). The neutrino. *Nature, 178*, 446–449.

105. Riordan, E. M. et al. (1974). Extraction of $R = \sigma_L/\sigma_T$ from deep inelastic $e-p$ and $e-d$ cross sections. *Physical Review Letters, 33*, 561–564.

106. Salam, A. (1957). Parity consevation and a two-component theory of the neutrino. *Nuovo Cimento, 5*, 299–301.

107. Salam, A. (1968). Weak and electromagnetic interactions. In N. Svartholm (Ed.), *Elementary Particle Theory, Proceedings of the 8th Nobel Symposium, Almqvist and Wiksell*, Stockholm.

108. Salam, A., & Ward, J. (1964). Electromagnetic and weak interactions. *Physics Letters, 13*, 168–170.

109. Schwinger, J. (1957). A theory of the fundamental interactions. *Annals of Physics, 2*, 407–434.

110. Slavnov, A. A. (1972). Ward identities in gauge theories. *Theoretical and Mathematical Physics, 10*, 152–160. English Translation: *Theoretical and Mathematical Physics, 10*, 99–108 (1972).

111. Sterman, G., & Weinberg, S. (1977). Jets from quantum chromodynamics. *Physical Review Letters, 39*, 1436–1439.

112. Stevenson, P. M. (1978). Comments on the Sterman-Weinberg jet formula. *Physics Letters, 78B*, 451–454.

113. Sudarshan, E. C. G., & Marshak, R. (1958). Chirality invariance and the universal fermi interaction. *Physical Review, 109*, 1860–1862.

114. 't Hooft, G. (1971a). Renormalizable of massless Yang-Mills fields. *Nuclear Physics, B33*, 173–199.

115. 't Hooft, G. (1971b). Renormalizable Lagrangians for massive Yang-Mills fields. *Nuclear Physics, B35*, 167–188.

116. 't Hooft, G. (1971c). Prediction for neutrino-electron scattering cross sections in weinberg's model of electroweak interaction. *Physics Letters, B37*, 195–196.

117. Taylor, J. C. (1971). Ward identities and charge renormalization. *Nuclear Physics, B33*, 436–444.

118. Veltman, M. J. G. (1981). The infrared-ultraviolet connection. *Acta Physica Polonica, B12*, 437.
119. Vilain, P. et al. (1994). Precision Measurement of Electroweak Parameters from the Scattering of Muon-Neutrinos on Electrons. *Physics Letters B, 335*, 246–252.
120. Weinberg, S. (1967). A Model of Leptons. *Physical Review Letters, 19*, 1264–1266.
121. Weinberg, S. (1996). *The Quantum Theory of Fields, II: Modern Applications*. Cambridge University Press, Cambridge.
122. Wu, C. S. et al. (1957). Experimental tests of parity conservation in beta decay. *Physical Review, 105*, 1413–1415.
123. Yanagida, Y. (1980). Horizontal symmetry and masses of neutrinos. *Progress of Theoretical Physics, 64*, 1103–1105.
124. Zee, A. (1973). Electron-positron annihilation in stagnant field theories. *Physical Review, D8*, 4038–4041.
125. Zweig, G. (1964). An $SU_3$ Model for strong interaction symmetry and its breaking. CERN Preprint, TH-401, 1–24.

# Recommended Reading

1. Beringer, J. et al. (2012). Particle data group. *Physical Review D, 86*, 010001.
2. DeWitt, B. (2014). *The global approach to quantum field theory*. Oxford: Oxford University Press.
3. Manoukian, E. B. (1981). Generalized decoupling theorem in quantum field theory. *Journal of Mathematics and Physics, 22*, 2258–2262.
4. Manoukian, E. B. (1984). Proof of the decoupling theorem in Minkowski space. *Journal of Mathematics and Physics, 25*, 1519–1523.
5. Manoukian, E. B. (1986a). Generalized conditions for the decoupling theorem of quantum field theory in Minkowski space with particles of vanishingly small masses. *Journal of Mathematics and Physics, 27*, 1879–1882.
6. Manoukian, E. B. (1986b). Action principle and quantization of gauge fields. *Physical Review, D34*, 3739–3749.
7. Manoukian, E. B. (1987). Functional differential equations for gauge theories. *Physical Review, D35*, 2047–2048.
8. Olive, K. A. et al. (2014). Particle Data Group. *Chinese Physics C, 38*, 090001.
9. Ross, G. G. (1985). *Grand Unified Theories*. Reading: Benjamin/Cummings Publishing.
10. Weinberg. S. (1996). *The Quantum Theory of Fields. II: Modern Applications*. Cambridge: Cambridge University Press.
11. Yongram, N., Manoukian, E. B., & Siranan, S. (2006). Polarization correlations in muon pair production in the electroweak model. *Modern Physics Letters A, 21*, 979–984.

# General Appendices

# Appendix I
# The Dirac Formalism

For the convenience of the reader we gather here some important equations dealing with the Dirac formalism for the description of spin $1/2$ particles. For derivations and detailed presentation, we refer the reader to Chapter 16 of Manoukian [6].

The Dirac equation is given by

$$\left(\frac{\gamma^\mu \partial_\mu}{i} + m\right)\psi = 0, \quad \{\gamma^\mu, \gamma^\nu\} = -2\eta^{\mu\nu}, \quad [\eta^{\mu\nu}] = \text{diag}[-1, 1, 1, 1] = [\eta_{\mu\nu}]. \tag{I.1}$$

In the presence of an external electromagnetic field, the Dirac equation reads

$$\left[\gamma^\mu\left(\frac{\partial_\mu}{i} - eA_\mu(x)\right) + m\right]\psi(x) = 0, \tag{I.2}$$

from which one obtains the equation $(\overline{\psi} = \psi^\dagger \gamma^0)$

$$\left[\gamma^\mu\left(\frac{\partial_\mu}{i} + eA_\mu(x)\right) + m\right]\psi^{\mathscr{C}}(x) = 0, \quad \psi^{\mathscr{C}}(x) = \mathscr{C}\overline{\psi}^{\mathsf{T}}(x), \quad \mathscr{C} = i\gamma^2\gamma^0, \tag{I.3}$$

thus introducing, in turn, the charge conjugation matrix $\mathscr{C}$, and the charge conjugate spinor $\psi^{\mathscr{C}}$.

Under a homogeneous Lorentz transformation which may include a 3D rotation (see (2.2.1), (2.2.3), (2.2.6))

$$x'^\mu = \Lambda^\mu{}_\nu x^\nu, \qquad \partial_\nu = \Lambda^\mu{}_\nu \partial'_\mu, \qquad \partial'_\mu = \Lambda_\mu{}^\nu \partial_\nu. \tag{I.4}$$

the Dirac equation reads

$$\left[\frac{\gamma^\nu \partial'_\nu}{i} + m\right]K\psi(x) = 0, \qquad K\psi(x) = \psi'(x'), \tag{I.5}$$

© Springer International Publishing Switzerland 2016
E.B. Manoukian, *Quantum Field Theory I*, Graduate Texts in Physics,
DOI 10.1007/978-3-319-30939-2

where the matrix $K$ satisfies the relations

$$K^\dagger \gamma^0 K = \gamma^0, \qquad \Lambda^\mu{}_\nu \gamma^\nu = K^{-1}\gamma^\mu K. \tag{I.6}$$

For infinitesimal transformations, we recall from (2.2.18), that $\Lambda^\mu{}_\nu \simeq \delta^\mu{}_\nu + \delta\omega^\mu{}_\nu$, $\delta\omega^{\mu\nu} = -\delta\omega^{\nu\mu}$, and we may set $K \simeq I + (i/2)\,\delta\omega^{\mu\nu} S_{\mu\nu}$, where $S_{\mu\nu}$ is to be determined. By substituting this expression in the last equation in (I.6) gives

$$[S^{\lambda\mu}, \gamma^\nu] = i\left(\eta^{\lambda\nu}\gamma^\mu - \eta^{\mu\nu}\gamma^\lambda\right), \quad \text{with solution} \quad S^{\lambda\mu} = \frac{i}{4}[\gamma^\lambda, \gamma^\mu]. \tag{I.7}$$

Some of the properties of the gamma matrices, based on their anti-commutations relations in (I.1) are given in Box I.1.

Box I.1: Some properties of the gamma matrices

$$\gamma^\mu\gamma^\nu = -\eta^{\mu\nu}I + \frac{1}{2}[\gamma^\mu, \gamma^\nu], \quad \text{Tr}[\gamma^\mu] = 0,$$

$$(\gamma^0)^2 = I, \quad (\gamma^i)^2 = -I, \quad i = 1, 2, 3.$$

$$\eta_{\mu\nu}\gamma^\mu\gamma^\nu = -4I,$$

$$\eta_{\mu\nu}\gamma^\mu(\gamma^\sigma)\gamma^\nu = 2\gamma^\sigma,$$

$$\eta_{\mu\nu}\gamma^\mu(\gamma^\sigma\gamma^\lambda)\gamma^\nu = 4\eta^{\sigma\lambda},$$

$$\eta_{\mu\nu}\gamma^\mu(\gamma^\sigma\gamma^\lambda\gamma^\rho)\gamma^\nu = 2\gamma^\rho\gamma^\lambda\gamma^\sigma,$$

$$\left[\gamma^\mu, [\gamma^\sigma, \gamma^\rho]\right] = 4\left(\gamma^\sigma\eta^{\mu\rho} - \gamma^\rho\eta^{\mu\sigma}\right),$$

$$\text{Tr}[\gamma^\mu\gamma^\nu] = -4\eta^{\mu\nu},$$

$$\text{Tr}[\gamma^\alpha\gamma^\beta\gamma^\mu\gamma^\nu] = 4(\eta^{\alpha\beta}\eta^{\mu\nu} - \eta^{\alpha\mu}\eta^{\beta\nu} + \eta^{\alpha\nu}\eta^{\beta\mu}),$$

$$\text{Tr}[\gamma^5\gamma^\alpha\gamma^\beta\gamma^\mu\gamma^\nu] = -4i\varepsilon^{\alpha\beta\mu\nu},$$

$\varepsilon^{\alpha\beta\mu\nu}$ totally anti-symmetric with $\varepsilon^{0123} = +1$.

$\text{Tr}[\text{odd number of } \gamma\text{'s}] = 0.$

$$\gamma^5 = i\gamma^0\gamma^1\gamma^2\gamma^3, \quad \text{Tr}[\gamma^5] = 0, \quad \text{Tr}[\gamma^5\gamma^\mu] = 0,$$

$$(\gamma^5)^2 = I, \quad \{\gamma^5, \gamma^\mu\} = 0,$$

$$(\gamma^\mu a_\mu)^2 = -I[\mathbf{a}^2 - (a^0)^2], \quad (\boldsymbol{\gamma}\cdot\mathbf{a})^2 = -I\mathbf{a}^2,$$

$$\mathbf{a} = (a_1, a_2, a_3), \quad a_0 = -a^0, \quad a_i = a^i, \quad i = 1, 2, 3.$$

The Dirac representation of the $\gamma^\mu$ matrices, in particular, is defined by

$$\gamma^0 = \begin{pmatrix} I & 0 \\ 0 & -I \end{pmatrix}, \quad \boldsymbol{\gamma} = \begin{pmatrix} 0 & \boldsymbol{\sigma} \\ -\boldsymbol{\sigma} & 0 \end{pmatrix}, \quad \gamma^5 \equiv i\gamma^0\gamma^1\gamma^2\gamma^3 = \begin{pmatrix} 0 & I \\ I & 0 \end{pmatrix}, \quad \text{(I.8)}$$

and $\sigma^1, \sigma^2, \sigma^3$ are the Pauli matrices

$$\sigma^1 = \begin{pmatrix} 0 & 1 \\ 1 & 0 \end{pmatrix}, \quad \sigma^2 = \begin{pmatrix} 0 & -i \\ i & 0 \end{pmatrix}, \quad \sigma^3 = \begin{pmatrix} 1 & 0 \\ 0 & -1 \end{pmatrix}, \quad \text{(I.9)}$$

satisfying, the relations $\{\sigma^i, \sigma^j\} = 2\delta^{ij}$.

In the momentum description, on the mass shell $p^0 = \sqrt{\mathbf{p}^2 + m^2}$, where $m$ is the mass of a particle, we have two sets of solutions: $u(\mathbf{p}, \sigma)$, $v(\mathbf{p}, \sigma)$, with $\sigma = \pm 1$ specifying spin states, satisfying the equations

$$(\gamma p + m)u(\mathbf{p}, \sigma) = 0, \qquad \bar{u}(\mathbf{p}, \sigma)(\gamma p + m) = 0, \qquad \text{(I.10)}$$

$$(-\gamma p + m)v(\mathbf{p}, \sigma) = 0, \qquad \bar{v}(\mathbf{p}, \sigma)(-\gamma p + m) = 0, \qquad \text{(I.11)}$$

and $u(\mathbf{p}, \sigma)$ may be taken as

$$u(\mathbf{p}, \sigma) = \sqrt{\frac{p^0 + m}{2m}} \begin{pmatrix} \xi_\sigma \\ \frac{\boldsymbol{\sigma} \cdot \mathbf{p}}{p^0 + m}\xi_\sigma \end{pmatrix}, \qquad \sigma = \pm 1, \qquad \text{(I.12)}$$

involving two normalized two component spinors satisfying $\xi_\sigma^\dagger \xi_{\sigma'} = \delta_{\sigma\sigma'}$, which, for an arbitrary unit vector $\mathbf{N} = (\cos\phi \sin\theta, \sin\phi \sin\theta, \cos\theta)$, may be taken as

$$\xi_{+\mathbf{N}} = \begin{pmatrix} e^{-i\phi/2} \cos\frac{\theta}{2} \\ e^{i\phi/2} \sin\frac{\theta}{2} \end{pmatrix}, \quad \xi_{-\mathbf{N}} = \begin{pmatrix} -e^{-i\phi/2} \sin\frac{\theta}{2} \\ e^{i\phi/2} \cos\frac{\theta}{2} \end{pmatrix}, \qquad \text{(I.13)}$$

$$\boldsymbol{\sigma} \cdot \mathbf{N} \xi_{\pm\mathbf{N}} = \pm\xi_{\pm\mathbf{N}}, \qquad i\sigma^2 \xi_{\pm\mathbf{N}}^* = \mp\xi_{\mp\mathbf{N}}. \qquad \text{(I.14)}$$

We note that the spin matrix is defined by

$$S^i = \frac{1}{2}\varepsilon^{ijk}S^{jk} = \frac{i}{8}\varepsilon^{ijk}[\gamma^j, \gamma^k], \qquad \mathbf{S} = \frac{1}{2}\boldsymbol{\Sigma} = \frac{1}{2}\begin{pmatrix} \boldsymbol{\sigma} & 0 \\ 0 & \boldsymbol{\sigma} \end{pmatrix}. \qquad \text{(I.15)}$$

In the Dirac representation, the charge conjugation matrix $\mathscr{C} = i\gamma^2\gamma^0$ is

$$\mathscr{C}|_{\mathrm{D}} = i\gamma^2\gamma^0\Big|_{\mathrm{D}} = \begin{pmatrix} 0 & -i\sigma^2 \\ -i\sigma^2 & 0 \end{pmatrix}, \tag{I.16}$$

with the charge conjugate of $u(\mathbf{p},\sigma)$ given by $\mathscr{C}\bar{u}^{\mathsf{T}}(\mathbf{p},\pm1) = \pm\gamma^5 u(\mathbf{p},\mp1)$. Thus we may define the second set of spinors in the momentum description by

$$v(\mathbf{p},\pm1) = \gamma^5 u(\mathbf{p},\mp1), \qquad \bar{v}(\mathbf{p},\pm1) = -\bar{u}(\mathbf{p},\mp1)\gamma^5, \tag{I.17}$$

up to phase factors, with $v(\mathbf{p},\sigma)$ as charge conjugate of $u(\mathbf{p},\sigma)$.

Under space reflection: $x' = (x^0, -\mathbf{x})$ the Dirac equation reads

$$\left[\frac{\gamma^\mu\partial'_\mu}{i} + m\right]\gamma^0\psi(x) = 0, \qquad \psi'(x') = \gamma^0\psi(x), \tag{I.18}$$

up to a phase factor for the latter. Hence under space reflection, $u(\mathbf{p},\sigma) \to \gamma^0 u(-\mathbf{p},\sigma) = +u(-\mathbf{p},\sigma)$, $v(\mathbf{p},\sigma) \to \gamma^0 v(-\mathbf{p},\sigma) = -v(-\mathbf{p},\sigma)$, having opposite (intrinsic) parities.

We have the following normalization conditions

$$\bar{u}(\mathbf{p},\sigma)u(\mathbf{p},\sigma') = \delta_{\sigma\sigma'}, \quad \bar{v}(\mathbf{p},\sigma)v(\mathbf{p},\sigma') = -\delta_{\sigma\sigma'}, \quad \bar{u}(\mathbf{p},\sigma)v(\mathbf{p},\sigma') = 0, \tag{I.19}$$

$$u^\dagger(\mathbf{p},\sigma)u(\mathbf{p},\sigma') = v^\dagger(\mathbf{p},\sigma)v(\mathbf{p},\sigma') = \frac{p^0}{m}\delta_{\sigma\sigma'}, \quad u^\dagger(-\mathbf{p},\sigma)v(\mathbf{p},\sigma') = 0, \tag{I.20}$$

and the completeness relations

$$\sum_\sigma u(\mathbf{p},\sigma)\bar{u}(\mathbf{p},\sigma) = \frac{(-\gamma p + m)}{2m} \equiv \mathbb{P}_+(p), \tag{I.21}$$

$$-\sum_\sigma v(\mathbf{p},\sigma)\bar{v}(\mathbf{p},\sigma) = \frac{(\gamma p + m)}{2m} \equiv \mathbb{P}_-(p). \tag{I.22}$$

Note that for

$$\mathbf{N} = \mathbf{p}/|\mathbf{p}|, \tag{I.23}$$

in (I.13),

$$u(\mathbf{p}, \pm) = \sqrt{\frac{p^0 + m}{2m}} \begin{pmatrix} \xi_{\pm} \\ \pm \frac{|\mathbf{p}|}{p^0 + m} \xi_{\pm} \end{pmatrix}. \tag{I.24}$$

For a massless Dirac particle, it is convenient to work in the *chiral* representation (see (2.3.3)), and with the unit vector $\mathbf{N}$ chosen along the momentum of the particle, as given in (I.23), we have spinors[1]

$$u(\mathbf{p}, +1) = \begin{pmatrix} \xi_+ \\ 0 \end{pmatrix} = \frac{I + \gamma^5}{2} \begin{pmatrix} \xi_+ \\ \xi_- \end{pmatrix}, \quad u(\mathbf{p}, -1) = \begin{pmatrix} 0 \\ \xi_- \end{pmatrix} = \frac{I - \gamma^5}{2} \begin{pmatrix} \xi_+ \\ \xi_- \end{pmatrix}, \tag{I.25}$$

$u^{\dagger}(\mathbf{p}, \pm 1) \, u(\mathbf{p}, \pm 1) = 1$. We may conveniently write

$$u(\pm \mathbf{p}, \pm 1) = \begin{pmatrix} \xi_{\pm} \\ 0 \end{pmatrix}, \quad u(\pm \mathbf{p}, \mp 1) = \begin{pmatrix} 0 \\ \xi_{\mp} \end{pmatrix}, \tag{I.26}$$

where $\pm$ with $\pm$ means spin projection in the same direction as of the momentum, while $\pm$ with $\mp$ means in the opposite directions. With an appropriately chosen phases, the latter are related by a parity transformation: $-\gamma^0 u(-\mathbf{p}, +1) \rightarrow u(+\mathbf{p}, +1)$, $-\gamma^0 u(+\mathbf{p}, -1) \rightarrow u(-\mathbf{p}, -1)$, with spin and momentum in opposite directions to spin and momentum in the same direction, and vice versa. Accordingly, if parity is not conserved, nature picks up only one of the helicities.

The interest in the equations in (I.25) is that $(I + \gamma^5)/2$ projects out a state corresponding to a right-handed particle, with spin along its momentum, while $(I - \gamma^5)/2$ projects out a left-handed one, with spin in opposite direction to its momentum. The corresponding particles are referred to, respectively, as right-handed and left-handed. The situation may be demonstrated as shown in Fig. I.1

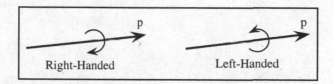

Right-Handed      Left-Handed

**Fig. I.1** Diagrams defining the handedness of a particle, where $\mathbf{p}$ is its 3-momentum

---

[1]For the relevant details, see Manoukian [6], p. 912.

The eigenvectors $u(\mathbf{p}, \pm 1)$ then satisfy the simultaneous eigenvalue equations:

$$\gamma^0 \boldsymbol{\gamma} \cdot \mathbf{p}\, u(\mathbf{p}, \pm 1) = |\mathbf{p}|\ u(\mathbf{p}, \pm 1), \tag{I.27}$$

$$\gamma^5 u(\mathbf{p}, \pm 1) = \pm\ u(\mathbf{p}, \pm 1), \tag{I.28}$$

$$\mathbf{S} \cdot \mathbf{N} u(\mathbf{p}, \pm 1) = \pm (1/2)\, u(\mathbf{p}, \pm 1), \tag{I.29}$$

where $\{\gamma^5,\ \mathbf{S} \cdot \mathbf{N},\ \gamma^0 \boldsymbol{\gamma} \cdot \mathbf{p}\}$ is a commuting set of operators,[2] and specify the state of a particle. The eigenvalues of $\gamma^5$ are referred to as the chiralities (handedness) of a particle, and $\mathbf{S} \cdot \mathbf{N} \equiv (\boldsymbol{\Sigma}/2) \cdot \mathbf{N}$ defines the helicity operator, with $u(\mathbf{p}, \pm)$ corresponding to a massless particle with two-helicity states. As seen above, if parity is not conserved a massless particle is to be considered to have only one helicity.

---

[2]Note that $\gamma^5$ does not commute with $\gamma^0(\boldsymbol{\gamma} \cdot \mathbf{p} + m)$ for $m \neq 0$.

# Appendix II
# Doing Integrals in Field Theory

For $a$ real, consider the following two integrals

$$\int_{-\infty}^{\infty} dy \, \cos(a y^2) = \sqrt{\frac{\pi}{2|a|}}, \qquad \int_{-\infty}^{\infty} dy \, \sin(a y^2) = \sqrt{\frac{\pi}{2|a|}} \, \text{sgn} \, a. \qquad \text{(II.1)}$$

Using the fact that $\sin \pi/4 = \cos \pi/4 = 1/\sqrt{2}$, they may be combined into an exponential form as

$$\int_{-\infty}^{\infty} dy \, e^{i a y^2} = \sqrt{\frac{\pi}{|a|}} \, \exp\left[ i\frac{\pi}{4} \, \text{sgn} \, a \right]. \qquad \text{(II.2)}$$

In view of applications in $n$-dimensional spacetime, we apply this integral to a typical expression of the denominator of a propagator, which may be first rewritten as

$$\frac{1}{(k^2 + M^2 - i\epsilon)} = i \int_0^{\infty} ds \, \exp\left[-is(k^2 + M^2 - i\epsilon)\right], \quad k^2 = \sum_{i=1}^{n-1} k_i^2 - (k^0)^2. \qquad \text{(II.3)}$$

In $n$-dimensional spacetime, our metric is defined by $[\eta_{\mu\nu}] = \text{diag}[-1, 1, \ldots, 1]$.

From (II.2) we obtain the following integrals , with $s > 0$,

$$\int_{-\infty}^{\infty} dk^0 \exp\left[i s (k^0)^2\right] = \sqrt{\frac{\pi}{s}} \, e^{i\frac{\pi}{4}}, \qquad \text{(II.4)}$$

$$\int_{-\infty}^{\infty} dk^1 \int_{-\infty}^{\infty} dk^2 \ldots \int_{-\infty}^{\infty} dk^{n-1} \exp\left[-i s \left((k^1)^2 + \cdots + (k^{n-1})^2\right)\right] = \left( \sqrt{\frac{\pi}{s}} \, e^{-i\frac{\pi}{4}} \right)^{n-1}, \qquad \text{(II.5)}$$

© Springer International Publishing Switzerland 2016
E.B. Manoukian, *Quantum Field Theory I*, Graduate Texts in Physics,
DOI 10.1007/978-3-319-30939-2

and hence

$$\int_{-\infty}^{\infty} dk^0 \int_{-\infty}^{\infty} dk^1 \dots \int_{\infty}^{\infty} dk^{n-1} \, e^{-isk^2} = \left(\frac{1}{i}\right)^{\frac{n-2}{2}} \left(\frac{\pi}{s}\right)^{\frac{n}{2}}, \quad k^2 = \sum_{i=1}^{n-1} (k^i)^2 - (k^0)^2; \quad s > 0.$$

(II.6)

In the remaining part of this appendix, we work only with $n = 4$, i.e., in 4-dimensional spacetime. We also suppress the $-i\epsilon$ factor in (II.3).

If we integrate the expression in (II.3) over $k$, we encounter a singularity in the $s$-integral at $s = 0$. We may, however, differentiate (II.3) twice with respect to $M^2$ to obtain

$$\frac{1}{(k^2 + M^2)^3} = -\frac{i}{2} \int_0^{\infty} ds \, s^2 \exp\left[-is(k^2 + M^2)\right],$$

(II.7)

leading from (II.6) to the useful integral

$$\int \frac{(dk)}{(k^2 + M^2)^3} = \frac{i\pi^2}{2} \frac{1}{M^2}, \quad (dk) = dk^0 dk^1 dk^2 dk^3,$$

(II.8)

where we have used the fact that $\int_0^{\infty} ds \exp - is(M^2 - i\epsilon) = -i/(M^2)$.

A further differentiation of (II.8) with respect to $M^2$, for example, leads to

$$\int \frac{(dk)}{(k^2 + M^2)^4} = \frac{i\pi^2}{6} \frac{1}{(M^2)^2}, \quad \int \frac{k^2 (dk)}{(k^2 + M^2)^4} = \frac{i\pi^2}{3} \frac{1}{M^2}.$$

(II.9)

We note that quite generally one may write,

$$\int (dk) \frac{k^\mu k^\nu}{(k^2 + M^2)^4} = A \eta^{\mu\nu},$$

(II.10)

$$\int (dk) \frac{k^{\mu_1} k^{\mu_2} k^{\mu_3} k^{\mu_4}}{(k^2 + M^2)^5} = B \left(\eta^{\mu_1\mu_2} \eta^{\mu_3\mu_4} + \eta^{\mu_1\mu_3} \eta^{\mu_2\mu_4} + \eta^{\mu_1\mu_4} \eta^{\mu_2\mu_3}\right).$$

(II.11)

To determine the coefficient $A$, for example, we may make a contraction over $\mu$, $\nu$, to obtain from the second equation in (II.9), $A = i\pi^2/12M^2$, using the fact that $\eta^\mu{}_\mu = 4$. Similarly, contractions over $\mu_1$, $\mu_2$, and $\mu_3$, $\mu_4$, give $24B$ to the right-hand side of (II.10), leading to $B = i\pi^2/96M^2$. These values are worth recording here:

$$A = \frac{i\pi^2}{12M^2}, \quad B = \frac{i\pi^2}{96M^2}.$$

(II.12)

Thus we are bound to encounter the following readily evaluated integrals:

$$\int (dk) \frac{(k^2)^{m-2}}{(k^2 + M^2)^n} = \frac{i\pi^2}{(M^2)^{n-m}} \frac{\Gamma(m)\,\Gamma(n - m)}{\Gamma(n)}, \tag{II.13}$$

which obviously exist for $n > m > 0$, where $\Gamma(z)$ is the gamma function.

Now we consider more complicated integrals involving an additional (external) momentum. For convenience, we enumerate the type of integrals considered.

1.

$$I(p^2) = \int (dk) \left( \frac{1}{[(k-p)^2 + M^2(p^2)]^2} - \frac{1}{[k^2 + M^2(p^2)]^2} \right) = 0, \tag{II.14}$$

where, as indicated, $M^2(p^2)$ may be a function of $p^2$. To show that this integral is indeed zero, we take its derivative with respect to $p_\sigma$. It is easily checked that the derivative of the integrand is zero giving $\partial I(p^2)/\partial p_\sigma = 0$. With the boundary condition that $I(0) = 0$, gives $I(p^2) = 0$. Here we note that if we define the *degree of divergence of an integral as the of power of k in the numerator minus the power of k in the denominator plus four*, then the degree of divergence of the above integral restricted to the first term is zero. This then allows one to make a shift of the integration variable $k \to k + p$ in the first part of the integral just mentioned leading to the net result zero for the integral $I(p^2)$. Of course such a shift of an integration variable is obviously valid if the degree of divergence of an integral is negative.

The degree of divergence of the following integrals are negative, thus by making a shift of the integration variable lead to the stated results by using, in the process, (II.8):

$$\int \frac{(dk)}{[(k-p)^2 + M^2(p^2)]^3} = \int \frac{(dk)}{[k^2 + M^2(p^2)]^3} = \frac{i\pi^2}{2M^2}, \tag{II.15}$$

$$\int (dk) \frac{k^\mu}{[(k-p)^2 + M^2(p^2)]^3} = p^\mu \int \frac{(dk)}{[k^2 + M^2(p^2)]^3} = \frac{i\pi^2 p^\mu}{2M^2}, \tag{II.16}$$

where in the latter equation, we have used the property that an integral which is odd in $k^\mu$ is zero. Another integral, where the shift of the integration variable in the first term is obviously permissible is

$$\int (dk) \left( \frac{k^\mu k^\nu}{[(k-p)^2 + M^2(p^2)]^3} - \frac{k^\mu k^\nu}{[k^2 + M^2(p^2)]^3} - \frac{p^\mu p^\nu}{[k^2 + M^2(p^2)]^3} \right) = 0. \tag{II.17}$$

2. To obtain the following integral

$$\int (dk) \left( \frac{k^\mu k^\nu k^\sigma}{[(k-p)^2 + M^2(p^2)]^3} - \frac{1}{4} \frac{\eta^{\mu\nu} p^\sigma + \eta^{\mu\sigma} p^\nu + \eta^{\nu\sigma} p^\mu}{[k^2 + M^2(p^2)]^2} \right)$$

$$= -\frac{5i\pi^2}{24} (\eta^{\mu\nu} p^\sigma + \eta^{\mu\sigma} p^\nu + \eta^{\nu\sigma} p^\mu) + \frac{i\pi^2}{2M^2(p^2)} p^\mu p^\nu p^\sigma, \qquad \text{(II.18)}$$

we denote the result of the integration by

$$C(p^2)(\eta^{\mu\nu} p^\sigma + \eta^{\mu\sigma} p^\nu + \eta^{\nu\sigma} p^\mu) + D(p^2) p^\mu p^\nu p^\sigma, \qquad \text{(II.19)}$$

and determine $C(p^2)$ and $D(p^2)$. To this end, note that the integral is zero for $p = 0$ since the resulting integrand is an odd function of $k$. Secondly, note that the degree of divergence of the integral restricted to the first term involving $k^\mu k^\nu k^\sigma$ is positive and hence we cannot simply make a shift of the integration variable $k$ to $k + p$ in evaluating this term. What we do instead is take the derivative of the integral with respect to $p_\lambda$ as shown below.

The derivative of the first term in the integrand, for example, is

$$(-3) \left( 2p^\lambda + \frac{\partial M^2(p^2)}{\partial p_\lambda} \right) \frac{k^\mu k^\nu k^\sigma}{[(k-p)^2 + M^2(p^2)]^4} + 6 \frac{k^\mu k^\nu k^\sigma k^\lambda}{[(k-p)^2 + M^2(p^2)]^4}.$$

The first term leads to an integral of degree of divergence $-1$, and the second to a degree of divergence 0. Hence we can make a shift of the integration variable $p$ to $p + k$ here. Continuing in this manner and using any of the integrations obtained before (II.18), we obtain for the derivative of the integral on the left-hand side of (II.18) explicitly

$$-\frac{5i\pi^2}{24} (\eta^{\mu\nu}\eta^{\sigma\lambda} + \eta^{\mu\sigma}\eta^{\nu\lambda} + \eta^{\nu\sigma}\eta^{\mu\lambda}) - \frac{i\pi^2}{2(M^2(p^2))^2} \frac{\partial M^2(p^2)}{\partial p_\lambda} p^\mu p^\nu p^\sigma$$

$$+ \frac{i\pi^2}{2M^2(p^2)} (\eta^{\mu\lambda} p^\nu p^\sigma + \eta^{\nu\lambda} p^\mu p^\sigma + \eta^{\sigma\lambda} p^\mu p^\nu). \qquad \text{(II.20)}$$

This is to be compared with the derivative of the expression in (II.19) with respect to $k_\lambda$,

$$\frac{\partial C(p^2)}{\partial p_\lambda} (\eta^{\mu\nu} p^\sigma + \eta^{\mu\sigma} p^\nu + \eta^{\nu\sigma} p^\mu) + C(p^2)(\eta^{\mu\nu}\eta^{\sigma\lambda} + \eta^{\mu\sigma}\eta^{\nu\lambda} + \eta^{\nu\sigma}\eta^{\mu\lambda})$$

$$+ \frac{\partial D(p^2)}{\partial p_\lambda} p^\mu p^\nu p^\sigma + D(p^2)(\eta^{\mu\lambda} p^\nu p^\sigma + \eta^{\nu\lambda} p^\mu p^\sigma + \eta^{\sigma\lambda} p^\mu p^\nu), \qquad \text{(II.21)}$$

from which we obtain

$$C(p^2) = -\frac{5 i \pi^2}{24}, \qquad D(p^2) = \frac{i \pi^2}{2 M^2(p^2)}, \qquad \text{(II.22)}$$

as given on the right-hand side of (II.18).

The following integrals is obtained in a similar manner

$$\int (dp) \left( \frac{p^\mu p^\nu}{[(p-k)^2 + M^2(k^2)]^2} - \frac{k^\mu k^\nu}{[p^2 + M^2(k^2)]^2} - \frac{1}{4} \frac{p^2 \eta^{\mu\nu}}{[p^2 + M^2(k^2)]^2} \right)$$

$$= -\frac{5 i \pi^2}{6} k^\mu k^\nu - \frac{i \pi^2}{6} k^2 \eta^{\mu\nu}. \qquad \text{(II.23)}$$

Here we have used $p$ as the *integration* variable for a direct application of this in Appendix $\Lambda$ of Chap. 3. The following integral is also quite useful

$$\int (dk) \left( \frac{k^\mu}{[(k-p)^2 + M^2(p^2)]^2} - \frac{p^\mu}{[k^2 + M^2(p^2)]^2} \right) = -\frac{i \pi^2}{2} p^\mu. \qquad \text{(II.24)}$$

The reader is strongly encouraged to go through the above details and realize the simplicity in evaluating these integrals by the method just developed.

Integrands appearing in (II.8)...(II.24) arise when one combines the product of two (or more) factors such as

$$\frac{1}{(k-p)^2 + M^2} \frac{1}{k^2 + m^2}, \qquad \text{(II.25)}$$

into one factor in the following manner. To this end, for two c-numbers $A$, $B$, consider the integral

$$\int_0^1 dx \, \frac{d}{dx} \frac{1}{[Ax + B(1-x)]} = \frac{1}{A} - \frac{1}{B}. \qquad \text{(II.26)}$$

By carrying out the differentiation with respect to $x$ and by dividing the resulting integral by $(B-A)$ gives the useful formula

$$\frac{1}{AB} = \int_0^1 dx \, \frac{1}{[Ax + B(1-x)]^2}. \qquad \text{(II.27)}$$

This is referred to as a Feynman parameter representation of the product on the left-hand side of (II.27).

As an interesting example for applying this formula and comparing it with the method developed above for evaluating some integrals, we will evaluate the following

$$I(p^2) = \int (dk) \left( \frac{1}{(k-p)^2 + m^2} - \frac{1}{k^2 + m^2} \right), \quad I(p^2)|_{p=0} = 0. \quad \text{(II.28)}$$

Let us apply the method developed earlier first to evaluate this integral by differentiating it with respect to $p_\mu$. This gives

$$\frac{\partial}{\partial p_\mu} I(p^2) = -2 \int (dk) \left( \frac{p^\mu}{[(k-p)^2 + m^2]^2} - \frac{k^\mu}{[(k-p)^2 + m^2]^2} \right) = -i\pi^2 p^\mu, \quad \text{(II.29)}$$

$$I(p^2) = -\frac{i\pi^2}{2} p^2, \quad \text{(II.30)}$$

where we have used the fact that the degree of divergence of the integral in (II.29) *restricted* to the *first* term is zero to make a shift of the integration variable in the first, and then used (II.24), to finally obtain $I(p^2)$. Now follow the *explicit* evaluation of (II.28).

Let us use (II.27). This leads from the latter to the evaluation of the integral

$$I(p^2) = -\int_0^1 dx \int (dk) \frac{p^2 - 2kp}{[(k-px)^2 + p^2 x(1-x) + m^2]^2}$$

$$= -\frac{i\pi^2}{2} p^2 + p^2 \int (dk) \int_0^1 \frac{(2x-1)\, dx}{[k^2 + k^2 x(1-x) + m^2]^2}, \quad \text{(II.31)}$$

where in obtaining the latter equality, we have, in the process, used (II.24). By noting that $p^2(1-2x)\, dx = d(p^2 x(1-x))$, we may carry out the integration over $x$ in (II.31) by parts, and then carry out the resulting integral over $k$ using (II.8). The integral on the right-hand side of (II.31), as the coefficient of $p^2$, then becomes

$$-i\pi^2 \int_0^1 dx \, \frac{x(1-x) p^2 (1-2x)}{[m^2 + p^2 x(1-x)]}$$

$$= -i\pi^2 \int_0^1 dx \, x(1-x) \frac{d}{dx} \ln \left[ 1 + \frac{p^2}{m^2} x(1-x) \right]$$

$$= +i\pi^2 \int_0^1 dx \, (1-2x) \ln \left[ 1 + \frac{p^2}{m^2} x(1-x) \right]. \quad \text{(II.32)}$$

The last integral is zero since the change of variable $x \to 1-x$ gives minus the same integral. That is, the integral on the right-hand side of (II.31) is zero and we obtain the result given in (II.30).

Another example that is worth considering is

$$\int (dk) \left( \frac{k^\mu}{(k-p)^2 + m^2} - \frac{1}{2} \frac{p^\mu k^2}{(k^2 + m^2)^2} \right) = -\frac{i \pi^2}{3} p^2 p^\mu,$$  (II.33)

and we leave it as an exercise for the reader to verify it by both methods given above.

A useful formula that follows from (II.27) is obtained by differentiating the latter with respect to $A$. This gives

$$\frac{1}{A^2 B} = 2 \int_0^1 x \, dx \, \frac{1}{[Ax + B(1-x)]^3}.$$  (II.34)

More generally,

$$\frac{1}{ABC} = 2 \int_0^1 dx \int_0^x dz \, \frac{1}{[A(1-x) + Bz + C(x-z)]^3}.$$  (II.35)

The latter may be generalized for the product of $N$, not necessarily identical, factors, and more conveniently written, as follows

$$\frac{1}{A_1 A_2 \cdots A_N} = (N-1)! \int_0^1 dx_1 \int_0^{x_1} dx_2 \cdots \int_0^{x_{N-2}} dx_{N-1}$$

$$\times \frac{1}{[A_1 x_{N-1} + A_2 (x_{N-2} - x_{N-1}) + \cdots + A_N (1 - x_1)]^N},$$  (II.36)

which, for $N = 3$, coincides with (II.35) upon setting $A_1 = B$, $A_2 = C$, $A_3 = A$, $x_1 = x$, $x_2 = z$.

# Appendix III
# Analytic Continuation in Spacetime Dimension and Dimensional Regularization

Upon differentiation of the expression in (II.3)

$$\frac{1}{(k^2 + M^2 - i\epsilon)} = i \int_0^\infty ds \exp[-is(k^2 + M^2 - i\epsilon)], \quad k^2 = \sum_{i=1}^{n-1} k_i^2 - (k^0)^2.$$

(III.1)

$(\nu - 1)$ number of times, we obtain

$$\frac{1}{(k^2 + M^2 - i\epsilon)^\nu} = \frac{(i)^\nu}{\Gamma(\nu)} \int_0^\infty ds\, s^{\nu-1} \exp[-is(k^2 + M^2 - i\epsilon)].$$

(III.2)

Hence from (II.6)

$$\int \frac{d^n k}{(2\pi)^n} \frac{1}{(k^2 + M^2 - i\epsilon)^\nu} = \frac{(i)^{\nu-\frac{n}{2}+1}}{(4\pi)^{\frac{n}{2}} \Gamma(\nu)} \int_0^\infty ds\, s^{\nu-\frac{n}{2}-1}\, e^{-s(iM^2+\epsilon)}.$$

(III.3)

Upon using the integral

$$\int_0^\infty ds\, s^{\nu-\frac{n}{2}-1}\, e^{-s(iM^2+\epsilon)} = \frac{1}{(iM^2 + \epsilon)^{\nu-\frac{n}{2}}} \Gamma\left(\nu - \frac{n}{2}\right),$$

(III.4)

we obtain

$$\int \frac{d^n k}{(2\pi)^n} \frac{1}{(k^2 + M^2 - i\epsilon)^\nu} = \frac{i}{(4\pi)^{\frac{n}{2}}} \frac{1}{(M^2 - i\epsilon)^{\nu-\frac{n}{2}}} \frac{\Gamma(\nu - \frac{n}{2})}{\Gamma(\nu)}, \quad \nu > \frac{n}{2}.$$

(III.5)

© Springer International Publishing Switzerland 2016
E.B. Manoukian, *Quantum Field Theory I*, Graduate Texts in Physics,
DOI 10.1007/978-3-319-30939-2

From the above integral it easily follows that

$$\int \frac{d^n k}{(2\pi)^n} \frac{k^2}{(k^2 + M^2 - i\epsilon)^\nu} = \frac{i}{(4\pi)^{\frac{n}{2}}} \frac{n}{2} \frac{1}{(M^2 - i\epsilon)^{\nu - \frac{n}{2} - 1}} \frac{\Gamma(\nu - \frac{n}{2} - 1)}{\Gamma(\nu)},$$

(III.6)

valid for $\nu > (n/2) + 1$.

One then defines integrals by an analytic extension of spacetime dimension to $D$ as follows

$$\int \frac{d^D k}{(2\pi)^D} \frac{1}{(k^2 + M^2 - i\epsilon)^\nu} = \frac{i}{(4\pi)^{\frac{D}{2}}} \frac{1}{(M^2 - i\epsilon)^{\nu - \frac{D}{2}}} \frac{\Gamma(\nu - \frac{D}{2})}{\Gamma(\nu)},$$

(III.7)

$$\int \frac{d^D k}{(2\pi)^D} \frac{k^2}{(k^2 + M^2 - i\epsilon)^\nu} = \frac{iD}{(4\pi)^{\frac{D}{2}}} \frac{(M^2 - i\epsilon)}{(M^2 - i\epsilon)^{\nu - \frac{D}{2}}} \frac{\Gamma(\nu - \frac{D}{2})}{(2\nu - D - 2)\Gamma(\nu)}.$$

(III.8)

*Dimensional regularization* consists of choosing

$$D = 4 - \varepsilon, \qquad \varepsilon > 0,$$

(III.9)

with $\varepsilon$ not to be confused with $\epsilon$ earlier. The so-called ultraviolet divergences in quantum field theory, i.e., at high energies, are equivalently expressed in terms of singularities in $\varepsilon$, with the latter *kept* $> 0$.

We may use Feynman parameters representation, as given in (II.27), to write

$$\frac{1}{(k - p)^2 k^2} = \int_0^1 dx \frac{1}{[(k - px)^2 + p^2 x(1 - x)]^2},$$

(III.10)

which from (III.7) leads to

$$\int_0^1 dx \int \frac{d^D k}{(2\pi)^D} \frac{1}{(k - p)^2 k^2} = \frac{i}{(4\pi)^{\frac{D}{2}}} \int_0^1 dx \frac{1}{[p^2 x(1 - x)]^{\varepsilon/2}} \Gamma(\varepsilon/2).$$

(III.11)

Upon using the integral

$$\int_0^1 dx \, x^{\alpha - 1}(1 - x)^{\beta - 1} = \frac{\Gamma(\alpha)\Gamma(\beta)}{\Gamma(\alpha + \beta)}, \quad \alpha > 0, \ \beta > 0,$$

(III.12)

we obtain

$$\int \frac{d^D k}{(2\pi)^D} \frac{1}{(k - p)^2 k^2} = \frac{i}{(4\pi)^{\frac{D}{2}}} \left(\frac{1}{p^2}\right)^{\varepsilon/2} \frac{\Gamma(\frac{\varepsilon}{2}) \Gamma^2(1 - \frac{\varepsilon}{2})}{\Gamma(2 - \varepsilon)}.$$

(III.13)

By shifting the variable $k$: $k \rightarrow k + px$, using the expression on the right-hand side of (III.10), and (III.12), in the process, we also obtain

$$\int \frac{d^D k}{(2\pi)^D} \frac{k^\mu}{(k-p)^2 k^2} = \frac{p^\mu}{2} \frac{i}{(4\pi)^{\frac{D}{2}}} \left(\frac{1}{p^2}\right)^{\varepsilon/2} \frac{\Gamma(\frac{\varepsilon}{2}) \Gamma^2(1-\frac{\varepsilon}{2})}{\Gamma(2-\varepsilon)}. \tag{III.14}$$

Here we have also used the property that the integral of an odd function of $k$ is zero. Another useful integral is obtained again by shifting the variable $k$: $k \rightarrow k + px$, by using in the process (II.8) and noting that in $D$ dimensions $\eta^\mu{}_\mu = D$ to obtain

$$\int \frac{d^D k}{(2\pi)^D} \frac{k^\mu k^\nu}{(k-p)^2 k^2} = \frac{1}{4(3-\varepsilon)} \frac{i}{(4\pi)^{\frac{D}{2}}} \left(\frac{1}{p^2}\right)^{\varepsilon/2} \frac{\Gamma(\frac{\varepsilon}{2}) \Gamma^2(1-\frac{\varepsilon}{2})}{\Gamma(2-\varepsilon)}$$
$$\times [-\eta^{\mu\nu} p^2 + (4-\varepsilon) p^\mu p^\nu]. \tag{III.15}$$

In particular, using the fact that $\eta^\mu{}_\mu = (4-\varepsilon)$, dimensional regularization gives from the above equation the interesting result

$$\int \frac{d^D k}{(2\pi)^D} \frac{1}{(k-p)^2} = 0. \tag{III.16}$$

The following properties of the $\Gamma$-function should be noted

$$\Gamma(z+1) = z\,\Gamma(z), \qquad \Gamma\left(1-\frac{\varepsilon}{2}\right) = 1 + \gamma_E \frac{\varepsilon}{2} + \mathcal{O}(\varepsilon^2),$$

$$\Gamma\left(\frac{\varepsilon}{2}\right) = \frac{2}{\varepsilon} - \gamma_E + \mathcal{O}(\varepsilon), \tag{III.17}$$

where $\gamma_E = 0.5772157\ldots$ is Euler's constant.

Finally, we note that due to dimensional reasons, when the dimension of spacetime is analytically continued to $D$, an integral in four dimensions is then defined by introducing an arbitrary mass parameter $\mu_D$, and by making, in the process, the replacement $(dk) \rightarrow (\mu_D)^\varepsilon d^D k$. Therefore, a four dimensional regularized integral may be defined as

$$\int \frac{(dk)}{(2\pi)^4} F(k,p) \Big|_{\text{Reg}} = (\mu_D)^\varepsilon \int \frac{d^D k}{(2\pi)^D} F(k,p). \tag{III.18}$$

We note that the integrals (III.13), (III.14), (III.15), using the definition in (III.18), all involve the following factor, when expanded in powers of $\varepsilon$,

$$\frac{1}{(4\pi)^{\frac{D}{2}}} \left(\frac{\mu_D^2}{p^2}\right)^{\varepsilon/2} \Gamma\left(\frac{\varepsilon}{2}\right) = \frac{1}{16\pi^2} \left[\frac{2}{\varepsilon} - \gamma_E - \ln(p^2/\mu_D^2) + \ln(4\pi) + \mathcal{O}(\varepsilon)\right], \tag{III.19}$$

up to overall multiplicative finite constant terms coming from the other gamma functions occurring in their factors, for $\varepsilon \to 0$. The parameter $\varepsilon$ has now replaced the ultraviolet cut-off.

It is also worth knowing that the following two integrals obtained by the same method as above, and by power counting, are ultraviolet finite (UV finite):

$$\int \frac{d^D k}{(2\pi)^D} \frac{1}{(p_1 - k)^2 (p_2 - k)^2 k^2} = \text{UV finite}, \qquad (\text{III.20})$$

$$\int \frac{d^D k}{(2\pi)^D} \frac{k^\mu}{(p_1 - k)^2 (p_2 - k)^2 k^2} = \text{UV finite}, \qquad (\text{III.21})$$

while with $Q = p_2 - p_1 \neq 0$,

$$\int \frac{d^D k}{(2\pi)^D} \frac{k^\mu k^\nu}{(p_1 - k)^2 (p_2 - k)^2 k^2} = \frac{i \eta^{\mu\nu}}{(4\pi)^{\frac{D}{2}}} \left(\frac{1}{Q^2}\right)^{\varepsilon/2} \frac{\Gamma(\frac{\varepsilon}{2})\Gamma^2(1 - \frac{\varepsilon}{2})}{\Gamma(3)\Gamma(3 - \varepsilon)} + \text{UV finite.}$$

$$(\text{III.22})$$

In $D = 4 - \varepsilon$ dimensions, the gamma matrices are defined to satisfy the following relations:

**Box III.1:** Some properties of the gamma matrices in $D = 4 - \varepsilon$ dimensions

$$\{\gamma^\mu, \gamma^\nu\} = -2\eta^{\mu\nu}I, \quad \eta^\mu{}_\mu = (4 - \varepsilon), \quad \gamma^\mu \gamma_\mu = -(4 - \varepsilon)I,$$

$$\gamma^\mu \gamma^\sigma \gamma_\mu = (2 - \varepsilon)\gamma^\sigma, \quad \text{Tr}\, I = 4, \quad \text{Tr}\,[\gamma^\mu \gamma^\nu] = -4\eta^{\mu\nu},$$

$$\text{Tr}\,[\gamma^\alpha \gamma^\beta \gamma^\mu \gamma^\nu] = 4(\eta^{\alpha\beta}\eta^{\mu\nu} - \eta^{\alpha\mu}\eta^{\beta\nu} + \eta^{\alpha\nu}\eta^{\beta\mu}),$$

$$\text{Tr}\,[\text{odd number of } \gamma\text{'s}] = 0.$$

# Appendix IV
# Schwinger's Point Splitting Method of Currents: Arbitrary Orders

The vacuum-to-vacuum transition amplitude corresponding to Dirac's equation $[\gamma^\mu(\partial_\mu/i - eA_\mu) + m]\psi = 0$, in the presence of an external electromagnetic field $A_\mu$, is given in (3.6.26) to be given by

$$\langle\, 0_+ \mid 0_-\rangle^{(e)} = \exp i\, W, \tag{IV.1}$$

which is normalized to unity for $e = 0$. Here $(x_\pm = x \pm \epsilon/2)$

$$i\, W = -\int_0^e \mathrm{d}e'\, (\mathrm{d}x)\, \left(\mathrm{Tr}\,[\,S(x_-, x_+; e'A)\gamma A\,]\, e^{\mathrm{i}e'\int_{x-}^{x+} \mathrm{d}\xi^\mu A_\mu(\xi)}\right), \quad \epsilon \to 0, \tag{IV.2}$$

as given in (3.6.27), with a priori dependence on $\epsilon$ arising from Schwinger's point splitting of the current as given in (3.6.24) and spelled out to be

$$\frac{\langle\, 0_+ \mid j^\mu(x) \mid 0_-\rangle^{(e)}}{\langle\, 0_+ \mid 0_-\rangle^{(e)}} = i\,\mathrm{Tr}\,\mathrm{Av}\left([\,S_+(x_-, x_+; eA)\gamma^\mu\,]\exp i\, e\int_{x-}^{x+}\mathrm{d}\xi^\mu A_\mu(\xi)\right), \tag{IV.3}$$

where Av stands for an average over $\epsilon^0 > 0$ and $\epsilon^0 < 0$, with $\epsilon$ in a space-like direction. Here Tr denotes a trace over the spinor indices of gamma matrices. We also note that $S^A(x_-, x_+) \equiv S_+(x_-, x_+; eA)$.

The purpose of this appendix, is to develop a perturbation expansion of (IV.2) to arbitrary orders in $eA_\mu$ in the limit $\epsilon \to 0$. We will see, in particular, that the perturbative expansion, due to the presence of Schwinger's line-integral, is modified from its naïve expression, only up to fourth order in the external electromagnetic field.

We first consider the expression

$$\mathrm{Tr}\,[\,S(x_-, x_+; eA)\gamma^\mu\,]A_\mu(x) \equiv \mathrm{Tr}\,[\gamma^\mu S(x_-, x_+; eA)\,]A_\mu(x), \tag{IV.4}$$

© Springer International Publishing Switzerland 2016
E.B. Manoukian, *Quantum Field Theory I*, Graduate Texts in Physics,
DOI 10.1007/978-3-319-30939-2

multiplying the Schwinger line-integral in (IV.2) for a coupling e. From the integral equation for $S(x_-, x_+ ; eA) \equiv S_+^A(x_-, x_+)$, in (3.2.12), we may carry out an expansion in powers of the external potential $A_\mu$ as

$$\text{Tr}\,[\,S(x_-, x_+ ; eA)\gamma^\mu\,]\,A_\mu(x) = A_{\mu_1}(x)\,\text{Tr}\,[S_+(x_- - x_+)\gamma^{\mu_1}] +$$

$$+ \sum_{N\geq 2} (e)^{N-1} \int (dx_2)\cdots(dx_N)\,A_{\mu_1}(x)A_{\mu_2}(x_2)\cdots A_{\mu_N}(x_N) \times$$

$$\times \text{Tr}\,[\,S_+(x_- - x_N)\gamma^{\mu_N}S_+(x_N - x_{N-1})\gamma^{\mu_{N-1}}\cdots\gamma^{\mu_2}S_+(x_2 - x_+)\gamma^{\mu_1}\,], \qquad \text{(IV.5)}$$

where $S_+(x - x')$ is the fermion propagator in (3.1.9). Equivalently for the expression on the right-hand side of (IV.5), we have

$$\text{Tr}\,[\,\gamma^\mu S(x_-, x_+ ; eA)\,]\,A_\mu(x) = A_{\mu_1}(x)\,\text{Tr}\,[\,\gamma^{\mu_1}S_+(x_- - x_+)] +$$

$$+ \sum_{N\geq 2} (e)^{N-1} \int (dx_2)\cdots(dx_N)\,A_{\mu_1}(x)A_{\mu_2}(x_2)\ldots A_{\mu_N}(x_N) \times$$

$$\times \text{Tr}\,[\,\gamma^{\mu_1}S_+(x_- - x_2)\gamma^{\mu_2}\ldots\gamma^{\mu_{N-1}}S_+(x_{N-1} - x_N)\gamma^{\mu_N}S_+(x_N - x_+)\,]. \qquad \text{(IV.6)}$$

We may rewrite the $\text{Tr}[(\,.\,)]$ term within the multiple integral in (IV.5) as $\text{Tr}\,[\,(\,.\,)^\top\,] \equiv \text{Tr}[\mathscr{C}^{-1}(\,.\,)^\top \mathscr{C}]$, where $\mathscr{C}$ is the charge conjugation matrix in (I.3), satisfying, in particular,

$$\mathscr{C}^{-1}(\gamma^\mu)^\top \mathscr{C} = -\gamma^\mu, \qquad \mathscr{C}^{-1}(S_+(x - x'))^\top \mathscr{C} = S_+(x' - x). \qquad \text{(IV.7)}$$

Accordingly, (IV.5) may be re-expressed as

$$\text{Tr}\,[\,S(x_-, x_+ ; eA)\gamma^\mu\,]\,A_\mu(x) = -A_{\mu_1}(x)\,\text{Tr}\,[\,\gamma^{\mu_1}S_+(x_+ - x_-)] +$$

$$+ \sum_{N\geq 2} (-1)^N (e)^{N-1} \int (dx_2)\cdots(dx_N)\,A_{\mu_1}(x)A_{\mu_2}(x_2)\cdots A_{\mu_N}(x_N) \times$$

$$\times \text{Tr}\,[\,\gamma^{\mu_1}S_+(x_+ - x_2)\,\gamma^{\mu_2}\cdots\gamma^{\mu_{N-1}}S_+(x_{N-1} - x_N)\gamma^{\mu_N}S_+(x_N - x_-)\,]. \qquad \text{(IV.8)}$$

Eqs. (IV.6), (IV.8) allow us to write (IV.5) as the average of the just mentioned two equivalent expressions. That is

$$\text{Tr}\,[\,S(x_-, x_+ ; eA)\gamma^\mu\,]\,A_\mu(x) = \frac{1}{2}A_{\mu_1}(x)\,\text{Tr}\left[\,\gamma^{\mu_1}[\,S_+(x_- - x_+) - S_+(x_+ - x_-)]\,\right]$$

$$+ \sum_{N\geq 2} (e)^{N-1} \int (dx_2)\ldots(dx_N)\,A_{\mu_1}(x)A_{\mu_2}(x_2)\cdots A_{\mu_N}(x_N) \times$$

$$\times \frac{1}{2}\,[\,F^{\mu_1\mu_2\cdots\mu_N}(x_-, x_2, \ldots, x_N, x_+) + (-1)^N\,F_N^{\mu_1\mu_2\cdots\mu_N}(x_+, x_2, \cdots, x_N, x_-)\,],$$

$$\text{(IV.9)}$$

where

$$F^{\mu_1\mu_2\cdots\mu_N}(x, x_2, \ldots, x_N, x') = \text{Tr}\,[\,\gamma^{\mu_1}S_+(x-x_2)\gamma^{\mu_2}S_+(x_2-x_3)\cdots\gamma^{\mu_N}S_+(x_N-x')\,].$$
(IV.10)

Note that $S_+(x_--x_+)$, $S_+(x_+-x_-)$, are *independent* of $x$.

By carrying out Fourier transforms,

$$A_\mu(x) = \int \frac{(dQ)}{(2\pi)^4}\, e^{iQx}A_\mu(Q), \qquad S_+(x-x') = \int \frac{(dp)}{(2\pi)^4}\, e^{ip(x-x')}S_+(p), \quad \text{(IV.11)}$$

(IV.9) becomes

$$\text{Tr}\,[\,S(x_-, x_+; eA)\gamma^{\mu}\,]A_\mu(x)$$

$$= \int \frac{(dp)}{(2\pi)^4}\frac{(dQ_1)}{(2\pi)^4}\, e^{iQ_1x}\,\frac{1}{2}\,[\,e^{-ip\epsilon}e^{iQ_1\epsilon/2} - e^{ip\epsilon}e^{-iQ_1\epsilon/2}\,]A_{\mu_1}(Q_1)\text{Tr}\,[\,\gamma^{\mu_1}S_+(p)\,]\,+$$

$$+ \sum_{N\geq 2} \int \frac{(dp)}{(2\pi)^4}\frac{(dQ_1)}{(2\pi)^4}\cdots\frac{(dQ_N)}{(2\pi)^4}\, e^{i(Q_1+\cdots+Q_N)x}\,\times$$

$$\times\,\frac{1}{2}\,e^{N-1}\,[\,e^{-ip\epsilon}\,e^{i\Sigma' Q_i\epsilon/2} + (-1)^N e^{ip\epsilon}\,e^{-i\Sigma' Q_i\epsilon/2}\,]A_{\mu_1}(Q_1)\cdots A_{\mu_N}(Q_N)\times$$

$$\times\,\text{Tr}\big\lfloor\gamma^{\mu_1}S_+(p-\frac{Q_1}{2})\gamma^{\mu_2}S_+(p-\frac{Q_1}{2}-Q_2)\ldots\gamma^{\mu_N}S_+(p-\frac{Q_1}{2}-Q_2-\ldots-Q_N)\big].$$
(IV.12)

Now we consider the contribution of the gauge compensating factor, i.e., of the Schwinger line-integral in (IV.2), for a given coupling e. Since we will eventually take the limits of $\epsilon$ going to zero, we may carry out an expansion as follows

$$\exp[\,i\,e\int_{x_-}^{x_+}d\xi^\mu A_\mu(\xi)\,] = 1 + I_1 + I_2 + I_3 + \cdots. \qquad \text{(IV.13)}$$

To carry out the $\xi^\mu$-integration, we make a change of variable $\xi^\mu \to \lambda : \xi^\mu = x^\mu + (\epsilon^\mu/2)\lambda$, $-1 \leq \lambda \leq +1$, make a Fourier transform of $A_\mu(x)$, and expand in powers of $\epsilon$ to obtain for $\epsilon \simeq 0$

$$I_1 = 2\left(\frac{ie}{2}\right)\int\frac{(dQ_1)}{(2\pi)^4}\,e^{iQ_1x}A_{\mu_1}(Q_1)\Big[1 - \frac{(Q_1\epsilon)^2}{24} + \cdots\Big]\epsilon^{\mu_1}, \qquad \text{(IV.14)}$$

$$I_2 = 4\left(\frac{ie}{2}\right)^2\frac{1}{2!}\int\frac{(dQ_1)}{(2\pi)^4}\frac{(dQ_2)}{(2\pi)^4}\,e^{i(Q_1+Q_2)x}A_{\mu_1}(Q_1)A_{\mu_2}(Q_2)[1 + \cdots]\epsilon^{\mu_1}\epsilon^{\mu_2}, \qquad \text{(IV.15)}$$

$$I_3 = 8\left(\frac{ie}{2}\right)^3\frac{1}{3!}\int\frac{(dQ_1)}{(2\pi)^4}\frac{(dQ_2)}{(2\pi)^4}\frac{(dQ_3)}{(2\pi)^4}\,e^{i(Q_1+Q_2+Q_3)x}A_{\mu_1}(Q_1)A_{\mu_2}(Q_2)\times$$

$$\times A_{\mu_3}(Q_3)[1 + \cdots]\epsilon^{\mu_1}\epsilon^{\mu_2}\epsilon^{\mu_3}. \qquad \text{(IV.16)}$$

We will show that all the terms represented by ... in (IV.14), (IV.15), (IV.16), as well as $I_4$, $I_5$, ... in (IV.13), will not contribute in the limiting procedure.

We multiply (IV.12) by the expression $[1 + I_1 + I_2 + I_3 + \ldots]$ in (IV.13), extract the $(N-1)$th order terms in e of the product, integrate over e as indicated in (IV.2), and integrate over $x$ as well. Here we note the basic property of the factor $[e^{-ip\epsilon}e^{i\sum' Q_i\epsilon/2} + (-1)^N e^{ip\epsilon}e^{-i\sum' Q_i\epsilon/2}]/2$ in (IV.12), in reference to (IV.13), (IV.14), (IV.15), (IV.16):

$$\epsilon^{\mu_1}\ldots\epsilon^{\mu_k}\frac{1}{2}\Big[e^{-ip\epsilon}e^{i\sum' Q_i\epsilon/2} + (-1)^{N-k}e^{ip\epsilon}e^{-i\sum' Q_i\epsilon/2}\Big]$$

$$= \Big(i\frac{\partial}{\partial p_{\mu_1}}\Big)\ldots\Big(i\frac{\partial}{\partial p_{\mu_k}}\Big)\frac{1}{2}\Big[e^{-ip\epsilon}e^{i\sum' Q_i\epsilon/2} + (-1)^N e^{ip\epsilon}e^{-i\sum' Q_i\epsilon/2}\Big]. \qquad \text{(IV.17)}$$

We may then integrate by parts over $p$ to apply these derivatives with respect to the $p_{\mu_j}$, replacing the $\epsilon^{\mu_j}$ in (IV.14), (IV.15), (IV.16), as indicated above, to the Tr[ . ] factor in (IV.12), and take the limit $\epsilon \to 0$. The process is straightforward, and we obtain for the $N$th- order term in e for $iW$ in (IV.2)

$$iW\Big|_{(N)} = -\frac{(e)^N}{N}\int(dx)\frac{(dQ_1)}{(2\pi)^4}\cdots\frac{(dQ_N)}{(2\pi)^4}e^{i(Q_1 + \cdots + Q_N)x}$$

$$\times A_{\mu_1}(Q_1)\cdots A_{\mu_N}(Q_N)L^{\mu_1\cdots\mu_N}(Q_1,\ldots,Q_N), \qquad \text{(IV.18)}$$

where

$$L^{\mu_1\cdots\mu_N}(Q_1,\ldots,Q_N) = \frac{1}{2}\Big[1 + (-1)^N\Big]\int\frac{(dp)}{(2\pi)^4}\Pi^{\mu_1\cdots\mu_N}(p\,;Q_1,\ldots,Q_N), \qquad \text{(IV.19)}$$

and we immediately note that only terms even in $N$, i.e., for $N = 2, 4, \ldots$, contribute. This is known as Furry's Theorem . It is a consequence of charge conjugation invariance of electrodynamics (if $A_\mu$ is also made to transform to $-A_\mu$ consistently with Maxwell's equations).

Before spelling out the explicit expressions for $\Pi^{\mu_1\cdots\mu_N}(p\,;Q_1,\ldots,Q_N)$, for the various $N$, we first recall Gauss' Theorem in 4D conveniently written in the way it presents itself in the above formulae for a function $f(p)$ involving a product of Dirac propagators. It reads

$$\int(dp)\frac{\partial}{\partial p_\mu}f(p) = \int d\sigma^\mu f(p), \qquad \text{(IV.20)}$$

with the right-hand side defined as a surface integral, where $d\sigma^\mu$ is an element of surface. With $p^0 \to ip^0, d\sigma^\mu = n^\mu d\sigma$, $n^\mu$ a unit vector along $p^\mu$, if $d\sigma$ scaled by a parameter $\lambda$, it grows like $\lambda^3$. Accordingly, if $f(\lambda p)$ vanishes faster than $\lambda^{-3}$, then the integral on the right-hand side is zero.

According to the rule in (IV.17), the gauge compensating terms $I_n$ in (IV.13), each of them, lead to $n$ or more derivatives $\partial/\partial p_\mu$, corresponding to the $\epsilon^\mu$ appearing in the $I_n$,[3] and *acting* on the Tr$[\,.\,]$ factor in (IV.12). That is, a gauge compensating term $I_n$ leads explicitly to integrals of the form

$$\int (\mathrm{d}p) \frac{\partial}{\partial p_{\nu_1}} \left( \frac{\partial}{\partial p_{\nu_2}} \cdots \frac{\partial}{\partial p_{\nu_n}} \right)$$

$$\times \mathrm{Tr}\Big[ \gamma^{\mu_1} S_+(p - \frac{Q_1}{2}) \gamma^{\mu_2} S_+(p - \frac{Q_1}{2} - Q_2) \ldots \gamma^{\mu_k} S_+(p - \frac{Q_1}{2} - Q_2 - \cdots - Q_k) \Big],$$

$$(IV.21)$$

and may involve even more derivatives with respect to $p$. By Gauss' Theorem, such integrals would vanish for $n - 1 + k \geq 4$, since $S(\lambda p) = \mathcal{O}(\lambda^{-1})$. Accordingly, for $k = 1$, only, $I_1, I_2, I_3$ may contribute, and for $k = 2$, only $I_1, I_2$ may contribute, and finally for $k = 3$ only $I_1$ may contribute. Hence we may infer that only $\Pi^{\mu_1\mu_2}$, and $\Pi^{\mu_1\mu_2\mu_3\mu_4}$ would involve an addition to the expressions in (IV.2) coming from gauge compensating terms. Needless to say $\Pi^{\mu_1\mu_2\mu_3}$ is zero, corresponding to $N = 3$.

Accordingly for $N = 6, 8, \ldots,$

$$\Pi^{\mu_1 \ldots \mu_N}(p; Q_1, \ldots, Q_N)$$

$$= \mathrm{Tr}\Big[ \gamma^{\mu_1} S_+(p - \frac{Q_1}{2}) \gamma^{\mu_2} S_+(p - \frac{Q_1}{2} - Q_2) \ldots \gamma^{\mu_N} S_+(p - \frac{Q_1}{2} - Q_2 - \cdots - Q_N) \Big]_{\mathrm{sym}},$$

$$(IV.22)$$

where $[\,.\,]_{\mathrm{sym}}$ means to symmetrize over the $(N - 1)$ pairs of "generalized indices" $(\mu_2, Q_2), \ldots, (\mu_N, Q_N)$.

For $N = 4$, we have

$$\Pi^{\mu_1\mu_2\mu_3\mu_4}(p; Q_1, \ldots, Q_4) =$$

$$\mathrm{Tr}\Big[ \gamma^{\mu_1} S_+(p - \frac{Q_1}{2}) \gamma^{\mu_2} S_+(p - \frac{Q_1}{2} - Q_2) \ldots \gamma^{\mu_4} S_+(p - \frac{Q_1}{2} - Q_2 - Q_3 - Q_4) \Big]$$

$$+ \frac{\partial}{\partial p_{\mu_4}} \mathrm{Tr}\Big[ \gamma^{\mu_1} S_+(p - \frac{Q_1}{2}) \gamma^{\mu_2} S_+(p - \frac{Q_1}{2} - Q_2) \gamma^{\mu_3} S_+(p - \frac{Q_1}{2} - Q_2 - Q_3) \Big]$$

$$+ \frac{1}{2!} \frac{\partial}{\partial p_{\mu_3}} \frac{\partial}{\partial p_{\mu_4}} \mathrm{Tr}\Big[ \gamma^{\mu_1} S_+(p - \frac{Q_1}{2}) \gamma^{\mu_2} S_+(p - \frac{Q_1}{2} - Q_2) \Big]$$

$$+ \frac{1}{3!} \frac{\partial}{\partial p_{\mu_2}} \frac{\partial}{\partial p_{\mu_3}} \frac{\partial}{\partial p_{\mu_4}} \mathrm{Tr}\Big[ \gamma^{\mu_1} S_+(p) \Big]. \qquad (IV.23)$$

---

[3] See, e.g., (IV.14), (IV.15), (IV.16).

A symmetrization is necessary. However much simplification of this expression may be done first. To this end, note that for any given $q$, we may write

$$\frac{1}{\gamma(p-Q)+m} = \frac{1}{\gamma p+m} + \frac{1}{\gamma(p-Q)+m}\,\gamma Q\,\frac{1}{\gamma p+m}. \tag{IV.24}$$

Clearly, the second term introduces one more power of the momentum $p$ in the denominator, and hence invoking Gauss' Theorem, we may infer that we may set the $Q$ momenta in the second and third terms on the right-hand side of (IV.23) equal to zero. On the other hand symmetrizing the second term on the right-hand side of (IV.23) over $\mu_2$, $\mu_3$, after setting the $Q$ momenta equal to zero, it takes the simple form

$$\frac{1}{2}\frac{\partial}{\partial p_{\mu_4}}\mathrm{Tr}\big[\gamma^{\mu_1}S_+(p)\gamma^{\mu_2}S_+(p)\gamma^{\mu_3}S_+(p)\big]+\frac{1}{2}\frac{\partial}{\partial p_{\mu_4}}\mathrm{Tr}\big[\gamma^{\mu_1}S_+(p)\gamma^{\mu_3}S_+(p)\gamma^{\mu_2}S_+(p)\big]$$

$$\equiv -\frac{1}{2}\frac{\partial}{\partial p_{\mu_3}}\frac{\partial}{\partial p_{\mu_4}}\mathrm{Tr}\big[\gamma^{\mu_1}S_+(p)\gamma^{\mu_2}S_+(p)\big], \tag{IV.25}$$

which exactly cancels the third term on the right-hand side of (IV.23) (after setting the $Q$'s equal to zero). Hence (IV.23) finally reduces to

$$\Pi^{\mu_1\mu_2\mu_3\mu_4}(p;Q_1,\ldots,Q_4)$$

$$= \mathrm{Tr}\big[\gamma^{\mu_1}S_+(p-\tfrac{Q_1}{2})\gamma^{\mu_2}S_+(p-\tfrac{Q_1}{2}-Q_2)\ldots\gamma^{\mu_4}S_+(p-\tfrac{Q_1}{2}-Q_2-Q_3-Q_4)\big]_{\mathrm{sym}}$$

$$+\frac{1}{3}\frac{\partial}{\partial p_{\mu_2}}\frac{\partial}{\partial p_{\mu_3}}\frac{\partial}{\partial p_{\mu_4}}\mathrm{Tr}\big[\gamma^{\mu_1}S_+(p)\big]. \tag{IV.26}$$

For $N=2$, we simply have ($Q_2 = -Q_1$)

$$\Pi^{\mu_1\mu_2}(p;Q_1,-Q_1) = \mathrm{Tr}\big[\gamma^{\mu_1}S_+(p-\tfrac{Q_1}{2})\gamma^{\mu_2}S_+(p+\tfrac{Q_1}{2})\big]$$

$$+Q_1^{\nu_1}Q_1^{\nu_2}\frac{1}{24}\frac{\partial}{\partial p^{\nu_1}}\frac{\partial}{\partial p^{\nu_2}}\frac{\partial}{\partial p_{\mu_2}}\mathrm{Tr}\big[\gamma^{\mu_1}S_+(p)\big]+\frac{\partial}{\partial p_{\mu_2}}\mathrm{Tr}\big[\gamma^{\mu_1}S_+(p)\big]. \tag{IV.27}$$

Thus we have obtained all the terms in (IV.18), (IV.19), for the $N$th order contribution to $iW$ in (IV.2). After integrating (IV.18) over $x$, the following expression emerges for $iW$

$$iW = -\sum_{N=2,4,\ldots}\frac{(e)^N}{N}\int\frac{(dQ_1)}{(2\pi)^4}\ldots\frac{(dQ_N)}{(2\pi)^4}(2\pi)^4\delta^4(Q_1+\cdots+Q_N)$$

$$\times A_{\mu_1}(Q_1)\ldots A_{\mu_N}(Q_N)\int\frac{(dp)}{(2\pi)^4}\,\Pi^{\mu_1\ldots\mu_N}(p;Q_1,\ldots,Q_N), \tag{IV.28}$$

where $\Pi^{\mu_1 \cdots \mu_N}(p\,;Q_1,\ldots,Q_N)$ is given in (IV.22) for $N = 6,8,\ldots,$ and in (IV.26) for $N = 4$, and in (IV.27) for $N = 2$. The expression in (IV.27) for $N = 2$ was possibly first derived in details by K. Johnson[4], and by Schwinger himself.

---

[4]See, e.g., K. Johnson (1965). Brandeis University Summer Institute in Theoretical Physics. In *Lectures on particles and fields*. Englewood Cliffs: Prentice Hall.

# Appendix V
# Renormalization and the Underlying Subtractions

In this appendix, we learn how to define the renormalized Feynman integral, associated with a graph, given its unrenormalized Feynman integrand.

One defines a graph by specifying a set of vertices $\mathcal{V} = (v_1, \ldots, v_r)$, a set of lines $\mathcal{L} = (\ell_1, \ldots, \ell_s)$, together with a rule describing which of these lines join which of the vertices. By removing all the external lines of a graph, one generates an amputated graph. Vertices $\{v_i\}$ in an amputated graph, to which external lines were attached are referred to as external vertices, while their remaining vertices are referred to as internal vertices. From now on we will consider only amputated graphs and amputated subdiagrams with the latter to be defined below. By specifying a subset $\mathcal{V}' \subset \mathcal{V}$ of the vertices and all the lines in $\mathcal{L}$ of a graph $G$ that join these vertices, but not those lines joining vertices not in $\mathcal{V}'$, one defines a subgraph $G'$ of $G$. By specifying a subset $\mathcal{V}' \subset \mathcal{V}$ of the vertices and not necessarily all the lines in $\mathcal{L}$ of the original graph that join these vertices, one defines a subdiagram $g$ of $G$. Examples of an (amputated) graph, together with a subgraph and a subdiagram are given in Fig. V.1.

A vertex of an amputated subdiagram at which an external line was attached previous to the removal of the latter is referred to as an external vertex, while the other vertices are referred to as internal vertices. All the lines of an amputated subdiagram are, by definition, internal lines.

A subdiagram is called disconnected if it is connected out of two or more subdiagrams in which any two of them have no vertices and no lines in common. If upon cutting or removing an internal line of a subdiagram, the number of the connected parts of the latter is increased, it is called an improper line. A subdiagram is called a proper connected subdiagram, if it is connected and has no improper lines. A subdiagram is called proper but not connected, if it is constructed out of two or more subdiagrams none of which have improper lines. Two subdiagrams $g_1$, $g_2$ specified by the pairs $(\mathcal{V}_1, \mathcal{L}_1)$, $(\mathcal{V}_2, \mathcal{L}_2)$, for which $\mathcal{V}_1 \subset \mathcal{V}_2$, and $\mathcal{L}_1 \subset \mathcal{L}_2$, we write $g_1 \subset g_2$. In this work, the symbol $\subset$ may include equality, while $\subsetneq$ is used to exclude an equality.

© Springer International Publishing Switzerland 2016
E.B. Manoukian, *Quantum Field Theory I*, Graduate Texts in Physics,
DOI 10.1007/978-3-319-30939-2

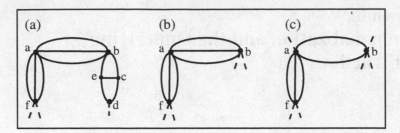

**Fig. V.1** A graph is introduced in part (**a**). The *dashed lines* stand for external lines that have been removed defining an amputated graph. Vertices at $f$ and $d$ represent external vertices for the graph, while the remaining ones represent internal vertices. A subgraph is introduced in part (**b**) with four external lines that have been removed. In this case $f$ and $b$ represent external vertices of the subgraph, while $a$ represents an internal vertex. A subdiagram is introduced in part (**c**) where one of the lines joining the vertices at $a$ and $b$ has been cut with the resulting two external lines removed. The vertices at $a$, $b$ and $f$ represent external vertices for the subdiagram in part (**c**)

Since the subtractions of renormalization involve a sequence of Taylor expansions in the momenta associated with the external lines of a proper graph and in the momenta of the external momenta of proper subdiagrams, we have to spell out how such momenta are defined for the proper graph and the proper subdiagrams. This is considered next.

Suppose that a proper and connected graph $G$ has $m$ external vertices with total external momenta $\{q_1, \ldots, q_m\} \equiv q^G$, $q_r = (q_r^0, q_r^1, q_r^2, q_r^3)$. The integration variables associated with the graph are denoted by $\{k_1^0, k_1^1, \ldots, k_n^3\} \equiv k^G$. A line labeled $\ell$, joining a vertex $\upsilon_i$ to a vertex $\upsilon_j$, carries a momentum denoted by $Q_{ij\ell}$ which may be written as

$$Q_{ij\ell} = k_{ij\ell} + q_{ij\ell}, \quad k_{ij\ell} = \sum_{s=1}^{n} a_{ij\ell}^s k_s, \quad q_{ij\ell} = \sum_{r=1}^{m} b_{ij\ell}^r q_r, \quad Q_{ij\ell} = -Q_{ji\ell}.$$

$$(V.1)$$

At each external vertex $\upsilon_j$ of $G$,

$$\sum_{i\ell}{}^{G} Q_{ij\ell} = q_j^G, \quad \text{and by momentum conservation} \quad \sum_{j} q_j^G = 0, \qquad (V.2)$$

where the sum is over all $i$ corresponding to all the vertices with lines $\ell$ joining them to the external vertex $\upsilon_j$. For an internal vertex, the right-hand side of the above first equality is set equal to zero.

A line joining a vertex $\upsilon_i$ to a vertex $\upsilon_j$ in $G$, will be represented by

$$\Delta_{+ij\ell}(Q_{ij\ell}, \mu_{ij\ell}) = f_{ij\ell}(Q_{ij\ell}, \mu_{ij\ell})[Q_{ij\ell}^2 + \mu_{ij\ell}^2]^{-1}, \qquad (V.3)$$

involving a familiar multiplicative factor $f_{ij\ell}(Q_{ij\ell}, \mu_{ij\ell})$ when one is dealing with particles of non-zero spin. Let $j$ be fixed, and consider the set $\{\upsilon_{i(j)}\}_{1 \leq i \leq r_j}$ of vertices attached by lines to the vertex $v_j$ in $G$, and consider the set $\mathscr{L}^G(v_j)$ of all lines joining the vertices $v_{i(j)}$ to the vertex $v_j$. Moreover, let $\{Q_{ij\ell}\}_{1\leq\ell\leq s_{ij}}$ be the set of momenta carried by these lines. Then we assign to the veretx $v_j$ a polynomial $\mathscr{P}_j = \mathscr{P}_j(Q_{ij1},\ldots,Q_{njs_{nj}})$. The unrenormalized Feynman integrand associated with the proper and connected graph $G$, up to an overall multiplicative constants (involving couplings, etc.), is of the form

$$I_G = \prod_{ij\ell,i<j}^{G} \mathscr{P}_j \, \Delta_{+ij\ell}. \tag{V.4}$$

We introduce canonical variables,[4] by choosing

$$\sum_{i\ell}^{G} q_{ij\ell} = q_j^G, \quad \text{and} \quad q_{ij\ell} = u_i - u_j, \tag{V.5}$$

where $u_i$, $u_j$ are four vectors, and the second equality above means that the external variables of the lines joining a vertex $v_i$ to the vertex $v_j$ are all chosen to be equal. In particular, we also note that (V.2), (V.5) imply that

$$\sum_{i\ell}^{G} k_{ij\ell} = 0, \qquad \sum_{ij\ell}^{G} k_{ij\ell} = 0, \tag{V.6}$$

with the latter defining a constraint. Equation (V.5) provides $(\#\mathscr{V} - 1)$ independent solutions of the $(\#\mathscr{V} - 1)$ independent differences $(u_j - u_i)$, where $\#\mathscr{V}$ denotes the number of vertices. We may write the variables $q_{ij\ell}$ as a linear combination of the elements in $\{q_j^G\} \equiv q^G$: $q_{ij\ell} = q_{ij\ell}(q^G)$.

If $4n$ denotes the number of independent integration variables associated with $G$ then $n = \#\mathscr{L}^G - \#\mathscr{V}^G + 1$. Eqs. (V.6) imply that only $n$ of the $\#\mathscr{L}^G k_{ij\ell}$ are independent, and we may write $k_{ij\ell} = k_{ij\ell}(k)$. We provide examples of canonical decompositions of some graphs and some diagrams obtained from them.

*Example V.1* Consider the graph in Fig. V.2a, with external vertices at 1 and 2. At the vertex $v_1$

$$q = q_{211} + q_{212} + q_{213}, \quad \text{or} \quad u_2 - u_1 = \frac{1}{3}q, \quad \text{i.e.} \quad q_{21\ell} = \frac{1}{3}q, \quad \ell = 1, 2, 3. \tag{V.7}$$

At vertex $v_1$, we also have: $k_{211} + k_{212} + k_{213} = 0$. Let $k_{211} = k_1$, $k_{212} = k_2$, then $k_{213} = -k_1 - k_2$. Therefore a canonical choice of variables are

$$Q_{211} = k_1 + \frac{1}{3}q, \quad Q_{212} = k_2 + \frac{1}{3}q, \quad Q_{213} = -k_1 - k_2 + \frac{1}{3}q. \quad \Diamond \tag{V.8}$$

---

[4]Canonical variables were introduced by Zimmermann [9].

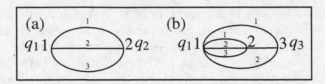

**Fig. V.2** (a) A graph with external vertices at 1 and 2, and external momenta $q_1 \equiv q$, $q_2 \equiv -q$. (b) A graph with external vertices at 1 and 3 and an internal vertex at 2

*Example V.2* Consider the graph in Fig. V.2b with $q_1 \equiv q$, $q_3 \equiv -q$. At the vertices, $\upsilon_1$, $\upsilon_2$, $\upsilon_3$, we have, respectively, with $\upsilon_2$ an internal vertex

$$q = q_{211} + q_{212} + q_{213} + q_{311} + q_{312} = 3(u_2 - u_1) + 2(u_3 - u_1), \qquad (V.9)$$

$$0 = q_{121} + q_{122} + q_{123} + q_{32} = 3(u_1 - u_2) + (u_3 - u_2), \qquad (V.10)$$

$$-q = q_{131} + q_{132} + q_{23} = 2(u_1 - u_3) + (u_2 - u_3). \qquad (V.11)$$

The unique solutions of the differences are then: $u_1 - u_2 = -q/11$, $u_1 - u_3 = -4q/11$, $u_2 - u_3 = -3q/11$. That is, $q_{12\ell} = -q/11$, $\ell = 1, 2, 3$; $q_{13\ell} = -4q/11$, $\ell = 1, 2$; $q_{23} = -3q/11$. For 16 integration variables, we may write for the four vectors: $k_{121} = k_1$, $k_{122} = k_2$, $k_{123} = k_3$, $k_{131} = k_4$. For the remaining integration variables, we note that at $\upsilon_2$, and $\upsilon_3$, respectively, we have

$$k_{121} + k_{122} + k_{123} + k_{32} = 0, \qquad k_{131} + k_{132} + k_{23} = 0, \qquad (V.12)$$

from which $k_{32} = -k_1 - k_2 - k_3$, and $k_{132} = -k_1 - k_2 - k_3 - k_4$. Accordingly a canonical set of variables are

$$Q_{121} = k_1 - \frac{1}{11}q, \; Q_{122} = k_2 - \frac{1}{11}q, \; Q_{123} = k_3 - \frac{1}{11}q \; Q_{131} = k_4 - \frac{4}{11}q, \qquad (V.13)$$

$$Q_{132} = -k_1 - k_2 - k_3 - k_4 - \frac{4}{11}q, \; Q_{32} = -k_1 - k_2 - k_3 + \frac{3}{11}q. \; \diamond \qquad (V.14)$$

*Example V.3* Consider the graph $G$ in Fig. V.3 below, with $q_1 = -q$, $q_4 = q$. Let $k_{12} = 3(k_1 + 3k_2)/8$, $k_{34} = -3(3k_1 + k_2)/8$. Following the procedure in the above two examples gives rise to the following canonical decompositions:

$$Q_{12} = \frac{3}{8}k_1 + \frac{9}{8}k_2 + \frac{1}{2}q, \quad Q_{13} = -\frac{3}{8}k_1 - \frac{9}{8}k_2 + \frac{1}{2}q, \quad Q_{32} = \frac{3}{4}k_1 - \frac{3}{4}k_2, \qquad (V.15)$$

$$Q_{34} = -\frac{9}{8}k_1 - \frac{3}{8}k_2 + \frac{1}{2}q, \quad Q_{24} = \frac{9}{8}k_1 + \frac{3}{8}k_2 + \frac{1}{2}q. \; \diamond \qquad (V.16)$$

Canonical decomposition corresponding to a proper and connected subdiagram $g$ of $G$, will be defined in a similar manner. We write the momentum associated with a line in $g$ as

$$Q_{ij\ell} = k^G_{ij\,\ell} + q^G_{ij\,\ell} = k^g_{ij\,\ell} + q^g_{ij\,\ell}, \tag{V.17}$$

and at each external vertex $v_j$ of $g$,

$$\sum_{i\ell}{}^g q^g_{ij\ell}(k^G, q^G) = q^g_j(k^G, q^G), \quad \sum_j{}^g q^g_j(k^G, q^G) = 0 \quad \text{and} \quad q^g_{ij\ell} = w_i - w_j, \tag{V.18}$$

where $w_i$, $w_j$ are four vectors, and the sum over $i\ell$ pertains to the subdiagram $g$. If $v_j$ is an internal vertex of $g$ then we set $q^g_j = 0$. Once the $q^g_{ij\ell}$ are determined, then $k^g_{ij\ell} = Q_{ij\ell} - q^g_{ij\ell}$. We show that the $k^g_{ij\ell}$ are linear combination of the elements in $k^G$ only and are independent of the elements in $q^G$. To see this, set the integration variables in $k^G$, equal to zero, then

$$\sum_{i\ell}{}^g q^g_{ij\ell}(0, q^G) = q^g_j(0, q^G), \qquad \sum_j^g \left( \sum_{i\ell}^g q^g_{ij\ell}(0, q^G) \right) = 0,$$

$$q^g_{ij\ell}(0, q^G) = (w_i - w_j)\big|_{k^G - 0} = w_i' - w_j'. \tag{V.19}$$

But for $k^G = 0$, $Q_{ij\ell} = q^G_{ij\ell} = k^g_{ij\ell}(0, q^G) + q^g_{ij\ell}(0, q^G)$ and

$$\sum_{i\ell}{}^g q^G_{ij\ell} = \sum_{i\ell}{}^g Q_{ij\ell} = q^g_j(0, q^G), \quad \left( \sum_{i\ell}{}^g q^G_{ij\ell} \right) = 0, \quad q^G_{ij\ell} = u_i' - u_j', \tag{V.20}$$

for $i, j, \ell$ pertaining to the subdiagram $g$. Uniqueness of the solutions of the $(\#\mathscr{V}^g - 1)$ conditions of the $(\#\mathscr{V}^g - 1)$ independent differences $u_i' - u_j'$ or $w_i' - w_j'$, implies that $u_i' - u_j' = w_i' - w_j'$. That is $k^g_{ij\ell}(0, q^G) = 0$ and hence $k^g_{ij\ell} = k^g_{ij\ell}(k^G)$.

Similarly for two proper connected subdiagram $g' \subsetneqq g$, $k^{g'}_{ij\ell} + q^{g'}_{ij\ell} = k^g_{ij\ell} + q^g_{ij\ell}$, and $k^{g'}_{ij\ell} = k^{g'}_{ij\ell}(k^g)$, $q^{g'}_{ij\ell} = q^{g'}_{ij\ell}(k^g, q^g)$. In particular, the $q^{g'}_j$ are linear combinations of the $k^g_{ij\ell}$ in $g/g'$. Here we note that once the unrenormalized Feynman integrands $I_g$ and $I_{g'}$ have been defined, $I_{g/g'}$ is defined through $I_g = I_{g/g'} I_{g'}$. Note that

$$\sum_{i\ell}{}^{g'} q^G_{ij\ell} = q^{g'}(k^g, k^g), \quad \text{at each vertex } v_j \text{ of } g', \tag{V.21}$$

with the sum over $i\ell$ pertaining to the subdiagram $g'$. Also

$$q_j^g = \sum_{i\ell}{}^g q_{ij\ell}^G = \sum_{i\ell}{}^{g'} Q_{ij\ell} + \sum_{i\ell}{}^{g/g'} Q_{ij\ell}, \qquad (V.22)$$

and hence the above two equations imply that

$$q_j^{g'}(k^g, q^g) = q_j^g - \sum_{i\ell}{}^{g/g'}\left(k_{ij\ell}^g + q_{ij\ell}^g\right), \qquad (V.23)$$

which establishes the stated result.

*Example V.4* Consider the graph $G$ and the two subdiagrams $g_1$ and $g_2$ in Fig. V.3. From (V.15), (V.16), (V.17), at the vertices 2 and of 3 of subdiagram $g_1$, respectively,

$$q_2^{g_1} = q_{42}^{g_1} + q_{32}^{g_1} = Q_{42} + Q_{32} = w_4 - w_2 + w_3 - w_2 = -\frac{3}{8}k_1 - \frac{9}{8}k_2 - \frac{1}{2}q, \qquad (V.24)$$

$$q_3^{g_1} = q_{43}^{g_1} + q_{23}^{g_1} = Q_{43} + Q_{23} = w_4 - w_3 + w_2 - w_3 = +\frac{3}{8}k_1 + \frac{9}{8}k_2 - \frac{1}{2}q. \qquad (V.25)$$

These lead to $(q_4^{g_1} = q)$

$$q_{34}^{g_1} = -\frac{1}{8}k_1 - \frac{3}{8}k_2 + \frac{1}{2}q, \quad q_{24}^{g_1} = \frac{1}{8}k_1 + \frac{3}{8}k_2 + \frac{1}{2}q, \quad q_{32}^{g_1} = -\frac{1}{4}k_1 - \frac{3}{4}k_2. \qquad (V.26)$$

For the internal variables we then have

$$k_{34}^{g_1} = Q_{34} - q_{34}^{g_1} = -k_1, \quad k_{24}^{g_1} = Q_{24} - q_{24}^{g_1} = k_1, \quad k_{32}^{g_1} = Q_{32} - q_{32}^{g_1} = k_1. \qquad (V.27)$$

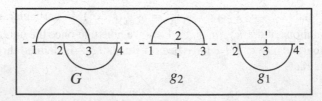

**Fig. V.3** A graph $G$ with external vertices at 1 and 4, showing only two of its proper subdiagrams $g_1$ and $g_2$ having each three external vertices. The *dashed lines* are just to remind us of the *external lines* that have been removed.

Similarly for the subdiagram $g_2$ we have $\left(q_1^{g_2} = -q,\ q_2^{g_2} = (9/8)\,k_1 + (3/8)\cdot k_2 + (q/2),\ q_3^{g_2} = -(9/8)\,k_1 - (3/8)\,k_2 + (q/2)\right)$

$$q_{32}^{g_2} = \frac{3}{4}\,k_1 + \frac{1}{4}\,k_2,\quad q_{13}^{g_2} = -\frac{3}{8}\,k_1 - \frac{1}{8}\,k_2 + \frac{1}{2}\,q,\quad q_{12}^{g_2} = \frac{3}{8}\,k_1 + \frac{1}{8}\,k_2 + \frac{1}{2}\,q,$$
(V.28)

$$k_{32}^{g_2} = -k_2,\qquad k_{13}^{g_2} = -k_2,\qquad k_{12}^{g_2} = k_2. \qquad \diamond \qquad \text{(V.29)}$$

Once the external momentum variables $q_{ij\ell}^{G}$ of the graph $G$ in question, as well as the external variables $q_{ij\ell}^{g}$ of all proper and connected subdiagrams of $G$ have been defined, one may define Taylor operations $T_G$, $T_g$ in their external momenta about the origin up to their (superficial) degree of divergences. The (superficial) degree of divergence $d(g)$ of a proper connected subdiagram $g$ involving $4\,t$ integration variables is defined by

$$d(g) = \deg I_g + 4\,t. \tag{V.30}$$

where "deg" is here defined by scaling the integration variables in $I_g$ by a parameter $\lambda$ and by considering the power of $\lambda$ for $\lambda \to \infty$. For a proper not necessarily connected subdiagram $g$ involving proper and connected subdiagrams $g_1, g_2, \ldots, g_s$, the (superficial) degree of divergence of $g$ is defined by $d(g) = \sum_{i=1}^{s} d(g_i)$.

We are particularly interested in the consecutive application of two or more Taylor operations. To this end, consider two proper subdiagrams $g' \subsetneqq g$, we define $T_g\, T_{g'} I_g$ by the following procedure. We first note that

$$T_{g'}\, I_{g'} = F(Q, k^{g'}, q^{g'}), \tag{V.31}$$

where the function $I_{g/g'} F(Q, k^{g'}, q^{g'})$ should be now expressed in terms the internal $k^g$, and external $q^g$ variables of $g$. To do this, we use (V.17), and the functional relations $k^{g'} = k^{g'}(k^g)$, $q^{g'} = q^{g'}(k^g, q^g)$. The application of consecutive several Taylor operation $T_{g'_s} \ldots T_{g'_1} I_{g'_s}$ for proper subdiagrams $g'_1 \subsetneqq \ldots \subsetneqq g'_s$ is similarly handled.

Now we are ready to define the renormalized $R$ Feynman integrand involving Taylor operations, about the origin, applied to the external variables of a proper and connected graph $G$, and to its proper, not necessarily connected subdiagrams. The external variables of such subdiagrams constitute the totality of the external variables of its connected parts.

The renormalized Feynman integrand $R$, involving subtractions, is defined as follows:

$$R = \left[1 + \sum_{D} \prod_{g \in D} (-T_g)\right] I_G, \tag{V.32}$$

where the sum is over *all* non-empty sets $D$ such that

(i) If $g \in D$, then $g$ is a proper, but not necessarily connected, subdiagram of $G$ with $d(g) \geq 0$. If $d(G) \geq 0$, then one of the elements of $D$ may be $G$ itself.

(ii) If $g_1$, $g_2 \in D$, then either $g_1 \subsetneq g_2$ or $g_2 \subsetneq g_1$. If $g_1 \subsetneq g_2$, then the ordering of the Taylor operations in (V.32) is as $\ldots (-T_{g_2}) \ldots (-T_{g_1}) \ldots$.

Note that the Taylor operations are applied directly to the integrand $I_G$ in the momentum representation, and no questions of divergences arise in (V.32). The sets $D$ are called renormalization sets.

We see that an important task is to find the renormalization sets $D$ associated with a graph $G$. For example, for the graph $G$ in Fig. V.3, the normalization sets, say for simplicity for scalar field self-coupling in 6 dimensional spacetime, are: $\{G\}$, $\{G, g_1\}$, $\{G, g_2\}$, $\{g_1\}$, $\{g_2\}$. Note that the subdiagram obtained from the graph $G$ by cutting (i.e, removing) line 23 has a degree of divergence -2 in 6 dimensions. Note also that $\{g_1, g_2\}$ is not a $D$ set, since neither $g_1 \subsetneq g_2$ nor $g_2 \subsetneq g_1$. The renormalized integrand then becomes

$$\left[ 1 + (-T_G) + (-T_G)(-T_{g_1}) + (-T_G)(-T_{g_2}) + (-T_{g_1}) + (-T_{g_2}) \right] I_G$$

$$= (1 - T_G) \left[ 1 + (-T_{g_1}) + (-T_{g_2}) \right] I_G,$$

$$(V.33)$$

with $d(G) = 2$, $d(g_1) = d(g_2) = 0$.

*Remarks*

(1) The key point in establishing the finiteness of the renormalized Feynman integral in Euclidean space for massive particles $\mu_{ij\ell}^2 \neq 0$ is to show that when any subset of the integration variables become large, one can always group the Taylor operations in a particular way, and show that the renormalized integrand has just right behavior and vanishes rapidly asymptotically as a function of these variables and guaranties its integrability. A proof of this is given in Manoukian [3]). For a proof of the finiteness of the renormalized in Minkowski space, in the sense of distributions, see also the latter reference, and for a simpler and direct proof see Manoukian [4, 5], including for zero-mass particles.

(2) Once the finiteness of the renormalized theory is established, with subtractions at the origin, it may be then normalized at other points.

(3) The above subtraction procedure may be reduced to the one referred to as the Bogoliubov-Parasiuk-Hepp-Zimmermann scheme, and for the proof of the underlying theorem, referred to as the "Unifying Theorem of Renormalization",[5] see Manoukian [3].

(4) The renormalization procedure with subtractions is equivalent to adding terms to the original Lagrangian, referred to as counter-terms, and adjusting their numerical coefficients in such a way that parameters appearing in the theory

---

[5] See also Zeidler [8, p. 972] and Figueroa and Gracia-Bondia [1].

(masses, couplings,...) are taken from experiments. For the equivalence of the subtraction scheme and the counter-term formalism, see Manoukian [2, 3]. This establishes the criterion of renormalizabilty and is expressed as follows. One, *a prioi*, may generalize the original Lagrangian density by adding to it new terms. If the counter-terms needed to make the (modified) theory finite and consistent, have the same structures as of terms in the (modified) Lagrangian density and are *finite* in number, then the (modified) theory is called renormalizable. In the latter case only a finite number of parameters are taken from experiments. For example in QED, the counter-terms have exactly the same structures of terms in the Dirac-Maxwell Lagrangian density and the values of the two parameters the (renormalized) mass and (renormalized) charge of the electron are taken from experiments. If the number of counter-terms needed are infinite in number, then one may need to fix an infinite number of parameters and the theory loses its predictive power.

(5) Asymptotic behavior of the *renormalized* theory, such as at high-energy, or for small masses, or for large masses, and other variations, as well as of the proof of the decoupling theorem consistently used in QCD, see Manoukian [3].

(6) Regarding the author's work in the completion of the renormalization program stemming that of Salam's, Streater [7] writes: *"It is the end of a long chapter in the history of physics"*. He also states: *"Physicists found Salam's [method] easier than the BPH one"*.

(7) When you write down a local Lagrangian density, the parameters (couplings, masses,...) are introduced at infinite energies (specified by large ultraviolet cut-offs) and are unattainable. Renormalization theory eliminates these parameters in favor of physically measurable parameters.

# Recommended Reading

1. Figueroa, H., & Gracia-Bondia, J. M. (2004). The uses of Connes and Kreimer's algebraic formulation of renormalization. *International Journal of Modern Physics, A19*, 2739–2754. hep–th/0301015v2.
2. Manoukian, E. B. (1979). Subtractions vs Counterterms. *Nuovo Cimento, 53A*, 345–358.
3. Manoukian, E. B. (1983a). *Renormalization*. New York/London/Paris: Academic Press.
4. Manoukian, E. B. (1983b). Elementary proof of $\epsilon \to +0$ limit of Renormalized Feynman amplitudes. *Journal of Physics: Mathematical and General, 16*, 4131–4133.
5. Manoukian, E. B. (1984b). Elementary proof of $\epsilon \to +0$ limit of Renormalized Feynman amplitudes. II: Theories involving zero mass particles. *Journal of Physics: Mathematical and General, 17*, 1931–1935.
6. Manoukian, E. B. (2006). *Quantum theory: A wide spectrum*. Dordrecht: Springer.
7. Streater, R. F. (1985). Review of Renormalization by E. B. Manoukian. *Bulletin of London Mathematical Society, 17*, pp. 509–510.
8. Zeidler, E. (2009). *Quantum field theory II: Quantum electrodynamics*. Berlin: Springer. pp. 972–975.
9. Zimmermann, W. (1969). Convergence of Bogoliubov's method of renormalization in momentum space. *Communications in Mathematical Physics, 15*, 208–234.

# Solutions to the Problems

## Chapter 2

2.1. Note that $\cos 2\theta = 1/\sqrt{|\mathbf{a}|^2 + 1}$, $\sin 2\theta = |\mathbf{a}|/\sqrt{\mathbf{a}^2 + 1}$, where we have used $\{\gamma^0, \boldsymbol{\gamma}\} = 0$, $(\boldsymbol{\gamma} \cdot \mathbf{a})^2 = -|\mathbf{a}|^2$. Also note that $(|\mathbf{a}| \equiv a)$

$$G\gamma^0 G^{-1} = \gamma^0 \left[ \cos 2\theta - \frac{\boldsymbol{\gamma} \cdot \mathbf{a}}{|\mathbf{a}|} \sin 2\theta \right]$$

$$G\gamma^0 \boldsymbol{\gamma} \cdot \mathbf{a} \, G^{-1} = \gamma^0 \left[ \boldsymbol{\gamma} \cdot \mathbf{a} \cos 2\theta + a \sin 2\theta \right].$$

Hence $G\gamma^0 (\boldsymbol{\gamma} \cdot \mathbf{a} + 1) G^{-1} = \gamma^0 \Big[ \boldsymbol{\gamma} \cdot \mathbf{a} \big( \cos 2\theta - (\sin 2\theta/|\mathbf{a}|) \big) + (\cos 2\theta + a \sin 2\theta) \Big]$.

The statement of the problem then follows from the expressions of $\cos 2\theta$ and $\sin 2\theta$ given above.

2.2. Let $\mathbf{a} = \mathbf{p}/m$, $\cos \theta = \sqrt{(p^0 + m)/2p^0}$, $\sin \theta = |\mathbf{p}|/\sqrt{2p^0(p^0 + m)}$ in Problem 2.1. Then $GHG^{-1} = m\gamma^0 \sqrt{\mathbf{p}^2/m^2 + 1} = \gamma^0 \sqrt{\mathbf{p}^2 + m^2}$.

2.3. The integral in question is given by: $\int d\rho_j \, (\rho_j - \beta_j)(\alpha_0 + c_1 \rho_j)$ which reduces to $\int d\rho_j \, \rho_j (\alpha_0 + c_1 \beta_j) = \alpha_0 + c_1 \beta_j = f(\beta_j)$, where $c_1$ is a c number.

2.4. The integral is given by $\int d\rho \, (\alpha_0 + c_1 \rho) = \int d\rho \, c_1 \rho = c_1$, since $c_1$ is a c number. On the other hand, $(\partial/\partial\rho)(\alpha_0 + c_1 \rho) = c_1 (\partial/\partial\rho)\rho = c_1$.

2.5. (i) $\int d\rho_R \rho_R = 1 = \left( \int d\rho_R \rho_R \right)^*$. The latter, in turn, is equal to $\int \rho_R \, (d\rho_R)^* = -\int (d\rho_R)^* \, \rho_R$. That is, $(d\rho_R)^* = -d\rho_R$, and similarly $(d\rho_I)^* = -d\rho_I$. From the definition $d\rho = (d\rho_R + i \, d\rho_I)$ we obtain $(d\rho)^* = (-d\rho_R^* - (i)(-1)d\rho_I^*)$, from which the condition $(d\rho)^* = -d\rho^*$ follows. (ii) Using the latter result, we have $\left( \int d\rho \, \rho \right)^* = \int \rho^* (-d\rho^*) = \int d\rho^* \rho^*$. Recall that complex conjugation reverses the order in a product.

© Springer International Publishing Switzerland 2016
E.B. Manoukian, *Quantum Field Theory I*, Graduate Texts in Physics,
DOI 10.1007/978-3-319-30939-2

2.6. The integral in question may be rewritten as:

$$\int d\bar{\eta}_1 \eta_1 [(i)^2 (\rho_1 - \alpha_1)(\bar{\rho}_1 - \bar{\alpha}_1)] \, \eta_1 d\eta_1 \dots d\bar{\eta}_n \eta_n [(i)^2 (\rho_n - \alpha_n)(\bar{\rho}_n - \bar{\alpha}_n)] \, \eta_n d\eta_n.$$

The result then follows from the integrals $\int d\bar{\eta}\,\bar{\eta} = 1$, $\int \eta \, d\eta = -1$. Note that every factor like $\bar{\eta}_j(\rho_j - \alpha_j)$, $(\bar{\rho}_j - \bar{\alpha}_j)\eta_j$, $(\rho_j - \alpha_j)(\bar{\rho}_j - \bar{\alpha}_j)$ commutes with everything for any $j$.

2.7. $[\delta/\delta\bar{\eta}_a(x)] \exp[i\bar{\eta}A\,\eta] = i \exp[i\bar{\eta}A\,\eta] \int (dx')A_{ab}(x,x')\eta_b(x')$. A further application of $[\delta/\delta\eta_c(y)]$ to the latter gives: $\exp[i\bar{\eta}A\,\eta]\times$

$$\times \left[ \left( (i)(-i)\int (dx'')\bar{\eta}_d(x'')A_{dc}(x'',y) \right) \left( \int (dy)A_{ad}(x,y)\eta_d(y) \right) + iA_{ac}(x,y) \right],$$

which coincides with (2.6.15), where we note that $\exp[i\bar{\eta}A\eta]$ commutes with everything.

2.8. Using the identities in (2.6.11), (2.6.12), we may write to the leading order

$$Z[\bar{\eta}, \eta] = \left[ 1 + i e \int (dx)(i)\frac{\delta}{\delta\eta(x)}\gamma^\mu A_\mu(x)(-i)\frac{\delta}{\delta\bar{\eta}(x)} \right]$$

$$\times \int \mathscr{D}\bar{\rho}\,\mathscr{D}\rho \exp\left( i\left[ -\bar{\rho}\left(\gamma^\mu\frac{\partial_\mu}{i} + m\right)\rho + \bar{\rho}\eta + \bar{\eta}\rho \right] \right).$$

But the functional integral is from (2.6.27) is equal to $C\exp[i\bar{\eta}S_+\eta]$, where the multiplicative factor $C$ is independent of $(\bar{\eta}, \eta)$ and, of course, independent of $e$ as well. Hence from (2.6.15), we have

$$Z[\bar{\eta}, \eta] = C\left[ 1 + i e \int (dx)(i)\frac{\delta}{\delta\eta(x)}\gamma^\mu A_\mu(x)(-i)\frac{\delta}{\delta\bar{\eta}(x)} \right]\exp[i\bar{\eta}S_+\eta]$$

$$= C\left[ 1 - e\int(dx)\text{Tr}[\gamma^\mu A_\mu(x)S_+(x,x)] + i e \int(dx)(dx')(dy)\times\right.$$

$$\left.\times \bar{\eta}(x)\,S_+(x,x')\gamma^\mu A_\mu(x')S_+(x',y)\,\eta(y)\right]\exp[i\bar{\eta}S_+\eta].$$

Therefore to the leading order, $\exp[i\bar{\eta}S_+^A\eta] = Z[\bar{\eta}, \eta]/Z[0,0]$ is equal to

$$\left[ 1 + i e \int(dx)(dx')(dy)\bar{\eta}(x)\,S_+(x,x')\gamma^\mu A_\mu(x')S_+(x',y)\,\eta(y) \right]\exp[i\bar{\eta}S_+\eta].$$

In particular carrying the functional derivatives: $(-i)[\delta/\delta\eta(y)][\delta/\delta\bar\eta(x)]$ of $\exp[i\bar\eta S_+^A \eta]$ and then setting $\bar\eta = 0$, $\eta = 0$, we obtain from the above equation, to the leading order in $A_\mu$,

$$S_+^A(x,y) = S_+(x,y) + e \int (dx')\, S_+(x,x')\gamma^\mu A_\mu(x') S_+(x',y),$$

where $S_+^{-1}(x,x') = (\gamma^\mu \partial_\mu/i + m)\,\delta^{(4)}(x-x')$. $S_+$ is worked out in Sect. 3.1 with an appropriate boundary condition.

## Chapter 3

3.1. At $t = -a$, the second expression is $(1/2)\exp(-\infty) = 0$, and $f(t)$ is continuous at this point and vanishes. The continuity at $t = 0$ is obvious too. For $t \to a$ from below, write $t = a - \epsilon$, $\epsilon \to +0$. Then the third expression is
$[1 - (1/2)\exp(-2(a/\epsilon)-1)] \to 1$ for $\epsilon \to +0$, and $f(t)$ is continuous at $t = a$ as well and is equal to 1.

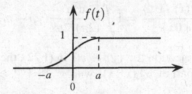

The derivative is given by

$$f'(t) \overset{\cdot}{=} \begin{cases} 0, & t < -a, \\ \frac{a}{(t+a)^2} \exp\left[-2\left(\frac{a}{t+a} - 1\right)\right], & -a \le t < 0, \\ \frac{a}{(t-a)^2} \exp\left[-2\left(\frac{a}{-t+a} - 1\right)\right], & 0 \le t < a, \\ 0, & a \le t. \end{cases}$$

The continuity of $f'(t)$ is immediate by noting, in the process, that at $t = +a$ we have $(1/\epsilon)^2\exp(-1/\epsilon) \to 0$ for $\epsilon \to +0$, and is obviously continuous at $t = 0$. It is easy to see that $f'(t)$ provides a continuous representation of a Dirac delta for $a \to 0$, which peaks at $t = 0$ and vanishes for $t < -a$, $t \ge a$.

3.2. The left-hand side of the equation in question is given by

$$\int \frac{(dp)}{(2\pi)^4} \frac{-\gamma p + m}{p^2 + m^2 - i\epsilon} e^{ip(x'-x)} \left(-\frac{\gamma \overleftarrow{\partial}}{i} + m\right) = \delta^{(4)}(x' - x),$$

where we have used the identity $(-\gamma p + m)(\gamma p + m) = (p^2 + m^2)$.

3.3. For a weak external electromagnetic potential we have seen in (3.2.14) that

$$S_+^A(x, x') \simeq S_+(x - x') + e \int (dx'') S_+(x - x'') \gamma^\mu A_\mu(x'') S_+^A(x'', x').$$

Using the expression for $S_+(x - x')$ in (3.1.10) for $x^0 > x'^0$, and the Fourier transform of the field $A_\mu(x)$ in (3.2.18), we obtain for the e-dependent part of $\langle \psi(x) \rangle_A$ in (3.2.13)

$$e^2(i)^2 \int \frac{d^3\mathbf{p}'}{(2\pi)^3 2p'^0} \frac{d^3\mathbf{p}}{(2\pi)^3 2p^0} (-\gamma p' + m) \gamma^\mu (-\gamma p + m) A_\mu(p' - p) e^{ip'x} e^{-ipx},$$

for $x'^0 \ll x''^0 \ll x^0$. Using the projections in (I.21) for $(-\gamma p' + m)$, $(-\gamma p + m)$, as well as the contribution to $\langle \psi(x) \rangle_A$ coming from the first term on the right-hand side of the equation for $S_+^A(x, x')$ above, as given in (3.2.17), the result in (3.2.15)/(3.2.16) follows.

3.4. We explicitly have

$$\frac{\langle 0_+ \mid \psi(x) \mid 0_- \rangle}{\langle 0_+ \mid 0_- \rangle} = i \int \frac{d^3\mathbf{p}}{(2\pi)^3 2p^0} (\gamma p + m) \, \eta(-p) \, e^{-ipx}.$$

Upon using the projection for $(\gamma p + m)$ in (I.22), the result follows.

3.5. The e-dependent part of $S_+^A(x', x)$, with $x^0 \gg x''^0 \gg x'^0$ is explicitly given by

$$e(i)^2 \int \frac{d^3\mathbf{p}'}{(2\pi)^3 2p'^0} \frac{d^3\mathbf{p}}{(2\pi)^3 2p^0} (\gamma p + m) \gamma^\mu (\gamma p' + m) \, \delta A_\mu(p' - p) \, e^{-ipx'} e^{ip'x}$$

$$= \int \sum_{\sigma\sigma'} \frac{d^3\mathbf{p}'}{(2\pi)^3 2p'^0} \frac{d^3\mathbf{p}}{(2\pi)^3 2p^0} [-i \, v(\mathbf{p}, \sigma) \, e^{-ipx'}] [-i e \, \overline{v}(\mathbf{p}, \sigma) \, \gamma^\mu \, v(\mathbf{p}', \sigma')]$$

$$\times v(\mathbf{p}', \sigma') \, e^{ip'x} \, \delta A_\mu(p' - p),$$

where we have used the projection in (I.22). Moreover by writing the e-independent part of $S_+^A(x', x)$, i.e., $S_+(x', x)$, as

$$S_+(x', x) = \int \sum_{\sigma, \sigma'} \frac{m \, d^3\mathbf{p}'}{p'^0 (2\pi)^3} \frac{m \, d^3\mathbf{p}}{p^0 (2\pi)^3} (2\pi)^3 \frac{p^0}{m} \delta_{\sigma'\sigma} \delta^{(3)}(\mathbf{p}' - \mathbf{p})$$

$$\times [-i \, v(\mathbf{p}, \sigma) \, e^{-ipx'}] \overline{v}(\mathbf{p}', \sigma') \, e^{ip'x},$$

the result stated for $\langle 0_+ \mid \overline{\psi}(x) \mid 0_- \rangle_{\delta A} / \langle 0_+ \mid 0_- \rangle_{\delta A}$ in (3.3.14) follows, by multiplying $S_+^A(x', x)$ by $\overline{\eta}(x')$, and integrating over $x'$ as indicated in (3.3.13).

3.6. This is simply obtained by taking the absolute value squared of (3.3.41), and restrict the **p**-integration as indicated in the region specified by $\Delta$, picking up the spin $\sigma$, and dividing by the normalization factor $N$.

3.7. By using the projection given in (I.21), we have

$$\frac{1}{2} \sum_{\sigma\sigma'} |\bar{u}(\mathbf{p}',\sigma')\, \gamma^0 u(\mathbf{p},\sigma)|^2 = \frac{1}{8m^2}\, \mathrm{Tr}\left[ (-\gamma p' + m)\gamma^0(-\gamma p + m)\gamma^0 \right]$$

$$= \frac{1}{8m^2}\{p'_\mu p_\nu\, \mathrm{Tr}[\gamma^\mu \gamma^0 \gamma^\nu \gamma^0] + m^2\, \mathrm{Tr}[\gamma^0 \gamma^0]\},$$

where we have used the fact that the trace of an odd number of gamma matrices is zero. Finally using the identities

$$\mathrm{Tr}[\gamma^\mu \gamma^0 \gamma^\nu \gamma^0] = 4(\eta^{\mu 0}\eta^{\nu 0} - \eta^{\mu\nu}\eta^{00} + \eta^{\mu 0}\eta^{\nu 0}), \quad \mathrm{Tr}[\gamma^0 \gamma^0] = 4,$$

and $pp' = |\mathbf{p}|^2 \cos\theta - (p^0)^2$, $|\mathbf{p}|^2 = (p^0)^2 - m^2$, $(1+\cos\theta) = 2\cos^2\theta/2$, $(1-\cos\theta) = 2\sin^2(\theta/2)$, the result follows.

3.8. $\mathbb{P}_+(p)$, $\mathbb{P}_-(p)$, defined in (I.21), (I.22), satisfy, in particular, the following equations:

$$(\mathbb{P}_\pm(p)\gamma^0)^\dagger = \mathbb{P}_\pm(p)\gamma^0, \qquad (2m)^2\mathbb{P}_+(p)\gamma^0\mathbb{P}_+(p) = 2p^0(-\gamma p + m),$$

$$(2m)^2\mathbb{P}_-(p)\gamma^0\mathbb{P}_-(p) = -2p^0(\gamma p + m), \qquad \mathbb{P}_\pm(p^0,\mathbf{p}))\gamma^0\mathbb{P}_\mp(p^0,-\mathbf{p}) = 0.$$

Using the anti-commutation relation $\{\psi_a(y), \psi_b^\dagger(y')\} = \delta_{ab}\delta^3(\mathbf{y} - \mathbf{y}')$, for $y^0 = y'^0 (= 0)$, in (3.5.11), we note that due to the last identity in the above set of equations that the cross term in $\{\psi_a(x), \bar{\psi}_b(x')\}$, for arbitrary $x, x'$, vanishes and we obtain from the remaining identities above

$$\{\psi_a(x), \bar{\psi}_b(x')\} = \left(-\frac{\gamma\partial}{i} + m\right)_{ab} \int \frac{d^3\mathbf{p}}{(2\pi)^3 2p^0}\left[e^{ip(x-x')} - e^{-ip(x-x')}\right].$$

Upon changing the variable $\mathbf{p} \to -\mathbf{p}$ in the second term in the above integrand, the statement in the problem follows.

3.9. Note that $(-\gamma\partial/i + m)_{ab} \exp[i\mathbf{p}\cdot(\mathbf{x} - \mathbf{x}')][\sin(p^0 - p'^0)]$ becomes simply
$-\gamma^0_{ab}p^0/i$ at equal times, and the $-p^0/i$ factor cancels the product of $-i$ and $1/p^0$ in the integrand in $\Delta(x - x')$, giving $\gamma^0_{ab}\delta^3(\mathbf{x} - \mathbf{x}')$. The result then follows by multiplying by $\gamma^0$.

3.10. The $\xi$ integral may be expressed in terms of the sine function, or equivalently

$$I = \frac{\varepsilon^\mu}{2} \int \frac{(dQ)}{(2\pi)^4} A_\mu(Q)\, e^{iQx} \int_{-1}^{+1} d\lambda\, e^{i(\epsilon Q/2)\lambda}$$

$$= \epsilon^\mu \int \frac{(dQ)}{(2\pi)^4} A_\mu(Q) \left[ 1 - \frac{1}{3!}\left(\frac{\epsilon Q}{2}\right)^2 + \frac{1}{5!}\left(\frac{\epsilon Q}{2}\right)^4 - \cdots \right] e^{iQx},$$

and is an odd function in $\epsilon$.

3.11. From (I.15) $\frac{1}{4}[\gamma^i, \gamma^j]\varepsilon_{ijk} = \Sigma_k$. Moreover $[\gamma^0, \gamma^i] = 2\begin{pmatrix} 0 & \sigma^i \\ \sigma^i & 0 \end{pmatrix}$. Hence the expressions $F_{0i} = -F^{0i} = -E^i$, $F_{ij} = \varepsilon_{ijk}B^k$, lead from

$$\frac{i}{4}[\gamma^\mu, \gamma^\nu]F_{\mu\nu} = 2\frac{i}{4}[\gamma^0, \gamma^i]F_{0i} + \frac{i}{4}[\gamma^i, \gamma^j]F_{ij},$$

to the stated result, consistent with (3.7.13) for **E**, **B** along the third axis.

3.12. The formal substitution $s \to is$ amounts in replacing (3.8.5) by

$$\frac{2\,\mathrm{Im}\,W^{(e)}}{VT} = -\frac{1}{4\pi^2} \mathrm{Re}\, i \int_0^\infty \frac{ds}{s^3} e^{-ism^2} \left[ (s|eE|\coth s|eE|) - \frac{(seE)^2}{3} - 1 \right].$$

The expression within the square brackets is real. Accordingly, we may make the replacement $\mathrm{Re}[i\,e^{-ism^2}] = \sin(sm^2)$, leading to an integrand which is an even function of $s$. Thus the above integral may be rewritten as

$$\frac{2\,\mathrm{Im}\,W^{(e)}}{VT} = -\frac{1}{8\pi^2} \int_{-\infty}^\infty \frac{ds}{s^3} \sin(sm^2) \left[ (s|eE|\coth s|eE|) - \frac{(seE)^2}{3} - 1 \right],$$

$$= \frac{1}{8\pi^2} \mathrm{Im} \int_{-\infty}^\infty \frac{ds}{s^3} e^{-ism^2} \left[ (s|eE|\coth s|eE|) - \frac{(seE)^2}{3} - 1 \right].$$

It is precisely because of the $(seE)^2/3$, term within the square brackets that the point $s = 0$ is *not* a pole of the integrand. By closing the integral from below as shown in the c.w. direction in the complex $s$-plane, we enclose all the poles $-in\pi//|eE|, n = 1, 2, \ldots$, and note that $-i(i(\mathrm{Im}s))m^2 = m^2(\mathrm{Im}s)$, thus $e^{m^2(\mathrm{Im}s)} \to 0$ for $(\mathrm{Im}s) \to -\infty$.

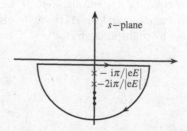

For $s \simeq -in\pi/|eE|$, with $n = 1, 2, \ldots$, the integrand approaches: $e^{-ism^2}/\left[s^2\left(s + \frac{in\pi}{|eE|}\right)\right]$, and the residue theorem gives

$$\frac{2\operatorname{Im}W^{(e)}}{VT} = \frac{1}{8\pi^2}\operatorname{Im}(-2\pi i)\sum_{n=1}^{\infty}\frac{|eE|^2}{-n^2\pi^2}e^{-n\pi m^2/|eE|} = \frac{\alpha E^2}{\pi^2}\sum_{n=1}^{\infty}\frac{e^{-n\pi m^2/|eE|}}{n^2},$$

coinciding with the one in (3.8.8).

3.13. The identity follows by considering the chain of equalities

$$\frac{1}{(\gamma(p + \delta p) + m)} - \frac{1}{(\gamma p + m)}$$

$$= \frac{1}{(\gamma(p + \delta p) + m)}\Big[(\gamma p + m) - (\gamma(p + \delta p) + m)\Big]\frac{1}{(\gamma p + m)}$$

$$= \frac{1}{(\gamma(p + \delta p) + m)}(-\gamma\delta p)\frac{1}{(\gamma p + m)},$$

and the identity follows upon taking the limit $\delta p_\nu \to 0$.

3.14. Due to the matrix nature of $U(x)$, $\partial_\mu U(x)$ and $U(x)$ do not commute. From the gauge transformation in (3.10.3) $A_\nu \to A'_\nu = UA_\nu U^{-1} + (i/g)U\partial_\nu U^{-1}$, hence using the fact that $\partial_\mu(U^{-1}U) = 0$ i.e., $U^{-1}(\partial_\mu U) = -(\partial_\mu U^{-1})U$, then following expression emerges for $\partial_\mu\left(UA_\nu U^{-1} + (i/g)U\partial_\nu U^{-1}\right) =$

$$U(\partial_\mu A_\nu)U^{-1} - \Big[U\partial_\mu U^{-1}, UA_\nu U^{-1}\Big] - \frac{i}{g}(U\partial_\mu U^{-1})(U\partial_\nu U^{-1}) + \frac{i}{g}U\partial_\mu\partial_\nu U^{-1}.$$

Accordingly,

$$\partial_\mu A'_\nu - \partial_\nu A'_\mu = U(\partial_\mu A_\nu - \partial_\nu A_\mu)U^{-1} - \Big[U\partial_\mu U^{-1}, UA_\nu U^{-1}\Big]$$

$$+ \Big[U\partial_\nu U^{-1}, UA_\mu U^{-1}\Big] - \frac{i}{g}\Big[U\partial_\mu U^{-1}, U\partial_\nu U^{-1}\Big],$$

$$-ig[A'_\mu, A'_\nu] = -ig\, U[A_\mu, A_\nu]U^{-1} - \Big[U\partial_\nu U^{-1}, UA_\mu U^{-1}\Big]$$

$$+ \Big[U\partial_\mu U^{-1}, UA_\nu U^{-1}\Big] + \frac{i}{g}\Big[U\partial_\mu U^{-1}, U\partial_\nu U^{-1}\Big].$$

The last two equations give

$$\partial_\mu A'_\nu - \partial_\nu A'_\mu - ig[A'_\mu, A'_\nu] = U\Big(\partial_\mu A_\nu - \partial_\nu A_\mu - ig[A_\mu, A_\nu]\Big)U^{-1},$$

which provides the gauge transformation of $G_{\mu\nu} = \partial_\mu A_\nu - \partial_\nu A_\mu - ig[A_\mu, A_\nu]$.

3.15. $(\gamma^\mu(\partial/i\partial z^\mu) + m)S_+(z-y) = \delta^{(4)}(z-y)$. Upon multiplying the latter, from the right, by $S_+^{-1}(y-x)$, and integrating with respect to $y$, we obtain: $S_+^{-1}(z-x)$ $= (\gamma^\mu(\partial/i\partial z^\mu) + m)\delta^{(4)}(z-x) = \delta^{(4)}(z-x)\left(-\gamma^\mu \overleftarrow{\partial}/i\overleftarrow{\partial}x^\mu + m\right)$. Hence

$$\exp[iF(z)]\left[S_+^{-1}(z-x) + \delta^{(4)}(z-x)\,\gamma^\mu\frac{\partial}{\partial x^\mu}F(x)\right]$$

$$= \exp[iF(x)]\,\delta^{(4)}(z-x)\left(\gamma^\mu\left[-\frac{\overleftarrow{\partial}_\mu}{i} + \overrightarrow{\partial}_\mu F(x)\right] + m\right)$$

$$= \delta^{(4)}(z-x)\left[-\frac{\gamma^\mu \overleftarrow{\partial}_\mu}{i} + m\right]\exp[iF(x)] = S_+^{-1}(z-x)\,\exp[iF(x)],$$

which is the statement in the problem.

## Chapter 4

4.1. Using the facts that $\langle 0\sigma|\mathbf{P} = \mathbf{0}$, and $\langle 0\sigma|P^0 = m\langle 0\sigma|$, we may infer from (4.2.23)–(4.2.31) that

$$\langle 0\sigma|\exp[-i\alpha J_{03}]\exp[-i\theta\mathbf{n}\cdot\mathbf{J}]P^i = \dot{R}^{ij}\langle 0\sigma|\,Y^j\exp[-i\alpha J_{03}]\exp[-i\theta\mathbf{n}\cdot\mathbf{J}]$$

$$= R^{ij}m\,\sinh\alpha\,\delta_{j3}\langle 0\sigma|\exp[-i\alpha J_{03}]\exp[-i\theta\mathbf{n}\cdot\mathbf{J}].$$

With $\mathbf{n}$ given in (4.2.24), we may refer to (2.2.11) to infer that $R^{33} = \cos\theta$, $R^{23} = \sin\phi\sin\theta$, $R^{13} = \cos\phi\sin\theta$, and use the fact that $\sinh\alpha = |\mathbf{p}|/m$, to conclude that the state in (4.2.23) becomes multiplied by $\mathbf{p}$ by the action of $\mathbf{P}$, where $\mathbf{p}$ is given in (4.2.33). On the other hand, $Y^0 = P^0\cosh\alpha + P^3\sinh\alpha$, as given in (4.2.30), and $m\cosh\alpha = \sqrt{|\mathbf{p}|^2 + m^2}$, which verifies (4.2.32) for the application of $P^0$ as well.

4.2. From (4.2.14) we note that $W^0$ commutes with $\mathbf{J}$. On the other hand, consider the expression: $K(\alpha) = \exp[-i\alpha J_{03}]W^0\exp[i\alpha J_{03}]$, with B.C. $K(0) = W^0$. We explicitly have $K'(0) = W^3$, $K''(\alpha) = K(\alpha)$, with the latter two equalities following from (4.2.14), from which we obtain $K(\alpha) = W^3\sinh\alpha + W^0\cosh\alpha$. This leads to the statement of the problem by finally noting that $\langle 0\sigma|W^0 = 0$, $\langle 0\sigma|W^3 = m\sigma\langle 0,\sigma|$.

4.3. If $W^\mu = 0$, then we may write $W^\mu = 0 \times P^\mu$, and there is nothing to prove. Accordingly suppose that $W^\mu \neq 0$, that is, $|W^0| = |\mathbf{W}| > 0$. From the orthogonality of the two vectors: $P^0W^0 = |\mathbf{P}||\mathbf{W}|\cos\theta$ or $|W^0||P^0| = |\mathbf{P}||\mathbf{W}||\cos\theta|$ from which $|\cos\theta| = 1$. That is, $\mathbf{W} = \lambda\mathbf{P}$. On the other hand, the relation $W^0P^0 = \mathbf{W}\cdot\mathbf{P} = \lambda|\mathbf{P}|^2$ also implies that we also have $W^0 = \lambda P^0$, since $P^0 = |\mathbf{P}| > 0$, which completes the proof.

4.4.  $W^0 = (1/2)\,\varepsilon^{ijk}P_i J_{jk}$. But $(1/2)\,\varepsilon^{ijk}J_{jk} = J^i$, hence $W^0 = \mathbf{P}\cdot\mathbf{J}$. On the other hand, $W^i = (1/2)\varepsilon^{i\nu\sigma\lambda}P_\nu J_{\sigma\lambda}$. The latter may be rewritten as $P^0 J^i + \varepsilon^{0ijk}P_j J_{0k}$. But $J_{0k} = -N_k$, from which $\mathbf{W} = P^0\mathbf{J} - \mathbf{P}\times\mathbf{N}$.

4.5.  Upon writing $A_\mu(x) = t_c A_{c\mu}(x)$, and using the commutation rule $[t_a, t_b] = \mathrm{i}f_{abc}t_c$, as given in (3.10.4), the result follows.

4.6.  We note that $\pi^{\alpha\mu} = \partial\mathscr{L}/\partial(\partial_\mu A_\alpha(x)) = F^{\alpha\mu}$. Hence from (4.4.6)

$$\Omega^{\mu\nu\lambda} = \frac{\mathrm{i}}{2}\Big(F^{\alpha\mu}(S^{\nu\lambda})_\alpha{}^\beta A_\beta + F^{\alpha\lambda}(S^{\mu\nu})_\alpha{}^\beta A_\beta + F^{\alpha\nu}(S^{\mu\lambda})_\alpha{}^\beta A_\beta\Big) = F^{\mu\lambda}A^\nu.$$

Maxwell's equations: $\partial_\lambda F^{\lambda\mu} = 0$, give $\partial_\lambda \Omega^{\mu\nu\lambda} = F^{\mu\lambda}\,\partial_\lambda A^\nu$. Accordingly, from (4.4.16)

$$T^{\mu\nu} = \big(-(1/4)\eta^{\mu\nu}F^{\alpha\beta}F_{\alpha\beta} - F^{\alpha\mu}\partial^\nu A_\alpha\big) - F^{\mu\lambda}\partial_\lambda A^\nu$$
$$= -(1/4)\eta^{\mu\nu}F^{\alpha\beta}F_{\alpha\beta} + F^{\mu\alpha}F^\nu{}_\alpha,$$

which is obviously symmetric and gauge invariant, i.e., under the transformation $A_\alpha \to A_\alpha + \partial_\alpha\lambda$, for the latter. Using the elementary identity $\partial^\nu F_{\alpha\beta} = \partial_\alpha F^\nu{}_\beta - \partial_\beta F^\nu{}_\alpha$, and the asymmetry of $F^{\alpha\beta}$ the conservation law follows $\partial_\mu T^{\mu\nu} = -\frac{1}{2}F^{\alpha\beta}\partial^\nu F_{\alpha\beta} + F^{\mu\alpha}\partial_\mu F^\nu{}_\alpha = 0$. $T^{\mu\nu}$ is traceless.

4.7.  $\big(\partial\mathscr{L}/\partial(\partial_\mu\psi)\big)_r - \mathrm{i}\,\overline{\psi}\,\gamma^\mu$. Recall from (1.7) that $S^{\nu\lambda} = (\mathrm{i}/4)[\gamma^\nu, \gamma^\lambda]$. Hence from the definition of $\Omega^{\mu\nu\lambda}$ in (4.4.6), taking into account the adjoint field contribution,

$$\frac{1}{2}\Big(\Omega^{\mu\nu\lambda} + (\Omega^{\mu\nu\lambda})^\dagger\Big) = -\frac{1}{4}\overline{\psi}\Big(\{\gamma^\mu, S^{\nu\lambda}\} + \{\gamma^\lambda, S^{\mu\nu}\} + \{\gamma^\nu, S^{\mu\lambda}\}\Big)\psi$$

$$= -\frac{\mathrm{i}}{8}\overline{\psi}\Big(\gamma^\mu\gamma^\nu\gamma^\lambda - \gamma^\lambda\gamma^\nu\gamma^\mu\Big)\psi.$$

Upon using the Dirac equations: $\partial_\lambda\overline{\psi}\,\gamma^\lambda = \mathrm{i}m\overline{\psi}$, $\gamma^\lambda\partial_\lambda\psi = -\mathrm{i}m\psi$, we obtain

$$\frac{1}{2}\partial_\lambda\Big(\Omega^{\mu\nu\lambda} + (\Omega^{\mu\nu\lambda})^\dagger\Big) = -\frac{\mathrm{i}}{4}\overline{\psi}\Big(\gamma^\mu\overleftrightarrow{\partial^\nu} - \gamma^\nu\overleftrightarrow{\partial^\mu}\Big)\psi,$$

$\mathscr{L} = 0$, and from (4.4.16)

$$T^{\mu\nu} = -\frac{1}{2}\Big(\mathrm{i}\,\overline{\psi}\,\gamma^\mu\partial^\nu\psi + (\mathrm{i}\,\overline{\psi}\,\gamma^\mu\partial^\nu\psi)^\dagger\Big) + \frac{\mathrm{i}}{4}\overline{\psi}\Big(\gamma^\mu\overleftrightarrow{\partial^\nu} - \gamma^\nu\overleftrightarrow{\partial^\mu}\Big)\psi$$

$$= \frac{1}{4\mathrm{i}}\overline{\psi}\Big(\gamma^\mu\overleftrightarrow{\partial^\nu} + \gamma^\nu\overleftrightarrow{\partial^\mu}\Big)\psi,$$

which is obviously symmetric.

4.8.   The general expression for the energy-momentum tensor, given in the previous problem is $T^{\mu\nu} = (1/4\,\mathrm{i})\,\overline{\psi}\big(\gamma^\mu \overleftrightarrow{\partial}{}^\nu + \gamma^\nu \overleftrightarrow{\partial}{}^\mu\big)\psi$. In particular,

$$T^{00} = \frac{1}{2\,\mathrm{i}}\,\psi^\dagger \overleftrightarrow{\partial}{}^0 \psi, \qquad T^{0k} = \frac{1}{4\,\mathrm{i}}\,\overline{\psi}\big(\gamma^0 \overleftrightarrow{\partial}{}^k + \gamma^k \overleftrightarrow{\partial}{}^0\big)\psi.$$

The Dirac equations give $\partial^0 \psi = \gamma^0(\boldsymbol{\gamma}\cdot\overrightarrow{\nabla}+\mathrm{i}m)\psi$, $\partial^0\overline{\psi} = \overline{\psi}(\boldsymbol{\gamma}\cdot\overleftarrow{\nabla}-\mathrm{i}m)\gamma^0$. The right-hand sides of these two equations together with the identity $\gamma^k \gamma^j = (1/2)[\gamma^k, \gamma^j] - \eta^{kj}$, lead to the following expressions

$$T^{00} = \overline{\psi}\Big(m + \frac{1}{2\,\mathrm{i}}\,\boldsymbol{\gamma}\cdot\overleftrightarrow{\nabla}\Big)\psi, \qquad T^{0k} = \frac{1}{8\,\mathrm{i}}\,\partial_j\big(\psi^\dagger\,[\gamma^j, \gamma^k]\,\psi\big) + \frac{1}{2\,\mathrm{i}}\,\psi^\dagger \overleftrightarrow{\partial}{}^k \psi.$$

4.9.   This commutation relation follows by the application of the general equal commutation rule of the commutator of bilinear forms:

$$\Big[\,[\psi_a^\dagger(x), \psi_b(x)],\, [\psi_c^\dagger(x'), \psi_d(x')]\,\Big]$$

$$= 2\,\delta_{bc}\,\delta^3(\mathbf{x} - \mathbf{x}')\,[\psi_a^\dagger(x), \psi_d(x')] - 2\,\delta_{ad}\,\delta^3(\mathbf{x} - \mathbf{x}')\,[\psi_c^\dagger(x'), \psi_b(x)]. \quad (*)$$

Note that the equal-time anti-commutation relations in (4.3.41) imply that since $\pi_b(x) = \mathrm{i}\,\overline{\psi}_b\gamma^0$, $\{\psi_a(x), \psi_b^\dagger(x')\} = \delta^3(\mathbf{x} - \mathbf{x}')$, $\{\psi_a(x), \psi_b(x')\} = 0$. To prove the identity (*), note that the following ones, in turn

$$\psi_a^\dagger(x)\psi_b(x)\psi_c^\dagger(x')\psi_d(x') = \psi_c^\dagger(x')\psi_d(x')\psi_a^\dagger(x)\psi_b(x)$$

$$- \delta_{ad}\,\delta^3(\mathbf{x} - \mathbf{x}')\,\psi_c^\dagger(x')\psi_b(x) + \delta_{bc}\,\delta^3(\mathbf{x} - \mathbf{x}')\,\psi_a^\dagger(x)\psi_d(x'),$$

$$\psi_a^\dagger(x)\psi_d(x')\psi_b^\dagger(x)\psi_c(x') = -\psi_d(x')\psi_c^\dagger(x')\psi_a^\dagger(x)\psi_b(x)$$

$$- \delta_{bc}\,\delta^3(\mathbf{x} - \mathbf{x}')\,\psi_d(x')\psi_a^\dagger(x) + \delta_{ad}\,\delta^3(\mathbf{x} - \mathbf{x}')\,\psi_b(x)\psi_c^\dagger(x'),$$

are sufficient to establish the identity in question, by a mere relabeling of the spinor indices, and by the fact that

$$\partial_k \delta^3(\mathbf{x} - \mathbf{x}')\Big[\psi_c^\dagger(x')\psi_b(x) + \psi_c^\dagger(x)\psi_b(x')\Big]$$

$$= \partial_k \delta^3(\mathbf{x} - \mathbf{x}')\Big[\psi_c^\dagger(x')\psi_b(x') + \psi_c^\dagger(x)\psi_b(x)\Big].$$

4.10.  The right-hand side of (4.7.129) is given by:

$$-(p^0 p^i)/m^2 + \mathrm{e}_0^{0*}\mathrm{e}_0^i = 0, \quad \text{for} \quad \mu = 0, \nu = i,$$

$$-p^0 p^0/m^2 + \mathrm{e}_0^{0*}\mathrm{e}_0^0 = -(\mathbf{p}^2 + m^2)/m^2 + \mathbf{p}^2/m^2 = -1, \quad \text{for} \quad \mu = \nu = 0,$$

$$-p^i p^j/m^2 + \mathrm{e}_+^{i*}\mathrm{e}_+^j + \mathrm{e}_-^{i*}\mathrm{e}_-^j + (p^0)^2 p^i p^j/(m^2 |\mathbf{p}|^2) = \delta^{ij}, \text{ for } \mu = i, \nu = j,$$

using, in the process, the 3D completeness relation for the last relation.

4.11. By multiplying the given equation, in turn, by $\gamma_\mu$ and $\partial_\mu$ and by considering the two resulting equations *simultaneously*, we are led to the following equivalent two equations $(\gamma\partial \equiv \gamma^\mu\partial_\mu,\ \partial K = \partial_\mu K^\mu,\ \gamma K = \gamma_\mu K^\mu)$

$$m\,\partial\psi = (\gamma\partial\,\gamma K) + \partial K, \quad \gamma\psi = -2\,(\gamma\partial\,\gamma K + \partial K)/(\mathrm{i}m^2) - 3\,\gamma K/m,$$

which when are substituted back in the initial equation in the problem give

$$\left(\frac{\gamma\partial}{\mathrm{i}} + m\right)\psi^\mu = \left[\eta^{\mu\nu} + \frac{1}{3}\left(\gamma^\mu\gamma^\nu + \gamma^\mu\frac{\partial^\nu}{\mathrm{i}m} - \gamma^\nu\frac{\partial^\mu}{\mathrm{i}m} + \frac{2}{m^2}\frac{\partial^\mu}{\mathrm{i}}\frac{\partial^\nu}{\mathrm{i}}\right)\right]K_\nu$$

$$-\frac{2}{3\,m^2}\left(\frac{\gamma\partial}{\mathrm{i}} + m\right)\left[\frac{\partial^\mu}{\mathrm{i}}\gamma^\nu - \frac{\partial^\nu}{\mathrm{i}}\gamma^\mu + \left(\frac{\gamma\partial}{\mathrm{i}} - m\right)\gamma^\mu\gamma^\nu\right]K_\nu.$$

Upon multiplying this equation by $(-\gamma\partial/\mathrm{i}+m)$, the statement in the problem then follows from an analysis similar to the one leading to (4.7.143), and by making note that the first term in the above equation multiplied by $(-\gamma\partial/\mathrm{i}+m)$ is nothing but $\rho^{\mu\nu}K_\nu$, where $\rho^{\mu\nu}$ is given in (4.7.145) in the momentum description, while the coefficient of $(-\Box+m^2)$ of the second term is the non-propagating, non-singular term, mentioned in the statement of the problem. The equations in (4.7.137), (4.7.138) are also satisfied for $K^\mu = 0$.

4.12. Consider the equation for $\psi^0$ in (4.7.137) $(\gamma\partial/\mathrm{i} + m)\psi^0 = 0$ for $K^\mu = 0$. It may be rewritten as

$$0 = \frac{\gamma^0\partial_0\psi^0}{\mathrm{i}} + \left(\frac{\boldsymbol{\gamma}\cdot\boldsymbol{\nabla}}{\mathrm{i}} + m\right)\psi^0 = -\gamma^0\frac{\partial_i\psi^i}{\mathrm{i}} + \left(\frac{\boldsymbol{\gamma}\cdot\boldsymbol{\nabla}}{\mathrm{i}} + m\right)\gamma^0\gamma^i\psi^i,$$

where we have used, in turn, the constraint in (4.7.137) and (4.7.138) to eliminate $\psi^0$. The constraint equation in (4.7.148) follows upon multiplying the above equation by $-\gamma^0$.

4.13. The following two equalities are explicitly verified

$$\gamma^i\beta^{ij} = \frac{\partial^j}{\mathrm{i}m} + \frac{\boldsymbol{\gamma}\cdot\boldsymbol{\nabla}}{3\,\mathrm{i}m}\left(\gamma^j + 2\frac{\partial^j}{\mathrm{i}m}\right),$$

$$\frac{\partial^i}{\mathrm{i}}\beta^{ij} = \frac{\partial^j}{\mathrm{i}} + \frac{1}{3}\frac{\boldsymbol{\gamma}\cdot\boldsymbol{\nabla}}{\mathrm{i}}\left(\gamma^j + 2\frac{\partial^j}{\mathrm{i}m}\right) - \frac{\boldsymbol{\gamma}\cdot\boldsymbol{\nabla}}{\mathrm{i}}\frac{\partial^j}{\mathrm{i}m} - \frac{\nabla^2}{3m}\left(\gamma^j + 2\frac{\partial^j}{\mathrm{i}m}\right).$$

The equation in question follows directly from combining these two identities.

4.14. By working in the chiral representation of the gamma matrices, given in (2.3.3), $u(\pm 1)$ may be taken (see (I.25)) and conveniently normalized, as

$$u(+1) = \sqrt{|\mathbf{p}|}\begin{pmatrix}\xi_+\\0\end{pmatrix}, \quad u(-1) = \sqrt{|\mathbf{p}|}\begin{pmatrix}0\\\xi_-\end{pmatrix}, \quad \sum_\sigma u(\sigma)\bar{u}(\sigma) = \gamma p,$$

where $\xi_\pm$ are defined in (I.13). We may then write

$$P_+{}^{ij} = -\frac{1}{2} \sum_{\lambda,\lambda'} e_\lambda^i e_\lambda^{k*} \gamma^\ell \left( u(+)\bar{u}(+) + u(-)\bar{u}(-) \right) \gamma^k e_{\lambda'}^\ell e_{\lambda'}^{j*}. \qquad (*)$$

By considering, for example, the vector $\mathbf{p}$ to be along the 3-axis, and $\mathbf{e}_+ = (1,-i,0)/\sqrt{2}$, $\mathbf{e}_- = (-1,-i,0)/\sqrt{2}$, the following are easily established

$$\sum_{\lambda'} e_{\lambda'}^{j*} \mathbf{e}_{\lambda'} \cdot \boldsymbol{\gamma} = \delta^{j1}\gamma^1 + \delta^{j2}\gamma^2,$$

$$\sum_{\lambda'} e_{\lambda'}^{j*} \mathbf{e}_{\lambda'} \cdot \boldsymbol{\gamma}\, u(+) = \sqrt{2}\, e_-^j u(-),$$

$$\bar{u}(+) \sum_\lambda e_\lambda^i \mathbf{e}_\lambda^* \cdot \boldsymbol{\gamma} = -\sqrt{2}\, e_-^i \bar{u}(-),$$

and with an almost identical analysis carried out for the $u(-)\bar{u}(-)$ in $(*)$ completes the verification of (4.7.193).

4.15. In the presence of the external source, (4.7.103), for $\nu = 0$ becomes replaced by $-\partial_k F^{k0} + m^2 V^0 = K^0$, and with $F^{k0} = \pi^k$, denoting the canonical conjugate momenta of the components $V^k$, the *dependent* field $V^0$ satisfies the equation

$$V^0 = \frac{1}{m^2}\left(K^0 + \partial_k \pi^k\right).$$

By keeping the canonical conjugate momentum $\pi^k$ and its *space* derivative fixed, we have

$$\frac{\delta}{\delta K_\mu(x)} V^0(x') = \frac{1}{m^2} \eta^{0\mu} \delta^{(4)}(x'-x), \quad \text{or equivalently}$$

$$\frac{\delta}{\delta K_\mu(x)} V^\nu(x') = \frac{1}{m^2} \delta^\nu{}_0 \eta^{0\mu} \delta^{(4)}(x'-x).$$

On the other hand, $(-i)\delta/\delta K_\nu(x')\langle 0_+ | 0_-\rangle = \langle 0_+ | V^\nu(x') | 0_-\rangle$, and hence from (4.6.38),

$$(-i)\frac{\delta}{\delta K_\mu(x)}(-i)\frac{\delta}{\delta K_\nu(x')} \langle 0_+ | 0_-\rangle = \langle 0_+ | \left(V^\mu(x)V^\nu(x')\right)_+ 0_-\rangle$$

$$- i\langle 0_+ | \frac{\delta}{\delta K_\mu(x)} V^\nu(x') | 0_-\rangle$$

$$= \langle 0_+ | \left(V^\mu(x)V^\nu(x')\right)_+ | 0_-\rangle - \frac{i}{m^2}\delta^\nu{}_0 \eta^{0\mu}\delta^{(4)}(x'-x)\langle 0_+ | 0_-\rangle,$$

from which the statement in the problem follows upon multiplying by i, and is a consequence of the presence of a dependent field.

4.16. Although $V_0(x)$ is a dependent field we note that

$$\lambda\,(-i)\frac{\delta}{\delta K^\mu(x)}\,(i)\frac{\delta}{\delta\eta(x)}\,\gamma^\mu\,(-i)\frac{\delta}{\delta\bar\eta(x)}\,\langle\,0_+|\,0_-\rangle$$

$$=\lambda\,(-i)\frac{\delta}{\delta K^\mu(x)}\,\langle\,0_+|\,(\bar\psi(x)\gamma^\mu\psi(x))_+\,|\,0_-\rangle$$

$$=\lambda\,\langle\,0_+|\,(\bar\psi(x)\gamma^\mu\psi(x)V_\mu(x))_+\,|\,0_-\rangle=\langle\,0_+|\,\mathscr{L}_I(x)\,|\,0_-\rangle$$

where the functional derivative $\delta/\delta K^0(x)$ does not generate an additional term as in (4.6.38) because $\bar\psi(x)\gamma^\mu\psi(x)$ in the first equality consists only of independent fields.

## Chapter 5

5.1. The solution immediately follows by writing down the rotation matrix $[R^{ik}]$ explicitly and the polarization vectors as $3\times 1$ (column) matrices. The expression for $[R^{ik}]$ follows from (5.2.6) to be

$$[R^{ik}]=\begin{pmatrix}\cos^2\phi\,\cos\theta+\sin^2\phi & \sin\phi\,\cos\phi\,(\cos\theta-1) & \cos\phi\,\sin\theta\\ \sin\phi\,\cos\phi\,(\cos\theta-1) & \sin^2\phi\,\cos\theta+\cos^2\phi & \sin\phi\,\sin\theta\\ -\cos\phi\,\sin\theta & -\sin\phi\,\sin\theta & \cos\theta\end{pmatrix}.$$

5.2. $\partial_\mu\int(dx')D^{\mu\nu}(x-x')J_\nu(x')=\lambda\int(dx')D_+(x-x')\partial'_\nu J^\nu(x')=\lambda\chi(x).$

5.3. Going through these various steps, we have: $E_n(K,R)\exp[-iE_n(K,R)T]$

$$=i\frac{d}{dT}\exp[-iE_n(K,R)T]=-\frac{d^2}{dT^2}\left(\frac{\exp[-iE_n(K,R)T]}{E_n(K,R)}\right),$$

$$\frac{\exp[-iF_n(K,R)T]}{E_n(K,R)}=\frac{2i}{T}\frac{\partial}{\partial\mathbf{K}^2}\exp\left[-i\sqrt{\mathbf{K}^2+\frac{n^2\pi^2}{R^2}}\,T\right].$$

Therefore

$$E_n(K,R)\exp[-iE_n(K,R)T]=-2i\frac{d^2}{dT^2}\left\{\frac{1}{T}\frac{\partial}{\partial\mathbf{K}^2}\exp\left[-i\sqrt{\mathbf{K}^2+\frac{n^2\pi^2}{R^2}}\,T\right]\right\}.$$

An elementary application of $\partial/\partial a$ to the above equation gives the stated result, where recall that $R$ is expressed in terms of $a$.

5.4. Clearly

$$f_{ka}(\mathbf{n}_1,\mathbf{n}_2)=\frac{1}{C}\left|e^{-i|k|\mathbf{n}_1\cdot\mathbf{R}_1}\,e^{-i|k|\mathbf{n}_2\cdot\mathbf{R}_2}+e^{-i|k|\mathbf{n}_1\cdot\mathbf{R}_2}\,e^{-i|k|\mathbf{n}_2\cdot\mathbf{R}_1}\right|^2.$$

independently of $f(|\mathbf{k}|)$ where

$$C = \int d\Omega_1 \int d\Omega_2 \left| e^{-i|\mathbf{k}|\mathbf{n}_1 \cdot \mathbf{R}_1} e^{-i|\mathbf{k}|\mathbf{n}_2 \cdot \mathbf{R}_2} + e^{-i|\mathbf{k}|\mathbf{n}_1 \cdot \mathbf{R}_2} e^{-i|\mathbf{k}|\mathbf{n}_2 \cdot \mathbf{R}_1} \right|^2 .$$

Upon using the elementary integrals,

$$\int_{-1}^{1} dx \, x \, e^{-ikax} = 2i \left[ \frac{\cos ka}{ka} - \frac{\sin ka}{k^2 a^2} \right], \quad C = 32\pi \left[ 1 + \frac{\sin^2 ka}{k^2 a^2} \right],$$

we obtain the following expression

$$\langle c \rangle = \left( \frac{\cos ka}{ka} - \frac{\sin ka}{k^2 a^2} \right)^2 \Big/ \left( 1 + \frac{\sin^2 ka}{k^2 a^2} \right).$$

5.5. For a matrix $S = 1/[A - e_0 B]$, we have

$$\frac{\partial}{\partial e_0} \frac{1}{A - e_0 B} = \frac{1}{\delta e_0} \left[ \frac{1}{A - (e_0 + \delta e_0) B} - \frac{1}{A - e_0 B} \right], \quad \delta e_0 \to 0$$

$$= \frac{1}{A - (e_0 + \delta e_0) B} \left[ \frac{(A - e_0 B) - (A - (e_0 + \delta e_0) B)}{\delta e_0} \right] \frac{1}{A - e_0 B}$$

$$\to \frac{1}{A - e_0 B} B \frac{1}{A - e_0 B},$$

and the result follows by matrix multiplication with matrix elements indices specified by spacetime variables.

5.6. From (5.7.3) and (5.7.4), we have

$$\partial_\mu \langle 0_+ | j^\mu | 0_- \rangle = i e_0 \left[ \langle 0_+ | \overline{\psi} | 0_- \rangle \eta - \overline{\eta} \langle 0_+ | \psi | 0_- \rangle \right],$$

upon taking matrix element between vacuum states. The first result then follows from the explicit equations for $\langle 0_+ | \psi(x) | 0_- \rangle$ in (5.7.13), $\langle 0_+ | \overline{\psi}(x) | 0_- \rangle$ in (5.7.14), and (5.7.7). Now write $\langle 0_+ | 0_- \rangle = F[\delta/\delta J]\langle 0_+ | 0_- \rangle_{0\gamma}$ in (5.7.27). Then

$$\partial_\mu \langle 0_+ | A^\mu(x) | 0_- \rangle = \lambda F[\delta/\delta J] \int (dx') D_+(x - x') \partial'^\nu J_\nu(x') \langle 0_+ | 0_- \rangle_{0\gamma}.$$

Upon using the identity: $F[\delta/\delta J] J_\nu(x) = \left[ J_\nu(x) + \delta F[T]/\delta T^\nu(x) \right]\big|_{T^\nu = \delta/\delta J_\nu}$, the expansion

$$S_+(y, y'; e_0 T) = S_+(x - x') - ie_0 \int (dy) \, S_+(y - y_1) \gamma_\mu T^\mu(y_1) S_+(y_1 - y') + \cdots ,$$

and the identity

$$\int (dx')(dy_1) D_+(x-x') \partial_\mu^{x'} \frac{\delta}{\delta T_\mu(x')} S_+(y-y_1) \gamma_\mu T^\mu(y_1) S_+(y_1-y')$$

$$= -i [D_+(x-y) S_+(y-y') - D_+(x-y') S_+(y-y')],$$

which applied to the $\exp[i\,\overline{\eta} S_+(.; e_0 \hat{A})\,\eta]$ factor leads to the stated result. By the same method just developed, the other factor $\exp[\int de'\, \mathrm{Tr}[\gamma \hat{A}(.) S_+ (.; e'\hat{A})]$ gives no contribution (see also Appendix IV).

5.7. Write the exponential in (5.8.2) involving $(-i)\delta/\delta J^\mu$, acting on $\langle 0_+ \mid 0_- \rangle_{0\gamma}$, as

$$\exp i [A_0 + e_0 A_1 + e_0^2(A_2 + iB_2) + e_0^3 A_3 + \cdots] =$$

$$\left[ 1 + i e_0 A_1 + i e_0^2 \left( A_2 + i B_2 + \frac{i}{2} A_1^2 \right) + i e_0^3 \left( A_3 - \frac{1}{3!} A_1^3 + i A_1 (A_2 + i B_2) \right) + \cdots \right] e^{iA_0},$$

up to third order, and where $B_2$ is the coefficient of $e_0^2$ in (5.8.6). Obviously, $a_0 = A_0$ as the zeroth order involving no functional differentiations. Also $a_1 = [\overline{\eta} S \gamma^\mu S \eta] D_{\mu\nu} J^\nu$ in a matrix notation. Clearly, $A_1^3$ involves only disconnected parts as it involves only three photon propagators since to *third* order they cannot connect together three different expressions of the form $[\overline{\eta} S \gamma^\mu S \eta]$. $B_2$ leads to $(1/2)[(-i) D_{\mu_1\mu_2} + D_{\mu_1\nu_1} J^{\nu_1} D_{\mu_2\nu_2} J^{\nu_2}] K^{\mu_1\mu_2}$, and the first term in the latter $(1/2)(-i) D_{\mu_1\mu_2} K^{\mu_1\mu_2}$ involves no external lines and should be omitted due to the normalization condition of $\langle 0_+ \mid 0_- \rangle$. For $A_1^2$, the only connected term is $[\overline{\eta} S \gamma^{\mu_1} S \eta][\overline{\eta} S \gamma^{\mu_2} S \eta](-i) D_{\mu_1\mu_2}$. The application of $A_2$ is straightforward and leads only to connected parts, and so is for $A_3$. The application of $iA_1 A_2$ leads to

$$i [\overline{\eta} S \gamma^{\mu_1} S \eta][\overline{\eta} S \gamma^{\mu_2} S \gamma^{\mu_3} \eta] \times \Big( (-i) D_{\mu_1\mu_3} D_{\mu_2\nu_2} J^{\nu_2} +$$

$$+ D_{\mu_1\nu_1} J^{\nu_1} D_{\mu_2\nu_2} J^{\nu_2} D_{\mu_3\nu_3} J^{\nu_3} + (-i) D_{\mu_1\mu_2} D_{\mu_3\nu_3} J^{\nu_3} + (-i) D_{\mu_2\mu_3} D_{\mu_1\nu_1} J^{\nu_1} \Big).$$

Clearly the second and the fourth terms within the round brackets give rise to disconnected parts and are to be omitted. Finally, $-A_1 B_2$ leads to

$$\frac{i}{2} [ [\overline{\eta} S \gamma^{\mu_3} S \eta] K^{\mu_1\mu_2} \Big[ D_{\mu_1\mu_3} D_{\mu_2\nu_2} J^{\nu_2} + D_{\mu_2\mu_3} D_{\mu_1\nu_1} J^{\nu_1} \Big],$$

retaining only connected parts.

5.8. This involves four terms. The term obtained by multiplying the first term in (5.9.23) by its complex conjugate is given by

$$\frac{1}{(p_2 - p_2')^4} [\bar{u}(\mathbf{p}_2', \sigma_2') \gamma^\mu u(\mathbf{p}_2, \sigma_2) \bar{u}(\mathbf{p}_1', \sigma_1') \gamma_\mu u(\mathbf{p}_1, \sigma_1)$$

$$\times \bar{u}(\mathbf{p}_1, \sigma_1) \gamma^\sigma u(\mathbf{p}_1', \sigma_1') \bar{u}(\mathbf{p}_2, \sigma_2) \gamma_\sigma u(\mathbf{p}_2', \sigma_2')]$$

$$= \frac{1}{(2m)^4} \frac{1}{(p_2 - p_2')^4} \text{Tr}[\gamma^\mu(-\gamma p_1' + m) \gamma^\sigma(-\gamma p_1 + m)]$$

$$\times \text{Tr}[\gamma_\mu(-\gamma p_2' + m) \gamma_\sigma(-\gamma p_2 + m)],$$

where we have used the relation $\sum_\sigma u_a(\mathbf{p}, \sigma) \bar{u}_b(\mathbf{p}, \sigma) = (-\gamma p + m)_{ab}/2m$. Application of the properties of the gamma matrices readily gives the first term in (5.9.32). It is easy to see that the two cross terms are identical, and direct applications of the method just given for the first term leads to the other two.

5.9. The four-spinors of the ingoing electrons are given by

$$u(\mathbf{p}_1, +) = \sqrt{\frac{p^0 + m}{2m}} \begin{pmatrix} \begin{pmatrix} 1 \\ 0 \end{pmatrix} \\ i\rho \begin{pmatrix} 1 \\ 0 \end{pmatrix} \end{pmatrix}, \ u(\mathbf{p}_2, -) = \sqrt{\frac{p^0 + m}{2m}} \begin{pmatrix} \begin{pmatrix} 0 \\ 1 \end{pmatrix} \\ i\rho \begin{pmatrix} 0 \\ 1 \end{pmatrix} \end{pmatrix},$$

where $\rho = \gamma\beta/(\gamma + 1) = \beta/(1 + \sqrt{1 - \beta^2})$. The four spinors of the outgoing electrons may be written as

$$u(\mathbf{p}_1', \sigma_1') = \sqrt{\frac{p^0 + m}{2m}} \begin{pmatrix} \xi_1 \\ \frac{\sigma \cdot \mathbf{p}_1'}{p^0 + m} \xi_1 \end{pmatrix}, \ u(\mathbf{p}_2', \sigma_2') = \sqrt{\frac{p^0 + m}{2m}} \begin{pmatrix} \xi_2 \\ -\frac{\sigma \cdot \mathbf{p}_1'}{p^0 + m} \xi_2 \end{pmatrix},$$

where

$$\mathbf{n}_1 = (0, \sin\chi_1, \cos\chi_1), \ \mathbf{n}_2 = (0, \sin\chi_2, \cos\chi_2), \ \mathbf{p}_1' = \gamma m \beta (1, 0, 0) = -\mathbf{p}_2',$$

$$\xi_j = (e^{-i\pi/4} \cos(\chi_j/2) \ e^{i\pi/4} \sin(\chi_j/2))^\top, \ j = 1, 2.$$

From (5.9.23), we have for the amplitude of the process

$$A \propto \xi_1^\dagger \xi_2^\dagger \left[ \begin{pmatrix} 0 \\ 1 \end{pmatrix}_1 \begin{pmatrix} 1 \\ 0 \end{pmatrix}_2 - \begin{pmatrix} 1 \\ 0 \end{pmatrix}_1 \begin{pmatrix} 0 \\ 1 \end{pmatrix}_2 \right].$$

This leads to $|A|^2/2 = P[\chi_1, \chi_2]$ as given in the problem, satisfying the completeness relation:

$$P[\chi_1, \chi_2] + P[\chi_1 + \pi, \chi_2] + P[\chi_1, \chi_2 + \pi] + P[\chi_1 + \pi, \chi_2 + \pi] = 1.$$

If only one spin is measured: $P[\chi_1, -] = P[\chi_1, \chi_2] + P[\chi_1, \chi_2 + \pi] = 1/2$, and similarly $P[-, \chi_2] = 1/2$. The obvious correlation between the two electron spins, that is the measurement of the spin of one is correlated with the spin value of the other, is expressed by the fundamental general relation $P[\chi_1, \chi_2] \neq P[\chi_1, -] P[-, \chi_2]$. It is interesting to note that the above probabilities are independent of the speed $\beta$ of the particles. This is due to the special choice of axes of measurements and spin orientations chosen in the problem. More generally polarization correlations depend on the speed of the underlying particles (see, e.g., Sect. 5.9.2 and references therein).

5.10. $(\gamma p + m)^2 = -p^2 + 2m\gamma p + m^2 = -(p^2 + m^2) + 2m(\gamma p + m)$ from which the identity follows.

5.11. From the method of the Feynman parameter representation, as given in (II.34), in Appendix I at the end of the book,

$$\int (dk) \frac{1}{(k^2 + m^2)^2} \frac{\Lambda^2}{(k^2 + \Lambda^2)} = 2\Lambda^2 \int_0^1 x \, dx \int \frac{(dk)}{\left[k^2 + \Lambda^2 + (m^2 - \Lambda^2)x\right]^3}$$

$$= i\pi^2 \int_0^1 \frac{\Lambda^2 x \, dx}{(\Lambda^2 + (m^2 - \Lambda^2)x)} = \frac{\Lambda^2}{m^2 - \Lambda^2} \int_0^1 dx \left[1 - \frac{\Lambda^2}{\Lambda^2 + (m^2 - \Lambda^2)x}\right],$$

using the integral (II.8). The stated result then follows upon carrying out the elementary $x$-integral, and then considering the limit $\Lambda^2 \to \infty$.

5.12. Using the integral

$$\int \frac{x \, dx}{x^2 - ax + a} = \frac{1}{2} \ln[x^2 - ax + a] + \frac{a}{2} \frac{1}{\sqrt{a - (a/2)^2}} \arctan\left(\frac{x - a/2}{\sqrt{a - (a/2)^2}}\right),$$

for $a > 0$, we obtain with $a = \mu^2/m^2 \to 0$, $C_{ir} = -(1/2)\ln(\mu^2/m^2) + (\mu/2m) \times [(\pi/2) + \mathscr{O}(\mu/m)]$. On the other hand $\mu^2 \partial C_{ir}/\partial \mu^2 = -C_{ir} + D_{ir}$.

5.13.

$$I \equiv \int \frac{(dk)}{[(p-k)^2 + m^2][k^2 + \mu^2]} = \int (dk) \int_0^1 \frac{dx}{F(k^2, x)},$$

where $F(k^2, x) = \left[k^2 + (p^2 + m^2)x(1 - x) + m^2 x^2 + \mu^2(1 - x)\right]^2$, after having made a shift of the (integration) variable $k \to k + px$. Upon integration over $x$ by parts, and using the elementary integral (II.8), we obtain $I =$

$i\pi^2[C_{\mathrm{uv}} + I_2]$, where $C_{\mathrm{uv}}$ is given in (5.10.21), and

$$I_2 = (p^2 + m^2) \int_0^1 \frac{x\,(1 - 2x)\,\mathrm{d}x}{[(p^2 + m^2)\,x(1 - x) + m^2x^2 + \mu^2(1 - x)]}$$

$$+ \int_0^1 \frac{x\,(2\,m^2x - \mu^2)\,\mathrm{d}x}{[(p^2 + m^2)\,x(1 - x) + m^2x^2 + \mu^2(1 - x)]}.$$

In the first integral $I_2^{(1)}$, we may set $(p^2 + m^2) = 0$, in the denominator, obtaining, $I_2^{(1)} = [(p^2 + m^2)/m^2][C_{\mathrm{ir}} - 2]$, for $\mu^2/m^2 \to 0$, where $C_{\mathrm{ir}}$ is given in (5.10.22). The second integral may be written, up to first order in $(p^2 + m^2)$, as

$$I_2^{(2)} = -(p^2 + m^2)\int_0^1 \frac{x^2(1 - x)(2\,m^2x - \mu^2)\,\mathrm{d}x}{[m^2x^2 + \mu^2(1 - x)]^2} + \int_0^1 \frac{x\,(2\,m^2x - \mu^2)\,\mathrm{d}x}{[m^2x^2 + \mu^2(1 - x)]}.$$

Both integrals are elementary except, perhaps, the integral

$$\mu^2 \int_0^1 \frac{x^2(1 - x)\,\mathrm{d}x}{[m^2x^2 + \mu^2(1 - x)]^2} \to \mu^2 \int_0^1 \frac{x^2\,\mathrm{d}x}{[m^2x^2 + \mu^2(1 - x)]^2},$$

where we have used, in the process, the value of the second integral $D_{\mathrm{ir}}$ in Problem 5.12, to set the second term to zero. The above integral, multiplied by $\mu^2$, vanishes like $\mathscr{O}(\mu/m)$. Accordingly, $I_2^{(2)} = (2 + [(p^2 + m^2)/m^2][3 - 2\,C_{\mathrm{ir}}])$. Hence $I_2 = 2 + [(p^2 + m^2)/m^2][C_{\mathrm{ir}} - 2 + 3 - 2\,C_{\mathrm{ir}}]$,

$$\int \frac{(\mathrm{d}k)}{[(p - k)^2 + m^2][k^2 + \mu^2]} = i\pi^2\Big[(C_{\mathrm{uv}} + 2) - 2\frac{\gamma p + m}{m}(C_{\mathrm{ir}} - 1) + \mathscr{O}\big((\gamma p + m)^2\big)\Big],$$

where we have used the identity in Problem 5.10. The statement of the problem follows upon multiplying this integral by $-ie^2[-(\lambda + 1)(\gamma p + m) - 2m]/(2\pi)^4$.

5.14. From (5.10.57), the modified Coulomb potential, to second order may be written as $U'(\mathbf{x}) = q^2\Big(Z_3 + (\alpha/3\pi)\,\Delta U'(\mathbf{x})\Big)/(4\pi|\mathbf{x}|)$, where

$$\Delta U'(\mathbf{x}) = \int_{(2m)^2}^{\infty} \frac{\mathrm{d}M^2}{M^2}\,e^{-M|\mathbf{x}|} + \int_{(2m)^2}^{\infty} \frac{\mathrm{d}M^2}{M^2}\Big[\Big(1 + \frac{2m^2}{M^2}\Big)\sqrt{1 - \frac{4m^2}{M^2}} - 1\Big]e^{-M|\mathbf{x}|}.$$

The first integral is readily expressed in terms of the exponential integral function,[6] and it is given by the equivalent integral with asymptotic

---

[6]See, e.g., I. S. Gradshteyn and I. M. Ryzhik (2000). Tables of Integrals, Series and Products (6th ed.). San Diego/San Francisco: Academic Press. pp. 875–877.

behavior

$$I_1 = 2\int_{2m|\mathbf{x}|}^{\infty} dt\, \frac{e^{-t}}{t} = -2\,\mathrm{Ei}(-2m|\mathbf{x}|) \simeq -2\,\gamma_E + 2\ln\Big(\frac{1}{2m|\mathbf{x}|}\Big) + \mathcal{O}(m|\mathbf{x}|),$$

where $\gamma_E = 0.5772157\ldots$ denotes Euler's constant. On the other hand, it is justifiable to take the limit $m|\mathbf{x}| \to 0$ inside the second integral. This amounts to evaluate the integral

$$I_2' \equiv \int_{(2m)^2}^{\infty} \frac{dM^2}{M^2}\Big[\Big(1 + \frac{2\,m^2}{M^2}\Big)\sqrt{1 - \frac{4\,m^2}{M^2}} - 1\Big].$$

Upon introducing the integration variable $z = 1 - (1 - 4m^2/M^2)^{1/2}$, the above integral simplifies to

$$I_2' = \int_0^1 dz\,\Big[-1 - 2z + z^2 - \frac{2}{z-2}\Big] = -2\Big[\ln\Big(\frac{1}{2}\Big) + \frac{5}{6}\Big].$$

All told, we obtain for $m|\mathbf{x}| \ll 1$,

$$U'(\mathbf{x}) - \frac{q^2}{4\pi|\mathbf{x}|}\Big(Z_3 + \frac{2\alpha}{3\pi}\Big[\ln\Big(\frac{1}{m|\mathbf{x}|}\Big) - \gamma_E - \frac{5}{6}\Big]\Big).$$

5.15. From (5.10.42), $\bar{u}(\mathbf{p}',\sigma)\gamma^\mu \Pi_{\mu\nu}(k)\,(1/k^2)u(\mathbf{p},\sigma)$ is equal to

$$\bar{u}(\mathbf{p}',\sigma)\,\gamma_\nu\,u(\mathbf{p},\sigma)\,\Pi(k^2) - \bar{u}(\mathbf{p}',\sigma)(\gamma p' - \gamma p)u(\mathbf{p},\sigma)\frac{k_\nu}{k^2}.$$

The Dirac spinors satisfy the equations $(\gamma p + m)u(\mathbf{p},\sigma) = 0$, $\bar{u}(\mathbf{p}',\sigma)(\gamma p' + m) = 0$. That is the last expression above is zero. The statement of the problem then follows from (5.10.44), (5.10.49), by taking the limit $k^2 \to 0$ in this order, where $\Pi(0) = 1 - Z_3$.

5.16. Using the normalization condition of $\rho(\mathbf{x})$, the change in potential energy is

$$\Delta U(\mathbf{x}) = -\alpha\Big[\int d^3\mathbf{x}'\,\frac{\rho(\mathbf{x}')}{|\mathbf{x}-\mathbf{x}'|} - \frac{1}{|\mathbf{x}|}\Big] = -\alpha\int d^3\mathbf{x}'\rho(\mathbf{x}')\Big[\frac{1}{|\mathbf{x}-\mathbf{x}'|} - \frac{1}{|\mathbf{x}|}\Big].$$

leading to an approximate energy shift

$$\delta E = -\alpha|\varphi_{n0}(0)|^2\int d^3\mathbf{x}'\rho(\mathbf{x}')\int d^3\mathbf{x}\Big[\frac{1}{|\mathbf{x}-\mathbf{x}'|} - \frac{1}{|\mathbf{x}|}\Big].$$

The $\mathbf{x}'$-integrand, multiplying $\rho(\mathbf{x}')$, is easily worked out, for example, by a Legendre polynomial expansion,[7] to be

$$\int d^3\mathbf{x}\left[\frac{1}{|\mathbf{x}-\mathbf{x}'|}-\frac{1}{|\mathbf{x}|}\right] = 2\pi\int_{|\mathbf{x}|<|\mathbf{x}'|} d|\mathbf{x}|\,|\mathbf{x}|^2\left[\frac{2}{|\mathbf{x}'|}-\frac{2}{|\mathbf{x}|}\right]+0 = -2\pi|\mathbf{x}'|^2/3,$$

and $\int d^3\mathbf{x}'\rho(\mathbf{x}')\,|\mathbf{x}'|^2 = 12\gamma^2$. This gives

$$\delta E = \frac{8\,\alpha^4\,m}{n^3}\,m^2\gamma^2.$$

5.17. $F^{k\nu}A_\nu = F^{k0}A_0 + F^{ki}A_i$. Using the notation $(-i)\delta/\delta J^\mu \equiv \widehat{A}_\mu$, note that $\widehat{A}_0(y)\,\widehat{F}^{k0}(x)\langle\,0_+\mid 0_-\rangle$ is equal to

$$\widehat{A}_0(y)\langle\,0_+|F^{k0}(x)\,|\,0_-\rangle = \langle\,0_+|\big(F^{k0}(x)A_0(y)\big)_+\,|\,0_-\rangle+i\frac{\partial^k}{\nabla^2}\,\delta^{(4)}(x-y)\langle\,0_+\mid 0_-\rangle,$$

where we have used (5.14.27) and (4.6.38). Hence

$$\langle\,0_+|\big(F^{k\nu}(x)A_\nu(y)\big)_+\,|\,0_-\rangle = \left[\widehat{F}^{k\nu}(x)\widehat{A}_\nu(y) - i\frac{\partial^k}{\nabla^2}\,\delta^{(4)}(x-y)\right]\langle\,0_+\mid 0_-\rangle.$$

5.18. We note from the just mentioned equation that

$$\pi^a = \partial_3^{-1}\partial^a\partial^0 A^3 - \partial^0 A^a, \qquad a = 1,2,$$

$$\partial_a\pi^a = \partial_3^{-1}\partial^0(\partial_a\partial^a A^3 - \partial^3\partial_a A^a) = \partial_3^{-1}\partial^0\nabla^2 A^3.$$

where we have used the relation $-\partial^a A^a = \partial_3 A^3$, in writing the last equality. From the equal-time commutation relations:

$$[A^a(x^0,\mathbf{x}'), \pi^b(x^0,\mathbf{x})] = i\,\delta^{ab}\delta^{(3)}(\mathbf{x}'-\mathbf{x}), \qquad a,b = 1,2,$$

and the above two equations in turn, lead to

$$[A^a(x^0,\mathbf{x}'), \partial_3^{-1}\partial^b\partial^0 A^3(x^0,\mathbf{x})] - [A^a(x^0,\mathbf{x}'), \partial^0 A^b(x^0,\mathbf{x})] = i\,\delta^{ab}\delta^{(3)}(\mathbf{x}'-\mathbf{x}),$$

$$[A^a(x^0,\mathbf{x}'), \partial_3^{-1}\partial^0 A^3(x^0,\mathbf{x})] = i\frac{\partial^a}{\nabla^2}\,\delta^{(3)}(\mathbf{x}'-\mathbf{x}),$$

---

[7]Recall: $(|\mathbf{x}-\mathbf{x}'|)^{-1} = (1/r_>)\sum_{n=0}^\infty (r_</r_>)^n P_n(\cos\theta)$, in a standard notation, $P_0(\cos\theta) = 1$, $P_1(\cos\theta) = \cos\theta$, $\int_{-1}^1 d\cos\theta\,P_n(\cos\theta)P_{n'}(\cos\theta) = 2\,\delta_{nn'}/(2n+1)$.

from which the following key equal-time commutation relations emerges

$$[A^a(x^0, \mathbf{x}'), \partial_0 A^b(x^0, \mathbf{x})] = i\left(\delta^{ab} - \frac{\partial^a \partial^b}{\nabla^2}\right)\delta^{(3)}(\mathbf{x}' - \mathbf{x}), \qquad a, b = 1, 2.$$

The expression $A^3 = -(\partial^a/\partial_3)A^a$ then leads to the equal time commutation relations in question.

5.19. The Fourier transform of the left-hand side of the identity reads $[\eta^{\alpha\beta} - (\eta^{\alpha j} k^j k^\beta/\mathbf{k}^2) - (\eta^{\beta j} k^j k^\alpha/\mathbf{k}^2) + (k^\alpha k^\beta/\mathbf{k}^2)]/k^2$, and coincides with $D_C^{\alpha\beta}(k)$ (see (5.14.13)).

5.20. Denote the left-hand side by $K$. This gives

$$(-i)\frac{\delta}{\delta\bar{\eta}(x)}K = \exp\left(-i e_0 \frac{\delta}{\delta\eta}\gamma^\mu \frac{\delta}{\delta\bar{\eta}}\partial_\mu \Lambda\right)\left[S_+(x - .)\,\eta(.)\right]\exp\left[i\bar{\eta}S_+\eta\right]$$

$$= S_+(x - .)\left[\eta(.) + i e_0 \gamma^\mu \frac{\delta}{\delta\bar{\eta}(.)}\partial_\mu \Lambda(.)\right]K,$$

which upon multiplying by $S_+^{-1}(z - x)$ and integrating over $x$, gives

$$\int (dx)\,S_+^{-1}(z - x)(-i)\frac{\delta}{\delta\bar{\eta}(x)}K = \left[\eta(z) + i e_0 \gamma^\mu\,\partial_\mu^z \Lambda(z)\frac{\delta}{\delta\bar{\eta}(z)}\right]K$$

which, after multiplying it by $\exp[i e_0 \Lambda(z)]$, may be rewritten as

$$\int (dx)\exp[i e_0 \Lambda(z)]\left[S_+^{-1}(z - x) + \delta^{(4)}(z - x)\,e_0\gamma^\mu\,\partial_\mu \Lambda(x)\right](-i)\frac{\delta}{\delta\bar{\eta}(x)}K$$

$$= \exp[i e_0 \Lambda(z)]\,\eta(z)K.$$

But

$$\exp[i e_0 \Lambda(z)]\left[S_+^{-1}(z - x) + \delta^{(4)}(z - x)\,e_0\gamma^\mu(\partial_\mu \Lambda(x))\right]$$
$$= S_+^{-1}(z - x)\exp[i e_0 \Lambda(x)],$$

(see Problem 3.15). Hence upon multiplying the former equation by: $\exp[-i e_0 \Lambda(y)]\,S_+(y - z)$, and integrating over $z$, we obtain

$$(-i)\frac{\delta}{\delta\bar{\eta}(y)}K = \int (dz)\,\exp[-i e_0 \Lambda(y)]\,S_+(y - z)\,\exp[i e_0 \Lambda(z)]\,\eta(z)\,K.$$

Functionally integrating over $\bar{\eta}(y)$ leads to the right-hand side of the equation stated in the problem, and incidentally satisfies the appropriate boundary condition for $e_0 \to 0$.

5.21. Upon setting $e_0 \left(\delta/\delta\rho(z)\right)\gamma^\mu\left(\delta/\delta\overline{\rho}(z)\right) = \hat{f}^\mu(z)$, with $\eta \to \rho$, $\overline{\eta} \to \overline{\rho}$, $J^\mu \to K^\mu$, we have from (5.15.1),

$$\exp\left[\int (dx)\left[\left(\eta^{\mu\nu} - \frac{\partial^\mu\partial^\nu}{\Box}\right)J_\nu(x)\right]\frac{\delta}{\delta K^\mu(x)}\right]F[\rho,\overline{\rho},K^\mu,\lambda]\bigg|_{K^\mu=0}$$

$$= \exp\left[\frac{i}{2}(\hat{f}^\mu + J^\mu)D_{\mu\nu}(\lambda)(\hat{f}^\nu + J^\nu)\right]\exp\left[-\frac{i}{2}(\partial_\mu J^\mu)\,G\,(\partial_\nu J^\nu)\right]$$

$$\times \exp\left[-i\hat{f}^\mu\partial_\mu\left(\frac{\partial_\alpha}{\Box^2}J^\alpha\right)\right]\exp\left[i\overline{\rho}\,S_+\,\rho\right],$$

where $G$ is defined in (5.15.25). From Problem 5.20, we also have, with

$$\Lambda = e_0\left[\frac{\partial_\alpha}{\Box^2}\,J^\alpha\right],$$

$$\exp\left[-i\hat{f}^\mu\partial_\mu\left(\frac{\partial_\alpha}{\Box^2}J^\alpha\right)\right]\exp\left[i\overline{\rho}\,S_+\,\rho\right] = \exp[i(\overline{\rho}\,e^{-i\Lambda})\,S_+\,(e^{i\Lambda}\rho)].$$

Since we eventually have to set the external Fermi sources to zero, we may make a change of these source variables,

$$\rho \to e^{-i\Lambda}\rho, \quad \overline{\rho} \to e^{i\Lambda}\overline{\rho},$$

and use the invariance of $\hat{f}^\mu$, under such a transformation, to reach the statement made in the problem by finally using, in the process, (5.15.20)–(5.15.22).

5.22. Set $e_0\left(\delta/\delta\rho(z)\right)\gamma^\mu\left(\delta/\delta\overline{\rho}(z)\right) = \hat{f}^\mu(z)$. Then from (5.15.20),

$$(-i)\frac{\delta}{\delta\overline{\eta}(x)}(i)\frac{\delta}{\delta\eta(y)}F[\eta,\overline{\eta},J^\mu,\lambda=0]\bigg|_{\eta=0,\overline{\eta}=0}$$

$$= (-i)\frac{\delta}{\delta\overline{\rho}(x)}(i)\frac{\delta}{\delta\rho(y)}\exp[\hat{Q}]\,F[\rho,\overline{\rho},K^\mu,\lambda]\bigg|_{\rho=0,\overline{\rho}=0,K^\mu=0}, \quad (*)$$

$$\hat{Q} = e_0\left[\tilde{a}^\mu_y\frac{\delta}{\delta K^\mu(y)} - \tilde{a}^\mu_x\frac{\delta}{\delta K^\mu(x)}\right] + \int (dx')\left[\left(\eta^{\mu\nu} - \tilde{a}'^\mu\partial'^\nu\right)J_\nu(x')\right]\frac{\delta}{\delta K^\mu(x')}.$$

The first term in $\hat{Q}$ may be more conveniently rewritten as

$$e_0\int (dx')\,\tilde{a}'^\mu[\delta^{(4)}(x'-x) - \delta^{(4)}(x'-y)]\frac{\delta}{\delta K^\mu(x')}.$$

With $\exp[\hat{Q}]$ generating translations in $K^\nu$, the right-hand side of the former equation $(*)$ is given, in matrix multiplication notation in spacetime, by

$$\exp[i\phi(J^\mu)](-i)\frac{\delta}{\delta\bar{\eta}(x)}(i)\frac{\delta}{\delta\eta(y)}\exp\left[\frac{i}{2}(\hat{f}^\mu + J^\mu)D_{\mu\nu}(\hat{f}^\nu + J^\nu)\right]$$

$$\times \exp\left[-i\hat{f}^\mu\partial_\mu\Lambda\right]\exp\left[i\bar{\rho}S_+\rho\right]\Big|_{\rho=0,\bar{\rho}=0,K^\mu=0},$$

$$\Lambda(z;x,y) = \int(dz')\,G(z-z')\left[e_0\left(\delta^{(4)}(z'-x) - \delta^{(4)}(z'-y)\right) - \partial^\mu J_\mu(z')\right],$$

$$\phi(J^\mu) = -J^\mu(.)\,\partial^{\cdot}_\mu\,\Lambda(.;x,y) + \frac{1}{2}\,g(.;x,y)\,G(.-.)\,g(.;x,y),$$

$$g(z;x,y) = e_0\left(\delta^{(4)}(z-x) - \delta^{(4)}(z-y)\right) - \partial^z_\mu J^\mu(z).$$

where $G(z-z')$ is defined in (5.15.25) with an ultraviolet cut-off. From Problem 5.20: $\exp\left[-i\hat{f}^\mu\partial_\mu\Lambda\right]\exp\left[i\bar{\rho}S_+\rho\right] = \exp[i(\bar{\rho}e^{-i\Lambda})S_+(e^{i\Lambda}\rho)]$. Hence upon defining sources

$$T = e^{i\Lambda}\rho, \quad \bar{T} = \bar{\rho}e^{i\Lambda},$$

using the chain rule: $(\delta/\delta\rho) = e^{i\Lambda}(\delta/\delta T)$, and simply evaluating the functional $\phi(J^\mu)$ above, we obtain

$$(i)(-i)\frac{\delta}{\delta\bar{\eta}(x)}(i)\frac{\delta}{\delta\eta(y)}F[\eta,\eta,J^\mu,\lambda = 0]\Big|_{\eta=0,\bar{\eta}=0,}$$

$$= e^{i\Psi[J^\mu]}e^{-ie_0^2\left[G(0)-G(x-y)\right]}(i)(-i)\frac{\delta}{\delta\bar{T}(x)}(i)\frac{\delta}{\delta T(y)}F[T,\bar{T},J^\mu,\lambda]\Big|_{T=0,\bar{T}=0,} \quad (**)$$

$$\Psi[J^\mu] = -\frac{1}{2}(\partial^\mu J_\mu)\,G\,(\partial^\nu J_\nu) - e_0\int(dz)J^\mu(z)\partial^z_\mu[G(z-x)-G(z-y)].$$

Upon dividing $(**)$ by $F[0,0,J^\mu,\lambda = 0]$, as given in (5.15.24), the statement of the problem follows.

## Chapter 6

6.1. For infinitesimal transformations $V(x) \simeq I + ig_0\,\theta_c(x)t_c$, and the transformation rule $A^\mu \to (VA^\mu V^{-1} + i(V)\partial^\mu V^{-1}/g_0)$, defined in (6.2.4), gives $A^\mu \xrightarrow{} A^\mu + ig_0[t_c,t_b]\theta_c A_b^\mu + t_c\partial^\mu\theta_c$, where we have used the relation $A^\mu = t_b A_b^\mu$. This leads to the following infinitesimal transformation, upon using the antisymmetric nature of the structure constants,

$$A_a^\mu \to A_a^\mu + \nabla_{ac}^\mu\,\theta_c, \quad \nabla_{ac}^\mu = \delta_{ac}\,\partial^\mu + g_0 f_{abc}A_b^\mu.$$

6.2. We explicitly have

$$[\nabla^\mu, \nabla^\nu]_{cb} = g_0 f_{cab}(\partial^\mu A_a^\nu - \partial^\nu A_a^\mu) + g_0^2 (f_{cda} f_{aeb} - f_{cea} f_{adb}) A_d^\mu A_e^\nu.$$

Using the identity $f_{cda} f_{aeb} - f_{cea} f_{adb} = -f_{cba} f_{ade} = f_{cab} f_{ade}$, the result follows upon factoring out $g_0 f_{cab}$, and using the definition of $G_a^{\mu\nu}$.

6.3. From the commutation relation in (6.2.7), established in the previous problem, we have

$$\nabla_{ab\mu} \nabla_{bc\nu} G_c^{\mu\nu} = \nabla_{ab\nu} \nabla_{bc\mu} G_c^{\mu\nu} + g_0 f_{abc} G_{b\mu\nu} G_c^{\mu\nu}$$

$$= \nabla_{ab\nu} \nabla_{bc\mu} G_c^{\mu\nu} = -\nabla_{ab\mu} \nabla_{bc\nu} G_c^{\mu\nu},$$

where in going from the first line to the second we have used the antisymmetry of $f_{abc}$. In the last equality we have used the fact that $G_c^{\mu\nu} = -G_c^{\nu\mu}$, and relabeled $\mu \leftrightarrow \nu$, thus establishing the equality.

6.4. By an integral representation of the delta functional, up to an unimportant multiplicative constant, the left-hand side becomes

$$\int \Pi_{bx} \mathscr{D}\phi_b(x) \exp i[\int (dx)(\phi_a(x)\partial_\mu \mathscr{A}_a^\mu + \frac{\lambda}{2}\phi_a(x)\phi_a(x))] = \int \Pi_{bx} \mathscr{D}\phi_b(x)$$

$$\times \exp \frac{i\lambda}{2} [\int (dx)\phi_a(x)\phi_a(x)] \exp -\frac{i}{2\lambda} [\int (dx)\partial_\mu \mathscr{A}_a^\mu(x)\,\partial_\nu \mathscr{A}_a^\nu(x)],$$

upon completing the squares in the exponential, and shifting the variable $\phi_a$. The result follows after integration over the latter variable.

6.5. From Problem 6.1: $A_a^{(\theta)\mu} \simeq A_a^\mu + (\delta_{ab}\partial^\mu + g_0 f_{acb}A_c^\mu)\theta_b$. The constraint: $\partial_k A_a^{(\theta)k} = 0$ gives $\theta_a \simeq (-\partial_k/\partial^2)A_a^k + \mathcal{O}(A^2)$. When the latter is substituted back in the expression for $A_a^{(\theta)\mu}$, we obtain

$$A_a^{(\theta)\mu} \simeq (\eta^{\mu\nu} - [(\partial^\mu \eta^{\nu k}\partial_k)/\partial^2])A_{a\nu} + \mathcal{O}(A^2).$$

6.6. This expression is obtained from the corresponding differential cross section for $e^-\mu^- \to e^-\mu^-$ in Sect. 5.9.3, by replacing the expression

$$\frac{1}{(p^0/M)(p'^0/M)} M^2 \sum_{\text{spins}} \text{Tr}[(\bar{u}(\mathbf{p}',\sigma')\gamma^\mu u(\mathbf{p},\sigma))(\bar{u}(\mathbf{p},\sigma)\gamma^\nu u(\mathbf{p}',\sigma'))] \text{ in it by}$$

$$\frac{1}{(p^0/M_p)(p'^0/M_p)} M_p^2 \sum_{\text{spins}} \text{Tr}[\langle p',\sigma'|j^\mu(0)|p,\sigma\rangle\langle p\sigma|j^\nu(0)|p',\sigma'\rangle], \quad \text{where}$$

$$\langle p',\sigma'|j^\mu(0)|p\sigma\rangle = \bar{u}(\mathbf{p}',\sigma')[\gamma^\mu F_1(Q^2) + \frac{[\gamma^\mu, \gamma^\alpha]}{4M_p}Q_\alpha \kappa F_2(Q^2)]u(\mathbf{p},\sigma),$$

as readily follows by invariance arguments and application of the Dirac equation $(\gamma p + M_p)u(\mathbf{p}, \sigma) = 0$. $M_p$ denotes the mass of the proton. This gives

$$\frac{d\sigma}{d\Omega}\bigg|_{TF} = \frac{\alpha^2}{4E^2 \sin^4\frac{\vartheta}{2}} \frac{E'}{E}\left[\left(F_1^2 + \frac{Q^2}{4M_p^2}\kappa^2 F_2^2\right)\cos^2\frac{\vartheta}{2} + \frac{Q^2}{2M_p^2}\left(F_1 + \kappa F_2\right)^2 \sin^2\frac{\theta}{2}\right].$$

The result follows upon setting: $G_E = [F_1 - (Q^2/4M_p^2)\kappa F_2]$, $G_M = [F_1 + \kappa F_2]$.

6.7.  Let $(k, k')$, denote the momenta of $(e^-, e^+)$, and $(p, p')$ denote the momenta of $q, \bar{q}$. Using the fact that

$$m_e^2 \sum_{\text{spins}} \text{Tr}\left[\left(\bar{v}(\mathbf{k}', \sigma')\gamma^\mu u(\mathbf{k}, \sigma)\right)\left(\bar{u}(\mathbf{k}, \sigma)\gamma^\nu v(\mathbf{k}', \sigma')\right)\right] \rightarrow [k'^\mu k^\nu + k'^\nu k^\mu - \eta^{\mu\nu} kk'],$$

for $m_e \rightarrow 0$. The corresponding expression for the quarks is then

$$[p'^\mu p^\nu + p'^\nu p^\mu - \eta^{\mu\nu} pp'] \quad \text{and hence,}$$

$$[k'^\mu k^\nu + k'^\nu k^\mu - \eta^{\mu\nu} kk'][p'_\mu p_\nu + p'_\nu p_\mu - \eta_{\mu\nu} pp']$$

$$= 2[kp'\ kp + kp\ kp'] \propto (1 + \cos^2\vartheta),$$

in the CM frame, where $\mathbf{k} \cdot \mathbf{p}/|\mathbf{k} \cdot \mathbf{p}| = \cos\vartheta$. That is, $\vartheta$ is the angle made by the momentum of an emerging quark $q$ relative to that of the electron. Hence

$$d\sigma/d\Omega \propto 3\, e^4 \sum_f (e_f^2/c^2)(1 + \cos^2\vartheta),$$

where $e_f$ is the charge of the quark of a given flavor, and the factor 3 is for the three different colors. The cross section then works out to be $\sigma \propto 3\, e^4 \sum_f (e_f^2/c^2)(16\,\pi)/3$. Upon comparison of this expression with the cross section for $e^+ e^- \rightarrow \mu^+ \mu^-$, with masses set equal to zero in (6.5.7), we obtain

$$\frac{d\sigma}{d\Omega} = \alpha^2 \frac{(1 + \cos^2\vartheta)}{4s} 3\sum_f (e_f^2/e^2), \qquad \sqrt{s} = \text{CM energy.}$$

6.8.  (i) By using a Feynman parameter representation and shifting the variable of integration $k$, the integrand becomes replaced by:

$$2\int_0^1 dx \int_0^x dz\, 1/[k^2 + Q^2(1-x)z]^3,$$

which by using the integral representation over $k$ in (II.7) in Appendix II, at the end of the book, gives

$$\frac{i}{(4\pi)^{D'/2}}\left(\frac{1}{Q^2}\right)^{(1-\delta/2)}\frac{\Gamma\left(1-\frac{\delta}{2}\right)}{\Gamma(3)}\,2\int_0^1 dx\,(1-x)^{(-1+\delta/2)}\int_0^x dz\,z^{(-1+\delta/2)}.$$

Finally carrying out the $z$-integral, followed by the use of the integral (III.12), involving gamma functions, the stated result follows.

(ii) As in part (i), except the $x-z$ integrands become simply multiplied by $[p_1^\mu(1-x)+p_2^\mu z]$, after the shift of the integration variables $k$ and setting, in the process, an odd integral in $k$ equal to zero. Finally the $x-z-$ integrals are readily carried out as above leading to the stated result.

6.9. By using the Feynman parameter representation in Problem 6.8 above, and shifting the $k$-integration variable again, the denominator of the integrand becomes simply

$$k^\mu k^\nu + (p_1^\mu(1-x)+p_2^\mu z)(p_1^\nu(1-x)+p_2^\nu z),$$

after setting an odd integral in $k$ equal to zero. The integral involving the $k^\mu k^\nu$ part may be ultraviolet-regularized using the integral in (III.8), while the integral involving $(p_1^\mu(1-x)+p_2^\mu z)(p_1^\nu(1-x)+p_2^\nu z)$ may be infrared-regularized as in Problem 6.8. Finally, the $(x,z)$ – integrations yield the stated result in a straightforward manner as in Problem 6.8.

6.10. It is sufficient to spell out, the general infra-red singular structure of the function $h_{IR}(\mu_{D'}^2/Q^2,\delta)$. To this end, we refer to the right-hand sides of the integrals in Box 6.2 of the regularized integrals in Sect. 6.6. If an integral depends on $p_1^\mu$, then multiplying it, by $p_1^\mu$ gives zero, and if multiplied by either, $p_{2\mu}$, or $Q_\mu$, give a factor $Q^2$ which cancels out the factor $1/Q^2$ multiplying $(\mu_{D'}^2/Q^2)^{-\delta/2}$. Similar statements follow if the integral depends on $p_2^\mu$. On the other hand, if we multiply the first integral by either $p_1 p_2$ or $Q^2$, these terms cancel out again the $1/Q^2$ factor just mentioned. That is, in all the terms contributing to the fermion-gluon vertex, the $1/Q^2$ factor multiplying $(\mu_{D'}^2/Q^2)^{-\delta/2}$ is canceled out in the infra-red regularized part. The most infra-red singular part in evaluating the vertex function comes from the first integral in the Table now involving the factor $(\mu_{D'}^2/Q^2)^{-\delta/2}\Gamma^2(\delta/2)$. Therefore the infra-red singular structure of $h_{IR}(\mu_{D'}^2/Q^2,\delta)$ is given by a linear combination of the following terms: $1/\delta^2$, $1/\delta$, $(1/\delta)\ln(Q^2/\mu_{D'}^2)$, $\ln^2(Q^2/\mu_{D'}^2)$.

**6.11.**

$$\sum_{nP_n} (2\pi)^4 \delta^4(p + Q - P_n) \langle P, \sigma | j_\mu(0) | nP_n \rangle \langle nP_n | j_\mu(0) | P, \sigma \rangle$$

$$= \sum_{nP_n} \int (\mathrm{d}y) \, e^{i(p_n - P - Q)y} \langle P, \sigma | j_\mu(0) | nP_n \rangle \langle nP_n | j_\mu(0) | P, \sigma \rangle$$

$$= \int (\mathrm{d}y) \, e^{-iQy} \langle P, \sigma | j_\mu(y/2) j_\mu(-y/2) | P, \sigma \rangle$$

where we have used the fact $\langle P, \sigma | j_\mu(0) | nP_n \rangle$

$$= e^{i(y/2)P} e^{-i(y/2)P_n} \langle P, \sigma | e^{[-i(y/2)\mathrm{Mom.Op.}]} j_\mu(0) \, e^{[i(y/2)\mathrm{Mom.Op.}]} | P_n \rangle,$$

and a similar expression for the other factor, and we finally summed over $(n, P_n)$.

**6.12.** Since $P_i = 0, Q_i = Q_3 \delta_{i3}$, we explicitly have $W_{11} = W_{22} = W_1$. Hence for a transversal photon

$$\sigma_\mathrm{T} \sim \epsilon_\lambda^\mu W_{\mu\sigma} \epsilon_\lambda^\sigma = W_1 \geq 0,$$

for $\lambda = 1, 2$. On the other hand, $\epsilon_0^\mu W_{\mu\sigma} \epsilon_0^\sigma$, involves the following three terms:

$$[(Q^2 + v^2)/Q^2]^2 [(Q^2 + v^2) W_2 / Q^2 - W_1],$$

$$[2v^2(Q^2 + v^2)/Q^2][W_1 - (Q^2 + v^2) W_2 / Q^2],$$

$$(v^2/Q^2)^2 [(Q^2 + v^2) W_2 / Q^2 - W_1].$$

Their sum gives

$$\sigma_\mathrm{L} \sim \epsilon_0^\mu W_{\mu\sigma} \epsilon_0^\sigma = [W_2(Q^2 + v^2)/Q^2 - W_1] \geq 0,$$

which establishes (6.9.13). Equations (6.9.14), (6.9.16) follow upon multiplying (6.9.13) by $2xM$, with $x = Q^2/2Mv$, and finally using, in the process, the definitions in (6.9.15).

**6.13.** We explicitly have (see also (6.9.7))

$$W_i^{\mu\nu} = \frac{1}{2\pi\xi M} \frac{e_i^2}{e^2} \int \frac{\mathrm{d}^3\mathbf{p}'}{2p'^0 (2\pi)^3} [.]^{\mu\nu} (2\pi)^4 \delta^{(4)}(\xi P + Q - p'),$$

where $[.]^{\mu\nu}$ is obtained from (6.9.5) by making the substitutions: $k \to \xi P$, $k' \to p'$, and finally using the conservation law $p' = \xi P + Q$.

6.14. The integral on the left-hand side is equal to

$$\int (dp')\,\Theta(p'^{0})\,\delta(p'^{2}+\xi^{2}M^{2})\,\delta^{(4)}(\xi P+Q-p')$$

$$= \Theta(\xi P^{0}+Q^{0})\delta((\xi P+Q)^{2}+\xi^{2}M^{2}) = \delta(2\xi PQ+Q^{2}).$$

The result in question then follows upon taking $2QP$ outside the argument of $\delta(2\xi PQ+Q^{2})$.

6.15. The vertex function $V^{\mu}$ for a spin 0 boson going from momentum $p$ to $p'$ after interacting with the virtual photon must be of the form $p^{\mu}+p'^{\mu}$, with equal coefficients due to gauge invariance: $Q^{\mu}(p_{\mu}+p'_{\mu}) = 0$, where $Q = p'-p$. On the other hand, from the definition of $Q$, we may rewrite

$$(p^{\mu}+p'^{\mu}) = 2(p^{\mu}+Q^{\mu}/2).$$

Also $pQ = -Q^{2}/2$, i.e., $Q^{\mu}/2 = -p\,Q\,Q^{\mu}/Q^{2}$. Thus the vertex function for the spin 0 boson, consistent with gauge invariance, is simply proportional to $(p^{\mu}-pQ\,Q^{\mu}/Q^{2})$. This in turn gives rise to a structure function contribution proportional to

$$(p^{\mu}-pQ\,Q^{\mu}/Q^{2})(p^{\nu}-p\,Q\,Q^{\nu}/Q^{2}),$$

involving *no* $(\eta^{\mu\nu}-Q^{\mu}Q^{\nu}/Q^{2})$ term. This leads to $W_{1} = 0$ and the results stated in the problem follow.

6.16. We may write

$$\sum_{i} e_{i}^{2}\,xf_{i}(x) = xf(x)\sum_{i}e_{i}^{2} = (2/3)\,xf(x) = (2/3)(1/3)\sum_{i}xf_{i}(x),$$

where we have used the fact that $\sum_{i}e_{i}^{2} = (4/9)+(1/9)+(1/9) = 2/3$. Upon integration over $x$, this gives the relation stated in the problem.

6.17. For $A_{qG}^{n}$, we note that $n \geq 3$ implies that $n(n+1) \geq 3\times 4 = 12$, and hence $0 < 1/n(n+1) \leq 1/12$. Also, we may write

$$\sum_{j=2}^{n}1/j = 1/2+1/3+\sum_{j=4}^{n}1/j.$$

The inequality for $A_{qG}^{n}$, then follows. The lower bound is easy to obtain, just omit the positive part. The same reasoning leads to the inequalities in (6.11.18) for $A_{GG}^{n}$, and $A_{Gq}^{n}$, $A_{q\bar{q}}^{n}$ in (6.11.17), where note, for example, that $A_{Gq}^{n}$ in (6.11.12) may be rewritten as: $(4/3)[(1/(n-1))+(2/n(n^{2}-1))]$.

6.18. Let $t = \ln(Q^{2}/\Lambda^{2})$, then $\tau = (1/b_{0})\ln\left[\ln(Q^{2}/\Lambda^{2})/\ln(Q_{o}^{2}/\Lambda^{2})\right]$, or $\tau = (1/b_{0})\ln(t/t_{o})$. This gives $(d\tau/dt) = (1/b_{0}t) = \alpha_{s}(Q^{2})/(2\pi)$, where we have used the relation $\alpha_s(Q^2) = 1/(\beta_0 t)$, to lowest order. From the chain rule $d/dt = (d\tau/dt)(d/d\tau)$, the relation follows.

6.19. (i) This directly follows by noting that $(A_n D_n - B_n C_n) = \lambda_n^+ \lambda_n^-$, and that $A_n + D_n = \lambda_n^+ + \lambda_n^-$, upon carrying out the multiplication of the three matrices on the left-hand side of (6.11.32). (ii) From (6.11.17), $0 < 8n_f A_{qq}^n A_{Gq}^n \equiv 4 B_n C_n \leq (98/135)n_f$. Thus $\lambda_n^+ - \lambda_n^- = \sqrt{(A_n - D_n)^2 + 4B_n C_n}$ is real and positive. Also from (6.11.16), (6.11.18) we establish the positivity of $(A_n - D_n) \equiv (A_{qG}^n - A_{GG}^n)$:

$$\frac{59}{45} + \frac{n_f}{3} + \frac{10}{3}\vartheta_4^n < (A_n - D_n) < \frac{49}{18} + \frac{n_f}{3} + \frac{10}{3}\vartheta_4^n.$$

Hence using the fact that for any two positive numbers $a, b$: $\sqrt{a^2 + b^2} \leq (a + b)$, we obtain from (6.11.34), (6.11.16)

$$\lambda_n^+ \leq \frac{1}{2}[A_n + D_n + (A_n - D_n) + \sqrt{(98/135)n_f}],$$

$$= \frac{1}{2}[2A_n + \sqrt{(98/135)n_f}] < \frac{1}{2}\left[-\frac{50}{9} + \sqrt{(98/135)n_f}\right] < 0,$$

with the upper bound, as shown, is strictly negative for unusually large $n_f < 43$. On the other hand $(A_n - D_n) > 4B_n C_n$, and $(A_n + D_n) > \sqrt{2}(A_n + D_n)$ with the latter being negative. Hence

$$\lambda_n^- = (1/2)[A_n + D_n - \sqrt{(A_n - D_n)^2 + 4B_n C_n}]$$

$$> (\sqrt{2}/2)[2D_n] > -\sqrt{2}[(11/2) + (n_f/3) + 6\vartheta_4^n],$$

as follows from (6.11.18). (iv) Finally

$$(\lambda_n^+ - A_n) = (1/2)[-(A_n - D_n) + \sqrt{(A_n - D_n)^2 + 4B_n C_n}] > 0,$$

$$(\lambda_n^- - A_n) = -(1/2)[(A_n - D_n) + \sqrt{(A_n - D_n)^2 + 4B_n C_n}] < 0.$$

6.20. For $x^0 > y^0$, $\left(\exp[+ig\int_{y^0}^{x^0} d\xi \, A_0(\xi, \mathbf{x})]\right)_+$

$$= 1 + (ig)\int_{y^0}^{x^0} d\xi_1 A_0(\xi_1, \mathbf{x}) + (ig)^2 \int_{y^0}^{x^0} d\xi_2 \int_{x'^0}^{\xi_2} d\xi_1 A_0(\xi_2, \mathbf{x})A_0(\xi_1, \mathbf{x}) + \cdots,$$

and $\partial_0(.)_+ = igA_0(x^0, \mathbf{x})[1 + (ig)\int_{y^0}^{x^0} d\xi_1 A_0(\xi_1, \mathbf{x}) + \cdots]$.

6.21. For $T > 0$, consider the following expression, as a function of $t \geq 0$:

$$Q_-(t) = h_-(t, a, 0, a)V^{-1}(t, a)$$

with condition $Q_-(0) = V^{-1}(0, a)$. Using the fact that

$$dh_-(t, a, 0, a)/dt = -igh_-(t, a, 0, a)A_0(t, a),$$

we obtain $dQ_-(t)/dt$

$$= -igh_-(t, a; 0, a)V^{-1}(t, a)V(t, a)\big[A_0(t, a) - (1/ig)(d/dt)\big]V^{-1}(t, a),$$

which is just $-ig\, Q_-(t)A_0^V(t, a)$. Upon integration from $0$ to $T$, gives:

$$h_-(T, a; 0, a)V^{-1}(T, a) = V^{-1}(0, a)h_-^V(T, a; 0, a),$$

from which the transformation rule in question follows. The transformation rules of the last two are almost identical to the first two by simply exchanging time variables with space variables.

6.22. If a priori $\phi^+$ is zero, then by a specific choice of the transformation in (6.14.23), as a phase factor, we may remove any phase that $\phi^0$ may have upon the transformation in. Otherwise, suppose that $\phi^+ \neq 0$. Using the identity

$$\exp[i\varphi\mathbf{n} \cdot \boldsymbol{\sigma}/2] = \cos(\varphi/2) + i\,\mathbf{n} \cdot \boldsymbol{\sigma} \sin(\varphi/2),$$

the transformation in (6.14.22) gives

$$\begin{pmatrix} \cos(\varphi/2) + i n_3 \sin(\varphi/2) & (i n_1 + n_2) \sin(\varphi/2) \\ (i n_1 - n_2) \sin(\varphi/2) & \cos(\varphi/2) - i n_3 \sin(\varphi/2) \end{pmatrix} \begin{pmatrix} \phi^+ \\ \phi^0 \end{pmatrix}.$$

Upon considering the expression $\phi^0/\phi^+$, and writing the latter as $(\phi^0/\phi^+) = (a + ib)$, with $a$ and $b$ real, it is easily checked, by equating the resulting upper entry to zero, i.e., by setting

$$\big(\cos(\varphi/2) + i n_3 \sin(\varphi/2)\big) + (i n_1 + n_2)\sin(\varphi/2)(a + ib) = 0,$$

that this equation has always a solution for all real $a$ and $b$, by appropriate choices of $n_1$, $n_2$, $n_3$, and $\varphi$. This makes the resulting upper entry equal to zero in vacuum expectation value. Any phase that may arise from the second row of the transformation above may be removed by the appropriate choice of the transformation in (6.14.23), giving finally a real non-negative field for the lower entry in vacuum expectation value. Thus such transformations give rise to a field as given on the right-hand side of (6.14.26), with its components satisfying Eq. (6.14.27).

6.23. Using the anti-commutativity of the fermion fields, we have

$$[\bar{e}_L \gamma^\rho \nu_L [[\bar{\nu}_L \gamma_\rho e_L]] = -(1/4)\bar{e}_a e_D \bar{\nu}_c \nu_B \big[[\gamma^\rho(1 - \gamma^5)]_{aB} [\gamma_\rho(1 - \gamma^5)]_{cD}\big].$$

Multiplying the Fierz identity[8]

$$(\gamma^\rho)_{ab}(\gamma_\rho)_{cd} = -\delta_{ad}\delta_{cb} - \frac{1}{2}(\gamma^\rho)_{ad}(\gamma_\rho)_{cb} - \frac{1}{2}(\gamma^5\gamma^\rho)_{ad}(\gamma^5\gamma_\rho)_{cb} + (\gamma^5)_{ad}(\gamma^5)_{cb},$$

by $(1 - \gamma^5)_{bB}(1 - \gamma^5)_{dD}$ and using the identities $\{\gamma^5, \gamma^\mu\} = 0, (\gamma^5)^2 = 1,$ give $-[\gamma^\rho(1 - \gamma^5)]_{aD}[\gamma_\rho(1 - \gamma^5)]_{cB}$, and the identity immediately follows.

6.24. Following the method in Sects. 5.9.3, 5.9.1, and averaging over the spin of the muon and summing over the spins of the product particles, we have

$$(2M_\mu)(2m_e)(2m_{\nu_e})(2m_{\nu_\mu})\frac{1}{2}\sum_{\text{spins}}|\mathscr{A}|^2\Big|_{m_e, m_{\nu_e}, m_{\nu_\mu} \to 0} = 64\,G_F^2\,(p_\mu\,k_\nu)(k_1^\mu\,k_2^\nu),$$

which we conveniently denote by $X$. The decay rate is then given by $(p = (M_\mu, \mathbf{0}))$

$$d\Gamma = \frac{1}{16M_\mu}\int X\frac{d^3\mathbf{k}}{(2\pi)^3|\mathbf{k}|}\frac{d^3\mathbf{k}_1}{(2\pi)^3|\mathbf{k}_1|}\frac{d^3\mathbf{k}_2}{(2\pi)^3|\mathbf{k}_2|}(2\pi)^4\delta^{(4)}(p-k-k_1-k_2).$$

Also note that

$$p_\mu k_\nu\int\left(d^3\mathbf{k}_1 d^3\mathbf{k}_2/|\mathbf{k}_1||\mathbf{k}_2|\right)k_1^\mu k_2^\nu\delta^{(4)}(p-k-k_1-k_2) = (\pi M_\mu^2/6)[3M_\mu|\mathbf{k}|-4|\mathbf{k}|^2].$$

Accordingly, using the fact that $d^3\mathbf{k}/|\mathbf{k}| = E_e\,dE_e\,d\Omega$, and noting by conservation of energy and momentum that the maximum value of the energy $E_e$ attained by the electron corresponds to the neutrinos moving in the same direction leading to $E_e|_{max} = M_\mu/2$, we readily obtain, upon carrying the $\mathbf{k}$-integration, the decay rate stated in the problem.

---

[8]For many details on Fierz identity and some of its generalizations, see Appendix A to Chapter 2 in Volume II.

# Index

© Springer International Publishing Switzerland 2016
E.B. Manoukian, *Quantum Field Theory I*, Graduate Texts in Physics,
DOI 10.1007/978-3-319-30939-2

Printed in the United States
By Bookmasters